VOLUME FOUR HUNDRED AND SEVENTY-FIVE

METHODS IN ENZYMOLOGY

Single Molecule Tools, Part B: Super-Resolution, Particle Tracking, Multiparameter, and Force Based Methods

METHODS IN ENZYMOLOGY

Editors-in-Chief

JOHN N. ABELSON AND MELVIN I. SIMON

Division of Biology
California Institute of Technology
Pasadena, California

Founding Editors

SIDNEY P. COLOWICK AND NATHAN O. KAPLAN

VOLUME FOUR HUNDRED AND SEVENTY-FIVE

Methods in
Enzymology

Single Molecule Tools, Part B: Super-Resolution, Particle Tracking, Multiparameter, and Force Based Methods

EDITED BY

NILS G. WALTER
Department of Chemistry
Single Molecule Analysis Group
University of Michigan, Ann Arbor
Michigan, USA

AMSTERDAM • BOSTON • HEIDELBERG • LONDON
NEW YORK • OXFORD • PARIS • SAN DIEGO
SAN FRANCISCO • SINGAPORE • SYDNEY • TOKYO
Academic Press is an imprint of Elsevier

Academic Press is an imprint of Elsevier
525 B Street, Suite 1900, San Diego, CA 92101-4495, USA
30 Corporate Drive, Suite 400, Burlington, MA 01803, USA
32 Jamestown Road, London NW1 7BY, UK

First edition 2010

Copyright © 2010, Elsevier Inc. All Rights Reserved.

No part of this publication may be reproduced, stored in a retrieval system or transmitted in any form or by any means electronic, mechanical, photocopying, recording or otherwise without the prior written permission of the publisher

Permissions may be sought directly from Elsevier's Science & Technology Rights Department in Oxford, UK: phone (+44) (0) 1865 843830; fax (+44) (0) 1865 853333; email: permissions@elsevier.com. Alternatively you can submit your request online by visiting the Elsevier web site at http://elsevier.com/locate/permissions, and selecting *Obtaining permission to use Elsevier material*

Notice
No responsibility is assumed by the publisher for any injury and/or damage to persons or property as a matter of products liability, negligence or otherwise, or from any use or operation of any methods, products, instructions or ideas contained in the material herein. Because of rapid advances in the medical sciences, in particular, independent verification of diagnoses and drug dosages should be made

For information on all Academic Press publications
visit our website at elsevierdirect.com

ISBN: 978-0-12-381482-1
ISSN: 0076-6879

Printed and bound in United States of America
10 11 12   10 9 8 7 6 5 4 3 2 1

Working together to grow
libraries in developing countries

www.elsevier.com | www.bookaid.org | www.sabre.org

ELSEVIER   BOOK AID International   Sabre Foundation

Contents

Contributors	*xiii*
Preface	*xxi*
Volumes in Series	*xxiii*

1. Super-Accuracy and Super-Resolution: Getting Around the Diffraction Limit — 1
Erdal Toprak, Comert Kural, and Paul R. Selvin

1. Overview: Accuracy and Resolution — 2
2. Getting Super-Accuracy — 4
3. Calculating Super-Accuracy — 7
4. Reaching Super-Resolution — 13
5. Future Directions — 22
References — 22

2. Molecules and Methods for Super-Resolution Imaging — 27
Michael A. Thompson, Julie S. Biteen, Samuel J. Lord, Nicholas R. Conley, and W. E. Moerner

1. Introduction — 28
2. Molecules for Super-Resolution Imaging — 30
3. Selected Methods for Super-Resolution Imaging — 46
Acknowledgments — 55
References — 55

3. Tracking Single Proteins in Live Cells Using Single-Chain Antibody Fragment-Fluorescent Quantum Dot Affinity Pair — 61
Gopal Iyer, Xavier Michalet, Yun-Pei Chang, and Shimon Weiss

1. Introduction — 62
2. The Method: Targeting QDs via a Single-Chain Variable Fragment-Hapten Pair — 64
3. Functionalization of QDs — 66
4. Quantification of the Number of FL Molecules per FL-pc-QD — 69
5. Binding of FL-QDs to Anti-scFv Fusion Constructs — 71
6. DNA Constructs for Single FL-QD Imaging in Live Cells — 73

7. Single-Molecule Imaging of Live Mammalian Cells	74
Acknowledgments	75
References	77

4. Recording Single Motor Proteins in the Cytoplasm of Mammalian Cells — 81

Dawen Cai, Neha Kaul, Troy A. Lionberger,
Diane M. Wiener, Kristen J. Verhey, and Edgar Meyhofer

1. Introduction	82
2. Basic Principles	84
3. Labeling Molecular Motors for *In Vivo* Observations	85
4. Instrumentation for Tracking Single Motors *In Vivo*	89
5. Detailed Experimental Procedures	95
6. Summary and Conclusions	103
Acknowledgments	103
References	104

5. Single-Particle Tracking Photoactivated Localization Microscopy for Mapping Single-Molecule Dynamics — 109

Suliana Manley, Jennifer M. Gillette,
and Jennifer Lippincott-Schwartz

1. Introduction	110
2. Description of the sptPALM Method	111
3. Labeling with Photoactivatable Fluorescent Probes	113
4. Tracking Single Molecules	114
5. Experimental Example: sptPALM on a Membrane Protein	116
6. Conclusions	118
References	118

6. A Bird's Eye View: Tracking Slow Nanometer-Scale Movements of Single Molecular Nano-Assemblies — 121

Nicole Michelotti, Chamaree de Silva, Alexander E. Johnson-Buck,
Anthony J. Manzo, and Nils G. Walter

1. Introduction	122
2. DNA-Based Nanowalkers	125
3. Considerations for Fluorescence Imaging of Slowly Moving Particles	129
4. Single-Molecule Fluorescence Tracking of Nanowalkers	131
5. Extracting Super-Resolution Position Information	136
6. Concluding Remarks	143
Acknowledgments	144
References	144

7. Anti-Brownian Traps for Studies on Single Molecules — 149
Alexander P. Fields and Adam E. Cohen

1. Theoretical Overview	150
2. Anti-Brownian Trapping Systems	155
3. The ABEL Trap	161
4. Applications	169
5. Future Work: *En Route* to Single Fluorophores	170
Acknowledgments	171
References	171

8. Plasmon Rulers as Dynamic Molecular Rulers in Enzymology — 175
Björn M. Reinhard, Jaime M. Yassif, Peter Vach, and Jan Liphardt

1. Introduction	176
2. The Basic Idea: Distance Dependence of Plasmon Coupling	177
3. Hardware Needed for Single Particle Rayleigh Scattering Spectroscopy	179
4. Which Readout—Intensity, Polarization, or Color?	181
5. Ruler Calibration?	182
6. Plasmon Ruler Assembly and Purification	184
7. Example 1: Dynamics of DNA Bending and Cleavage by Single EcoRV Restriction Enzymes	186
8. Example 2: Spermidine Modulated Ribonuclease Activity Probed by RNA Plasmon Rulers	192
9. Outlook	194
References	196

9. Quantitative Analysis of DNA-Looping Kinetics from Tethered Particle Motion Experiments — 199
Carlo Manzo and Laura Finzi

1. Introduction	200
2. Change-Point Algorithm	201
3. Data Clustering and Expectation-Maximization Algorithm	203
4. Adaptation of the Method to the Case of TPM Data Analysis	204
5. Performance of the Method	209
6. Comparison with the Threshold Method	212
7. Application to TPM Experiments: CI-Induced Looping in λ-DNA	215
8. Conclusions	217
Acknowledgments	218
References	218

10. Methods in Statistical Kinetics — 221
Jeffrey R. Moffitt, Yann R. Chemla, and Carlos Bustamante

1. Introduction — 222
2. The Formalism of Statistical Kinetics — 223
3. Characterizing Fluctuations — 232
4. Extracting Mechanistic Constraints from Moments — 240
5. Conclusions and Future Outlook — 248
References — 255

11. Visualizing DNA Replication at the Single-Molecule Level — 259
Nathan A. Tanner and Antoine M. van Oijen

1. Introduction — 260
2. Observing Replication Loops with Tethered Bead Motion — 261
3. Fluorescence Visualization of DNA Replication — 271
Acknowledgments — 277
References — 277

12. Measurement of the Conformational State of F_1-ATPase by Single-Molecule Rotation — 279
Daichi Okuno, Mitsunori Ikeguchi, and Hiroyuki Noji

1. Introduction — 280
2. Sample Preparation — 282
3. Single-Molecule Cross-Link Experiment — 287
4. Pausing with AMP-PNP or/and N_3^- — 293
Acknowledgments — 295
References — 295

13. Magnetic Tweezers for the Study of DNA Tracking Motors — 297
Maria Manosas, Adrien Meglio, Michelle M. Spiering,
Fangyuan Ding, Stephen J. Benkovic, François-Xavier Barre,
Omar A. Saleh, Jean François Allemand,
David Bensimon, and Vincent Croquette

1. Introduction — 298
2. Experimental Setup — 299
3. Methods and Protocols — 300
4. Application to the Study of FtsK — 308
5. Application to the Study of the GP41 Helicase — 312
6. Conclusions — 318
Acknowledgments — 319
References — 319

14. Single-Molecule Dual-Beam Optical Trap Analysis of Protein Structure and Function   321

Jongmin Sung, Sivaraj Sivaramakrishnan, Alexander R. Dunn, and James A. Spudich

1. Introduction   322
2. Insights into Myosin Function Using a Dual-Beam Optical Trap   322
3. Optical Trap Instrumentation   324
4. Optical Trapping Experiment   358
5. Data Analysis   361
6. Conclusion   369
Acknowledgments   372
References   372

15. An Optical Apparatus for Rotation and Trapping   377

Braulio Gutiérrez-Medina, Johan O. L. Andreasson, William J. Greenleaf, Arthur LaPorta, and Steven M. Block

1. Introduction   378
2. Optical Trapping and Rotation of Microparticles   379
3. The Instrument   385
4. Fabrication of Anisotropic Particles   390
5. Instrument Calibration   395
6. Simultaneous Application of Force and Torque Using Optical Tweezers   400
7. Conclusions   402
Acknowledgments   402
References   403

16. Force–Fluorescence Spectroscopy at the Single-Molecule Level   405

Ruobo Zhou, Michael Schlierf, and Taekjip Ha

1. Introduction   406
2. Setup   407
3. Optical Trapping   410
4. Fluorescence Detection   413
5. Coalignment of Confocal and Optical Trapping   416
6. Sample Preparation Protocols   418
7. Applications to Biological Systems   423
8. Outlook   423
Acknowledgments   424
References   424

17. **Combining Optical Tweezers, Single-Molecule Fluorescence Microscopy, and Microfluidics for Studies of DNA–Protein Interactions** — 427

Peter Gross, Géraldine Farge, Erwin J. G. Peterman, and Gijs J. L. Wuite

1. Introduction	428
2. Instrumentation	430
3. Preparation of Reagents	443
4. Combining Optical Trapping, Fluorescence Microscopy, and Microfluidics: Example Protocols	445
5. Conclusions	450
Acknowledgments	451
References	451

18. **Accurate Single-Molecule FRET Studies Using Multiparameter Fluorescence Detection** — 455

Evangelos Sisamakis, Alessandro Valeri, Stanislav Kalinin, Paul J. Rothwell, and Claus A. M. Seidel

1. Introduction	456
2. FRET Theory	461
3. Fluorescence Properties and Measurement Techniques	466
4. Qualitative Description of smFRET	474
5. Quantitative Description of smFRET	489
6. Discussion	503
Acknowledgments	505
References	506

19. **Atomic Force Microscopy Studies of Human Rhinovirus: Topology and Molecular Forces** — 515

Ferry Kienberger, Rong Zhu, Christian Rankl, Hermann J. Gruber, Dieter Blaas, and Peter Hinterdorfer

1. Introduction	516
2. Results and Discussion	517
References	536

20. High-Speed Atomic Force Microscopy Techniques for Observing Dynamic Biomolecular Processes — 541

Daisuke Yamamoto, Takayuki Uchihashi, Noriyuki Kodera, Hayato Yamashita, Shingo Nishikori, Teru Ogura, Mikihiro Shibata, and Toshio Ando

1. Introduction — 542
2. Survey of Requirements for High-Speed Bio-AFM Imaging — 543
3. Substrate Surfaces — 544
4. Control of Diffusional Mobility — 554
5. Protein 2D Crystals as Targets to Study — 555
6. Low-Invasive Imaging — 558
7. UV Flash-Photolysis of Caged Compounds — 559
8. Cantilever Tip — 561
References — 562

21. Nanopore Force Spectroscopy Tools for Analyzing Single Biomolecular Complexes — 565

Olga K. Dudko, Jérôme Mathé, and Amit Meller

1. Introduction — 566
2. The Nanopore Method — 567
3. DNA Unzipping Kinetics Studied Using Nanopore Force Spectroscopy — 577
4. Conclusions and Summary — 585
Acknowledgments — 587
References — 587

22. Analysis of Single Nucleic Acid Molecules with Protein Nanopores — 591

Giovanni Maglia, Andrew J. Heron, David Stoddart, Deanpen Japrung, and Hagan Bayley

1. Background: Analysis of Nucleic Acids with Nanopores — 593
2. Electrical Recording with Planar Lipid Bilayers — 601
3. Nanopores — 610
4. Materials — 613
5. Data Acquisition and Analysis — 616
References — 619

Author Index — *625*
Subject Index — *649*

Contributors

Jean François Allemand
Laboratoire de Physique Statistique, Université Paris Diderot, and Département de Biologie, Ecole Normale Superieure; Institut Universitaire de France, Paris, France

Toshio Ando
Department of Physics, Kanazawa University, Kakuma-machi, Kanazawa, and Core Research for Evolutional Science and Technology (CREST) of the Japan Science and Technology Agency, Sanban-cho, Chiyoda-ku, Tokyo, Japan

Johan O. L. Andreasson
Department of Physics, Stanford University, Stanford, California, USA

François-Xavier Barre
CNRS, Centre de Génétique Moléculaire, Gif-sur-Yvette, and Université Paris-Sud, Orsay, Paris, France

Hagan Bayley
Department of Chemistry, University of Oxford, Oxford, United Kingdom

Stephen J. Benkovic
Department of Chemistry, The Pennsylvania State University, University Park, Pennsylvania, USA

David Bensimon
Laboratoire de Physique Statistique, Université Paris Diderot, and Département de Biologie, Ecole Normale Superieure, Paris, France; Department of Chemistry and Biochemistry, UCLA, Los Angeles, California, USA

Julie S. Biteen
Department of Chemistry, Stanford University, Stanford, California, USA

Dieter Blaas
Max F. Perutz Laboratories, Medical University of Vienna, Vienna, Austria

Steven M. Block
Department of Biology, and Department of Applied Physics, Stanford University, Stanford, California, USA

Carlos Bustamante
Department of Physics and Jason L. Choy Laboratory of Single-Molecule Biophysics, and Departments of Molecular and Cell Biology, and Chemistry, Howard Hughes Medical Institute, University of California, Berkeley, California, USA

Dawen Cai
Department of Cell and Developmental Biology, and Department of Biophysics, University of Michigan, Ann Arbor, Michigan, USA

Yun-Pei Chang
Department of Chemistry and Biochemistry, California NanoSystems Institute, University of California, Los Angeles, California, USA

Yann R. Chemla
Department of Physics and Center for Biophysics and Computational Biology, University of Illinois at Urbana-Champaign, Urbana, Illinois, USA

Adam E. Cohen
Department of Chemistry and Chemical Biology, and Department of Physics, Harvard University, Cambridge, Massachusetts, USA

Nicholas R. Conley
Department of Chemistry, Stanford University, Stanford, California, USA

Vincent Croquette
Laboratoire de Physique Statistique, Université Paris Diderot, and Département de Biologie, Ecole Normale Superieure, Paris, France

Chamaree de Silva
Single Molecule Analysis Group, Department of Chemistry, University of Michigan, Ann Arbor, Michigan, USA

Fangyuan Ding
Laboratoire de Physique Statistique, Université Paris Diderot, and Département de Biologie, Ecole Normale Superieure, Paris, France

Olga K. Dudko
Department of Physics and Center for Theoretical Biological Physics, University of California, San Diego, La Jolla, California, USA

Alexander R. Dunn
Department of Chemical Engineering, Stanford University, Stanford, California, USA

Géraldine Farge
Department of Physics and Astronomy and Laser Centre, VU University, De Boelelaan, Amsterdam, The Netherlands

Alexander P. Fields
Department of Biophysics, Harvard University, Cambridge, Massachusetts, USA

Laura Finzi
Physics Department, Emory University, Atlanta, Georgia, USA

Jennifer M. Gillette
Section on Organelle Biology, Cell Biology and Metabolism Program, NICHD, National Institutes of Health, Bethesda, Maryland, USA

William J. Greenleaf
Department of Applied Physics, Stanford University, Stanford, California, USA

Peter Gross
Department of Physics and Astronomy and Laser Centre, VU University, De Boelelaan, Amsterdam, The Netherlands

Hermann J. Gruber
Institute for Biophysics, Johannes Kepler University of Linz, Linz, Austria

Braulio Gutiérrez-Medina
Department of Biology, Stanford University, Stanford, California, USA

Taekjip Ha
Department of Physics and Center for the Physics of Living Cells, University of Illinois at Urbana-Champaign, and Howard Hughes Medical Institute, Urbana, Illinois, USA

Andrew J. Heron
Department of Chemistry, University of Oxford, Oxford, United Kingdom

Peter Hinterdorfer
Institute for Biophysics, Johannes Kepler University of Linz, Linz, Austria

Mitsunori Ikeguchi
Graduate School of Nanobioscience, Yokohama City University, Yokohama, Japan

Gopal Iyer
Department of Chemistry and Biochemistry, California NanoSystems Institute, University of California, Los Angeles, California, USA

Deanpen Japrung
Department of Chemistry, University of Oxford, Oxford, United Kingdom

Alexander E. Johnson-Buck
Single Molecule Analysis Group, Department of Chemistry, University of Michigan, Ann Arbor, Michigan, USA

Stanislav Kalinin
Institut für Physikalische Chemie, Lehrstuhl für Molekulare Physikalische Chemie, Heinrich-Heine-Universität, Universitätsstraße 1, Düsseldorf, Germany

Neha Kaul
Department of Mechanical Engineering, University of Michigan, Ann Arbor, Michigan, USA

Ferry Kienberger
Institute for Biophysics, Johannes Kepler University of Linz, Linz, Austria

Noriyuki Kodera
Department of Physics, Kanazawa University, Kakuma-machi, Kanazawa, and Core Research for Evolutional Science and Technology (CREST) of the Japan Science and Technology Agency, Sanban-cho, Chiyoda-ku, Tokyo, Japan

Comert Kural
Department of Biophysics, University of Illinois, Urbana, Illinois, and Immune Disease Institute, Harvard Medical School, Boston, Massachusetts, USA

Arthur LaPorta
Department of Physics, Biophysics Program, Institute for Physical Science and Technology, University of Maryland, College Park, Maryland, USA

Troy A. Lionberger
Department of Mechanical Engineering, and Program in Cellular and Molecular Biology, University of Michigan, Ann Arbor, Michigan, USA

Jan Liphardt
Department of Physics, Biophysics Graduate Group, and Bay Area Physical Sciences Oncology Center, University of California at Berkeley, and Physical Biosciences Division, Lawrence Berkeley National Laboratory, Berkeley, California, USA

Jennifer Lippincott-Schwartz
Section on Organelle Biology, Cell Biology and Metabolism Program, NICHD, National Institutes of Health, Bethesda, Maryland, USA

Samuel J. Lord
Department of Chemistry, Stanford University, Stanford, California, USA

Giovanni Maglia
Department of Chemistry, University of Oxford, Oxford, United Kingdom

Suliana Manley
Institute of Physics of Biological Systems, Swiss Federal Institute of Technology (EPFL), Lausanne, Switzerland

Maria Manosas
Laboratoire de Physique Statistique, Université Paris Diderot, and Département de Biologie, Ecole Normale Superieure, Paris, France

Anthony J. Manzo
Single Molecule Analysis Group, Department of Chemistry, University of Michigan, Ann Arbor, Michigan, USA

Carlo Manzo
Physics Department, Emory University, Atlanta, Georgia, USA

Jérôme Mathé
Laboratoire LAMBE (UMR 8587—CNRS-CEA-UEVE), Université d'Evry-val d'Essonne, Evry, France

Adrien Meglio
Laboratoire de Physique Statistique, Université Paris Diderot, and Département de Biologie, Ecole Normale Superieure, Paris, France

Amit Meller
Department of Biomedical Engineering and Department of Physics, Boston University, Boston, Massachusetts, USA

Edgar Meyhofer
Department of Biophysics, and Department of Mechanical Engineering; Program in Cellular and Molecular Biology, University of Michigan, Ann Arbor, Michigan, USA

Xavier Michalet
Department of Chemistry and Biochemistry, California NanoSystems Institute, University of California, Los Angeles, California, USA

Nicole Michelotti
Single Molecule Analysis Group, Department of Chemistry, and Department of Physics, University of Michigan, Ann Arbor, Michigan, USA

W. E. Moerner
Department of Chemistry, Stanford University, Stanford, California, USA

Jeffrey R. Moffitt
Department of Physics and Jason L. Choy Laboratory of Single-Molecule Biophysics, University of California, Berkeley, California, USA

Shingo Nishikori
Department of Molecular Cell Biology, Institute of Molecular Embryology and Genetics, Kumamoto University, Kumamoto, Japan

Hiroyuki Noji
The Institute of Scientific and Industrial Research, Osaka University, Osaka, Japan

Teru Ogura
Core Research for Evolutional Science and Technology (CREST) of the Japan Science and Technology Agency, Sanban-cho, Chiyoda-ku, Tokyo, and

Department of Molecular Cell Biology, Institute of Molecular Embryology and Genetics, Kumamoto University, Kumamoto, Japan

Daichi Okuno
The Institute of Scientific and Industrial Research, Osaka University, Osaka, Japan

Erwin J. G. Peterman
Department of Physics and Astronomy and Laser Centre, VU University, De Boelelaan, Amsterdam, The Netherlands

Christian Rankl
Institute for Biophysics, Johannes Kepler University of Linz, Linz, Austria

Björn M. Reinhard
Department of Chemistry, The Photonics Center, Boston University, Boston, Massachusetts, USA

Paul J. Rothwell
Institut für Physikalische Chemie, Lehrstuhl für Molekulare Physikalische Chemie, Heinrich-Heine-Universität, Universitätsstraße 1, Düsseldorf, Germany

Omar A. Saleh
Materials Department and BMSE Program, University of California, Santa Barbara, California, USA

Michael Schlierf
Department of Physics and Center for the Physics of Living Cells, University of Illinois at Urbana-Champaign, Urbana, Illinois, USA

Claus A. M. Seidel
Institut für Physikalische Chemie, Lehrstuhl für Molekulare Physikalische Chemie, Heinrich-Heine-Universität, Universitätsstraße 1, Düsseldorf, Germany

Paul R. Selvin
Department of Biophysics, and Department of Physics, University of Illinois, Urbana, Illinois, USA

Mikihiro Shibata
Department of Physics, Kanazawa University, Kakuma-machi, Kanazawa, Japan

Evangelos Sisamakis
Institut für Physikalische Chemie, Lehrstuhl für Molekulare Physikalische Chemie, Heinrich-Heine-Universität, Universitätsstraße 1, Düsseldorf, Germany, and Department of Applied Physics, Experimental Biomolecular Physics, Royal Institute of Technology, Stockholm, Sweden

Sivaraj Sivaramakrishnan
Department of Biochemistry, Stanford University School of Medicine, Stanford, California, USA

Michelle M. Spiering
Department of Chemistry, The Pennsylvania State University, University Park, Pennsylvania, USA

James A. Spudich
Department of Biochemistry, Stanford University School of Medicine, Stanford, California, USA

David Stoddart
Department of Chemistry, University of Oxford, Oxford, United Kingdom

Jongmin Sung
Department of Biochemistry, Stanford University School of Medicine, and Department of Applied Physics, Stanford University, Stanford, California, USA

Nathan A. Tanner
Department of Biological Chemistry and Molecular Pharmacology, Harvard Medical School, Boston, Massachusetts, USA

Michael A. Thompson
Department of Chemistry, Stanford University, Stanford, California, USA

Erdal Toprak
Department of Biophysics, University of Illinois, Urbana, Illinois, and Department of Systems Biology, Harvard Medical School, Boston, Massachusetts, USA

Takayuki Uchihashi
Department of Physics, Kanazawa University, Kakuma-machi, Kanazawa, and Core Research for Evolutional Science and Technology (CREST) of the Japan Science and Technology Agency, Sanban-cho, Chiyoda-ku, Tokyo, Japan

Peter Vach
Department of Physics, Biophysics Graduate Group, and Bay Area Physical Sciences Oncology Center, University of California at Berkeley, California, USA

Alessandro Valeri
Institut für Physikalische Chemie, Lehrstuhl für Molekulare Physikalische Chemie, Heinrich-Heine-Universität, Universitätsstraße 1, Düsseldorf, Germany

Antoine M. van Oijen
Department of Biological Chemistry and Molecular Pharmacology, Harvard Medical School, Boston, Massachusetts, USA

Kristen J. Verhey
Department of Cell and Developmental Biology, and Department of Biophysics; Program in Cellular and Molecular Biology, University of Michigan, Ann Arbor, Michigan, USA

Nils G. Walter
Single Molecule Analysis Group, Department of Chemistry, University of Michigan, Ann Arbor, Michigan, USA

Shimon Weiss
Department of Chemistry and Biochemistry, California NanoSystems Institute, University of California, Los Angeles, California, USA

Diane M. Wiener
Department of Mechanical Engineering, University of Michigan, Ann Arbor, Michigan, USA

Gijs J. L. Wuite
Department of Physics and Astronomy and Laser Centre, VU University, De Boelelaan, Amsterdam, The Netherlands

Daisuke Yamamoto
Department of Physics, Kanazawa University, Kakuma-machi, Kanazawa, and Core Research for Evolutional Science and Technology (CREST) of the Japan Science and Technology Agency, Sanban-cho, Chiyoda-ku, Tokyo, Japan

Hayato Yamashita
Department of Physics, Kanazawa University, Kakuma-machi, Kanazawa, Japan

Jaime M. Yassif
Department of Physics, Biophysics Graduate Group, and Bay Area Physical Sciences Oncology Center, University of California at Berkeley, and Physical Biosciences Division, Lawrence Berkeley National Laboratory, Berkeley, California, USA

Ruobo Zhou
Department of Physics and Center for the Physics of Living Cells, University of Illinois at Urbana-Champaign, Urbana, Illinois, USA

Rong Zhu
Christian Doppler Laboratory for Nanoscopic Methods in Biophysics, Johannes Kepler University of Linz, Linz, Austria

Preface

Ever since Feynman's suggestion in the early 1960s that "there's plenty of room at the bottom," single-molecule tools have seen an exponential rise in popularity (note that exponentially increasing rates are characteristic of explosions!). One can hardly go to a Biophysical Society meeting these days without being impressed by the literally thousands of posters and seminars that show data exploiting the unique capabilities of single-molecule probing techniques. Among their benefits are that they: (i) can directly observe events at the molecular level; (ii) reveal rare and/or transient species and heterogeneities along a reaction pathway, which are often lost in ensemble averages; (iii) can directly access the low copy numbers (typically 1–1000) of any specific biopolymer in a single cell; (iv) afford counting and nanometer-accuracy localization of molecules in spatially distributed samples such as a cell; (v) enable the ultimate miniaturization and multiplexing of biological assays such as DNA sequencing; (vi) allow of the direct measurement of the mechanical forces affecting and enacted by biopolymers; and (vii) yield standard population-averaged information from the statistics of many single-molecule observations. A half-century of single-molecule tool development has yielded technical advances that have demonstrated each of these advantages, and more are sure to emerge.

Yet in any field enjoying increasing popularity, there inevitably comes a crossroads, which inspired MIE volumes 472 and 475. To advance beyond being used or studied only by a limited (and eventually vanishing) group of specialists, a set of tools or area of research needs to find more widespread appreciation. Many methods that are commonplace in labs today – such as gel electrophoresis, PCR, and sequencing – made that transition from specialist's art to general practitioner's basic tool by a combination of being very appealing and becoming easy to master. The two MIE volumes are aimed to facilitate this transition by, often for the first time, revealing for a broad selection of single-molecule tools those details that pioneering specialists rarely have the space to cover in their research publications.

Compiling methods from an emerging field is a daunting task, since new tools are developed nearly daily. The resulting selection is by necessity incomplete, limited by both the availability of contributors and my gaps in knowledge. Yet through the vigorous response to my solicitation of articles, what was planned as one volume became two, somewhat loosely organized by theme. While editing each of these works, I became increasingly impressed by the consistently superb quality of the contributions, in terms of both style and substance. I am therefore very grateful to John Abelson for convincing me to take on the job as editor, and trusting me with it, to the phenomenal group of authors (some of which even made the deadline), and to the staff at Elsevier for allowing me to divide the contributions into two volumes and supporting me in numerous other ways. My hope is that the hard work by everyone involved bears fruit and helps spread the word and enthusiasm about the power of single-molecule tools.

Nils G. Walter

Methods in Enzymology

VOLUME I. Preparation and Assay of Enzymes
Edited by SIDNEY P. COLOWICK AND NATHAN O. KAPLAN

VOLUME II. Preparation and Assay of Enzymes
Edited by SIDNEY P. COLOWICK AND NATHAN O. KAPLAN

VOLUME III. Preparation and Assay of Substrates
Edited by SIDNEY P. COLOWICK AND NATHAN O. KAPLAN

VOLUME IV. Special Techniques for the Enzymologist
Edited by SIDNEY P. COLOWICK AND NATHAN O. KAPLAN

VOLUME V. Preparation and Assay of Enzymes
Edited by SIDNEY P. COLOWICK AND NATHAN O. KAPLAN

VOLUME VI. Preparation and Assay of Enzymes *(Continued)*
Preparation and Assay of Substrates
Special Techniques
Edited by SIDNEY P. COLOWICK AND NATHAN O. KAPLAN

VOLUME VII. Cumulative Subject Index
Edited by SIDNEY P. COLOWICK AND NATHAN O. KAPLAN

VOLUME VIII. Complex Carbohydrates
Edited by ELIZABETH F. NEUFELD AND VICTOR GINSBURG

VOLUME IX. Carbohydrate Metabolism
Edited by WILLIS A. WOOD

VOLUME X. Oxidation and Phosphorylation
Edited by RONALD W. ESTABROOK AND MAYNARD E. PULLMAN

VOLUME XI. Enzyme Structure
Edited by C. H. W. HIRS

VOLUME XII. Nucleic Acids (Parts A and B)
Edited by LAWRENCE GROSSMAN AND KIVIE MOLDAVE

VOLUME XIII. Citric Acid Cycle
Edited by J. M. LOWENSTEIN

VOLUME XIV. Lipids
Edited by J. M. LOWENSTEIN

VOLUME XV. Steroids and Terpenoids
Edited by RAYMOND B. CLAYTON

VOLUME XVI. Fast Reactions
Edited by KENNETH KUSTIN

VOLUME XVII. Metabolism of Amino Acids and Amines (Parts A and B)
Edited by HERBERT TABOR AND CELIA WHITE TABOR

VOLUME XVIII. Vitamins and Coenzymes (Parts A, B, and C)
Edited by DONALD B. MCCORMICK AND LEMUEL D. WRIGHT

VOLUME XIX. Proteolytic Enzymes
Edited by GERTRUDE E. PERLMANN AND LASZLO LORAND

VOLUME XX. Nucleic Acids and Protein Synthesis (Part C)
Edited by KIVIE MOLDAVE AND LAWRENCE GROSSMAN

VOLUME XXI. Nucleic Acids (Part D)
Edited by LAWRENCE GROSSMAN AND KIVIE MOLDAVE

VOLUME XXII. Enzyme Purification and Related Techniques
Edited by WILLIAM B. JAKOBY

VOLUME XXIII. Photosynthesis (Part A)
Edited by ANTHONY SAN PIETRO

VOLUME XXIV. Photosynthesis and Nitrogen Fixation (Part B)
Edited by ANTHONY SAN PIETRO

VOLUME XXV. Enzyme Structure (Part B)
Edited by C. H. W. HIRS AND SERGE N. TIMASHEFF

VOLUME XXVI. Enzyme Structure (Part C)
Edited by C. H. W. HIRS AND SERGE N. TIMASHEFF

VOLUME XXVII. Enzyme Structure (Part D)
Edited by C. H. W. HIRS AND SERGE N. TIMASHEFF

VOLUME XXVIII. Complex Carbohydrates (Part B)
Edited by VICTOR GINSBURG

VOLUME XXIX. Nucleic Acids and Protein Synthesis (Part E)
Edited by LAWRENCE GROSSMAN AND KIVIE MOLDAVE

VOLUME XXX. Nucleic Acids and Protein Synthesis (Part F)
Edited by KIVIE MOLDAVE AND LAWRENCE GROSSMAN

VOLUME XXXI. Biomembranes (Part A)
Edited by SIDNEY FLEISCHER AND LESTER PACKER

VOLUME XXXII. Biomembranes (Part B)
Edited by SIDNEY FLEISCHER AND LESTER PACKER

VOLUME XXXIII. Cumulative Subject Index Volumes I–XXX
Edited by MARTHA G. DENNIS AND EDWARD A. DENNIS

VOLUME XXXIV. Affinity Techniques (Enzyme Purification: Part B)
Edited by WILLIAM B. JAKOBY AND MEIR WILCHEK

Volume XXXV. Lipids (Part B)
Edited by John M. Lowenstein

Volume XXXVI. Hormone Action (Part A: Steroid Hormones)
Edited by Bert W. O'Malley and Joel G. Hardman

Volume XXXVII. Hormone Action (Part B: Peptide Hormones)
Edited by Bert W. O'Malley and Joel G. Hardman

Volume XXXVIII. Hormone Action (Part C: Cyclic Nucleotides)
Edited by Joel G. Hardman and Bert W. O'Malley

Volume XXXIX. Hormone Action (Part D: Isolated Cells, Tissues, and Organ Systems)
Edited by Joel G. Hardman and Bert W. O'Malley

Volume XL. Hormone Action (Part E: Nuclear Structure and Function)
Edited by Bert W. O'Malley and Joel G. Hardman

Volume XLI. Carbohydrate Metabolism (Part B)
Edited by W. A. Wood

Volume XLII. Carbohydrate Metabolism (Part C)
Edited by W. A. Wood

Volume XLIII. Antibiotics
Edited by John H. Hash

Volume XLIV. Immobilized Enzymes
Edited by Klaus Mosbach

Volume XLV. Proteolytic Enzymes (Part B)
Edited by Laszlo Lorand

Volume XLVI. Affinity Labeling
Edited by William B. Jakoby and Meir Wilchek

Volume XLVII. Enzyme Structure (Part E)
Edited by C. H. W. Hirs and Serge N. Timasheff

Volume XLVIII. Enzyme Structure (Part F)
Edited by C. H. W. Hirs and Serge N. Timasheff

Volume XLIX. Enzyme Structure (Part G)
Edited by C. H. W. Hirs and Serge N. Timasheff

Volume L. Complex Carbohydrates (Part C)
Edited by Victor Ginsburg

Volume LI. Purine and Pyrimidine Nucleotide Metabolism
Edited by Patricia A. Hoffee and Mary Ellen Jones

Volume LII. Biomembranes (Part C: Biological Oxidations)
Edited by Sidney Fleischer and Lester Packer

VOLUME LIII. Biomembranes (Part D: Biological Oxidations)
Edited by SIDNEY FLEISCHER AND LESTER PACKER

VOLUME LIV. Biomembranes (Part E: Biological Oxidations)
Edited by SIDNEY FLEISCHER AND LESTER PACKER

VOLUME LV. Biomembranes (Part F: Bioenergetics)
Edited by SIDNEY FLEISCHER AND LESTER PACKER

VOLUME LVI. Biomembranes (Part G: Bioenergetics)
Edited by SIDNEY FLEISCHER AND LESTER PACKER

VOLUME LVII. Bioluminescence and Chemiluminescence
Edited by MARLENE A. DELUCA

VOLUME LVIII. Cell Culture
Edited by WILLIAM B. JAKOBY AND IRA PASTAN

VOLUME LIX. Nucleic Acids and Protein Synthesis (Part G)
Edited by KIVIE MOLDAVE AND LAWRENCE GROSSMAN

VOLUME LX. Nucleic Acids and Protein Synthesis (Part H)
Edited by KIVIE MOLDAVE AND LAWRENCE GROSSMAN

VOLUME 61. Enzyme Structure (Part H)
Edited by C. H. W. HIRS AND SERGE N. TIMASHEFF

VOLUME 62. Vitamins and Coenzymes (Part D)
Edited by DONALD B. MCCORMICK AND LEMUEL D. WRIGHT

VOLUME 63. Enzyme Kinetics and Mechanism (Part A: Initial Rate and Inhibitor Methods)
Edited by DANIEL L. PURICH

VOLUME 64. Enzyme Kinetics and Mechanism
(Part B: Isotopic Probes and Complex Enzyme Systems)
Edited by DANIEL L. PURICH

VOLUME 65. Nucleic Acids (Part I)
Edited by LAWRENCE GROSSMAN AND KIVIE MOLDAVE

VOLUME 66. Vitamins and Coenzymes (Part E)
Edited by DONALD B. MCCORMICK AND LEMUEL D. WRIGHT

VOLUME 67. Vitamins and Coenzymes (Part F)
Edited by DONALD B. MCCORMICK AND LEMUEL D. WRIGHT

VOLUME 68. Recombinant DNA
Edited by RAY WU

VOLUME 69. Photosynthesis and Nitrogen Fixation (Part C)
Edited by ANTHONY SAN PIETRO

VOLUME 70. Immunochemical Techniques (Part A)
Edited by HELEN VAN VUNAKIS AND JOHN J. LANGONE

VOLUME 71. Lipids (Part C)
Edited by JOHN M. LOWENSTEIN

VOLUME 72. Lipids (Part D)
Edited by JOHN M. LOWENSTEIN

VOLUME 73. Immunochemical Techniques (Part B)
Edited by JOHN J. LANGONE AND HELEN VAN VUNAKIS

VOLUME 74. Immunochemical Techniques (Part C)
Edited by JOHN J. LANGONE AND HELEN VAN VUNAKIS

VOLUME 75. Cumulative Subject Index Volumes XXXI, XXXII, XXXIV–LX
Edited by EDWARD A. DENNIS AND MARTHA G. DENNIS

VOLUME 76. Hemoglobins
Edited by ERALDO ANTONINI, LUIGI ROSSI-BERNARDI, AND EMILIA CHIANCONE

VOLUME 77. Detoxication and Drug Metabolism
Edited by WILLIAM B. JAKOBY

VOLUME 78. Interferons (Part A)
Edited by SIDNEY PESTKA

VOLUME 79. Interferons (Part B)
Edited by SIDNEY PESTKA

VOLUME 80. Proteolytic Enzymes (Part C)
Edited by LASZLO LORAND

VOLUME 81. Biomembranes (Part H: Visual Pigments and Purple Membranes, I)
Edited by LESTER PACKER

VOLUME 82. Structural and Contractile Proteins (Part A: Extracellular Matrix)
Edited by LEON W. CUNNINGHAM AND DIXIE W. FREDERIKSEN

VOLUME 83. Complex Carbohydrates (Part D)
Edited by VICTOR GINSBURG

VOLUME 84. Immunochemical Techniques (Part D: Selected Immunoassays)
Edited by JOHN J. LANGONE AND HELEN VAN VUNAKIS

VOLUME 85. Structural and Contractile Proteins (Part B: The Contractile Apparatus and the Cytoskeleton)
Edited by DIXIE W. FREDERIKSEN AND LEON W. CUNNINGHAM

VOLUME 86. Prostaglandins and Arachidonate Metabolites
Edited by WILLIAM E. M. LANDS AND WILLIAM L. SMITH

VOLUME 87. Enzyme Kinetics and Mechanism (Part C: Intermediates, Stereo-chemistry, and Rate Studies)
Edited by DANIEL L. PURICH

VOLUME 88. Biomembranes (Part I: Visual Pigments and Purple Membranes, II)
Edited by LESTER PACKER

VOLUME 89. Carbohydrate Metabolism (Part D)
Edited by WILLIS A. WOOD

VOLUME 90. Carbohydrate Metabolism (Part E)
Edited by WILLIS A. WOOD

VOLUME 91. Enzyme Structure (Part I)
Edited by C. H. W. HIRS AND SERGE N. TIMASHEFF

VOLUME 92. Immunochemical Techniques (Part E: Monoclonal Antibodies and General Immunoassay Methods)
Edited by JOHN J. LANGONE AND HELEN VAN VUNAKIS

VOLUME 93. Immunochemical Techniques (Part F: Conventional Antibodies, Fc Receptors, and Cytotoxicity)
Edited by JOHN J. LANGONE AND HELEN VAN VUNAKIS

VOLUME 94. Polyamines
Edited by HERBERT TABOR AND CELIA WHITE TABOR

VOLUME 95. Cumulative Subject Index Volumes 61–74, 76–80
Edited by EDWARD A. DENNIS AND MARTHA G. DENNIS

VOLUME 96. Biomembranes [Part J: Membrane Biogenesis: Assembly and Targeting (General Methods; Eukaryotes)]
Edited by SIDNEY FLEISCHER AND BECCA FLEISCHER

VOLUME 97. Biomembranes [Part K: Membrane Biogenesis: Assembly and Targeting (Prokaryotes, Mitochondria, and Chloroplasts)]
Edited by SIDNEY FLEISCHER AND BECCA FLEISCHER

VOLUME 98. Biomembranes (Part L: Membrane Biogenesis: Processing and Recycling)
Edited by SIDNEY FLEISCHER AND BECCA FLEISCHER

VOLUME 99. Hormone Action (Part F: Protein Kinases)
Edited by JACKIE D. CORBIN AND JOEL G. HARDMAN

VOLUME 100. Recombinant DNA (Part B)
Edited by RAY WU, LAWRENCE GROSSMAN, AND KIVIE MOLDAVE

VOLUME 101. Recombinant DNA (Part C)
Edited by RAY WU, LAWRENCE GROSSMAN, AND KIVIE MOLDAVE

VOLUME 102. Hormone Action (Part G: Calmodulin and Calcium-Binding Proteins)
Edited by ANTHONY R. MEANS AND BERT W. O'MALLEY

VOLUME 103. Hormone Action (Part H: Neuroendocrine Peptides)
Edited by P. MICHAEL CONN

VOLUME 104. Enzyme Purification and Related Techniques (Part C)
Edited by WILLIAM B. JAKOBY

VOLUME 105. Oxygen Radicals in Biological Systems
Edited by LESTER PACKER

VOLUME 106. Posttranslational Modifications (Part A)
Edited by FINN WOLD AND KIVIE MOLDAVE

VOLUME 107. Posttranslational Modifications (Part B)
Edited by FINN WOLD AND KIVIE MOLDAVE

VOLUME 108. Immunochemical Techniques (Part G: Separation and Characterization of Lymphoid Cells)
Edited by GIOVANNI DI SABATO, JOHN J. LANGONE, AND HELEN VAN VUNAKIS

VOLUME 109. Hormone Action (Part I: Peptide Hormones)
Edited by LUTZ BIRNBAUMER AND BERT W. O'MALLEY

VOLUME 110. Steroids and Isoprenoids (Part A)
Edited by JOHN H. LAW AND HANS C. RILLING

VOLUME 111. Steroids and Isoprenoids (Part B)
Edited by JOHN H. LAW AND HANS C. RILLING

VOLUME 112. Drug and Enzyme Targeting (Part A)
Edited by KENNETH J. WIDDER AND RALPH GREEN

VOLUME 113. Glutamate, Glutamine, Glutathione, and Related Compounds
Edited by ALTON MEISTER

VOLUME 114. Diffraction Methods for Biological Macromolecules (Part A)
Edited by HAROLD W. WYCKOFF, C. H. W. HIRS, AND SERGE N. TIMASHEFF

VOLUME 115. Diffraction Methods for Biological Macromolecules (Part B)
Edited by HAROLD W. WYCKOFF, C. H. W. HIRS, AND SERGE N. TIMASHEFF

VOLUME 116. Immunochemical Techniques
(Part H: Effectors and Mediators of Lymphoid Cell Functions)
Edited by GIOVANNI DI SABATO, JOHN J. LANGONE, AND HELEN VAN VUNAKIS

VOLUME 117. Enzyme Structure (Part J)
Edited by C. H. W. HIRS AND SERGE N. TIMASHEFF

VOLUME 118. Plant Molecular Biology
Edited by ARTHUR WEISSBACH AND HERBERT WEISSBACH

VOLUME 119. Interferons (Part C)
Edited by SIDNEY PESTKA

VOLUME 120. Cumulative Subject Index Volumes 81–94, 96–101

VOLUME 121. Immunochemical Techniques (Part I: Hybridoma Technology and Monoclonal Antibodies)
Edited by JOHN J. LANGONE AND HELEN VAN VUNAKIS

VOLUME 122. Vitamins and Coenzymes (Part G)
Edited by FRANK CHYTIL AND DONALD B. MCCORMICK

VOLUME 123. Vitamins and Coenzymes (Part H)
Edited by FRANK CHYTIL AND DONALD B. MCCORMICK

VOLUME 124. Hormone Action (Part J: Neuroendocrine Peptides)
Edited by P. MICHAEL CONN

VOLUME 125. Biomembranes (Part M: Transport in Bacteria, Mitochondria, and Chloroplasts: General Approaches and Transport Systems)
Edited by SIDNEY FLEISCHER AND BECCA FLEISCHER

VOLUME 126. Biomembranes (Part N: Transport in Bacteria, Mitochondria, and Chloroplasts: Protonmotive Force)
Edited by SIDNEY FLEISCHER AND BECCA FLEISCHER

VOLUME 127. Biomembranes (Part O: Protons and Water: Structure and Translocation)
Edited by LESTER PACKER

VOLUME 128. Plasma Lipoproteins (Part A: Preparation, Structure, and Molecular Biology)
Edited by JERE P. SEGREST AND JOHN J. ALBERS

VOLUME 129. Plasma Lipoproteins (Part B: Characterization, Cell Biology, and Metabolism)
Edited by JOHN J. ALBERS AND JERE P. SEGREST

VOLUME 130. Enzyme Structure (Part K)
Edited by C. H. W. HIRS AND SERGE N. TIMASHEFF

VOLUME 131. Enzyme Structure (Part L)
Edited by C. H. W. HIRS AND SERGE N. TIMASHEFF

VOLUME 132. Immunochemical Techniques (Part J: Phagocytosis and Cell-Mediated Cytotoxicity)
Edited by GIOVANNI DI SABATO AND JOHANNES EVERSE

VOLUME 133. Bioluminescence and Chemiluminescence (Part B)
Edited by MARLENE DELUCA AND WILLIAM D. MCELROY

VOLUME 134. Structural and Contractile Proteins (Part C: The Contractile Apparatus and the Cytoskeleton)
Edited by RICHARD B. VALLEE

VOLUME 135. Immobilized Enzymes and Cells (Part B)
Edited by KLAUS MOSBACH

VOLUME 136. Immobilized Enzymes and Cells (Part C)
Edited by KLAUS MOSBACH

VOLUME 137. Immobilized Enzymes and Cells (Part D)
Edited by KLAUS MOSBACH

VOLUME 138. Complex Carbohydrates (Part E)
Edited by VICTOR GINSBURG

VOLUME 139. Cellular Regulators (Part A: Calcium- and Calmodulin-Binding Proteins)
Edited by ANTHONY R. MEANS AND P. MICHAEL CONN

VOLUME 140. Cumulative Subject Index Volumes 102–119, 121–134

VOLUME 141. Cellular Regulators (Part B: Calcium and Lipids)
Edited by P. MICHAEL CONN AND ANTHONY R. MEANS

VOLUME 142. Metabolism of Aromatic Amino Acids and Amines
Edited by SEYMOUR KAUFMAN

VOLUME 143. Sulfur and Sulfur Amino Acids
Edited by WILLIAM B. JAKOBY AND OWEN GRIFFITH

VOLUME 144. Structural and Contractile Proteins (Part D: Extracellular Matrix)
Edited by LEON W. CUNNINGHAM

VOLUME 145. Structural and Contractile Proteins (Part E: Extracellular Matrix)
Edited by LEON W. CUNNINGHAM

VOLUME 146. Peptide Growth Factors (Part A)
Edited by DAVID BARNES AND DAVID A. SIRBASKU

VOLUME 147. Peptide Growth Factors (Part B)
Edited by DAVID BARNES AND DAVID A. SIRBASKU

VOLUME 148. Plant Cell Membranes
Edited by LESTER PACKER AND ROLAND DOUCE

VOLUME 149. Drug and Enzyme Targeting (Part B)
Edited by RALPH GREEN AND KENNETH J. WIDDER

VOLUME 150. Immunochemical Techniques (Part K: *In Vitro* Models of B and T Cell Functions and Lymphoid Cell Receptors)
Edited by GIOVANNI DI SABATO

VOLUME 151. Molecular Genetics of Mammalian Cells
Edited by MICHAEL M. GOTTESMAN

VOLUME 152. Guide to Molecular Cloning Techniques
Edited by SHELBY L. BERGER AND ALAN R. KIMMEL

VOLUME 153. Recombinant DNA (Part D)
Edited by RAY WU AND LAWRENCE GROSSMAN

VOLUME 154. Recombinant DNA (Part E)
Edited by RAY WU AND LAWRENCE GROSSMAN

VOLUME 155. Recombinant DNA (Part F)
Edited by RAY WU

VOLUME 156. Biomembranes (Part P: ATP-Driven Pumps and Related Transport: The Na, K-Pump)
Edited by SIDNEY FLEISCHER AND BECCA FLEISCHER

VOLUME 157. Biomembranes (Part Q: ATP-Driven Pumps and Related Transport: Calcium, Proton, and Potassium Pumps)
Edited by SIDNEY FLEISCHER AND BECCA FLEISCHER

VOLUME 158. Metalloproteins (Part A)
Edited by JAMES F. RIORDAN AND BERT L. VALLEE

VOLUME 159. Initiation and Termination of Cyclic Nucleotide Action
Edited by JACKIE D. CORBIN AND ROGER A. JOHNSON

VOLUME 160. Biomass (Part A: Cellulose and Hemicellulose)
Edited by WILLIS A. WOOD AND SCOTT T. KELLOGG

VOLUME 161. Biomass (Part B: Lignin, Pectin, and Chitin)
Edited by WILLIS A. WOOD AND SCOTT T. KELLOGG

VOLUME 162. Immunochemical Techniques (Part L: Chemotaxis and Inflammation)
Edited by GIOVANNI DI SABATO

VOLUME 163. Immunochemical Techniques (Part M: Chemotaxis and Inflammation)
Edited by GIOVANNI DI SABATO

VOLUME 164. Ribosomes
Edited by HARRY F. NOLLER, JR., AND KIVIE MOLDAVE

VOLUME 165. Microbial Toxins: Tools for Enzymology
Edited by SIDNEY HARSHMAN

VOLUME 166. Branched-Chain Amino Acids
Edited by ROBERT HARRIS AND JOHN R. SOKATCH

VOLUME 167. Cyanobacteria
Edited by LESTER PACKER AND ALEXANDER N. GLAZER

VOLUME 168. Hormone Action (Part K: Neuroendocrine Peptides)
Edited by P. MICHAEL CONN

VOLUME 169. Platelets: Receptors, Adhesion, Secretion (Part A)
Edited by JACEK HAWIGER

VOLUME 170. Nucleosomes
Edited by PAUL M. WASSARMAN AND ROGER D. KORNBERG

VOLUME 171. Biomembranes (Part R: Transport Theory: Cells and Model Membranes)
Edited by SIDNEY FLEISCHER AND BECCA FLEISCHER

VOLUME 172. Biomembranes (Part S: Transport: Membrane Isolation and Characterization)
Edited by SIDNEY FLEISCHER AND BECCA FLEISCHER

VOLUME 173. Biomembranes [Part T: Cellular and Subcellular Transport: Eukaryotic (Nonepithelial) Cells]
Edited by SIDNEY FLEISCHER AND BECCA FLEISCHER

VOLUME 174. Biomembranes [Part U: Cellular and Subcellular Transport: Eukaryotic (Nonepithelial) Cells]
Edited by SIDNEY FLEISCHER AND BECCA FLEISCHER

VOLUME 175. Cumulative Subject Index Volumes 135–139, 141–167

VOLUME 176. Nuclear Magnetic Resonance (Part A: Spectral Techniques and Dynamics)
Edited by NORMAN J. OPPENHEIMER AND THOMAS L. JAMES

VOLUME 177. Nuclear Magnetic Resonance (Part B: Structure and Mechanism)
Edited by NORMAN J. OPPENHEIMER AND THOMAS L. JAMES

VOLUME 178. Antibodies, Antigens, and Molecular Mimicry
Edited by JOHN J. LANGONE

VOLUME 179. Complex Carbohydrates (Part F)
Edited by VICTOR GINSBURG

VOLUME 180. RNA Processing (Part A: General Methods)
Edited by JAMES E. DAHLBERG AND JOHN N. ABELSON

VOLUME 181. RNA Processing (Part B: Specific Methods)
Edited by JAMES E. DAHLBERG AND JOHN N. ABELSON

VOLUME 182. Guide to Protein Purification
Edited by MURRAY P. DEUTSCHER

VOLUME 183. Molecular Evolution: Computer Analysis of Protein and Nucleic Acid Sequences
Edited by RUSSELL F. DOOLITTLE

VOLUME 184. Avidin-Biotin Technology
Edited by MEIR WILCHEK AND EDWARD A. BAYER

VOLUME 185. Gene Expression Technology
Edited by DAVID V. GOEDDEL

VOLUME 186. Oxygen Radicals in Biological Systems (Part B: Oxygen Radicals and Antioxidants)
Edited by LESTER PACKER AND ALEXANDER N. GLAZER

VOLUME 187. Arachidonate Related Lipid Mediators
Edited by ROBERT C. MURPHY AND FRANK A. FITZPATRICK

VOLUME 188. Hydrocarbons and Methylotrophy
Edited by MARY E. LIDSTROM

VOLUME 189. Retinoids (Part A: Molecular and Metabolic Aspects)
Edited by LESTER PACKER

VOLUME 190. Retinoids (Part B: Cell Differentiation and Clinical Applications)
Edited by LESTER PACKER

VOLUME 191. Biomembranes (Part V: Cellular and Subcellular Transport: Epithelial Cells)
Edited by SIDNEY FLEISCHER AND BECCA FLEISCHER

VOLUME 192. Biomembranes (Part W: Cellular and Subcellular Transport: Epithelial Cells)
Edited by SIDNEY FLEISCHER AND BECCA FLEISCHER

VOLUME 193. Mass Spectrometry
Edited by JAMES A. MCCLOSKEY

VOLUME 194. Guide to Yeast Genetics and Molecular Biology
Edited by CHRISTINE GUTHRIE AND GERALD R. FINK

VOLUME 195. Adenylyl Cyclase, G Proteins, and Guanylyl Cyclase
Edited by ROGER A. JOHNSON AND JACKIE D. CORBIN

VOLUME 196. Molecular Motors and the Cytoskeleton
Edited by RICHARD B. VALLEE

VOLUME 197. Phospholipases
Edited by EDWARD A. DENNIS

VOLUME 198. Peptide Growth Factors (Part C)
Edited by DAVID BARNES, J. P. MATHER, AND GORDON H. SATO

VOLUME 199. Cumulative Subject Index Volumes 168–174, 176–194

VOLUME 200. Protein Phosphorylation (Part A: Protein Kinases: Assays, Purification, Antibodies, Functional Analysis, Cloning, and Expression)
Edited by TONY HUNTER AND BARTHOLOMEW M. SEFTON

VOLUME 201. Protein Phosphorylation (Part B: Analysis of Protein Phosphorylation, Protein Kinase Inhibitors, and Protein Phosphatases)
Edited by TONY HUNTER AND BARTHOLOMEW M. SEFTON

VOLUME 202. Molecular Design and Modeling: Concepts and Applications (Part A: Proteins, Peptides, and Enzymes)
Edited by JOHN J. LANGONE

VOLUME 203. Molecular Design and Modeling: Concepts and Applications (Part B: Antibodies and Antigens, Nucleic Acids, Polysaccharides, and Drugs)
Edited by JOHN J. LANGONE

VOLUME 204. Bacterial Genetic Systems
Edited by JEFFREY H. MILLER

VOLUME 205. Metallobiochemistry (Part B: Metallothionein and Related Molecules)
Edited by JAMES F. RIORDAN AND BERT L. VALLEE

VOLUME 206. Cytochrome P450
Edited by MICHAEL R. WATERMAN AND ERIC F. JOHNSON

VOLUME 207. Ion Channels
Edited by BERNARDO RUDY AND LINDA E. IVERSON

VOLUME 208. Protein–DNA Interactions
Edited by ROBERT T. SAUER

VOLUME 209. Phospholipid Biosynthesis
Edited by EDWARD A. DENNIS AND DENNIS E. VANCE

VOLUME 210. Numerical Computer Methods
Edited by LUDWIG BRAND AND MICHAEL L. JOHNSON

VOLUME 211. DNA Structures (Part A: Synthesis and Physical Analysis of DNA)
Edited by DAVID M. J. LILLEY AND JAMES E. DAHLBERG

VOLUME 212. DNA Structures (Part B: Chemical and Electrophoretic Analysis of DNA)
Edited by DAVID M. J. LILLEY AND JAMES E. DAHLBERG

VOLUME 213. Carotenoids (Part A: Chemistry, Separation, Quantitation, and Antioxidation)
Edited by LESTER PACKER

VOLUME 214. Carotenoids (Part B: Metabolism, Genetics, and Biosynthesis)
Edited by LESTER PACKER

VOLUME 215. Platelets: Receptors, Adhesion, Secretion (Part B)
Edited by JACEK J. HAWIGER

VOLUME 216. Recombinant DNA (Part G)
Edited by RAY WU

VOLUME 217. Recombinant DNA (Part H)
Edited by RAY WU

VOLUME 218. Recombinant DNA (Part I)
Edited by RAY WU

VOLUME 219. Reconstitution of Intracellular Transport
Edited by JAMES E. ROTHMAN

VOLUME 220. Membrane Fusion Techniques (Part A)
Edited by NEJAT DÜZGÜNEŞ

VOLUME 221. Membrane Fusion Techniques (Part B)
Edited by NEJAT DÜZGÜNEŞ

VOLUME 222. Proteolytic Enzymes in Coagulation, Fibrinolysis, and Complement Activation (Part A: Mammalian Blood Coagulation Factors and Inhibitors)
Edited by LASZLO LORAND AND KENNETH G. MANN

VOLUME 223. Proteolytic Enzymes in Coagulation, Fibrinolysis, and Complement Activation (Part B: Complement Activation, Fibrinolysis, and Nonmammalian Blood Coagulation Factors)
Edited by LASZLO LORAND AND KENNETH G. MANN

VOLUME 224. Molecular Evolution: Producing the Biochemical Data
Edited by ELIZABETH ANNE ZIMMER, THOMAS J. WHITE, REBECCA L. CANN, AND ALLAN C. WILSON

VOLUME 225. Guide to Techniques in Mouse Development
Edited by PAUL M. WASSARMAN AND MELVIN L. DEPAMPHILIS

VOLUME 226. Metallobiochemistry (Part C: Spectroscopic and Physical Methods for Probing Metal Ion Environments in Metalloenzymes and Metalloproteins)
Edited by JAMES F. RIORDAN AND BERT L. VALLEE

VOLUME 227. Metallobiochemistry (Part D: Physical and Spectroscopic Methods for Probing Metal Ion Environments in Metalloproteins)
Edited by JAMES F. RIORDAN AND BERT L. VALLEE

VOLUME 228. Aqueous Two-Phase Systems
Edited by HARRY WALTER AND GÖTE JOHANSSON

VOLUME 229. Cumulative Subject Index Volumes 195–198, 200–227

VOLUME 230. Guide to Techniques in Glycobiology
Edited by WILLIAM J. LENNARZ AND GERALD W. HART

VOLUME 231. Hemoglobins (Part B: Biochemical and Analytical Methods)
Edited by JOHANNES EVERSE, KIM D. VANDEGRIFF, AND ROBERT M. WINSLOW

VOLUME 232. Hemoglobins (Part C: Biophysical Methods)
Edited by JOHANNES EVERSE, KIM D. VANDEGRIFF, AND ROBERT M. WINSLOW

VOLUME 233. Oxygen Radicals in Biological Systems (Part C)
Edited by LESTER PACKER

VOLUME 234. Oxygen Radicals in Biological Systems (Part D)
Edited by LESTER PACKER

VOLUME 235. Bacterial Pathogenesis (Part A: Identification and Regulation of Virulence Factors)
Edited by VIRGINIA L. CLARK AND PATRIK M. BAVOIL

VOLUME 236. Bacterial Pathogenesis (Part B: Integration of Pathogenic Bacteria with Host Cells)
Edited by VIRGINIA L. CLARK AND PATRIK M. BAVOIL

VOLUME 237. Heterotrimeric G Proteins
Edited by RAVI IYENGAR

VOLUME 238. Heterotrimeric G-Protein Effectors
Edited by RAVI IYENGAR

VOLUME 239. Nuclear Magnetic Resonance (Part C)
Edited by THOMAS L. JAMES AND NORMAN J. OPPENHEIMER

VOLUME 240. Numerical Computer Methods (Part B)
Edited by MICHAEL L. JOHNSON AND LUDWIG BRAND

VOLUME 241. Retroviral Proteases
Edited by LAWRENCE C. KUO AND JULES A. SHAFER

VOLUME 242. Neoglycoconjugates (Part A)
Edited by Y. C. LEE AND REIKO T. LEE

VOLUME 243. Inorganic Microbial Sulfur Metabolism
Edited by HARRY D. PECK, JR., AND JEAN LEGALL

VOLUME 244. Proteolytic Enzymes: Serine and Cysteine Peptidases
Edited by ALAN J. BARRETT

VOLUME 245. Extracellular Matrix Components
Edited by E. RUOSLAHTI AND E. ENGVALL

VOLUME 246. Biochemical Spectroscopy
Edited by KENNETH SAUER

VOLUME 247. Neoglycoconjugates (Part B: Biomedical Applications)
Edited by Y. C. LEE AND REIKO T. LEE

VOLUME 248. Proteolytic Enzymes: Aspartic and Metallo Peptidases
Edited by ALAN J. BARRETT

VOLUME 249. Enzyme Kinetics and Mechanism (Part D: Developments in Enzyme Dynamics)
Edited by DANIEL L. PURICH

VOLUME 250. Lipid Modifications of Proteins
Edited by PATRICK J. CASEY AND JANICE E. BUSS

VOLUME 251. Biothiols (Part A: Monothiols and Dithiols, Protein Thiols, and Thiyl Radicals)
Edited by LESTER PACKER

VOLUME 252. Biothiols (Part B: Glutathione and Thioredoxin; Thiols in Signal Transduction and Gene Regulation)
Edited by LESTER PACKER

VOLUME 253. Adhesion of Microbial Pathogens
Edited by RON J. DOYLE AND ITZHAK OFEK

VOLUME 254. Oncogene Techniques
Edited by PETER K. VOGT AND INDER M. VERMA

VOLUME 255. Small GTPases and Their Regulators (Part A: Ras Family)
Edited by W. E. BALCH, CHANNING J. DER, AND ALAN HALL

VOLUME 256. Small GTPases and Their Regulators (Part B: Rho Family)
Edited by W. E. BALCH, CHANNING J. DER, AND ALAN HALL

VOLUME 257. Small GTPases and Their Regulators (Part C: Proteins Involved in Transport)
Edited by W. E. BALCH, CHANNING J. DER, AND ALAN HALL

VOLUME 258. Redox-Active Amino Acids in Biology
Edited by JUDITH P. KLINMAN

VOLUME 259. Energetics of Biological Macromolecules
Edited by MICHAEL L. JOHNSON AND GARY K. ACKERS

VOLUME 260. Mitochondrial Biogenesis and Genetics (Part A)
Edited by GIUSEPPE M. ATTARDI AND ANNE CHOMYN

VOLUME 261. Nuclear Magnetic Resonance and Nucleic Acids
Edited by THOMAS L. JAMES

VOLUME 262. DNA Replication
Edited by JUDITH L. CAMPBELL

VOLUME 263. Plasma Lipoproteins (Part C: Quantitation)
Edited by WILLIAM A. BRADLEY, SANDRA H. GIANTURCO, AND JERE P. SEGREST

VOLUME 264. Mitochondrial Biogenesis and Genetics (Part B)
Edited by GIUSEPPE M. ATTARDI AND ANNE CHOMYN

VOLUME 265. Cumulative Subject Index Volumes 228, 230–262

VOLUME 266. Computer Methods for Macromolecular Sequence Analysis
Edited by RUSSELL F. DOOLITTLE

VOLUME 267. Combinatorial Chemistry
Edited by JOHN N. ABELSON

VOLUME 268. Nitric Oxide (Part A: Sources and Detection of NO; NO Synthase)
Edited by LESTER PACKER

VOLUME 269. Nitric Oxide (Part B: Physiological and Pathological Processes)
Edited by LESTER PACKER

VOLUME 270. High Resolution Separation and Analysis of Biological Macromolecules (Part A: Fundamentals)
Edited by BARRY L. KARGER AND WILLIAM S. HANCOCK

VOLUME 271. High Resolution Separation and Analysis of Biological Macromolecules (Part B: Applications)
Edited by BARRY L. KARGER AND WILLIAM S. HANCOCK

VOLUME 272. Cytochrome P450 (Part B)
Edited by ERIC F. JOHNSON AND MICHAEL R. WATERMAN

VOLUME 273. RNA Polymerase and Associated Factors (Part A)
Edited by SANKAR ADHYA

VOLUME 274. RNA Polymerase and Associated Factors (Part B)
Edited by SANKAR ADHYA

VOLUME 275. Viral Polymerases and Related Proteins
Edited by LAWRENCE C. KUO, DAVID B. OLSEN, AND STEVEN S. CARROLL

VOLUME 276. Macromolecular Crystallography (Part A)
Edited by CHARLES W. CARTER, JR., AND ROBERT M. SWEET

VOLUME 277. Macromolecular Crystallography (Part B)
Edited by CHARLES W. CARTER, JR., AND ROBERT M. SWEET

VOLUME 278. Fluorescence Spectroscopy
Edited by LUDWIG BRAND AND MICHAEL L. JOHNSON

VOLUME 279. Vitamins and Coenzymes (Part I)
Edited by DONALD B. MCCORMICK, JOHN W. SUTTIE, AND CONRAD WAGNER

VOLUME 280. Vitamins and Coenzymes (Part J)
Edited by DONALD B. MCCORMICK, JOHN W. SUTTIE, AND CONRAD WAGNER

VOLUME 281. Vitamins and Coenzymes (Part K)
Edited by DONALD B. MCCORMICK, JOHN W. SUTTIE, AND CONRAD WAGNER

VOLUME 282. Vitamins and Coenzymes (Part L)
Edited by DONALD B. MCCORMICK, JOHN W. SUTTIE, AND CONRAD WAGNER

VOLUME 283. Cell Cycle Control
Edited by WILLIAM G. DUNPHY

VOLUME 284. Lipases (Part A: Biotechnology)
Edited by BYRON RUBIN AND EDWARD A. DENNIS

VOLUME 285. Cumulative Subject Index Volumes 263, 264, 266–284, 286–289

VOLUME 286. Lipases (Part B: Enzyme Characterization and Utilization)
Edited by BYRON RUBIN AND EDWARD A. DENNIS

VOLUME 287. Chemokines
Edited by RICHARD HORUK

VOLUME 288. Chemokine Receptors
Edited by RICHARD HORUK

VOLUME 289. Solid Phase Peptide Synthesis
Edited by GREGG B. FIELDS

VOLUME 290. Molecular Chaperones
Edited by GEORGE H. LORIMER AND THOMAS BALDWIN

VOLUME 291. Caged Compounds
Edited by GERARD MARRIOTT

VOLUME 292. ABC Transporters: Biochemical, Cellular, and Molecular Aspects
Edited by SURESH V. AMBUDKAR AND MICHAEL M. GOTTESMAN

VOLUME 293. Ion Channels (Part B)
Edited by P. MICHAEL CONN

VOLUME 294. Ion Channels (Part C)
Edited by P. MICHAEL CONN

VOLUME 295. Energetics of Biological Macromolecules (Part B)
Edited by GARY K. ACKERS AND MICHAEL L. JOHNSON

VOLUME 296. Neurotransmitter Transporters
Edited by SUSAN G. AMARA

VOLUME 297. Photosynthesis: Molecular Biology of Energy Capture
Edited by LEE MCINTOSH

VOLUME 298. Molecular Motors and the Cytoskeleton (Part B)
Edited by RICHARD B. VALLEE

VOLUME 299. Oxidants and Antioxidants (Part A)
Edited by LESTER PACKER

VOLUME 300. Oxidants and Antioxidants (Part B)
Edited by LESTER PACKER

VOLUME 301. Nitric Oxide: Biological and Antioxidant Activities (Part C)
Edited by LESTER PACKER

VOLUME 302. Green Fluorescent Protein
Edited by P. MICHAEL CONN

VOLUME 303. cDNA Preparation and Display
Edited by SHERMAN M. WEISSMAN

VOLUME 304. Chromatin
Edited by PAUL M. WASSARMAN AND ALAN P. WOLFFE

VOLUME 305. Bioluminescence and Chemiluminescence (Part C)
Edited by THOMAS O. BALDWIN AND MIRIAM M. ZIEGLER

VOLUME 306. Expression of Recombinant Genes in Eukaryotic Systems
Edited by JOSEPH C. GLORIOSO AND MARTIN C. SCHMIDT

VOLUME 307. Confocal Microscopy
Edited by P. MICHAEL CONN

VOLUME 308. Enzyme Kinetics and Mechanism (Part E: Energetics of Enzyme Catalysis)
Edited by DANIEL L. PURICH AND VERN L. SCHRAMM

VOLUME 309. Amyloid, Prions, and Other Protein Aggregates
Edited by RONALD WETZEL

VOLUME 310. Biofilms
Edited by RON J. DOYLE

VOLUME 311. Sphingolipid Metabolism and Cell Signaling (Part A)
Edited by ALFRED H. MERRILL, JR., AND YUSUF A. HANNUN

VOLUME 312. Sphingolipid Metabolism and Cell Signaling (Part B)
Edited by ALFRED H. MERRILL, JR., AND YUSUF A. HANNUN

VOLUME 313. Antisense Technology
(Part A: General Methods, Methods of Delivery, and RNA Studies)
Edited by M. IAN PHILLIPS

VOLUME 314. Antisense Technology (Part B: Applications)
Edited by M. IAN PHILLIPS

VOLUME 315. Vertebrate Phototransduction and the Visual Cycle (Part A)
Edited by KRZYSZTOF PALCZEWSKI

VOLUME 316. Vertebrate Phototransduction and the Visual Cycle (Part B)
Edited by KRZYSZTOF PALCZEWSKI

VOLUME 317. RNA–Ligand Interactions (Part A: Structural Biology Methods)
Edited by DANIEL W. CELANDER AND JOHN N. ABELSON

VOLUME 318. RNA–Ligand Interactions (Part B: Molecular Biology Methods)
Edited by DANIEL W. CELANDER AND JOHN N. ABELSON

VOLUME 319. Singlet Oxygen, UV-A, and Ozone
Edited by LESTER PACKER AND HELMUT SIES

VOLUME 320. Cumulative Subject Index Volumes 290–319

VOLUME 321. Numerical Computer Methods (Part C)
Edited by MICHAEL L. JOHNSON AND LUDWIG BRAND

VOLUME 322. Apoptosis
Edited by JOHN C. REED

VOLUME 323. Energetics of Biological Macromolecules (Part C)
Edited by MICHAEL L. JOHNSON AND GARY K. ACKERS

VOLUME 324. Branched-Chain Amino Acids (Part B)
Edited by ROBERT A. HARRIS AND JOHN R. SOKATCH

VOLUME 325. Regulators and Effectors of Small GTPases
(Part D: Rho Family)
Edited by W. E. BALCH, CHANNING J. DER, AND ALAN HALL

VOLUME 326. Applications of Chimeric Genes and Hybrid Proteins
(Part A: Gene Expression and Protein Purification)
Edited by JEREMY THORNER, SCOTT D. EMR, AND JOHN N. ABELSON

VOLUME 327. Applications of Chimeric Genes and Hybrid Proteins
(Part B: Cell Biology and Physiology)
Edited by JEREMY THORNER, SCOTT D. EMR, AND JOHN N. ABELSON

VOLUME 328. Applications of Chimeric Genes and Hybrid Proteins (Part C: Protein–Protein Interactions and Genomics)
Edited by JEREMY THORNER, SCOTT D. EMR, AND JOHN N. ABELSON

VOLUME 329. Regulators and Effectors of Small GTPases (Part E: GTPases Involved in Vesicular Traffic)
Edited by W. E. BALCH, CHANNING J. DER, AND ALAN HALL

VOLUME 330. Hyperthermophilic Enzymes (Part A)
Edited by MICHAEL W. W. ADAMS AND ROBERT M. KELLY

VOLUME 331. Hyperthermophilic Enzymes (Part B)
Edited by MICHAEL W. W. ADAMS AND ROBERT M. KELLY

VOLUME 332. Regulators and Effectors of Small GTPases (Part F: Ras Family I)
Edited by W. E. BALCH, CHANNING J. DER, AND ALAN HALL

VOLUME 333. Regulators and Effectors of Small GTPases (Part G: Ras Family II)
Edited by W. E. BALCH, CHANNING J. DER, AND ALAN HALL

VOLUME 334. Hyperthermophilic Enzymes (Part C)
Edited by MICHAEL W. W. ADAMS AND ROBERT M. KELLY

VOLUME 335. Flavonoids and Other Polyphenols
Edited by LESTER PACKER

VOLUME 336. Microbial Growth in Biofilms (Part A: Developmental and Molecular Biological Aspects)
Edited by RON J. DOYLE

VOLUME 337. Microbial Growth in Biofilms (Part B: Special Environments and Physicochemical Aspects)
Edited by RON J. DOYLE

VOLUME 338. Nuclear Magnetic Resonance of Biological Macromolecules (Part A)
Edited by THOMAS L. JAMES, VOLKER DÖTSCH, AND ULI SCHMITZ

VOLUME 339. Nuclear Magnetic Resonance of Biological Macromolecules (Part B)
Edited by THOMAS L. JAMES, VOLKER DÖTSCH, AND ULI SCHMITZ

VOLUME 340. Drug–Nucleic Acid Interactions
Edited by JONATHAN B. CHAIRES AND MICHAEL J. WARING

VOLUME 341. Ribonucleases (Part A)
Edited by ALLEN W. NICHOLSON

VOLUME 342. Ribonucleases (Part B)
Edited by ALLEN W. NICHOLSON

VOLUME 343. G Protein Pathways (Part A: Receptors)
Edited by RAVI IYENGAR AND JOHN D. HILDEBRANDT

VOLUME 344. G Protein Pathways (Part B: G Proteins and Their Regulators)
Edited by RAVI IYENGAR AND JOHN D. HILDEBRANDT

VOLUME 345. G Protein Pathways (Part C: Effector Mechanisms)
Edited by RAVI IYENGAR AND JOHN D. HILDEBRANDT

VOLUME 346. Gene Therapy Methods
Edited by M. IAN PHILLIPS

VOLUME 347. Protein Sensors and Reactive Oxygen Species (Part A: Selenoproteins and Thioredoxin)
Edited by HELMUT SIES AND LESTER PACKER

VOLUME 348. Protein Sensors and Reactive Oxygen Species (Part B: Thiol Enzymes and Proteins)
Edited by HELMUT SIES AND LESTER PACKER

VOLUME 349. Superoxide Dismutase
Edited by LESTER PACKER

VOLUME 350. Guide to Yeast Genetics and Molecular and Cell Biology (Part B)
Edited by CHRISTINE GUTHRIE AND GERALD R. FINK

VOLUME 351. Guide to Yeast Genetics and Molecular and Cell Biology (Part C)
Edited by CHRISTINE GUTHRIE AND GERALD R. FINK

VOLUME 352. Redox Cell Biology and Genetics (Part A)
Edited by CHANDAN K. SEN AND LESTER PACKER

VOLUME 353. Redox Cell Biology and Genetics (Part B)
Edited by CHANDAN K. SEN AND LESTER PACKER

VOLUME 354. Enzyme Kinetics and Mechanisms (Part F: Detection and Characterization of Enzyme Reaction Intermediates)
Edited by DANIEL L. PURICH

VOLUME 355. Cumulative Subject Index Volumes 321–354

VOLUME 356. Laser Capture Microscopy and Microdissection
Edited by P. MICHAEL CONN

VOLUME 357. Cytochrome P450, Part C
Edited by ERIC F. JOHNSON AND MICHAEL R. WATERMAN

VOLUME 358. Bacterial Pathogenesis (Part C: Identification, Regulation, and Function of Virulence Factors)
Edited by VIRGINIA L. CLARK AND PATRIK M. BAVOIL

VOLUME 359. Nitric Oxide (Part D)
Edited by ENRIQUE CADENAS AND LESTER PACKER

VOLUME 360. Biophotonics (Part A)
Edited by GERARD MARRIOTT AND IAN PARKER

VOLUME 361. Biophotonics (Part B)
Edited by GERARD MARRIOTT AND IAN PARKER

VOLUME 362. Recognition of Carbohydrates in Biological Systems (Part A)
Edited by YUAN C. LEE AND REIKO T. LEE

VOLUME 363. Recognition of Carbohydrates in Biological Systems (Part B)
Edited by YUAN C. LEE AND REIKO T. LEE

VOLUME 364. Nuclear Receptors
Edited by DAVID W. RUSSELL AND DAVID J. MANGELSDORF

VOLUME 365. Differentiation of Embryonic Stem Cells
Edited by PAUL M. WASSAUMAN AND GORDON M. KELLER

VOLUME 366. Protein Phosphatases
Edited by SUSANNE KLUMPP AND JOSEF KRIEGLSTEIN

VOLUME 367. Liposomes (Part A)
Edited by NEJAT DÜZGÜNEŞ

VOLUME 368. Macromolecular Crystallography (Part C)
Edited by CHARLES W. CARTER, JR., AND ROBERT M. SWEET

VOLUME 369. Combinational Chemistry (Part B)
Edited by GUILLERMO A. MORALES AND BARRY A. BUNIN

VOLUME 370. RNA Polymerases and Associated Factors (Part C)
Edited by SANKAR L. ADHYA AND SUSAN GARGES

VOLUME 371. RNA Polymerases and Associated Factors (Part D)
Edited by SANKAR L. ADHYA AND SUSAN GARGES

VOLUME 372. Liposomes (Part B)
Edited by NEJAT DÜZGÜNEŞ

VOLUME 373. Liposomes (Part C)
Edited by NEJAT DÜZGÜNEŞ

VOLUME 374. Macromolecular Crystallography (Part D)
Edited by CHARLES W. CARTER, JR., AND ROBERT W. SWEET

VOLUME 375. Chromatin and Chromatin Remodeling Enzymes (Part A)
Edited by C. DAVID ALLIS AND CARL WU

VOLUME 376. Chromatin and Chromatin Remodeling Enzymes (Part B)
Edited by C. DAVID ALLIS AND CARL WU

VOLUME 377. Chromatin and Chromatin Remodeling Enzymes (Part C)
Edited by C. DAVID ALLIS AND CARL WU

VOLUME 378. Quinones and Quinone Enzymes (Part A)
Edited by HELMUT SIES AND LESTER PACKER

VOLUME 379. Energetics of Biological Macromolecules (Part D)
Edited by JO M. HOLT, MICHAEL L. JOHNSON, AND GARY K. ACKERS

VOLUME 380. Energetics of Biological Macromolecules (Part E)
Edited by JO M. HOLT, MICHAEL L. JOHNSON, AND GARY K. ACKERS

VOLUME 381. Oxygen Sensing
Edited by CHANDAN K. SEN AND GREGG L. SEMENZA

VOLUME 382. Quinones and Quinone Enzymes (Part B)
Edited by HELMUT SIES AND LESTER PACKER

VOLUME 383. Numerical Computer Methods (Part D)
Edited by LUDWIG BRAND AND MICHAEL L. JOHNSON

VOLUME 384. Numerical Computer Methods (Part E)
Edited by LUDWIG BRAND AND MICHAEL L. JOHNSON

VOLUME 385. Imaging in Biological Research (Part A)
Edited by P. MICHAEL CONN

VOLUME 386. Imaging in Biological Research (Part B)
Edited by P. MICHAEL CONN

VOLUME 387. Liposomes (Part D)
Edited by NEJAT DÜZGÜNEŞ

VOLUME 388. Protein Engineering
Edited by DAN E. ROBERTSON AND JOSEPH P. NOEL

VOLUME 389. Regulators of G-Protein Signaling (Part A)
Edited by DAVID P. SIDEROVSKI

VOLUME 390. Regulators of G-Protein Signaling (Part B)
Edited by DAVID P. SIDEROVSKI

VOLUME 391. Liposomes (Part E)
Edited by NEJAT DÜZGÜNEŞ

VOLUME 392. RNA Interference
Edited by ENGELKE ROSSI

VOLUME 393. Circadian Rhythms
Edited by MICHAEL W. YOUNG

VOLUME 394. Nuclear Magnetic Resonance of Biological Macromolecules (Part C)
Edited by THOMAS L. JAMES

VOLUME 395. Producing the Biochemical Data (Part B)
Edited by ELIZABETH A. ZIMMER AND ERIC H. ROALSON

VOLUME 396. Nitric Oxide (Part E)
Edited by LESTER PACKER AND ENRIQUE CADENAS

VOLUME 397. Environmental Microbiology
Edited by JARED R. LEADBETTER

VOLUME 398. Ubiquitin and Protein Degradation (Part A)
Edited by RAYMOND J. DESHAIES

VOLUME 399. Ubiquitin and Protein Degradation (Part B)
Edited by RAYMOND J. DESHAIES

VOLUME 400. Phase II Conjugation Enzymes and Transport Systems
Edited by HELMUT SIES AND LESTER PACKER

VOLUME 401. Glutathione Transferases and Gamma Glutamyl Transpeptidases
Edited by HELMUT SIES AND LESTER PACKER

VOLUME 402. Biological Mass Spectrometry
Edited by A. L. BURLINGAME

VOLUME 403. GTPases Regulating Membrane Targeting and Fusion
Edited by WILLIAM E. BALCH, CHANNING J. DER, AND ALAN HALL

VOLUME 404. GTPases Regulating Membrane Dynamics
Edited by WILLIAM E. BALCH, CHANNING J. DER, AND ALAN HALL

VOLUME 405. Mass Spectrometry: Modified Proteins and Glycoconjugates
Edited by A. L. BURLINGAME

VOLUME 406. Regulators and Effectors of Small GTPases: Rho Family
Edited by WILLIAM E. BALCH, CHANNING J. DER, AND ALAN HALL

VOLUME 407. Regulators and Effectors of Small GTPases: Ras Family
Edited by WILLIAM E. BALCH, CHANNING J. DER, AND ALAN HALL

VOLUME 408. DNA Repair (Part A)
Edited by JUDITH L. CAMPBELL AND PAUL MODRICH

VOLUME 409. DNA Repair (Part B)
Edited by JUDITH L. CAMPBELL AND PAUL MODRICH

VOLUME 410. DNA Microarrays (Part A: Array Platforms and Web-Bench Protocols)
Edited by ALAN KIMMEL AND BRIAN OLIVER

VOLUME 411. DNA Microarrays (Part B: Databases and Statistics)
Edited by ALAN KIMMEL AND BRIAN OLIVER

VOLUME 412. Amyloid, Prions, and Other Protein Aggregates (Part B)
Edited by INDU KHETERPAL AND RONALD WETZEL

VOLUME 413. Amyloid, Prions, and Other Protein Aggregates (Part C)
Edited by INDU KHETERPAL AND RONALD WETZEL

VOLUME 414. Measuring Biological Responses with Automated Microscopy
Edited by JAMES INGLESE

VOLUME 415. Glycobiology
Edited by MINORU FUKUDA

VOLUME 416. Glycomics
Edited by MINORU FUKUDA

VOLUME 417. Functional Glycomics
Edited by MINORU FUKUDA

VOLUME 418. Embryonic Stem Cells
Edited by IRINA KLIMANSKAYA AND ROBERT LANZA

VOLUME 419. Adult Stem Cells
Edited by IRINA KLIMANSKAYA AND ROBERT LANZA

VOLUME 420. Stem Cell Tools and Other Experimental Protocols
Edited by IRINA KLIMANSKAYA AND ROBERT LANZA

VOLUME 421. Advanced Bacterial Genetics: Use of Transposons and Phage for Genomic Engineering
Edited by KELLY T. HUGHES

VOLUME 422. Two-Component Signaling Systems, Part A
Edited by MELVIN I. SIMON, BRIAN R. CRANE, AND ALEXANDRINE CRANE

VOLUME 423. Two-Component Signaling Systems, Part B
Edited by MELVIN I. SIMON, BRIAN R. CRANE, AND ALEXANDRINE CRANE

VOLUME 424. RNA Editing
Edited by JONATHA M. GOTT

VOLUME 425. RNA Modification
Edited by JONATHA M. GOTT

VOLUME 426. Integrins
Edited by DAVID CHERESH

VOLUME 427. MicroRNA Methods
Edited by JOHN J. ROSSI

VOLUME 428. Osmosensing and Osmosignaling
Edited by HELMUT SIES AND DIETER HAUSSINGER

VOLUME 429. Translation Initiation: Extract Systems and Molecular Genetics
Edited by JON LORSCH

VOLUME 430. Translation Initiation: Reconstituted Systems and Biophysical Methods
Edited by JON LORSCH

VOLUME 431. Translation Initiation: Cell Biology, High-Throughput and Chemical-Based Approaches
Edited by JON LORSCH

VOLUME 432. Lipidomics and Bioactive Lipids: Mass-Spectrometry–Based Lipid Analysis
Edited by H. ALEX BROWN

VOLUME 433. Lipidomics and Bioactive Lipids: Specialized Analytical Methods and Lipids in Disease
Edited by H. ALEX BROWN

VOLUME 434. Lipidomics and Bioactive Lipids: Lipids and Cell Signaling
Edited by H. ALEX BROWN

VOLUME 435. Oxygen Biology and Hypoxia
Edited by HELMUT SIES AND BERNHARD BRÜNE

VOLUME 436. Globins and Other Nitric Oxide-Reactive Protiens (Part A)
Edited by ROBERT K. POOLE

VOLUME 437. Globins and Other Nitric Oxide-Reactive Protiens (Part B)
Edited by ROBERT K. POOLE

VOLUME 438. Small GTPases in Disease (Part A)
Edited by WILLIAM E. BALCH, CHANNING J. DER, AND ALAN HALL

VOLUME 439. Small GTPases in Disease (Part B)
Edited by WILLIAM E. BALCH, CHANNING J. DER, AND ALAN HALL

VOLUME 440. Nitric Oxide, Part F Oxidative and Nitrosative Stress in Redox Regulation of Cell Signaling
Edited by ENRIQUE CADENAS AND LESTER PACKER

VOLUME 441. Nitric Oxide, Part G Oxidative and Nitrosative Stress in Redox Regulation of Cell Signaling
Edited by ENRIQUE CADENAS AND LESTER PACKER

VOLUME 442. Programmed Cell Death, General Principles for Studying Cell Death (Part A)
Edited by ROYA KHOSRAVI-FAR, ZAHRA ZAKERI, RICHARD A. LOCKSHIN, AND MAURO PIACENTINI

VOLUME 443. Angiogenesis: *In Vitro* Systems
Edited by DAVID A. CHERESH

VOLUME 444. Angiogenesis: *In Vivo* Systems (Part A)
Edited by DAVID A. CHERESH

VOLUME 445. Angiogenesis: *In Vivo* Systems (Part B)
Edited by DAVID A. CHERESH

VOLUME 446. Programmed Cell Death, The Biology and Therapeutic Implications of Cell Death (Part B)
Edited by ROYA KHOSRAVI-FAR, ZAHRA ZAKERI, RICHARD A. LOCKSHIN, AND MAURO PIACENTINI

VOLUME 447. RNA Turnover in Bacteria, Archaea and Organelles
Edited by LYNNE E. MAQUAT AND CECILIA M. ARRAIANO

VOLUME 448. RNA Turnover in Eukaryotes: Nucleases, Pathways and Analysis of mRNA Decay
Edited by LYNNE E. MAQUAT AND MEGERDITCH KILEDJIAN

VOLUME 449. RNA Turnover in Eukaryotes: Analysis of Specialized and Quality Control RNA Decay Pathways
Edited by LYNNE E. MAQUAT AND MEGERDITCH KILEDJIAN

VOLUME 450. Fluorescence Spectroscopy
Edited by LUDWIG BRAND AND MICHAEL L. JOHNSON

VOLUME 451. Autophagy: Lower Eukaryotes and Non-Mammalian Systems (Part A)
Edited by DANIEL J. KLIONSKY

VOLUME 452. Autophagy in Mammalian Systems (Part B)
Edited by DANIEL J. KLIONSKY

VOLUME 453. Autophagy in Disease and Clinical Applications (Part C)
Edited by DANIEL J. KLIONSKY

VOLUME 454. Computer Methods (Part A)
Edited by MICHAEL L. JOHNSON AND LUDWIG BRAND

VOLUME 455. Biothermodynamics (Part A)
Edited by MICHAEL L. JOHNSON, JO M. HOLT, AND GARY K. ACKERS (RETIRED)

VOLUME 456. Mitochondrial Function, Part A: Mitochondrial Electron Transport Complexes and Reactive Oxygen Species
Edited by WILLIAM S. ALLISON AND IMMO E. SCHEFFLER

VOLUME 457. Mitochondrial Function, Part B: Mitochondrial Protein Kinases, Protein Phosphatases and Mitochondrial Diseases
Edited by WILLIAM S. ALLISON AND ANNE N. MURPHY

VOLUME 458. Complex Enzymes in Microbial Natural Product Biosynthesis, Part A: Overview Articles and Peptides
Edited by DAVID A. HOPWOOD

VOLUME 459. Complex Enzymes in Microbial Natural Product Biosynthesis, Part B: Polyketides, Aminocoumarins and Carbohydrates
Edited by DAVID A. HOPWOOD

VOLUME 460. Chemokines, Part A
Edited by TRACY M. HANDEL AND DAMON J. HAMEL

VOLUME 461. Chemokines, Part B
Edited by TRACY M. HANDEL AND DAMON J. HAMEL

VOLUME 462. Non-Natural Amino Acids
Edited by TOM W. MUIR AND JOHN N. ABELSON

VOLUME 463. Guide to Protein Purification, 2nd Edition
Edited by RICHARD R. BURGESS AND MURRAY P. DEUTSCHER

VOLUME 464. Liposomes, Part F
Edited by NEJAT DÜZGÜNEŞ

VOLUME 465. Liposomes, Part G
Edited by NEJAT DÜZGÜNEŞ

VOLUME 466. Biothermodynamics, Part B
Edited by MICHAEL L. JOHNSON, GARY K. ACKERS, AND JO M. HOLT

VOLUME 467. Computer Methods Part B
Edited by MICHAEL L. JOHNSON AND LUDWIG BRAND

VOLUME 468. Biophysical, Chemical, and Functional Probes of RNA Structure, Interactions and Folding: Part A
Edited by DANIEL HERSCHLAG

VOLUME 469. Biophysical, Chemical, and Functional Probes of RNA Structure, Interactions and Folding: Part B
Edited by DANIEL HERSCHLAG

VOLUME 470. Guide to Yeast Genetics: Functional Genomics, Proteomics, and Other Systems Analysis, 2nd Edition
Edited by GERALD FINK, JONATHAN WEISSMAN, AND CHRISTINE GUTHRIE

VOLUME 471. Two-Component Signaling Systems, Part C
Edited by MELVIN I. SIMON, BRIAN R. CRANE, AND ALEXANDRINE CRANE

VOLUME 472. Single Molecule Tools, Part A: Fluorescence Based Approaches,
Edited by NILS G. WALTER

VOLUME 473. Thiol Redox Transitions in Cell Signaling, Part A Chemistry and Biochemistry of Low Molecular Weight and Protein Thiols
Edited by ENRIQUE CADENAS AND LESTER PACKER

VOLUME 474. Thiol Redox Transitions in Cell Signaling, Part B Cellular Localization and Signaling
Edited by ENRIQUE CADENAS AND LESTER PACKER

VOLUME 475. Single Molecule Tools, Part B: Super-Resolution, Particle Tracking, Multiparameter, and Force Based Methods
Edited by NILS G. WALTER

CHAPTER ONE

SUPER-ACCURACY AND SUPER-RESOLUTION: GETTING AROUND THE DIFFRACTION LIMIT

Erdal Toprak,[*,†] Comert Kural,[*,‡] and Paul R. Selvin[*,§]

Contents

1. Overview: Accuracy and Resolution 2
2. Getting Super-Accuracy 4
3. Calculating Super-Accuracy 7
   3.1. Kinesin walks hand-over-hand 8
   3.2. Kinesin and dynein *in vivo* 10
   3.3. FIONA without fluorescence 10
   3.4. FIONA in a live organism 12
4. Reaching Super-Resolution 13
   4.1. Single-molecule high-resolution imaging with photobleaching 14
   4.2. Single-molecule high-resolution colocalization 15
   4.3. PALM and STORM 17
   4.4. Stimulated emission depletion microscopy (STED) 19
5. Future Directions 22
References 22

Abstract

In many research areas such as biology, biochemistry, and biophysics, measuring distances or identifying and counting objects can be of great importance. To do this, researchers often need complicated and expensive tools in order to have accurate measurements. In addition, these measurements are often done under nonphysiological settings. X-ray diffraction, for example, gets Angstrom-level structures, but it requires crystallizing a biological specimen. Electron microscopy (EM) has about 10 Å resolution, but often requires frozen (liquid nitrogen) samples. Optical microscopy, while coming closest to physiologically relevant conditions, has been limited by the minimum distances to be measured, typically about the diffraction limit, or ~200 nm. However, most

---

[*] Department of Biophysics, University of Illinois, Urbana, Illinois, USA
[†] Department of Systems Biology, Harvard Medical School, Boston, Massachusetts, USA
[‡] Immune Disease Institute, Harvard Medical School, Boston, Massachusetts, USA
[§] Department of Physics, University of Illinois, Urbana, Illinois, USA

biological molecules are <5–10 nm in diameter, and getting molecular details requires imaging at this scale. In this chapter, we will describe some of the experimental approaches, from our lab and others, that push the limits of localization accuracy and optical resolution in fluorescence microscopy.

1. OVERVIEW: ACCURACY AND RESOLUTION

There are actually two fundamentally different types of limits, namely, *accuracy* and *resolution*. In general, the historic limit on how well you can do has been the Rayleigh diffraction limit of light, or about 200 nm. Modern techniques reach up to approximately a nanometer or two in accuracy, and 20 nm in the minimum distance between identical points, which is resolution.

Accuracy is the minimum distance or volume that one can locate a particle's position within a certain time period. Our lab has generally focused on enhancing the localization accuracy of single-molecule microscopy, often called super-accuracy. Our work on fluorescence imaging with one nanometer accuracy (FIONA) achieves ~ 1 nm accuracy in 1–500 ms. FIONA has been applied to *in vitro* and *in vivo* samples, and to living organisms (Kural *et al.*, 2005, 2009; Yildiz *et al.*, 2003). It can be extended to nonfluorescent (absorbing) probes as well (Kural *et al.*, 2007). Naturally, FIONA is just the culmination of a long string of work in an effort to reach super-accuracy (Bobroff, 1986; Gelles *et al.*, 1988; Ghosh and Webb, 1994; Schmidt *et al.*, 1996).

Resolution is the minimum distance or volume that can be measured *between* two (identical) particles in a given period of time. For visible fluorescence in the far-field, it is $\sim \lambda/2$ or ~ 200–300 nm. However, with modern super-resolution methods, the optical resolution is brought down to ~ 8–25 nm. The general idea is to have two fluorophores where they can be separately imaged, because they either emit or are excited at different wavelengths, or can be temporarily or permanently turned off. As long as there is only a single emitting fluorophore left at one time within a diffraction-limited spot, it can be localized to high precision via a FIONA type of measurement. The process is then repeated many times, and the full high-resolution image is reconstructed (more on super-accuracy and super-resolution later).

Most super-accuracy and super-resolution techniques rely on imaging single molecules. The technology for single-molecule biophysics was pioneered in the late 1970s by a handful of researchers. In 1976, Sakmann and Neher developed a patch clamp technique that allowed them to measure the electrical currents across single ion channels (Neher and Sakmann, 1976). In 1986, Ashkin *et al.* invented optical tweezers that enabled several scientists to characterize many biomolecules at the single-molecule level by applying forces as small as several piconewtons (pN) and measuring nanometer-to-Angstrom-sized movements (Abbondanzieri *et al.*, 2005; Ashkin *et al.*, 1986, 1990; Block *et al.*, 1990; Bustamante *et al.*, 1994; Mehta *et al.*, 1999; Moffitt *et al.*, 2006; Vale *et al.*, 1985). In 1989, Moerner and Kador reported the first detection of single molecules in a solid at the cryo temperatures (Moerner and Kador, 1989). This was quickly followed in 1990 by Orrit and Bernard who dramatically improved the signal-to-noise ratio by using fluorescence (Orrit and Bernard, 1990). This was then followed by Soper *et al.*'s (1991) observation of single fluorophores at room temperature using far-field microscopy and Betzig *et al.*'s near-field optical microscopy (Betzig and Chichester, 1993; Betzig and Trautman, 1992). These discoveries and the invention (Axelrod, 1989) and use (Funatsu *et al.*, 1995; Vale *et al.*, 1996) of total internal reflection fluorescence microscopy (TIRF) led to imaging of single fluorophores and fluorescent proteins (FPs) using wide-field microscopy at room temperature. Since then, there has been a dramatic increase in the number of single-molecule experiments.

Modern super-resolution techniques such as single-molecule high-resolution imaging with photobleaching (SHRIMP) or single-molecule high-resolution colocalization (SHREC) rely on sequentially and accurately localizing single molecules (Balci *et al.*, 2005; Gordon *et al.*, 2004). SHRIMP achieves 10 nm resolution and uses two (or more) identical fluorophores (or FPs). One of the identical dyes is turned off by a permanent photobleaching, which occurs stochastically upon excitation. SHREC (Churchman *et al.*, 2005; Toprak and Selvin, 2007) and similar techniques (Michalet *et al.*, 2001) reach 8–10 nm resolution using two photostable organic fluorophores, or quantum dots (Lagerholm *et al.*, 2006; Lidke *et al.*, 2005), that have spectrally distinct wavelengths. Photoactivated localization microscopy (PALM) (Betzig *et al.*, 2006; Shroff *et al.*, 2007) and fluorescence photoactivated localization microscopy (FPALM) (Hess *et al.*, 2006), achieve about 25 nm resolution using photoactivatable fluorescent proteins (FAPs). FAPs are variants of FPs that can be turned on and off by a UV laser pulse, and then made to emit with a second longer-wavelength laser pulse. Stochastic optical reconstruction microscopy (STORM) is another super-resolution technique which essentially uses the same algorithm as (F)PALM. Although STORM formally includes FAPs, it mostly uses a wide variety of photoswitchable organic probes (Bates *et al.*, 2007; Huang *et al.*, 2008a). It also achieves about 25 nm resolution.

Finally, there are two methods for getting super-resolution that do not need single fluorophores. The first one is structured-illumination microscopy (SIM) (Gustafsson, 2005, 2008; Kner *et al.*, 2009). SIM can currently achieve twofold better (∼120 nm) optical resolution than classical optical microscopy, although using nonlinear techniques, the resolution is formally unbounded. The second method is called stimulated emission depletion microscopy (STED; Hell, 2007; Hell and Wichmann, 1994; Klar *et al.*, 2000; Willig *et al.*, 2006). STED works by transiently turning off all fluorophores in the focal region, except those located at the zero of a second, overlapping beam (the STED beam). It has achieved 16 nm resolution with organic fluorophores (Westphal and Hell, 2005) and 65 nm resolution at video rates (Westphal et al., 2008). STED's ability to acquire data rapidly is one of the unique aspects of the technique, compared to the single-molecule methods.

Finally, improved optics and microscopy is only one part of the solution for achieving high-resolution images. In the case of fluorescence, there is the nontrivial problem of specifically labeling biomolecules, particularly in the cell. Fortunately, there are FPs, that is, green fluorescent protein (GFP) and its many different colored derivatives (Giepmans et al., 2006; Shaner et al., 2008). These are genetically encodable probes that enable perfect alignment with the biological molecule of interest. Of course, FPs have their own problems. They are not compatible with staining of tissue samples that are not already genetically encodable. There can also be problems of functionality because of the large size of FPs ($\sim$ 5 nm and larger for oligomers; Ormo et al., 1996). More significantly, FPs are *not* sufficiently photostable and bright enough to see a single one (Shaner et al., 2008). This is especially true if one wants to look for an extended period of time, that is, more than a few seconds. There are examples of single GFPs detected, particularly *in vitro* (Pierce et al., 1997), and for six GFPs *in vivo* (Cai et al., 2007), but photostability of FPs remains a major problem (Shaner et al., 2008). Chemically synthesized organic fluorophores also have major problems, particularly for getting past the cell membrane and staining their targets specifically. The fluorophores are generally negatively charged to prevent them from aggregating, but this makes it very difficult to penetrate into the cell. Fluorescein derivatives, including fluorescein acetates/esters (Rotman and Papermaster, 1966) and FLASH (for fluorescein-based arsenical hairpin binder) and ReASH (for resorufin-based arsenical hairpin binder) (Griffin et al., 1998), get through the membrane, but are not sufficiently photostable for single-molecule detection (Park et al., 2004). Other dyes, such as tetramethylrhodamine, pass through the membrane but label nonspecifically. Finally quantum dots, which are very promising colloidal fluorescent particles, are very big (>3–20 nm) and therefore neither easily get into cells nor label organelles specifically (Michalet et al., 2005).

2. Getting Super-Accuracy

Super-accuracy was first shown in a biological setting in 1996 when Schmidt et al. achieved 30 nm accuracy in single-molecule diffusion (Schmidt et al., 1996). The *accuracy* with which you can determine the position of the particle is limited by the number of photons emitted (or absorbed), and a few other parameters. Figure 1.1 shows a (fluorescent) image of a single dye. The accuracy is determined by how well you can

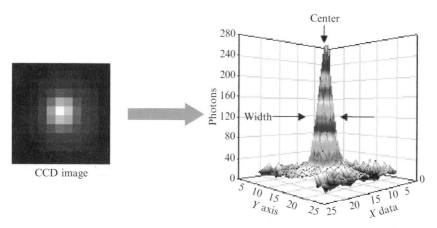

Figure 1.1 CCD image (left) of a fluorescent dye and corresponding emission pattern (right). The width (w) is ~ 250 nm, but the center can be located to within $w/\sqrt{N}$, where $N = 10^4$, is the total number of photons, making the center's uncertainty ± 1.3 nm.

locate the center of the fluorescence distribution. Thompson et al. showed experimentally that they could localize a fluorescent bead with up to ~ 2 nm accuracy and derived a relationship for the accuracy (see Thompson et al., 2002, Box 1). The accuracy is approximately equal to the width, or the diffraction limit of light, divided by the square root of the number of photons. The diffraction limit is typically $\lambda/2\text{NA}$ where λ is the wavelength of the emitted light and NA is the numerical aperture of the collecting lens. For visible light ($\lambda \sim 500$ nm), NA = 1.4, and 14,000 photons collected in Fig. 1.1, one can locate the centroid of the distribution with ~ 1.3 nm accuracy. Obviously, this is much better than the diffraction limit of light.

There are a few other parameters that come into determining the accuracy. One is the size of the detection pixel with which the fluorescence is imaged. Obviously, if the pixel is too large, you cannot resolve the point spread function (PSF). If the pixel is too small, the signal-to-noise ratio diminishes and the PSF may become anisotropic (Enderlein et al., 2006). In most cases, using ~ 85–120 nm (or a 150–250× magnification onto the typically 16–24 μm CCD (charge coupled devices) pixels) works well. The other issue is the background noise, either stemming from the detector, or from the sample in the form of autofluorescence. The former can be essentially eliminated with the advent of super-quiet, back-thinned, electron multiplying charge coupled devices (EMCCDs). Autofluorescence can be greatly minimized if your sample can be excited by a thin strip of light at the water–glass interface (Fig. 1.2). This is common, for example, in *in vitro* measurements where the sample is stuck down to the glass coverslip or slide, or *in vivo* measurements, where one is interested in looking at the cell

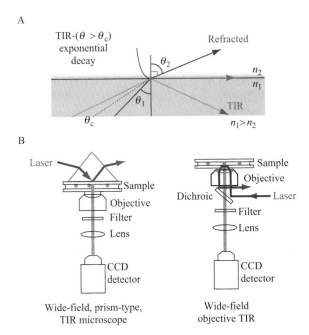

Figure 1.2 Schematics of total internal reflection. (A) In TIR, an excitation light enters a sample at a large angle such that it totally reflects at an glass (n_1)–water (n_2) interface. However, a small amount of energy enters the water, which exponentially dies away with about a 60 nm decay length. Consequently, fluorophores at the interface will be excited, but those beyond ~ 100 nm will not be excited. (B) There are two ways of exciting TIR. One is a prism type, and the other is an objective type. A prism type was initially the most popular, and still has certain advantage, but with objective lenses that are quieter, this has become the most popular and sensitive. Adapted from Selvin and Ha (2007).

membrane. The thin strip of light is created by exciting the sample with TIRF (Fig. 1.2).

In TIRF, the excitation light comes in at a large angle such that the light is totally internal reflected at the interface between the glass coverslip (or slide) and the water immersing the sample. Despite the complete reflection, there is a small amount of energy that reaches the sample. This energy dies away with a decay length ($1/e$) of ~ 60 nm, depending on the refractive indices of the two adjacent media. Hence, it excites your sample, but not the fluorophores above it. Of course, this is only good when the fluorescence is near the surface. A detailed description of building a custom-made TIRF setup for single fluorophore imaging can be found in the third chapter of the *Single Molecule Techniques* book edited by Selvin and Ha (2007).

Finally, there is a problem of how long the fluorescence lasts. We have found that adding the enzymes protocatechuic acid (PCA)/protocatechuate-3,4-dioxygenase (PCD) makes the fluorophores very stable, lasting from 30 s to several minutes (Aitken et al., 2008). We initially used glucose oxidase and catalase, but found that the enzyme underwent some formulation changes (in the United States) and was no longer available in its original formulation (Yildiz et al., 2003). (The catalase was made by Roche, from horses. In the United States, they switched to a molecular biology formulation, which we found did not work as well.)

3. CALCULATING SUPER-ACCURACY

The fundamental goal is to determine the center, or mean value of the distribution, $\mu = (x_0, y_0)$, and its uncertainty, the standard error of the mean (σ_μ). σ_μ tells you how well you can localize the fluorophore. The relation between σ_μ and the number of collected photons (N), the pixel size of the imaging detector (a), the standard deviation of the background (b, which includes background fluorescence and detector noise), and the width of the distribution (standard deviation, s_i, in direction i) was derived by Thompson et al. (2002) in two dimensions:

$$\sigma_{\mu_i} = \sqrt{\left(\frac{s_i^2}{N} + \frac{a^2/12}{N} + \frac{8\pi s_i^4 b^2}{a^2 N^2}\right)} \qquad (1.1)$$

where the index i refers to the x or y direction. The first term (s_i^2/N) is the photon noise, the second term is the effect of finite pixel size of the detector, and the last term is the effect of background.

Control experiments were done to demonstrate the ability to localize a translationally immobile dye. Figure 1.1 shows the PSF of an individual Cy3 dye attached to a coverslip via a DNA–biotin–streptavidin linkage, immersed in an aqueous buffer, and obtained with objective-type TIRF with an integration time of 0.5 s (Woehlke and Schliwa, 2000). For the highlighted PSF, $N = 14,200$ photons, $a = 86$ nm, $b = 11$, $s_y = 122$ nm, $s_x = 125$ nm (the full-width-half-max of the distribution, $FWHM_i = .354$ $s_i \approx 287$ nm). Based on Eq. (1.1), the expected σ_μ is 1.24 nm in each direction. Photon noise only (first term, Eq. (1.1)) leads to $\sigma_\mu = 1.02$ nm, pixelation (second term, Eq. (1.1)) increases σ_μ to 1.04 nm, and background noise (third term, Eq. (1.1)) increases σ_μ to 1.24 nm, showing that photon noise is the dominant contributor to σ_μ. A 2-D Gaussian yields an excellent fit ($r^2 = 0.994$; $\chi_r^2 = 1.48$; Woehlke and Schliwa, 2000, Fig. S2), with $\sigma_\mu = 1.3$ nm, in excellent agreement with the expected value. We note that fitting Fig. 1.1 with a 2-D Airy disk, which is formally the proper function to use, rather than a 2-D Gaussian fit, does not lead to a significantly improved fit (Cheezum et al., 2001). We now turn to various examples of FIONA, namely *in vitro*, *in vivo*, in living organism, and with nonfluorescent tags.

3.1. Kinesin walks hand-over-hand

Kinesin is a dimeric motor protein that walks processively along microtubules for several microns (Block et al., 1990; Vale et al., 1985; Yildiz et al., 2004a). Kinesin is in charge of moving organelles and lipid vesicles toward the cellular membrane and has a step size of 8.3 nm (Svoboda et al., 1993). This step size was measured using optical tweezers by attaching a micron-sized bead to the cargo-binding domain. Until 2004, there was an uncertainty about the walking mechanism of kinesin. One of the proposed models, known as the inchworm model, suggested that one of the two heads of kinesin was taking an 8.3 nm step and the other head followed this by another concerted 8.3 nm movement (Hua et al., 2002; Yildiz et al., 2004a). The other proposed model was a hand-over-hand model where each head was taking 16.6 nm steps and becoming the leading and trailing heads alternatingly. In 2004, two elegant back-to-back publications clearly showed that kinesin walks hand-over-hand. The first one was done using optical tweezers with a kinesin mutant that had a truncated stalk (Asbury et al., 2003). In this work, Asbury et al. observed alternating fast and slow steps that suggested that both of the kinesin heads were equally responsible for generating steps via ATP hydrolysis. The second paper, published by Yildiz et al. (2004a), used FIONA for directly observing head movements. In this work, only one of the heads was labeled with a Cy3 dye. The expected head movements for the hand-over-hand mechanism were alternating ~ 16.6 and ~ 0 nm head movements. Indeed, this was the actual step size shown. Note that, in particular, the 0 nm step size could be inferred from this experiment. By plotting a histogram of the time between steps, one gets a function that looks like: $y(t) = k_1 k_2 (t e^{-k_1 t} + t e^{-k_2 t})/(k_1 - k_2)$, with each step occurring with a rate constant of k_i. With the two rates equal, this reduces to $y(t) = tk^2 e^{-kt}$. This result is clearly different from the exponential decay function expected if the underlying process is a single step and is Poissonian. The data agree very well with this model (Fig. 1.3D). Now we know that myosin V (Yildiz et al., 2003), myosin VI (Yildiz et al., 2004b), kinesin, and cytoplasmic dynein (Reck-Peterson et al., 2006) all use the hand-over-hand mechanism. This is in contrast to an inchworm mechanism for certain nuclear (DNA) motors (Myong et al., 2007).

The time resolution of early FIONA experiments was limited to 0.3–0.5 s since Cy3 or bifunctional rhodamine dyes were used, and CCD cameras were slower at that time. Now, we can accurately track single organic fluorophores using 20 ms exposure times (unpublished results), at

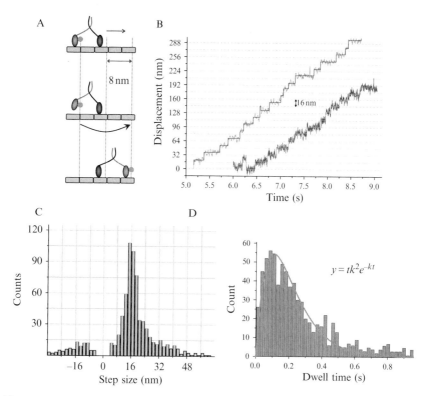

Figure 1.3 Stepping dynamics of kinesin molecules. (A) A kinesin molecule labeled with a quantum dot (red filled circle) on one of the microtubule-binding heads (brown). (B) Two sample traces of kinesin motility. Abrupt 16 nm steps were observed. (C) The step size distribution of quantum dot labeled kinesin molecules. (D) The hidden 0 nm step size can be detected because it leads to an initial linear rise. Adapted from Toprak et al. (2009).

the cost of less total time until photobleaching. A time resolution down to 1 ms when tracking quantum dots is possible. This advance is important for showing how kinesin acts *in vitro* at higher ATP concentrations (Toprak et al., 2009) and for *in vivo* measurements (Kural et al., 2005; also see later). As an *in vitro* example, we labeled kinesin's microtubule-binding head with a streptavidin-coated quantum dot (QS655, Molecular Probes; Fig. 1.3B). We imaged kinesin molecules using 2 or 4 ms exposure times at 5 μM ATP. This ATP concentration is 16-fold higher and the time resolution 125–250-fold faster than for our original experiments using Cy3 dyes. We clearly observed abrupt $\sim$16 nm steps (Fig. 1.3B) (Toprak et al., 2009). The average step size was 16.3 $\pm$ 4.4 nm (Fig. 1.3C).

Also, with quantum dots, because they are so bright, enough photons can be collected to detect small step sizes. The center-of-mass steps of kinesin, for example, has been measured with FIONA to be 8.4 ± 0.7 nm (Yardimci et al., 2008), in good agreement with the optical trap value of 8.3 nm (Svoboda et al., 1993). We note that the smallest step size that can be measured is often not limited by the amount of photons, but rather by the noise sources such as drift and vibrations.

3.2. Kinesin and dynein in vivo

It is of course desirable to look at molecular motors inside a cell (Kural et al., 2005). For this goal, we had two problems: how to label them with a photostable dye, and how to reach high enough of a time resolution given that the [ATP] is about 2 mM within cells, greatly accelerating the motor. As mentioned before, it is very difficult to use single fluorophores inside the cell. Hence, we tracked movements of fluorescently tagged organelles, such as peroxisomes, for studying the dynamics of motor proteins. Because there were hundreds of GFPs per peroxisome, but they all moved together, the criteria held for determining the position of a single molecule with high precision. Furthermore, the GFPs were surprisingly photostable. Fortunately, the time resolution we needed was about 1 ms, which was just accessible with our fastest (Andor) EMCCD camera.

Figure 1.4 is the result. We found that the peroxisomes moved in 8.3 nm steps, which was the value for kinesin and dynein step sizes *in vitro*. Even more interestingly, we saw that peroxisomes undergo a saltatory up-and-back motion, as is common among cargos. Presumably kinesin-1 is responsible for the forward motion, toward the positive-end of the microtubule, and dynein is responsible for the backwards motion, towards the negative-end. (Some more recent results suggest that dynein may be able to undergo forward and backward motion even in the absence of kinesin, possibly due to the presence of microtubule associated proteins (Dixit et al., 2008) or dynactin (Vendra et al., 2007).) How the transitions between the forward and backward motion occur remains unclear since they were faster than our time resolution.

3.3. FIONA without fluorescence

If sufficient contrast can be achieved in bright-field illumination, there is no reason to use fluorescence to accurately determine the centroid position of tiny organelles or vesicles. For tracking *Xenopus* melanosomes, which consist of dark melanin granules, bright-field illumination is sufficient. One advantage is that a simple microscope setup can be used, although we did use a rather sophisticated camera (a back-thinned EMCCD, Andor).

Figure 1.5 is one such example. An image is taken, inverted, and a 2-D Gaussian fit to the resulting PSF, which yields an accuracy of 2.4 nm with a 1 ms acquisition time. This result is verified by piezo-stepper experiments

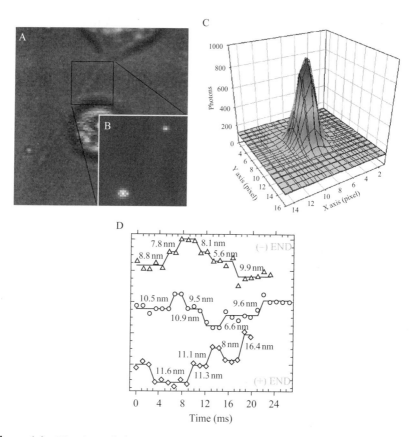

Figure 1.4 Kinesin and dynein steps can be detected *in vivo* using FIONA. (A) A bright-field image of a Drosophila S2 cell, with a process (long thick object) sticking out of it due to actin depolymerization. (B) Looking at the rectangular area via fluorescence microscopy, two GFP-peroxisomes are visible. The time resolution was 1.1 ms. (C) The PSF of the GFP-peroxisomes are plotted and the center was determine with an accuracy of be 1.5 nm. (D) The movement of a cargo being moved *in vivo* by either dynein, going away from the nucleus (the (−)-ve direction), or kinesin, going towards the nucleus (the (+)-ve direction), or switching between the two motors. It walks with 8 nm center-of-mass. Adapted from Kural *et al.* (2005).

performed on fixed melanophores where steps of 8 nm can be easily resolved by using this tracking scheme. We named this technique bright-field imaging with one nanometer accuracy (bFIONA) (Kural *et al.*, 2007). Since bFIONA tracking does not depend on fluorescence, high spatial resolution is not limited by photobleaching. Furthermore, the absence of toxic by-products of photobleaching, such as oxygen radicals, makes bFIONA an appropriate tool to track organelles as long as desired, without severely harming cells.

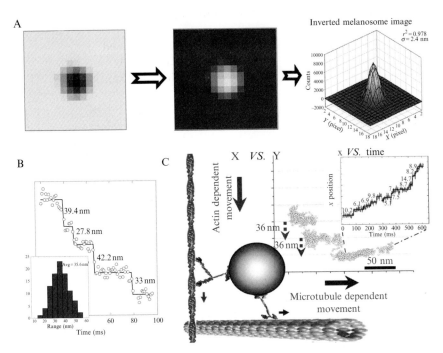

Figure 1.5 bFIONA. (A) The image of a dark melanosome is inverted and then plotted and fit to 2.4 nm accuracy. (B) Adding nocodozole depolymerizes the microtubules, leaving actin. The 36 nm center-of-mass steps are visible. (C) Both of the dyes can be tracked very accurately and one can see the transfer from actin to microtubules. Adapted from Kural et al. (2007).

By tracking melanosomes with bFIONA, we have shown that heterotrimeric kinesin-2 (KIF3) takes 8 nm steps (as in the case of conventional kinesin, KIF5). In cells where microtubules are depolymerized, we could resolve myosin V stepping as well. Our results also imply that intermediate filaments can hinder the movement of organelles in the cytoplasm since the uninterrupted run lengths of melanosome transport increased 1.5-fold in the absence of intermediate filaments.

3.4. FIONA in a live organism

Is it practical to use FIONA in a live organism, as opposed to *in vitro* or in cultured cells? To be able to address such a question one needs to make sure that the target organism can easily be labeled, which often means via genetic engineering, and be optically clear. The tiny nematode *Caenorhabditis elegans* was shown to be a perfect model organism for this purpose (Kural et al., 2009). The subcellular compartments that belong to the neuronal system of this worm can be tracked with sub-10 nm localization accuracy in a ∼5 ms time resolution by FIONA.

Fluorescence imaging can be successfully performed on *C. elegans* because of its transparent body. Since the mechanosensory nerves are contiguous and aligned parallel to the cuticle, the fluorophores within these cells can be excited with TIR illumination if the worms can be immobilized adjacent to the coverslip. We have found that the ELKS proteins, which plays a regulatory role in synaptic development, can be imaged in a punctuate pattern in mechanosensory neurons when it is genetically labeled with GFP or photoconvertible DENDRA2 probes (Fig. 1.6A). 2-D Gaussian fits to the emission patterns of the fluorescent ELKS spots let us to localize the peak within 4.6 nm in 4 ms (Fig. 1.6B). The immobilization of the worms is performed by sandwiching them between a glass support and a coverslip. This is a very straightforward and effective way of performing stable measurements on simultaneous multiple worms without using anesthetic reagents that can alter the metabolism of the nematode. We have verified the vitality of the worms by making sure that their movement is restored after releasing the pressure between the glass layers.

Our results suggested that ELKS punctae are motionless in general, but are occasionally carried by molecular motors (Fig. 1.6C,D). In addition, we replaced the 15-protofilament microtubules with an 11-protofilament mutation. Even though this eliminates the touch sensitivity of the worms, it does not affect the organization and the punctuate pattern of ELKS. Other mutations that disrupt degenerin/epithelial Na^+ channel formation or interrupt their association with the epidermis (cuticle) also did not have an effect on the ELKS punctae distribution.

4. REACHING SUPER-RESOLUTION

Optical resolution is the minimum distance necessary to distinguish two light emitting particles. If two objects are closer than the diffraction limit ($\lambda/2NA$), their PSFs overlap and you cannot tell that they are, in fact, two separate emitting objects (Fig. 1.7). If they are separated by a distance larger than the diffraction limit, their PSFs are well separated. To get improved resolution, one can try to decrease the width of individual PSFs, for example, by using a shorter wavelength of light; or one of the PSFs can be transiently or permanently photobleached (in SHRIMP, FPALM, STORM, or STED), or one can minimize the overlap of the PSFs, by, for example, making them spectrally distinct (in SHREC). We now go through several of these techniques, listed and summarized in section 1.

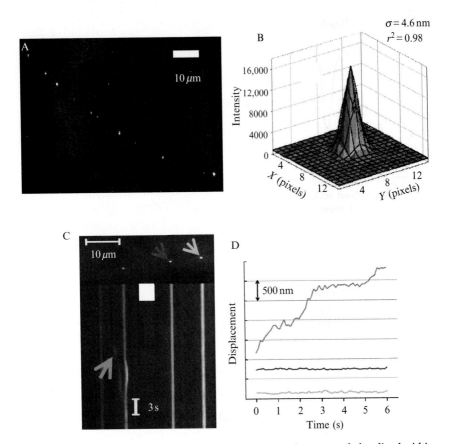

Figure 1.6 Fluorescently labeled protein complexes can be accurately localized within a live *C. elegans*. (A) DENDRA2::ELKS spots decorate the mechanosensory neurons along the *C. elegans* body. (B) The peak of the two dimensional Gaussian fit to the emission of a DENDRA::ELKS puncta can be localized within less than 5 nm in 4 ms. (C) A kymograph showing the moving and stationary GFP::ELKS spots in neurons. The displacement of spots pointed with red, blue and green arrows are shown in (D) with red, blue and green colors respectively. Adapted from Kural *et al.* (2009). (See Color Insert.)

4.1. Single-molecule high-resolution imaging with photobleaching

If there are many fluorescent probes that are within a diffraction-limited spot, they appear to be a single bright spot. One solution is to sequentially switch off these fluorophores, and localizing them one by one. If the fluorophores are excited simultaneously, they will have a stochastic lifetime until photobleaching. Let us say there are two fluorophores, called A and B. If A photobleaches first, then B can subsequently be located with FIONA-type accuracy. Furthermore, A can then be localized in earlier images by

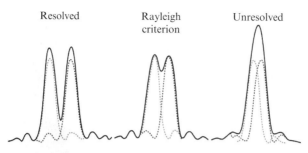

Figure 1.7 Two point particles at differing distances from each other create a well-resolved (left), a barely resolved (middle) or an unresolved image. If, however, the two PSF's can be resolved such that one particle at a time is imaged, then FIONA can be applied to each of them and their distances can theoretically be found to a few nanometers accuracy.

subtracting the photons of B from those of A + B, the former coming from the time following photobleaching. This technique was called single-molecule high-resolution imaging with photobleaching (SHRIMP). Figure 1.8 is an example of two dyes placed apart 329.7 ± 2.2 nm, where they can just be resolved by fitting a double-peaked Gaussian function to the two PSFs. They can also be resolved by SHRIMP, where we find 324.6 ± 1.6 nm. Smaller distances were successfully resolved by SHRIMP. However, smaller distances cannot be resolved by fitting multipeaked Gaussian functions. Gordon et al. (2004) labeled both ends of short DNA pieces (10, 17, and 51 nm) with two identical fluorophores in order to test the limits of SHRIMP. All of these distances were successfully resolved. Balci et al. (2005) later applied the same approach to measure the inter-head distance of myosin VI molecules labeled with GFPs on the actin-binding domains.

Qu et al. (2004) independently used the technique (which they called nanometer-localized multiple single-molecule (NALMS) fluorescence microscopy) to cover up to five separate fluorophores on DNA. We have recently applied this technique to study many—up to several hundred—fluorophores (P.R.S. Paul Simonson, unpublished data). One of the great advantages of SHRIMP is the convenience of using only one type of fluorophore for imaging. This feature increases fluorescent labeling efficiency and also makes imaging straightforward without having to correct any chromatic aberrations. A regular TIRF microscope can be used, although one may want a reasonably powerful laser to make the photobleaching time shorter.

4.2. Single-molecule high-resolution colocalization

SHREC was first introduced by Churchman et al. (2005) and later by Yildiz et al. (Toprak and Selvin, 2007). Using SHREC, one can measure distances larger than 10 nm that are out of the range of single-molecule fluorescence

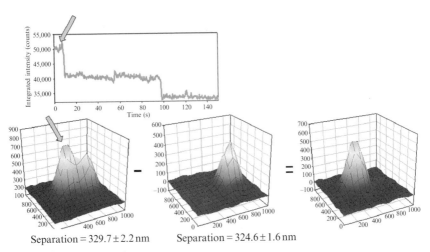

Figure 1.8 SHRIMP. Two fluorophores can be imaged (left). When one stochastically is photobleached, (arrow) the position of the other one can be fit to FIONA accuracy (middle). Subtracting this from the left-most image yields and fitting with FIONA, yields the position of the first dye. The agreement between SHRIMP technique, and normal fitting for (when the dyes are sufficiently far apart to be fit in the normal way), is excellent. The SHRIMP technique, however, has been shown to extend down to 10 nm. Adapted from Reference (Gordon et al., 2004).

resonance energy transfer (FRET) and conventional light microscopy (Ha et al., 1996). SHREC works by differentially labeling two sites of a biological molecule with two fluorophores that have spectrally different fluorescence emissions. Unlike FRET, both of the dyes are simultaneously excited with one or two lasers depending on the dyes. The resulting fluorescence emission is spectrally split using a homemade or commercial (e.g., a Dual-view, by Optical Insights) optical apparatus, and imaged with the same CCD (two separate CCD cameras can also be used if necessary). One common problem of the technique is the chromatic aberration in the imaging apparatus. That is, light emanating from a single spot consisting of two different wavelengths gets focused to different x, y, z spots. This interferes with the colocalization efficiency. After obtaining the images of both of the dyes, the images are localized on the CCD coordinates and these positions are mapped using a second (or higher) order polynomial. The standard way of assigning the mapping polynomial is imaging multicolor beads and colocalizing the spectrally split images so that the spectrally split emission of a fluorescent bead is mapped onto itself.

Churchman et al. used SHREC for measuring the length of short DNA segments that had Cy3 and Cy5 dyes on both ends. They successfully demonstrated that distances larger than 10 nm can be measured. They also monitored the movements of both of the two actin-binding heads of

Super-Accuracy and Super-Resolution: Getting Around the Diffraction Limit 17

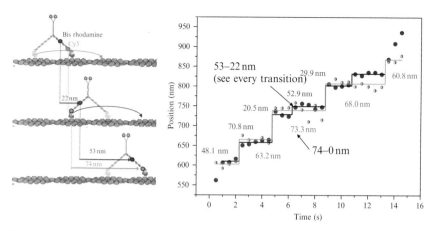

Figure 1.9 SHREC. Two fluorophores which emit at different wavelength can be imaged separately and the position of each dye fit with FIONA-like accuracy. The minimum distance between the two dyes, that is, the resolution, is approximately 10 nm. In this particular case, a myosin V, which moves with 37 nm center-of-mass, is labeled with a Bis-rhodamine (blue) and a Cy5 (red) (left). With the Cy5 labeled very close to the motor domain, a "hand-over-hand" motion causes the Cy5 to move 0-74-0-74 nm. The other dye, bis-rhodamine, labeled on the "leg," translocates by $(37-x)$ and $(37+x)$ nm, where x is the distance from the center-of-mass. Adapted from Toprak and Selvin (2007). (See Color Insert.)

myosin V simultaneously. One of the heads was labeled with Cy3 and the other head was labeled with Cy5. Churchman *et al.* showed that each head was taking a 74 nm step after hydrolyzing an ATP molecule and the heads were alternating between trailing and leading positions. Similar experiments were later independently conducted by Yildiz *et al.* for exploring the stepping mechanism of myosin V further (Toprak and Selvin, 2007). As is shown in Fig. 1.9, two CaMs were differentially labeled with Cy3 and Cy5. The results were confirming the experiments of Churchman *et al.* except that a larger variety of events was observed since CaMs were labeled instead of the heads. Similar dynamics were also observed by Warshaw *et al.* (2005), where they used quantum dots instead of dyes. Similar approaches have been applied to photo-blinking quantum dots and fluorophores as well (Lidke *et al.*, 2005; Rosenberg *et al.*, 2005).

4.3. PALM and STORM

SHRIMP and SHREC are extremely effective when the number of fluorophores is limited. However, they are not sufficient to be able to resolve the structure of a complex molecule that is labeled with a high density of fluorophores. In order to overcome this difficulty, a number of groups used a laser to activate a dye, and then another different-wavelength laser to excite the activated dye. Because only a small number of dyes were activated at any one time, very good resolution could be achieved. Two groups used PALM and FPALM based on a photoswitchable GFP (Patterson and Lippincott-Schwartz, 2002; often related to EosGFP; Wiedenmann et al., 2004). They used intense irradiation with 413-nm wavelength, which increased fluorescence 100-fold upon excitation with 488-nm light. They showed that they could get ~25 nm resolution in fixed cells, although the initial observations took 24 h of microscopy time (Fig. 1.10; Betzig et al., 2006; Hess et al., 2006). One group has extended this to a second, differently colored FP and shortened the total time to 5–30 min (Shroff et al., 2007). Unfortunately, this has not been achieved (yet) in living cells.

To get more colors in a photoswitchable way, STORM uses conjugates of two different organic fluorophores (Bates et al., 2005, 2007; Huang et al., 2008a, 2009). Formally, STORM includes all aspects of (F)PALM, including FAP, but often what is used are photoswitchable dyes. These had the surprising and favorable characteristic of being turned on and off several times (Bates et al., 2005), as well as having many color combinations. The dye pair, called the "activator" and the "reporter" and both fluorescent probes, have to be in close proximity—a few nanometers—in order to facilitate photoswitching. (The photophysics is not really understood at this point.) For this purpose, the reporter and activator pairs can be conjugated to antibodies, double-stranded DNA molecules, or can be linked covalently. The reporter fluorophore is often a cyanine dye (generally Cy5, Cy5.5, or Cy7), and a proximal fluorophore with a shorter

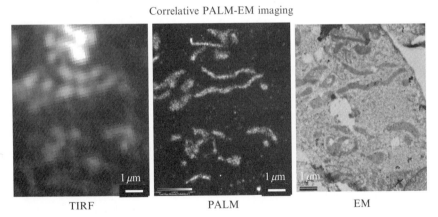

Correlative PALM-EM imaging

TIRF   PALM   EM

Figure 1.10 Images taken with TIRF, PALM, and by EM. The increased resolution of PALM is evident, and the structure correlates well with EM. Adapted from Betzig et al. (2006).

wavelength (e.g., Cy2 or Cy3) is used as the activator. Because a wide variety of activator–reporter pairs are available, it is possible to simultaneously use several different pairs, which give differing excitation or emission wavelengths. For example, the pairs Cy3–Cy5 and Cy3–Cy5.5 can both be used, where Cy3 requires a single wavelength to activate, but Cy5 and Cy5.5 emit at distinct wavelengths. Or Cy2–Cy5 and Cy3–Cy5 can be used where Cy2 and Cy3 require different excitation wavelengths but in both pairs the final signal (coming from Cy5 in both cases) is detected at the same wavelength. The latter has the advantage that chromatic aberrations do not have to be accounted for.

Molecules that are labeled with a high density of activator–reporter pairs can be imaged with sub-diffraction-limit resolution (up to 20 nm) with STORM (Fig. 1.11A). Only one dye pair is turned on in a diffraction-limited spot, and a FIONA-type image is taken. The process is then repeated many times, developing a high-resolution image. The Zhuang group managed to perform STORM on fixed cells where different intracellular molecules could be imaged in multiple colors (Bates et al., 2007). They also could extend the high-resolution localization to the third (z) dimension with up to ~ 50 nm axial resolution by making use of the ellipticity of individual PSFs introduced by cylindrical lenses inserted in the emission path (Fig. 1.11B; Huang et al., 2008a).

A potentially important new derivative has been added to the STORM repertoire called direct STORM (dSTORM). It has the advantage of *not* using dimers of fluorophores. Rather it relies on a single fluorophore, sometimes in the presence of reducing agents, sometimes in the presence of air-saturated solution (van de Linde et al., 2008, 2009). The molecules go into a metastable dark state and are hence turned off, while STORM is acquired. Approximately 20 nm resolution is achieved. Because only a single dye is used, there are no problems with chromatic aberrations, and it is very simple because only one excitation laser is used.

4.4. Stimulated emission depletion microscopy (STED)

Invented by Stefan Hell, STED is unlike other techniques in that it is not limited to single molecules. It also uses "normal," albeit bright, fluorophores, rather than tandems of fluorophores. It also does not require any chromatic (or other) correction. In STED, as in conventional fluorescence techniques, a fluorophore is driven into its excited state and then proceeds to relax within a few picoseconds into the vibrationally lowest electronically excited state. At this point, it can make a transition to the ground state by either emission of a photon, that is, fluorescence, or nonradiative decay. So far, this is all the "normal" part of the fluorescence process. However, Hell applies a second beam that stimulates the excited state to spontaneously revert back down to a vibrationally upper state of the electronic ground state

z-direction, only the light going towards the objective is collected, whereas in the x–y plane, all of the light is collected. The solution is to add another lens in the z-direction to collect more of the light. In addition, and most importantly, the sample is excited from both directions and the two beams interfere. The technique has been called 4Pi for scanning microscopy (Hell and Stelzer, 1992; Hell *et al.*, 2009), or I5M for wide-field illumination (Gustafsson *et al.*, 1999).

5. Future Directions

High accuracy and high-resolution microscopy has progressed tremendously in the past 10 or so years. A challenge that remains is to examine live cells, as the advances so far have been limited to *in vitro* samples and fixed cells. Looking at living cells, living specimens, and ultimately actual human tissues are the next steps. But here, the problem of autofluorescence and scattering is challenging. Perhaps nonlinear microscopy with its low scattering cross-sections is a favorable technique (Denk *et al.*, 1990; Larson *et al.*, 2003). New ways of specifically targeting fluorophores, especially smaller quantum dots (Michalet *et al.*, 2005; Smith and Nie, 2009), are badly needed. Perhaps then we can truly talk about seeing molecular changes in live samples in the optical microscope.

REFERENCES

Abbondanzieri, E. A., Greenleaf, W. J., Shaevitz, J. W., Landick, R., and Block, S. M. (2005). Direct observation of base-pair stepping by RNA polymerase. *Nature* **438**(7067), 460–465.

Aitken, C. E., Marshall, R. A., and Puglisi, J. D. (2008). An oxygen scavenging system for improvement of dye stability in single-molecule fluorescence experiments. *Biophys. J.* **94**(5), 1826–1835.

Asbury, C. L., Fehr, A. N., and Block, S. M. (2003). Kinesin moves by an asymmetric hand-over-hand mechanism. *Science* **302**(5653), 2130–2134.

Ashkin, A., Dziedzic, J. M., Bjorkholm, J. E., and Chu, S. (1986). Observation of a single-beam gradient force optical trap for dielectric particles. *Optics Lett.* **11**, 288–290.

Ashkin, A., Schutze, K., Dziedzic, J. M., Euteneuer, U., and Schliwa, M. (1990). Force generation of organelle transport measured in vivo by an infrared laser trap. *Nature* **348**(6299), 346–348.

Axelrod, D. (1989). Total internal reflection fluorescence microscopy. *Meth. Cell Biol.* **30**, 245–270.

Balci, H., Ha, T., Sweeney, H. L., and Selvin, P. R. (2005). Interhead distance measurements in myosin VI via SHRImP support a simplified hand-over-hand model. *Biophys. J.* **89**(1), 413–417.

Bates, M., Blosser, T. R., and Zhuang, X. (2005). Short-range spectroscopic ruler based on a single-molecule optical switch. *Phys. Rev. Lett.* **94**(10), 108101.

Bates, M., Huang, B., Dempsey, G. T., and Zhuang, X. (2007). Multicolor super-resolution imaging with photo-switchable fluorescent probes. *Science* **317**(5845), 1749–1753.

Betzig, E., and Chichester, R. J. (1993). Single molecules observed by near-field scanning optical microscopy. *Science* **262**(5138), 1422–1425.
Betzig, E., and Trautman, J. K. (1992). Near-field optics: Microscopy, spectroscopy, and surface modification beyond the diffraction limit. *Science* **257**(5067), 189–195.
Betzig, E., Patterson, G. H., Sougrat, R., Lindwasser, O. W., Olenych, S., Bonifacino, J. S., Davidson, M. W., Lippincott-Schwartz, J., and Hess, H. F. (2006). Imaging intracellular fluorescent proteins at nanometer resolution. *Science* **313**(5793), 1642–1645.
Block, S. M., Goldstein, L. S., and Schnapp, B. J. (1990). Bead movement by single kinesin molecules studied with optical tweezers. *Nature* **348**(6299), 348–352.
Bobroff, N. (1986). Position measurement with a resolution and noise-limited instrument. *Rev. Sci. Inst.* **57**(6), 1152–1157.
Bustamante, C., Marko, J. F., Siggia, E. D., and Smith, S. (1994). Entropic elasticity of lambda-phage DNA. *Science* **265**(5178), 1599–1600.
Cai, D., Verhey, K., and Meyhofer, E. (2007). Tracking single kinesin molecules in the cytoplasm of mammalian cells. *Biophys. J.* **92**(12), 4137–4144.
Cheezum, M. K., Walker, W. F., and Guilford, W. H. (2001). Quantitative comparison of algorithms for tracking single fluorescent particles. *Biophys. J.* **81**(4), 2378–2388.
Churchman, L. S., Okten, Z., Rock, R. S., Dawson, J. F., and Spudich, J. A. (2005). Single molecule high-resolution colocalization of Cy3 and Cy5 attached to macromolecules measures intramolecular distances through time. *Proc. Natl. Acad. Sci. USA* **102**(5), 1419–1423.
Denk, W., Strickler, J. H., and Webb, W. W. (1990). Two-photon laser scanning fluorescence microscopy. *Science* **248**(4951), 73–76.
Dixit, R., Ross, J. L., Goldman, Y. E., and Holzbaur, E. L. (2008). Differential regulation of dynein and kinesin motor proteins by tau. *Science* **319**, 1086–1089.
Enderlein, J., Toprak, E., and Selvin, P. R. (2006). Polarization effect on position accuracy of fluorophore localization. *Optic. Express* **14**(18), 8111–8120.
Funatsu, T., Harada, Y., Tokunaga, M., Saito, K., and Yanagida, T. (1995). Imaging of single fluorescent molecules and individual ATP turnovers by single myosin molecules in aqueous solution. *Nature* **374**, 555–559.
Gelles, J., Schnapp, B. J., and Sheetz, M. P. (1988). Tracking kinesin-driven movements with nanometre-scale precision. *Nature* **331**(6155), 450–453.
Ghosh, R. N., and Webb, W. W. (1994). Automated detection and tracking of individual and clustered cell surface low density lipoprotein receptor molecules. *Biophys. J.* **66**(5), 1301–1318.
Giepmans, B. N. G., Adams, S. R., Ellisman, M. H., and Tsien, R. Y. (2006). Review – The fluorescent toolbox for assessing protein location and function. *Science* **312**(5771), 217–224.
Gordon, M. P., Ha, T., and Selvin, P. R. (2004). Single-molecule high-resolution imaging with photobleaching. *Proc. Natl. Acad. Sci. USA* **101**(17), 6462–6465.
Griffin, B. A., Adams, S. R., and Tsien, R. Y. (1998). Specific covalent labeling of recombinant protein molecules inside live cells. *Science* **281**, 269–272.
Gustafsson, M. G. (2005). Nonlinear structured-illumination microscopy: Wide-field fluorescence imaging with theoretically unlimited resolution. *Proc. Natl. Acad. Sci. USA* **102**(37), 13081–13086.
Gustafsson, M. G., Agard, D. A., and Sedat, J. W. (1999). I5M: 3D widefield light microscopy with better than 100 nm axial resolution. *J. Microsc.* **195**(Pt 1), 10–16.
Gustafsson, M. G., Shao, L., Carlton, P. M., Wang, C. J., Golubovskaya, I. N., Cande, W. Z., Agard, D. A., and Sedat, J. W. (2008). Three-dimensional resolution doubling in wide-field fluorescence microscopy by structured illumination. *Biophys. J.* **94**(12), 4957–4970.

Ha, T., Enderle, T., Ogletree, D. F., Chemla, D. S., Selvin, P. R., and Weiss, S. (1996). Probing the interaction between two single molecules: Fluorescence resonance energy transfer between a single donor and a single acceptor. *Proc. Natl. Acad. Sci. USA* **93**(13), 6264–6268.

Hell, S. W. (2007). Far-field optical nanoscopy. *Science* **316**(5828), 1153–1158.

Hell, S., and Stelzer, E. H. K. (1992). Fundamental improvement of resolution with a 4Pi-confocal fluorescence microscope using two-photon excitation. *Optics Comm.* **93**, 277–282.

Hell, S. W., and Wichmann, J. (1994). Breaking the diffraction resolution limit by stimulated emission: Stimulated emission depletion microscopy. *Optic. Lett.* **19**(11), 780–782.

Hell, S. W., Schmidt, R., and Egner, A. (2009). Diffraction-unlimited three-dimensional optical nanoscopy with opposing lenses. *Nat. Photonics* **3**, 381–387.

Hess, S. T., Girirajan, T. P., and Mason, M. D. (2006). Ultra-high resolution imaging by fluorescence photoactivation localization microscopy. *Biophys. J.* **91**(11), 4258–4272.

Hua, W., Chung, J., and Gelles, J. (2002). Distinguishing inchworm and hand-over-hand processive kinesin movement by neck rotation measurements. *Science* **295**(5556), 844–848.

Huang, B., Wang, W., Bates, M., and Zhuang, X. (2008a). Three-dimensional super-resolution imaging by stochastic optical reconstruction microscopy. *Science* **319**(5864), 810–813.

Huang, B., Jones, S. A., Brandenburg, B., and Zhuang, X. (2008b). Whole-cell 3D STORM reveals interactions between cellular structures with nanometer-scale resolution. *Nat. Meth.* **5**, 1047–1052.

Huang, B., Bates, M., and Zhuang, X. (2009). Super-resolution fluorescence microscopy. *Annu. Rev. Biochem.* **78**, 993–1016.

Klar, T. A., Jakobs, S., Dyba, M., Egner, A., and Hell, S. W. (2000). Fluorescence microscopy with diffraction resolution limit broken by stimulated emission. *Proc. Natl. Acad. Sci. USA* **97**(15), 8206–8210.

Kner, P., Chhun, B. B., Griffis, E. R., Winoto, L., and Gustafsson, M. G. (2009). Super-resolution video microscopy of live cells by structured illumination. *Nat. Meth.* **6**(5), 339–342.

Kural, C., Kim, H., Syed, S., Goshima, G., Gelfand, V. I., and Selvin, P. R. (2005). Kinesin and dynein move a peroxisome in vivo: A tug-of-war or coordinated movement? *Science* **308**, 1469–1472.

Kural, C., Serpinskaya, A. S., Chou, Y. H., Goldman, R. D., Gelfand, V. I., and Selvin, P. R. (2007). Tracking melanosomes inside a cell to study molecular motors and their interaction. *Proc. Natl. Acad. Sci. USA* **104**, 5378–5382.

Kural, C., Nonet, M. L., and Selvin, P. R. (2009). FIONA on the living worm, *Caenorhabditis elegans*. *Biochemistry* **48**(22), 4663–4665.

Lagerholm, B. C., Averett, L., Weinreb, G. E., Jacobson, K., and Thompson, N. L. (2006). Analysis method for measuring submicroscopic distances with blinking quantum dots. *Biophys. J.* **91**(8), 3050–3060.

Larson, D. R., Zipfel, W. R., Williams, R. M., Clark, S. W., Bruchez, M. P., Wise, F. W., and Webb, W. W. (2003). Water-soluble quantum dots for multiphoton fluorescence imaging in vivo. *Science* **300**(5624), 1434–1436.

Lidke, K., Rieger, B., Jovin, T., and Heintzmann, R. (2005). Superresolution by localization of quantum dots using blinking statistics. *Optic. Express* **13**(18), 7052–7062.

Mehta, A. D., Rief, M., Spudich, J. A., Smith, D. A., and Simmons, R. M. (1999). Single-molecule biomechanics with optical methods. *Science* **283**(5408), 1689–1695.

Michalet, X., Lacoste, T. D., and Weiss, S. (2001). Ultrahigh-resolution colocalization of spectrally separable point-like fluorescent probes. *Methods* **25**(1), 87–102.

Michalet, X., Pinaud, F. F., Bentolila, L. A., Tsay, J. M., Doose, S., Li, J. J., Sundaresan, G., Wu, A. M., Gambhir, S. S., and Weiss, S. (2005). Quantum dots for live cells, in vivo imaging, and diagnostics. *Science* **307**(5709), 538–544.

Moerner, W. E., and Kador, L. (1989). Optical detection and spectroscopy of single molecules in a solid. *Phys. Rev. Lett.* **62**, 2535–2538.

Moffitt, J. R., Chemla, Y. R., Izhaky, D., and Bustamante, C. (2006). Differential detection of dual traps improves the spatial resolution of optical tweezers. *Proc. Natl. Acad. Sci. USA* **103**(24), 9006–9011.

Myong, S., Bruno, M. M., Pyle, A. M., and Ha, T. (2007). Spring-loaded mechanism of DNA unwinding by hepatitis C virus NS3 helicase. *Science* **317**, 513–516.

Neher, E., and Sakmann, B. (1976). Single-channel currents recorded from membrane of denervated frog muscle fibres. *Nature* **260**(5554), 799–802.

Ormo, M., Cubitt, A. B., Kallio, K., Gross, L. A., Tsien, R. Y., and Remington, S. J. (1996). Crystal structure of the *Aequorea victoria* green fluorescent protein. *Science* **273**(5280), 1392–1395.

Orrit, M., and Bernard, J. (1990). Single pentacene molecules detected by fluorescence excitation in a p-terphenyl crystal. *Phys. Rev. Lett.* **65**(21), 2716–2719.

Park, H., Hanson, G., Duff, S., and Selvin, P. (2004). Nanometer localization of single ReAsH molecules. *J. Microsc.* **216**, 199–205.

Patterson, G. H., and Lippincott-Schwartz, J. (2002). A photoactivatable GFP for selective photolabeling of proteins and cells. *Science* **297**(5588), 1873–1877.

Pierce, D. W., Hom-Booher, N., and Vale, R. D. (1997). Imaging individual green fluorescent proteins. *Nature* **388**(6640), 338.

Qu, X., Wu, D., Mets, L., and Scherer, N. F. (2004). Nanometer-localized multiple single-molecule fluorescence microscopy. *Proc. Natl. Acad. Sci. USA* **101**(31), 11298–11303.

Reck-Peterson, S. L., Yildiz, A., Carter, A. P., Gennerich, A., Zhang, N., and Vale, R. D. (2006). Single-molecule analysis of dynein processivity and stepping behavior. *Cell* **126**(2), 335–348.

Rosenberg, S. A., Quinlan, M. E., Forkey, J. N., and Goldman, Y. E. (2005). Rotational motions of macro-molecules by single-molecule fluorescence microscopy. *Accounts Chem. Res.* **38**(7), 583–593.

Rotman, B., and Papermaster, B. W. (1966). Membrane properties of living mammalian cells as studied by enzymatic hydrolysis of fluorogenic esters. *Proc. Natl. Acad. Sci. USA* **55**(1), 134–141.

Schmidt, T., Schutz, G. J., Baumgartner, W., Gruber, H. J., and Schindler, H. (1996). Imaging of single molecule diffusion. *Proc. Natl. Acad. Sci. USA* **93**(7), 2926–2929.

Schmidt, R., Wurm, C. A., Jakobs, S., Engelhardt, J., Egner, A., and Hell, S. W. (2008). Spherical nanosized focal spot unravels the interior of cells. *Nat. Meth.* **5**(6), 539–544.

Selvin, P. R., and Ha, T. (2007). Single molecule techniques: A laboratory manual. Cold Spring Harbor, New York.

Shaner, N. C., Lin, M. Z., McKeown, M. R., Steinbach, P. A., Hazelwood, K. L., Davidson, M. W., and Tsien, R. Y. (2008). Improving the photostability of bright monomeric orange and red fluorescent proteins. *Nat. Meth.* **5**(6), 545–551.

Shroff, H., Galbraith, C. G., Galbraith, J. A., White, H., Gillette, J., Olenych, S., Davidson, M. W., and Betzig, E. (2007). Dual-color superresolution imaging of genetically expressed probes within individual adhesion complexes. *Proc. Natl. Acad. Sci. USA* **104**(51), 20308–20313.

Smith, A. M., and Nie, S. (2009). Next-generation quantum dots. *Nat. Biotechnol.* **27**(8), 732–733.

Soper, S. A., Shera, E. B., Martin, J. C., Jett, J. H., Hahn, H., Nutter, L., and Keller, R. A. (1991). Single-molecule detection of rhodamine-6G in ethanolic solutions using continuous wave laser excitation. *Anal. Chem.* **63**(5), 432–437.

Svoboda, K., Schmidt, C. F., Schnapp, B. J., and Block, S. M. (1993). Direct observation of kinesin stepping by optical trapping interferometry. *Nature* **365**(6448), 721–727.
Thompson, R. E., Larson, D. R., and Webb, W. W. (2002). Precise nanometer localization analysis for individual fluorescent probes. *Biophys. J.* **82**(5), 2775–2783.
Toprak, E., and Selvin, P. R. (2007). New fluorescent tools for watching nanometer-scale conformational changes of single molecules. *Annu. Rev. Biophys. Biomol. Struct.* **36**, 349–369.
Toprak, E., Yildiz, A., Hoffman, M. T., Rosenfeld, S. S., and Selvin, P. R. (2009). Why kinesin is so processive. *Proc. Natl. Acad. Sci. USA* **106**, 12717–12722.
Vale, R. D., Reese, T. S., and Sheetz, M. P. (1985). Identification of a novel force-generating protein, kinesin, involved in microtubule-based motility. *Cell* **42**(1), 39–50.
Vale, R. D., Funatsu, T., Pierce, D. W., Romberg, L., Harada, Y., and Yanagida, T. (1996). Direct observation of single kinesin molecules moving along microtubules. *Nature* **380** (6573), 451–453.
van de Linde, S., Kasper, R., Heilemann, M., and Sauer, M. (2008). Photoswitching microscopy with standard fluorophores. *Appl. Phys. B* **93**, 725–773.
van de Linde, S., Endesfelder, U., Mukherjee, A., Schuttpelz, M., Wiebusch, G., Wolter, S., Heilemann, M., and Sauer, M. (2009). Multicolor photoswitching microscopy for subdiffraction-resolution fluorescence imaging. *Photochem. Photobiol. Sci.* **8**(4), 465–469.
Vendra, G., Hamilton, R. S., and Davis, I. (2007). Dynactin suppresses the retrograde movement of apically localized mRNA in *Drosophila* blastoderm embryos. *RNA* **13**, 1860–1867.
Warshaw, D. M., Kennedy, G. G., Work, S. S., Krementsova, E. B., Beck, S., and Trybus, K. M. (2005). Differential labeling of myosin V heads with quantum dots allows direct visualization of hand-over-hand processivity. *Biophys. J.* **88**(5), L30–L32.
Westphal, V., and Hell, S. W. (2005). Nanoscale resolution in the focal plane of an optical microscope. *Phys. Rev. Lett.* **94**, 143903.
Westphal, V., Rizzoli, S. O., Lauterbach, M. A., Kamin, D., Jahn, R., and Hell, S. W. (2008). Video-rate far-field optical nanoscopy dissects synaptic vesicle movement. *Science* **320**, 246–249.
Wiedenmann, J., Ivanchenko, S., Oswald, F., Schmitt, F., Rocker, C., Salih, A., Spindler, K. D., and Nienhaus, G. U. (2004). EosFP, a fluorescent marker protein with UV-inducible green-to-red fluorescence conversion. *Proc. Natl. Acad. Sci. USA* **101**(45), 15905–15910.
Willig, K. I., Rizzoli, S. O., Westphal, V., Jahn, R., and Hell, S. W. (2006). STED microscopy reveals that synaptotagmin remains clustered after synaptic vesicle exocytosis. *Nature* **440**(7086), 935–939.
Woehlke, G., and Schliwa, M. (2000). Walking on two heads: The many talents of kinesin. *Nat. Rev. Mol. Cell. Biol.* **1**(1), 50–58.
Yardimci, H., van Duffelen, M., Mao, Y., Rosenfeld, S. S., and Selvin, P. R. (2008). The mitotic kinesin CENP-E is a processive transport motor. *Proc. Natl. Acad. Sci. USA* **105**(16), 6016–6021.
Yildiz, A., Forkey, J. N., McKinney, S. A., Ha, T., Goldman, Y. E., and Selvin, P. R. (2003). Myosin V walks hand-over-hand: Single fluorophore imaging with 1.5-nm localization. *Science* **300**, 2061–2065.
Yildiz, A., Tomishige, M., Vale, R. D., and Selvin, P. R. (2004a). Kinesin walks hand-over-hand. *Science* **303**, 676–678.
Yildiz, A., Park, H., Safer, D., Yang, Z., Chen, L. Q., Selvin, P. R., and Sweeney, H. L. (2004b). Myosin VI steps via a hand-over-hand mechanism with its lever arm undergoing fluctuations when attached to actin. *J. Biol. Chem.* **279**(36), 37223–37226.

CHAPTER TWO

MOLECULES AND METHODS FOR SUPER-RESOLUTION IMAGING

Michael A. Thompson, Julie S. Biteen, Samuel J. Lord, Nicholas R. Conley, *and* W. E. Moerner

Contents

1. Introduction	28
2. Molecules for Super-Resolution Imaging	30
2.1. Strategies for designing and characterizing photoactivatable fluorophores	30
2.2. The azido-DCDHF class of photoswitchable fluorophores	34
2.3. Synthesis and characterization of covalent Cy3–Cy5 heterodimers	40
2.4. EYFP as a photoswitchable emitter	43
3. Selected Methods for Super-Resolution Imaging	46
3.1. PALM in live *C. crescentus* bacterial cells	46
3.2. Three-dimensional single-molecule imaging using double-helix photoactivated localization microscopy	50
Acknowledgments	55
References	55

Abstract

By looking at a fluorescently labeled structure one molecule at a time, it is possible to side-step the optical diffraction limit and obtain "super-resolution" images of small nanostructures. In the Moerner Lab, we seek to develop both molecules and methods to extend super-resolution fluorescence imaging. Methodologies and protocols for designing and characterizing fluorophores with switchable fluorescence required for super-resolution imaging are reported. These fluorophores include azido-DCDHF molecules, covalently linked Cy3–Cy5 dimers, and also the first example of a photoswitchable fluorescent protein, enhanced yellow fluorescent protein (EYFP). The imaging of protein superstructures in living *Caulobacter crescentus* bacteria is used as an example of the power of super-resolution imaging by single-molecule photoswitching to extract information beyond the diffraction limit. Finally, a new method is described for obtaining three-dimensional super-resolution information using a double-helix point-spread function.

Department of Chemistry, Stanford University, Stanford, California, USA

Methods in Enzymology, Volume 475
ISSN 0076-6879, DOI: 10.1016/S0076-6879(10)75002-3

© 2010 Elsevier Inc.
All rights reserved.

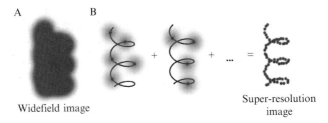

Widefield image   Super-resolution image

Scheme 2.1 *Schematic showing the key idea of super-resolution imaging of a structure by PALM.* (A) It is not possible to resolve the underlying structure in a conventional widefield fluorescence image because the fluorescent labels are in high concentration and the images overlap. (B) Using controllable fluorophores, it is possible to turn on and image a sparse subset of molecules which then can be localized with nanometer precision (black line is the underlying structure being sampled). Once the first subset of molecules photobleaches, another subset is turned on and localized. This process is repeated and the resulting localizations summed to give a super-resolution image of the underlying structure.

authors' research. The molecules include two classes of small-molecule labels that can be photoactivated or photoswitched between emissive and dark states, as well as the first-reported photoswitchable fluorescent protein, enhanced yellow fluorescent protein (EYFP). The methods described include protocols for super-resolution studies in live bacteria and a novel method for obtaining three-dimensional super-resolution image information using a microscope with a double-helix PSF.

2. MOLECULES FOR SUPER-RESOLUTION IMAGING

2.1. Strategies for designing and characterizing photoactivatable fluorophores

Taking into account both the localization precision (Ober *et al.*, 2004; Thompson *et al.*, 2002) and the Nyquist–Shannon sampling theorem (Nyquist, 1928; Shannon, 1949), the best emitters for photoactivation and localization-based super-resolution imaging will maximize the number of well-localized unique molecules per area per time (Shroff *et al.*, 2008). To achieve this, good photoactivatable fluorophores must be bright, emit many photons, densely label the sample, and have a high contrast between on and off states. In addition, the probe must be easily photoactivated to avoid cell damage from short-wavelength illumination.

One of the most important parameters for single-molecule fluorophores is the number of photons each molecule emits before photobleaching ($N_{tot,e}$). Scaling inversely with $N_{tot,e}$ is the photobleaching quantum yield (Φ_B), or the probability of bleaching with each photon absorbed (see below for equations). A very low value of Φ_B corresponds to not only a long-lived fluorophore but also to very high precision in localizing the point emitter (Ober *et al.*, 2004; Thompson *et al.*, 2002).

Besides the number of photons emitted, fluorophore labeling density is another important variable that determines the ultimate resolution. Because super-resolution imaging by switching point sources is effectively a sampling of the true underlying structure, there are well-known restrictions on the labeling density (or spatial sampling frequency) for a correct reproduction of the structure from the samples at a given resolution. For instance, the Nyquist–Shannon sampling theorem (Biteen *et al.*, 2008; Nyquist, 1928; Shannon, 1949; Shroff *et al.*, 2007, 2008) requires that the fluorophores label the structure of interest at a frequency (number per spatial distance) that is at least two molecules per the desired spatial scale to be resolved. This requirement extends further as well, in that labels must actually be localized to the same average density. This criterion adds a further restriction on the emitters in that the turn-on ratio (i.e., the contrast between the bright and dark states of the molecule) must be very high, lest the many weakly emitting "off" molecules in a diffraction-limited spot drown out the signal from the one "on" molecule.

2.1.1. Photobleaching and photoconversion quantum yields

The photobleaching quantum yield may be measured on a bulk sample of the activated emitters, and is defined as the probability of photobleaching per photon absorbed or the ratio of the bleaching rate R_B to the rate of absorbing photons R_{abs}:

$$\Phi_{B(P)} = \frac{R_{B(P)}}{R_{abs}} = \frac{1}{\tau_{B(P)} R_{abs}} = \frac{1}{\tau_{B(P)} \sigma_\lambda I_\lambda \left(\frac{\lambda}{hc}\right)}, \quad (2.1)$$

where $\tau_{B(P)}$ is the decay constant in an exponential fit to the decay curve, the absorption cross-section σ_λ is related to the molar absorption coefficient ε_λ by the equation $\sigma_\lambda = 2303\varepsilon_\lambda/N_A \approx 10^{-16}$ cm^2 for good emitters, I_λ is the irradiance (intensity) at the sample, λ is the excitation wavelength, h is Planck's constant, and c is the speed of light. Bulk photobleaching decay curves are often not single exponential, and the average decay constant for a two-exponential fit, $F = \sum_{i=1}^{n=2} \alpha_i e^{(-t/\tau_i)}$, is given by

$$\bar{\tau} = f_1 \tau_1 + f_2 \tau_2 = \frac{\alpha_1 \tau_1^2 + \alpha_2 \tau_2^2}{\alpha_1 \tau_1 + \alpha_2 \tau_2}, \quad (2.2)$$

where $f_i = \alpha_i \tau_i / \sum_j \alpha_j \tau_j$ is the fractional area under the multiexponential curve (Lakowicz, 2006). (Some other authors use $t_{90\%}$, the irradiation time in seconds for 90% conversion to product, as a more practical measure than the decay constant $\bar{\tau}$(Adams *et al.*, 1989); so values should be compared carefully.)

Photoconversion from a precursor fluorogen to the emissive form can be monitored in bulk by measuring changes over time in absorbance or emission of the reactant or photoproduct of interest. The quantum yield of photoconversion, Φ_P, is defined in Eq. (2.1), with τ_P as the average decay constant from the exponential fit of the decaying absorption values for the starting material. Note that Φ_P is the probability that the starting material will react for each photon absorbed. A fraction of those molecules then become fluorescent because the photoreaction yield to fluorescent product is usually less than unity.

2.1.2. Effective turn-on ratio

This section describes an accurate and experimentally convenient method for measuring the turn-on ratio of photoactivatable fluorophores. The goal of this experiment is to measure how many times brighter an activated molecule is than the preactivated fluorogen. Thus, the limit we care about is when the intensity from one bright molecule I_{on} equals the intensity from n_{off} dark fluorogens (i.e., when $I_{on} = n_{off}I_{off}$):

$$R = \frac{I_{on}}{I_{off}} = \frac{n_{off}I_{off}}{I_{off}} = n_{off}. \quad (2.3)$$

Assuming that every dark molecule becomes fluorescent is rarely correct (i.e., $n_{on} < n_{off}$ is the common situation). One could measure R by averaging over many single molecules; however, this would select only the fluorogens that become fluorescent, and would be artificially inflated.

Alternatively, it is more accurate to measure an effective turn-on ratio that takes into account the reaction yield. In a bulk experiment,[1] integrate the background-subtracted intensities over a large region before activation (S_{off}) and after activation to steady-state (S_{on}). Not all copies of the fluorogen convert to the fluorescent species, as the simple ratio R assumes above; the overall reaction yield p is almost always less than unity, because there are often nonfluorescent photoproducts. Therefore, the total number of emitters that will turn on is the reaction yield times the number of precursor molecules: $n_{on} = pn_{off}$. The ratio of the background-subtracted signals in a

---

[1] For bulk experiments, the fluorogens are doped into a film (e.g., polymer, gelatin, agarose) at approximately 1–2 orders of magnitude higher concentration than in single-molecule experiments, but otherwise are imaged under similar conditions. This measurement assumes that one is working in a concentration regime where the emitters are dense enough to obtain a sufficient statistical sampling of the population but separated enough to avoid self-quenching or excimer behavior.

bulk experiment gives the effective turn-on ratio R_{eff}, which is the experimentally relevant parameter:

$$R_{\text{eff}} = \frac{S_{\text{on}}}{S_{\text{off}}} = \frac{n_{\text{on}} I_{\text{on}}}{n_{\text{off}} I_{\text{off}}} = \frac{p n_{\text{off}} I_{\text{on}}}{n_{\text{off}} I_{\text{off}}} = p \frac{I_{\text{on}}}{I_{\text{off}}} = pR = p n_{\text{off}} = n_{\text{on}}. \quad (2.4)$$

The value R_{eff} corresponds directly to the maximum number of molecules n_{on} that one could localize in a diffraction-limited spot before the aggregate signal ($n_{\text{off}} I_{\text{off}}$) of all the dark fluorogens required for that number of localizations equals the signal from one emitting molecule I_{on}.

Resolution on the order of nanometers or tens of nanometers requires labels with densities of many thousands of localizations per μm^2 and therefore turn-on ratios in the hundreds or thousands. For example, if we assume the diffraction limit to be approximately 250 nm, the area of the diffraction-limited spot is about 50,000 nm^2. If $R_{\text{eff}} = 325$, there is a maximum of 325 localizations in each diffraction-limited spot, so the distance between each localization is approximated by $\sqrt{50,000\,\text{nm}^2/325} = 12$ nm. The Nyquist–Shannon theorem (Nyquist, 1928; Shannon, 1949) requires a sampling at least of twice the desired resolution, limiting the resolution to about 25 nm (in two dimensions). For three dimensions, the excitation volume in z is much larger than in x–y. Therefore, much higher contrast ratios and labeling densities are required for high-resolution imaging in three dimensions. For more details, see the supplemental material of Shroff *et al.* (2008).

As a side note, the measured value of R_{eff} should be considered a lower limit, because any molecules already in the "on" state before activation (preactivated molecules) contribute to the signal in the frames before activation, thus lowering the measured value of the parameter. The fraction q of preactivated molecules should be kept low by protecting the fluorophore stock solution and samples from room lights and by pretreating the sample with the imaging wavelength to return preactivated molecules to the "off" state (prebleaching the sample) if possible. Regardless, some preactivation will inevitably occur. We can calculate the effect preactivation has on measuring R_{eff} by including signal from preactivated molecules in the dark measurement:

$$\begin{aligned} R_{\text{eff,preact}} &= \frac{S_{\text{on}}}{S_{\text{off,preact}}} = \frac{n_{\text{on}} I_{\text{on}}}{n_{\text{off}} I_{\text{off}} + q n_{\text{off}} I_{\text{on}}} \\ &= \frac{n_{\text{on}} I_{\text{on}}}{n_{\text{off}} I_{\text{off}} \left(1 + \frac{q n_{\text{off}} I_{\text{on}}}{n_{\text{off}} I_{\text{off}}}\right)} = \frac{pR}{1 + qR} = \frac{R_{\text{eff,true}}}{1 + qR}. \end{aligned} \quad (2.5)$$

From Eq. (2.5), the multiplicative correction factor to convert from measured to true effective turn-on ratio is $(1 + qR)$. Even 0.1% preactivation could artificially deflate the measured value by half (assuming the $R = I_{\text{on}}/I_{\text{off}}$ of one isolated molecule is 1000). Therefore, minimizing preactivation (or, alternatively, maximizing prebleach) before measuring the effective turn-on ratio can increase the value of R_{eff} such that it approaches the true ratio.

Table 2.1 provides a summary of data for some of the photoactivatable and photoswitchable fluorophores in current use for super-resolution imaging. Where possible, estimated values were extracted from the literature, but this was not possible in many cases. We kindly encourage researchers to report values determined as described above to enable consistent, fair comparisons between the systems.

2.1.3. Single-molecule photon-count analysis

Single-molecule movies can be used to extract the total number of detected photons before photobleaching, where all the photons (minus background) contributing to a single-molecule spot are spatially and temporally integrated. The camera's electron-multiplication gain and conversion gain defined as the number of photoelectrons per A to D converter count are used to convert digital A-to-D counts back to photoelectrons, which are detected photons. Results are plotted using the probability distribution of photobleaching: $P_N = m_N/M$, the ratio of the number of single molecules m_N surviving after emitting a given number of photons N, relative to the total number of molecules M in the measurement set (Molski, 2001). In other words, the probability for any value of photons emitted is determined by counting the number of single molecules that emitted that number of photons or more divided by the total number of molecules in the population: if 50 molecules emitted at least 500,000 photons before photobleaching, and the other 150 bleached before emitting 500,000, the probability $P_{500,000} = m_{500,000}/M = 50/200 = 0.25$. The value of P_N is plotted for each value of N, and fitted using one or two exponential decays. The decay constant can be extracted from a fit as in Eq. (2.2). The probability-distribution approach for determining average photons detected avoids any artifact from choice of bin size, and gives comparable results to histogramming.

To calculate the number of *emitted* photons $N_{tot,e}$, the measured value of detected photons is corrected using the collection efficiency of the setup ($D = \eta_Q F_{coll} F_{opt} F_{filter}$), which is the product of the camera quantum efficiency η_Q, the angular collection factor F_{coll} determined by the objective NA, the transmission factor through the objective and microscope optics F_{opt}, and the transmission factor through the various filters F_{filter}, respectively (Moerner and Fromm, 2003). For wide-field imaging, D values are typically 5–15%.

2.2. The azido-DCDHF class of photoswitchable fluorophores

For several years, in collaboration with the laboratory of Robert J. Twieg at Kent State University, we have been exploring the properties of push–pull fluorophores containing an amine donor covalently linked to an electron

Table 2.1 Photophysical properties of various photoswitchable molecules, including whether the fluorophore can be cycled between bright and dark states multiple times, absorption and emission peaks, and molar absorption coefficient, fluorescence quantum yield, photoconversion quantum yield, turn-on ratio, photobleaching quantum yield, and total photons emitted

	Reversible?[a]	$\lambda_{abs}/\lambda_{em}$ (nm)	ε_{max} (M^{-1} cm^{-1})	Φ_F	Φ_P	Turn-on ratio[b]	Φ_B	$N_{tot,e}$
DCDHF-V-P-azide (Lord et al., 2008, 2010)	No	570/613	54,100	0.025–0.39[c]	Good (0.0059)	Excellent (325–1270)[d]	4.1×10^{-6}	2.3×10^6
DCDHF-V-PF$_4$-azide (Lord et al., 2010; Pavani et al., 2009)	No	463/578	20,000	0.0062+	Very good (0.017)		9.2×10^{-6}	
DCM-azide (Lord et al., 2010)	No	456/599	31,100	0.18	Excellent (0.085)		6.2×10^{-6}	
Cy3/Cy5 + thiol (Bates et al., 2005; Conley et al., 2008; Huang et al., 2008b; Schmidt et al., 2002)[e]	Yes	647/662	200,000	0.18	Very good	Excellent (≤ 1000)[f]		~670,000
PC-RhB[g]	Yes	552/580	110,000	0.65	Moderate	Moderate		~600,000
EYFP (Biteen et al., 2008; Tsien, 1998; Dickson et al., 1997; Harms et al., 2001; Patterson et al., 2001; Schmidt et al., 2002; Biteen et al., 2009)[h,i]	Yes	514/527	83,400	0.61	Moderate (1.6×10^{-6})		5.5×10^{-5}	~140,000

(continued)

acceptor group, such as a dicyanomethylenedihydrofuran (DCDHF). Recently, we created a novel class of photoactivatable single-molecule fluorophores by replacing the amine with an azide, which is not a donor. With long-wavelength pumping, the azido fluorogenic molecules are dark, but applying low-intensity activating blue light photochemically converts the azide to an amine, which restores the donor–acceptor character, the red-shifted absorption, and the bright fluorescent emission.

2.2.1. Design
Photoactivatable (or "photocaged") donor-conjugated network-acceptor (or "push–pull") chromophores can be designed by disrupting the charge-transfer band (Doub and Vandenbelt, 1947; Stevenson, 1965), and therefore significantly blue-shifting the absorption to the extent that it is no longer resonant with the imaging laser. In these cases, photoactivation requires a photoreaction that converts the disrupting component to a substituent that is capable of donating electrons into the chromophore's conjugated network.

An azide can be the disrupting substituent since azides are weakly electron-withdrawing (Hansch et al., 1991). Recovering fluorescence is possible because aryl azides are known to be photolabile. The photochemistry of aryl azides has been studied extensively (Schriven, 1984), and the photoreaction most often reported involves the loss of dinitrogen and rearrangement to a seven-membered azepine heterocycle, which is unlikely to act as a strong electron-donating group. Fortunately, Soundararajan and Platz (1990) demonstrated that electron-withdrawing substituents on the benzene can stabilize the nitrene intermediate and promote the formation of amine and azo groups instead of rearranging to the azepine. Because push–pull chromophores inherently contain a strong electron-withdrawing substituent, an azido push–pull molecule should photoconvert to the fluorescent amino version upon irradiation with activating light that is resonant with the blue-shifted absorption. Note that the photoactivation of these azido fluorogens is an irreversible chemical reaction, so reversible photoswitching is not possible. This can be a drawback in cases which require cycling between bright and dark forms, such as in STED or when using PALM to image dynamics. However, when imaging static structures, a probe that activates, emits millions of photons, then disappears permanently is desired. Otherwise, fluorophores turning back on means that some portions of the structure are localized over and over, thus complicating the subsequent image analysis and reconstruction. Photoactivation of azido DCDHFs as well as other push–pull chromophores has been previously demonstrated (Lord et al., 2008, 2010; Pavani et al., 2009). Detailed syntheses by colleagues in the Twieg laboratory have been reported in previous papers and in supporting online material (Lord et al., 2008, 2010; Pavani et al., 2009).

2.2.2. Photoreaction characterization

Bulk chemical studies revealed the photoproducts of DCDHF-V-P-azide after illumination with blue or UV light. Full experimental details have been published elsewhere (Lord et al., 2008). Photoproducts identified as DCDHF-V-P-amine and DCDHF-V-P-nitro were confirmed using column chromatography, NMR, and HPLC-MS; the spectra of the photoproducts matched those of pure, synthesized versions. Figure 2.1 enumerates several possible photoreactions of the azido push–pull fluorogens.

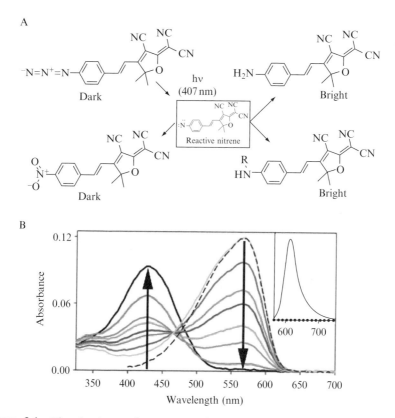

Figure 2.1 *Photochemistry and spectroscopy of an azido DCDHF fluorogen.* (A) Various products resulting from photochemical conversion of an azido DCDHF fluorogen. (B) Absorption curves in ethanol (bubbled with N_2) showing photoactivation of 1 (λ_{abs} = 424 nm) over time to fluorescent product 2 (λ_{abs} = 570 nm). Different colored curves represent 0, 10, 90, 150, 240, 300, 480, and 1320 s of illumination by 3.1 mW cm^{-2} of diffuse 407-nm light, where the arrows show the direction of increasing time. The sliding isosbestic point may indicate a build-up of reaction intermediates. Dashed line is the absorbance of pure, synthesized 2. (Inset) Dotted line is weak preactivation fluorescence of 1 excited at 594 nm; solid line is strong postactivation fluorescence resulting from exciting 2 at 594 nm, showing >100-fold turn-on ratio. (Adapted and reproduced with permission from Lord et al., 2008. Copyright 2008 American Chemical Society.)

Previous studies have reported that the DCDHF class of single-molecule fluorophores emits millions of photons and have low values for Φ_B (Lord et al., 2009; Willets et al., 2005). Therefore, azido DCDHFs are attractive as photoactivatable probes for PALM. For instance, single molecules of DCDHF-V-PF$_4$-azide can be photoactivated and localized to less than 20 nm standard deviation in all three dimensions (Pavani et al., 2009).

As shown in Table 2.1, adding electron-withdrawing fluorine substituents to the phenyl group (DCDHF-V-PF and DCDHF-V-PF$_4$) has little effect on the photostability parameter Φ_B. DCM is also a strong single-molecule fluorophore, with a Φ_B of 6.2×10^{-6}, which is only slightly worse than most DCDHFs. Moreover, DCM has a higher fluorescence quantum yield in solution, and thus is more likely to be bright throughout a sample (not just in rigid environments, which is a feature of DCDHFs).

We determined that the lower limit to the effective turn-on ratio for DCDHF-V-P-azide in PMMA is $R_{eff} = 325 \pm 15$. This effective turn-on ratio corresponds to thousands of localizations per μm^2 and a Nyquist–Shannon limit on the resolution of approximately 25 nm in two dimensions. The upper limit is the turn-on ratio of a single molecule that does indeed become fluorescent, $R = 1270 \pm 500$ (Lord et al., 2010).

A common trait among the azido push–pull fluorogens is their high sensitivity to photoactivating illumination (as measured by the photoconversion quantum yield Φ_P). For some fluorogens, only tens or hundreds of photons need be absorbed before the fluorogen converts to a fluorescent product. This is important because high doses of blue or UV light can kill cells or alter phenotype. This benefit comes with the requirement that sample preparation be carried out in the dark or under red light to avoid preactivation. Azido DCDHF fluorogens can be activated and form bright fluorophores in a live-cell environment. Figure 2.2 shows azido-DCDHF fluorogens activated with low amounts of blue light in live Chinese Hamster Ovary (CHO) cells. The resulting fluorophores are bright in the aqueous environment of the cell. In the data from Fig. 2.2, the cells were incubated with azido DCDHF fluorogens which penetrate the cell membrane and nonspecifically label the interior of the cell. The azido push–pull chromophores meet many of the critical requirements for high-resolution PALM: Several emit millions of photons before irreversibly photobleaching, are photoconverted with high quantum efficiency, exhibit high turn-on ratios, and possess moderate molar absorption coefficients and quantum yields.

2.3. Synthesis and characterization of covalent Cy3–Cy5 heterodimers

In its original implementation, the STORM method utilized the Cy3/Cy5 photoswitching system, which requires thiol and oxygen scavenger additives (Rust et al., 2006). In that system, a Cy5 dye molecule is optically

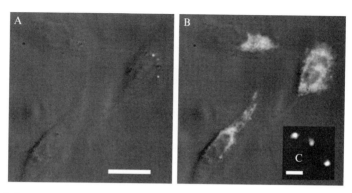

Figure 2.2 *Photoactivation of the azido DCDHF fluorogen in live mammalian cells.* (A) Three Chinese Hamster Ovary cells incubated with azido DCDHF fluorogen are dark before activation. (B) The fluorophore lights up in the cells after activation with a 10-s flash of diffuse, low-irradiance (0.4 W cm^{-1}) 407-nm light. The white-light transmission image is merged with the fluorescence images, excited at 594 nm (~1 kW cm^{-1}). Scalebar: 20 µm. (C) Single molecules of the activated fluorophore in a cell under higher magnification. Scalebar: 800 nm. (Adapted with permission from Lord et al., 2008. Copyright 2008 American Chemical Society.)

excited in the presence of a thiol until it enters into a meta-stable dark state. Recovery of the Cy5 fluorescent state is induced by low-intensity excitation of a proximal Cy3 fluorophore; the percentage of Cy5 emitters restored to the fluorescent state can be controlled by the intensity and duration of the Cy3 excitation pulse. Because the Cy5 reporter molecule must be located within ~2 nm of the Cy3 activator molecule for efficient reactivation to occur, structures are typically labeled using heterolabeled antibodies (step 1, *vide infra*). The labeling ratio of activator to reporter molecules on each antibody (~3:1) is controlled during labeling by employing different concentrations of the reactive activator and reactive reporter fluorophores (Bates et al., 2007). The probabilistic nature of this labeling strategy results in an ill-defined photoswitching system that exhibits undesirable heterogeneity in switching rates. To remedy this shortcoming, we recently reported the preparation and bulk- and single-molecule characterization of a Cy3–Cy5 covalent heterodimer (Conley et al., 2008).

2.3.1. Synthesis

For preparation of Cy3–Cy5 covalent heterodimers **4** and **5** (Fig. 2.3A), commercially available, reactive cyanine dyes **1–3** (Conley et al., 2008) are used; hydrazides and NHS esters are known to be cross-reactive (Al Jammaz et al., 2006). Cy3-NHS ester **1** and Cy5-hydrazide **2** are coupled in DMSO/triethylamine at 50 °C to produce the unreactive Cy3–Cy5 heterodimer **4**. Similarly, the amine-reactive Cy3–Cy5 NHS ester **5** is prepared by coupling Cy3-bis(NHS ester) **3** with **2**. Both **4** and **5** are readily purified by silica gel column chromatography using a dichloromethane/methanol eluent, and isolated in yields of approximately 75% and 25%, respectively.

2.3.2. Bulk photophysical properties

As shown in Fig. 2.3, the absorption spectra of **4** and **5** contain the characteristic Cy3 and Cy5 absorbance peaks in a ratio that corresponds approximately to the molar absorptivity ratio of the two dyes (0.6:1). Optical excitation of the Cy3 in **4** or **5** produces considerable Förster resonance energy transfer (FRET), confirming the covalent linkage in

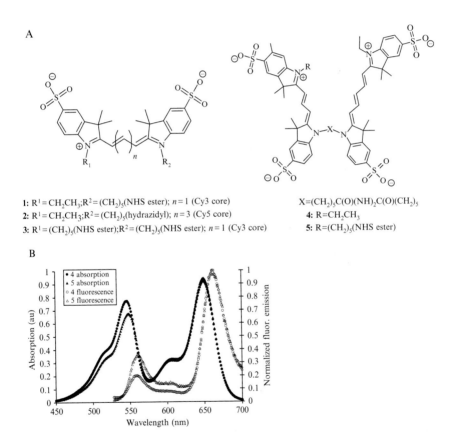

Figure 2.3 *Synthesis and bulk characterization of covalently linked Cy3–Cy5 dimers.* (A) Structures of reactive cyanine dyes and covalent heterodimers. (B) Absorption (solid) and fluorescence emission (hollow, λ_{ex}= 516 nm) spectra of Cy3–Cy5 covalent heterodimers 4 and 5 (in water; 3.7 μM for absorption; 37 nM for fluorescence) before photodarkening. (Adapted and reproduced with permission from Conley *et al.*, 2008. Copyright 2008 American Chemical Society.)

each heterodimer. The fluorescence lifetime of the Cy3 donor in **4** (0.15 ± 0.03 ns) is shorter than the lifetime of monomeric Cy3 (0.254 ± 0.007 ns), whose measured value reproduces those found in the literature (Los and Wood, 2007; Los et al., 2008). From these data, the FRET efficiency of **4** is determined to be 0.41, where the relatively low value is possibly due to nonoptimal orientation of the transition dipoles.

2.3.3. Single-molecule behavior

Bovine serum albumin (BSA) was sparsely labeled with NHS ester–Cy3–Cy5 **5** through its lysine residues. After immobilization onto a glass cover slip, single-molecule photoswitching of **5**-labeled BSA was achieved in the presence of 2-mercaptoethanesulfonate and a glucose oxidase oxygen scavenging system, as shown in Fig. 2.4A and B. The single-molecule photoswitching properties of **5** are characterized by controllable reactivation, varying "on" times, and occasional spontaneous recovery from a long-lived dark state.

The utility of **5** in STORM super-resolution imaging was demonstrated by covalently attaching it to free amines on the surface of *Caulobacter crescentus* cells, which contain a sub-diffraction-limited appendage, known as a stalk. While the stalk cannot be easily observed in white-light microscopy, it was successfully imaged with 30-nm resolution using STORM (Fig. 2.4C). We anticipate that Cy3–Cy5 covalent heterodimers will eventually replace more cumbersome methods for achieving Cy3/Cy5 proximity in the super-resolution imaging of biological systems; indeed, since our initial report, other reactive AlexaFluor and CyDye covalent heterodimers have been described (Huang et al., 2008b).

2.4. EYFP as a photoswitchable emitter

The use of EYFP as a photoswitchable emitter vastly expands the number of biological specimens immediately available for super-resolution imaging. In most reported PALM imaging, the photoactivatable fluorescent protein has been selected from various sophisticated constructs such as PA-GFP, Dronpa, Kaede, tdEosFP, Dendra2, rsFastLime, PA-mCherry, and rsCherryRev (Betzig et al., 2006; Geisler et al., 2007; Niu and Yu, 2008; Stiel et al., 2008; Subach et al., 2009). However, single immobilized and apparently bleached yellow FPs (S65G, S72A, T203Y or S65G, S72A, T203F) were shown to reactivate with violet light more than 10 years ago (Dickson et al., 1997), and the closely related enhanced yellow fluorescent protein EYFP (S65G, V68L, S72A, T203Y) was recently used for live-cell PALM imaging of the *C. crescentus* structural protein MreB (Biteen et al., 2008). EYFP is widely used for routine fluorescent protein fusions due to the absence of physiological perturbations such as agglomeration and mislocalization, and N-terminal EYFP-MreB fusions have been previously shown to be

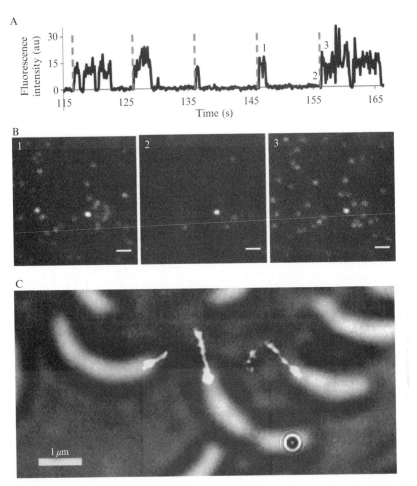

Figure 2.4 *Photoswitching behavior of and super-resolution imaging using the Cy3–Cy5 covalent dimer.* (A) Representative single-molecule fluorescence time trace of 5-labeled bovine serum albumin showing reactivation cycles 12–16, denoted by the dashed lines. (B) Fluorescence images at times 1, 2, and 3 corresponding to the times labeled in panel (A). Scale bar, 1 μm. (C) Super-resolution fluorescence image of *C. crescentus* stalks with 30 nm resolution superimposed on a white-light image of the cells. The *C. crescentus* cells were incubated in 4 μM of Cy3–Cy5 NHS ester for 1 h and then washed five times before imaging to remove free fluorophores. The data were acquired over 2048 100-ms imaging frames with 633 nm excitation at 400 W cm^{-2}. After initial imaging and photobleaching of the Cy3–Cy5 dimers, the molecules were reactivated every 10 s for 0.1 s with 532-nm light at 10 W cm^{-2}. (Adapted and reproduced with permission from Conley *et al.*, 2008. Copyright 2008 American Chemical Society.)

functional in *C. crescentus* and other bacteria, making this a physiologically relevant system (Carballido-López and Errington, 2003; Figge et al., 2004; Gitai et al., 2004). Furthermore, single-molecule imaging of EYFP-MreB with 514-nm excitation has previously shown that this fluorescent protein is a bright single-molecule emitter for live-cell imaging (Kim et al., 2006).

Figure 2.5 shows EYFP photoreactivation in live cells (Biteen et al., 2008). Figure 2.5A shows a single imaging frame where the fluorescence image is superimposed over a negative-contrast white-light transmission image of the *C. crescentus* cell. Two single EYFP-MreB molecules can be identified in panel A, which was acquired after the initial bleaching step. After further imaging with 514-nm light, all fluorophores had bleached, as observed in panel B. EYFP reactivation was achieved after this initial bleaching step by administering a 2-s dose of 407-nm laser illumination.

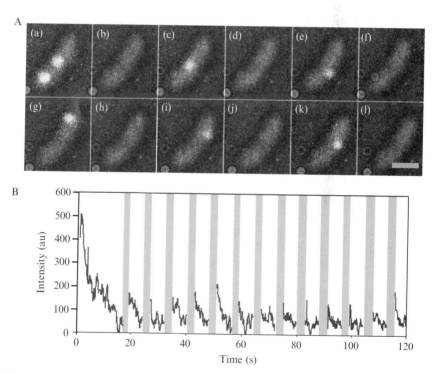

Figure 2.5 *Reactivation of EYFP-MreB fusions in live C. crescentus cells.* (A) Single 100-ms acquisition frames showing isolated EYFP-MreB molecules (a, c, and e) upon photoactivation and no single-molecule emission (b, d, f) after photobleaching. The spot in the bottom left of each image is an imaging artifact. (B) Bulk reactivation of a sample of 22 cells. The grey lines indicate 2-s pulses of 407 nm light. (Adapted and reproduced with permission from Biteen et al., 2008. Copyright 2008 Nature Publishing Group.)

This pulse length, as well as the reactivation intensity of 10^3–10^4 W cm^{-2}, was chosen such that, at most, one EYFP molecule was reactivated in each diffraction-limited region and a different subensemble of EYFP molecules was activated in each cycle of this process.

In Fig. 2.5B, the total emission intensity from 22 live *C. crescentus* cells expressing EYFP-MreB is displayed as a function of time. After an initial bleaching period, flashes of 407-nm activation energy were applied between imaging cycles. These reactivation cycles were used to calculate the photoreactivation quantum yield of the EYFP fluorophore. The measured relationship between activation time and percent reactivation can be plotted and is quasi-linear. In measurements of EYFP-MreB in live *C. crescentus* cells, 370 photons are absorbed by each EYFP molecule per second of the activation pulse. From the slope of the plot, a reactivation quantum yield of 1.6×10^{-6} was found for EYFP (Biteen *et al.*, 2009). This number is on the same order of magnitude as the activation quantum yield for PA-GFP, and only 1 order of magnitude smaller than the photoswitching quantum yield of the highly engineered protein, tdEos (see Table 2.1). EYFP can therefore be viewed as a useful photoswitchable fluorophore for super-resolution imaging.

3. SELECTED METHODS FOR SUPER-RESOLUTION IMAGING

3.1. PALM in live *C. crescentus* bacterial cells

Recently, photoreactivation and single-molecule imaging of EYFP were applied to image superstructures of the bacterial actin protein MreB in live *C. crescentus* cells with sub-40-nm resolution (Biteen *et al.*, 2008). The experiments were unique in that they used the natural treadmilling motion of MreB monomers to increase the number of localizations and thus improve the resolution of the measurements. This is an example of the general principle that prudent use of live-cell dynamics can overcome limitations on labeling concentrations to obtain higher resolution. While EYFP is used in the example shown here, the techniques and analysis can be applied to any photoswitchable or photoactivatable fluorophore.

3.1.1. Sample preparation

C. crescentus cells expressing 100–1000 copies of an N-terminal EYFP-MreB fusion in a background of unlabeled MreB molecules are prepared on an agarose pad for imaging as described in references (Biteen *et al.*, 2008; Kim *et al.*, 2006). The following is a protocol for preparing samples on an agarose pad that helps alleviate axial drift resulting from drying and shifting of the agarose pad.

1. A solid medium consisting of 2.5% agarose in M2G buffer is heated until mobile, but before reaching a rolling boil.
2. 1 ml of the agarose solution is sandwiched between two plasma-etched 30 mm × 50 mm #1 glass coverslips (typically 175 μm thick).
3. 1 ml of cells growing in log phase are spun at 13,400 rpm for 90 s.
4. The cell suspension is washed 3× with clean M2G buffer.
5. The cells are pelleted and resuspended in 0.1 ml of clean M2G.
6. The top coverslip is removed from the agarose pad and 1 μl of cells with 0.1 μl of Tetraspeck bead solution (diluted 20× from commercial stock) is deposited onto the pad.
7. A plasma etched glass coverslip is placed on the cells, forming the imaging interface.
8. The nonimaging coverslip is removed and replaced with a small 25 mm×25 mm #1 coverslip, excess agarose is cut away
9. The edge of the agarose and glass is sealed with melted paraffin wax.

Preparing the sample this way minimizes axial drift caused by drying and shifting of the agarose. The wide-field, single-molecule epifluorescence microscopy setup has been fully described previously (Biteen et al., 2008, 2009; Moerner and Fromm, 2003).

3.1.2. Image reconstruction

Since a major part of the super-resolution imaging process consists of the PSF fitting and the algorithm used to display the final image, details of this process are presented here. For each frame, the emitters are identified, their positions localized, and their fit evaluated. Once this is accomplished, all of the localized molecule positions are summed to produce a final super-resolution image. For each imaging frame, putative single-molecule emitters are identified by: (a) setting a threshold equal to the image average intensity plus the standard deviation of intensities, (b) recording the position of the highest point above the threshold, (c) excluding a 1 × 1-μm region about the point, and (d) repeating steps (b) and (c) until no points are identified above the threshold. Each putative emitter image is fit with a symmetric 2D Gaussian function using the MATLAB nonlinear least squares regression function, nlinfit. The use of a Gaussian rather than an Airy disk is justified in these experiments because the background noise in the image makes the tails of the Airy disk hard to quantify. To compensate for $x-y$ stage drift during imaging, the positions of the EYFP-MreB molecules are determined relative to the position of fixed fiduciaries (Tetraspeck fluorescent microspheres). Here, it is important to note that each localization event, and thus each determination of a position along the polymeric MreB structure, comes from a single 100-ms frame. Thus, multiple position determinations can hail from a single MreB-EYFP molecule as it treadmills through an MreB filament, allowing one to use a smaller real concentration of fluorescent protein fusions to obtain a large number of localizations (up to several hundred per cell).

Localizations are rejected if the fit returns a negative amplitude, a standard deviation greater than 320 nm, or a 95% confidence interval in the fit of the molecular position which is greater than the standard diffraction limit. The first and third conditions reflect a failed fit. The second condition requires that the subimage is a single diffraction-limited spot and therefore rejects both out-of-focus fits and subimages that likely contain more than one EYFP molecule. Furthermore, the MATLAB statistics toolbox function nlparci is used to determine the confidence intervals for all fit parameters, and the localization are thrown out if the 96% confidence intervals of the center positions are greater than x_0 or y_0 themselves. All remaining (successful) fitted positions are summed to output an image, where each position is plotted as a unit-area symmetric 2D Gaussian profile centered at the position found by the fit and with the standard deviation of each Gaussian equal to the average 96% confidence interval for the center position of all good fits. (The theoretical photon-limited localization accuracy for a point emitter can also be determined analytically for the symmetric 2D Gaussian PSF as a function of number of photons recorded, the background level, and the pixel size (Thompson et al., 2002).) We found the 96% confidence interval to be a conservative value relative to this theoretical limit. As long as the theoretical limit is not violated, the selection of the confidence interval can be altered somewhat to prevent excessively punctate images when the labeling or sampling concentration is limited.

Though most fitted positions come from a single localization event, especially in the case of the time-lapse images in Fig. 2.6, oversampling occurs when the one fluorophore is localized in the same confidence-interval-limited region in multiple frames. Such oversampling can be treated in a postprocessing step in which localizations within a set radius (e.g., 20 nm) are considered to correspond to the same molecule. In these cases, the photons from the neighboring fits are combined to yield a single molecule in the output image with position and fit accuracy corresponding to that of the reconstituted emission spot. Again, each single molecule is plotted as a unit-area Gaussian profile with fixed standard deviation given by the average statistical error in localization (96% confidence interval) of the center of all successful fits.

Figure 2.6 shows super-resolution images of the EYFP-MreB protein superstructure in *C. crescentus* (Biteen et al., 2009), where we have identified the presence of two different structures at distinct stages in the cell cycle: a regularly spaced band-like arrangement of MreB molecules that suggests a helix in the stalked cell in Fig. 2.6B–D, and a ring of MreB molecules at the cell mid-plane in the predivisional cell in Fig. 2.6A and E. The *C. crescentus* cells are ~ 2 μm in length and ~ 0.5 μm in diameter. Since the depth of field in our system is similar to the thickness of the cells, the super-resolution images represent a 2D projection of the MreB structures. It is worth noting

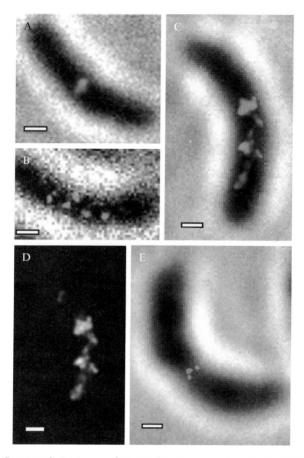

Figure 2.6 *Super-resolution images of MreB in live* C. crescentus *cells*. (A–B) Images taken using standard PALM (C–D) images taken using time-lapse imaging to obtain higher labeling density using. (A) Image of MreB forming a midplane ring in a predivisional cell. (B) Banded MreB structure in a stalk cell. (C) Quasi-helical MreB structure at 40 nm resolution observed using time-lapse PALM. (D) Structure in panel (C) displayed without white-light image in order to highlight the continuity of the structure. (E) Time-lapse PALM image of MreB midplane ring in a predivisional cell. (Adapted and reproduced with permission from Biteen *et al.*, 2008. Copyright 2008 Nature Publishing Group.)

that these images show a much smaller field of view than most images of mammalian cells. In addition, bulk epifluorescence images of EYFP-MreB under these conditions in stalked cells showed no structure whatsoever.

3.1.3. Time-lapse imaging

Understanding super-resolution features derived from many image acquisitions requires careful consideration of the emitter photophysics and the dynamics of the underlying structure. Indeed, this is a general problem with PALM methods, how to distinguish static from dynamic structure, and must be considered carefully in each case. Figure 2.6A and B was attained with 15 cycles of a 2-s 407-nm activation pulse followed by 5 s of fluorescence image acquisition (fifty 100-ms frames). MreB filaments have been previously found in single-molecule studies to have an average length of 390 nm and to consist of single MreB units treadmilling along polymeric MreB filaments at a rate of 6.0 nm (1.2 monomer additions) per second (Kim et al., 2006). Since the MreB molecules travel slowly along polymer chains, the treadmilling motion of each activated molecule during a 5-s acquisition is only 30 nm, that is, on the order of the resolution limit of the live-cell super-resolution technique. Because the fluorescence of EYFP can be switched on multiple times, it can suffer from being oversampled and result in apparent distortions of the underlying structure.

Analytically removing the oversampling, as described above, wastes a large number of these critical labels. However, since the cells are alive, the dynamics of MreB monomer treadmilling along filaments in the present experiment were exploited to acquire more positional information for the same number of fluorophores (Kim et al., 2006). For this purpose, time-lapse imaging is employed (Biteen et al., 2009). In time-lapse imaging, a delay is introduced between each imaging frame, and the sample is only illuminated with imaging light during the short acquisition time. Here, only a small number of mobile EYFP-MreB molecules are activated, but due to the dark periods between imaging frames, they are imaged for tens of seconds or minutes. Over this longer time, each activated molecule traces out up to ∼300 nm of path along the filaments. Here, a large number of localizations can be obtained from this small population of fluorophores, and the distinct localization events elucidate more of the underlying superstructure.

In time-lapse, the effective labeling concentration is increased. In Fig. 2.6C–E, results from time-lapse imaging of EYFP-MreB in *C. crescentus* are presented. As in continuous-acquisition super-resolution imaging, two distinct MreB superstructures are identified in the cells: a quasi-helical arrangement in a stalked cell (6C, 6D) and a midplane ring in the predivisional cell (6E). Clearly, the images obtained in this manner are more continuous than those shown in Fig. 2.6A and B, and provide more information on the protein localization patterns. In particular, in the case of the stalked cell in Fig. 2.6B, the improved resolution of time-lapse imaging makes visible strands that join the bands observed in Fig. 2.6B, showing that such bands are likely part of a continuous structure.

3.2. Three-dimensional single-molecule imaging using double-helix photoactivated localization microscopy

The methods described thus far have been limited to the two dimensions of the image plane. The standard PSF cannot be used effectively for three-dimensional localization for two major reasons: (1) the standard PSF contains very little information about the emitter's axial position for several hundred nanometers above and below the focal plane, meaning that it is difficult to axially localize an emitter in and around the focal plane; and (2) the standard PSF is symmetric above and below the focal plane, which implies that a molecule located 500 nm above the focal plane cannot be distinguished from a molecule which is 500 nm below the focal plane. For this reason several groups have proposed methods for axial localization that both increase the axial information in the focal plane and break the symmetry of the standard PSF by using interferometry (Shtengel et al., 2009), astigmatism (Huang et al., 2008a), and biplane defocusing (Juette et al., 2008).

Recently, we have pursued a new strategy for three-dimensional super-resolution imaging using an engineered PSF that resembles a double helix in three dimensions (Pavani et al., 2009). Each z position throughout the depth of field of the instrument is represented by a unique orientation of two spots that are roughly diffraction limited in size on the detector. This PSF shape is termed a double-helix PSF, or DH-PSF, which was developed and provided by R. Piestun and S.R.P. Pavani at the University of Colorado (Pavani and Piestun, 2008). The DH-PSF is generated by a superposition of many Gauss–Laguerre modes derived through an optimization procedure designed to maximize diffraction efficiency and to localize most of the intensity into the two main lobes (Pavani and Piestun, 2008). Our work combining the DH-PSF with single-molecule PALM imaging yields double-helix photoactivated localization microscopy (DH-PALM) (Pavani et al., 2009). DH-PALM should be superior to other methods for three-dimensional imaging because the DH-PSF has high and uniform Fisher information in x, y, and z, implying excellent localization precision in all three dimensions. The lobes of the DH-PSF maintain approximately the same size throughout a wide range of axial positions, leading to a wider depth of field than simple defocusing methods.

The setup is simple to implement on any standard microscope setup with single-molecule sensitivity. The setup essentially involves adding a $4f$ image processing section in the collection path of the microscope between the image plane and the EMCCD detector. The required image transformation convolves the standard PSF with the DH-PSF. The simplest setup consists of two achromatic lenses and a spatial light (phase) modulator (SLM). Although an SLM is a simple and flexible way to generate the DH-PSF, it can only modulate vertically polarized light, and thus discards half of the photons that would normally be used in analysis. This is not a fundamental limitation, in fact, other phase modulator or mask designs can be used. The phase modulator is placed in the Fourier plane and a phase pattern corresponding to the Fourier transform of the DH-PSF is multiplied with the Fourier transform of the image. Inverse Fourier transformation by the second lens of the setup onto the camera gives the DH-PSF. Fig. 2.7A

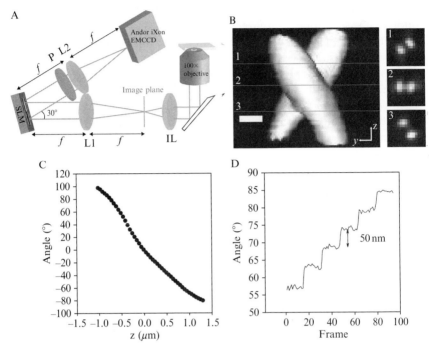

Figure 2.7 *Localization of 200-nm fluorescent beads using the DH-PALM microscope.* (A) Schematic illustrating the collection path in a DH-PALM setup, IL is imaging lens, L1 and L2 are 150 mm focal length achromat lenses, P is a linear polarizer, and SLM is a phase-only spatial light modulator. (B) Three-dimensional representation of the experimentally observed DH-PSF (created with VolumeJ (Abràmoff and Viergever, 2002) with slices taken at z positions of approximately (1) −450 nm, (2) 0 nm, and (3) 500 nm where 0 is taken to be the designed focal plane of the microscope. (C) Calibration of angle between the two lobes as a function of the distance between the objective surface and the bead with 0 being the position when the lobes are horizontal with respect to one another. (D) Plot of angle between the lobes versus frame for a fluorescent bead as the objective is scanned through 50 nm steps showing clear steps in the angle with a low standard deviation in each step.

shows a schematic of the collection path. The excitation path of the microscope is identical to that used in more conventional single-molecule experiments (Moerner and Fromm, 2003).

Although the experimental setup is relatively simple, the analysis of three-dimensional super-resolution images is far more complicated than the equivalent two-dimensional analysis, for many reasons. One is that no imaging system is free from aberration and thus no experimental PSF is identical to a theoretical model PSF of the system in three dimensions (Mlodzianoski et al., 2009). The axial position determined from the image of a molecule will always be based on interpolation from a calibration of the system. For DH-PALM imaging, calibration consists of measuring the angle between the two lobes as a function of the z position of the emitter. The distance between the object and the sample is controlled by a piezo-driven nanopositioner. The angle as a function of the z position of the emitter follows a quasi-linear shape as shown in Fig. 2.7C using a 200 nm fluorescent bead as a calibration source. With a bright fluorescent bead it is possible to obtain 1–2 nm precision in all three dimensions making it a useful calibration sample. Because even the slightest alignment change can affect accurate z position estimation, calibrations should be done frequently, and calibration several times throughout a day of experiments was found to be quite helpful. Also the calibration needs to be done with a sample that has the same refractive index characteristics as the sample of interest. Index mismatches can cause faulty z position estimates because of aberrations and therefore need to always be carefully considered (Deng and Shaevitz, 2009). For the measurement in Fig. 2.7, the bead was immobilized on a glass coverslip with index matching immersion oil above and below the glass to minimize index mismatch.

Another reason for increased complexity in analysis is that, because experimental PSFs always deviate appreciably from theory, it can be ineffective to derive an equation for localization precision as a function of the number of photons collected as was done in Thompson *et al.* (2002). This makes it difficult to assess how well one can localize a molecule based on a single measurement. A more effective strategy to measure localization precision is to find the position of a fixed point-like emitter (single molecule or a fluorescent bead) many times as to derive a population of measurements from which a standard deviation (localization precision) can be obtained for a specific number of photons detected, background noise, pixel size, etc. (as shown in Fig. 2.8). When collecting these data it is important to remove deterministic drifts such that the localization precision measurements reflect solely the uncertainty due to the Poisson nature of the photon detection process and not stage drift, a systematic error that can be a serious problem in commercial inverted microscopes. Deterministic noise can be removed by the use of a bright fiduciary (Pavani *et al.*, 2009) or by removing correlated drift through autocorrelation analysis of all molecular positions (Huang *et al.*, 2008a). It is best to generate data like Fig. 2.8 multiple times to form a look-up table that gives the localization precision as a function of the number of photons detected and the background noise. For any single acquisition in an actual image, an estimate of the localization precision can then be obtained from the lookup table.

To extract as much information as possible from the PSF, one must use a maximum likelihood estimator, or some other statistically efficient estimator that is based on the observed PSF (Mlodzianoski *et al.*, 2009), but this can be computationally prohibitive. Rather, our initial studies have used two computationally efficient but statistically inefficient estimators.

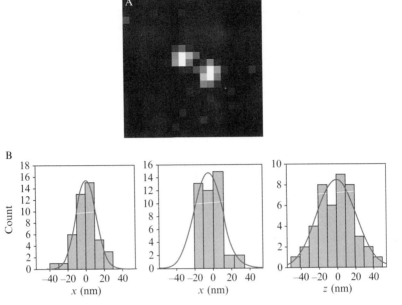

Figure 2.8 *Single-molecule superlocalization using the DH-PSF.* (A) Image of a single DCDHF-V-PF$_4$-azide molecule coming through the DH-PALM imaging system. (B) Histograms of 44 positions of the single molecule in (A) with standard deviations of 12.8 nm in x, 12.1 nm in y, and 19.5 nm in z. For these nonoptimized measurements, the average number of photons detected was 9300 with background noise of 48 photons/pixel and a pixel size of 160 nm. (Adapted and reproduced with permission from Pavani et al., 2009. Copyright 2009 National Academy of Sciences.)

The simplest technique is to find the centroid of each DH-PSF lobe and extract the lateral (x–y) position from the midpoint between the lobes and the z position from the angle. A second, superior estimator fits each lobe to a Gaussian using the function nlinfit in MATLAB. Image analysis consists of breaking the image into small regions of interest around each molecule to be analyzed. The small box image is then fit to the sum of two Gaussian functions using the MATLAB function nlinfit. The lateral position of the molecule is the midpoint between the centers of the two lobes calculated from the fit. Then the axial position of the emitter is found by interpolation of the angle found between the two lobes with a calibration curve of the type shown in Fig. 2.7C. This estimator was chosen because it gives more precise results than a simple centroid estimator, but is not as computationally expensive or complicated to derive and implement as a maximum likelihood estimator.

Proof-of-principle experiments that demonstrate super-resolution imaging of two DCDHF-V-PF$_4$-azide molecules separated by only 36 nm in three-dimensional space doped in PMMA have been performed. It was found that DH-PALM can precisely localize single molecules over a 2 μm depth of field, which is twice as large as previously published methods (Juette et al., 2008; Pavani et al., 2009). A detailed analysis of localization precision *versus* number of detected photons has recently appeared (Thompson, et al., 2010). With additional analysis of images using Fisher information concepts, additional improvements in PSF design and optimal use of the available photons, further gains in localization precision are to be expected in future work. It will be an important step to also apply the DH-PALM scheme to the study of super-resolution dynamics and structure in biological systems.

ACKNOWLEDGMENTS

We warmly thank Hsiao-lu Lee for assisting in cell culture for the data shown in Fig. 2.2, the laboratory of Prof. Lucy Shapiro for *C. crescentus* cell lines, and the laboratory of Rafael Piestun for collaboration regarding DH-PALM. The work on DCDHF molecules would not have been possible without Prof. Robert J. Twieg's group at Kent State University who synthesized the molecules. This work was supported in part by the National Institutes of Health Roadmap for Medical Research Grant No. 1P20-HG003638, and by Award Number R01GM085437 from the National Institute of General Medical Sciences. MAT and NRC were supported by National Science Foundation Graduate Research Fellowships. MAT was also supported by a Bert and DeeDee McMurtry Stanford Graduate Fellowship.

REFERENCES

Abràmoff, M. D., and Viergever, M. A. (2002). Computation and visualization of three dimensional motion in the orbit. *IEEE Trans. Med. Imag.* **21,** 61–78.
Adams, S. R., Kao, J. P. Y., and Tsien, R. Y. (1989). Biologically useful chelators that take up Ca^{2+} upon illumination. *J. Am. Chem. Soc.* **111,** 7957–7968.
Al Jammaz, I., Al-Otaibi, B., Okarvi, S., and Amartey, J. (2006). Novel synthesis of [^{18}F]-fluorobenzene and pyridinecarbohydrazide-folates as potential PET radiopharmaceuticals. *J. Labelled Compd. Rad.* **49,** 125–137.
Ambrose, W. P., Basché, T., and Moerner, W. E. (1991). Detection and spectroscopy of single pentacene molecules in a p-terphenyl crystal by means of fluorescence excitation. *J. Chem. Phys.* **95,** 7150–7163.
Ando, R., Hama, H., Yamamoto-Hino, M., Mizuno, H., and Miyawaki, A. (2002). An optical marker based on the UV-induced green-to-red photoconversion of a fluorescent protein. *Proc. Natl. Acad. Sci. USA* **99,** 12651–12656.
Ando, R., Mizuno, H., and Miyawaki, A. (2004). Regulated fast nucleocytoplasmic shuttling observed by reversible protein highlighting. *Science* **306,** 1370–1373.
Bates, M., Blosser, T. R., and Zhuang, X. (2005). Short-range spectroscopic ruler based on a single-molecule switch. *Phys. Rev. Lett.* **94,** 108101-1–108101-4.
Bates, M., Huang, B., Dempsey, G. T., and Zhuang, X. (2007). Multicolor super-resolution imaging with photo-switchable fluorescent probes. *Science* **317,** 1749–1753.
Betzig, E. (1995). Proposed method for molecular optical imaging. *Opt. Lett.* **20,** 237–239.

Betzig, E., Patterson, G. H., Sougrat, R., Lindwasser, O. W., Olenych, S., Bonifacino, J. S., Davidson, M. W., Lippincott-Schwartz, J., and Hess, H. F. (2006). Imaging intracellular fluorescent proteins at nanometer resolution. *Science* **313**, 1642–1645.

Biteen, J. S., Thompson, M. A., Tselentis, N. K., Bowman, G. R., Shapiro, L., and Moerner, W. E. (2008). Super-resolution imaging in live *Caulobacter Crescentus* cells using photoswitchable EYFP. *Nat. Meth.* **5**, 947–949.

Biteen, J. S., Thompson, M. A., Tselentis, N. K., Shapiro, L., and Moerner, W. E. (2009). Superresolution imaging in live *Caulobacter crescentus* cells using photoswitchable enhanced yellow fluorescent protein. *SPIE Proc.* **7185**, 71850I.

Carballido-López, R., and Errington, J. (2003). The bacterial cytoskeleton: In vivo dynamics of the actin-like protein Mbl of *Bacillus subtilis*. *Dev. Cell* **4**, 19–28.

Chudakov, D. M., Lukyanov, S., and Lukyanov, K. A. (2007). Tracking intracellular protein movements using photoswitchable fluorescent proteins PS-CFP2 and Dendra2. *Nat. Protocols* **2**, 2024–2032.

Conley, N. R., Biteen, J. S., and Moerner, W. E. (2008). Cy3-Cy5 covalent heterodimers for single-molecule photoswitching. *J. Phys. Chem. B* **112**, 11878–11880.

Deng, Y., and Shaevitz, J. W. (2009). Effect of aberration on height calibration in three-dimensional localization-based microscopy and particle tracking. *Appl. Opt.* **48**, 1886–1890.

Dickson, R. M., Cubitt, A. B., Tsien, R. Y., and Moerner, W. E. (1997). On/off blinking and switching behavior of single green fluorescent protein molecules. *Nature* **388**, 355–358.

Doub, L., and Vandenbelt, J. M. (1947). The ultraviolet absorption spectra of simple unsaturated compounds. I. mono- and p-disubstituted benzene derivatives. *J. Am. Chem. Soc.* **69**, 2714–2723.

Figge, R. M., Divakaruni, A. V., and Gober, J. W. (2004). MreB, the cell shape-determining bacterial actin homologue, co-ordinates cell wall morphogenesis in *Caulobacter crescentus*. *Mol. Microbiol.* **51**, 1321–1332.

Fölling, J., Belov, V., Kunetsky, R., Medda, R., Schönle, A., Egner, A., Eggeling, C., Bossi, M., and Hell, S. W. (2007). Photochromic rhodamines provide nanoscopy with optical sectioning. *Angew. Chem. Int. Ed.* **46**, 6266–6270.

Geisler, C., Schönle, A., von Middendorff, C., Bock, H., Eggeling, C., Egner, A., and Hell, S. W. (2007). Resolution of $\lambda/10$ in fluorescence microscopy using fast single molecule photo-switching. *Appl. Phys. A* **88**, 223–226.

Gelles, J., Schnapp, B. J., and Sheetz, M. P. (1988). Tracking Kinesin-driven movements with nanometre-scale precision. *Nature* **4**, 450–453.

Gitai, Z., Dye, N., and Shapiro, L. (2004). An actin-like gene can determine cell polarity in bacteria. *Proc. Natl. Acad. Sci. USA* **101**, 8643–8648.

Gunewardene, M. S., and Hess, S. T. (2008). Annual meeting of the biophysical society: Photoactivation yields and bleaching yield measurements for PA-FPs. *Biophys. J.* **94**, 840–848.

Gustafsson, M. G. L. (2005). Nonlinear structured-illumination microscopy: Wide-field fluorescence imaging with theoretically unlimited resolution. *Proc. Natl. Acad. Sci. USA* **102**, 13081–13086.

Güttler, F., Irngartinger, T., Plakhotnik, T., Renn, A., and Wild, U. P. (1994). Fluorescence microscopy of single molecules. *Chem. Phys. Lett.* **217**, 393.

Habuchi, S., Ando, R., Dedecker, P., Verheijen, W., Mizuno, H., Miyawaki, A., and Hofkens, J. (2005). Reversible single-molecule photoswitching in the GFP-like fluorescent protein Dronpa. *Proc. Natl. Acad. Sci. USA* **102**, 9511–9516.

Hansch, C., Leo, A., and Taft, R. W. (1991). A survey of hammett substituent constants and resonance and field parameters. *Chem. Rev.* **91**, 165–195.

Harms, G. S., Cognet, L., Lommerse, P. H. M., Blab, G. A., and Schmidt, T. (2001). Autofluorescent proteins in single-molecule research: Applications to live cell imaging microscopy. *Biophys. J.* **80,** 2396–2408.

Heintzmann, R., Jovin, T. M., and Cremer, C. (2002). Saturated patterned excitation microscopy—A concept for optical resolution improvement. *J. Opt. Soc. Am. A* **19,** 1599–1609.

Hell, S. W. (2007). Far-field optical nanoscopy. *Science* **316,** 1153–1158.

Hell, S. W., and Wichmann, J. (1994). Breaking the diffraction resolution limit by stimulated emission: Stimulated-emission-depletion fluorescence microscopy. *Opt. Lett.* **19,** 780–782.

Hess, S. T., Girirajan, T. P. K., and Mason, M. D. (2006). Ultra-high resolution imaging by fluorescence photoactivation localization microscopy. *Biophys. J.* **91,** 4258–4272.

Huang, B., Wang, W., Bates, M., and Zhuang, X. (2008a). Three-dimensional super-resolution imaging by stochastic optical reconstruction microscopy. *Science* **319,** 810–813.

Huang, B., Jones, S. A., Brandenburg, B., and Zhuang, X. (2008b). Whole-cell 3D STORM reveals interactions between cellular structures with nanometer-scale resolution. *Nat. Meth.* **5,** 1047–1052.

Juette, M. F., Gould, T. J., Lessard, M. D., Mlodzianoski, M. J., Nagpure, B. S., Bennett, B. T., Hess, S. T., and Bewersdorf, J. (2008). Three-dimensional sub-100 nm resolution fluorescence microscopy of thick samples. *Nat. Meth.* **5,** 527–529.

Kim, S. Y., Gitai, Z., Kinkhabwala, A., Shapiro, L., and Moerner, W. E. (2006). Single molecules of the bacterial actin MreB undergo directed treadmilling motion in *Caulobacter crescentus*. *Proc. Natl. Acad. Sci. USA* **103,** 10929–10934.

Klar, T. W., and Hell, S. W. (1999). Subdiffraction resolution in far-field fluoresence microscopy. *Opt. Lett.* **24,** 954–956.

Kremers, G., Hazelwood, K. L., Murphy, C. S., Davidson, M. W., and Piston, D. W. (2009). Photoconversion in orange and red fluorescent proteins. *Nat. Methods* **6,** 355–358.

Lakowicz, J. R. (2006). Principles of Fluorescence Spectroscopy. Springer Science, New York.

Lord, S. J., Conley, N. R., Lee, H. D., Samuel, R., Liu, N., Twieg, R. J., and Moerner, W. E. (2008). A photoactivatable push–pull fluorophore for single-molecule imaging in live cells. *J. Am. Chem. Soc.* **130,** 9204–9205.

Lord, S. J., Conley, N. R., Lee, H. D., Nishimura, S. Y., Pomerantz, A. K., Willets, K. A., Lu, Z., Wang, H., Liu, N., Samuel, R., Weber, R., Semyonov, A. N., *et al.* (2009). DCDHF fluorophores for single-molecule imaging in cells. *ChemPhysChem* **10,** 55–65.

Lord, S. J., *et al.* (2010). Azido push–pull fluorogens photoactivate into bright fluorescent labels. *J. Phys. Chem. B* 10.1021/jp907080r, (in press).

Los, G. V., and Wood, K. (2007). The HaloTag: A novel technology for cell imaging and protein analysis. *Methods Mol. Biol.* **356,** 195–208.

Los, G. V., Encell, L. P., McDougall, M. G., Hartzell, D. D., Karassina, N., Zimprich, C., Wood, M. G., Learish, R., Ohana, R. F., Urh, M., Simpson, D., Mendez, J., *et al.* (2008). HaloTag: A novel protein labeling technology for cell imaging and protein analysis. *ACS Chem. Biol.* **3,** 373–382.

Michalet, X., and Weiss, S. (2006). Using photon statistics to boost microscopy resolution. *Proc. Natl. Acad. Sci. USA* **103,** 4797–4798.

Mlodzianoski, M. J., Juette, M. F., Beane, G. L., and Bewersdorf, J. (2009). Experimental characterization of 3D localization techniques for particle-tracking and super-resolution microscopy. *Opt. Express* **17,** 8264–8277.

Moerner, W. E. (2006). Single-molecule mountains yield nanoscale images. *Nat. Methods* **3,** 781–782.

Moerner, W. E. (2007). New directions in single-molecule imaging and analysis. *Proc. Natl. Acad. Sci. USA* **104,** 12596–12602.
Moerner, W. E., and Fromm, D. P. (2003). Methods of single-molecule fluorescence spectroscopy and microscopy. *Rev. Sci. Instrum.* **74,** 3597–3619.
Molski, A. (2001). Statistics of the bleaching number and the bleaching time in single-molecule fluorescence spectroscopy. *J. Chem. Phys.* **114,** 1142–1147.
Niu, L., and Yu, P. (2008). Investigating intracellular dynamics of FtsZ cytoskeleton with photoactivation single-molecule tracking. *Biophys. J.* **95,** 2009–2016.
Nyquist, H. (1928). Certain topics in telegraph transmission theory. *Trans AIEE* **47,** 617–644.
Ober, R. J., Ram, S., and Ward, E. S. (2004). Localization accuracy in single-molecule microscopy. *Biophys. J.* **86,** 1185–1200.
Patterson, G., Day, R. N., and Piston, D. (2001). Fluorescent protein spectra. *J. Cell Sci.* **114,** 837–838.
Pavani, S. R. P., and Piestun, R. (2008). High-efficiency rotating point spread functions. *Opt. Express* **16,** 3484–3489.
Pavani, S. R. P., Thompson, M. A., Biteen, J. S., Lord, S. J., Liu, N., Twieg, R. J., Piestun, R., and Moerner, W. E. (2009). Three-dimensional, single-molecule fluorescence imaging beyond the diffraction limit by using a double-helix point spread function. *Proc. Natl. Acad. Sci. USA* **106,** 2995–2999.
Rust, M. J., Bates, M., and Zhuang, X. (2006). Sub-diffraction-limit imaging by stochastic optical reconstruction microscopy (STORM). *Nat. Methods* **3,** 793–795.
Schmidt, T., Kubitscheck, U., Rohler, D., and Nienhaus, U. (2002). Photostability data for fluorescent dyes: An update. *Single Mol.* **3,** 327.
Schriven, E.F.V. (1984). Azides and nitrenes: Reactivity and utility. Academic Press, Orlando, FL.
Shaner, N. C., Campbell, R. E., Steinbach, P. A., Giepmans, B. N. G., Palmer, A. E., and Tsien, R. Y. (2004). Improved monomeric red, orange and yellow fluorescent proteins derived from *Discosoma* sp. red fluorescent protein. *Nat. Biotech.* **22,** 1567–1572.
Shannon, C. E. (1949). Communication in the presence of noise. *Proc. IRE* **37,** 10–21.
Shroff, H., Galbraith, C. G., Galbraith, J. A., White, H., Gillette, J., Olenych, S., Davidson, M. W., and Betzig, E. (2007). Dual-color superresolution imaging of genetically expressed probes within individual adhesion complexes. *Proc. Natl. Acad. Sci. USA* **104,** 20308–20313.
Shroff, H., Galbraith, C. G., Galbraith, J. A., and Betzig, E. (2008). Live-cell photoactivated localization microscopy of nanoscale adhesion dynamics. *Nat. Methods* **5,** 417–423.
Shtengel, G., Galbraith, J. A., Galbraith, C. G., Lippincott-Schwartz, J., Gillette, J. M., Manley, S., Sougrat, R., Waterman, C. M., Kanchanawong, P., Davidson, M. W., Fetter, R. D., and Hess, H. F. (2009). Interferometric fluorescent super-resolution microscopy resolves 3D cellular ultrastructure. *Proc. Natl. Acad. Sci.* **106,** 3125–3130.
Soper, S. A., Nutter, H. L., Keller, R. A., Davis, L. M., and Shera, E. B. (1993). The photophysical constants of several fluorescent dyes pertaining to ultrasensitive fluorescence spectroscopy. *Photochem. Photobiol.* **57,** 972–977.
Soundararajan, N., and Platz, M. S. (1990). Descriptive photochemistry of polyfluorinated azide derivatives of methyl benzoate. *J. Org. Chem.* **55,** 2034–2044.
Stevenson, P. E. (1965). Effects of chemical substitution on the electronic spectra of aromatic compounds: Part I. The effects of strongly perturbing substituents on benzene. *J. Mol. Spectrosc.* **15,** 220–256.
Stiel, A. C., Andresen, M., Bock, H., Hillbert, M., Schilde, J., Schönle, A., Eggeling, C., Egner, A., Hell, S. W., and Jakobs, S. (2008). Generation of monomeric reversibly switchable red fluorescent proteins for far-field nanoscopy. *Biophys. J.* **95,** 2989–2997.

Subach, F. V., Patterson, G. H., Manley, S., Gillette, J. M., Lippincott-Schwartz, J., and Verkhusha, V. V. (2009). Photoactivatable mCherry for high-resolution two-color fluorescence microscopy. *Nat. Methods* **6**, 153–159.

Thompson, M. A., Lew, M. D., Badieirostami, M., and Moerner, W. E. (2010). Localizing and tracking single nanoscale emitters in three dimensions with high spatio-temporal resolution using a double-helix point spread function. *Nano Lett.* **10**, 211–218.

Thompson, R. E., Larson, D. R., and Webb, W. W. (2002). Precise nanometer localization analysis for individual fluorescent probes. *Biophys. J.* **82**, 2775–2783.

Tsien, R. Y. (1998). The green fluorescent protein. *Annu. Rev. Biochem.* **67**, 509–544.

van Oijen, A. M., Köhler, J., Schmidt, J., Müller, M., and Brakenhoff, G. J. (1998). 3-dimensional super-resolution by spectrally selective imaging. *Chem. Phys. Lett.* **292**, 183–187.

van Oijen, A. M., Köhler, J., Schmidt, J., Müller, M., and Brakenhoff, G. J. (1999). Far-field fluorescence microscopy beyond the diffraction limit. *J. Opt. Soc. Am. A* **16**, 909–915.

Wiedenmann, J., Ivanchenko, S., Oswald, F., Schmitt, F., Röcker, C., Salih, A., Spindler, K., and Nienhaus, G. U. (2004). EosFP, a fluorescent marker protein with UV-inducible green-to-red fluorescence conversion. *Proc. Natl Acad. Sci. USA* **101**, 15905–15910.

Willets, K. A., Nishimura, S. Y., Schuck, P. J., Twieg, R. J., and Moerner, W. E. (2005). Nonlinear optical chromophores as nanoscale emitters for single-molecule spectroscopy. *Acc. Chem. Res.* **38**, 549–556.

CHAPTER THREE

TRACKING SINGLE PROTEINS IN LIVE CELLS USING SINGLE-CHAIN ANTIBODY FRAGMENT-FLUORESCENT QUANTUM DOT AFFINITY PAIR

Gopal Iyer, Xavier Michalet, Yun-Pei Chang, *and* Shimon Weiss

Contents

1. Introduction	62
2. The Method: Targeting QDs via a Single-Chain Variable Fragment-Hapten Pair	64
3. Functionalization of QDs	66
3.1. Reagents for peptide coating	66
3.2. Peptide coating of QDs	66
3.3. Varying the stoichiometry of FL-pc-QDs	68
4. Quantification of the Number of FL Molecules per FL-pc-QD	69
5. Binding of FL-QDs to Anti-scFv Fusion Constructs	71
5.1. Quantification of the binding affinity of FL-QD to α-FL scFv	71
6. DNA Constructs for Single FL-QD Imaging in Live Cells	73
6.1. Labeling of live mammalian cells	74
7. Single-Molecule Imaging of Live Mammalian Cells	74
Acknowledgments	75
References	77

Abstract

Quantum dots (QDs) are extremely bright fluorescent imaging probes that are particularly useful for tracking individual molecules in living cells. Here, we show how a two-component system composed of a high-affinity single-chain fragment antibody and its cognate hapten (fluorescein) can be utilized for tracking individual proteins in various cell types. The single-chain fragment antibody against fluorescein is genetically appended to the protein of interest, while the hapten fluorescein is attached to the end of the peptide that is used to coat the QDs. We describe (i) the method used to functionalize QDs with

Department of Chemistry and Biochemistry, California NanoSystems Institute, University of California, Los Angeles, California, USA

Methods in Enzymology, Volume 475
ISSN 0076-6879, DOI: 10.1016/S0076-6879(10)75003-5

© 2010 Elsevier Inc.
All rights reserved.

fluorescein peptides; (ii) the method used to control the stoichiometry of the hapten on the surface of the QD; and (iii) the technical details necessary to observe single molecules in living cells.

1. INTRODUCTION

Fluorescence microscopy of live cells is a central tool in modern biological studies. The ability to adapt recombinant DNA methods for the fusion of fluorescent proteins (FPs) to target proteins in the cell, combined with the development of advanced light microscopy, has allowed researchers to explore trafficking of proteins, their turnover and localization (Lippincott-Schwartz and Patterson, 2003), cellular movement during embryogenesis (Kwon and Hadjantonakis, 2007), protein–protein interactions (Villalobos et al., 2007), and cell lineage development in live cells (Wacker et al., 2007) and whole organisms (Sakaue-Sawano et al., 2008). Recent progress in use of fluorescence imaging in biology has gained from developments in optical instruments and imaging techniques (Betzig et al., 2006; Michalet et al., 2003; Westphal et al., 2008), expansion of the FP repertoire (Chudakov et al., 2005; Giepmans et al., 2006), and introduction of quantum dots (QDs) (Bruchez et al., 1998; Michalet et al., 2005; Pons and Mattoussi, 2009). These advances, however, critically depend on the ease, versatility, and efficiency of fluorescently labeling macromolecules *in vitro* and *in vivo*.

Genetic encoding of proteins of interest with a FP is the most popular labeling method in cells (Tsien, 1998). Other modes of attachment of exogenous fluorophores to cellular proteins are based on (i) enzymatic strategies such as acyl carrier protein/phosphopantetheine transferase (George et al., 2004); Q-tag/transglutaminase (Lin and Ting, 2006), biotin acceptor peptide/biotin ligase (Howarth et al., 2005), prokaryotic hydrolase/Halo tag (Los et al., 2008), formylglycine-generating enzyme/aldehyde tag (Carrico et al., 2007), and Farnesyltransferase/CVIA peptide tag (Duckworth et al., 2007); and (ii) noncovalent affinity labeling methods such as the tetracysteine/biarsenical system (Andresen et al., 2004; Hoffmann et al., 2005), dihydrofolate reductase (DHFR) (Miller et al., 2004, 2005), the bis-arsenical/SplAsH (spirolactam Arsenical Hairpin binder) system (Bhunia and Miller, 2007), biotin/streptavidin (Weber et al., 1989), and barnase/barstar (Wang et al., 2004). QDs have emerged as a very promising class of fluorophores for multiple biological applications (Michalet et al., 2005), including live cell imaging. Labeling cellular proteins with QDs depends on the way they are functionalized. Covalent and noncovalent attachment of proteins and small molecules to QDs can be achieved using standard conjugation chemistry (Michalet et al., 2005).

One of the most interesting approaches has been to functionalize QDs with peptides. Peptide-coated QDs (pc-QDs) are advantageous (Iyer *et al.*, 2007; Pinaud *et al.*, 2004; Zhou and Ghosh, 2007) for several reasons: (i) The peptide coat is a natural cloak protecting the cell from the potentially toxic inorganic materials the QD is comprised of, (ii) exposed native functional groups such as $-NH_2$ and $-COOH$ can be manipulated by standard chemistries to react with the target of interest, and (iii) the stoichiometry of peptides with various reactive handles can be adjusted by varying the number and type of peptides displayed on the large surface of QDs. In particular, by tuning the functional peptide stoichiometry to one or few peptides per QD, it is possible to prevent or minimize the unwanted attachment of several target proteins to each QD.

A new frontier for fluorescence microscopy has recently emerged with the ability to observe single molecules in live cells, opening the way for a true molecular understanding of cellular processes (Hibino *et al.*, 2009; Joo *et al.*, 2008; Wennmalm and Simon, 2007). However, organic dyes and FPs are unfortunately prone to rapid photobleaching, limiting the time span of single-molecule observations. In addition, fluorescence saturation (asymptotic saturation of fluorescence at large excitation rates due to triplet blinking) and the finite fluorescence lifetime of fluorophores limit the temporal resolution of these single-molecule observations due to limited signal-to-noise ratio. Brighter and more stable fluorophores such as QDs have the potential to further our ability to explore fast as well as long-term single-molecule behaviors in live cells, by circumventing these two limitations. But since they are exogenous probes and cannot be genetically encoded, other means of targeting cellular proteins have to be developed. Furthermore, in order to harness QDs' tremendous potential for multicolor imaging, the development of several orthogonal targeting schemes will be necessary. The more conventional approach to address the QDs' targeting issue has been to directly attach primary (Watanabe and Higuchi, 2007) or secondary (Dahan *et al.*, 2003) antibodies to QDs. Although extremely versatile and simple (not requiring any modification of the target protein), this approach has several pitfalls. First, antibodies are large (as large as the QDs themselves), further increasing the size of the exogenous label, potentially interfering with the protein target's function. Second, antibodies are divalent, and since multiple antibodies could be bound to a single large QD, this raises the issue of multivalency of the probe and therefore of aggregation of the target. Third, antibody availability and specificity need to be addressed for each new protein of interest.

Direct attachment of the protein of interest to QDs by chemical crosslinking to an existing cysteine is a possible strategy, although it is easier to perform *in vitro* than *in vivo*. All other alternatives require modification of the protein sequence to add a chemical moiety that will allow binding to the functionalized QD. Several examples of such modifications already exist in the literature. For instance, we recently showed how fusion of the chicken avidin sequence to a protein-moiety of interest allows its targeting by biotinylated pc-QDs at the single-molecule level (Pinaud et al., 2009). Ting's group (Howarth and Ting, 2008) showed how enzymatic biotinylation of proteins allowed streptavidin-functionalized QDs to target these proteins. Dahan's group (Roullier et al., 2009) designed, characterized and applied QD-trisNTA, which integrates tris-nitrilotriacetic acid, a small and high-affinity recognition unit for the ubiquitous polyhistidine protein tag. Direct visualization of protein–protein interactions at the single-molecule level was made possible for type-1 interferon subunits using a two-color QD tracking approach (Roullier et al., 2009). Although quite powerful, these methods have their share of problems. In particular, avidin or streptavidin are tetramers with four binding sites for biotin, raising the possibility of multivalency. Also, these systems can only be used to target a single protein species at a time. Orthogonal binding schemes are still needed if several different proteins need to be targeted simultaneously. All of the approaches described earlier have merits; each can be used for a single protein target at a time, necessitating the mastery of different chemistries and strategies for multicolor labeling of several different species.

2. The Method: Targeting QDs via a Single-Chain Variable Fragment-Hapten Pair

To address these issues, we proposed a general strategy for targeting proteins with single QDs, based on the use of single-chain variable fragment (scFv) antibodies (Boder et al., 2000) against small molecules (or haptens). The principle of our approach consists in fusing the scFv sequence to the cellular protein of interest and targeting it with QDs functionalized with the corresponding hapten. For a successful single-molecule tracking experiment, the QDs need to stay attached to the target protein for a long enough time, requiring a high-affinity interaction between the scFv and the hapten (in the range of a 10^{-12} to 10^{-15} M dissociation equilibrium constant).

Other hurdles in live cell experiments are nonspecific cellular labeling and a large molecular weight "penalty" of the fused affinity reagents (streptavidin: ~60 kDa and immunoglobulin IgG: 150 kDa). To circumvent these hurdles, we adapted a high-affinity scFv evolved against monovalent fluorescein (FL) ligand (Boder et al., 2000). The anti-FITC scFv dubbed 4M5.3 (Boder et al., 2000) is the highest known engineered affinity antibody to an antigen (FL) and has an affinity of 270 fM, a dissociation half-time of >5 days and a low molecular weight of 30 kDa. The advantage of using anti-FITC scFv antibody for QD targeting (in lieu of GFP) is the possibility for long-term imaging (due to QDs' high brightness and excellent photostability) together with such a high-affinity interaction. We therefore synthesized an N-terminal fluorescein isothiocyanate (FITC)-labeled peptide and used it to solubilize QDs. These coated QDs emit light in two colors (green from FITC and red from QD; Iyer et al., 2008). The ability to fuse the scFv's DNA sequence to genes expressed on the surface of a variety of cell types offers the possibility to simultaneously target several different-color QDs to different proteins by using (and further evolving) other high-affinity, orthogonal, scFv–hapten pairs. This approach is therefore well suited for monitoring and long-term tracking of individual proteins in cells and can be combined with multicolor labeling experiments with other color QDs and FPs. As an example, we generated hamster prion (PrP) DNA construct fused to this scFv and demonstrated that fusion of scFv to PrP does not impair its localization to the cell surface; we were also able to monitor its membrane diffusion by single-particle tracking of attached FITC-QDs (Chang et al., 2008), further showcasing the versatility of FITC-QDs as high-affinity binding partners.

A typical scFv has a molecular weight of 27 kDa which is comparable in size to a FP (e.g., eGFP: 30 kDa) and will therefore, in favorable cases, not interfere with the function of the protein to which it is fused. Genetic engineering of the protein fusion and its expression are in this respect similar to the standard approach used for the creation of FP–protein constructs, with similar advantages and inconveniences. However, used in combination with sparse labeling with hapten-functionalized QDs, it allows overcoming a major problem exhibited in FP–protein systems, which is the control of expression levels and thus, the amount of expressed-labeled proteins. By separating expression and labeling, our approach allows for external control of the degree of labeling of the target protein, by simply adjusting the concentration and incubation time of hapten-functionalized QDs.

Functionalization of the QDs with haptens is performed using our previously described peptide-coating technique (Pinaud et al., 2004). In this technique, a simple exchange reaction replaces the hydrophobic trioctyl-phosphine oxide (TOPO) molecules solubilizing QDs in organic solvent, by short (~ 20 amino acid (aa)) amphipathic peptides solubilizing QDs in aqueous buffers. By chemically grafting haptens to the peptide sequence (either before or after coating), the pc-QDs are equipped with the desired hapten. If needed, alternate methods for functionalizing QDs with a small hapten could be used instead of our peptide-coating approach.

The generality of this method comes from the availability of a growing number of scFvs against various haptens developed for various purposes and characterized by very small (tight) dissociation constants (Brichta et al., 2005). The availability of many such "affinity pairs" allows for orthogonal labeling of multiple targets with different-color QDs. Here, we illustrate this approach with the particular case of the 4M5.3 scFv (α-FL scFv) developed against FL by the Wittrup group. This affinity pair is characterized by a dissociation constant of 270 fM in phosphate-buffered saline (PBS) and 48 fM in low salt buffer (Boder and Wittrup, 2000). We call the resulting QDs FL-pc-QDs, for FL-functionalized pc-QDs. As will become clear later, an advantage of using FL as hapten is the ability to quantify the presence and stoichiometry of the hapten on QDs. In the following, we fully characterize this system, demonstrate targeting of single FL-pc-QDs to scFv displayed on the surface of live yeast cells and, finally, show an example of long-term tracking of individual scFv–PrP fusion proteins in live neuronal cells. The fluorophore of choice needs to be a good single-molecule fluorophore, with good photostability, high quantum yield (which results in a better signal-to-noise ratio using less laser pumping power), large extinction coefficient, and low residency time in the triplet state (which produces interruptions of fluorescence emission).

3. Functionalization of QDs

Synthesis of QDs is reviewed elsewhere (Murray et al., 1993; Peng et al., 1997). In short, it is usually carried out at high temperatures from organometallic precursors in the presence of coordinating solvents. At the end of the reaction QDs are capped with a mixture of TOPO and alkyl amines, rendering them hydrophobic and not amenable for biological applications. Several methods exist to solubilize them in water, including lipid micelles, encapsulating TOPO with silica coating, or exchanging the hydrophobic coating with amphiphilic ligands. QDs can either be synthesized in house with the appropriate equipment, or purchased from various vendors such as Invitrogen Corporation, Evident Technologies, Ocean Nanotech, and others.

3.1. Reagents for peptide coating

Crude peptides with sequence FITC–GSESGGSESGFCCFCCFCCF–CONH$_2$ (FL-peptide, FL-P) and KGSESGGSESGFCCFCCFCCF–CONH$_2$ (K-peptide or K-P) were purchased from Anaspec (San Jose, CA). Polyethylene glycol reagents (PEG 330) were obtained from Pierce Biotechnology (Rockford, IL). All chemicals used for QD solubilization were purchased from Sigma-Aldrich (St. Louis, MO).

3.2. Peptide coating of QDs

1. TOPO-coated CdSe/ZnS QDs are synthesized as described by Dabbousi et al. (1997).

2. 20 μl of 1 μM TOPO-coated QDs are pelleted in a table top centrifuge at 10,000 rpm for 5 min in 10 volumes of methanol. The resulting pellet is resuspended in 450 μl of anhydrous pyridine and transferred to a 1.5 ml microcentrifuge tube labeled A. The pellet can be refluxed for few seconds at 60 °C on a heating top if the pellet is difficult to dissolve. Discard the QD solution if the pellet is particulate during the process of dissolution.
3. 4 mg of FL-P is weighed out and added to 50 μl of anhydrous DMSO in a 1.5 ml microcentrifuge tube labeled B. It is important to ensure that the peptide is dissolved completely. Place the tube in a water bath set at 60 °C with continuous monitoring if the pellet does not dissolve easily.
4. Pipette the pyridine containing QDs from tube A to tube B, immediately add 11.6 μl of tetramethyl ammonium hydroxide 25% (w/v) in methanol (Sigma), and vigorously mix the tube for few seconds.
5. Spin the tube at 12,000 rpm for 2–3 min in a high-speed microcentrifuge. Pour off the supernatant into a designated solvent waste container. Resuspend the pellet in 300 μl of anhydrous DMSO. Before adding DMSO, remove residual pyridine from the walls of the tube.
6. Prepare an Illustra NAP-25 gel filtration column (GE Healthcare) by equilibrating the column with distilled water 4–5 times per the manufacturer's instructions.
7. Load the column after the pellet has completely resuspended in DMSO. Do not vortex the pellet. Tube containing the QD pellet should be left standing at room temperature to dissolve slowly. The presence of aggregates during the process of dissolution indicates improper solubilization and should be discarded. Allow the DMSO containing QDs to settle into the column bed. Top the column with distilled water and monitor the band of QDs with a hand-held UV lamp.
8. Collect 1 ml of the QDs in a low-binding microcentrifuge tube. At this stage, QDs are solubilized with FL-P and can be used in a biological environment.
9. Measure the concentration of column-eluted FL-QDs at 493 and 607 nm with a scanning spectrophotometer using the extinction coefficients of FL 85,200 M^{-1} cm^{-1} and QDs 353,762.8 M^{-1} cm^{-1}, respectively.
10. Dialyze the FL-QDs using a 300,000 MWCO cellulose ester or PVDF membrane (spectra/Por, Spectrum Labs, Rancho Dominguez, CA) against 4 l of PBS with frequent (3–4 times) change in the span of 6 h and overnight once. Dialysis should be carried out at room temperature protected from light.
11. Measure the concentration of column-eluted FL-QDs at 493 and 607 nm with a scanning spectrophotometer. Measuring before and after dialysis

will determine the final yield of the FL-QDs and will help in determining the scale of FL-QD synthesis required for all experiments.
12. To check colloidal properties, perform gel electrophoresis of the FL-QDs on a 1% agarose gel. Evaluate the presence of excess FL-P using a fluorescent gel scanner equipped with laser excitation at 488 nm or an imaging station equipped with a simple CCD camera with appropriate excitation and emission filters. If free FL-P are present at this stage, either attempt to repurify using a NAP-25 column or start over. A schematic representation of an FL-QD is shown in Fig. 3.1.

3.3. Varying the stoichiometry of FL-pc-QDs

Single-molecule experiments that involve affinity pairs should meet the following requirements: (a) The binding pair should have an affinity constant in the high picomolar to femtomolar range (see later), (b) affinity pairs should be orthogonal to each other (for multiplexing purposes) and otherwise inert (to minimize nonspecific labeling cross-reactivity), (c) should be small (nonperturbative to the cell), and (d) display single avidity and precise stoichiometry (to avoid cross-linking of targets inside the cell). We have

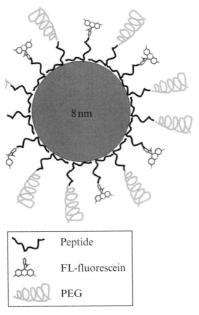

Figure 3.1 Schematic representation of QD functionalized with FL-P. Addition of polyethylene glycol to the pc-QDs minimizes nonspecific binding and sticking to surface during imaging.

been able to control the labeling density of the small hapten FL on the surface of QDs by varying the ratio of FL-P to NH_2-peptide in the following way:

1. To obtain 100% FL-P coverage, follow the protocol as described earlier.
2. For 50% coverage, weigh out 2 mg each of FL-P and NH_2-peptide, while for 10% coverage, weigh out 0.4 and 3.6 mg of FL-P and NH_2-peptide, respectively, and proceed as described in Section 3.2.
3. After dialysis into PBS, pH 7.4, unreacted amine groups can be blocked with small molecule PEG 330-NHS ester (Pierce, USA).
4. PEGylation can be carried out in PBS, pH 7.4, in 10–100-fold excess for 1 h at room temperature. The reaction is terminated by addition of 100 mM Glycine, pH 7.0, or any amine containing buffer. It is not necessary to remove the unreacted PEG as this does not interfere with downstream labeling applications.

4. Quantification of the Number of FL Molecules per FL-pc-QD

It is important to quantify the number of hapten molecules that cover the QD after conjugation. We have previously shown (Iyer *et al.*, 2007) that attaching a fluorescent label to the bioconjugated molecule can aid quantification of the resulting stoichiometry. A typical absorbance spectrum of FL-QDs is depicted in Fig. 3.2.

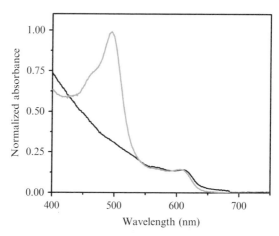

Figure 3.2 Normalized absorbance spectra of pc-QDs emitting at 620 nm. FL pc-QDs have absorption peaks at 493 and 610 nm, while pc-QDs have an absorption peak at 610 nm only.

A simple way to compute the number of FL-P coating each QD is by light absorption measurement. Absorption of light at a wavelength λ by a FL-pc-QD can have several causes: absorption by the QD if $\lambda < \lambda_{QD}$ (where λ_{QD} is the first exciton peak of the QD absorption), (ii) absorption by FL if $\lambda < \lambda_{FL}$ (where λ_{FL} is the upper limit of the absorption of FL), and (iii) absorption by the peptides themselves. The latter is negligible at visible wavelengths, and therefore it is easy to take measurements in two separate situations: one in which, first, only the QDs are absorbing ($\lambda_{FL} < \lambda < \lambda_{QD}$) and, second, where both FL and QDs are absorbing ($\lambda < \lambda_{FL} < \lambda_{QD}$). As we show next, these two measurements are sufficient to extract the number n of FL-P per QD using some simple assumptions and reference samples.

Let n be the unknown average number of FL-P per QD. The extinction coefficient of FL at 493 nm, $\varepsilon_{FL}(493)$, is provided by the manufacturer ($\varepsilon_{FL}(493) = 85{,}200$ cm^{-1} M^{-1}). The first exciton peak of the QD used in this experiment is 610 nm. If we measure the absorption of a FL-pc-QD at both 493 and 610 nm, we will obtain data corresponding to the two different situations described earlier. Assuming that the wavelength dependence of the QD and FL absorption spectra (i.e., the extinction coefficients) are not affected by the presence of the other species in close proximity, the total absorption at a wavelength λ will be:

$$A_{FL-pc-QD}\lambda = (\varepsilon_{QD}\lambda + n\varepsilon_{FL}\lambda)c_{QD}L, \quad (3.1)$$

where c_{QD} and L are the QD concentration and the excitation path length, respectively. The extinction coefficient of CdSe QDs (at their first exciton peak wavelength) has been experimentally measured by Peng and collaborators (Yu *et al.*, 2003) to depend on the first exciton peak wavelength according to:

$$\varepsilon_{QD} = 5857 D^{2.65}$$
$$D = (1.6122 \times 10^{-9})\lambda^4 - (2.6575 \times 10^{-6})\lambda^3 \quad (3.2)$$
$$+ (1.6242 \times 10^{-3})\lambda^2 - (0.4277\lambda + 41.57)$$

where D is the diameter of the QD core in nm, and ε is expressed in M^{-1} cm^{-1}.

From Eq. (3.2), we can calculate $\varepsilon_{QD}(610\text{ nm})$, and by substituting it in Eq. (3.1) (written for $\lambda = 610$ nm), obtain the QD concentration:

$$c_{QD} = \frac{A_{FL-pc-QD}(610)}{\varepsilon_{QD}(610)L}. \quad (3.3)$$

Writing Eq. (3.1) for $\lambda = 493$ nm, we obtain the following equation for n:

$$n = \frac{\left[\frac{A_{FL-pc-QD}(493)}{A_{FL-pc-QD}(610)}\varepsilon_{QD}(610) - \varepsilon_{QD}(493)\right]}{\varepsilon_{FL}(493)}. \tag{3.4}$$

The only unknown in this expression is $\varepsilon_{QD}(493$ nm$)$, which can be easily obtained from $\varepsilon_{QD}(610$ nm$)$ and the measurement of the QD-only absorption spectrum:

$$\varepsilon_{QD}(493) = \varepsilon_{QD}(610)\frac{A_{QD}(493)}{A_{QD}(610)}. \tag{3.5}$$

The number of FL-P per QD is thus given by the following equation:

$$n = \left[\frac{A_{FL-pc-QD}(493)}{A_{FL-pc-QD}(610)} - \frac{A_{QD}(493)}{A_{QD}(610)}\right]\frac{\varepsilon_{QD}(610)}{\varepsilon_{FL}(493)}. \tag{3.6}$$

The uncertainty in this number can be calculated by the standard rule of error propagation and can be kept below 1 if all parameters in Eq. (3.6) are affected by a relative error of a few percentages only.

5. Binding of FL-QDs to Anti-scFv Fusion Constructs

After coating of the QDs with FL-P, the positive control yeast strain EBY 100 (*MATa ura3-52 trp1 leu2Δ1 his3Δ200 pep4::HIS3 prb1Δ1.6R can1 GAL* (pIU211:*URA*3) harboring plasmid pCT-4M5.3 (Boder *et al.*, 2000) can be used to confirm binding of FL-QDs to α-FL scFv (using both green and red fluorescence to detect FL and QDs, respectively). Growth and induction conditions of yeast cells have been described in Chao *et al.* (2006). Plasmids and strains can be requested from Dane Wittrup's lab, MIT, for testing individual preparation of FL-QDs.

5.1. Quantification of the binding affinity of FL-QD to α-FL scFv

Individual FL-QD preparations can be tested for binding affinity using yeast strain EBY 100 expressing α-FL scFv. This step is optional as our lab has determined that K_d values obtained with FL-QDs are sufficient to carry out single-molecule imaging experiments (Iyer *et al.*, 2008).

The method for measuring and determining the binding constant for FL-QDs is briefly described here. Although the α-FL scFv developed by the Wittrup group has a measured dissociation constant of 270 fM in PBS and 48 fM in low salt buffer (Boder et al., 2000), this affinity concerns the free FL molecule. In our system, FL is cross-linked to a peptide, which is itself bound to a QD and is surrounded by other peptides and PEG molecules. This crowded environment could potentially reduce the affinity of α-FL scFv for the cross-linked FL species.

Briefly, incubate α-FL scFv expressing yeast cells with varying concentrations of FL-QDs and measure their total FL fluorescence. Plotting total FL as function of concentration should result in a sigmoidal curve, with $[Fl-QD]_{1/2}$ being a characteristic concentration at which half of the change is observed (as expected for a simple single-step binding process). Growth, induction, labeling, and titration of FL-pc-QDs of 4M5.3 surface-displaying yeast are carried out as described previously (Chao et al., 2006):

1. Incubate α-FL scFv-induced yeast cells with 1 pM to 600 nM FL-pc-QDs for 30 min at room temperature or 4 °C in low-binding microcentrifuge tubes on a rotator. The rotator should be incubated in the dark or the tubes be covered with aluminum foil.
2. Spin cells at 13,000 rpm for 30 s and wash pellet three times with 1× PBS.
3. Transfer labeled yeast cells to FACS tubes.
4. Turn on BD FACSAria (Becton Dickinson, Franklin Lakes, NJ) flow cell sorter equipped with BD FACSDiVa software for at least 30 min. The cytometer has three lasers emitting at 488, 633, and 405 nm. The 488 nm laser excites FITC, whose emission is detected using a 530/30 filter. The 405 nm laser excites QDs, whose emission is detected using a 610/20 filter.
5. Obtain a forward versus side scatter profile for unlabeled yeast cells. If necessary, set compensation, although in our hands, a single control—cells stained for FITC indicated that no compensation was necessary. Compensation (PMT settings) may be required when using multiple fluorophores whose emission signals overlap.
6. Import FACS data in ASCII or txt format and analyze using software written in LabView (National Instruments, Austin, TX) to provide better control of the cell selection criteria (or gates).
7. Plot the (arithmetic) mean fluorescence signal of the gated cells in each channel (FL or QD). Make sure to reject cells with fluorescence signal <0 (in either the FL or QD channel), which should correspond to cells with signal below background level when titrating with 10%, 50%, and 100% FL-QDs as described in Iyer et al. (2008).
8. Fit the percentage of cells retained for analysis with the same simple binding model (Chao et al., 2006) used for the fluorescence signal itself. Obtain dissociation constants K_d for the FL and QD signal, respectively.

In summary, these measurements of the dissociation constant of the α-FL scFv/FL-QD pair using yeast as a model system can be optimized by the experimenter for further single-molecule imaging experiments in cells. We have observed that the increased dissociation constant measured (6.9 ± 5.1 nM) did not prevent single-molecule observation (as reported subsequently), showing that relatively large dissociation constants (in the nM) range are not an obstacle for this kind of application (Iyer *et al.*, 2008).

6. DNA Constructs for Single FL-QD Imaging in Live Cells

A typical application for the FL-QD/4M5.3 affinity pair is to fuse 4M5.3 to the N- or C-terminus of a membrane or glycophosphatidyl (GPI)-anchored protein. Fusions to cytoplasmic proteins have not been attempted especially since delivery of QDs into the cytoplasm is not straightforward and appropriate controls for folding of the 4M5.3 fusion protein has to be tested on a case-by-case basis. A typical DNA fusion construct for mammalian cells is illustrated in Fig. 3.3. To test if the FITC pc-QDs are suitable for single-particle tracking experiments in live mammalian cells, we designed the construct by fusing the anti-scFv 4M5.3 to a GPI-anchored protein so that the single-chain antibody is well exposed on the outer membrane. To construct anti-scFv 4M5.3–PrP fusion protein, the coding cDNA sequence of anti-scFv 4M5.3 was inserted to the immediate distal end of the signal peptide cleavage site (aa-23) of full length mouse PrP protein-bearing epitope 3F4 (Iyer *et al.*, 2008). The resulting construct is still anchored via the GPI anchor to the outer leaflet of the plasma membrane. The chimeric scFv–PrP construct is then cloned into mammalian expression vector pCDNA3 (Invitrogen) and transfected into mouse neuroblastoma cell line Neuro-2a (N2a, ATCC) using lipofectamine 2000 reagent (Invitrogen) following the manufacturer's protocol. Single stable clones are selected, isolated, and maintained by growth medium supplemented with 0.8 mg ml^{-1} G418 (Invitrogen).

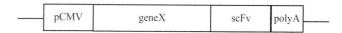

Figure 3.3 Schematic map of a mammalian DNA construct composed of gene X fused to single-chain fragment antibody anti-scFv 4M5.3 at the C-terminus. Similar fusions can be engineered to the N-terminus and also to GPI anchored protein. Reprinted with permission from Gopal Iyer *et al.* High affinity scFv- Hapten Pair as a Tool for Quantum Dot Labelling and Tracking of Single Proteins in Live Cells, Nano letters, 8 (12). Copyright 2008, American Chemical Society.

6.1. Labeling of live mammalian cells

This procedure can be adapted for most mammalian cell lines. Here, we describe the procedure for labeling mouse neuroblastoma cell line Neuro-2a (N2a from ATCC):

1. 35 mm culture dishes with No 1.0 or 1.5 glass bottom (Mattek Corp., MA) are coated with 0.1% poly-L-lysine for several hours, and then washed with PBS before cell plating.
2. Two days before microscopy experiments, N2a cells are plated at a density of 2×10^5 cells on the coated glass-bottom culture dish and maintained in DMEM supplemented with 10% (v/v) fetal bovine serum and 1× pen/strep (penicillin and streptomycin—100× stock). The cells are allowed to grow for 48 h at 37 °C with 5% CO_2.
3. Wash the cells once with Hank's buffer salt solution (HBSS, Invitrogen) to remove residual medium.
4. Block cells with 1% (w/v) bovine serum albumin (BSA) in HBSS for 1 h at 37 °C.
5. For ensemble labeling, incubate 1–10 nM FL-QDs with 1% (w/v) BSA in HBSS for 10 min at room temperature. For single-molecule imaging, use 2–10 pM FL-QDs under identical conditions.
6. Wash 4–5 times with HBSS buffer and if necessary, image cells using an epifluorescence microscope to optimize the wash conditions.
7. Uniform membrane staining should be visible in the green and red channels.

7. SINGLE-MOLECULE IMAGING OF LIVE MAMMALIAN CELLS

Single-molecule imaging experiments are carried out using total internal reflectance (TIRF) excitation (IX71, Olympus) and a high numerical aperture (NA) oil immersion objective (Olympus Planapo 60X, NA 1.45). The detection path is home-built and permits dual-color imaging, as described in Pinaud et al. (2009). Images are acquired by a back-illuminated EMCCD camera (Cascade 512B, Photometrics) controlled by the WinView software (Photometrics). Since a custom magnification is used, the size of the camera pixel in the object plane is calibrated using a reticle with 10 μm pitch ruling. In our experiments, the pixel size was 95 nm per pixel. The lower limit on time resolution depends on the number of collected photons required for good signal-to-noise ratio. Typically, single fluorophores are imaged at rates of 200 Hz or less, with a maximum number of emitted photons before photobleaching near 10^6:

1. N2a cells expressing 4M5.3-prp chimera are grown in glass-bottom culture dish (Mattek Corp.) for room temperature imaging. Alternatively, cells are imaged using the 6-well glass-bottom chamber plates (Wafergen, CA) that can be programmed for imaging at 37 °C.
2. Switch to bright-field imaging to focus onto a region of interest and record this image. If desired, record a DIC image of the cells.
3. To image FL-QDs, the samples are excited by a fiber-coupled 488-nm laser line at 2–5 mW from an argon laser. Images are further magnified and filtered through a band pass filter (620BP40 Semrock) for FL-QD imaging. In the case of dual-color imaging, FL-QDs and GFPs are imaged simultaneously by a custom-built dual-view system with dichroic mirror 560LP and filters 628BP40 and 525BP50 for the red and green channel, respectively.
4. Using the fluorescence excitation, focus on the cells and record a fluorescence image.
5. Switch to TIRF angle mode and adjust the focus to the basal membrane of individual adherent cells. Individual fluorescent spots can be detected with the camera or through the eye pieces. Acquire a TIRF image if desired.
6. A sequence of images can then be acquired to follow the movement of individual fluorescent spots. In our experiments, a minimum of 1000 frames at 100 ms per frame is typically recorded for single-particle tracking as seen in Fig. 3.4A and B.
7. Single-particle tracking is performed using custom in-house software (AsteriX) written in LabVIEW (Pinaud et al., 2009). Briefly, individual QD images are fitted by a Gaussian intensity profile, resulting in sub-pixel determination of the center position. Single-particle trajectories can then be analyzed by different approaches, as described (Pinaud et al., 2009).
8. Fluorescent intensity along a single-time trace (camera photo counts per frame) along the trajectory is the determining factor to exclude aggregate from single FL-pc-QDs. A blinking intensity time trace is characteristic of a single QD, illustrated in Fig. 3.4C and D, but is not a necessary condition. Absolute time trace intensity significantly larger than the average may indicate the observation of an aggregate, especially if no blinking is observed.

ACKNOWLEDGMENTS

We thank Professor Wittrup for providing the 4M5.3 scFv plasmid, Professor Lindquist for the mouse PrP plasmid, and Dr. Thomas Dertinger for help with FACS data import. We acknowledge the help of Dr. Ingrid Schmid, supervisor of the UCLA Jonsson Comprehensive Cancer Center (JCCC) and Center for AIDS Research Flow Cytometry Core Facility that is supported by National Institutes of Health awards CA-16042 and AI-28697, and by the JCCC, the UCLA AIDS Institute, and the David Geffen School of Medicine at UCLA.

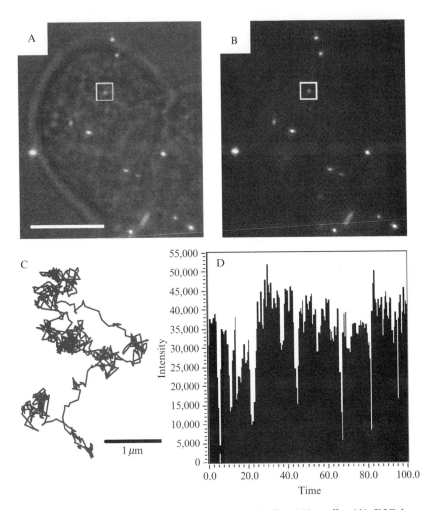

Figure 3.4 Imaging of scFv–PrP with FL-QDs in live N2a cells. (A) DIC image overlaid with QD signal (white dots), scale bar = 10 μm. (B) Fluorescent image. (C) 1000 frames single-QD trajectory of QD indicated in white square. (D) Intensity time trace (camera counts per frame) along the trajectory exhibiting a blinking pattern of single QD. Iyer G, Pinaud F, Tsay J, Weiss S : Solubilization of quantum dots with a recombinant peptide from *Escherichia coli*. Small Vol 3. Issue 5 Pages 793–798. Copyright Wiley-VCH Verlag GmbH & Co. KGaA. Reproduced with permission.

Ensemble confocal imaging was performed at the UCLA/CNSI Advanced Light Microscopy Shared Facility. This work was supported by NIH/NIBIB BRP Grant 5-R01-EB000312 and the NSF, The Center for Biophotonics, an NSF Science and Technology Center managed by the University of California, Davis, under Cooperative Agreement PHY0120999.

REFERENCES

Andresen, M., Schmitz-Salue, R., and Jakobs, S. (2004). Short tetracysteine tags to β-tubulin demonstrate the significance of small labels for live cell imaging. *Mol. Biol. Cell.* **15**, 5616–5622.

Betzig, E., Patterson, G. H., Sougrat, R., Lindwasser, O. W., Olenych, S., Bonifacino, J. S., Davidson, M. W., Lippincott-Schwartz, J., and Hess, H. F. (2006). Imaging intracellular fluorescent proteins at nanometer resolution. *Science* **313**, 1642–1645.

Bhunia, A. K., and Miller, S. C. (2007). Labeling tetracysteine-tagged proteins with a SplAsH of color: A modular approach to bis-arsenical fluorophores. *Chem. Bio. Chem.* **8**, 1642–1645.

Boder, E. T., and Wittrup, K. D. (2000). Yeast surface display for directed evolution of protein expression, affinity, and stability. *Meth. Enzymol.* **328**, 430–444.

Boder, E. T., Midelfort, K. S., and Wittrup, K. D. (2000). Directed evolution of antibody fragments with monovalent femtomolar antigen-binding affinity. *Proc. Natl. Acad. Sci. USA* **97**, 10701–10705.

Brichta, J., Hnilova, M., and Viskovic, T. (2005). Generation of hapten-specific recombinant antibodies: antibody phage display technology: A review. *Vet. Med. Czech.* **50**, 231–252.

Bruchez, M., Jr., Moronne, M., Gin, P., Weiss, S., and Alivisatos, A. P. (1998). Semiconductor nanocrystals as fluorescent biological labels. *Science* **281**, 2013–2016.

Carrico, I. S., Carlson, B. L., and Bertozzi, C. R. (2007). Introducing genetically encoded aldehydes into proteins. *Nat. Chem. Biol.* **3**, 321–322.

Chang, Y. P., Pinaud, F., Antelman, J., and Weiss, S. (2008). Tracking bio-molecules in live cells using quantum dots. *J. Biophoton* **1**, 287–298.

Chao, G., Lau, W. L., Hackel, B. J., Sazinsky, S. L., Lippow, S. M., and Wittrup, K. D. (2006). Isolating and engineering human antibodies using yeast surface display. *Nat. Protoc.* **1**, 755–768.

Chudakov, D. M., Lukyanov, S., and Lukyanov, K. A. (2005). Fluorescent proteins as a toolkit for in vivo imaging. *Trends Biotechnol.* **23**, 605–613.

Dabbousi, R. O., Rodriguez-Viejo, J., Mikulec, F. V., Heine, J. R., Mattoussi, H., Ober, R., Jensen, K. F., and Bawendi, M. G. (1997). (CdSe)ZnS core-shell quantum dots: synthesis and characterization of a size series of highly luminescent nanocrystallites. *J. Phys. Chem. B* **101**, 9463–9475.

Dahan, M., Levi, S., Luccardini, C., Rostaing, P., Riveau, B., and Triller, A. (2003). Diffusion dynamics of glycine receptors revealed by single-quantum dot tracking. *Science* **302**, 442–445.

Duckworth, B. P., Zhang, Z., Hosokawa, A., and Distefano, M. D. (2007). Selective labeling of proteins by using protein farnesyltransferase. *Chem. Bio. Chem.* **8**, 98–105.

George, N., Pick, H., Vogel, H., Johnsson, N., and Johnsson, K. (2004). Specific labeling of cell surface proteins with chemically diverse compounds. *J. Am. Chem. Soc.* **126**, 8896–8897.

Giepmans, B. N., Adams, S. R., Ellisman, M. H., and Tsien, R. Y. (2006). The fluorescent toolbox for assessing protein location and function. *Science* **312**, 217–224.

Hibino, K., Hiroshima, M., Takahashi, M., and Sako, Y. (2009). Single-molecule imaging of fluorescent proteins expressed in living cells. *Meth. Mol. Bio.* **544**, 451–460.

Hoffmann, C., Gaietta, G., Bünemann, M., Adams, S. R., Oberdorff-Maass, S., Behr, B., Vilardaga, J. P., Tsien, R. Y., Ellisman, M. H., and Lohse, M. J. (2005). A FlAsH-based FRET approach to determine G protein – Coupled receptor activation in living cells. *Nat. Meth.* **2**, 171–176.

Howarth, M., and Ting, A. Y. (2008). Imaging proteins in live mammalian cells with biotin ligase and monovalent streptavidin. *Nat. Protocols* **3**, 534–545.

Howarth, M., Takao, K., Hayashi, Y., and Ting, A. Y. (2005). Targeting quantum dots to surface proteins in living cells with biotin ligase. *Proc. Natl. Acad. Sci. USA* **102**(21), 7583–7588.

Iyer, G., Pinaud, F., Tsay, J., and Weiss, S. (2007). Solubilization of quantum dots with a recombinant peptide from *Escherichia coli*. *Small* **3**, 793–798.

Iyer, G., Michalet, X., Chang, Y. P., Pinaud, F. F., Matyas, S. E., Payne, G., and Weiss, S. (2008). High affinity scFv-hapten pair as a tool for quantum dot labeling and tracking of single proteins in live cells. *Nano Letters* **8**, 4618–4623.

Joo, C., Balci, H., Ishitsuka, Y., Buranachai, C., and Ha, T. (2008). Advances in single-molecule fluorescence methods for molecular biology. *Ann. Rev. Biochem.* **77**, 51–76.

Kwon, G. S., and Hadjantonakis, A.-K. (2007). Eomes::GFP – A tool for live imaging cells of the trophoblast, primitive streak, and telencephalon in the mouse embryo. *Genesis* **45**, 208–217.

Lin, C. W., and Ting, A. Y. (2006). Transglutaminase-catalyzed site-specific conjugation of small-molecule probes to proteins in vitro and on the surface of living cells. *J. Am. Chem. Soc.* **128**, 4542–4543.

Lippincott-Schwartz, J., and Patterson, G. (2003). Development and use of fluorescent protein markers in living cells. *Science* **300**, 87–91.

Los, G. V., Encell, L. P., McDougall, M. G., Hartzell, D. D., Karassina, N., Zimprich, C., Wood, M. G., Learish, R., Ohana, R. F., Urh, M., Simpson, D., Mendez, J., *et al.* (2008). HaloTag: A novel protein labeling technology for cell imaging and protein analysis. *ACS Chem. Biol.* **3**(6), 373–382.

Michalet, X., Michalet, X., Kapanidis, A. N., Laurence, T., Pinaud, F., Doose, S., Pflughoefft, M., and Weiss, S. (2003). The power and prospects of fluorescence microscopies and spectroscopies. *Ann. Rev. Biophys. Biomol. Struct.* **32**, 161–182.

Michalet, X., Pinaud, F. F., Bentolila, L. A., Tsay, J. M., Doose, S., Li, J. J., Sundaresan, G., Wu, A. M., Gambhir, S. S., and Weiss, S. (2005). Quantum dots for live cells, in vivo imaging, and diagnostics. *Science* **307**, 538–544.

Miller, L. W., Sable, J., Goelet, P., Sheetz, M. P., and Cornish, V. W. (2004). Methotrexate conjugates: A molecular in vivo protein tag. *Angew Chem. Intl. Ed.* **43**, 1672–1675.

Miller, L. W., Cai, Y., Sheetz, M. P., and Cornish, V. W. (2005). In vivo protein labeling with trimethoprim conjugates: a flexible chemical tag. *Nat. Meth.* **2**, 255–257.

Murray, C. B., Norris, D. J., and Bawendi, M. G. (1993). Synthesis and characterization of nearly monodisperse CdE (E = S, Se, Te) semiconductor nanocrystallites. *J. Am. Chem. Soc.* **115**, 8706–8715.

Peng, X., Schlamp, M. C., Kadavanich, A. V., and Alivisatos, A. P. (1997). Epitaxial growth of highly luminescent CdSe/CdS core/shell nanocrystals with photostability and electronic accessibility. *J. Am. Chem. Soc.* **119**, 7019–7029.

Pinaud, F., King, D., Moore, H.-P., and Weiss, S. (2004). Bioactivation and cell targeting of semiconductor CdSe/ZnS nanocrystals with phytochelatin-related peptides. *J. Am. Chem. Soc.* **126**, 6115–6123.

Pinaud, F., Michalet, X., Iyer, G., Margeat, E., Moore, H. P., and Weiss, S. (2009). Dynamic partitioning of a glycosyl-phosphatidylinositol-anchored protein in glycosphingolipid-rich microdomains imaged by single-quantum dot tracking. *Traffic* **10**, 691–712.

Pons, T., and Mattoussi, H. (2009). Investigating biological processes at the single molecule level using luminescent quantum dots. *Ann. Biomed. Eng.* **37**(10), 1934–1959.

Roullier, V., Clarke, S., You, C., Pinaud, F., Gouzer, G. G., Schaible, D., Marchi-Artzner, V., Piehler, J., and Dahan, M. (2009). High-affinity labeling and tracking of individual histidine-tagged proteins in live cells using Ni2+ tris-nitrilotriacetic acid quantum dot conjugates. *Nano Letters* **9**, 1228–1234.

Sakaue-Sawano, A., Kurokawa, H., Morimura, T., Hanyu, A., Hama, H., Osawa, H., Kashiwagi, S., Fukami, K., Miyata, T., Miyoshi, H., Imamura, T., Ogawa, M., *et al.*

(2008). Visualizing spatiotemporal dynamics of multicellular cell-cycle progression. *Cell* **132**, 487–498.
Tsien, R. Y. (1998). The green fluorescent protein. *Annu. Rev. Biochem.* **67**, 509–544.
Villalobos, V., Naik, S., and Piwnica-Worms, D. (2007). Current state of imaging protein–protein interactions in vivo with genetically encoded reporters. *Annu. Rev. Biomed. Eng.* **9**, 321–349.
Wacker, S. A., Oswald, F., Wiedenmann, J., and Knöchel, W. (2007). A green to red photoconvertible protein as an analyzing tool for early vertebrate development. *Dev. Dyn.* **236**, 473–480.
Wang, T., Tomic, S., Gabdoulline, R. R., and Wade, R. C. (2004). How optimal are the binding energetics of barnase and barstar? *Biophys. J.* **87**, 1618–1630.
Watanabe, T. M., and Higuchi, H. (2007). Stepwise movements in vesicle transport of HER2 by motor proteins in living cells. *Biophys. J.* **92**, 4109–4120.
Weber, P., Ohlendorf, D. H., Wendoloski, J. J., and Salemme, F. R. (1989). Structural origins of high-affinity biotin binding to streptavidin. *Science* **243**, 85–88.
Wennmalm, S., and Simon, S. M. (2007). Studying individual events in biology. *Annu. Rev. Biochem.* **76**, 419–446.
Westphal, V., Rizzoli, S. O., Lauterbach, M. A., Kamin, D., Jahn, R., and Hell, S. W. (2008). Video-rate far-field optical nanoscopy dissects synaptic vesicle movement. *Science* **320**, 246–249.
Yu, W. W., Qu, L., Guo, W., and Peng, X. (2003). Experimental determination of the extinction coefficient of CdTe, CdSe, and CdS nanocrystals. *Chem. Mater.* **15**, 2854–2860.
Zhou, M., and Ghosh, I. (2007). Quantum dots and peptides: A bright future together. *Pept. Sci.* **88**, 325–339.

CHAPTER FOUR

RECORDING SINGLE MOTOR PROTEINS IN THE CYTOPLASM OF MAMMALIAN CELLS

Dawen Cai,[*,†] Neha Kaul,[‡] Troy A. Lionberger,[‡,§] Diane M. Wiener,[‡] Kristen J. Verhey,[*,†,§] and Edgar Meyhofer[†,‡,§]

Contents

1. Introduction	82
2. Basic Principles	84
3. Labeling Molecular Motors for *In Vivo* Observations	85
3.1. Organelles as labels for cytoplasmic motors	85
3.2. Quantum dots	87
3.3. Organic fluorophores and genetic tags	87
4. Instrumentation for Tracking Single Motors *In Vivo*	89
4.1. Multicolor TIRF illumination	90
4.2. Dark-field excitation to track nanoparticles	92
4.3. Recording single-molecule events	93
4.4. Dual channel image recording	94
4.5. Postimage processing	94
4.6. Microscope accessories	95
5. Detailed Experimental Procedures	95
5.1. Fluorescent protein-labeled motors	96
5.2. Fluorescent protein fusion plasmids	96
5.3. Cell culture	97
5.4. Transient expression of fluorescent protein-labeled kinesin motors and microtubule markers	99
5.5. Recording and analyzing single-molecule *in vivo* events	99
5.6. Recording image sequences	100
5.7. Identifying individual single motor events in live cells	100
5.8. SD maps	101
5.9. High-resolution tracking	102
6. Summary and Conclusions	103
Acknowledgments	103
References	104

[*] Department of Cell and Developmental Biology, University of Michigan, Ann Arbor, Michigan, USA
[†] Department of Biophysics, University of Michigan, Ann Arbor, Michigan, USA
[‡] Department of Mechanical Engineering, University of Michigan, Ann Arbor, Michigan, USA
[§] Program in Cellular and Molecular Biology, University of Michigan, Ann Arbor, Michigan, USA

Methods in Enzymology, Volume 475
ISSN 0076-6879, DOI: 10.1016/S0076-6879(10)75004-7

© 2010 Elsevier Inc.
All rights reserved.

Abstract

Biomolecular motors are central to the function and regulation of all cellular transport systems. The molecular mechanisms by which motors generate force and motion along cytoskeletal filaments have been mostly studied *in vitro* using a variety of approaches, including several single-molecule techniques. While such studies have revealed significant insights into the chemomechanical transduction mechanisms of motors, important questions remain unanswered as to how motors work in cells. To understand how motor activity is regulated and how motors orchestrate the transport of specific cargoes to the proper subcellular domain requires analysis of motor function *in vivo*. Many transport processes in cells are believed to be powered by single or very few motor molecules, which makes it essential to track, in real time and with nanometer resolution, individual motors and their associated cargoes and tracks. Here we summarize, contrast, and compare recent methodological advances, many relying on advanced fluorescent labeling, genetic tagging, and imaging techniques, that lay the foundation for groundbreaking approaches and discoveries. In addition, to illustrate the impact and capabilities for these methods, we highlight novel biological findings where appropriate.

1. Introduction

Biomolecular motors play crucial roles in a large number of cellular functions and are essential components of a carefully orchestrated transport system that maintains the structural and functional properties of cells. Prominent examples of these cellular functions include muscle contraction, vesicular transport in the cytoplasm, formation of the spindle apparatus, segregation of chromosomes during mitosis and meiosis, and powering the beating motion of cilia and flagella. The principal mechanism by which molecular motors accomplish these diverse functions is binding to and moving along cytoskeletal filaments (either microtubules or actin filaments) by transducing the chemical energy available from the hydrolysis of ATP into mechanical work (Block, 2007; Gennerich and Vale, 2009; Schliwa and Woehlke, 2003). Because of the central importance of these mechanisms to biology, a major research effort has been directed toward understanding the molecular mechanism underpinning biomolecular motor function. Spurred by the discovery that the motility of single kinesin motors along microtubules can be reconstituted from purified proteins in cell-free assays (Howard *et al.*, 1989; Vale *et al.*, 1985), a large number of progressively more sophisticated single-molecule techniques, including laser trapping and fluorescence tracking experiments, have been developed to precisely characterize the stepwise movements and forces that molecular motors generate as they translocate along their respective cytoskeletal filaments (Meyhofer and Howard, 1995; Ray et al., 1993; Svoboda et al., 1993; Visscher et al., 1999; Yildiz et al., 2004). For example, such studies have demonstrated that Kinesin-1 generates maximum forces of about 6 pN and moves processively by taking 8 nm steps. These studies have also shown that kinesin steps over the microtubule lattice parallel to the protofilaments by coordinating the alternating catalytic activity of its two motor domains. While attention of such biophysical work is now shifting toward deciphering the molecular details of how motion and forces are generated and how coordination between the heads is accomplished (Block, 2007), critical questions need to be addressed with regard to the function and regulation of molecular motors in cells in order to understand how different kinesin motors function *in vivo*. Key questions, for instance, include how biomolecular motors move in the crowded cellular environment, how many motors are involved in specific transport processes, how different motors interact in the transport of specific cargoes, how individual motors are directed to specific cellular targets, and how motor activity is regulated within cells. Significant evidence supports the view that relatively few motors, perhaps even single molecules, power many of the transport processes (Laib et al., 2009; Shubeita et al., 2008). Therefore, addressing these challenging problems not only requires analysis of motor function *in vivo*, but also makes it essential to track, in real time and with nanometer resolution, individual motor molecules and their associated cargoes, cytoskeletal filaments, and other proteins in living cells.

Over the last few years, significant technical progress has been made in a number of closely related approaches that make it now possible to follow individual motor molecules in living cells (Cai et al., 2007, 2009; Courty et al., 2006; Kulic et al., 2008; Kural et al., 2005; Nan et al., 2005, 2008). Generally, these techniques rely on labeling motors with one of a rapidly expanding variety of fluorescent labels or with bright, scattering nanoparticles, and on imaging these labels in living cells using low-background confocal, total internal reflection fluorescence, or dark-field microscopy. With the availability of highly sensitive digital cameras, it has also become readily possible to acquire even very faint signals with video resolution, enabling the localization of diffraction-limited spots with subpixel accuracy by fitting appropriate functions to the acquired intensity distributions. As long as it can be assured that individual particles or molecules are being detected and tracked, the position of these spots can be determined with a resolution that is, in principle, only limited by the number of photons that are available to estimate their precise location (Thompson et al., 2002). Using these new approaches, many of which were initially applied to the detailed analysis of motility *in vitro*, recent work has demonstrated the stepwise movement of motors *in vivo*, presented significant support that Kinesin-1 behaves *in vitro* similarly as *in vivo*, and revealed that kinesin motors differ in their ability to select specific subsets of microtubule tracks in cells (Cai et al., 2007, 2009; Kural et al., 2005, 2007; Levi and Gratton, 2007; Nan et al., 2005, 2008).

In this chapter, we review the methodological foundations, with regard to both the required instruments and biological techniques that have been successfully applied to record single or few motors in living cells. In addition to emphasizing the approaches we have taken in our own work (Cai et al., 2007, 2009), we contrast the advantages and drawbacks of different single-molecule imaging and recording techniques and evaluate how the different characteristics meet the requirements of biological studies of *in vivo* motor function.

2. Basic Principles

Imaging of single biomolecules in live cells is based on a single particle tracking approach where individual molecules are labeled with high contrast or bright, fluorescent particles or labels such that their centroids can be determined with high resolution from image sequences. Beginning with relatively large latex particles *in vitro* (Gelles et al., 1988), the techniques were first successfully applied to *in vivo* studies by characterizing biological processes at the plasma membrane (Barak and Webb, 1982; Cherry, 1992; Kusumi et al., 2005; Sako and Yanagida, 2003; Saxton and Jacobson, 1997; Schmidt et al., 1996). Tracking single molecules inside intact cells has been technically more challenging because of the relatively large cellular autofluorescence and the need to introduce labeled components into cells. By leveraging recent improvements in various (fluorescent) labels, microscopy techniques and CCD detectors, the basic particle tracking approaches have now been successfully adapted to image intracellular events with nanometer resolution (Cai et al., 2007; Courty et al., 2006; Michalet et al., 2005; Nan et al., 2005). Total internal reflection fluorescence microscopy (TIRFM) has played a crucial role in realizing these measurements because of the associated reduction in background fluorescence. In addition, the availability of reliable, low-cost solid-state lasers at multiple wavelengths as efficient and flexible light sources for fluorophore excitation and particle scattering, and sensitive electron-multiplication CCD (EMCCD) cameras have facilitated precise image recording and quantitative analysis of microscopy data that are essential for nanometer-resolution tracking of single molecules. Intriguingly, these same basic single particle tracking principles are now being further extended and developed in a number of super-resolution imaging techniques (like photoactivated location microscopy (PALM) and stochastic optical reconstruction microscopy (STORM)) to circumvent the diffraction limited performance of the conventional light microscope (see e.g., Huang et al., 2008).

3. LABELING MOLECULAR MOTORS FOR IN VIVO OBSERVATIONS

Perhaps the most critical and challenging step in experimentally analyzing the cellular function and molecular mechanisms of motors in living cells is that light-emitting or light-scattering molecules or particles have to be attached to the motors of interest to track their movement and characterize their interactions with other cellular components. For high-resolution tracking (temporal and spatial) these particles need to be optically bright and photostable, yet neither the label nor the labeling procedure should alter or influence the functional or physiological properties of the motor or other relevant cellular constituents. The labeling should be highly specific to individual molecules and not induce oligomerization or clustering of labels and molecular motors. Also, the simultaneous labeling of molecular motors, cytoskeletal tracks, cargoes, scaffolding proteins, linkers or regulatory components with multiple, distinguishable (e.g., spectroscopically distinct) fluorophores is highly desirable to study the molecular mechanisms of biomolecular motor-based transport. Clearly, these are conflicting criteria that need to be carefully balanced, given the biological question at hand. To date, there is no single best label, and not surprisingly, the development of new labels is attracting much attention (see e.g., Chang *et al.*, 2008; Johnsson, 2009; Shaner *et al.*, 2005, 2008). Below, we discuss various labeling and single particle tracking approaches (ranging from organelle labels to single organic fluorophores) that have been employed for studying (single) motors in live cells.

3.1. Organelles as labels for cytoplasmic motors

The discovery of kinesin (Kinesin-1) as a motor for fast axonal transport is explicitly linked to the study of organelle transport (Brady, 1985; Vale *et al.*, 1985). In fact, organelle transport had been investigated for many years prior to the discovery of kinesins, leading to important technical advances like video-enhanced differential interference contrast (DIC) (Allen *et al.*, 1981; Salmon and Tran, 1998). It is therefore not surprising that organelles, which are relatively large in size and are known to be transported by various motors, have been utilized to analyze the movement of motor molecules in live cells. A relatively simple, but powerful approach is to take advantage of the natural labeling of melanosomes (melanin-containing, pigmented organelles in melanophores), which can be visualized directly in bright-field microscopy due to the high contrast between the melanosomes and the background of the cytoplasm (Gross *et al.*, 2002; Kural *et al.*, 2007; Levi *et al.*, 2006). In a recent study, Kural *et al.* (2007) were able to use this approach and track the position of individual melanosomes with ~2 nm spatial and 1.1 ms temporal resolution by fitting a 2D Gaussian function to the reversed (or negative) bright-field image of melanosomes. This method (bFIONA or bright-field FIONA) is an extension of the previously developed, fluorescence-based FIONA (fluorescence imaging with one-nanometer accuracy, see Section 4.5) (Yildiz et al., 2003). Since it is now quite well established that myosin V, cytoplasmic dynein, and Kinesin-2 motors are all located on and involved in the movement of these organelles in melanophores, selective depolymerization of either actin filaments or microtubules (with latrunculin B or nacodazole, respectively) reduces motor interactions to microtubule-based dynein and kinesin motors or actin filament-based myosin V. Using this approach, the authors were able to characterize the anterograde and retrograde stepping ($\sim$8 nm) of Kinesin-2 and dynein along microtubules, and they showed that myosin V transported organelles along actin filaments in 35-nm increments. Most intriguingly, in the presence of both microtubules and actin filaments, the melanosomes moved along actin filaments and microtubules nearly simultaneously, indicating that clearly distinct diffusion events for switching from one cytoskeletal filament to the other are not necessary.

To track the motion of molecular motors and organelles in nonmelanophore cells, approaches have been developed to intensely label organelles for single particle tracking. The movement of peroxisomes by kinesin and dynein motors was recorded by Kural et al. (2005) and subsequently Kulic et al. (2008) by constitutively expressing EGFP (enhanced green fluorescent protein) with a peroxisome targeting sequence in cultured *Drosophila* S2 cells, which leads to the accumulation of a large number of EGFP molecules in these organelles. Because of the large number of photons that can be collected from the labeled peroxisomes, it is possible to determine the centroid location of labeled, single peroxisome particles with about 1 ms and 1.5 nm accuracy. Nan et al. (2005) followed the same basic strategy and introduced bright, fluorescent quantum dots (QDs, see Section 3.2) into endocytotic vesicles of human lung cancer (A549) cells by exposing the cells directly to streptavidin and biotin-polyarginine-coated QDs. Typically, 5–30 QDs were taken up into individual endocytotic vesicles. Because of their bright fluorescence signals, these labeled endocytotic vesicles can be tracked with high temporal ($\sim$0.3 ms) and spatial ($\sim$1.5 nm) resolution following the same single particle or centroid tracking approach outlined above. The same group has now extended this work (Nan et al., 2008) to even higher resolution, by replacing the QD labels with 100–150-nm-diameter gold nanoparticles. Such large gold particles generate very strong scattering signals when excited by laser excitation and objective-based dark-field microscopy, and they are also very photostable. Using this gold nanoparticle system, it is now possible to follow organelles *in vivo* in the cytoplasm at 25-μs and 1.5-nm resolution. With this improved methodology, Nan et al. present evidence that cytoplasmic dynein takes 8, 12, 16, 20, and 24-nm steps over the full range of physiologic velocities, while kinesin steps consistently in 8-nm intervals.

3.2. Quantum dots

QDs are semiconductor nanocrystals with highly desirable fluorescence properties for tracking motors inside live cells (see e.g., Alivisatos et al., 2005; Michalet et al., 2005). QDs are about 20-fold brighter, and vastly more photostable (by orders of magnitude) than organic dye molecules and fluorescent proteins. To ensure solubilization and functionality, the nanocrystal core of commercially available QDs is encapsulated by a shell and polymer-coat to link a variety of bioconjugation molecules (like streptavidin or antibodies) to the QD surface. Thus, the size of these particles typically falls into the range of 15–20 nm. The disadvantages to motor-QD couplings are that the linked molecules must be generated *in vitro* and then introduced into cells and that the stoichiometry of motors on an individual QD cannot be controlled. The physical size of the QD, which is often larger than the actual motor domain under investigation, can also be a concern.

Courty et al. (2006) directly attached truncated, biotinylated *Drosophila* Kinesin-1 motor domains to QDs at two stoichiometries, confirmed motility of the kinesin-QD conjugates in *in vitro* motility assays, and then loaded the kinesin-QDs into the cytoplasm of HeLa cells with a cell loading technique that is based on osmotic lysis of pinocytotic vesicles. Alternatively, microinjection of the QDs into the cells was used, but Courty et al. (2006) report that the pinocytotic loading technique proved to be more reliable. Kinesin-labeled QDs located uniformly in the cytoplasm of cells but were excluded from the nucleus. Moreover, Courty et al. showed that the labeled particles were not aggregating, by using fluorescence intermittency and mixtures of QDs emitting at 605 and 655 nm as signatures for possible aggregation of the QDs. By means of conventional fluorescence microscopy, they observed (as expected from the labeling ratios) both diffusing QDs and QDs exhibiting linear trajectories. Analysis of the linear trajectories showed that single QDs were transported at a velocity of ~ 0.6 μm/s and over distances of about 1.7 μm, in agreement with single-molecule *in vitro* properties of Kinesin-1 from *Drosophila*. More recently, Nelson et al. (2009) developed a similar QD-based method and successfully tracked the movement of myosin Va motors in COS cells. Interestingly, their work implies that myosin Va motors move processively and at the same time undergo a random walk through the dense and randomly oriented cortical actin network.

3.3. Organic fluorophores and genetic tags

With the systematic development of an array of fluorescent proteins and genetic tags, we now have available a large library of fluorescent labels that offer the advantage of being genetically (and therefore precisely) attached to the protein of interest with exactly known stoichiometries (Davidson and Campbell, 2009; Shaner et al., 2005, 2008; Snapp, 2009). In addition, these and deflect the excitation wavelengths in fast sequence toward the sample while maintaining broadband spectral transmission suitable for all fluorophore emission profiles (Ross et al., 2007).

4.2. Dark-field excitation to track nanoparticles

Alternative single particle tracking methods capable of enabling both high temporal bandwidths and high spatial resolution in live cells include using gold nanoparticle labels or plasmon coupling between two gold nanoparticles. However, both techniques require specialized excitation illumination. Utilizing the scattering signal from gold nanoparticles is highly advantageous because their photostability allows high excitation intensities and consequently high temporal resolution and long observation times. Molecular motors labeled with gold nanoparticles have been used in tracking experiments *in vitro* (Dunn and Spudich, 2007) and *in vivo* (Nan et al., 2008), using dark-field illumination with laser excitation to achieve 25-μs temporal resolution and 2-nm spatial resolution (Braslavsky et al., 2001; Nan et al., 2008; Nishiyama et al., 2001). The optical path of objective dark-field illumination is similar to objective-based TIRF microscopy (Fig. 4.1B). However, the dichroic mirror in TIRF microscopy is replaced by a small mirror (Fig. 4.1B, M) that is $\sim$5 mm diameter and attached to a glass substrate. By carefully aligning the excitation source, this mirror reflects the incoming light into the back-focal plane of the microscope objective. The collected, scattered light passes through the glass substrate and is measured with a CCD camera or photodiode. Because the imaging planes of the scattered light (which is necessarily divergent as it is emitted from the specimen) and reflected light (which is collimated) are different, the reflected illumination can be effectively blocked by an appropriately placed aperture (Fig. 4.1A and B). Another experimental approach to measuring distances within live cells, plasmon coupling (Jun et al., 2009;

Ultima U100) and an appropriate dichroic mirror (D3, multiband laser dichroic from Chroma Technology) into the back focal plane of the TIRF microscope objective. The TTM allows precise alignment of the laser excitation sources for TIRF. Fluorescence emission signals (here shown as light and dark gray paths) are transmitted through the dichroic mirror (D3) and imaged by the tube lens (TL). A custom dual-view system splits (via dichroic mirror D4) the image into two-color components and projects them side-by-side (with mirrors (M) and lens (IL) onto the sensor of a single EMCCD camera (Photometrics 512B Cascade). (B) Objective Dark-field. A small mirror (M) attached to an antireflection-coated glass substrate (G) reflects the illumination laser beam (solid line) into the back-focal plane of the microscope objective lens (OL). The scattered light from gold nanoparticles (dashed line) is collected by the objective, passed through the glass substrate, and is imaged by the TL and a magnifying imaging telescope (IT) onto a CCD. Back-reflected illumination light from the glass coverslip is blocked by an aperture (A) to increase the signal-to-background ratio. (See Color Insert.)

Rong et al., 2008), maintains the advantages of dark-field illumination and is additionally capable of reporting the distance between two gold nanoparticles (to within 15 nm *in vivo*) by measuring their spectral response (Reinhard et al., 2005). Despite the advantages of gold nanoparticle tracking, large gold nanoparticles must be used (40–200 nm in diameter) to achieve the desired temporal and spatial resolution, which limits their applicability to many biological problems.

4.3. Recording single-molecule events

The most versatile and widely used devices for capturing images of single molecules *in vivo* are digital CCD cameras. They offer high quantum efficiency over a broad range of wavelengths, pixel-limited resolution, and a wide dynamic signal range. However, the readout noise from CCD chips is significant when dim, single-molecule events are imaged. Moreover, following the above discussion on labels for observing single motor events in cells, these measurements will always be limited by the number of photons emitted from a given label and collected by the CCD. Specifically, the spatial and temporal resolutions are both highly dependent on the number of collected photons, and therefore are inversely related. During recent years, the development of EMCCD sensors by E2V and Texas Instruments has led to significant technical improvements that enabled the development of cameras with >90% quantum efficiency in the wavelength range from 500 to 700 nm (now available from a number of camera manufacturers, including Andor, Roper Scientific, and Hamamatsu). The on-board gain function of these EMCCD cameras circumvents previous limitations in the readout noise of CCD devices to achieve single photon sensitivity. The cameras also feature high readout rates (10–12 MHz) leading to full frame transfer rates of >30 Hz for 512 × 512 images and frame rates in excess of 1 kHz for smaller frame transfer chips and subarrays. In our recent work, we used a Photometrics 512B Cascade model from Roper Scientific with success (Cai et al., 2007, 2009). This camera is based on a back-illuminated, frame transfer CCD device (CCD97 from E2V) with a 512 × 512-pixel imaging array, on-chip electron multiplication gain, and the highest available quantum efficiency of >90%. The camera is thermoelectrically cooled to −30 °C, supports fast readout at a rate of 10 MHz, and is a considerable improvement over intensified CCD cameras that we used in past work. In general, EMCCD cameras have played a critical role in enabling high-resolution single particle tracking experiments (highlighted by FIONA, Yildiz et al., 2003, and recent super-resolution microscopy; see e.g., Betzig et al., 2006; Huang et al., 2008).

To further increase the temporal resolution in motor tracking experiments with large gold nanoparticles (100–150 nm), Nan et al. (2008) developed a novel detection strategy that uses dark-field microscopy and a high-bandwidth quadrant photodiode to determine the centroid of the gold particles (Nan et al.,

2008). Any displacement of a single gold nanoparticle detected (in 2D) with the photodiode is immediately offset via a feedback loop-based adjustment of a piezoelectric sample stage. Because of the high bandwidth of the photodiode and the bright scattering signal from the gold nanoparticle, this unique detection method achieves a spatial (1.5 nm) and temporal (40 kHz) resolution that is not possible with current EMCCD cameras.

4.4. Dual channel image recording

Dual-color imaging for truly simultaneous acquisition of images from two fluorophores can be achieved by splitting the emitted fluorescence signals for side-by-side capture by a single camera and image processor. This can be achieved through either an internal configuration, wherein the beam splitter is inserted between the objective and eyepiece, or an external configuration, wherein the beam splitter is placed outside the microscope before the CCD camera (Kinosita et al., 1991). Commercial dual-view attachments for microscopes (such as those offered by Hamamatsu, Optical Insights, and Photometrics) are designed based on the above-mentioned configurations. Although they are equipped with exchangeable filter cubes, the position of the lenses is not adjustable to independently focus each channel. Since the microscope objectives used for TIRFM typically have extremely high numerical apertures (NAs), it is impossible to achieve practical, apochromatic designs over the full visible spectrum (i.e., the two images may not be parfocal). For example, when we simultaneously imaged mCit and mCherry fusion proteins in our TIRF system with a 1.45 NA, 100× α-Plan-Fluar (Zeiss) and an Optical Insights Dual-View system, the two images were noticeably focused to different imaging planes. As a solution to this problem, we inserted a long focal distance "contact" lens (focal length $\sim$150 mm) into the light path of the longer emission wavelength channel to further converge the beam and achieve critical focus of both images in the same plane.

4.5. Postimage processing

Originally, particle tracking methods were developed for the analysis of imaging data from *in vitro* motility experiments. By employing contrast-enhancing techniques such as DIC microscopy, the centroid of a reporter particle (submicron latex beads) could be determined with subpixel resolution using methods such as image cross-correlation to achieve a resolution of 1–2 nm with an acquisition rate of 30 Hz (Gelles et al., 1988). With the use of new fluorescent labels and imaging methods, techniques emerged to also track fluorophore labeled molecules with high spatial and temporal resolution. Most current localization techniques take advantage of the fact that the point spread function (PSF) of a diffraction-limited fluorescent particle in a conventional fluorescence microscope can be well approximated by a 2D Gaussian. Fitting single, diffraction-limited spots produced by fluorescent molecules with a 2D Gaussian function, popularly known as FIONA, has been used widely to characterize the single-molecule behavior of molecular motors both *in vitro* (Yildiz et al., 2003, 2004) and *in vivo* (Courty et al., 2006; Kural et al., 2005, 2007, 2009). FIONA has proved capable of localizing a target molecule to within 1.5 nm with millisecond resolution, and has recently been extended to colocalize two fluorescent dyes, each with unique spectral characteristics, labeling a single DNA. By spectrally discriminating the dyes prior to imaging onto a single CCD camera, this two-color FIONA, also known as SHREC (single-molecule high-resolution colocalization), has been demonstrated to localize two fluorophores within 10 nm of each other (Churchman et al., 2005).

4.6. Microscope accessories

Many standard mechanical microscope stages and stepper designs for sample manipulation employ gears for x–y positioning. Because such systems exhibit significant backlash, it is highly advisable to replace such systems with a precision, flexure hinge-based, piezoelectric design as available from several manufacturers (e.g., Physik Instrumente, Mad City Labs, Queensgate, Jenoptik). For studies aimed at retrospectively correlating live-cell imaging with immunolabeling of fixed cells, the ability to reproducibly locate the same cell in the same orientation is desirable. We have achieved this by designing a specialized, compact mounting frame to which tissue culture dishes with optical quality cover glasses are temporarily glued. The two perpendicular edges of the frame are accurately mounted on the microscope stage with the aid of three alignment pins that allow reproducible mounting with ~ 1-μm accuracy. The mounting frame and culture dish are held in place with microscope sample clips.

A critical aspect for live-cell imaging is to maintain physiological conditions for cells on the microscope stage. Temperature, humidity, CO_2 and O_2 tension, and the pH of the buffer should be sustained once cells are transferred to the microscope stage. This may be best achieved through the use of a micro CO_2 incubator and a small, objective-based, electrical heating system (Ince et al., 1983) during high resolution, vibration, and drift-sensitive measurements, rather than a large, commercial incubator system.

5. Detailed Experimental Procedures

As outlined above, the specificity of labeling and the nondisruptive introduction of motor molecules into the cytoplasm are critical to the experimental analysis of single motor molecules in live cells. Courty et al.

(2006), and more recently Nelson et al. (2009), followed the general approach used in in vitro studies and conjugated purified motor proteins to QDs (Courty et al., 2006; Nelson et al., 2009) and introduced them into the cells by pinocytosis and osmotic lysis (Okada and Rechsteiner, 1982). We prefer to use genetic tagging with fluorescent proteins to achieve the most physiological in vivo imaging conditions. Here we outline the methodological details that allowed us to track single kinesin motors in cells with fluorescent protein-based genetic tags.

5.1. Fluorescent protein-labeled motors

Transient expression of fluorescent protein-labeled molecules is used extensively in cell biology to study protein functions in living mammalian cells (Giepmans et al., 2006). This approach also offers the distinct advantage that expression and synthesis via the cell's own molecular machineries ensures proper folding and modifications of the kinesin motor and other cytoskeletal proteins. To generate sufficiently bright labeling, we engineered a three-tandem copy of mCit as label and fused it to Kinesin-1 (Cai et al., 2007). This method was subsequently leveraged for two-color TIRF experiments (Cai et al., 2009) in which we simultaneously imaged single kinesin motors and cytoskeletal microtubule tracks (Fig. 4.2).

5.2. Fluorescent protein fusion plasmids

As a bright variant of the jellyfish *Aequorea victoria* green fluorescent protein (GFP), mCit was chosen as fluorescence tag. Its longer emission wavelength (peak emission 527 nm) allows for high-efficiency emission collection while

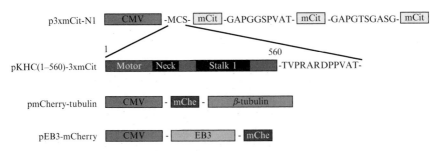

Figure 4.2 Design of plasmids for transfection of mammalian cells with fluorescent fusion proteins. The vector p3×mCit-N1 was constructed based on pEGFP-N1 backbone (Clontech) and contains three-tandem mCit fluorescent proteins proceeded by a multiple cloning site (MCS). A truncated, constitutively active, dimeric Kinesin-1 motor domain (heavy chain amino acids 1-560, KHC1-560) was then inserted in frame into the (MCS). The β-tubulin and EB3 genes were amplified by PCR and inserted in frame into the pmCherry-C1 and pmCherry-N1 vectors, respectively. Mammalian expression of these constructs is under the control of the CMV promoter.

using a 488-nm argon ion laser for excitation. To increase labeling brightness, a vector containing three-tandem copies of mCit (p3×mCit-N1) was created by PCR using the EGFP-N1 vectors (Clontech, Palo Alto, CA) as a backbone. Ten amino acid linkers (GAPGGSPVAT and GAPGTSGASG) connect the three copies of mCit (Fig. 4.2). In our initial studies, the three mCit polypeptides were encoded by the same nucleotide sequence. Occasional and random mutation or recombination of this sequence, presumably due to the three identical coding sequences, led us to redesign the 3×mCit tag such that the three mCit polypeptides are coded by different nucleotide sequences. Constitutively active Kinesin-1 motors are generated by deletion of the autoinhibitory tail domain of the kinesin heavy chain (KHC) subunit. These motors, KHC(1-891) or KHC(1-560), are fused to the N'-terminus of p3×mCit. Our initial studies utilized a 12-amino acid linker (Fig. 4.2, TVPRARDPPVAT) between KHC and the mCits, although shorter linker sequences are sufficient. We used the CMV promoter because its high expression efficiency made it possible to achieve proper expression levels for single-molecule imaging in just a few hours after transfection (Fig. 4.3). However, if single-molecule imaging over days is preferred, a truncated CMV promoter (Watanabe and Mitchison, 2002) or β-actin promoter (Dieterlen *et al.*, 2009) can be used for driving expression at lower levels over long periods. For two-color TIRF experiments, other polypeptides of interest can be labeled with other fluorescent proteins. In our experiments, we used mCherry as a second fluorescent protein (Shaner *et al.*, 2005). mCherry was fused to the α-tubulin subunit to label all microtubules or to end-binding protein 3 (EB3) to label dynamic microtubules. mCherry-tubulin and EB3-mCherry were generated by inserting α-tubulin and EB3 PCR products in frame to the C'- and N'-termini of mCherry, respectively (Fig. 4.2).

5.3. Cell culture

Fibroblast cells (e.g., COS cells) are ideally suited for live-cell imaging due to their hardiness and tolerance of exogenous gene expression. They attach readily to glass coverslips and their flat morphology is particularly suited for single-molecule microscopy because of the reduced background fluorescence. Other cells lines with flat morphology, such as PtK2, MRC-5, and LLC-PK1, are also well suited for live-cell imaging. COS cells are routinely cultured in 10-cm plates (Falcon or Corning) in growth medium (10% (v/v) FBS, 1% (w/v) L-glutamine, in DMEM medium) at 37 ° C in an incubator with 5% (v/v) CO_2. Normally, COS cells divide every 24 h when cultured at 10–90% confluence. For passaging or splitting cells for transfection, 70–90% confluent cultures are used. The media is first aspirated and the cells are washed once with 2 ml of sterile PBS buffer (pH 7.4). After removing the PBS wash buffer, 2 ml of 0.25% (w/v) trypsin is added to

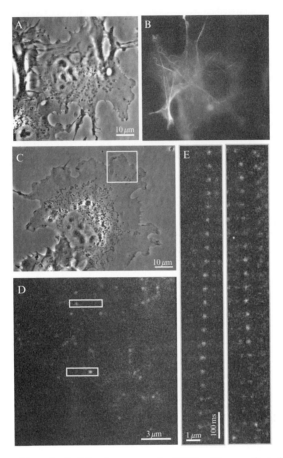

Figure 4.3 Tracking single kinesin motors in COS cells transfected with kinesin-3×mCit fusion protein. (A) Phase-contrast image of a cultured COS cells. (B) Epifluorescence image of a transfected COS cell expressing kinesin-3×mCit fusion protein at a high level. At a low expression level (C) the cell's periphery is well-suited for tracking single kinesin motors (white box, magnified view in (D)). A kymograph analysis of two single molecule events (white boxes in (D)) is shown in (E), indicating that both motors move processively at physiological rates (0.6 μm/s).

the plate. When the cells start to detach from the culture dish (2–10 min, depending on temperature), 2 ml of fresh culture medium is added to quench the trypsin. Cells should be completely separated from each other by pipetting before counting with a standard hemocytometer. 10^5 cells are added with 2 ml of fresh culture medium to a 35-mm cell culture dish containing a 25 mm × 25 mm, #1.5 cover glass (MatTek Corporation). The cells are returned to the incubator and allowed to settle for 6–20 h. Before transfection, the cells should be checked by phase contrast microscopy to confirm that most cells are attached to the cover glass and exhibit the typical flat morphology. Cells should reach ~50% confluence before transfection and should not be over 90% confluent at the time of imaging.

5.4. Transient expression of fluorescent protein-labeled kinesin motors and microtubule markers

To initiate the transient expression of labeled kinesin motors, we prefer TransIT-LTI (Takara Mirus Bio, Madison, WI) or Fugene6 (Roche) and follow the manufacturer's protocols.

1. Briefly, the culture medium (2 ml) in the 35-mm culture dish is exchanged 2–4 h before transfection. The transfection reagent is then used following the manufacturer's protocol.
2. The transfection reagent (stored at $-20\,^\circ$C) is warmed to room temperature 10 min before use. To transfect the kinesin-FP motor alone, 3 μl of gently vortexed transfection reagent is suspended thoroughly in 100 μl of serum-free Opti-Mem media (Invitrogen), followed by 0.5 μg of KHC-3×mCit plasmid DNA (DNA concentration should be ~0.5 μg/μl).
3. The transfection mixture is then incubated at room temperature for 20 min before adding it evenly to culture dish. The transfected cells are returned to the incubator and allowed to express for 4–10 h to achieve low expression levels for single-molecule imaging (Fig. 4.3).
4. To simultaneously image kinesin motors and dynamic plus ends of microtubules, 0.5–1 μg of EB3-mCherry and 0.5 μg of KHC-3×mCit plasmid DNA are mixed prior to adding them to the Opti-MEM transfection reagent mixture.
5. To simultaneously image kinesin motors and all microtubules, 1 μg of mCherry-tubulin is transfected into COS cells and expressed for 24–30 h to allow mCherry-tubulin to be incorporated into all the microtubules. The same cells are then retransfected with 0.5 μg of Kinesin-3×mCit plasmid DNA and allowed to express for an additional 4–10 h.

5.5. Recording and analyzing single-molecule *in vivo* events

To image single molecules *in vivo* in transfected COS cells, we proceed as follows:

1. Transfected cells are carefully rinsed with 37 $^\circ$C Ringers solution (10 mM HEPES/KOH, 155 mM NaCl, 5 mM KCl, 2 mM CaCl$_2$, 1 mM MgCl$_2$, 2 mM NaH$_2$PO$_4$, 10 mM glucose, pH 7.2). Cells are maintained in Ringers buffer during imaging.

2. Immersion oil should be applied to the high NA TIRF objective before the 35 mm culture dish is mounted on the heated microscope stage. A temperature-controlled objective collar is preferred to keep the cells at 37 °C.
3. Cells with proper expression levels for imaging are then searched for under conditions meant to minimize excessive photobleaching. To do this, low excitation laser power (~ 0.03 mW) is used at this stage. Cells with high expression levels usually show significant labeling (decorating) of microtubules with kinesin motors, which can be identified easily by direct visual examination with the wide-field binocular tube of the TIRF microscope (Fig. 4.3B). Although such high levels of expression and crowding of motors on microtubules hinders the observation of individual motility events, these cells are useful for finding the exact focal plane and fine-tuning TIRF excitation. Cells with lower expression levels are more difficult to identify under low-power excitation conditions, but imaging with the EMCCD camera at 200 ms integration time clearly reveals transfected kinesin motors within the cellular background fluorescence (Fig. 4.3C and D). Notice that the autofluorescence background of the cells helps identify cell boundaries (Fig. 4.3B).

5.6. Recording image sequences

1. Once a cell with proper expression level is identified, the beam diameter of the incident laser illumination is adjusted to the desired imaging area (~ 30 μm diameter) with the TIRF field-iris. This further limits unnecessary photobleaching and cell damage.
2. To accurately track single motor motility events with sufficient resolution and to clearly distinguish diffusion events, higher laser powers (0.1–0.5 mW) are used and long image sequences (> 300 frames) at frame rates between 10 and 100 Hz are recorded with the EMCCD camera.
3. Depending on the expression level, multiple image sequences can be taken from the same area of a cell. Typically we wait a few minutes between image series to allow enough unbleached motor molecules to move/diffuse into the field of view.

5.7. Identifying individual single motor events in live cells

Single motor motility events are identified as diffraction-limited fluorescent spots (~ 250 nm in diameter) that undergo unidirectional processive movement. In live-cell imaging, a large number of motility events can be used to identify microtubule tracks. We find such microtubule tracks by calculating standard deviation maps (SD maps) of the image series (see below). Overlaying individual frames from the image series on the SD map allows the

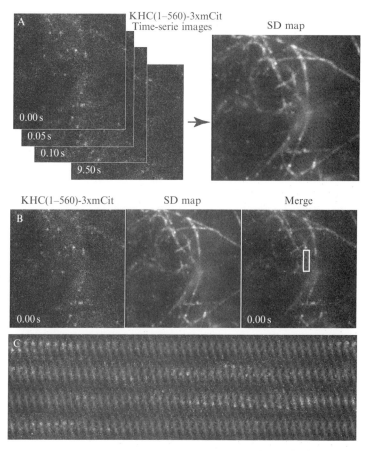

Figure 4.4 Standard deviation map. (A) A 9.5 s long image series containing 190 frames was calculated as described in the text. (B) Overlaying raw images with a SD map helps identify individual motility events. (C) Kymograph of the boxed region in (B) shows that numerous motility events occur on a highlighted track in the SD map.

alignment of motility events with microtubule tracks and provides additional criteria to define single kinesin motility events (Fig. 4.4B and C).

5.8. SD maps

The purpose of the standard deviation map (SD map) is to determine the location of microtubule tracks and to calculate how frequently motors bind to and move along particular tracks (relative to other tracks) (Fig. 4.4). In live-cell imaging sequences, four main noise sources degrade the motor's fluorescence signal: camera (readout) noise, cellular autofluorescence, diffusing fluorescent motors, and out-of-focus background fluorescence. These noise sources fluctuate randomly and therefore generate background fluorescence whereas the fluorescence intensity of the motor should be significant relative to these noise sources. Thus, motor motility events should contribute significantly to the fluorescence intensity fluctuation (or standard deviation) of a particular pixel of interest. We calculated SD maps (Fig. 4.4A) by determining the statistical intensity variation of each pixel location from the raw images of a video sequence (an image stack) and plotting them in the form of an image. In brief, for an image stack containing Z slices of images with $M \times N$ pixels, the intensity (I) of each pixel in the standard deviation map was calculated with ImageJ (ZProjector_StandardDeviation) as

$$I_{\text{std. dev},[m][n]} = \sqrt{\frac{1}{Z-1}\sum_{k=1}^{Z}\left(I_{[m][n][k]} - \frac{1}{Z}\sum_{k=1}^{z}I_{[m][n][k]}\right)^2}$$

In our work on the possible role of posttranslational modifications of microtubules in regulating cellular transport, analysis of imaging data with SD maps proved extremely helpful in identifying subpopulations of microtubules that are distinguished by different kinesin families to enable targeting of transport events to distinct subcellular domains (Cai et al., 2009).

5.9. High-resolution tracking

To analyze the movement of kinesins with subdiffraction-limited resolution, we use single particle tracking methods (see FIONA in Section 4.5). Given the limited brightness of our fluorescent protein tags, we obtain a spatial resolution of ~20 nm when tracking kinesins at frame rates of 30 Hz (Cai et al., 2007). We implemented FIONA-based particle tracking using 2D Gaussian fitting routines available in MATLAB (The MathWorks, Natick, MA) and ImageJ (NIH, http://rsbweb.nih.gov/ij/).

The accuracy of FIONA depends on a number of experimental parameters, including reduced image quality due to fluorescence background (cellular autofluorescence, freely diffusing fluorescent label) and undesirable photophysical properties of the fluorescent marker (quantum dot blinking, fluorescence photobleaching, self-quenching of multiply-labeled proteins). The fitting routine used to analyze images of a moving, diffraction-limited spot may also be prone to localization errors due to inherent optical aberrations (spherical and chromatic aberration, coma, astigmatism, image distortion, etc.), unclean optical components (especially dust in conjugated imaging planes, such as the CCD chip), and mechanical and thermal drift of the stage and other critical microscope components. For a quantitative comparison of the accuracies and errors associate with various fitting methods, including 2D Gaussian fitting and image cross-correlation, refer to Cheezum et al. (2001). In general, however, experimental methods are now well developed such that the primary limitation for any FIONA-based measurement capable of localizing single, fluorescent molecules is not the diffraction limit imposed by the Rayleigh criterion, but rather the strength (brightness above background) of the signal. Under such conditions photon shot noise dominates all other noise sources (including background fluorescence, pixel effects, and CCD readout noise) limiting the localization analysis (Thompson et al., 2002). Thus, the uncertainty of localizing the center of diffraction-limited spots, ε, is simply a function of the NA, the wavelength of light (λ), and the number of photons collected (N) and well approximated by Greenleaf et al. (2007):

$$\varepsilon \cong \frac{\lambda}{2NA\sqrt{N}}$$

6. Summary and Conclusions

Recent advances in imaging and labeling techniques have made it possible to routinely track the movement and distribution of single motor molecules and cytoskeletal proteins in the cytoplasm of mammalian cells. These methods are based on single particle tracking approaches where the centroid position of a label is determined with sub-diffraction-limited resolution. Due to their relatively high brightness, methods based on organelle tracking, gold particles, or aggregates of larger numbers of QDs enable particle tracking with superior spatial and temporal (nanometer and sub-millisecond) resolution, but lack precisely defined labeling stoichiometries and molecular definition that characterize fluorescent protein tags and genetic labels. We expect that the large number of problems remaining in cell biology that hinge on our ability to study the interactions and functions of multiprotein complexes at or close to the single-molecule level *in vivo* will be facilitated by new genetic labeling methods to address remaining limitations, including brightness, photostability, and restricted dye selection. Nonetheless, the existing methods presented in this chapter, which utilize multicolor fluorescence imaging techniques, are bound to impact on many important aspects of modern cell biology and will define the complex interactions of molecular motors and the cytoskeleton.

ACKNOWLEDGMENTS

We apologize that many original contributions could not be cited because of space limitations. Our original work in the area is supported by National Institutes of Health (NIH) grants to KJV (GM070862 and GM083254) and EM (GM076476 and GM083254).

REFERENCES

Alivisatos, A. P., Gu, W. W., and Larabell, C. (2005). Quantum dots as cellular probes. *Ann. Rev. Biomed. Eng.* **7,** 55–76.
Allen, R. D., Allen, N. S., and Travis, J. L. (1981). Video-enhanced contrast, differential interference contrast (AVEC-DIC) microscopy: A new method capable of analyzing microtubule-related motility in the reticulopodial network of Allogromia laticollaris. *Cell Motil.* **1,** 291–302.
Axelrod, D. (2003). Total internal reflection fluorescence microscopy in cell biology. *Methods Enzymol.* **361,** 1–33.
Barak, L. S., and Webb, W. W. (1982). Diffusion of low density lipoprotein–receptor complex on human fibroblasts. *J. Cell Biol.* **95,** 846–852.
Betzig, E., Patterson, G. H., Sougrat, R., Lindwasser, O. W., Olenych, S., Bonifacino, J. S., Davidson, M. W., Lippincott-Schwartz, J., and Hess, H. F. (2006). Imaging intracellular fluorescent proteins at nanometer resolution. *Science* **313,** 1642–1645.
Block, S. M. (2007). Kinesin motor mechanics: Binding, stepping, tracking, gating, and limping. *Biophys. J.* **92,** 2986–2995.
Brady, S. T. (1985). A novel brain ATPase with properties expected for the fast axonal-transport motor. *Nature* **317,** 73–75.
Braslavsky, I., Amit, R., Ali, B. M. J., Gileadi, O., Oppenheim, A., and Stavans, J. (2001). Objective-type dark-field illumination for scattering from microbeads. *Appl. Optics* **40,** 5650–5657.
Cai, D. W., Verhey, K. J., and Meyhofer, E. (2007). Tracking single kinesin molecules in the cytoplasm of mammalian cells. *Biophys. J.* **92,** 4137–4144.
Cai, D., McEwen, D. P., Martens, J. R., Meyhofer, E., and Verhey, K. J. (2009). Single molecule imaging reveals differences in microtubule track selection between Kinesin motors. *PLoS Biol.* **7,** e1000216.
Chang, Y. P., Pinaud, F., Antelman, J., and Weiss, S. (2008). Tracking bio-molecules in live cells using quantum dots. *J. Biophoton.* **1,** 287–298.
Cheezum, M. K., Walker, W. F., and Guilford, W. H. (2001). Quantitative comparison of algorithms for tracking single fluorescent particles. *Biophys. J.* **81,** 2378–2388.
Cherry, R. J. (1992). Keeping track of cell surface receptor. *Trends Cell Biol.* **2,** 242–244.
Churchman, L. S., Okten, Z., Rock, R. S., Dawson, J. F., and Spudich, J. A. (2005). Single molecule high-resolution colocalization of Cy3 and Cy5 attached to macromolecules measures intramolecular distances through time. *Proc. Natl. Acad. Sci. USA* **102,** 1419–1423.
Courty, S., Luccardini, C., Bellaiche, Y., Cappello, G., and Dahan, M. (2006). Tracking individual kinesin motors in living cells using single quantum-dot imaging. *Nano Lett.* **6,** 1491–1495.
Davidson, M. W., and Campbell, R. E. (2009). Engineered fluorescent proteins: Innovations and applications. *Nat. Methods* **6,** 713–717.
Dieterlen, M. T., Wegner, F., Schwarz, S. C., Milosevic, J., Schneider, B., Busch, M., Romuss, U., Brandt, A., Storch, A., and Schwarz, J. (2009). Non-viral gene transfer by nucleofection allows stable gene expression in human neural progenitor cells. *J. Neurosci. Methods* **178,** 15–23.
Dunn, A. R., and Spudich, J. A. (2007). Dynamics of the unbound head during myosin V processive translocation. *Nat. Struct. Mol. Biol.* **14,** 246–248.
Funatsu, T., Harada, Y., Tokunaga, M., Saito, K., and Yanagida, T. (1995). Imaging of single fluorescent molecules and individual ATP turnovers by single myosin molecules in aqueous solution. *Nature* **374,** 555–559.
Gaietta, G., Deerinck, T. J., Adams, S. R., Bouwer, J., Tour, O., Laird, D. W., Sosinsky, G. E., Tsien, R. Y., and Ellisman, M. H. (2002). Multicolor and electron microscopic imaging of connexin trafficking. *Science* **296,** 503–507.

Gelles, J., Schnapp, B. J., and Sheetz, M. P. (1988). Tracking kinesin-driven movements with nanometre-scale precision. *Nature* **331**, 450–453.
Gennerich, A., and Vale, R. D. (2009). Walking the walk: How kinesin and dynein coordinate their steps. *Curr. Opin. Cell Biol.* **21**, 59–67.
Giepmans, B. N., Adams, S. R., Ellisman, M. H., and Tsien, R. Y. (2006). The fluorescent toolbox for assessing protein location and function. *Science* **312**, 217–224.
Greenleaf, W. J., Woodside, M. T., and Block, S. M. (2007). High-resolution, single-molecule measurements of biomolecular motion. *Annu. Rev. Biophys. Biomol. Struct.* **36**, 171–190.
Griffin, B. A., Adams, S. R., and Tsien, R. Y. (1998). Specific covalent labeling of recombinant protein molecules inside live cells. *Science* **281**, 269–272.
Gross, S. P., Tuma, M. C., Deacon, S. W., Serpinskaya, A. S., Reilein, A. R., and Gelfand, V. I. (2002). Interactions and regulation of molecular motors in Xenopus melanophores. *J. Cell Biol.* **156**, 855–865.
Howard, J., Hudspeth, A. J., and Vale, R. D. (1989). Movement of microtubules by single kinesin molecules. *Nature* **342**, 154–158.
Huang, B., Jones, S. A., Brandenburg, B., and Zhuang, X. (2008). Whole-cell 3D STORM reveals interactions between cellular structures with nanometer-scale resolution. *Nat. Methods* **5**, 1047–1052.
Ince, C., Ypey, D. L., Diesselhoff-Den Dulk, M. M., Visser, J. A., De Vos, A., and Van Furth, R. (1983). Micro-CO2-incubator for use on a microscope. *J. Immunol. Methods* **60**, 269–275.
Johnsson, K. (2009). Visualizing biochemical activities in living cells. *Nat. Chem. Biol.* **5**, 63–65.
Jun, Y. W., Sheikholeslami, S., Hostetter, D. R., Tajon, C., Craik, C. S., and Alivisatos, A. P. (2009). Continuous imaging of plasmon rulers in live cells reveals early-stage caspase-3 activation at the single-molecule level. *Proc. Natl. Acad. Sci. USA* **106**, 17735–17740.
Keppler, A., Gendreizig, S., Gronemeyer, T., Pick, H., Vogel, H., and Johnsson, K. (2003). A general method for the covalent labeling of fusion proteins with small molecules in vivo. *Nat. Biotechnol.* **21**, 86–89.
Kinosita, K., Jr., Itoh, H., Ishiwata, S., Hirano, K., Nishizaka, T., and Hayakawa, T. (1991). Dual-view microscopy with a single camera: Real-time imaging of molecular orientations and calcium. *J. Cell Biol.* **115**, 67–73.
Kulic, I. M., Brown, A. E. X., Kim, H., Kural, C., Blehm, B., Selvin, P. R., Nelson, P. C., and Gelfand, V. I. (2008). The role of microtubule movement in bidirectional organelle transport. *Proc. Natl. Acad. Sci. USA* **105**, 10011–10016.
Kural, C., Kim, H., Syed, S., Goshima, G., Gelfand, V. I., and Selvin, P. R. (2005). Kinesin and dynein move a peroxisome in vivo: A tug-of-war or coordinated movement? *Science* **308**, 1469–1472.
Kural, C., Serpinskaya, A. S., Chou, Y. H., Goldman, R. D., Gelfand, V. I., and Selvin, P. R. (2007). Tracking melanosomes inside a cell to study molecular motors and their interaction. *Proc. Natl. Acad. Sci. USA* **104**, 5378–5382.
Kural, C., Nonet, M. L., and Selvin, P. R. (2009). FIONA on Caenorhabditis elegans. *Biochemistry* **48**, 4663–4665.
Kusumi, A., Nakada, C., Ritchie, K., Murase, K., Suzuki, K., Murakoshi, H., Kasai, R. S., Kondo, J., and Fujiwara, T. (2005). Paradigm shift of the plasma membrane concept from the two-dimensional continuum fluid to the partitioned fluid: High-speed single-molecule tracking of membrane molecules. *Annu. Rev. Biophys. Biomol. Struct.* **34**, 351–378.
Laib, J. A., Marin, J. A., Bloodgood, R. A., and Guilford, W. H. (2009). The reciprocal coordination and mechanics of molecular motors in living cells. *Proc. Natl. Acad. Sci. USA* **106**, 3190–3195.

Levi, V., and Gratton, E. (2007). Exploring dynamics in living cells by tracking single particles. *Cell Biochem. Biophys.* **48**, 1–15.

Levi, V., Gelfand, V. I., Serpinskaya, A. S., and Gratton, E. (2006). Melanosomes transported by myosin-V in Xenopus melanophores perform slow 35 nm steps. *Biophys. J.* **90**, L07–L09.

Meyhofer, E., and Howard, J. (1995). The force generated by a single kinesin molecule against an elastic load. *Proc. Natl. Acad. Sci. USA* **92**, 574–578.

Michalet, X., Pinaud, F. F., Bentolila, L. A., Tsay, J. M., Doose, S., Li, J. J., Sundaresan, G., Wu, A. M., Gambhir, S. S., and Weiss, S. (2005). Quantum dots for live cells, in vivo imaging, and diagnostics. *Science* **307**, 538–544.

Nan, X., Sims, P. A., Chen, P., and Xie, X. S. (2005). Observation of individual microtubule motor steps in living cells with endocytosed quantum dots. *J. Phys. Chem. B.* **109**, 24220–24224.

Nan, X. L., Sims, P. A., and Xie, X. S. (2008). Organelle tracking in a living cell with microsecond time resolution and nanometer spatial precision. *Chemphyschem* **9**, 707–712.

Nelson, S. R., Ali, M. Y., Trybus, K. M., and Warshaw, D. M. (2009). Random walk of processive, quantum dot-labeled myosin Va molecules within the actin cortex of COS-7 cells. *Biophys. J.* **97**, 509–518.

Nishiyama, M., Muto, E., Inoue, Y., Yanagida, T., and Higuchi, H. (2001). Substeps within the 8-nm step of the ATPase cycle of single kinesin molecules. *Nat. Cell Biol.* **3**, 425–428.

Okada, C. Y., and Rechsteiner, M. (1982). Introduction of macromolecules into cultured mammalian cells by osmotic lysis of pinocytic vesicles. *Cell* **29**, 33–41.

Park, H., Hanson, G. T., Duff, S. R., and Selvin, P. R. (2004). Nanometre localization of single ReAsH molecules. *J. Microsc.* **216**, 199–205.

Ray, S., Meyhofer, E., Milligan, R. A., and Howard, J. (1993). Kinesin follows the microtubule's protofilament axis. *J. Cell Biol.* **121**, 1083–1093.

Reinhard, B. M., Siu, M., Agarwal, H., Alivisatos, A. P., and Liphardt, J. (2005). Calibration of dynamic molecular rule based on plasmon coupling between gold nanoparticles. *Nano Lett.* **5**, 2246–2252.

Rong, G. X., Wang, H. Y., Skewis, L. R., and Reinhard, B. M. (2008). Resolving subdiffraction limit encounters in nanoparticle tracking using live cell plasmon coupling microscopy. *Nano Lett.* **8**, 3386–3393.

Ross, J., Buschkamp, P., Fetting, D., Donnermeyer, A., Roth, C. M., and Tinnefeld, P. (2007). Multicolor single-molecule spectroscopy with alternating laser excitation for the investigation of interactions and dynamics. *J. Phys. Chem. B.* **111**, 321–326.

Roy, R., Hohng, S., and Ha, T. (2008). A practical guide to single-molecule FRET. *Nat. Methods* **5**, 507–516.

Sako, Y., and Yanagida, T. (2003). Single-molecule visualization in cell biology. *Nat. Rev. Mol. Cell Biol.* SS1–SS5.

Salmon, E. D., and Tran, P. (1998). High-resolution video-enhanced differential interference contrast (VE-DIC) light microscopy. *Methods Cell Biol.* **56**, 153–184.

Saxton, M. J., and Jacobson, K. (1997). Single-particle tracking: Applications to membrane dynamics. *Annu. Rev. Biophys. Biomol. Struct.* **26**, 373–399.

Schliwa, M., and Woehlke, G. (2003). Molecular motors. *Nature* **422**, 759–765.

Schmidt, T., Schutz, G. J., Baumgartner, W., Gruber, H. J., and Schindler, H. (1996). Imaging of single molecule diffusion. *Proc. Natl. Acad. Sci. USA* **93**, 2926–2929.

Shaner, N. C., Steinbach, P. A., and Tsien, R. Y. (2005). A guide to choosing fluorescent proteins. *Nat. Methods* **2**, 905–909.

Shaner, N. C., Lin, M. Z., McKeown, M. R., Steinbach, P. A., Hazelwood, K. L., Davidson, M. W., and Tsien, R. Y. (2008). Improving the photostability of bright monomeric orange and red fluorescent proteins. *Nat. Methods* **5**, 545–551.

Shubeita, G. T., Tran, S. L., Xu, J., Vershinin, M., Cermelli, S., Cotton, S. L., Welte, M. A., and Gross, S. P. (2008). Consequences of motor copy number on the intracellular transport of Kinesin-1-driven lipid droplets. *Cell* **135,** 1098–1107.

Simon, S. M. (2009). Partial internal reflections on total internal reflection fluorescent microscopy. *Trends Cell Biol.* **19,** 661–668.

Snapp, E. L. (2009). Fluorescent proteins: A cell biologist's user guide. *Trends Cell Biol.* **19,** 649–655.

Svoboda, K., Schmidt, C. F., Schnapp, B. J., and Block, S. M. (1993). Direct observation of kinesin stepping by optical trapping interferometry. *Nature* **365,** 721–727.

Thompson, R. E., Larson, D. R., and Webb, W. W. (2002). Precise nanometer localization analysis for individual fluorescent probes. *Biophys. J.* **82,** 2775–2783.

Tokunaga, M., Imamoto, N., and Sakata-Sogawa, K. (2008). Highly inclined thin illumination enables clear single-molecule imaging in cells. *Nat. Methods* **5,** 159–161.

Vale, R. D., Reese, T. S., and Sheetz, M. P. (1985). Identification of a novel force-generating protein, kinesin, involved in microtubule-based motility. *Cell* **42,** 39–50.

Visscher, K., Schnitzer, M. J., and Block, S. M. (1999). Single kinesin molecules studied with a molecular force clamp. *Nature* **400,** 184–189.

Walter, N. G., Huang, C. Y., Manzo, A. J., and Sobhy, M. A. (2008). Do-it-yourself guide: How to use the modern single-molecule toolkit. *Nat. Methods* **5,** 475–489.

Watanabe, N., and Mitchison, T. J. (2002). Single-molecule speckle analysis of actin filament turnover in lamellipodia. *Science* **295,** 1083–1086.

Yildiz, A., Forkey, J. N., McKinney, S. A., Ha, T., Goldman, Y. E., and Selvin, P. R. (2003). Myosin V walks hand-over-hand: Single fluorophore imaging with 1.5-nm localization. *Science* **300,** 2061–2065.

Yildiz, A., Tomishige, M., Vale, R. D., and Selvin, P. R. (2004). Kinesin walks hand-over-hand. *Science* **303,** 676–678.

CHAPTER FIVE

SINGLE-PARTICLE TRACKING PHOTOACTIVATED LOCALIZATION MICROSCOPY FOR MAPPING SINGLE-MOLECULE DYNAMICS

Suliana Manley,* Jennifer M. Gillette,[†] *and* Jennifer Lippincott-Schwartz[†]

Contents

1. Introduction	110
2. Description of the sptPALM Method	111
3. Labeling with Photoactivatable Fluorescent Probes	113
3.1. Photoactivatable fluorescent proteins	113
3.2. Photocaged dyes	114
4. Tracking Single Molecules	114
4.1. Molecule identification	114
4.2. Tracking algorithms	115
5. Experimental Example: sptPALM on a Membrane Protein	116
6. Conclusions	118
References	118

Abstract

Recent developments in single-molecule localization techniques using photoactivatable fluorescent proteins have allowed the probing of single-molecule motion in a living cell with high specificity, millisecond time resolution, and nanometer spatial resolution. Analyzing the dynamics of individual molecules at high densities in this manner promises to provide new insights into the mechanisms of many biological processes, including protein heterogeneity in the plasma membrane, the dynamics of cytoskeletal flow, and clustering of

---

* Institute of Physics of Biological Systems, Swiss Federal Institute of Technology (EPFL), Lausanne, Switzerland
[†] Section on Organelle Biology, Cell Biology and Metabolism Program, NICHD, National Institutes of Health, Bethesda, Maryland, USA

Methods in Enzymology, Volume 475
ISSN 0076-6879, DOI: 10.1016/S0076-6879(10)75005-9

© 2010 Elsevier Inc.
All rights reserved.

receptor complexes in response to signaling cues. Here we describe the method of single-molecule tracking photoactivated localization microscopy (sptPALM) and discuss how its use can contribute to a quantitative understanding of fundamental cellular processes.

1. INTRODUCTION

Dissecting the molecular mechanisms for biological processes lends great insight into how these systems function, both in health and disease states. Likewise, fluorescence imaging of specific proteins allows for the identification of dynamic processes and causal relationships. Single-molecule measurements aim to exploit and expand upon both of these powerful avenues by elucidating the dynamics of specific proteins at the molecular scale. By following the motions of single molecular motors such as myosin or kinesin with high spatial and temporal resolution, the mechanics of their work cycle have been revealed (Svoboda et al., 1993; Yildiz et al., 2003). For many biological processes, including receptor trafficking, cytoskeletal dynamics, and cellular signaling, the collective behaviors of molecules may be important, and it would be interesting to extend these measurements to many molecules. Moreover, obtaining information on statistically rare events requires the acquisition of data on large numbers of single molecules, ideally under multiple conditions. However, traditional single-molecule measurements using fluorescence imaging require an extremely low labeling density. This is due to the diffraction of light when passing through a lens, which spreads the fluorescence of a single molecule into an Airy disk nearly 100 times the size of the molecule itself. Thus, in the case of closely packed membrane proteins, only one protein in thousands can be labeled to distinguish individual molecules. Since many small structures such as clathrin-coated pits and viruses typically contain only 1000 copies or fewer of a single protein, the extension of single-molecule imaging to their study was not evident until the recent advent of photoswitchable fluorescent probes (Ando et al., 2002; Patterson and Lippincott-Schwartz, 2002). In this era of large data sets, new advances in particle tracking combined with photoactivation of fluorescence have enabled the study of the dynamics of these previously inaccessible systems. Here, we describe the method of single-particle tracking combined with single-molecule localization using photoactivated localization microscopy (PALM) (Betzig et al., 2006), called single-particle tracking PALM (sptPALM) (Manley et al., 2008). This approach permits the study of protein dynamics in the context of living cells where proteins exist in dense populations.

2. Description of the sptPALM Method

The technological basis underlying sptPALM is the localization of the positions of specific proteins at near-molecular spatial resolution. This method further exploits the switchable properties of photoactivatable fluorescent proteins (PA-FPs; Ando *et al.*, 2002; Patterson and Lippincott-Schwartz, 2002; Wiedenmann *et al.*, 2004) used in PALM, although the same principle (termed STORM) has been applied to imaging with antibody-targeted photoswitchable synthetic labels (Rust *et al.*, 2006). In PALM (Fig. 5.1), PA-FPs are continuously activated, imaged, and bleached in order to temporally separate molecules that would otherwise be spatially indistinguishable.

Because PA-FPs are activated stochastically, it is possible to ensure that highlighted molecules are sparse enough so that in each image only a single molecule (or less) is activated within a given diffraction-limited region. This is achieved by using a low intensity illumination at the activation wavelength (typically in the ultraviolet), while adjusting the excitation intensity to maintain a sparse pool of molecules by photobleaching. The position of each molecule is localized by fitting the measured photon distribution with the point spread function of the microscope (typically approximated with a Gaussian function). This has been demonstrated to provide superior accuracy in molecular position finding, relative to cross-correlation, sum-absolute difference, and centroid finding (Cheezum *et al.*, 2001). The locations found during multiple imaging repetitions can then be summed to obtain a composite super-resolution

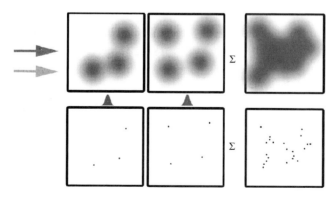

Figure 5.1 *Schematic of PALM analysis.* Raw data are acquired (top row) under continuous activation and excitation, then the image of each molecule is fit with a Gaussian to yield nanometric position information (bottom row). The summed information provides high-density, high-resolution molecular maps.

PALM image that contains information about the location of many single molecules, even those sharing a single diffraction-limited region. In a typical PALM experiment, thousands of raw images are captured over a period of a few minutes, from which hundreds of thousands of molecules are localized. When performed in fixed cells, this approach provides a high-density map of a protein's distribution at any particular time within the cell.

Proteins in living cells usually do not reside in a single place for very long, rather they are highly dynamic. Whether organized into multicomponent cytoskeletal structures, signaling ensembles or freely diffusing, proteins within cells are on the move. Understanding this motion is critical for providing a mechanistic basis for how proteins facilitate cell function. An approach for mapping the trajectories of individual molecules in living cells at very high densities is sptPALM (Manley et al., 2008). It combines the classic method of single-particle tracking with PALM. In sptPALM, fluorophores are stochastically activated (Fig. 5.2A) and imaged at video rate or slower, depending on the dynamics of the system. For slower processes, such as actin retrograde flow, slower imaging rates (commensurate with

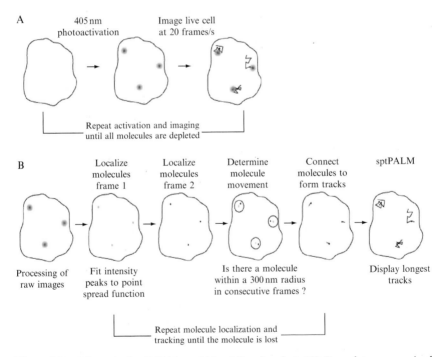

Figure 5.2 *Schematic of sptPALM acquisition (A) and analysis (B).* Raw data are acquired (A) under continuous activation and excitation, then the image of each molecule is fit with a Gaussian to yield nanometric position information (B). Molecular positions in consecutive frames are then associated based on their proximity.

lower excitation power or incorporating a pause between images) are desirable to maintain fluorescence over significant motions. Molecular trajectories can be reconstructed by connecting the positions of photoactivated fluorophores in consecutive images (Fig. 5.2B). Thus, the information obtained can be used to create spatially resolved maps of mobility (Manley *et al.*, 2008; Niu and Yu, 2008). In addition, maps of interacting proteins can be made by labeling different proteins with spectrally distinct fluorophores (Subach *et al.*, 2009), although this has yet to be extended to sptPALM.

3. Labeling with Photoactivatable Fluorescent Probes

There are two primary means of labeling proteins for measurements with sptPALM, these are genetic encoding and expression of a PA-FP chimera, or targeting with an antibody or small molecule coupled to photocaged or photoswitchable dyes. In both cases, precautions must be taken to ensure minimal perturbations due to the labels. This generally means selecting cells for imaging with low expression levels, and demonstrating preservation of protein function in the chimeric form.

3.1. Photoactivatable fluorescent proteins

The discovery and development of fluorescent proteins, which merited the 2008 Nobel Prize in chemistry, has proved to be a powerful tool for investigating the organization and dynamics of proteins in cells, tissue, and animals. The noninvasive nature of fluorescence imaging makes it ideal for studying living samples, and cells can readily be made to express DNA encoding proteins of interest labeled with PA-FP. Unlike affinity labeling with antibodies or small molecules, there is no background from nonspecific labeling. This can be of paramount importance when studying single molecules, whose fluorescence can easily be overcome by background fluorescence. Moreover, numerous PA-FPs have been developed in recent years (Shaner *et al.*, 2007) beginning with PAGFP (Patterson and Lippincott-Schwartz, 2002), offering many possibilities for sptPALM probes. Thus far, EosFP (Manley *et al.*, 2008), PAmCherry (Subach *et al.*, 2009), and Dendra2 (Gurskaya *et al.*, 2006) have been successfully used to map the dynamics of single molecules with sptPALM. Furthermore, single-molecule performance has now been used as a criterion for the selection of mutants to improve photon yields and signal-to-noise ratios (Subach *et al.*, 2009), directly targeting the weaknesses of fluorescent proteins for sptPALM measurements.

3.2. Photocaged dyes

With any single-molecule measurement, the number of photons translates directly into how well a molecule can be localized (Thompson *et al.*, 2002). Low photon yields can be an issue with molecular localization, and this problem is most severe for higher dimensional imaging, as in (x,y,z) three-dimensional imaging or (two-dimensional + time) live cell imaging (Shroff *et al.*, 2008). Chemical dyes generally have increased photon yields compared to fluorescent proteins, but they have their own drawbacks. They cannot be genetically encoded and so must be targeted to the relevant protein, typically using antibodies. Antibodies have limited targeting efficiency and their large size adds uncertainty to the position of the target molecule. Also, because they cannot penetrate the plasma membrane, membranes must be permeabilized before imaging molecules inside the cell, creating potential artifacts and making this approach unsuitable for live cell imaging of intracellular structures.

An alternative approach is to use small molecule labeling with chemical tags, such as the human DNA repair protein O6-alkylguanine-DNA (hAGT) and its derivatives (CLIP) (Gautier *et al.*, 2008; Keppler *et al.*, 2004) or dihydrofolate reductase (DHFR) (Miller *et al.*, 2005) conjugated with caged fluorophores. For a more complete list of small molecule targeting methods, see Fernandez-Suarez and Ting (2008). These labels have several advantages over FPs or antibody targeting. They are targeted to genetically expressed protein fusions with high affinity. There is the potential to make them membrane permeable and therefore compatible with live cell imaging. Finally, they can achieve the high photon yields of chemical fluorescent tags such as rhodamine.

4. Tracking Single Molecules

As indicated in Fig. 5.2B, there are two analyses involved in recreating multiple single-molecule trajectories. In the first, single molecules are identified and located. In the second, locations are connected to form trajectories.

4.1. Molecule identification

The image of a single molecule can be described by an Airy diffraction pattern, whose rings contain information about its vertical position. For two-dimensional imaging, as is frequently the case in membrane-bound molecular imaging and is enforced by total internal reflection fluorescence (TIRF) imaging, the molecular image can be fit with a Gaussian. This fit captures the central peak of the Airy diffraction pattern (Cheezum et al., 2001), and the associated errors with determining the molecule's centroid position have been well explored (Thompson et al., 2002).

Further criteria for peak intensities and shapes are set to discard peaks corresponding to multiple molecules, which are too bright, broad, or lacking radial symmetry. An alternative criterion to identify single molecules is single-step photobleaching, which could conceivably be implemented into an automated identification algorithm.

As molecules move, the photons they emit are spread by their motion. Thus, it is best to acquire frames with the shortest possible exposure time to minimize blurring. At the same time, exposure times and excitation intensities must be high enough to allow for nanometer-scale localization, requiring hundreds of photons in each image. In addition to minimizing the collection time for single frames, the frame rate for data acquisition must be selected carefully to capture the dynamics of interest. That is, individual molecules should move distances between consecutive frames that are larger than the uncertainties in their positions to avoid unnecessary oversampling. Another critical parameter is the degree of photoactivation at each time frame. If too many molecules are photoactivated simultaneously, as occurs with high (fast) photoactivation, their diffraction-limited images will overlap, making the localization of single molecules impossible.

4.2. Tracking algorithms

Tracking many single molecules requires identifying molecules, and determining their correspondence in consecutive frames. While this is straightforward for data with very sparse single molecules, and can be performed manually for few tracks, automation is necessary when tracking the hundreds or thousands of molecules that can be acquired in a single cell using sptPALM. For the purposes of sptPALM, data are typically acquired at high densities of molecules, such that a simple proximity-based algorithm that identifies each molecule with the nearest molecule in the subsequent frame is not sufficient. Numerous algorithms exist for particle tracking, with different probabilistic weightings or adjustable radial cutoffs designating molecular trajectories (Anderson et al., 1992; Crocker and Grier, 1996; Ghosh and Webb, 1994; Jaqaman et al., 2008). For a more complete review of the different multitarget tracking algorithms, see Kalaidzidis (2009). Automated tracking can be achieved by calculating the probability that identified molecules in one frame correspond to identified molecules in the next frame, on a frame-by-frame basis (Anderson et al., 1992; Crocker and Grier, 1996). Alternatively, a global probability can be calculated for all frames and optimized (Jaqaman et al., 2008). However, this method is computationally intensive and requires an underlying model to be introduced for the motions of molecules, to reduce the number of possible trajectories.

A useful criterion for diffusing molecules is that the probability for a molecule to diffuse a distance L or less in a time Δt is given by $P(L, \Delta t) = 1 - \exp(-L^2/4D\Delta t)$, where D is the diffusion coefficient (Saxton, 1993). Thus, given the typical range of diffusion coefficients for membrane proteins of 0.1–0.5 $\mu m^2/s$ (Kenworthy et al., 2004), L can be chosen to yield an appropriate probability of detecting molecules.

5. Experimental Example: sptPALM on a Membrane Protein

In the example presented here, we begin by plating cells onto a glass coverslip. Coverglass thickness and index of refraction should be chosen for compatibility with your objective. For the purposes of studying the motion of membrane proteins, use of TIRF is implemented to reduce background light. Use of a camera with single-photon sensitivity is also necessary to collect single-molecule fluorescence emission.

1. Plate cells on a glass coverslip.
2. Transfect cells ~18 h before measuring. Grow to 80–90% confluency.
3. Fasten coverslip securely to microscope stage. Cells should be immersed in CO_2 independent medium, unless a CO_2 chamber is used for the microscope. We have built custom TIRF setups using both Olympus and Zeiss microscopes. In principle, any TIRF microscope that includes a 405 nm laser and the appropriate excitation wavelength for your chosen fluorophore should work.
4. Find a transfected cell, and center it in the field of view.
5. Take a DIC image prior to and after excitation to serve as a reference point for cell viability.
6. Begin streaming live cell images while illuminating with excitation wavelength.
7. Once molecules are sparse (initial molecules have bleached), gradually increase activation power to maintain a constant density of single molecules. This typically occurs within the first 100 images.
8. Localize molecules using fitting algorithm. Molecules should be subjected to strict criteria regarding their localization and the goodness of fit to a radially symmetric Gaussian. Several groups are now offering free software to do this (Hedde et al., 2009; Henriques et al., 2010).
9. Perform automated tracking. Several groups offer free software to do this, including Crocker, Weeks, and Grier (http://www.physics.emory.edu/~weeks/idl/index.html) and Jaqaman and Danuser (http://lccb.hms.harvard.edu/software.html).

At this stage of the analysis, many statistical measures may be applied to determine the kinds of dynamics present, whether they are diffusive, subdiffusive, or directed. Importantly, sufficiently long trajectories must be used to distinguish between modes of motion or to calculate diffusion coefficients. In the case of determining the diffusion coefficients of single vesicular stomatitis virus G (VSVG) proteins, we required trajectories to contain at least 15 positions to keep fractional errors in diffusion coefficients below 0.49 (Manley et al., 2008). The single-molecule trajectories and the associated diffusion coefficients are shown for a single cell in Fig. 5.3. The error associated with trajectories should be calculated to determine whether inferred differences are statistically significant (Qian et al., 1991). In addition, checks for a nonrandom distribution can be useful in determining whether molecules are clustered. Typical measures of clustering include the pair correlation function (Subach et al., 2009) or Ripley's K-function (Appleyard et al., 1985), but caution must again be used to ensure that molecular blinking does not lead to artifactual identification of a single molecule as a cluster of distinct molecules. This is especially important if molecules are used that switch reversibly into dark states, such as Dronpa (Habuchi et al., 2005). An additional complication may arise in the case of TIRF microscopy, since undulations in the membrane may give the impression of fluctuations in molecular densities. This may be checked by using interference reflectance microscopy on the same cells, which will reveal where cell membranes are near or far from the coverslip (Verschuren, 1985). Likewise, in membrane sheet preparations, any patchiness in the lift-off procedure can give rise to an apparent clustering of remaining proteins.

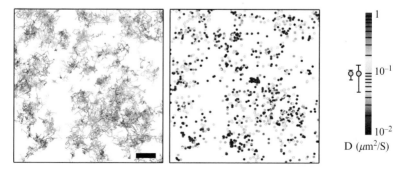

Figure 5.3 Example of sptPALM trajectories (left) and diffusion coefficient map (right) for the membrane protein VSVG. Color bar provides a scale for diffusion coefficients. Scale bar is 1 μm. (See Color Insert.)

6. Conclusions

The method of sptPALM represents a step toward a statistical description of single proteins in cells, where information on thousands of molecules provides a more complete picture of the wide variety of molecular motions taking place in living cells. Thus far, the method of sptPALM has been used to study the diffusion of membrane proteins (VSVG), membrane-binding proteins (Gag), and cytoskeletal proteins (actin and FtsZ). These studies have revealed new information about these systems. In the case of VSVG and Gag, these proteins were found to have strikingly different diffusional behaviors on the plasma membrane: VSVG diffuses freely, while many Gag molecules are immobilized in domains the size of a budding viral particle ($\sim$100–200 nm diameter) (Manley et al., 2008). In the case of the actin-related protein, FtsZ, sptPALM revealed that the remodeling of FtsZ molecules in Z-rings within bacteria occurs by local binding and dissociation rather than by active transport or treadmilling behavior (Niu and Yu, 2008). These findings provide a glimpse of the types of questions and diverse biological systems that may be investigated using sptPALM.

In the near future, it will be possible to combine sptPALM with additional imaging modalities, which will greatly enhance the kinds of biological information obtained by these kinds of experiments. This includes multicolor (Subach et al., 2009) and three-dimensional (Shtengel et al., 2009; Vaziri et al., 2008) imaging capabilities, as well as functional imaging.

REFERENCES

Anderson, C. M., Georgiou, G. N., Morrisoni, I. E. G., Stevenson, G. V. W., and Cherry, R. J. (1992). Tracking of cell surface receptors by fluorescence digital imaging microscopy using a charge-coupled device camera. Low-density lipoprotein and influenza virus receptor mobility at 4 degrees C. *J. Cell Sci.* **101,** 415–425.

Ando, R., Hama, H., Yamamoto-Hino, M., Mizuno, H., and Miyawaki, A. (2002). An optical marker based on the UV-induced green-to-red photoconversion of a fluorescent protein. *Proc. Natl. Acad. Sci. USA* **99,** 12651–12657.

Appleyard, S. T., Witkowski, J. A., and Ripley, B. D. (1985). A novel procedure for pattern analysis of features present on freeze-fractured plasma membranes. *J. Cell Sci.* **74,** 105–117.

Betzig, E., Patterson, G. H., Sougrat, R., Lindwasser, O. W., Olenych, S., Bonifacino, J. S., Davidson, M. W., Lippincott-Schwartz, J., and Hess, H. F. (2006). Imaging intracellular fluorescent proteins at nanometer resolution. *Science* **313,** 1642–1645.

Cheezum, M. K., Walker, W. F., and Guilford, W. H. (2001). Quantitative comparison of algorithms for tracking single fluorescent particles. *Biophys. J.* **81,** 2378–2388.

Crocker, J. C., and Grier, D. G. (1996). Methods of digital video microscopy for colloidal studies. *J. Colloid Interface Sci.* **179,** 298–310.

Fernandez-Suarez, M., and Ting, A. Y. (2008). Fluorescent probes for super-resolution imaging in living cells. *Nat. Rev. Mol. Cell Biol.* **9,** 929–943.

Gautier, A., Juillerat, A., Heinis, C., Correa, I. R., Jr, Kindermann, M., Beaufils, F., and Johnsson, K. (2008). An engineered protein tag for multiprotein labeling in living cells. *Chem. Biol.* **15,** 128–136.

Ghosh, R. N., and Webb, W. W. (1994). Automated detection and tracking of individual and clustered cell surface low density lipoprotein receptor molecules. *Biophys. J.* **66,** 1301–1318.

Gurskaya, N. G., Verkhusha, V. V., Shcheglov, A. S., Staroverov, D. B., Chepurnykh, T. V., Fradkov, A. F., Lukyanov, S., and Lukyanov, K. A. (2006). Engineering of a monomeric green-to-red photoactivatable fluorescent protein induced by blue light. *Nat. Biotechnol.* **24,** 461–465.

Habuchi, S., Ando, R., Dedecker, P., Verheijen, W., Mizuno, H., Miyawaki, A., and Hofkens, J. (2005). Reversible single-molecule photoswitching in the GFP-like fluorescent protein Dronpa. *Proc. Natl. Acad. Sci. USA* **102,** 9511–9516.

Hedde, P. N., Fuchs, J., Ozwald, F., Wiedenmann, J., and Nienhaus, G. U. (2009). Online image analysis software for photoactivation localization microscopy. *Nat. Methods* **6,** 689–690.

Henriques, R., Lelek, M., Fornasiero, E. F., Valtorta, F., Zimmer, C., and Mhlanga, M. M. (2010). QuickPALM: 3D real-time photoactivation nanoscopy image processing in ImageJ. *Nat. Methods* **7,** 339–340.

Jaqaman, K., Loerke, D., Mettlen, M., Kuwata, H., Grinstein, S., Schmid, S. L., and Danuser, G. (2008). Robust single-particle tracking in live-cell time-lapse sequences. *Nat. Methods* **5,** 695–702.

Kalaidzidis, Y. (2009). Multiple objects tracking in fluorescence microscopy. *J. Math. Biol.* **58,** 57–80.

Kenworthy, A. K., Nichols, B. J., Remmert, C. L., Hendrix, G. M., Kumar, M., Zimmerberg, J., and Lippincott-Schwartz, J. (2004). Dynamics of putative raft-associated proteins at the cell surface. *J. Cell Biol.* **165,** 735–746.

Keppler, A., Kindermann, M., Gendreizig, S., Pick, H., Vogel, H., and Johnsson, K. (2004). Labeling of fusion proteins of O6-alkylguanine-DNA alkyltransferase with small molecules in vivo and in vitro. *Methods* **32,** 437–444.

Manley, S., Gillette, J. M., Patterson, G. H., Shroff, H., Hess, H. F., Betzig, E., and Lippincott-Schwartz, J. (2008). High-density mapping of single-molecule trajectories with photoactivated localization microscopy. *Nat. Methods* **5,** 155–157.

Miller, L. W., Cai, Y., Sheetz, M. P., and Cornish, V. W. (2005). In vivo protein labeling with trimethoprim conjugates: A flexible chemical tag. *Nat. Methods* **2,** 255–257.

Niu, L., and Yu, J. (2008). Investigating intracellular dynamics of FtsZ cytoskeleton with photoactivation single-molecule tracking. *Biophys. J.* **95,** 2009–2016.

Patterson, G. H., and Lippincott-Schwartz, J. (2002). A photoactivatable GFP for selective photolabeling of proteins and cells. *Science* **297,** 1873–1877.

Qian, H., Sheetz, M. P., and Elson, E. L. (1991). Single particle tracking: Analysis of diffusion and flow in two-dimensional systems. *Biophys. J.* **60,** 910–921.

Rust, M. J., Bates, M., and Zhuang, X. (2006). Sub-diffraction-limit imaging by stochastic optical reconstruction microscopy (STORM). *Nat. Methods* **3,** 793–795.

Saxton, M. J. (1993). Lateral diffusion in an archipelago: Single-particle diffusion. *Biophys. J.* **64,** 1766–1780.

Shaner, N. C., Patterson, G. H., and Davidson, M. W. (2007). Advances in fluorescent protein technology. *J. Cell Sci.* **120,** 4247–4260.

Shroff, H., Galbraith, C. G., Galbraith, J. A., and Betzig, E. (2008). Live-cell photoactivated localization microscopy of nanoscale adhesion dynamics. *Nat. Methods* **5,** 417–423.

Shtengel, G., Galbraith, J. A., Galbraith, C. G., Lippincott-Schwartz, J., Gillette, J. M., Manley, S., Sougrat, R., Waterman, C. M., Kanchanawong, P., Davidson, M.,

Fetter, R., and Hess, H. F. (2008). Interferometric flourescent super-resolution microscopy resolves 3D cellular nano-architecture. *Proc. Natl. Acad. Sci. USA* **106**(9), 3125–3130.

Subach, F. V., Patterson, G. H., Manley, S., Gillette, J. M., Lippincott-Schwartz, J., and Verkhusha, V. V. (2009). Photoactivatable mCherry for high-resolution two-color fluorescence microscopy. *Nat. Methods* **6**, 153–159.

Svoboda, K., Schmidt, C. F., Schnapp, B. J., and Block, S. M. (1993). Direct observation of kinesin stepping by optical trapping interferometry. *Nature* **365**, 721–727.

Thompson, R. E., Larson, D. R., and Webb, W. W. (2002). Precise nanometer localization analysis for individual fluorescent probes. *Biophys. J.* **82**, 2775–2783.

Vaziri, A., Tang, J., Shroff, H., and Shank, C. V. (2008). Multilayer three-dimensional super resolution imaging of thick biological samples. *Proc. Natl. Acad. Sci. USA* **105**, 20221–20226.

Verschuren, H. (1985). Interference reflection microscopy in cell biology: Methodology and applications. *J. Cell Sci.* **75**, 279–301.

Wiedenmann, J., Ivanchenko, S., Oswald, F., Schmitt, F., Rocker, C., Salih, A., Spindler, K. D., and Nienhaus, G. U. (2004). EosFP, a fluorescent marker protein with UV-inducible green-to-red fluorescence conversion. *Proc. Natl. Acad. Sci. USA* **101**, 15905–15910.

Yildiz, A., Forkey, J. N., McKinney, S. A., Ha, T., Goldman, Y. E., and Selvin, P. R. (2003). Myosin V walks hand-over-hand: Single fluorophore imaging with 1.5-nm localization. *Science* **300**, 2061–2065.

CHAPTER SIX

A BIRD'S EYE VIEW: TRACKING SLOW NANOMETER-SCALE MOVEMENTS OF SINGLE MOLECULAR NANO-ASSEMBLIES

Nicole Michelotti,[1,*,†] Chamaree de Silva,[1,2,*] Alexander E. Johnson-Buck,* Anthony J. Manzo,* and Nils G. Walter*

Contents

1. Introduction	122
1.1. History of fluorescent single-particle tracking	123
2. DNA-Based Nanowalkers	125
2.1. The molecular "spider"	125
2.2. Behavior-based molecular robots: Programming tasks into the walker's environment	127
2.3. Methods for characterizing nanowalker behavior	129
3. Considerations for Fluorescence Imaging of Slowly Moving Particles	129
3.1. Prolonging fluorophore lifetime	129
3.2. Control for stage and focal drift	130
4. Single-Molecule Fluorescence Tracking of Nanowalkers	131
4.1. Slide preparation	131
4.2. Experimental procedure	133
4.3. Instrumentation for imaging	134
5. Extracting Super-Resolution Position Information	136
5.1. Mapping the two emission channels onto one another	136
5.2. Quality control	137
5.3. Presentation	139
5.4. Error analysis	143
6. Concluding Remarks	143
Acknowledgments	144
References	144

* Single Molecule Analysis Group, Department of Chemistry, University of Michigan, Ann Arbor, Michigan, USA
† Department of Physics, University of Michigan, Ann Arbor, Michigan, USA
[1] These authors contributed equally to this work.
[2] Current address: Department of Chemistry, The Ohio State University, Columbus, Ohio, USA

Methods in Enzymology, Volume 475   © 2010 Elsevier Inc.
ISSN 0076-6879, DOI: 10.1016/S0076-6879(10)75006-0   All rights reserved.

121

Abstract

Recent improvements in methods of single-particle fluorescence tracking have permitted detailed studies of molecular motion on the nanometer scale. In a quest to introduce these tools to the burgeoning field of DNA nanotechnology, we have exploited fluorescence imaging with one-nanometer accuracy (FIONA) and single-molecule high-resolution colocalization (SHREC) to monitor the diffusive behavior of synthetic molecular walkers, dubbed "spiders," at the single-molecule level. Here we discuss the imaging methods used, results from tracking individual spiders on pseudo-one-dimensional surfaces, and some of the unique experimental challenges presented by the low velocities ($\sim$3 nm/min) of these nanowalkers. These experiments demonstrate the promise of fluorescent particle tracking as a tool for the detailed characterization of synthetic molecular nanosystems at the single-molecule level.

1. INTRODUCTION

During the last few decades, there has been remarkable growth in the use of fluorescence spectroscopy in biophysical studies. Fluorescence-based tools are now being employed to understand the properties and dynamics of proteins (Giepmans et al., 2006; Min et al., 2005; Schuler and Eaton, 2008; Shi et al., 2008) and nucleic acids (Ditzler et al., 2007; Joo et al., 2008; Pljevaljcic and Millar, 2008; Zhao and Rueda, 2009). They are implemented in cutting-edge applications in medical and clinical chemistry for high-throughput screening and detection (Gribbon et al., 2004; Hintersteiner and Auer, 2008), as well as in cellular imaging for characterizing the localization and movement of intracellular components (Lippincott-Schwartz et al., 2003; Michalet et al., 2005; Moerner, 2007; Walter et al., 2008). Continued advances in fluorescence techniques and instrumentation have fueled applications of fluorescence spectroscopy to more detailed characterization of biomolecules. For example, there has been rapid expansion in the use of nucleotide analogs as fluorescent probes (Rist and Marino, 2002) to detect and characterize single nucleic acid molecules in real-time. Fluorescence resonance energy transfer (FRET) has emerged as a particularly powerful tool to probe distances (Deniz et al., 1999; Ha et al., 1996; Stryer and Haugland, 1967), conformational changes (Kim et al., 2002; Truong and Ikura, 2001; Weiss, 2000), and dynamics (Al-Hashimi and Walter, 2008; Zhuang et al., 2002) of macromolecules on the order of 1–10 nm. To measure larger distances of typically 10 nm or more, single-particle fluorescence tracking has proven useful (Barak and Webb, 1982; Churchman et al., 2005; Gordon et al., 2004; Yildiz et al., 2003).

The accuracy of particle tracking techniques is determined in part by the finite resolution of light microscopy. When light passes through a lens with a circular aperture as it does in a single-molecule fluorescence microscope, the focused light emitted from a point-like source forms a diffraction pattern known as an Airy pattern. The radius of the bright central region, called the Airy disk, can be approximated as $\lambda/(2 \times NA)$, where λ is the wavelength of the light source and NA is the numerical aperture of the lens. Within this radius, according to Rayleigh's criterion, no features may be resolved (Hecht, 2002). When imaging in the visible spectrum with a typical NA of 1.2, the radius of the Airy disk, and therefore the optical resolution limit, is ~250 nm. However, so-termed super-resolution (or super-accuracy) methods have been developed in the last few years that overcome this optical resolution barrier and bring the localization accuracy of a single particle down to the low nanometer range (Betzig and Chichester, 1993; Heintzmann et al., 2002; Hess et al., 2006; Hofmann et al., 2005; Huang et al., 2009; Moerner, 2007; Qu et al., 2004; Rust et al., 2006). It should be noted that we are using the term "super-resolution" here in its broader sense for all techniques that localize (and track over time) one or more single-molecule emitters at higher than the diffraction limit of accuracy (Moerner, 2007), while the more narrow sense of the term refers only to imaging techniques that resolve many closely spaced emitters by observing only few of them at a time over a time series of images (Huang et al., 2009).

Low-nanometer localization accuracy can allow for tracking slowly moving molecular devices in real-time. Several years ago, we suggested applying modern single-molecule fluorescence microscopy tools to nanotechnology as a way of monitoring and ultimately enabling control of the assembly and function of the desired structures and devices (Rueda and Walter, 2005). Here, we present in detail the implementation of nanometer-scale tracking of a novel type of autonomously walking DNA nano-assembly termed a "spider" (Lund et al., 2010).

1.1. History of fluorescent single-particle tracking

Fluorescent single-particle tracking entails using the fluorescence emission from a point-like source to accurately determine the emitting particle's location, typically over a span of time ranging from milliseconds to minutes that depends on the speed of particle motion. One of the first examples of fluorescent particle tracking was accomplished by Barak and Webb in their study of the diffusion of intensely fluorescent (~45 fluorophores) low density lipoprotein (LDL)–receptor complexes along human fibroblasts in which they were able to observe the movement of as few as one to three "molecules" in a given region (Barak and Webb, 1982). By using low temperatures, single chromophores were first optically detected in solids by Moerner and Kador (1989). Advancements in single-molecule fluorescence techniques allowed for single-molecule detection in more biologically relevant conditions with ever-improving localization accuracy.

For example, near-field scanning optical microscopy (NSOM) brought the tracking error down to ~14 nm in solution at room temperature (Betzig and Chichester, 1993). The high temporal as well as spatial resolution often required for single-particle tracking was accomplished by Schmidt *et al.* (1996) who successfully combined a low localization error of ~30 nm with a high time resolution of 40 ms, enabling them to study the diffusion of single phospholipids in a phospholipid membrane (Schmidt *et al.*, 1996).

Experimental fluorescent single-particle tracking called for the development of its theoretical counterpart. Using a maximum likelihood estimation analysis, Bobroff developed a quantitative method for analyzing the statistical error in position measurements made with light and particle signals, taking into consideration the measurement signal, noise distribution, and instrument resolution. This was done particularly for a Gaussian signal (Bobroff, 1986). Based on Bobroff's least-squares fitting approach, Webb and coworkers derived a simple equation for calculating the standard error of the mean of the position measurements (σ_μ) that depends on the instrumentation parameters and features of the Gaussian fit:

$$\sigma_{\mu_i} = \sqrt{\left(\frac{s_i^2}{N} + \frac{a^2/12}{N} + \frac{8\pi s_i^4 b^2}{a^2 N^2}\right)}, \qquad (6.1)$$

where s_i is the standard deviation of the Gaussian distribution of the *i*th index that indicates either the *x*- or *y*-direction, N is the number of photons, a is the pixel size, and b is the standard deviation of the background. The first term (s_i^2/N) arises from photon noise, the second term represents the effect of the finite pixel size of the camera, and the third term arises from the background signal of the sample (Thompson *et al.*, 2002; Yildiz *et al.*, 2003). Webb and coworkers were able to determine, according to this equation, the position of stationary beads with ~2 nm localization precision (Thompson *et al.*, 2002).

These advances in single-molecule fluorescence imaging and analysis laid a firm foundation for the development of fluorescence imaging with one-nanometer accuracy (FIONA) and related techniques. Developed by Selvin and coworkers, FIONA enabled the localization of singly fluorophore-labeled myosin V motor proteins along microtubules with typically 3 nm precision (<1.5 nm precision for the brighter molecules) and 0.5 s temporal resolution, using a total internal reflection fluorescence (TIRF) microscope at room temperature. This accomplishment was achieved by maximizing the number of photons collected (to ~5000–10,000 photons) while optimizing the camera pixel size (to 86 nm) and minimizing the background noise (to a standard deviation of ~11 photons) (Yildiz *et al.*, 2003). Complementing FIONA, Spudich and coworkers developed a technique, termed single-molecule high-resolution colocalization (SHREC), which utilizes two fluorophores of differing emission spectra mapped onto the same space and measures interfluorophore distances with 1-nm precision (based on 482 molecules). To demonstrate this technique, they verified the expected ∼36 nm distance between the two heads of myosin V (Churchman et al., 2005).

We adapted FIONA and SHREC, both of which have previously been used to study ATP-fueled biological motor proteins (Churchman et al., 2005; Yildiz et al., 2003), to study synthetic DNA-fueled nanowalkers termed spiders (Lund et al., 2010).

2. DNA-Based Nanowalkers

DNA nanotechnology, the study of constructing programmable nanometer-scale structures and devices based on the Watson–Crick base-pairing rules and catalytic properties of DNA, has continuously accelerated in pace since its conception in the early 1980s (Douglas et al., 2009; Rothemund, 2006; Seeman, 2007). DNA nanotechnology has recently yielded synthetic molecular machines that mimic naturally occurring bipedal nanowalkers (Bath et al., 2009; Green et al., 2008; Omabegho et al., 2009; Sherman and Seeman, 2004; Shin and Pierce, 2004; Yin et al., 2004). These DNA-based nanowalkers, consisting of two single-stranded DNA (ssDNA) "legs," traverse tracks composed of ssDNA strands complementary to the legs in an experimentally controlled direction using thermodynamically favored strand displacement (or exchange) by an ssDNA fuel strand as an energy source (Fig. 6.1A).

2.1. The molecular "spider"

Due to slow kinetics, strand displacement is not an ideal source of energy for molecular walkers. Biological motor proteins have velocities on the order of ∼1 μm/s *in vitro* under saturating ATP conditions (King and Schroer, 2000; Kural et al., 2005; Nishiura et al., 2004), while these synthetic nanowalkers are limited by the kinetics of unwinding one DNA duplex and forming another, both ∼15–50 base pairs in length (Bath et al., 2009; Green et al., 2008; Omabegho et al., 2009; Sherman and Seeman, 2004; Shin and Pierce, 2004; Yin et al., 2004), leading to estimated velocities of ∼10 nm/h (Shin and Pierce, 2004). In addition, while it has been predicted that they have the ability to traverse longer tracks, as yet they have only been shown to accomplish a few successive steps along short tracks approximated to be typically on the lower end of tens of nanometers (Green et al., 2008; Omabegho et al., 2009).

The catalytic power of deoxyribozymes, or DNAzymes, offers an attractive alternative to strand displacement for driving locomotion. DNAzymes

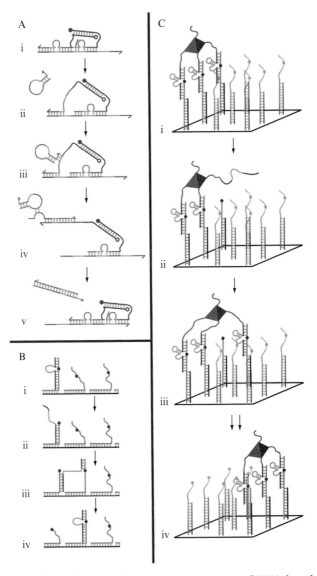

Figure 6.1 Example mechanisms for processive movement of DNA-based nanowalkers. (A) Biped nanowalker from Green *et al.* (2008) that utilizes fuel consisting of two complementary DNA hairpins. Colors (gray scales) represent complementary sequences. (i) The competition between identical feet to bind to the track permits the exposure of a toehold region in the left foot (ii). A hairpin hybridizes to the toehold region (iii), displacing the left foot from the track. A second, complementary hairpin hybridizes to the first hairpin (iv) to form a waste product, allowing the foot to rebind to the track with equal probability to the left or right. (B) Single-stranded deoxyribozyme-based nanowalker from Tian *et al.* (2005). (i) The 10–23 deoxyribozyme (red with orange active site) is able to cleave its substrate (green) at a specific site (purple) in the

are DNA sequences with the ability to site-specifically cleave chimeric DNA–RNA substrates in the presence of an appropriate divalent metal ion cofactor. Tian et al. (2005) were the first to incorporate DNAzymes into their nanowalker in the form of the 10–23 DNAzyme (Santoro and Joyce, 1997). As it cleaves an oligonucleotide on its track, the DNAzyme dissociates from the shorter cleavage product and is able to progress along the track via displacement of the still bound longer product portion by an adjacent substrate strand (Fig. 6.1B) (Tian et al., 2005). The speed of movement may still be limited by the kinetics of strand displacement despite the shorter cleavage product, and the processivity limited by the risk of complete dissociation of the single leg from its track.

To overcome these limitations, Stojanovic and coworkers recently developed a polypedal DNAzyme-based nanowalker dubbed a "spider" (Pei et al., 2006). Spiders consist of a streptavidin "body" bound to multiple biotinylated 8–17-based DNAzyme (Li et al., 2000) "legs." The spider's multivalent binding allows it to remain securely bound to the surface even as individual DNAzymes cleave their substrate and detach, and therefore allows a large number of substrate sites to be visited and cleaved by a single spider before it dissociates from the surface (Pei et al., 2006). Once a leg cleaves its bound substrate, it can more rapidly dissociate from the 10-nucleotide long products and bind another 18 nucleotide-long substrate in the vicinity. Mathematical modeling of this system suggests that these properties will result in spiders undergoing biased movement on a substrate-field, avoiding sites they have previously visited (Fig. 6.1C) (Antal and Krapivsky, 2007; Samaii et al., 2010).

2.2. Behavior-based molecular robots: Programming tasks into the walker's environment

Due to their ability to sense and respond to stimuli (for instance, leg substrates), nano-assemblies such as spiders can be considered behavior-based molecular robots (Brooks, 1991). Although they cannot themselves store complex programs or instructions, one can influence their behavior by exposing them to

---

presence of Mg^{2+}. The shorter end dissociates from its product (ii) and hybridizes to a neighboring strand (iii). Via displacement of the cleavage product by neighboring substrates, the deoxyribozyme progresses along the track (iv). (C) Spider moving along a three-substrate-wide origami track (Lund et al., 2010). The spider is composed of a streptavidin body, a capture leg, spacers, all shown in blue, and three 8–17 deoxyribozyme legs (red binding arms with orange active site). (i) The deoxyribozyme spider legs hybridize to substrates (green) that are attached to the origami scaffold via hybridization to staple overhangs (black). In the presence of Zn^{2+}, each leg of the spider cleaves its substrate, dissociates from its products (ii) and hybridizes to a neighboring strand (iii). The greater affinity of the leg for the substrate than the cleavage products makes it energetically favorable for the strand to bind to the full substrate, generating a biased-random walk from the cleaved toward the uncleaved substrate (iv). (See Color Insert.)

controlled environmental cues. It is possible to, for example, direct a spider to complete simple tasks such as "start," "stop," and "turn" by precisely controlling the position and sequence of the substrates to which the spider has access (Lund et al., 2010). Hence, tracks of substrate can define a program of movement consisting of local environmental cues (e.g., bends in the track, tight binding sites acting as "fly paper") which are executed by the spider, and represent an initial step toward a world of useful molecular robots.

To assess the capability of spiders to execute such programs, tracks with a feature resolution of 6 nm were engineered using DNA origami technology (Rothemund, 2006). A rectangular origami scaffold was constructed using the 7-kilobase single-stranded M13mp18 genomic DNA, which was shaped and held in place with the aid of 202 oligodeoxynucleotide staple strands that hybridize to complementary regions of the M13mp18 DNA. Specific staple strands in the array were extended on their 5′ end with specific sequences to position three types of surface features on the scaffold. First, a single staple near one corner of the origami tile was extended to contain the START sequence that is partially complementary to the noncatalytic, ssDNA "capture" leg of the spider, which positions the spider at the start of the track. In assembling the spider–origami complex, the spider was first allowed to bind to the START position before any other substrates were added in order to ensure specific binding at the START site. Second, the track was laid by hybridizing cleavable substrate to a specific overhang sequence that had been incorporated into particular staples. These substrates contain a single ribonucleotide (rA) upstream of G at the cleavage site to permit cleavage by the three DNAzyme legs (Fig. 6.1C). Six staples at the end of the track were extended with a unique sequence to similarly attach noncleavable all-DNA substrate analogs (STOP) intended to trap the spider once it reaches the end of the track ("fly paper"). The STOP strands and spiders were labeled with the cyanine fluorophores Cy5 and Cy3, respectively, to allow the spider to be tracked by super-resolution fluorescence microscopy relative to the STOP position. For purposes of immobilizing the origami–spider complexes on avidin-coated quartz slides for TIRF microscopy, four staples near the corners of the origami tile were biotinylated.

To initiate motion of the spider by displacement from the START, a TRIGGER DNA strand that is fully complementary to the START sequence was added ("start" command). Zinc(II) ions were then added to promote cleavage of bound substrates by the DNAzyme legs and thus walking ("walk" command). Spiders were predicted to graze the surface until they reach and become trapped at the STOP position ("stop" command). Scaffolded tracks were designed with either a linear shape or incorporating one left- or right-handed turn, resulting in a program of motion for the spider. All of the tracks were three substrates wide and at least 14 substrates long, thus actualizing a pseudo-one-dimensional path or program of motion for the spiders to follow.

2.3. Methods for characterizing nanowalker behavior

Surface plasmon resonance (SPR) was previously employed to detect movement of spiders with two to six legs in a polymer matrix containing a high density of DNAzyme substrate, yielding ensemble estimates of cleavage rate and processivity for spiders (Pei *et al.*, 2006). Native polyacrylamide gel electrophoresis (PAGE) is commonly used to detect procession of DNA-based nanowalkers along tracks, since the site at which the walker is bound to the track will influence the overall topology of the complex and, hence, its electrophoretic mobility (Omabegho *et al.*, 2009). While useful, these ensemble-averaging techniques do not directly report on movement and provide limited information concerning the distribution of possible behaviors for individual nano-assemblies.

One technique Lund *et al.* (2010) have used to study individual molecular spiders is atomic force microscopy (AFM), which yields detailed, high-resolution "snapshots" of trajectories followed by the nano-assemblies. While AFM yields insight into the behavior of individual spiders, real-time observation of the spider walk is not readily achieved. This is thought to be due to the inhibition of walking caused by the mica surface necessary for sample immobilization and the potential for mechanical disruption of the sample associated with repeated scanning by the AFM probe.

A complementary single-molecule technique is TIRF microscopy, which permits the visualization of individual fluorescent molecules on a microscope slide by limiting the excitation volume to a thin (~100–200 nm) sheet near the surface of the slide, thereby suppressing background noise. By fluorescently labeling the spider and origami track, we have used this technique to monitor the motion of spiders along prescriptive DNA origami tracks.

3. Considerations for Fluorescence Imaging of Slowly Moving Particles

While spiders have experimentally been shown to be faster and more processive than previous DNA-based nanowalkers, traversing a 100 nm track with speeds on the order of several nm/min, they are still significantly slower than myosin V, resulting in heightened challenges for fluorescence imaging due to fluorophore photobleaching and stage drift.

3.1. Prolonging fluorophore lifetime

Fluorophores are subject to permanent photobleaching: After a limited number of excitation events ($\sim 10^6$) (Willets *et al.*, 2003), they will remain in a permanent dark state, often induced by reaction of the excited state with molecular oxygen. Optimizing the fluorophore lifetime in order to track a specific fluorescently labeled particle for an extended period of time may be accomplished in multiple ways. The first intuitive way is by illuminating the sample only periodically, rather than continuously. However, this method has the often undesired consequence of reducing the experimental time resolution. The goal is, therefore, to strike a balance between the time resolution commensurate with the velocity of the moving particle, the camera exposure (or photon integration) time necessary to obtain super-resolution position accuracy, and fluorophore longevity. For our spider origami applications, the photon integration time is typically 2.5 s with a 12.5-s dark period between acquisitions of successive images, resulting in a time resolution of 15 s. With velocities of $\sim$3 nm/min, the spider moves $\sim$1 nm/frame, which constitutes sufficient time resolution considering the net travel distance of 100 nm.

A complementary means of extending fluorophore lifetime is to introduce an oxygen scavenging system (OSS), which reduces the rate of oxygen-dependent photobleaching by sequestering molecular oxygen from solution. A widely used OSS is a coupled enzyme system consisting of glucose oxidase and catalase that converts glucose and molecular oxygen into gluconic acid and water (Benesch and Benesch, 1953). However, Aitken et al. (2008) found an improved OSS that consists of 25 nM protocatechuate dioxygenase (PCD), 2.5 mM protocatechuate (PCA), and 1 mM Trolox. The PCD employs a nonheme iron center that catalyzes the conversion of PCA and molecular oxygen into β-carboxy-cis,cis-muconic acid (Aitken et al., 2008), while the antioxidant Trolox suppresses slow blinking and photobleaching of cyanine dyes (Rasnik et al., 2006). We use as much as fivefold the concentrations of the components specified above to further prolong fluorophore lifetime.

While sufficient fluorophore longevity is the primary concern for lengthy single-molecule fluorescence tracking experiments, other fluorophore properties should also be considered. The fluorophores should be photostable to reduce excessive blinking (access to reversible dark states) that will interrupt tracking. Also, it is preferable to choose fluorescent probes with high brightness (molar extinction coefficient times quantum yield) to maximize the number of photons emitted and, hence, position determination accuracy. The fluorescent signal may also be increased by multiply labeling a tracked particle, although it has been noted that Cy5 can self-quench when closely clustered (Gruber et al., 2000). We label the spider with two to three Cy3 molecules and each of the six STOP strands at the end of the DNA origami track with one Cy5 molecule. We do not have a detrimental self-quenching problem with these redundant probes.

3.2. Control for stage and focal drift

When imaging for long periods of time (several minutes to hours), stage and focal drift due to thermal fluctuations and mechanical instability become problematic. A practical approach to correct for such drift is to use a fiduciary marker(s) and determine the relative motion between the moving particle and stationary marker by subtracting the trajectory of the marker from that of the particle. We use the cluster of six Cy5 fluorescent probes on the STOP strands as a fiduciary marker for spider motion on the corresponding origami track. Controls should be performed in which the essential divalent metal ion cofactor is omitted to verify that the spider is stationary in its absence. In addition, since determination of spider movement is based on only its corresponding fiduciary marker, it is important to control for possible aberrant movement of the marker by comparing its trajectory with those of neighboring markers to ensure that the trajectories exhibit the same drift pattern.

4. SINGLE-MOLECULE FLUORESCENCE TRACKING OF NANOWALKERS

The following super-resolution fluorescence imaging protocol permits real-time tracking of fluorescently labeled, slowly moving nanowalkers along DNA origami paths.

4.1. Slide preparation

First, the origami tiles need to be immobilized so as to lie flat on the microscope slide's surface. This is accomplished via a strong noncovalent bond between the avidin-coated slide and four biotinylated staples located near the four corners of the underside of each origami tile. To maximize the probability that all biotins bind securely to the surface, the avidin coating on the slide surface must be dense, which we accomplish here by aminosilanization followed by reaction with a bifunctional diisothiocyanate to covalently anchor avidin (Fig. 6.2A).

1. Drill two 1-mm holes in each slide (Finkenbeiner, fused silica, 1 in.× 3 in. × 1 mm) to be used for buffer exchange. It is helpful to have a template to follow when drilling the holes to ensure that the coverslip will fit over both holes (Fig. 6.2B).
2. Fluorescent and other impurities must be removed from the slide surface prior to imaging. Submerge the slides in a boiling "piranha" solution (5% (v/v) ammonium hydroxide (Aldrich, 231-704-8) and 14% (v/v) hydrogen peroxide (Acros, 202465000)) for 20 min. Rinse well with deionized water, flame sides with a propane torch for 1 min each. Sonicate slides in 1 M potassium hydroxide (KOH; Aldrich, 215813) for 30 min. Rinse again thoroughly with deionized water, then rinse with acetone (Fisher, A949-4) and sonicate for $\sim$15–20 min in acetone.

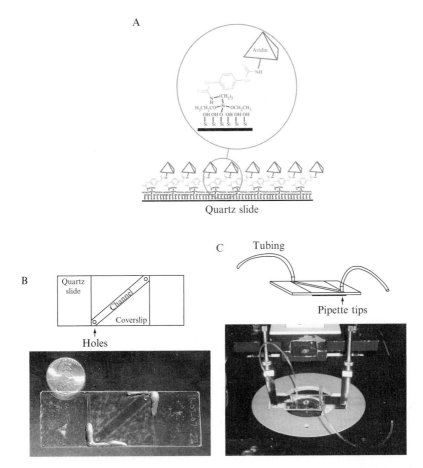

Figure 6.2 Microscope slide preparation. (A) The slide is densely coated with avidin (tetrahedron) covalently bound via a PDTIC linker (light gray) to an aminosilanized quartz slide surface. (B) A fluidic channel is created between the two 1-mm holes in the quartz slide by sandwiching two pieces of double-sided sticky tape between the slide and the coverslip. (C) Tubing for fluid introduction and disposal is attached to the holes in microscope slide by using pipette tips. Epoxy resin is added to generate a tight seal between the tubing and pipette tips as well as between the pipette tips and holes. Note that, when placed on the microscope stage, the coverslip is face-down, allowing the tubing on the microscope slide to be face-up.

3. To aminosilanize the slides, immerse clean slides in a 5% (v/v) 3-aminopropyltriethoxysilane (APTES; Sigma-Aldrich, A3648-100ML) acetone solution for 1 h. Rinse aminosilanized slides thoroughly with acetone. Dry at 80 °C for 1 h.
4. *para*-Diisothiocyanatobenzene (PDTIC; Acros, 417510050) covalently couples to the aminosilanized surface based on the approach developed

by Guo et al. (1994). Incubate the slides in 0.2% (w/v) PDTIC and 10% (v/v) pyridine (Fisher, P368-500) in dimethylformamide (Acros, spectroscopic grade, 40832-5000) for 2 h. Rinse thoroughly with acetone and methanol (Fisher, A452-4), and allow to dry.

5. Coat the slides with avidin. Pipette 70 μl of 0.5 mg/ml avidin (Sigma-Aldrich, A9275-2MG) on the center of each slide. Cover each slide with a glass coverslip (VWR, 48404-466) and incubate at room temperature for 2 h in a closed container above a water bath to maintain a humid environment. Remove the coverslips. Thoroughly rinse slides with deionized water. To quench the PDTIC not bound to avidin, apply 1 M NaCl in 40 mM NaOH to each slide for a few minutes. Thoroughly rinse the slides with deionized water. Dry with nitrogen.

6. Construct a flow channel between the drilled holes on each slide using two pieces of double-sided tape (Scotch, permanent, 1/2 in.) spaced 2–3 mm apart (Fig. 6.2B). Securely place a coverslip over the tape to cover the channel. Use epoxy glue (Hardman Adhesives 04001) along the edges of the coverslip to seal the channel. These avidin-coated slides may be stored at 4 °C in an evacuated desiccator for up to 4 weeks until mounting them on the microscope (Fig. 6.2C).

4.2. Experimental procedure

When choosing an appropriate imaging buffer, it is important to perform control experiments to ensure that the nano-assemblies are active under the buffer conditions. If a divalent transition metal cation such as Zn^{2+} is required to catalyze the reaction, be aware of possible chelation by buffering agents such as citrate in buffers such as saline–sodium citrate (SSC; 150 mM NaCl, 15 mM sodium citrate, pH 7.0). The imaging buffer used in our protocol is HEPES-buffered saline (HBS) (10 mM HEPES, 150 mM NaCl, pH 7.4). This protocol assumes that the origami–spider complex, with a Cy3-labeled spider and Cy5-labeled STOP position on the origami, has already been assembled as described in Lund et al. (2010).

1. Let 1 μl of 10 nM origami–spider complex incubate with 1 μl of 1–100 μM TRIGGER strand for at least 30 min to allow adequate time for the TRIGGER to hybridize to the START strand.
2. For a volume of 200 μl (the recommended minimum volume for any solution that will be flushed through the fluidic channel to ensure that the solution makes it all the way through the ~10 cm of tubing and to minimize the formation of air pockets), 1 × OSS is created by combining 5 μl 100 mM protocatechuate acid (PCA; Sigma, P5630-10G), 5 μl 1 μM protocatechuate-3,4-dioxygenase (PCD; Sigma, P8279-25UN), and 2 μl 100 mM Trolox (Acros, 218940050). We use concentrations as high as 5× OSS.

3. Add 1–5× OSS and 1× HBS to the sample to result in a final concentration of 100 pM spider–origami complex. This solution will henceforth be referred to as the sample.
4. Prepare the OSS buffer (1× HBS supplemented with OSS) as well as the reaction buffer (1× HBS supplemented with OSS and 1 mM ZnSO$_4$). If the zinc precipitates (the solution turns opaque), lower the concentration of ZnSO$_4$ or buffer pH accordingly.
5. Attach tubing (Tygon, 0.02′ ID, 63018-044) on both sides of the fluidic channel using pipette tips and epoxy. Plastic tubing should be just wide enough for a syringe needle (Fig. 6.2C).
6. Using a syringe with needle, flow 200 μl imaging buffer in the absence of OSS through the channel. Using excitation from the 532 and 638-nm lasers, image the slide with a TIRF microscope (described in detail in Section 5.3) prior to adding the sample. Inspect the slide to see if there is any fluorescent debris that may interfere with the signal emitted by the fluorophores. If so, photobleach these spots by exciting the location at a high laser intensity for several minutes until the background fluorescence has stabilized. If there are still many contaminants, use a new slide.
7. Flow the sample onto the slide in the dark and incubate for 2–10 min. To optimize the incubation time for an optimal density of nano-assemblies on the surface, excite the sample every few minutes with the lasers to see how many nano-assemblies have bound to the surface. Ideally, the nano-assemblies are dense enough to have a large number of point spread functions (PSFs) to analyze, but sparse enough that Gaussian functions may reliably be fit to the individual PSFs. We normally work with densities of ~ 0.03 molecules/μm^2.
8. Once the optimal density is achieved, flush out the extra unbound sample using the OSS buffer. Image the origami at room temperature using a 2.5-s camera integration time and 12.5-s delay during which a shutter is used to protect the fluorophores from excessive photobleaching.
9. Add the reaction buffer after imaging for 20 min. Omitting the divalent metal cofactor for the first several minutes provides an important control for movement in the absence of catalysis. Continue imaging for at least 60 min to allow ample time for the spider to reach the STOP position.

4.3. Instrumentation for imaging

We use a Newport ST-UT2 vibration isolation table in a temperature controlled room to minimize disturbances from vibrations and temperature fluctuations in the microscope's external environment. The table holds a home-built prism-based TIRF microscope equipped with a 1.2 NA 60× water objective (Olympus Uplanapo) for imaging (Fig. 6.3). A 532-nm ultracompact diode-pumped Nd:YAG laser (GCL-025-S, CrystaLaser) is

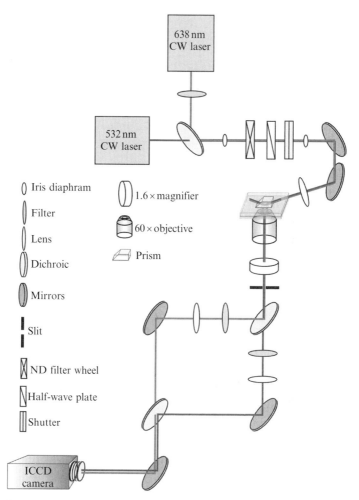

Figure 6.3 Single-molecule prism-based TIRF setup. Linearly polarized light from the 638-nm red diode laser passes through a 638 ± 10 nm clean-up filter and is reflected off a 610-nm cutoff dichroic. 532-nm light from the Nd:YAG passes through same dichroic to join the same beam path. Light from both lasers then passes through the following components, in order: an iris diaphragm, a neutral density filter, a $\lambda/2$-wave plate for 532-nm light, a shutter, an iris diaphragm, a series of mirrors (two are shown for ease of representation; three were used during our experiments), a focusing lens, the TIR prism, the microscope slide containing the sample, the 60× objective, a 1.6× magnifier (along with additional filters, mirrors, and image transferring lenses contained within the microscope), a slit, and a 610-nm cutoff dichroic mirror. The emission from Cy3 passes through a band-pass filter, while that of Cy5 passes through a long pass filter. These separated images are projected side-by-side onto an intensified charge-coupled device (ICCD) camera. (See Color Insert.)

used to excite Cy3, while a 638-nm red diode laser (Coherent CUBE 635-25C, Coherent Inc.) is used to excite Cy5.

The linearly polarized light from the red diode laser passes through a 638 ± 10 nm clean-up filter (Chroma, z640/20) before reflecting off a dichroic that the 532 nm light from the Nd:YAG laser passes through. The light from both lasers passes through an iris diaphragm that blocks excess stray light, followed by a neutral density filter to help regulate the laser power exciting the sample (Fig. 6.3). It continues through a $\lambda/2$-wave plate for 532-nm light and a shutter that is used to limit exposure time. As an extra measure for avoiding unwanted scattered light, a second iris diaphragm is used followed by a series of mirrors to redirect the light to the microscope where it passes through a focusing lens before entering the TIR prism and exciting the fluorescently labeled sample. The emitted light enters the 60× objective and passes through a 1.6× magnifier along with additional filters, mirrors, and image transferring lenses, resulting in an effective pixel size of 133 nm. The emission from the Cy3 and Cy5 fluorophores are then separated with a dichroic mirror with 610-nm cutoff (Chroma, 610DCXR). The Cy3 emission additionally passes through a band-pass filter (HQ580/60m, Chroma), while the Cy5 emission passes through a long pass filter (HQ655LP, Chroma). These spectrally separated images are projected side-by-side onto an intensified charge-coupled device (ICCD) camera (IPentamax HQ Gen III, Roper Scientific, Inc.). An aperture constructed of two razor blades and positioned at the microscope sideport image is used to adjust the size of the spectrally separated images on each half of the CCD chip. WinView32 software (Roper Scientific, Inc.) or suitable home-programmed software is used to immediately visualize and store the signal received from the camera. Camera gain values, or conversion values for converting the camera signals from arbitrary digital numbers (DN) to collected photoelectrons, are determined using the photon transfer method (Janesick, 1997).

5. Extracting Super-Resolution Position Information

After collection, the raw particle tracking data must be subjected to a series of analysis steps to yield and interpret super-resolution tracking data. A number of software routines custom-written in our laboratory implement these steps (Lund et al., 2010) and are available upon request.

5.1. Mapping the two emission channels onto one another

Since images of the Cy3 and Cy5 PSFs are projected onto separate halves of the ICCD camera and passed through different sets of optics, careful mapping of one channel onto the other is needed to determine which spider resides on a given origami. For this purpose we use a SHREC approach (Churchman et al., 2005) wherein fluorescent beads (FluoSpheres, F8810) visible in both channels are used as fiduciary markers to establish a locally weighted mean transformation between the channels in MATLAB (Mathworks, Natick, MA). Unlike global mapping routines, which attempt to establish a valid single transformation over the entire field, a locally weighted mean transformation accounts for local variations and distortions in the image due to, for example, aberrations in the optical path. Mapping the Cy5 channel onto the Cy3 channel enables us to pair each Cy3-labeled spider with its (most likely) Cy5-labeled STOP. Once this is performed, the motion of each spider relative to its STOP is determined by first applying a Gaussian fitting routine (Rust et al., 2006; Yildiz et al., 2003) to the individual Cy3 and Cy5 PSFs in MATLAB, followed by subtracting the Cy5 trajectory from that of the Cy3 (Fig. 6.4).

5.2. Quality control

In single-molecule microscopy, it is generally necessary to identify and reject traces that, due to poor tracking accuracy, fluorescent contamination, or failed assembly of the complexes, do not accurately represent the molecular behavior of interest. Suitable selection criteria are, however, challenging to implement without introducing some form of bias. To record spider motion along origami tracks, we employ a minimal yet essential set of selection criteria to analyze the data based on the following parameters (Lund et al., 2010):

1. *Intensity.* The signal-to-noise ratio must be sufficient to obtain high-resolution results from the Gaussian fitting routine (Thompson et al., 2002). In our current experiments, the minimum number of photons required for the peak intensity of a PSF is 1000.
2. *Ellipticity.* The ellipticity is calculated using the equation $\varepsilon = 1 - \sigma_{min}/\sigma_{max}$, where σ_{min} and σ_{max} are the standard deviations along the major and minor axes of the ellipse (Gordon et al., 2004) in a fixed plane. We demand that the maximum ellipticity be 0.3 for the included data points. Note that this cutoff may be dependent on the fluorophore distribution on the surface and may be modified to ensure that reliable data are not unintentionally removed. This approach filters out immobile dyes whose fixed polarization orientations result in elliptical PSFs and provides reassurance that we are tracking single molecules and not aggregates that rarely accumulate to form a perfectly circular PSF.
3. *Displacement.* The origami tracks are 100 nm in length. Position measurements that far exceed this net displacement do not likely represent a single spider moving along an intact origami track. A typical displacement cutoff is less than three standard deviations from the mean of all

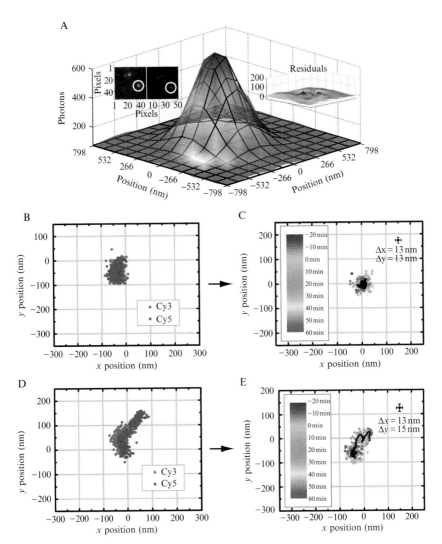

Figure 6.4 Extraction of trajectories from a single nanowalker on a linear origami track imaged by TIRF microscopy (Lund et al., 2010). (A) Point spread functions (PSFs) of Cy3-labeled spiders and Cy5-labeled origami are colocalized (inset, upper left) and fit separately to two-dimensional Gaussian functions in each movie frame to determine their coordinates over time. The fit has low residuals (inset, upper right). (B–E) Centroids from Gaussian fitting are plotted as a function of time to yield nanowalker trajectories. Even in the absence of the divalent metal ion cofactor, Cy3 and Cy5 coordinates show considerable drift (B), but when the trajectory of Cy5 is subtracted from that of Cy3 it becomes clear that the nanowalker is stationary on its track (C). In contrast, in the presence of 5 mM ZnSO$_4$, subtraction of the raw Cy5 and Cy3 trajectories (D) yields net movement of about 90 nm. In (C) and (E), HBS buffer containing either 0 or 5 mM ZnSO$_4$ was added to the sample at time $t = 0$ min. (See Color Insert.)

position measurements. However, to reduce the influence of extreme outliers, such as those caused by aberrant Gaussian fitting of transiently binding fluorescent contaminants, the median is used instead of the mean. To further reduce the influence of aberrant fitting, we also discard any position measurements greater than 500 nm from the spider's initial position.

4. *Colocalization.* SHREC analysis must confirm that each spider is within a reasonable distance of its STOP partner. Considering the track length of ~ 100 nm, we discard any Cy3 PSF that is more than 150 nm away from the nearest Cy5 PSF at any point in its trajectory.

5.3. Presentation

Single-molecule fluorescence experiments uniquely allow for a comparison of individual molecular behaviors, while averaging information from individual (diverse) traces is imperative to characterize the ensemble behavior and ensure statistical significance of individual trajectories (Lund *et al.*, 2010).

5.3.1. Trajectory plots

Trajectory plots consist of the spider position values (x, y) as a function of time, which are color-coded to ease visualization (Fig. 6.5A). We have also found it useful to incorporate an origami track representation drawn to scale into the background of these plots. Such plots provide intuitive insight concerning the range of behaviors exhibited by individual walkers; features such as start time, velocity, end time, total distance, and path traversed may be approximated across the samples at a glance. However, quantitative comparisons require additional representation tools.

5.3.2. Displacement plots

The displacement plot is generated by calculating the distance between the initial position and each subsequent position and plotting these displacements as a function of time (Fig. 6.5B). When determining the displacement of a slowly moving object in the presence of significant noise, it is important to carefully consider how to define the initial position since it will significantly affect the overall displacement calculated. A Monte Carlo simulation that compares the displacement calculated with and without the noise level of a typical experiment may be helpful in making this decision. Currently, we define the initial position as the arithmetic mean of the first 16 data points. Because noise artificially increases these displacements, the data are smoothed using a 16-point rolling average; this method is based on previous observations about measuring distances only slightly above the noise level (Churchman *et al.*, 2005). A bin size of 16 (4 min at 4 frames/min) is

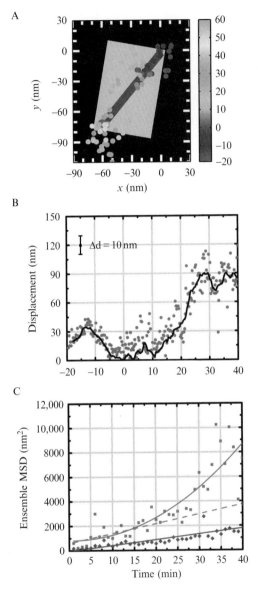

Figure 6.5 Characterization of nanowalker movement (Lund et al., 2010). (A) The two-dimensional trajectory of an individual spider is plotted as a function of time and compares reasonably well to a scale-drawn schematic of the origami track design (gray rectangle). HBS buffer containing 5 mM ZnSO$_4$ was added at time $t = 0$ min. (B) Displacement of the spider in panel (A) measured with respect to its position at the time of adding 5 mM ZnSO$_4$. Displacement plots calculated from both the raw trajectory (green dots) and a 4-min rolling average of the raw trajectory (black line) are shown. (C) Ensemble mean-square displacement (MSD) calculated from 16 spiders observed in HBS buffer containing 5 mM (green squares) or 0 mM (red diamonds) ZnSO$_4$ added at time $t = 0$ min. The former plot is fit with a power law function (green solid line), and the latter with a straight line (red line). To clearly illustrate the concave-up shape of the MSD in 5 mM ZnSO$_4$, a straight line is also fit to the first 15 min of this plot (green dotted line). (See Color Insert.)

appropriate because, with a typical velocity of 3 nm/min, a spider will move only 12 nm, which is generally within the error of our position measurements. The final or net displacement can be determined by heuristic or direct quantitative methods, or by inspection, and will depend on the system under study. Because many spiders on a 100-nm origami track will reach and become trapped on the STOP, we define the net displacement as the first local maximum in the rolling average to reach within 20 nm (a typical error value for displacement, calculated as described in Section 5.4) of the global maximum of the rolling average, and define the accompanying time coordinate as the stopping time. The average velocity of the spider, which provides a quantitative means for directly comparing individual molecules, is estimated by dividing the final net displacement by the stopping time.

5.3.3. Ensemble-averaged mean-square displacement plots

The characteristics of single particle motion can be determined from the sequence of positions and corresponding times made accessible by single particle tracking methods. The ability to observe the motions of individual particles allows one to sort trajectories into various modes of motion, and the stochastic nature of the particles call for statistical methods to be used for analysis (Qian et al., 1991; Saxton and Jacobson, 1997) by finding distributions of quantities characterizing the motion, such as diffusion coefficients, distances, velocities, anomalous diffusion exponents, corral sizes, etc. (Kusumi et al., 1993; Qian et al., 1991; Saxton and Jacobson, 1997). The mean-square displacement (MSD) is a particularly useful quantity for sorting and characterizing diffusive particle trajectories (Kusumi et al., 1993; Qian et al., 1991; Saxton and Jacobson, 1997), and can help uncover the spider walking mechanism. As the name would suggest, the MSD is produced by squaring the displacements between increasing time steps, averaging them over single trajectories for time-averaged MSD (TA MSD) or averaging over all the trajectories in a set (ensemble MSD), and plotting them as a function of time (Fig. 6.5C). For a trajectory in two dimensions, the MSD curve can be fit with the equation: $MSD = \langle r^2 \rangle = 4Dt^{\alpha}$, where the diffusion coefficient D and the α parameter help characterize the motion of the particle(s). A molecule exhibiting subdiffusive behavior, which is exemplified, for example, by confined particles, corresponds to the condition $0 < \alpha < 1$. Brownian diffusion corresponds to $\alpha = 1$. Superdiffusion corresponds to the case for $\alpha > 1$, where the MSD grows faster than it does in normal (Brownian) diffusion. Such processes are characterized by broad distributions of both waiting time and step size (with no meaningful mean), as opposed to subdiffusion, which is characterized by a broad distribution of only the waiting time but a narrow (Gaussian) step size distribution (Klafter and Sokolov, 2005). Diffusive processes with $\alpha \neq 1$ are generally referred to as anomalous diffusion and are commonplace in nature (Klafter and Sokolov, 2005). Finally, there are various modifications that the MSD fitting equation can have depending on the system under study, such as an additive velocity term for particles exposed to a continuous drift (directed transport) (Saxton and Jacobson, 1997).

A particle undergoing simple Brownian motion is an example of an ergodic system (where the time average of a single particle equals the ensemble average) and either TA MSD or ensemble MSD methods can be applied (Qian et al., 1991). By contrast, if applied to nonstationary systems (which implies nonergodicity), the TA MSD may yield misleading/incorrect results (Gross and Webb, 1988; He et al., 2008; Lubelski et al., 2008). The modification (cleavage) of a pseudo-one-dimensional origami track featuring START and STOP positions upon interaction with a spider allows the system to have a memory of previously visited locations and a bias to move toward new locations, which may be expected to result in the spider MSD increasing faster than that of a simple Brownian diffusion process (Antal and Krapivsky, 2007).

Complicated diffusion processes and noisy data can affect the MSD curves by further adding to the complexity of the underlying MSD model in various ways. For example, transitions between diffusive and nondiffusive segments of a trajectory (Saxton and Jacobson, 1997), crossover between anomalous dynamics on various time scales (Dieterich et al., 2008), and either noise inherent to the system (e.g., "biological noise") (Dieterich et al., 2008) or measurement noise can affect MSD curves at short time scales (Martin et al., 2002; Qian et al., 1991). Even for "perfect data" with no errors in the position measurements, the MSD and MSD-derived parameters will still have expected statistical variances due to the stochastic nature of a random walk; an estimate of these statistical variances is important to the analysis and should be assessed from the reproducibility of a series of corresponding measurements (Qian et al., 1991; Saxton and Jacobson, 1997). In addition, computer (e.g., Monte Carlo) simulations can generate random walks and provide a powerful tool to help assess the expected MSD curve and the processes and properties underlying a random walker's behavior (Dix and Verkman, 2008; Qian et al., 1991; Ritchie et al., 2005; Saxton and Jacobson, 1997). We use Monte Carlo simulations in MATLAB for this purpose (Lund et al., 2010). Because the proposed biased-walk mechanism is dependent on the difference in affinity of the spider's legs for the cleaved versus the intact substrate, a track composed completely of cleaved substrates should not exhibit this bias and would be predicted to undergo Brownian diffusion (Antal et al., 2007). To test for differences between the walking mechanism on cleaved and noncleaved substrate surfaces, a control origami track can be designed that contains cleaved substrate; the walking mechanism on the cleaved substrate is expected to be distinct from that on the intact substrate (Lund et al., 2010).

5.4. Error analysis

The theoretical error in tracking single particles may be determined from Eq. (6.1), which can be used to determine the accuracy of individual position determinations. For measurements that represent an average of multiple determinations for a stationary object over a span of time, a simple standard error of the mean would suffice. However, for an object such as a spider with a poorly defined velocity on short time scales, an alternative approach should be used. For the displacement plots, since a 16-point rolling average is used in smoothing the data, we use the following approach (Lund et al., 2010):

1. Define x- and y-errors, Δx_i and Δy_i, as the standard errors of the mean in the x- and y-dimension of all raw position measurements going into each averaged measurement; the resulting values for Δx_i and Δy_i can then be averaged (arithmetic mean) across all 16-point bins within the trajectory to get overall errors Δx and Δy as shown in Fig. 6.4, panels C and E.
2. Propagate these errors to the displacement measurement according to the equation:

$$\Delta d_i = \left|\frac{\partial d}{\partial x_i}\right| \Delta x_i + \left|\frac{\partial d}{\partial x_0}\right| \Delta x_0 + \left|\frac{\partial d}{\partial y_i}\right| \Delta y_i + \left|\frac{\partial d}{\partial y_0}\right| \Delta y_0$$
$$= \frac{|x_i - x_0|}{d_i}(\Delta x_i + \Delta x_0) + \frac{|y_i - y_0|}{d_i}(\Delta y_i + \Delta y_0), \quad (6.2)$$

where d_i, x_i, and y_i are the displacement and coordinate measurements for an individual point in the rolling average, and x_0, y_0, Δx_0, and Δy_0 are coordinate measurements and errors for the initial point defined as having $d = 0$.
3. Calculate the arithmetic mean of all individual displacement errors Δd_i to obtain an overall displacement error Δd for the entire trajectory. The result is a measure of the spread of individual displacement measurements around the 16-frame rolling average in a trajectory. An alternative approach is to sum the pixel intensities from individual images of a PSF in rolling 16-frame bins, fit these to 2-D Gaussian functions, and then calculate the error using Eq. (6.1). This may, however, underestimate the error as it assumes that all position determinations in each bin report on a single stationary object; that is, it ignores "biological noise."

6. Concluding Remarks

While autonomous deoxyribozyme-based walkers demonstrate the promise of molecular robotics, it is even more exciting to imagine applications such robots may see in the future. It has previously been shown how deoxyribozymes may be utilized for programming purposes (Stojanovic and Stefanovic, 2003). We now seek to expand the applications of the spider world while incorporating other single-molecule fluorescence techniques such as stochastic optical reconstruction microscopy (STORM) to monitor and characterize differing types of behavior with high spatial accuracy.

ACKNOWLEDGMENTS

This work was funded by the National Science Foundation (NSF) Collaborative Research: Chemical Bonding Center, award 0533019; and NSF Collaborative Research: EMT/MISC, award CCF-0829579. We thank Dr. Steven Taylor from Dr. Milan Stojanovic's lab at Columbia University for providing the spiders, Dr. Kyle Lund and Jeanette Nangreave from Dr. Hao Yan's lab at Arizona State University for preparing the origami tracks, and Dr. David Rueda for assembling the TIRF microscope.

REFERENCES

Aitken, C. E., Marshall, R. A., and Puglisi, J. D. (2008). An oxygen scavenging system for improvement of dye stability in single-molecule fluorescence experiments. *Biophys. J.* **94**, 1826–1835.

Al-Hashimi, H. M., and Walter, N. G. (2008). RNA dynamics: It is about time. *Curr. Opin. Struct. Biol.* **18**, 321–329.

Antal, T., and Krapivsky, P. L. (2007). Molecular spiders with memory. *Phys. Rev. E* **76**, 021121.

Antal, T., Krapivsky, P. L., and Mallick, K. (2007). Molecular spiders in one dimension. *J. Stat. Mech.—Theory Exp.* P08027.

Barak, L. S., and Webb, W. W. (1982). Diffusion of low-density lipoprotein-receptor complex on human-fibroblasts. *J. Cell Biol.* **95**, 846–852.

Bath, J., Green, S. J., Allen, K. E., and Turberfield, A. J. (2009). Mechanism for a directional, processive, and reversible DNA motor. *Small* **5**, 1513–1516.

Benesch, R. E., and Benesch, R. (1953). Enzymatic removal of oxygen for polarography and related methods. *Science* **118**, 447–448.

Betzig, E., and Chichester, R. J. (1993). Single molecules observed by near-field scanning optical microscopy. *Science* **262**, 1422–1425.

Bobroff, N. (1986). Position measurement with a resolution and noise-limited instrument. *Rev. Sci. Instrum.* **57**, 1152–1157.

Brooks, R. A. (1991). Intelligence without Representation. *Artif. Intell.* **47**, 139–159.

Churchman, L. S., Okten, Z., Rock, R. S., Dawson, J. F., and Spudich, J. A. (2005). Single molecule high-resolution colocalization of Cy3 and Cy5 attached to macromolecules measures intramolecular distances through time. *Proc. Natl. Acad. Sci. USA* **102**, 1419–1423.

Deniz, A. A., Dahan, M., Grunwell, J. R., Ha, T. J., Faulhaber, A. E., Chemla, D. S., Weiss, S., and Schultz, P. G. (1999). Single-pair fluorescence resonance energy transfer on freely diffusing molecules: Observation of Forster distance dependence and subpopulations. *Proc. Natl. Acad. Sci. USA* **96**, 3670–3675.

Dieterich, P., Klages, R., Preuss, R., and Schwab, A. (2008). Anomalous dynamics of cell migration. *Proc. Natl. Acad. Sci. USA* **105**, 459–463.

Ditzler, M. A., Aleman, E. A., Rueda, D., and Walter, N. G. (2007). Focus on function: Single molecule RNA enzymology. *Biopolymers* **87**, 302–316.

Dix, J. A., and Verkman, A. S. (2008). Crowding effects on diffusion in solutions and cells. *Annu. Rev. Biophys.* **37**, 247–263.

Douglas, S. M., Dietz, H., Liedl, T., Hogberg, B., Graf, F., and Shih, W. M. (2009). Self-assembly of DNA into nanoscale three-dimensional shapes. *Nature* **459**, 414–418.

Giepmans, B. N. G., Adams, S. R., Ellisman, M. H., and Tsien, R. Y. (2006). Review—The fluorescent toolbox for assessing protein location and function. *Science* **312**, 217–224.

Gordon, M. P., Ha, T., and Selvin, P. R. (2004). Single-molecule high-resolution imaging with photobleaching. *Proc. Natl. Acad. Sci. USA* **101**, 6462–6465.

Green, S. J., Bath, J., and Turberfield, A. J. (2008). Coordinated chemomechanical cycles: A mechanism for autonomous molecular motion. *Phys. Rev. Lett.* **101**, 238101.

Gribbon, P., Schaertl, S., Wickenden, M., Williams, G., Grimley, R., Stuhmeier, F., Preckel, H., Eggeling, C., Kraemer, J., Everett, J., Keighley, W. W., and Sewing, A. (2004). Experiences in implementing uHTS—cutting edge technology meets the real world. *Curr. Drug Discov. Technol.* **1**, 27–35.

Gross, D. J., and Webb, W. W. (1988). Cell surface clustering and mobility of the liganded LDL receptor measured by digital video fluorescence microscopy. *In* "Spectroscopic Membrane Probes," (L. M. Loew, ed.), pp. 19–45. CRC Press, Boca Raton, FL.

Gruber, H. J., Hahn, C. D., Kada, G., Riener, C. K., Harms, G. S., Ahrer, W., Dax, T. G., and Knaus, H. G. (2000). Anomalous fluorescence enhancement of Cy3 and cy3.5 versus anomalous fluorescence loss of Cy5 and Cy7 upon covalent linking to IgG and non-covalent binding to avidin. *Bioconjug. Chem.* **11**, 696–704.

Guo, Z., Guilfoyle, R. A., Thiel, A. J., Wang, R. F., and Smith, L. M. (1994). Direct fluorescence analysis of genetic polymorphisms by hybridization with oligonucleotide arrays on glass supports. *Nucleic Acids Res.* **22**, 5456–5465.

Ha, T., Enderle, T., Ogletree, D. F., Chemla, D. S., Selvin, P. R., and Weiss, S. (1996). Probing the interaction between two single molecules: Fluorescence resonance energy transfer between a single donor and a single acceptor. *Proc. Natl. Acad. Sci. USA* **93**, 6264–6268.

He, Y., Burov, S., Metzler, R., and Barkai, E. (2008). Random time-scale invariant diffusion and transport coefficients. *Phys. Rev. Lett.* **101**, 058101.

Hecht, E. (2002). Optics. Addison-Wesley, Reading, MA.

Heintzmann, R., Jovin, T. M., and Cremer, C. (2002). Saturated patterned excitation microscopy—A concept for optical resolution improvement. *J. Opt. Soc. Am. A* **19**, 1599–1609.

Hess, S. T., Girirajan, T. P. K., and Mason, M. D. (2006). Ultra-high resolution imaging by fluorescence photoactivation localization microscopy. *Biophys. J.* **91**, 4258–4272.

Hintersteiner, M., and Auer, M. (2008). Single-bead, single-molecule, single-cell fluorescence: Technologies for drug screening and target validation. *Ann. NY Acad. Sci.* **1130**, 1–11.

Hofmann, M., Eggeling, C., Jakobs, S., and Hell, S. W. (2005). Breaking the diffraction barrier in fluorescence microscopy at low light intensities by using reversibly photoswitchable proteins. *Proc. Natl. Acad. Sci. USA* **102**, 17565–17569.

Huang, B., Bates, M., and Zhuang, X. (2009). Super-resolution fluorescence microscopy. *Annu. Rev. Biochem.* **78**, 993–1016.

Janesick, J. (1997). CCD transfer method—Standard for absolute performance of CCDs and digital CCD camera systems. *Proc. SPIE* **3019**, 70–102.

Joo, C., Balci, H., Ishitsuka, Y., Buranachai, C., and Ha, T. (2008). Advances in single-molecule fluorescence methods for molecular biology. *Annu. Rev. Biochem.* **77**, 51–76.

Kim, H. D., Nienhaus, G. U., Ha, T., Orr, J. W., Williamson, J. R., and Chu, S. (2002). Mg^{2+}-dependent conformational change of RNA studied by fluorescence correlation and FRET on immobilized single molecules. *Proc. Natl. Acad. Sci. USA* **99**, 4284–4289.

King, S. J., and Schroer, T. A. (2000). Dynactin increases the processivity of the cytoplasmic dynein motor. *Nat. Cell Biol.* **2,** 20–24.

Klafter, J., and Sokolov, I. M. (2005). Anomalous diffusion spreads its wings. *Phys. World* **18,** 29–32.

Kural, C., Kim, H., Syed, S., Goshima, G., Gelfand, V. I., and Selvin, P. R. (2005). Kinesin and dynein move a peroxisome in vivo: A tug-of-war or coordinated movement? *Science* **308,** 1469–1472.

Kusumi, A., Sako, Y., and Yamamoto, M. (1993). Confined lateral diffusion of membrane-receptors as studied by single-particle tracking (Nanovid Microscopy)—Effects of calcium-induced differentiation in cultured epithelial-cells. *Biophys. J.* **65,** 2021–2040.

Li, J., Zheng, W., Kwon, A. H., and Lu, Y. (2000). In vitro selection and characterization of a highly efficient Zn(II)-dependent RNA-cleaving deoxyribozyme. *Nucleic Acids Res.* **28,** 481–488.

Lippincott-Schwartz, J., Altan-Bonnet, N., and Patterson, G. H. (2003). Photobleaching and photoactivation: Following protein dynamics in living cells. *Nat. Rev. Mol. Cell Biol.* S7–S14.

Lubelski, A., Sokolov, I. M., and Klafter, J. (2008). Nonergodicity mimics inhomogeneity in single particle tracking. *Phys. Rev. Lett.* **100,** 250602.

Lund, K., Manzo, A. J., Dabby, N., Michelotti, N., Johnson-Buck, A., Nangreave, J., Taylor, S., Pei, R., Stojanovic, M. N., Walter, N. G., Winfree, E., and Yan, H. (2010). Molecular robots guided by prescriptive landscapes. *Nature* (in press).

Martin, D. S., Forstner, M. B., and Kas, J. A. (2002). Apparent subdiffusion inherent to single particle tracking. *Biophys. J.* **83,** 2109–2117.

Michalet, X., Pinaud, F. F., Bentolila, L. A., Tsay, J. M., Doose, S., Li, J. J., Sundaresan, G., Wu, A. M., Gambhir, S. S., and Weiss, S. (2005). Quantum dots for live cells, in vivo imaging, and diagnostics. *Science* **307,** 538–544.

Min, W., English, B. P., Luo, G., Cherayil, B. J., Kou, S. C., and Xie, X. S. (2005). Fluctuating enzymes: Lessons from single-molecule studies. *Acc. Chem. Res.* **38,** 923–931.

Moerner, W. E. (2007). New directions in single-molecule imaging and analysis. *Proc. Natl. Acad. Sci. USA* **104,** 12596–12602.

Moerner, W. E., and Kador, L. (1989). Optical-detection and spectroscopy of single molecules in a solid. *Phys. Rev. Lett.* **62,** 2535–2538.

Nishiura, M., Kon, T., Shiroguchi, K., Ohkura, R., Shima, T., Toyoshima, Y. Y., and Sutoh, K. (2004). A single-headed recombinant fragment of Dictyostelium cytoplasmic dynein can drive the robust sliding of microtubules. *J. Biol. Chem.* **279,** 22799–22802.

Omabegho, T., Sha, R., and Seeman, N. C. (2009). A bipedal DNA Brownian motor with coordinated legs. *Science* **324,** 67–71.

Pei, R., Taylor, S. K., Stefanovic, D., Rudchenko, S., Mitchell, T. E., and Stojanovic, M. N. (2006). Behavior of polycatalytic assemblies in a substrate-displaying matrix. *J. Am. Chem. Soc.* **128,** 12693–12699.

Pljevaljcic, G., and Millar, D. P. (2008). Single-molecule fluorescence methods for the analysis of RNA folding and ribonucleoprotein assembly. *Methods Enzymol.* **450,** 233–252.

Qian, H., Sheetz, M. P., and Elson, E. L. (1991). Single particle tracking. Analysis of diffusion and flow in two-dimensional systems. *Biophys. J.* **60,** 910–921.

Qu, X. H., Wu, D., Mets, L., and Scherer, N. F. (2004). Nanometer-localized multiple single-molecule fluorescence microscopy. *Proc. Natl. Acad. Sci. USA* **101,** 11298–11303.

Rasnik, I., McKinney, S. A., and Ha, T. (2006). Nonblinking and long-lasting single-molecule fluorescence imaging. *Nat. Methods* **3,** 891–893.

Rist, M. J., and Marino, J. P. (2002). Fluorescent nucleotide base analogs as probes of nucleic acid structure, dynamics and interactions. *Curr. Org. Chem.* **6,** 775–793.

Ritchie, K., Shan, X. Y., Kondo, J., Iwasawa, K., Fujiwara, T., and Kusumi, A. (2005). Detection of non-Brownian diffusion in the cell membrane in single molecule tracking. *Biophys. J.* **88**, 2266–2277.

Rothemund, P. W. K. (2006). Folding DNA to create nanoscale shapes and patterns. *Nature* **440**, 297–302.

Rueda, D., and Walter, N. G. (2005). Single molecule fluorescence control for nanotechnology. *J. Nanosci. Nanotechnol.* **5**, 1990–2000.

Rust, M. J., Bates, M., and Zhuang, X. W. (2006). Sub-diffraction-limit imaging by stochastic optical reconstruction microscopy (STORM). *Nat. Methods* **3**, 793–795.

Samaii, L., Linke, H., Zuckermann, M. J., and Forde, N. R. (2010). Biased motion and molecular motor properties of bipedal spiders. *Phy. Rev. E.* **81**, 021106.

Santoro, S. W., and Joyce, G. F. (1997). A general purpose RNA-cleaving DNA enzyme. *Proc. Natl. Acad. Sci. USA* **94**, 4262–4266.

Saxton, M. J., and Jacobson, K. (1997). Single-particle tracking: Applications to membrane dynamics. *Annu. Rev. Biophys. Biomol. Struct.* **26**, 373–399.

Schmidt, T., Schutz, G. J., Baumgartner, W., Gruber, H. J., and Schindler, H. (1996). Imaging of single molecule diffusion. *Proc. Natl. Acad. Sci. USA* **93**, 2926–2929.

Schuler, B., and Eaton, W. A. (2008). Protein folding studied by single-molecule FRET. *Curr. Opin. Struct. Biol.* **18**, 16–26.

Seeman, N. C. (2007). An overview of structural DNA Nanotechnology. *Mol. Biotechnol.* **37**, 246–257.

Sherman, W. B., and Seeman, N. C. (2004). A precisely controlled DNA biped walking device. *Nano Lett.* **4**, 1203–1207.

Shi, J., Dertouzos, J., Gafni, A., and Steel, D. (2008). Application of single-molecule spectroscopy in studying enzyme kinetics and mechanism. *Methods Enzymol.* **450**, 129–157.

Shin, J. S., and Pierce, N. A. (2004). A synthetic DNA walker for molecular transport. *J. Am. Chem. Soc.* **126**, 10834–10835.

Stojanovic, M. N., and Stefanovic, D. (2003). A deoxyribozyme-based molecular automaton. *Nat. Biotechnol.* **21**, 1069–1074.

Stryer, L., and Haugland, R. P. (1967). Energy transfer: A spectroscopic ruler. *Proc. Natl. Acad. Sci. USA* **58**, 719–726.

Thompson, R. E., Larson, D. R., and Webb, W. W. (2002). Precise nanometer localization analysis for individual fluorescent probes. *Biophys. J.* **82**, 2775–2783.

Tian, Y., He, Y., Chen, Y., Yin, P., and Mao, C. D. (2005). Molecular devices—A DNAzyme that walks processively and autonomously along a one-dimensional track. *Angew. Chem. Int. Ed.* **44**, 4355–4358.

Truong, K., and Ikura, M. (2001). The use of FRET imaging microscopy to detect protein–protein interactions and protein conformational changes in vivo. *Curr. Opin. Struct. Biol.* **11**, 573–578.

Walter, N. G., Huang, C. Y., Manzo, A. J., and Sobhy, M. A. (2008). Do-it-yourself guide: How to use the modern single-molecule toolkit. *Nat. Methods* **5**, 475–489.

Weiss, S. (2000). Measuring conformational dynamics of biomolecules by single molecule fluorescence spectroscopy. *Nat. Struct. Biol.* **7**, 724–729.

Willets, K. A., Ostroverkhova, O., He, M., Twieg, R. J., and Moerner, W. E. (2003). Novel fluorophores for single-molecule imaging. *J. Am. Chem. Soc.* **125**, 1174–1175.

Yildiz, A., Forkey, J. N., McKinney, S. A., Ha, T., Goldman, Y. E., and Selvin, P. R. (2003). Myosin V walks hand-over-hand: Single fluorophore imaging with 1.5-nm localization. *Science* **300**, 2061–2065.

Yin, P., Yan, H., Daniell, X. G., Turberfield, A. J., and Reif, J. H. (2004). A unidirectional DNA walker that moves autonomously along a track. *Angew. Chem. Int. Ed.* **43**, 4906–4911.

Zhao, R., and Rueda, D. (2009). RNA folding dynamics by single-molecule fluorescence resonance energy transfer. *Methods* **49**, 112–117.

Zhuang, X. W., Kim, H., Pereira, M. J. B., Babcock, H. P., Walter, N. G., and Chu, S. (2002). Correlating structural dynamics and function in single ribozyme molecules. *Science* **296**, 1473–1476.

CHAPTER SEVEN

ANTI-BROWNIAN TRAPS FOR STUDIES ON SINGLE MOLECULES

Alexander P. Fields* and Adam E. Cohen[†,‡]

Contents

1. Theoretical Overview	150
1.1. Single-molecule spectroscopy yields important dynamical information but is hampered by Brownian motion	150
1.2. Brownian motion	152
1.3. Tracking and feedback	152
2. Anti-Brownian Trapping Systems	155
2.1. Tracking systems	156
2.2. Feedback systems	159
3. The ABEL Trap	161
3.1. Photon-by-photon feedback	163
3.2. Illumination system	164
3.3. Tracking and feedback system	165
3.4. Microfluidics	168
4. Applications	169
5. Future Work: *En Route* to Single Fluorophores	170
Acknowledgments	171
References	171

Abstract

Until recently, Brownian motion was seen as an immutable feature of small particles in room-temperature liquids. Molecules, viruses, organelles, and small cells jiggle incessantly due to countless collisions with thermally agitated molecules of solvent. Einstein showed in 1905 that this motion is intimately linked to the tendency of every system to relax toward thermal equilibrium.

In recent years, we and others have realized that Brownian motion is not as inescapable as one might think. By tracking the motion of a small particle and

---

* Department of Biophysics, Harvard University, Cambridge, Massachusetts, USA
† Department of Chemistry and Chemical Biology, Harvard University, Cambridge, Massachusetts, USA
‡ Department of Physics, Harvard University, Cambridge, Massachusetts, USA

applying correction forces to the particle or to the measurement apparatus, one can largely suppress the Brownian motion of particles as small as a few nanometers in diameter, in aqueous solution at room temperature. This new ability to stabilize single molecules has led to a host of studies on topics ranging from the conformational dynamics of DNA to the optical properties of metal nanoparticles.

In this review, we outline the physical principles behind suppression of Brownian motion. We discuss the relative merits of several systems that have been implemented. We give examples of studies performed with our anti-Brownian Electrokinetic trap (ABEL trap) as well as other anti-Brownian traps, and we discuss prospects for future research.

1. Theoretical Overview

1.1. Single-molecule spectroscopy yields important dynamical information but is hampered by Brownian motion

Single-molecule spectroscopy allows one to observe the inner workings of complex molecules, in a way that is not possible from bulk, ensemble-averaged measurements. Such studies provide information on the dynamics of nucleic acids (Abbondanzieri et al., 2008; Ha et al., 1999b) and proteins (Ha et al., 1999a; Lipman et al., 2003), both in vitro and in vivo (Cai et al., 2006; Kim et al., 2006). The chief difficulty in many single-molecule experiments is that the molecules do not hold still. For small particles this jiggling is so intense that, unless countermeasures are taken, the molecule is seen for only a fleeting moment. One can increase the diffusion-limited observation time by collecting photons from a larger volume, but such a strategy lowers the signal-to-background ratio.

Typical countermeasures against Brownian motion include immobilization on a surface (Ha, 2001), in the pores of a gel (Dickson et al., 1998; Lu et al., 1998), inside of a tethered lipid vesicle (Boukobza et al., 2001; Cisse et al., 2007), or in a liquid droplet (Katsura et al., 2001). In each case, however, there remains a persistent doubt whether the immobilized molecule behaves the same as its comrades in free solution.

Freely diffusing molecules can be studied via fluctuation spectroscopies such as fluorescence correlation spectroscopy (FCS; Meseth et al., 1999) or single-molecule burst analysis (Chen et al., 1999), but these techniques have typical observation times of 1 ms/molecule or less. Short observation times lead to two problems: few photons are collected from each molecule, so measurements of spectroscopic parameters are imprecise, and processes that last longer than the observation time are not resolved.

Many biological processes occur on the timescale of seconds to hours. As a concrete example of the problem facing single-molecule researchers, nobody has found a way to observe directly the catalytic cycle of the chaperonin GroEL, which takes 7–15 s, without disturbing this cycle (Kim et al., 2005).

Laser tweezers have led to a revolution in the fields of nanomanipulation and biophysics by allowing researchers to exert controlled forces on proteins bound to DNA or cytoskeletal elements (Bustamante et al., 2003). Although less widespread, magnetic tweezers (Gosse and Croquette, 2002) and AC dielectrophoresis (Voldman et al., 2001) have also been used to trap and manipulate micron-scale objects. Unfortunately, the forces generated by all of these techniques are too weak to trap objects smaller than ~ 100 nm in solution at room temperature, except under exceptional circumstances of resonant enhancement (Toussaint et al., 2007).

Optical trapping, dielectrophoresis, and magnetic tweezers exert very weak forces for two reasons. First, the force is proportional to the object's polarizability (electric or magnetic), which is proportional to the volume. So to trap a 10 nm object requires a million times as much input power as to trap a 1 μm object. Second, the force arises through a second-order interaction with the applied electric or magnetic field. That is, the field must first polarize the object, and then interact a second time to generate a force between the induced dipole and a gradient in the field. Electrophoretic forces, on the other hand, are much stronger than optical forces. Electrophoresis depends on charge rather than polarizability, and is first order in the field strength, rather than second order. This is why electrophoresis is commonly used to separate biomolecules, and optical forces are not. Electrophoresis is not typically used to trap biomolecules because an electrophoretic potential contains no minima away from the boundaries (although for particles whose momentum is significant, an AC electric field can dynamically trap a charged particle; Arnold and Hessel, 1985).

Feedback control is widely used to stabilize the motion of stochastic systems, where the stochasticity may arise from quantum, thermal, or manufacturing fluctuations. In particular, feedback may be used to cancel the Brownian motion of a single nanoscale object in solution, over some finite bandwidth. In contrast to passive trapping schemes, feedback trapping potentials need not have any local minima.

The performance of feedback traps scales more favorably for small particles than does that of laser tweezers. To trap objects smaller than 100 nm in diameter, laser tweezers require many watts of infrared light. This intense light runs the risk of cooking an object rather than trapping it. Feedback traps, in contrast, can hold objects as small as 15 nm in diameter with <1 mW of laser power, used solely for imaging the position of the object.

1.2. Brownian motion

Here we review the basic features of Brownian motion in a force field. A more detailed discussion can be found in Berg (1993). If a Brownian particle is released at rest at the origin at $t = 0$ and is free to diffuse in one dimension, the probability of finding the particle at position x at time t is:

$$P(x, t) = \frac{e^{-|x-vt|^2/4Dt}}{\sqrt{4\pi Dt}}, \quad (7.1)$$

where D is the diffusion coefficient of the particle and v is its average drift velocity due to external forces (Einstein, 1905). In two dimensions the motion along the two axes is statistically independent, so that the probability can be factored as: $P(x, y, t) = P(x, t)P(y, t)$, and similarly in three dimensions. Equation (7.1) describes a Gaussian distribution with average position $\langle x \rangle = vt$, and with variance $\langle x^2 \rangle - \langle x \rangle^2 = 2Dt$. It is often more convenient to work just with the mean and variance, rather than the full probability distribution of Eq. (7.1).

The diffusion coefficient of a small particle is given by the Stokes–Einstein relation:

$$D = \frac{k_B T}{6\pi\eta a}, \quad (7.2)$$

where k_B is Boltzmann's constant, T is the absolute temperature, η is the viscosity of the medium, and a is the hydrodynamic radius of the particle. The viscosity of pure water at room temperature is ~ 1 cP, or 10^{-3} kg/m s in SI units. Equation (7.2) implies that smaller particles diffuse faster. A decrease in the size of a particle can be partially compensated by an increase in the viscosity of the medium, but many biomolecules cease to function if the medium differs too much from the cellular milieu.

1.3. Tracking and feedback

Anti-Brownian traps use microscopy to track the position of a molecule, and apply feedback forces to keep the molecule close to a target location (Fig. 7.1). Here we review the general principles of any anti-Brownian trap. In Section 2 we discuss specific implementations.

How tightly can an anti-Brownian trap confine a particle? Several factors combine to limit the degree of confinement. These include: latency of the feedback loop, photophysics of the fluorophore, accuracy of the tracking, and strength of the restoring force.

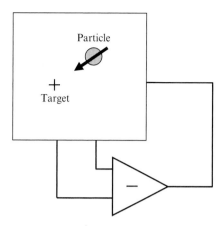

Figure 7.1 General scheme of any anti-Brownian motion trap.

Between iterations of the feedback, a molecule may wander a distance away from the target location. The displacement along each axis has RMS value $\delta x = \sqrt{2D\tau}$, where D is the diffusion coefficient of the particle and τ is the latency of the feedback. If this distance is comparable to the dimension of the observation volume, a large thermal fluctuation may knock the particle completely out of the trap. Thus, one wants to minimize the latency of the feedback, and this requirement becomes more stringent for smaller particles. Our current ABEL trap has a feedback latency as small as 10 μs.

Even with perfect feedback hardware, the feedback latency is limited by the finite rate at which photons reach the detector. Between photon detection events there is no information on the location of the particle, and thus no feedback. A fluorophore emitting photons that are detected at rate γ has a minimum average latency of γ^{-1}. The average value of γ may be as high as 50,000 s^{-1}, but many fluorophores have a tendency to blink. If a molecule is labeled with a single fluorophore, then the molecule may diffuse a considerable distance during an "off" event, and may even be lost from the trap. Thus, for single-fluorophore experiments it is important to use fluorophores with short triplet state lifetimes. Finally, most fluorophores photobleach, allowing the darkened particle to exit the trap.

The feedback must be accurate as well as fast. A single photon carries incomplete information about the location of a particle. Diffraction causes the photons coming from a single stationary molecule to appear to come from a blob whose shape is given by the point spread function of the microscope. Furthermore, background photons may originate from anywhere within the observation volume, leading to additional spurious information on the location of the molecule. Careful optical design may minimize the number of background photons, but the tracking uncertainty due to diffraction is much harder to eliminate.

One may decrease the uncertainty due to diffraction by averaging over many photons. If each photon is selected from a distribution of width w, then by averaging N photons one may localize a particle with accuracy $w/\sqrt{N}$. This fact is widely used to localize single fluorophores with sub-diffraction accuracy. If photons are detected at rate γ, then $\delta x = w/\sqrt{\gamma t}$ for a stationary particle.

For the case of freely diffusing particles, one cannot average indefinitely because the particle is a moving target. The increase in uncertainty due to diffusion eventually outweighs the decrease in uncertainty due to signal-averaging. The optimal integration time occurs when the uncertainties due to diffraction ($w/\sqrt{\gamma t}$) and diffusion ($\sqrt{2Dt}$) are equal (Fig. 7.2). This balance yields:

$$t_{opt} = \frac{w}{\sqrt{2D\gamma}}. \qquad (7.3)$$

Equation (7.3) is an important result for any particle-tracking system, not just an anti-Brownian trap. For instance, Eq. (7.3) gives an optimal frame-rate for video tracking of particles diffusing in solution or in a cell. We are not aware of other occurrences of Eq. (7.3) in the literature. Equation (7.3) can also be derived in a more rigorous manner using Bayesian statistics.

Any feedback system has a maximum velocity with which it can move the particle or the imaging system. This v_{max} may also limit the confinement of the molecule. The RMS velocity of a molecule diffusing a distance d is $v = 2D/d$, so the minimum trap confinement along each axis is $\delta x \geq 2D/v_{max}$. For electrokinetic feedback, v_{max} is relatively independent

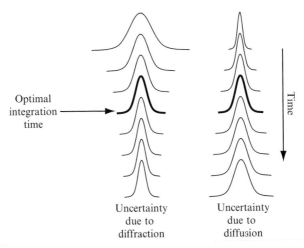

Figure 7.2 Balance of diffusion and diffraction in localization accuracy.

Table 7.1 Feedback parameters for commonly encountered substances

Object	D ($\mu m^2/s$)	Bandwidth (Hz)	Velocity ($\mu m/s$)
1 μm sphere	0.44	3.5	1.8
200 nm sphere	2.2	18	8.8
20 nm sphere	21.8	174	88
2 nm sphere	218	1740	870
Rhodamine 6G	280	2240	1120
Water[a]	2850	23,000	11,000
Macromolecules			
Ribosome 4S (*E. coli*)[b]	37	300	150
γ-Globulin[b]	37	300	150
BSA[c]	49	400	200
Hemoglobin[b]	69	550	275
Myoglobin[b]	100	800	400
Lysozyme[d]	111	890	445
Ribonuclease[b]	131	1050	525

Notes: Tyn and Gusek (1990) provide an expanded table with diffusion coefficients for 89 macromolecules and viruses. D is the diffusion coefficient. The feedback bandwidth required for successful trapping within $\delta x = 500$ nm is calculated from $1/t_r = 2D/\delta x^2$. To counter the Brownian motion, the trap must impose a velocity with RMS value $v = 2D/\delta x$.
[a] Mills (1973).
[b] Tyn and Gusek (1990).
[c] Meseth et al. (1999).
[d] Brune and Kim (1993).

of the size or nature of the particle, and in microfluidic geometries is $v_{max} \sim 10$ mm/s (Plenert and Shear, 2003). Table 7.1 shows that for most objects one might like to trap, the velocity required to confine the object to an area 500 nm in diameter is much less than 10 mm/s.

2. Anti-Brownian Trapping Systems

Anti-Brownian traps are capable of trapping objects far smaller than can be trapped by any other means under comparable conditions. They provide a way to study individual molecules in their native environment, bypassing the perils of surface attachment chemistry. The record of feedback forces also provides information along a dimension not usually accessible to single-molecule experiments. By analyzing the feedback forces one can extract time-dependent information about the diffusion coefficient, and possibly the electrokinetic mobility, of the trapped object.

Feedback traps consist of two subsystems: a tracking component and a feedback component. In each case there are multiple options available, with

Table 7.2 Anti-Brownian systems in the literature, and their performance specifications

Tracking scheme		Bandwidth (kHz)	Feedback scheme showing D_{max} ($\mu m^2/s$)		
			Laser translation	Stage translation	Electrokinetic
Camera		<0.3		<0.2 (3D)[a]	20 (2D)[b]
Multiple photodiodes		Fast		0.6 (3D)[c] 6 (3D)[d]	
Scanning	Galvo	<20	0.6 (3D)[e]	6.2 (2D)[f]	
	AOD	<50		20 (3D)[g]	30 (2D)[h]
	EOD	<1000			In progress
Multiple laser foci		<75,000			In progress[i]

Notes: All tracking systems except for photodiodes operate on a discrete clock, with maximum bandwidth given. Photodiodes operate quasi-continuously and rarely limit the feedback bandwidth. The numbers in the right half of the table indicate the maximum diffusion coefficient that has been trapped (in $\mu m^2/s$), and the number in parentheses is the number of dimensions of the trapping.
[a] Lu *et al.* (2007). No diffusion constant is listed in this reference. The particles trapped have 1.1 μm diameter, giving an estimate of 0.2 $\mu m^2/s$ in pure water, but the viscosity of the organic solvent used is unknown.
[b] Cohen and Moerner (2005c, 2006).
[c] Lessard *et al.* (2007).
[d] Xu *et al.* (2007).
[e] Levi *et al.* (2005).
[f] Berglund and Mabuchi (2005).
[g] McHale *et al.* (2007).
[h] Cohen and Moerner (2008).
[i] Davis *et al.* (2008).

different relative merits. We describe strategies used in our group and others to accomplish these goals. Some of these designs are also reviewed in Cang *et al.* (2008). Table 7.2 summarizes the combinations of tracking and feedback systems described thus far in the literature, and lists the maximum diffusion constant of the particles trapped in each.

2.1. Tracking systems

Fast tracking of molecules in solution requires an optical readout with high signal-to-background ratio. Here we consider some ways to track the position of molecules, and list some of their advantages and disadvantages.

2.1.1. Camera

Perhaps the most obvious way to determine the position of an object is to take a picture of it. The optimal camera for low-light tracking is a frame-transfer electron-multiplying (EM) charge-coupled device (CCD). Frame-transfer mode enables the camera to function at very high frame-rate, shortening the latency between emission of a photon and application of corresponding feedback, but also decreasing the number of photons per frame. High EM gain amplifies the signal due to photons above the electronic noise in the camera, improving the signal-to-noise ratio. However, the EM gain amplifies the signal from all photons equally—it does not distinguish between signal photons, originating from the molecule, and background photons originating elsewhere in the optical system. Camera-based imaging is compatible with wide field, total internal reflection (TIR), or scanned illumination, and was used in the original ABEL trap (Armani et al., 2005; Cohen and Moerner, 2005c). A fast image-fitting algorithm is necessary to convert each picture into a position estimate, without contributing excessively to the feedback latency. Details on one such algorithm are included in Cohen and Moerner (2005b).

The primary drawback of camera tracking is its low speed. Even in frame-transfer mode, EMCCD cameras typically cannot achieve a framerate above ~ 300 Hz. Even assuming instantaneous image fitting, feedback latency is several milliseconds. Such a system has been used to trap particles with diameters of 20 nm in aqueous solution, but smaller objects require higher viscosity or faster detection. On the other hand, cameras provide direct images of trapped particles, which can be useful when multiple objects are being observed or when the trapped object has significant internal structure (Cohen and Moerner, 2007a,b). For example, Lu et al. (2007) captured three-dimensional (3D) image stacks at a frame time of 6 s to track diffusion of multiple colloidal particles over a period of several hours, using a spinning-disk confocal microscope.

2.1.2. Multiphotodiode

Quadrant photodiodes retain some of the positional information provided by a camera but offer significantly improved bandwidth (often several megahertz). Such a detector can be substituted for a CCD camera in a wide-field imaging setup, and has been used for tracking (Cang et al., 2006). Unfortunately, commercially available quadrant photodiodes lack sensitivity and so are poorly suited to detection of dim objects. The optical detector with the best response properties for fast tracking of dim objects is the single-photon-sensitive avalanche photodiode (APD). When operating in Geiger mode, an APD generates a voltage pulse every time a photon is detected and provides unmatched sensitivity and time resolution. Photomultiplier tubes (PMTs) are also capable of low-light detection, but have lower quantum efficiency.

Several designs use multiple PMTs or APDs to track position. In 1971, Howard Berg used six PMTs to track a swimming bacterium (Berg, 1971). He coupled the input of each PMT to a fiber, and back-imaged the fiber inputs to the corners of an octahedron at the sample. The group of Haw Yang employs two prism mirrors reflecting light onto four APDs, two each for the x- and y-axes (Cang et al., 2007). 3D tracking is achieved using a defocused confocal pinhole so that the light intensity on a fifth APD is modulated in response to axial motion of the tracked particle. This design is susceptible to low-frequency drifts in illumination intensity and to fluctuations in the brightness of the trapped particle, which would be interpreted as changes in axial position. The Werner group stays closest to Berg's original design, back-imaging the photoreceptive areas of four APDs to the corners of a tetrahedron at the sample plane (Lessard et al., 2007). Such a design loses signal due to the splitting of the light prior to the confocal spatial filtering and due to the dead area between the confocal pinholes (i.e., the cladding of the optical fibers). In each of these designs, the position estimate is constructed using either analog electronics or a field-programmable gate array (FPGA) to scale and subtract the appropriate signals. The FPGA-based algorithm we describe in Section 3.3 could be adapted to fit a detection scheme of this style.

2.1.3. Scanning

An alternative way to detect position is to use a scanned laser beam and a single APD, as first proposed by Enderlein (2000). A confocal spot is rapidly rotated in a circle, with the radius of the circle approximately equal to the radius of the spot. If the molecule is in the center of the circle, the molecule emits a steady flux of fluorescent photons. If the molecule moves off-center, there is a modulation in the fluorescence intensity at the laser rotation frequency. The phase of this modulation, relative to the phase of the laser rotation, indicates the direction in which the molecule has moved. Enderlein's original proposal only tracks 2D diffusion, as he intended it for slowly diffusing molecules confined to two dimensions by cellular membranes ($D < 10^{-2}\ \mu m^2/s$).

Scanning of excitation light through the sample can be achieved using galvanometer scanners (galvos), acousto-optic beam deflectors (AODs), or electro-optic beam deflectors (EODs). The most important characteristic of these scanners for the purpose of fast tracking is their modulation bandwidth as this sets the maximum bandwidth of the feedback. Typical maximum scan rates are 20, 50, and 1000 kHz for galvos, AODs, and EODs, respectively. The limited bandwidth of the galvos makes them less suitable than AODs or EODs. EODs offer improved stability (Valentine et al., 2008) and lower wavelength dependence than AODs, but require high-voltage drive electronics. Either AODs or EODs offer sufficient scan speeds for fast tracking.

A simple way to calculate the phase of the fluorescence intensity modulation (and thus the position estimate) is to use a commercially available vector lock-in amplifier, with the photon signal as the input and the scanning signal as reference. The x and y output channels provide a direct readout of the position of the particle, provided the internal phase offset is adjusted correctly. Equation (7.3) specifies the optimal integration time for the lock-in, based on the diffusion constant and average count rate of the particle. Lock-in amplifiers often come with a minimum latency, which may inhibit trapping of smaller particles. Analog hardware circuits, such as the one described in Cohen and Moerner (2008), can shorten this delay, but are difficult to tune. The tracking scheme we outline in Section 3.3 for implementation on an FPGA allows near-optimal estimation while retaining flexibility.

The first implementation of scanning-based tracking is due to the group of Enrico Gratton, who used galvos to scan in 2D and a z-nanopositioner to move the objective between two focal planes for 3D tracking (Levi et al., 2003, 2005). Around the same time, Berglund and Mabuchi (2004, 2005) designed a 2D scanning trap, also using galvo scanners. The Mabuchi group has since made numerous improvements to their trap (Berglund and Mabuchi, 2006; Berglund et al., 2007); their most recently published design employs three AODs to vary laser light intensity sinusoidally between two focal planes in addition to the rotation within each plane, enabling fully 3D sensitivity (McHale et al., 2007). A hardware ABEL trap that tracks position by scanning a laser using an AOD has also been constructed (Cohen and Moerner, 2008; Jiang et al., 2008).

2.1.4. Multifocus

A clever alternative to direct scanning of a laser beam is to split pulsed excitation light into multiple beams that illuminate discrete points at the sample (Davis et al., 2008). By adjusting the distances traveled by each beam prior to hitting the sample, the illumination of each point can be separated in time. When a photon is detected, it is assigned with high probability to the position currently illuminated. The arrival-time information is obtained using multichannel gated detection. The resulting series of photon position measurements can then be used to estimate particle position using the scheme described in Section 3.3. Laser pulse frequencies are typically very high (~ 75 MHz), meaning that the system bandwidth is more likely to be limited by electronics and photon intensity than by the scanning itself. One concern with such a high scan rate is cross-talk due to the delay of the fluorescence excited state lifetime; this problem could be alleviated by using a pulse picker to lower the pulse frequency. Another concern is that tuning a multifocused illumination system (e.g., to change the number or position of scanned points) is more difficult than it is for the scanning systems described above.

2.2. Feedback systems

In addition to an optical system capable of tracking the motion of a particle in solution, a trap requires a feedback system to adjust the position of the particle relative to the tracking optics. Several strategies have been proposed and implemented.

2.2.1. Laser tracking

In Enderlein's original proposal for scanning laser tracking, the molecule is allowed to diffuse freely, and the center of the laser focus is adjusted to follow the molecule (Enderlein, 2000). Gratton and coworkers implemented this strategy for $x-y$ repositioning, and translated the objective directly using a nanopositioner along the z-axis (Levi et al., 2005). The problem with moving the laser to follow the position of a tracked particle is that both the scanning optics and the detection optics have limited range, and the spot will eventually fall off the edge of one of them. The Gratton group's design was aided in this regard by their use of galvos for scanning and a PMT for detection; AODs and EODs have much smaller maximum scan angles than galvos, and APDs have smaller photoreceptive areas than PMTs.

2.2.2. Stage feedback

Howard Berg's original tracking microscope translated the sample to keep bacteria within the field-of-view (Berg, 1971). This mechanism neatly avoids the issue of limited detector field-of-view. Beginning with the Mabuchi lab, several groups have adapted stage feedback to single-particle tracking (Berglund and Mabuchi, 2004; Cang et al., 2006; Lessard et al., 2007; Lu et al., 2007). Most mechanical translation stages are too slow to track very fast objects, and the finite travel of the stage limits the amount of time an object can be tracked. Additionally, tracking systems are easily confused by fixed pieces of dirt on a coverslip. As the system follows a particle diffusing over a surface, it is likely to encounter other particles stuck to the surface. Quite frequently the tracking system locks onto one of these fixed objects and loses the moving object of interest.

2.2.3. Electrokinetic feedback

Anti-Brownian Electrokinetic traps (ABEL traps) use electric fields to impose a drift on a particle that counteracts the Brownian motion. Two distinct electrokinetic effects are at play in an ABEL trap, called *electrophoresis* and *electroosmosis*. Electrophoretic forces arise from direct action of an applied electric field on the charge of a particle. Naturally, these forces apply only to charged particles.

To our surprise, we found that even nominally neutral particles move in an ABEL trap. The origin of this motion is the electroosmotic effect. When water is confined in a thin channel, the walls of the channel typically develop a charge due to preferential adsorption or desorption of ions of one charge. This fixed charge is screened by mobile charges of opposite sign within a nanometers-thick "Debye layer." An applied electric field exerts a force on these dissolved ions, and these ions exert a force on the water, leading to a net flow. This electroosmotic flow generates a hydrodynamic force which carries neutral particles at the same velocity as the flow.

By tailoring the surface charge of the channel, it is possible to augment or eliminate this electroosmotic flow. Thus, an ABEL trap can confine charged or neutral particles.

The applied electric field for trapping is only a weak perturbation to the molecule. Within a channel < 10 μm thick, a field of several hundred V/cm is sufficient to generate a large enough electroosmotic flow to cancel Brownian motion of most molecules. In contrast, an electric field $E > 25{,}000$ V/cm would be needed to impart a potential difference of order $k_B T$ to charges in a molecule of size < 10 nm.

The electrokinetic feedback applied in an ABEL trap overcomes many of the limitations of stage translation. Electrokinetic forces can move a particle far more quickly than can a piezo stage (Cohen and Moerner, 2005c). The tracking system is not thrown off by stuck dirt if there is none present at the start of the experiment, because the trapping is performed at a fixed position in the sample cell. An ABEL trap also allows one to position a trapped object relative to other fixed objects on the surface. On the other hand, it would be difficult or impossible to employ electrokinetic feedback to track particles inside cells or in other heterogeneous environments, while stage feedback performs this task with ease.

In addition to our own ABEL traps and ongoing efforts in the Moerner lab (Cohen, 2005; Cohen and Moerner, 2005a,b,c, 2008; Jiang *et al.*, 2008), Ben Shapiro's group has employed video tracking and an array of microelectrodes to steer multiple relatively large particles (e.g., yeast cells 4 μm in diameter) in solution (Armani *et al.*, 2005; Chaudhary and Shapiro, 2006). The group of Lloyd Davis has also described designs for both one- and three-dimensional ABEL traps (Davis *et al.*, 2008; Li *et al.*, 2008).

3. The ABEL Trap

Here we show in detail how the elements discussed above can be combined in a high-performance ABEL trap for trapping single molecules. The system described here has been constructed in our lab and, to our knowledge, traps smaller particles in solution than any comparable system.

The heart of our ABEL trap is a microfluidic cell which confines molecules to a thin pancake-shaped trapping region < 1 μm thick and ~10 μm wide. Feedback electric fields are applied to four electrodes, which are connected to the corners of the trapping region. The microfluidic cell is mounted in a fluorescence microscope, and an automated tracking and feedback system applies voltages to the electrodes to suppress the Brownian motion of a single particle.

In the early incarnations of our trap, we used a video camera and real-time image processing on a personal computer to follow the Brownian motion (Cohen, 2005; Cohen and Moerner, 2005a,b, 2006). Representative results obtained with this trap are shown in Fig. 7.3. This feedback system had a minimum latency of 4.5 ms, determined largely by the framerate of the camera. We subsequently switched to a hardware feedback scheme, in which the particle was illuminated with a rapidly rotating laser spot and the position of the particle was inferred from the timing of detected fluorescence photons. The signal processing was performed entirely in analog electronics. This lock-in scheme had a minimum latency of 25 μs, but could not implement the ideal trapping algorithm due to limitations of analog signal processing. We also switched from a polydimethylsiloxane (PDMS) trapping chamber first to one made of glass and later to one made of fused silica, decreasing the fluorescence background by a factor of 40 in the process. The result of all these improvements was the ability to trap biomolecules as small as 15 nm diameter in aqueous solution (Cohen and Moerner, 2008).

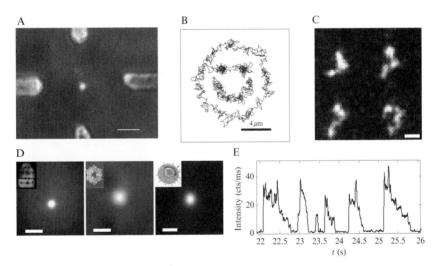

Figure 7.3 Trapping with the ABEL trap. (A) Single 200 nm diameter fluorescent polystyrene nanosphere. Scale bar = 5 μm. (B) Trajectory of a single 200 nm diameter nanosphere manipulated to draw out a smiley face over the course of 1 min. (C) Four images of a single fluorescently labeled molecule of λ-DNA held in the ABEL trap. Nearly 60,000 such images were acquired and analyzed to probe the internal molecular dynamics. Scale bar = 1 μm. (D) Images of a trapped molecule of GroEL, B-phycoerythrin, and a CdSe quantum dot. Scale bar = 2 μm. (E) Time-trace of the fluorescence intensity as a series of multiply labeled molecules of the chaperonin MmCpn entered the trap, photobleached, and were lost. This strategy has been used to count the number of fluorescently labeled ATP bound by the chaperonin.

3.1. Photon-by-photon feedback

Figure 7.4 shows a schematic of the electrical and optical components of our current ABEL trap. One can think of the tracking system as a high speed confocal microscope, with a frame rate of up to 100 kHz. These ultrafast scans are analyzed in real time by custom digital hardware, which locates the molecule within the scan pattern. The hardware calculates and applies feedback voltages based on every detected photon, with a latency of $< 10~\mu s$.

The key innovations over previous designs are:

1. Improved photon-by-photon feedback, with higher speed, lower noise, and greater flexibility than previous systems;

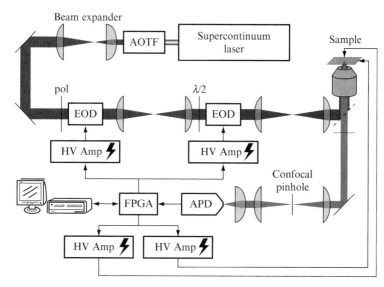

Figure 7.4 Optical layout of the ABEL trap. A 6 W supercontinuum laser emits light across the visible spectrum, up to eight spectral lines of which are selected by an acousto-optic tunable filter (AOTF). The beam is expanded, passed through a linear polarizer (pol), and scanned by two electrooptic beam deflectors (EODs). A half-wave Fresnel rhomb ($\lambda/2$) rotates the polarization by $90°$ between the EODs. Relay lenses are used to map the EOD deflections to pure 2D translations at the sample plane. The beam is reflected off a dichroic mirror (dashed line) and focused on the sample. Emitted photons are collected by the objective and filtered through the dichroic mirror and an emission filter (dotted line). The emission is reimaged through a confocal pinhole and focused onto an APD. Detected photons are reported to a field-programmable gate array (FPGA), which calculates appropriate feedback voltages to send to two high-voltage amplifiers (HV Amp) that apply feedback voltages to the sample. The FPGA also sets the deflection voltages applied by the EODs via two additional high-voltage amplifiers.

2. Adoption of EODs that allow higher speed laser scanning (up to 100 kHz), at multiple wavelengths simultaneously; and
3. A broadband supercontinuum light source that allows us to study any fluorophore that can be excited in the visible part of the spectrum.

3.2. Illumination system

Now we discuss in more detail the layout of Fig. 7.4. The light source is a 6 W supercontinuum laser (Fianium SC-450-6) that emits across the visible and near infrared spectrum. Up to eight spectral lines are selected by an acousto-optic tunable filter (AOTF, Crystal Technologies AODS 20160). This choice of illumination is crucial because it allows us to illuminate the sample with any wavelength or combination of wavelengths without realigning the optics. It is prohibitively complicated to align multiple beams from different sources onto a trapped particle, and the present approach obviates that challenge.

Two EODs (ConOptics, Danbury, CT) steer the position of the beam. These EODs allow much faster beam deflection compared to the AODs used in previous experiments (1 MHz vs. 50 kHz). Furthermore, the EODs work throughout the visible spectrum while AODs require realignment for each wavelength.

The optical system is designed to bring the illumination to a sharp focus and to convert the EOD deflections into pure translations of the beam in the plane of the sample (without any coupling between the EOD deflection and the *direction* of the beam at the sample). This design is achieved through a series of relay lenses that image the x-axis EOD onto the y-axis EOD, and then image both EODs onto the back-aperture of the objective. The EODs are polarization sensitive, so a polarizer is placed before the first EOD and a half-wave Fresnel rhomb is used to rotate the plane of polarization by 90° between the EODs. The excitation optics provides a laser beam that is precisely engineered spatially, spectrally, and temporally.

An objective lens (Olympus; 60×, PlanAPO, NA 1.45) illuminates the trapping region from below, and captures fluorescence from a molecule in the trap. This fluorescence is separated from back-scattered excitation light by a dichroic mirror and a high quality emission filter. The light is then passed through a pinhole and imaged onto an APD (Perkin Elmer SPCM-AQRH-14). Figure 7.5 illustrates a 12-pixel confocal scan pattern. Typical confocal microscopes scan the beam in a square lattice, for reasons of simplicity of design and analysis. However, a square lattice is not the optimal pattern for tracking a small particle. A hexagonal lattice maximizes the uniformity of the time-average illumination, while also maximizing the sensitivity of the tracking system to small displacements of the particle.

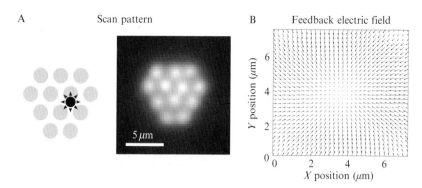

Figure 7.5 Confocal scan pattern. (A) Scan pattern for tracking single-molecule diffusion. (B) Calibration of tracking system obtained by scanning a fixed fluorescent bead through the tracking region and recording the pair of feedback voltages as a function of position.

The spacing of the pixels in Fig. 7.5A has been increased above the optimal value, to illustrate the discrete pixels in the scan pattern. The gray spots indicate the positions targeted by the laser beam. To test the accuracy of the tracking system, we placed a small fluorescent bead (represented as a star in Fig. 7.5A) on a piezoelectric scanning stage and moved the particle in a raster pattern through the illuminated region. We recorded the feedback fields that would have been applied to the particle had it been free in solution. Figure 7.5B shows these feedback fields, indicating that the tracking system is able to accurately generate a feedback field directed radially toward the origin.

3.3. Tracking and feedback system

The most important feature of the ABEL trap is a hardware-based tracking system implemented in an FPGA (National Instruments PCI-7831R). The FPGA uses the precise arrival time of every detected photon to form an estimate of the location of the particle. This estimate is processed, photon-by-photon, to generate feedback voltages.

Figure 7.6 shows a schematic of our tracking and feedback algorithm for a single axis (e.g., the x-axis). An identical algorithm operates in the orthogonal direction. A register stores coordinates of the laser beam for the scan pattern (top row). These coordinates are sampled at a rate of 1 MHz and fed to one of the EODs via a high-voltage amplifier. A second register records pulses from the APD (middle row). The occurrence of an APD pulse indicates that the molecule and the laser beam are likely to be in the same place. The FPGA records the position of the laser beam at the time of

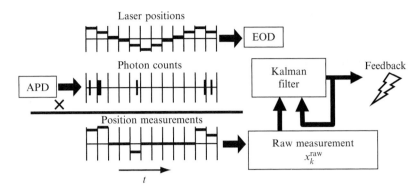

Figure 7.6 Tracking algorithm. The hardware controls the position of the laser (top) and records the arrival time of each photon (middle). On the basis of this information, the FPGA constructs a raw measurement of the particle position, which is fed into a recursive Kalman filter that outputs the feedback voltage. APD, avalanche photodiode; EOD, electro-optic beam deflector.

the APD pulse and stores this value in a third register. The positions of the photons detected in one scan cycle are averaged to form a raw measurement of the position of the particle.

3.3.1. Kalman filter signal-processing

The raw position estimates from each scan of the laser are sent through a Kalman filter (Hamilton, 1994) implemented in the FPGA hardware. The Kalman filter is a means to keep track of both the particle position, $\mathbf{x}$, and the uncertainty with which this position is known, $\sqrt{p}$. At each time-step, the Kalman filter performs two tasks:

1. It predicts the location of the particle at time-step k, conditional on all the measurements up to time $k-1$. This prediction involves moving the estimate of the particle position in accord with the applied feedback voltages, and increasing the uncertainty in the particle's position in accord with its diffusive motion:

$$\hat{\mathbf{x}}_{k|k-1} = \hat{\mathbf{x}}_{k-1|k-1} + \mathbf{E}_{k-1}\mu\Delta t,$$
$$\hat{p}_{k|k-1} = \hat{p}_{k-1|k-1} + 2D\Delta t. \qquad (7.4)$$

The subscripts $k|k-1$ indicate the prediction at time k, conditional on the information up to and including time $k-1$.

2. It adjusts the prediction from Step (1) based on the raw measurement, $\mathbf{x}_k^{\text{raw}}$, obtained from the photons detected during step k. In the limit of

fast scanning, the particle undergoes very little diffusive motion during a single laser scan. Then the uncertainty in the position of the particle is $w/\sqrt{n_k}$, where w is the width of the laser focus spot and n_k is the number of photons detected in scan k. The predicted position and the new data are combined with weighting factors determined by their relative uncertainty:

$$\hat{\mathbf{x}}_{k|k} = \frac{w^2 \hat{\mathbf{x}}_{k|k-1} + n_k \hat{p}_{k|k-1} \mathbf{x}_k^{\text{raw}}}{w^2 + n_k \hat{p}_{k|k-1}},$$

$$\hat{p}_{k|k} = \frac{w^2 \hat{p}_{k|k-1}}{w^2 + n_k \hat{p}_{k|k-1}}.$$
(7.5)

Equations (7.4) and (7.5) are simple enough to be implemented at high speed in the FPGA.

One concern with the tracking scheme outlined here is its requirement for prior knowledge of parameter values. The prediction step (Eq. (7.4)) requires knowledge of the diffusion constant (D) and electrokinetic mobility (μ) of the trapped object, and the update step (Eq. (7.5)) requires knowledge of the beam width (w) and the position of the beam at each point in the scan. The latter two parameters can be measured off-line by using a piezo stage to move an immobilized bead through the laser foci. The diffusion constant can be measured using FCS (Meseth et al., 1999). The electrokinetic mobility can be measured using multispot FCS in the presence of an applied electrical field (Dertinger et al., 2007). In our experience, the ability to trap is relatively insensitive to these parameter choices, so reasonable guesses can often substitute for precise values.

A key benefit of working with an FPGA is that the FPGA is connected to a computer so the user can set parameters in real time. The user interacts with the feedback circuit as though it were software, but the circuit runs with the performance of custom hardware. This real-time tunability is essential to finding the optimal parameters for trapping a small molecule. The previous version of the ABEL trap, with feedback implemented in analog hardware (Cohen and Moerner, 2008), was limited in the feedback algorithms it could use due to the difficulty of performing some kinds of analog computations.

Two high-voltage amplifiers (Model 7602, Krohn-Hite), generate the scan voltages for the EODs, and two more identical amplifiers generate the feedback voltages for the trap. The necessary feedback voltage strength is highly dependent on the electrokinetic mobility of the sample and on the geometry of the sample holder, but typically peak voltages less than 100 V are sufficient.

3.4. Microfluidics

The sample holder for the ABEL trap must be thin enough to confine molecules to within the focal depth of the objective (~1 μm) and must convey strong electric fields to the trap center. The sample holder we use (Fig. 7.7) consists of a shallow (~600 nm) central trapping region flanked by four deeper (~15 μm) channels etched within a 2.5 mm square piece of fused silica. To enclose the channels, we irreversibly bond the etched device to the center of a 1 in. square fused silica cover slip using sodium silicate (Wang et al., 1997). Fused silica is preferred due to its lower autofluorescence relative to glass or PDMS. The electrical resistance is much larger within the shallow trapping region than anywhere else, greatly increasing the electrical field strength experienced by trapped molecules.

We use a cast piece of PDMS to contain excess fluid around the sides of the fused silica device and to hold the feedback electrodes in place. Fluid

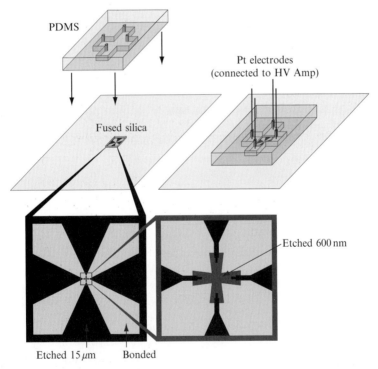

Figure 7.7 Sample holder for the ABEL trap. The fused silica sample cell contains a central trapping region 600 nm deep and deeper channels 15 μm deep as indicated. The etched 2.5 mm square piece of fused silica is bonded to a 1 in. square fused silica coverslip. A cast piece of PDMS (top) is reversibly bonded to the coverslip to contain fluid and to hold the platinum electrodes in place. PDMS, polydimethylsiloxane.

leakage around the corners of the device allows pressure equilibration without significant voltage loss. Damp tissue paper placed in an enclosed space containing the device alleviates sample evaporation. We use platinum electrodes to minimize electrochemical products.

4. APPLICATIONS

Anti-Brownian traps have been used in a wide range of experiments and applications. The initial demonstrations were mostly focused on physics applications. For instance, the Mabuchi lab observed the fluorescence emission from semiconductor quantum dots, and found that the photon statistics showed anti-bunching (McHale et al., 2007). This experiment was a conclusive proof that only a single quantum emitter was held in the trap. The Yang group at Berkeley used the feedback signal to reconstruct the diffusive trajectory of a single particle held in the trap. From the diffusion coefficient they determined the size of single particles (Xu et al., 2007). They also added a spectroscopic readout with which they observed, for the first time, orientation-dependent scattering from metal nanoparticles (Cang et al., 2006). In collaboration with Moerner, we generalized the feedback concept to include arbitrary position-dependent force fields (Cohen, 2005), and thereby studied the motion of nanoparticles in double-well and power law shaped potentials.

Due to the complexity of most biophysical experiments, these were not feasible until anti-Brownian traps became robust enough to operate reliably. Recently, there have been a number of exciting prospects on the biophysical front. The Shapiro lab demonstrated the ability to manipulate yeast cells using video tracking and electrokinetic feedback, and was even able to steer multiple particles simultaneously (Armani et al., 2006). We trapped single molecules of fluorescently labeled λ-DNA and recorded nearly 60,000 images of high-speed video of the internal conformational dynamics inside the trap (Cohen and Moerner, 2007a,b). These videos allowed us, for the first time, to measure the spectrum of internal conformational modes of a single relaxed polymer molecule. More recently, the Moerner lab has been using an ABEL trap to study individual chaperonin molecules (Jiang et al., 2008). By incubating the chaperonins with fluorescently labeled ATP, they can count, molecule-by-molecule, the number of bound ATP. This information provides a more detailed picture of cooperative binding of ATP than can be achieved from bulk measurements. The Gratton lab has used a laser scanning approach coupled with two-photon microscopy to follow the motion of single fluorescently labeled loci on interphase chromatin (Levi and Gratton, 2008). These experiments found previously unobserved "hops" of the fluorescent spot. Our early demonstration of trapping single virus particles, lipid vesicles, and proteins suggests that there are a great many applications waiting to be explored (Cohen and Moerner, 2006).

5. FUTURE WORK: *EN ROUTE* TO SINGLE FLUOROPHORES

An ultimate goal of anti-Brownian trapping is to immobilize a single fluorophore, with 10–15 carbon atoms and a hydrodynamic radius < 1 nm, in aqueous solution at room temperature. Once this goal is met, it is likely that one could trap any molecule that can be fluorescently labeled. The median size of human proteins is 375 amino acids, corresponding to a diameter of ~1.8 nm (Brocchieri and Karlin, 2005). With the current state-of-the-art ABEL trap technology, only the very largest proteins or protein complexes with diameters > 15 nm can be trapped. Small proteins have also been resistant to other single-molecule approaches because surface immobilization often disrupts their function. Indeed, the vast majority of proteins that have been studied by single-molecule techniques bind either to nucleic acids (Gore *et al.*, 2006; Wen *et al.*, 2008) or cytoskeletal elements (Svoboda *et al.*, 1993), and have been amenable to study precisely because their binding partners can be immobilized on a surface or on a bead. Many important cellular processes occur free-floating in the cytoplasm, and thus far have resisted single-molecule analysis.

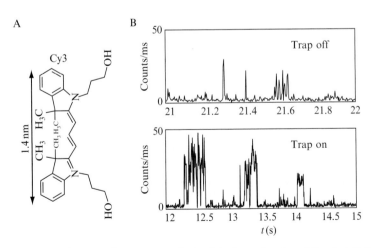

Figure 7.8 Confinement (but not trapping) of single molecules of the dye Cy3. (A) Molecular structure of Cy3, showing longest axis. (B) Representative fluorescence bursts detected in the absence or presence of electrokinetic trapping feedback.

Here we show results of some preliminary attempts to trap Cy3 using an early version of the photon-by-photon ABEL trap (Cohen and Moerner, 2008). A very dilute (150 pM) solution of Cy3 was loaded into an ABEL trap and fluorescence intensity time traces were recorded with the feedback on or off. Figure 7.8 shows representative time traces with the trap off (top), and on (bottom). Even with the trap off, molecules occasionally wandered into the trapping region and contributed a burst of fluorescence, but no molecule stayed in the trapping region for long. With the feedback on, the bursts lasted much longer, but molecules were still not stably trapped. Nonetheless, this result shows that it is possible to *suppress* the Brownian motion of a single fluorophore, without canceling this motion completely. This preliminary result makes us optimistic that anti-Brownian traps will reach the single-fluorophore limit in the near future.

ACKNOWLEDGMENTS

We gratefully acknowledge support of NSF grant CHE-0910824. This work was supported partially by the Materials Research Science and Engineering Center of the National Science Foundation under NSF Award Number DMR-02-13805. Fabrication was performed at the Harvard Faculty of Arts and Sciences (FAS) Center for Nanoscale Systems, a member of the National Nanotechnology Infrastructure Network, supported by NSF award ECS-0335765.

REFERENCES

Abbondanzieri, E. A., Bokinsky, G., Rausch, J. W., Zhang, J. X., Le Grice, S. F. J., and Zhuang, X. (2008). Dynamic binding orientations direct activity of HIV reverse transcriptase. *Nature* **453**, 184–189.
Armani, M., Chaudhary, S., Probst, R., Walker, S., and Shapiro, B. (2005). Control of microfluidic systems: Two examples, results, and challenges. *Int. J. Robust Nonlinear Control* **15**, 785.
Armani, M., Chaudhary, S., Probst, R., and Shapiro, B. (2006). Using feedback control of microflows to independently steer multiple particles. *J. Microelectromech. Syst.* **15**, 945–956.
Arnold, S., and Hessel, N. (1985). Photoemission from single electrodynamically levitated microparticles. *Rev. Sci. Instrum.* **56**, 2066.
Berg, H. C. (1971). How to track bacteria. *Rev. Sci. Instrum.* **42**, 868.
Berg, H. C. (1993). Random Walks in Biology. Princeton University Press, Princeton, NJ.
Berglund, A. J., and Mabuchi, H. (2004). Feedback controller design for tracking a single fluorescent molecule. *Appl. Phys. B* **78**, 653–659.
Berglund, A. J., and Mabuchi, H. (2005). Tracking-FCS: Fluorescence correlation spectroscopy of individual particles. *Opt. Express* **13**, 8069–8082.
Berglund, A. J., and Mabuchi, H. (2006). Performance bounds on single-particle tracking by fluorescence modulation. *Appl. Phys. B* **83**, 127–133.
Berglund, A. J., McHale, K., and Mabuchi, H. (2007). Feedback localization of freely diffusing fluorescent particles near the optical shot-noise limit. *Opt. Lett.* **32**, 145–147.

Boukobza, E., Sonnenfeld, A., and Haran, G. (2001). Immobilization in surface-tethered lipid vesicles as a new tool for single biomolecule spectroscopy. *J. Phys. Chem. B* **105**, 12165–12170.

Brocchieri, L., and Karlin, S. (2005). Protein length in eukaryotic and prokaryotic proteomes. *Nucleic Acids Res.* **33**, 3390.

Brune, D., and Kim, S. (1993). Predicting protein diffusion coefficients. *Proc. Natl. Acad. Sci. USA* **90**, 3835–3839.

Bustamante, C., Bryant, Z., and Smith, S. B. (2003). Ten years of tension: Single-molecule DNA mechanics. *Nature* **421**, 423–427.

Cai, L., Friedman, N., and Xie, X. S. (2006). Stochastic protein expression in individual cells at the single molecule level. *Nature* **440**, 358–362.

Cang, H., Wong, C. M., Xu, C. S., Rizvi, A. H., and Yang, H. (2006). Confocal three dimensional tracking of a single nanoparticle with concurrent spectroscopic readouts. *Appl. Phys. Lett.* **88**, 223901.

Cang, H., Xu, C. S., Montiel, D., and Yang, H. (2007). Guiding a confocal microscope by single fluorescent nanoparticles. *Opt. Lett.* **32**, 2729–2731.

Cang, H., Xu, C. S., and Yang, H. (2008). Progress in single-molecule tracking spectroscopy. *Chem. Phys. Lett.* **457**, 285–291.

Chaudhary, S., and Shapiro, B. (2006). Arbitrary steering of multiple particles independently in an electro-osmotically driven microfluidic system. *IEEE Trans. Control Syst. Technol.* **14**, 669–680.

Chen, Y., Müller, J. D., So, P. T. C., and Gratton, E. (1999). The photon counting histogram in fluorescence fluctuation spectroscopy. *Biophys. J.* **77**, 553–567.

Cisse, I., Okumus, B., Joo, C., and Ha, T. (2007). Fueling protein DNA interactions inside porous nanocontainers. *Proc. Natl. Acad. Sci. USA* **104**, 12646.

Cohen, A. E. (2005). Control of nanoparticles with arbitrary two-dimensional force fields. *Phys. Rev. Lett.* **94**, 118102.

Cohen, A. E., and Moerner, W. E. (2005a). An all-glass microfluidic cell for the ABEL trap: Fabrication and modeling. *Proc. SPIE* **5930**, 191–198.

Cohen, A. E., and Moerner, W. E. (2005b). The anti-Brownian electrophoretic trap (ABEL Trap): Fabrication and software. *Proc. SPIE* **5699**, 296.

Cohen, A. E., and Moerner, W. E. (2005c). Method for trapping and manipulating nanoscale objects in solution. *Appl. Phys. Lett.* **86**, 093109.

Cohen, A. E., and Moerner, W. E. (2006). Suppressing Brownian motion of individual biomolecules in solution. *Proc. Natl. Acad. Sci. USA* **103**, 4362–4365.

Cohen, A. E., and Moerner, W. E. (2007a). Internal mechanical response of a polymer in solution. *Phys. Rev. Lett.* **98**, 116001.

Cohen, A. E., and Moerner, W. E. (2007b). Principal-components analysis of shape fluctuations of single DNA molecules. *Proc. Natl. Acad. Sci. USA* **104**, 12622.

Cohen, A. E., and Moerner, W. E. (2008). Controlling Brownian motion of single protein molecules and single fluorophores in aqueous buffer. *Opt. Express* **16**, 6941–6956.

Davis, L., Sikorski, Z., Robinson, W., Shen, G., Li, X., Canfield, B., Lescano, I., Bomar, B., Hofmeister, W., Germann, J., King, J., White, Y., and Terekhov, A. (2008). Maximum-likelihood position sensing and actively controlled electrokinetic transport for single-molecule trapping. *Proc. SPIE* **6862**, 68620P.

Dertinger, T., Pacheco, V., von der Hocht, I., Hartmann, R., Gregor, I., and Enderlein, J. (2007). Two-focus fluorescence correlation spectroscopy: A new tool for accurate and absolute diffusion measurements. *ChemPhysChem* **8**, 433–443.

Dickson, R. M., Norris, D. J., and Moerner, W. E. (1998). Simultaneous imaging of individual molecules aligned both parallel and perpendicular to the optic axis. *Phys. Rev. Lett.* **81**, 5322–5325.

Einstein, A. (1905). On the movement of small particles suspended in stationary liquids required by the molecular-kinetic theory of heat. *Ann. Phys.* **17**, 549–560.

Enderlein, J. (2000). Tracking of fluorescent molecules diffusing within membranes. *Appl. Phys. B* **71,** 773–777.
Gore, J., Bryant, Z., Stone, M. D., Nollmann, M. N., Cozzarelli, N. R., and Bustamante, C. (2006). Mechanochemical analysis of DNA gyrase using rotor bead tracking. *Nature* **439,** 100–104.
Gosse, C., and Croquette, V. (2002). Magnetic tweezers: Micromanipulation and force measurement at the molecular level. *Biophys. J.* **82,** 3314–3329.
Ha, T. (2001). Single molecule fluorescence methods for the study of nucleic acids. *Curr. Opin. Struct. Biol.* **11,** 287–292.
Ha, T., Ting, A. Y., Liang, J., Caldwell, W. B., Deniz, A. A., Chemla, D. S., Schultz, P. G., and Weiss, S. (1999a). Single-molecule fluorescence spectroscopy of enzyme conformational dynamics and cleavage mechanism. *Proc. Natl. Acad. Sci. USA* **96,** 893.
Ha, T., Zhuang, X., Kim, H. D., Orr, J. W., Williamson, J. R., and Chu, S. (1999b). Ligand-induced conformational changes observed in single RNA molecules. *Proc. Natl. Acad. Sci. USA* **96,** 9077.
Hamilton, J. D. (1994). Time Series Analysis. Princeton University Press, Princeton, NJ.
Jiang, Y., Wang, Q., Cohen, A. E., Douglas, N., Frydman, J., and Moerner, W. E. (2008). Hardware-based anti-Brownian electrokinetic trap (ABEL Trap) for single molecules: Control loop simulations and application to ATP binding stoichiometry in multi-subunit enzymes. *Proc. SPIE* **7038,** 703807.
Katsura, S., Yamaguchi, A., Inami, H., Matsuura, S., Hirano, K., and Mizuno, A. (2001). Indirect micromanipulation of single molecules in water-in-oil emulsion. *Electrophoresis* **22,** 289–293.
Kim, S. Y., Semyonov, A. N., Twieg, R. J., Horwich, A. L., Frydman, J., and Moerner, W. E. (2005). Probing the sequence of conformationally induced polarity changes in the molecular chaperonin GroEL with fluorescence spectroscopy. *J. Phys. Chem. B* **109,** 24517–24525.
Kim, S. Y., Gitai, Z., Kinkhabwala, A., Shapiro, L., and Moerner, W. E. (2006). Single molecules of the bacterial actin MreB undergo directed treadmilling motion in *Caulobacter crescentus*. *Proc. Natl. Acad. Sci. USA* **103,** 10929–10934.
Lessard, G. A., Goodwin, P. M., and Werner, J. H. (2007). Three-dimensional tracking of individual quantum dots. *Appl. Phys. Lett.* **91,** 224106.
Levi, V., and Gratton, E. (2008). Chromatin dynamics during interphase explored by single-particle tracking. *Chromosome Res.* **16,** 439–449.
Levi, V., Ruan, Q., Kis-Petikova, K., and Gratton, E. (2003). Scanning FCS, a novel method for three-dimensional particle tracking. *Biochem. Soc. Trans.* **31,** 997–1000.
Levi, V., Ruan, Q., and Gratton, E. (2005). 3-D particle tracking in a two-photon microscope: Application to the study of molecular dynamics in cells. *Biophys. J.* **88,** 2919–2928.
Li, X., Hofmeister, W., Shen, G., Davis, L., and Daniel, C. (2008). Fabrication and characterization of nanofluidics device using fused silica for single protein molecule detection. In "Medical Device Materials IV (Proceedings of the Materials and Processes for Medical Devices Conference 2007)," (J. Gilbert, ed.), pp. 145–150. ASM International, Materials Park, OH.
Lipman, E. A., Schuler, B., Bakajin, O., and Eaton, W. A. (2003). Single-molecule measurement of protein folding kinetics. *Science* **301,** 1233.
Lu, H. P., Xun, L., and Xie, X. S. (1998). Single-molecule enzymatic dynamics. *Science* **282,** 1877–1882.
Lu, P. J., Sims, P. A., Oki, H., Macarthur, J. B., and Weitz, D. A. (2007). Target-locking acquisition with real-time confocal (TARC) microscopy. *Opt. Express* **15,** 8702–8712.
McHale, K., Berglund, A. J., and Mabuchi, H. (2007). Quantum dot photon statistics measured by three-dimensional particle tracking. *Nano Lett.* **7,** 3535–3539.

Meseth, U., Wohland, T., Rigler, R., and Vogel, H. (1999). Resolution of fluorescence correlation measurements. *Biophys. J.* **76,** 1619–1631.

Mills, R. (1973). Self-diffusion in normal and heavy water in the range 1–45°. *J. Phys. Chem.* **77,** 685–688.

Plenert, M. L., and Shear, J. B. (2003). Microsecond electrophoresis. *Proc. Natl. Acad. Sci. USA* **100,** 3853–3857.

Svoboda, K., Schmidt, C. F., Schnapp, B. J., and Block, S. M. (1993). Direct observation of Kinesin stepping by optical trapping interferometry. *Nature* **365,** 721–727.

Toussaint, K. C., Liu, M., Pelton, M., Pesic, J., Guffey, M. J., Guyot-Sionnest, P., and Scherer, N. F. (2007). Plasmon resonance-based optical trapping of single and multiple Au nanoparticles. *Opt. Express* **15,** 12017–12029.

Tyn, M. T., and Gusek, T. W. (1990). Prediction of diffusion coefficients of proteins. *Biotechnol. Bioeng.* **35,** 327–338.

Valentine, M. T., Guydosh, N. R., Gutiérrez-Medina, B., Fehr, A. N., Andreasson, J. O., and Block, S. M. (2008). Precision steering of an optical trap by electro-optic deflection. *Opt. Lett.* **33,** 599–601.

Voldman, J., Braff, R. A., Toner, M., Gray, M. L., and Schmidt, M. A. (2001). Holding forces of single-particle dielectrophoretic traps. *Biophys. J.* **80,** 531–541.

Wang, H. Y., Foote, R. S., Jacobson, S. C., Schneibel, J. H., and Ramsey, J. M. (1997). Low temperature bonding for microfabrication of chemical analysis devices. *Sens. Actuators B* **45,** 199–207.

Wen, J. D., Lancaster, L., Hodges, C., Zeri, A. C., Yoshimura, S. H., Noller, H. F., Bustamante, C., and Tinoco, I. (2008). Following translation by single ribosomes one codon at a time. *Nature* **452,** 598–603.

Xu, C. S., Cang, H., Montiel, D., and Yang, H. (2007). Rapid and quantitative sizing of nanoparticles using three-dimensional single-particle tracking. *J. Phys. Chem. C* **111,** 32–35.

CHAPTER EIGHT

PLASMON RULERS AS DYNAMIC MOLECULAR RULERS IN ENZYMOLOGY

Björn M. Reinhard,* Jaime M. Yassif,[†,‡] Peter Vach,[†] and Jan Liphardt[†,‡]

Contents

1. Introduction — 176
2. The Basic Idea: Distance Dependence of Plasmon Coupling — 177
3. Hardware Needed for Single Particle Rayleigh Scattering Spectroscopy — 179
4. Which Readout—Intensity, Polarization, or Color? — 181
   4.1. Color — 181
   4.2. Intensity — 181
   4.3. Polarization — 182
5. Ruler Calibration? — 182
6. Plasmon Ruler Assembly and Purification — 184
7. Example 1: Dynamics of DNA Bending and Cleavage by Single EcoRV Restriction Enzymes — 186
8. Example 2: Spermidine Modulated Ribonuclease Activity Probed by RNA Plasmon Rulers — 192
9. Outlook — 194
References — 196

Abstract

This chapter provides an introduction to the concept of "plasmon rulers," pairs of biopolymer-linked tethered nanoparticles which act as nonblinking, non-bleaching rulers for dynamic molecular distance measurements. Plasmon rulers utilize the distance dependence of the plasmon coupling between individual noble metal particles to measure distances. Although the plasmon ruler approach is still an emerging technology, proof-of-principle experiments have demonstrated that plasmon rulers can already be used to investigate structural fluctuations in nucleoprotein complexes, monitor nuclease catalyzed DNA or

* Department of Chemistry, The Photonics Center, Boston University, Boston, Massachusetts, USA
[†] Department of Physics, Biophysics Graduate Group, and Bay Area Physical Sciences Oncology Center, University of California at Berkeley, California, USA
[‡] Physical Biosciences Division, Lawrence Berkeley National Laboratory, Berkeley, California, USA

Methods in Enzymology, Volume 475 © 2010 Elsevier Inc.
ISSN 0076-6879, DOI: 10.1016/S0076-6879(10)75008-4 All rights reserved.

RNA cleavage reactions, and detect DNA bending. The physical concepts underlying plasmon rulers are summarized, and effective assembly approaches as well as recent applications are discussed. Plasmon rulers are a useful addition to the single molecule biophysics toolbox, since they allow single biomolecules to be continuously monitored for days at high temporal resolutions.

1. INTRODUCTION

The last 15 years have seen the rapid development of single-molecule biophysics. The intrinsic advantages of monitoring biological processes one molecule at a time are now widely appreciated. Single-molecule assays give insight into the molecular mechanisms of complex reactions by providing distributions of molecular properties and capturing transient intermediates without the need for synchronization (Xie and Lu, 1999). One achievement of single-molecule techniques in enzymology was the experimental verification of dynamic disorder: large fluctuations in the catalytic rate at a time scale comparable to or larger than the enzymatic cycle (van Oijen et al., 2003). These fluctuations reflect long-lived conformers interconverting over a broad range of time scales, which are typically concealed in ensemble measurements.

Most single-molecule assays do not directly monitor the catalyzed chemical reaction, but rather structural changes in the enzyme–substrate complex. During the catalytic cycle, many enzyme–substrate complexes undergo structural changes that involve the relative translation of individual components within the complex. To measure these enzyme mechanics, one needs tools capable of measuring nanometer distance changes within single molecules.

The method of choice for measuring distances and distance changes in dynamic biological systems has been Förster resonance energy transfer (FRET) between a donor and an acceptor dye (Ha et al., 1996, 1999). FRET exploits the $1/r^6$ dependence of energy transfer between two dyes on their separation r. By measuring the FRET efficiency, the dye separation (and thus the conformation of the labeled complex) can be inferred. FRET has been successfully applied to molecular motors (Smiley et al., 2006; Tomishige et al., 2006), RNA ribozymes (Zhuang et al., 2002), and DNA helicases (Rasnik et al., 2006) and binding proteins (Morgan et al., 2005). Like any measurement tool, FRET has limitations. When based on conventional organic dyes such as Cy3 and Cy5, the dynamic working range of FRET is limited to approximately 1–8 nm. Moreover, photobleaching of the dyes limits the maximum duration of single-molecule trajectories and the blinking of the dyes complicates the decomposition of single-molecule FRET traces into molecular trajectories and states.

These handicaps of FRET have motivated the development of alternative dynamic rulers, such as Raman based molecular rulers (Lal *et al.*, 2005), small-angle X-ray scattering interference between gold nanocrystal labels (Mastroianni *et al.*, 2009; Mathew-Fenn *et al.*, 2008), and nanometal surface energy transfer (Yun *et al.*, 2005). All of these techniques show great promise, but each of them has specific advantages and disadvantages, and each technique needs further refinement or characterization. In this review, we focus on another emerging technology that is not complicated by bleaching or blinking, allows higher temporal bandwidth, and has a significantly larger dynamic range than FRET: pairs of noble metal nanoparticles linked by biopolymers such as RNA, DNA, carbohydrates, and proteins (plasmon rulers).

Plasmon rulers exploit the distance dependence of the plasmon coupling between individual pairs of metal nanoparticles to measure distance and distance changes. Since the optical properties of the probes in plasmon rulers are based on light scattering, they do not blink and have an essentially unlimited lifetime. A drawback of plasmon rulers is their size, which is typically much larger than organic dyes. The nanoparticles used to build plasmon rulers typically have diameters of tens of nanometers, whereas conventional organic dyes have sizes of about 1 nm. Plasmon rulers are therefore primarily useful for applications where unlimited probe lifetime and a larger dynamic range outweigh the drawback of larger probe size.

In this chapter, we will review the current status of plasmon ruler technology and provide some examples of its applications. For example, these applications include monitoring enzyme induced DNA binding and bending as well as probing the kinetics of RNA digestion. Plasmon rulers are a young technology, and many groups are currently trying to improve them and to establish all their particularities and limitations. At the end of this chapter, we summarize some of these efforts and give an outlook.

2. The Basic Idea: Distance Dependence of Plasmon Coupling

The optical properties of noble metal nanoparticles are defined by particle plasmons, coherent oscillations of the particles' conduction band electrons (Kreibig and Vollmer, 1995). The resonance frequency of these plasmons depends on the dielectric function of the metal, the particle's size and shape, the presence of other nanoparticles, and the dielectric constant of the environment (Kelly *et al.*, 2003). At the resonance frequency, the optical cross section of the nanoparticles reaches its maximum and even weak incident fields induce a large response. Excellent reviews of the optical properties of nanoparticles are given in Halas (2005) and Kelly *et al.* (2003).

The near-fields of individual metal nanoparticles can interact when two or more nanoparticles approach (Prodan et al., 2003; Reinhard et al., 2005; Su et al., 2003), shifting the resonance wavelength. The distance-dependence of the resonance wavelength allows nanoparticle pairs to report their separation. In dimers of spherical particles, to which we confine our analysis here, two coupled plasmon modes can be excited using light: a longitudinal and a vertical plasmon mode (Fig. 8.1). In the longitudinal mode the coupled particle dipoles are aligned in a head-to-tail fashion along the long dimer axis, whereas in the vertical mode the dipoles have a parallel alignment perpendicular to the long interparticle axis. As a consequence of the relative orientation of the net dipole moments, the longitudinal plasmon mode redshifts with decreasing interparticle separation, whereas the vertical mode blueshifts. Under unpolarized white light illumination, the longitudinal mode is stronger and dominates the spectrum of the coupled dimer; the vertical mode can be resolved if it is selectively excited using linearly polarized light (Jain et al., 2007). An elegant theory to explain plasmon coupling is the plasmon hybridization method by Nordlander and coworkers (Prodan et al., 2003). The method, which is best described as an electromagnetic analog to molecular orbital theory, predicts the relative energies of the individual modes in complex plasmonic nanostructures.

The distance dependence of the plasmon coupling has been investigated at fixed interparticle distances with different nanostructures such as spherical (Jain et al., 2007; Maier et al., 2005), cylindrical (Haynes et al., 2003; Rechberger et al., 2003), and elliptical nanoparticles (Wei et al., 2004), trigonal prisms (Haynes et al., 2003), hollow nanoparticles (Yang et al., 2008), nanoshells (Lassiter et al., 2008), and opposing tip-to-tip Au triangle

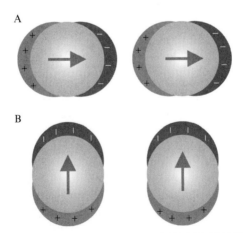

Figure 8.1 Charge distributions and dipole orientations in (A) longitudinal and (B) vertical plasmon modes.

(bowtie) nanostructures (Fromm et al., 2004). In these studies, nanostructures were fabricated with top-down fabrication techniques such as electron beam lithography and thus had fixed interparticle spacing. These studies are ideal for investigating the physical concepts of interparticle plasmon coupling at fixed separations.

While these experiments confirmed the feasibility of plasmon coupling based rulers in solutions of constant refractive index, nanofabricated nanostructures with fixed separation are inapplicable to many problems in single-molecule enzymology. Rather, single-molecule enzymologists require nanostructures that are free to move and this movement must somehow be coupled to an enzymatic reaction. The first experimental realization of plasmon rulers with clear biological applicability was achieved by Mirkin and coworkers on the bulk level in their colorimetric DNA sensing scheme (Elghanian et al., 1997). They observed a spectral red-shift in the bulk UV-Vis absorbance spectrum of a solution of DNA functionalized nanoparticles upon "polymerization" through a complementary probe strand. The first investigation of single-plasmon rulers, constructed by tethering two gold nanoparticle with single-stranded (ss)DNA, was reported in 2005 (Sonnichsen et al., 2005).

3. Hardware Needed for Single Particle Rayleigh Scattering Spectroscopy

Gold and silver nanoparticles with diameters of greater than 30 nm scatter certain colors of light very efficiently. The use of nanoparticle pairs as rulers requires detecting, imaging and, ideally, analyzing the spectrum of individual nanoparticle dimers. Microscopy techniques are needed that allow high contrast spectral imaging while eliminating the background. In fluorescence microscopy, one uses the Stokes' shift to separate excitation from emission, but that does not work with nanoparticles, since they are not fluorophores but rather light scatterers. This means that they "reflect" the incoming light, with some colors being scattered better than other colors, giving 40 nm gold nanoparticles a greenish hue when observed in the light microscope. The method of choice for investigating single nanoparticles is dark-field microscopy (Fig. 8.2) (Sonnichsen et al., 2000). In dark-field illumination, the excitation light is injected into the focal plane through a dark-field condenser at oblique angles from all directions. The oblique angles do not reach the objective unless they are scattered by an object in the specimen plane. This illumination scheme suppresses the background and enhances the imaging of the nanoparticles with high contrast, and spectral characterization of individual particles becomes possible. The recorded spectrum needs to be background and lineshape corrected by dividing the nanoparticle spectrum by a normalized white light spectrum of the excitation source.

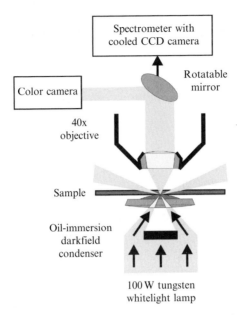

Figure 8.2 Schematic drawing of a dark-field microscope. The light is injected into the focal plane at oblique angles so that only the light scattered off individual particles in the specimen plane is collected by the objective.

Most conventional dark-field microscopes are suitable for single particle Rayleigh scattering spectroscopy. We typically perform measurements using an upright microscope (Zeiss Axioplan2) with a 100 W tungsten lamp, oil immersion dark-field condenser and a 40× objective. The scattered light is analyzed by a CCD detector (Andor iXonEM) attached to an imaging spectrometer (such as a SpectraPro 2300i).

Inverted microscopes are also suitable for taking measurements, but manufacturers typically do not sell dark-field condensers for these microscopes. However, all that is needed is a simple, inexpensive, cylindrical adapter that mates the dark-field optics to the condenser turret; any machine shop will be able to design such a mating adapter. The one we use is a 60 mm diameter aluminum tube (with a length of 30 mm), that has a Zeiss flange on one end and a Nikon flange on the other end. The only other consideration is that the numerical aperture (NA) of most dark-field condensers is not much bigger than 1.2, and thus, use of an expensive high NA objective will actually degrade, rather than enhance, signal-to-noise. A good starting point is a simple 40× objective with an NA of 1. There are also objectives with internal irises that allow the NA to be reduced (e.g., Zeiss Fluar 40×, max NA 1.3) that we have used with success.

4. Which Readout—Intensity, Polarization, or Color?

Distances and distance changes are encoded in both intensity and "color" (corresponding to the resonance wavelength) of the individual plasmon rulers.

4.1. Color

This is the "gold standard," since very few effects cause nanoparticle pairs to change their resonance wavelength—primarily, changes of interparticle separation. It is important to keep in mind that plasmon rulers are based on a different physical principle than the "surface plasmon" approaches with which many scientists are familiar. Surface plasmon approaches exploit the sensitivity of a metal's optical properties to its local refractive index. By contrast, plasmon rulers rely on the *coupling* of plasmons that arises when two particles are brought into proximity. Plasmon rulers are constructed from metal particles coated with thick stable claddings to minimize the (in this case) undesired sensitivity to local refractive index; since the material in direct contact with the metal does not change with time, spectral shifts due to changes of the buffer's refractive index are largely suppressed. Typical spectral shifts seen in surface plasmon approaches are on the order of 1–2 nm, whereas plasmon rulers can exhibit spectral shifts of greater than 60 nm. Color changes can be recorded with a spectrometer, but this limits the number of particles that can be observed simultaneously. This also reduces the temporal bandwidth. Ratiometric readout schemes (Rong *et al.*, 2008) help on both accounts, albeit with some loss of spectral resolution.

4.2. Intensity

Like the resonance wavelength, the total scattering cross section of pairs of coupled nanoparticles depends on the interparticle separation (Reinhard *et al.*, 2007). Intensities can generally be collected with higher temporal resolution than spectral information, because no dispersion is required. Intensity is a good readout for biological processes that involve large, sudden changes of interparticle separation, such as those resulting from nuclease cleavage of the tether that links two particles. However, measuring only intensity is less reliable than measuring both intensity and color, since there are imaging artifacts that change the amount of light collected from a sample, including changes of illumination intensity and drift of the microscope's focus. A good compromise may be to collect spectral information from some particle pairs in the field of view, and then to use this spectral information as an internal control or standard.

4.3. Polarization

There is one final consideration. The rotation of a plasmon ruler in the instrument frame will also change the amount of light scattered by it. This effect can be used to infer the orientation of a plasmon ruler (by varying the polarization of the illumination). However, this also means that intensity changes can arise from changes in interparticle separation, a rotation of the ruler, or a combination of the two. It is therefore theoretically ideal to measure both intensity and polarization, so that rotation can be distinguished from changes of interparticle separation.

5. RULER CALIBRATION?

One of the key questions to ask of any ruler is its accuracy and precision (repeatability). The biggest challenge in the plasmon ruler field is the heterogeneity of colloidal nanoparticle preparations. A solution of "40 nm gold nanoparticles" does not actually contain perfectly spherical nanoparticles, each with a diameter of 40 nm, but rather, a complicated mixture of nanoparticles with a range of sizes and shapes (Reinhard et al., 2005). Since the resonance wavelength of an individual particle depends both on its size and shape, a distribution in these parameters translates into a distribution of resonance wavelengths. Moreover, efficient coupling requires plasmons that are similar in energy, and so variation of plasmons also translates to variation of coupling coefficients.

In the early days, this particle heterogeneity raised doubts as to whether the signals collected from nanoparticle pairs could be interpreted in terms of interparticle separation. For example, transmission electron microscopy images of commercially available monodispersed nanoparticles (British Biocell International) reveal nonnegligible amounts of triangular and pentameric particles, such that sphere–sphere dimers likely constitute approximately 80% of all dimers (Reinhard et al., 2005). We synthesized a family of dsDNA-linked plasmon rulers spanning a range of interparticle separations and measured their resonance wavelengths. This work confirmed that there is indeed a correlation between resonance wavelength and spacer length and enabled a rough calibration of the plasmon resonance wavelength (λ_{res}) versus the interparticle separation (s) for dimers constructed from commercially available 42 and 87 nm gold nanoparticles. The resulting calibration curves are shown in Fig. 8.3. Two qualitative trends are apparent. First, the peak resonance of the 87 nm dimers is

Plasmon Rulers as Dynamic Molecular Rulers in Enzymology

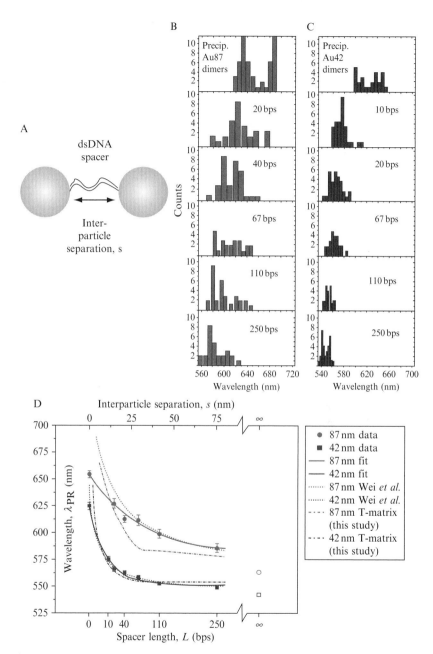

Figure 8.3 Plasmon resonance versus interparticle separation in dsDNA tethered gold nanoparticle dimers. (A) Schematic drawing of the plasmon rulers. Distributions of measured plasmon resonance wavelengths for selected dsDNA spacer lengths are plotted in (B) for 87-nm Au plasmon rulers and in (C) for 42-nm Au plasmon rulers. Data points for salt-precipitated dimers were included. (D) Plot of the average plasmon

consistently further in the red than for the 42 nm dimers, for all spacer lengths. In general, the larger the particle, the more red the resonant peak energy (and therefore the peak wavelength) will be. Second, the peak wavelength blue-shifts with increasing interparticle separation due to reduced near-field coupling, confirming a clear correlation between plasmon resonance wavelength and interparticle separation. These observations give credence to the notion that pairs of DNA and RNA tethered nanoparticles can be used to measure distance and distance changes. The measured resonance wavelength versus distance relationships were well-fit by simple exponentials allowing derivation of approximate $\lambda_{\text{res}}(s)$ relationships. A detailed investigation into the universal scaling behavior of interparticle coupling, combining both experimental and theoretical studies, was recently performed by Jain and el-Sayed (Jain et al., 2007).

The tentative conclusion regarding ruler calibration is that for a given nanoparticle pair, the relationship between interparticle separation and resonance wavelength is stable over time, but that slight shape and size variations from particle to particle lead to uncertainty in the estimate of the interparticle separation as judged by the resonance wavelength. This uncertainty should be experimentally quantified for each new sample of plasmon rulers; suggestions for how to do this are given in Reinhard et al. (2005). A simple test, for example, is to deposit single particles on a surface and analyze about 100 particles—the variation of intensity and $\lambda_{\max}$ immediately gives a sense of particle-to-particle homogeneity.

6. Plasmon Ruler Assembly and Purification

The specific experimental details of plasmon ruler assembly depend on the length and nature of the tether, but there are some general guidelines (Fig. 8.4). The first step is typically to replace the "proprietary" surface coating found on commercial gold nanoparticles with a known chemistry, such as phosphene groups. This exchange can be done, for example, by incubating gold nanoparticles in 1 mg/ml bis(p-sulphonatophenyl)phenylphosphine

---

resonance as a function of spacer length, L (bottom axis) and approximated interparticle distance x (top axis) for 42 nm (red squares) and 87 nm (blue circles) plasmon rulers. The plasmon resonance wavelengths for dimers with infinite separations (monomers) are included as open symbols. The reported errors are the standard errors of the mean. The continuous lines show fits (single exponentials $y(x) = A_0 \exp(-x/D_0) + C$) to the experimental data. Best-fit parameters for 42-nm Au particles: $C = 550.87$ nm, $A_0 = 73.48$ nm, $D_0 = 10.24$ nm; for 87-nm Au particles: $C = 579.66$ nm, $A_0 = 74.42$ nm, $D_0 = 30.23$ nm. The dotted lines represent T-matrix simulations by Wei et al. (2004) assuming illumination with light polarized along the interparticle axis and dot-dashed lines are T-matrix simulations assuming nonpolarized illumination. (See Color Insert.)

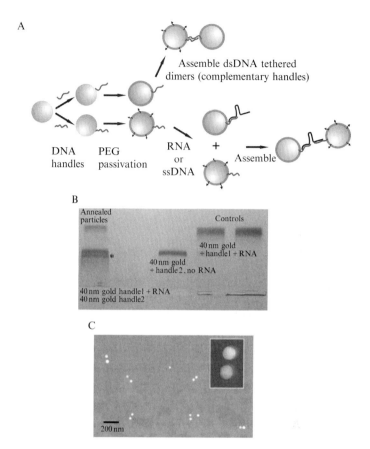

Figure 8.4 Plasmon ruler assembly. (A) Nucleic acid programmed self-assembly strategy for dsDNA-linked nanoparticle dimers and RNA, respectively. (B) Gel purification of 40 nm dimers connected by 155 nucleotide long RNA. The product band is marked with a ★. (C) TEM micrograph of dimers isolated from the ★ band. The isolated dimers are clearly separated by a gap, indicative that the particles are not sticking nonspecifically but are linked by the biopolymer.

(BSPP) at room temperature (RT) for at least 8 h. Then, these phosphene passivated particles are functionalized with tri-thiolated ssDNA in a ratio of 1:25 (gold nanoparticle/DNA). Finally, the ssDNA-gold nanoparticles are coated with a monolayer of short polyethylene glycol (PEG) ligands (typically, with a 4–8 h RT incubation in thiolated carboxy-terminated PEGs such as thiol-(EG)7-propionate at a ratio of 1:100,000). The PEG ligands are necessary to stabilize the particles in the buffer conditions (≥ 80 mM NaCl, 10 mM Tris–HCl, pH 7.0) ultimately required for efficient hybridization. For one of the two types of particles, small amounts of biotinylated PEGs are mixed in with the thiol-PEG–propionate (1:25 ratio) to facilitate immobilization of the assembled dimers to NeutrAvidin functionalized glass surfaces.

If ssDNA or RNA tethered nanoparticles are required, the ssDNA or RNA tether sequences need to contain 5′ and 3′ end sequences that are complementary to the respective DNA handles on the gold nanoparticles. The tether oligonucleotide is first annealed to one of the particles. After annealing overnight, the particles are thoroughly cleaned through repeated centrifugation and resuspension to remove excess DNA or RNA tether oligonucleotides. The cleaned particles are then combined with the second flavor of nanoparticles that were cleaned in a similar fashion and annealed at RT for several hours. To minimize multiple tether formation during annealing, the tether–particle ratio is limited to 30:1. Detailed reaction conditions for the synthesis of dsDNA tethered nanoparticle dimers can be found in Reinhard *et al.* (2007); for single-stranded RNA or DNA tethered dimers, see Skewis and Reinhard (2008).

After annealing, the assembled dimers should be purified and separated from monomers or larger nanoparticle assemblies. This can be conveniently achieved using gel-electrophoresis, which separates single particles from dimers based on size. In Fig. 8.4B a typical gel for the purification of RNA assembled dimers is shown. The gel contains particles with only handles (two outer right lanes), particles with the RNA annealed to one of the handles (middle lane), and the reaction mix (outer left lane). The reaction mix contains a new band that is absent in all the precursors; this band is cut out of the gel and the particles are then recovered from the gel using electroelution. The DNA-linked particles elute much like free DNA. Inspection of the recovered particles using TEM (Fig. 8.4C) shows that the band is highly enriched in dimers and that the individual particles are well separated, showing that the particles are not simply nonspecifically sticking together, but indeed are linked by a polymer tether.

In all these steps, it is absolutely critical to carefully optimize ionic strength, pH, and sample handling, to avoid undesired aggregation of the gold nanoparticles. Prior to adding PEG, particles should be suspended in a 25–40 mM NaCl buffer at pH 7.0. After PEG coating, the particles are less prone to aggregation when exposed to high ionic strength, and salt concentrations can be increased to 80 mM or higher. Avoiding excessive centrifugation is another means of preventing aggregation. If any aggregation occurs, the sample should be discarded.

7. Example 1: Dynamics of DNA Bending and Cleavage by Single EcoRV Restriction Enzymes

DNA bending plays a crucial role in determining the specificity in DNA–protein recognition, transcription regulation, and DNA packaging. DNA bending by proteins occurs typically on a millisecond timescale but it is desirable to be able to monitor particular DNAs for extended times, to explore the effects of enzyme concentration, ionic strength, and pH changes or presence of a cofactor. Plasmon rulers are almost ideal for this kind of experiment—different materials or buffers can be introduced into a flow chamber and the response of single nucleic acids can be measured, the buffer exchanged, and so forth, until a wide range of conditions have been explored with only one sample. Since there is no blinking or bleaching, experiments can be done over hours or days with continuous observation. Unlike with conventional fluorophores, experiment duration is only limited by clogging of the flow-chamber, degradation of the biological tether, or limited patience of the experimentalist.

One textbook example of a DNA-binding enzyme that bends its DNA substrate is the type II restriction endonuclease EcoRV. The homodimeric enzyme binds nonspecifically to DNA, translocates and binds to the target site, bends the DNA, cuts the DNA in a blunt-ended fashion by phosphoryl transfer and subsequently releases the product. The mechanism of the catalytic cycle has been intensively studied, and a wealth of structural and kinetic (both ensemble and single molecule) information is available, making this system an ideal test case to study the performance of plasmon rulers. EcoRV transiently bends its DNA substrate, and the bend angle is known from crystal structures to be $\sim 52°$. The detection of this bend angle and the associated small distance change in a single molecule in solution is a significant challenge. The plasmon rulers used in these studies contained DNA tethers 30, 40, and 60 base pairs (bp) in length and contained the EcoRV restriction site in the center. A kink of $52°$ in these DNAs leads to a change in the end-to-end distances of between 0.9 and 1.3 nm.

Before investigating DNA bending by EcoRV in detail, we first characterized the global cleavage kinetics. Figure 8.5 shows surface immobilized plasmon rulers as function of time; for better clarity only an extract of the total field of view at $40\times$ magnification of 100×150 μm^2 is shown. Five individual plasmon rulers are clearly discernible as green spots. The individual particles in the dimers are no longer resolvable in the optical microscope due to their subdiffraction limit separation. The total field of view of 100×150 μm^2 contained hundreds of plasmon rulers, which were all monitored in parallel.

At $t = 0$ s EcoRV restriction nuclease is flushed into the chamber. Individual cutting events catalyzed by the enzyme are clearly distinguishable by a sudden and irreversible drop in scattering intensity (marked by arrows). Control experiments using plasmon rulers without the EcoRV recognition site confirmed that these cleavage events are enzyme specific. Some of the monitored dimers ($<10\%$) exhibited more than one subsequent intensity drop, indicative of the existence of multiple tethers between the nanoparticle probes. In the vast majority of recorded trajectories, however, the scattering

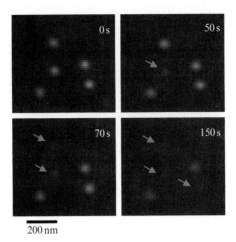

Figure 8.5 Plasmon rulers enable highly parallel single-molecule assays. Here DNA cleavage by EcoRV is shown for multiple individual particles. The image is an extract from a field of view with dimensions of 150×100 μm^2 containing hundreds of plasmon rulers. Individual dimers are shown as bright green dots. Dimer dissociation upon EcoRV-catalyzed DNA cleavage leads to a strong change in scattering intensities. Individual cleavage events are marked by red arrows.

intensity drops in one sudden step. This suggests that these plasmon rulers contain only a single DNA tether.

The strong signal change upon DNA cleavage facilitates kinetic studies of the enzyme cleavage reaction, for instance, as function of the divalent ion concentration. This is exemplified in Fig. 8.6. We monitored the plasmon rulers (40 bp DNA spacer) for 5 min after adding the enzyme, and then, plotted the cumulative fraction of rulers that were successfully cleaved within that time. Type II restriction endonucleases require divalent ions, and the cleavage rates depend on the nature of the divalent ions present; it is well known that Ca^{2+} in the presence of Mg^{2+} inhibits cleavage by EcoRV. The cleavage kinetics in Fig. 8.6 confirm the decelerating effect of increasing Ca^{2+} on the EcoRV cleavage kinetics in the presence of a fixed Mg^{2+} ion concentration (10 mM): The fitted first-order rate constants decrease from 0.036 s^{-1} (no Ca^{2+}) to 0.004 s^{-1} (2 mM Ca^{2+}).

In addition to quantifying the overall reaction kinetics, the plasmon ruler assay provides the opportunity to extract information about specific reaction steps in *individual tether* molecules. These details become accessible when the scattering intensity of individual plasmon rulers is analyzed as a function of time. Figure 8.7A shows an especially nice cleavage trajectory of a 40 bp DNA tether recorded at 85 Hz. Temporal resolutions of up to 240 Hz have already been achieved, albeit with lower spatial resolution.

The most noticeable feature in the scattering trajectories is the sudden intensity drop resulting from the dissociation of one of the two nanoparticles

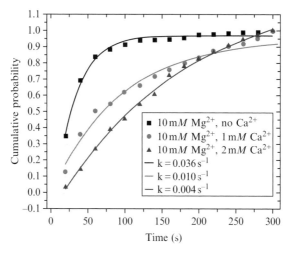

Figure 8.6 Percentage of plasmon rulers that have been cleaved as function of time for increasing Ca^{2+} concentrations. First-order kinetic fits are shown as continuous lines. EcoRV requires Mg^{2+} as a natural cofactor to catalyze DNA cleavage. Ca^{2+} can replace Mg^{2+} and inhibit the cleavage reaction in the presence of Mg^{2+}. A Mg^{2+} concentration of 10 mM was retained throughout.

upon DNA cleavage. This drop serves as fiduciary point for the detection of more subtle details in the trajectory that result from the interactions between the enzyme and the DNA. Many trajectories show a transient increase in scattering intensity, as shown for the scattering trajectory in Fig. 8.7A, right before the cleavage. This increase in scattering intensity is an indication of a transient decrease in the interparticle separation. To determine if this feature is really characteristic of the EcoRV-catalyzed DNA cleavage reaction, statistical analyses are required to test if the cleavage is correlated with the increase in scattering intensity. This can be achieved for instance by calculating the average intensity preceding the dimer dissociation for a statistically significant number of trajectories. This analysis is shown in Fig. 8.7B–D for 30-, 40-, and 60-bp plasmon rulers. In these plots the time of the plasmon ruler dissociation was set equal to 0 ms, and average intensities for shortening time intervals between [− 1770 to 0] and [− 24 to 0] ms were calculated. The total number of trajectories analyzed for Fig. 8.7B–D are 86 (30 bp) and 96 (40 bp) and 114 (60 bp).

For both the 30 and 40 bp spacers the average intensity increases before cleavage, which is indicative of a bending of the DNA through the enzyme. For the largest initial interparticle separation (60 bp spacer), no increase in intensity was observed. The differences among plasmon rulers with 30, 40, and 60 bp spacers are a consequence of the inverse relationship between sensitivity and spacer length; for the 60 bp spacer the coupling is too weak to

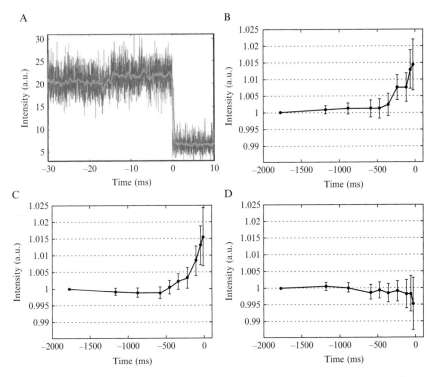

Figure 8.7 Single DNA cleavage events monitored by plasmon coupling. (A) Cleavage trajectory recorded at 85 Hz using an intensified CCD detector. A common feature observed in many trajectories is a slight increase in scattering intensity preceding dimer dissociation. (B–D) Average scattering intensities at defined intervals preceding the dimer dissociation at 0 ms for 30-, 40-, and 60-bp plasmon rulers. We set the time of the plasmon ruler dissociation equal to 0 ms and calculated average intensities for shortening time intervals between [−1770 to 0] and [−24 to 0] ms.

observe the bending. *A priori*, it cannot be excluded that the observed shift in the intensity is due to the binding but not the bending of the enzyme, since as mentioned in Section 4, the scattering intensity depends strongly on the interparticle distance but also very weakly on the dielectric constant of the surrounding medium. To quantify potential artifacts arising from environmental sensitivity, we mixed the enzyme with plasmon rulers in divalent free buffer in which the enzyme can bind but not bend the DNA. In these conditions, there was no detectable change of scattering intensity, suggesting that any effects due to environmental sensitivity (e.g., presence of an enzyme in the gap between the two particles) can be neglected in our experiments. Once divalent ions were added, we saw the expected increase in average scattering intensity immediately before cleavage, confirming our assignment of this signal to DNA bending.

After assigning the high-intensity state before DNA cleavage to the bent state, we analyzed the distribution of the durations of the high-intensity states (Fig. 8.8). In these plots the percentage of dimers that have been cleaved is plotted against the time they spend in the high-intensity state. Exponential fits to the decay curves in Fig. 8.8B give rate constants of $k = 0.50 \text{ s}^{-1}$ (30 bp DNA spacer) and $k = 0.46 \text{ s}^{-1}$ (40 bp spacer). These rate constants are very close to the first-order rate constant of 0.4 s^{-1} obtained for the later stages (DNA hydrolysis and product release) in the EcoRV catalytic cycle. Thus, although deviations from the bulk rates are to

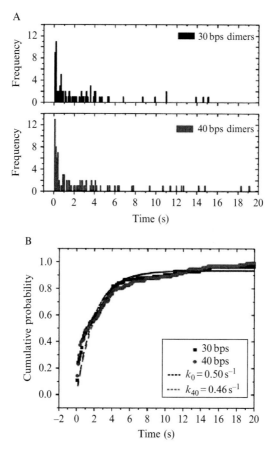

Figure 8.8 Kinetic and thermodynamic analysis of single-molecule data, in particular dwell-time analysis of plasmon rulers in high-intensity state. (A) Histograms with a bin size of 100 ms for 30- (upper) and 40-bp (lower) plasmon rulers. (B) Cumulative probability of plasmon ruler dissociation. Plots show the percentage of dimers with dwell times less than the indicated time. First-order kinetic fits are included as dashed lines ($k_{30} = 0.50 \text{ s}^{-1}$, $k_{40} = 0.46 \text{ s}^{-1}$). (See Color Insert.)

be expected in the plasmon ruler assay (the probes may slow the dynamics of the enzymatic reaction), such deviations are small (not more than $0.2\ \mathrm{s}^{-1}$).

8. EXAMPLE 2: SPERMIDINE MODULATED RIBONUCLEASE ACTIVITY PROBED BY RNA PLASMON RULERS

RNA plasmon rulers, in which the DNA tether is replaced by RNA, can be used to monitor dynamic distance changes in RNA molecules. As with DNA based rulers, the size and surface chemistry of the nanoparticles raises concerns about the potential perturbation of the catalytic activity of ribozymes or RNA-binding enzymes. On the bright side, the long dynamic working range of the plasmon rulers reduces the necessity of placing the probes close to the catalytically active center to be able to detect structural changes in the course of the catalytic cycle.

We used simple RNA–DNA hybridization based directly on the DNA protocol (Section 6) to generate RNA plasmon rulers in high yield and purity. The resulting rulers were used to monitor the influence of the endogenous trivalent cation spermidine on the cleavage kinetics of RNA by ribonuclease A (RNase A). In the absence of spermidine, efficient cleavage of the RNAs confirmed that the RNA tether was readily accessible for the RNase A. In addition, for most dimers the cleavage occurred in one rapid intensity drop, supporting the notion that only a single RNA molecule was tethered between the particles.

The main advantages that motivate the use of plasmon rulers in this example (RNA cleavage assays) are the high temporal resolution (hundreds of Hz) and the ability to watch many plasmon rulers at the same time. The latter enables investigations of the cleavage kinetics of RNase A on the single-molecule level, giving access to relative stabilities of weakly stabilized RNA subpopulations and their lifetimes. Specifically, the cleavage time Δt_{cl}, defined as the time gap between addition of the enzyme and observation of the cleavage event, is a useful measure of the cleavage kinetics of RNA-binding enzymes and can be used to quantify the relative stabilities of different RNA structures against enzymatic degradation. The cleavage times depend on all steps of the cleavage reaction, including the diffusion of the enzyme to the substrate, the binding of the enzyme, and the actual cleavage and subsequent product release.

In Fig. 8.9 cleavage time (Δt_{cl}) distributions for different spermidine concentrations are plotted (155-nt-long RNA tether). In the absence of spermidine, the cleavage time distribution is well fit by a single Gaussian centered at $\Delta t_{cl} = 3.8$ s. With increasing spermidine concentrations the cleavage times lengthen. Under physiological conditions, spermidine is

Plasmon Rulers as Dynamic Molecular Rulers in Enzymology 193

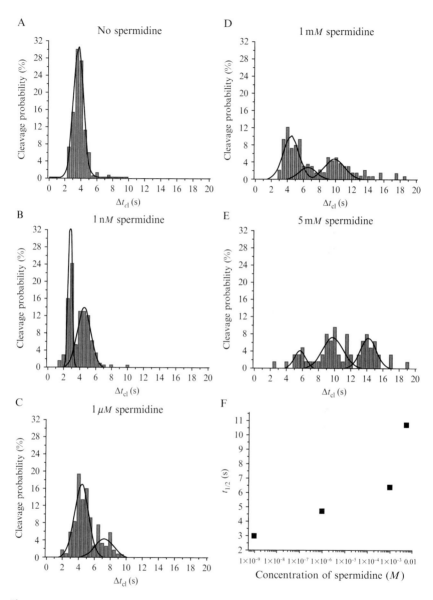

Figure 8.9 Cleavage time (Δt_{cl}) distributions as a function of spermidine concentration. (A) No spermidine; (B) 1 nM spermidine; (C) 1 μM spermidine; (D) 1 mM spermidine; (E) 5 mM spermidine. Gaussian fits mark potential subpopulations. Already a 1 nM spermidine concentration leads to a splitting of the Δt_{cl} distribution. With increasing spermidine concentrations, the cleavage events happen later and subpopulations with long lifetimes grow at the cost of fast subpopulations. (F) $t_{1/2}$ times (=time after which half of all cutting events have occurred) versus concentration.

positively charged and the RNA bound polycations reduce the effective negative charge of the polymer, reducing the nonspecific binding of the enzyme. In addition, spermidine stabilized double-stranded regions of the RNA are not amenable to cleavage through the ssRNA specific RNase A. Consequently, the number of cleavage sites available for RNase A decreases with increasing spermidine concentration.

Inspection of the Δt_{cl} distributions in Fig. 8.9 reveals that the cleavage kinetics not only decelerate but that addition of spermidine leads to clustering in the Δt_{cl} distribution. This finding points toward the existence of different RNA subpopulations with characteristic cleavage kinetics. In RNAs of identical length, similar cleavage rates are an indication of underlying structural similarities.

Another characteristic of Fig. 8.9 is that the number of subpopulations grows and shifts to later cleavage times with increasing spermidine concentration. The RNA subpopulations with longer lifetimes are *transiently* stabilized relative to RNAs with faster cleavage rates. Depending on the spermidine concentration, the lifetime of the later populations is extended by 1.8–8.5 s. The overall short lifetimes of these RNA species complicate their detection in conventional nuclease protection assays. The ability of plasmon rulers to detect short-lived, transient stabilizations makes them an exquisite tool for the detection of weak contacts in structural RNAs.

Overall, the plasmon ruler studies in this example reveal transient structural RNA stabilization in the presence of the polyamine spermidine. This finding indicates a potential regulation of RNA-binding enzyme activity through spermidine-induced changes in the charge and structure of the RNA substrate. Perhaps most importantly for this review, these studies recapitulate what is already known about endogenous polyamines such as spermine, spermidine, and putrescine with more established techniques (Hammann *et al.*, 1997; Koculi *et al.*, 2004; Tabor and Tabor, 1984).

9. Outlook

Plasmon rulers are an emerging technology. Many aspects of the technology remain to be refined. One known issue is the dependence of the optical response on both refractive index and interparticle separation (Jain and el-Sayed, 2008; Jain *et al.*, 2007). This effect complicates the detection of very small distance changes under conditions of simultaneous distance and refractive index fluctuations and makes careful control experiments necessary. As a rule of thumb, spectral shifts with magnitudes below 2 nm should be interpreted with great caution until suitable control experiments (see, e.g., Section 7) have been performed. To unambiguously discriminate between deformation of the biological tether and typically less interesting refractive index changes, additional information besides resonance wavelength or scattering cross section is desirable. As already mentioned, polarization sensitive detection facilitates separation of structural changes from refractive index changes (Wang and Reinhard, 2009). Rotations of the long dimer axis change the polarization of the scattered light and can easily be detected, providing for an additional indicator of structural changes in discrete particle dimers. The polarization anisotropy, the difference of the scattering cross sections along the long and short dimer axes divided by the total scattering cross section, depends on interparticle separation (Grecco and Martinez, 2006). Measuring the polarization anisotropy is thus an alternative approach for measuring distances and distance changes. Simulations in the dipolar coupling limit show that the polarization anisotropy is less sensitive toward refractive index than the resonance wavelength (Wang and Reinhard, 2009).

Another consideration for the quantitative analysis of plasmon ruler experiments results from the intrinsic heterogeneity of nanoparticle probes. The resonance wavelength of individual particles depends on the particle size and even more sensitively on the particle shape (Halas, 2005; Hu et al., 2006; Kelly et al., 2003). As a consequence, the plasmon resonance in individual plasmon rulers and the distance-dependent plasmon coupling efficiency can vary from one dimer to the next. One key challenge for further improvement of plasmon ruler technology for quantitative distance measurements will be to refine colloid synthesis and postsynthetic processing steps to obtain nanoparticles with extremely narrow size and shape distributions.

Clearly, refinement of plasmon ruler synthesis will also be welcome. The plasmon rulers used thus far were typically obtained through DNA or RNA annealing that did not include a separation step to eliminate dimers with multiple tethers. Cleavage experiments in which more than 90% of plasmon rulers were cleaved in a single step give credence to the notion that, under the chosen assembly conditions described above, the nanoparticles are connected by a single tether molecule. However, rational approaches to the synthesis of particles with defined numbers of tethers have obvious advantages. High-performance liquid chromatography (HPLC) can be used to separate 20 nm gold nanoparticles with different numbers of DNA molecules (Claridge et al., 2008). The resulting purified DNA functionalized nanoparticles are good inputs for plasmon ruler experiments.

Finally, plasmon ruler applications will also benefit from new detection approaches, such as simple ratiometric schemes (Rong et al., 2008). In this scheme the scattering intensities of individual particles are simultaneously monitored in two color channels. One wavelength is chosen to be on resonance with the particle plasmon (or slightly blue-shifted) and the other detection channel is red-shifted. This combination makes spectral red-shifts detectable as changes in the intensity ratio of the two color channels.

REFERENCES

Claridge, S. A., Liang, H. Y. W., Basu, S. R., Frechet, J. M. J., and Alivisatos, A. P. (2008). Isolation of discrete nanoparticle—DNA conjugates for plasmonic applications. *Nano Lett.* **8**(4), 1202–1206.

Elghanian, R., Storhoff, J. J., Mucic, R. C., Letsinger, R. L., and Mirkin, C. A. (1997). Selective colorimetric detection of polynucleotides based on the distance-dependent optical properties of gold nanoparticles. *Science* **277**(5329), 1078–1081.

Fromm, D. P., Sundaramurthy, A., Schuck, P. J., Kino, G., and Moerner, W. E. (2004). Gap-dependent optical coupling of single "Bowtie" nanoantennas resonant in the visible. *Nano Lett.* **4**(5), 957–961.

Grecco, H. E., and Martinez, O. E. (2006). Distance and orientation measurement in the nanometric scale based on polarization anisotropy of metallic dimers. *Opt. Express* **14**(19), 8716–8721.

Ha, T., Enderle, T., Ogletree, D. F., Chemla, D. S., Selvin, P. R., and Weiss, S. (1996). Probing the interaction between two single molecules: Fluorescence resonance energy transfer between a single donor and a single acceptor. *Proc. Natl. Acad. Sci. USA* **93**(13), 6264–6268.

Ha, T., Ting, A. Y., Liang, J., Caldwell, W. B., Deniz, A. A., Chemla, D. S., Schultz, P. G., and Weiss, S. (1999). Single-molecule fluorescence spectroscopy of enzyme conformational dynamics and cleavage mechanism. *Proc. Natl. Acad. Sci. USA* **96**(3), 893–898.

Halas, N. (2005). Playing with plasmons. Tuning the optical resonant properties of metallic nanoshells. *MRS Bull.* **30**, 362–367.

Hammann, C., Hormes, R., Sczakiel, G., and Tabler, M. (1997). A spermidine-induced conformational change of long-armed hammerhead ribozymes: Ionic requirements for fast cleavage kinetics. *Nucleic Acids Res.* **25**(23), 4715–4722.

Haynes, C. L., McFarland, A. D., Zhao, L. L., Van Duyne, R. P., Schatz, G. C., Gunnarsson, L., Prikulis, J., Kasemo, B., and Kall, M. (2003). Nanoparticle optics: The importance of radiative dipole coupling in two-dimensional nanoparticle arrays. *J. Phys. Chem. B* **107**(30), 7337–7342.

Hu, M., Chen, J. Y., Li, Z. Y., Au, L., Hartland, G. V., Li, X. D., Marquez, M., and Xia, Y. N. (2006). Gold nanostructures: Engineering their plasmonic properties for biomedical applications. *Chem. Soc. Rev.* **35**(11), 1084–1094.

Jain, P. K., and el-Sayed, M. A. (2008). Noble metal nanoparticle pairs: Effect of medium for enhanced nanosensing. *Nano Lett.* **8**, 4347.

Jain, P. K., Huang, W. Y., and El-Sayed, M. A. (2007). On the universal scaling behavior of the distance decay of plasmon coupling in metal nanoparticle pairs: A plasmon ruler equation. *Nano Lett.* **7**(7), 2080–2088.

Kelly, K. L., Coronado, E., Zhao, L. L., and Schatz, G. C. (2003). The optical properties of metal nanoparticles: The influence of size, shape, and dielectric environment. *J. Phys. Chem. B* **107**(3), 668–677.

Koculi, E., Lee, N. K., Thirumalai, D., and Woodson, S. A. (2004). Folding of the tetrahymena ribozyme by polyamines: Importance of counterion valence and size. *J. Mol. Biol.* **341**(1), 27–36.

Kreibig, U., and Vollmer, M. (1995). Optical Properties of Metal Clusters. Springer-Verlag, Berlinpp. 13–193.

Lal, S., Grady, N. K., Goodrich, G. P., and Halas, N. J. (2005). Profiling the near field of a plasmonic nanoparticle with Raman-based molecular rulers. *Nano Lett.* **6**(10), 2338–2343.

Lassiter, J. B., Aizpurua, J., Hernandez, L. I., Brandl, D. W., Romero, I., Lal, S., Hafner, J., Nordlander, H. P., and Halas, N. J. (2008). Close encounters between two nanoshells. *Nano Lett.* **8**(4), 1212–1218.

Maier, S. A., Friedman, M. D., Barclay, P. E., and Painter, O. (2005). Experimental demonstration of fiber-accessible metal nanoparticle plasmon waveguides for planar energy guiding and sensing. *Appl. Phys. Lett.* **86**(7), 071103.

Mastroianni, A. J., Sivak, D. A., Geissler, P. L., and Alivisatos, A. P. (2009). Probing the conformational distributions of subpersistence length DNA. *Biophys. J.* **97**(5), 1408–1417.

Mathew-Fenn, R. S., Das, R., and Harbury, P. A. (2008). Remeasuring the double helix. *Science* **322**(5900), 446–449.

Morgan, M. A., Okamoto, K., English, D. S., and Kahn, J. D. (2005). Single-molecule FRET studies of Lac repressor-DNA loop population distributions. *Biophys. J.* **88**(1), 382a.

Prodan, E., Radloff, C., Halas, N. J., and Nordlander, P. (2003). A hybridization model for the plasmon response of complex nanostructures. *Science* **302**(5644), 419–422.

Rasnik, I., Myong, S., and Ha, T. (2006). Unraveling helicase mechanisms one molecule at a time. *Nucleic Acids Res.* **34**(15), 4225–4231.

Rechberger, W., Hohenau, A., Leitner, A., Krenn, J. R., Lamprecht, B., and Aussenegg, F. R. (2003). Optical properties of two interacting gold nanoparticles. *Opt. Commun.* **220**(1–3), 137–141.

Reinhard, B. M., Siu, M., Agarwal, H., Alivisatos, A. P., and Liphardt, J. (2005). Calibration of dynamic molecular rulers based on plasmon coupling between gold nanoparticles. *Nano Lett.* **5**(11), 2246–2252.

Reinhard, B. M., Sheikholeslami, S., Mastroianni, A., Alivisatos, A. P., and Liphardt, J. (2007). Use of plasmon coupling to reveal the dynamics of DNA bending and cleavage by single EcoRV restriction enzymes. *Proc. Natl. Acad. Sci. USA* **104**(8), 2667–2672.

Rong, G., Wang, H., Skewis, L. R., and Reinhard, B. M. (2008). Resolving subdiffraction limit encounters in nanoparticle tacking using live cell plasmon coupling microscopy. *Nano Lett.* **8**, 3386.

Skewis, L. R., and Reinhard, B. M. (2008). Spermidine modulated ribonuclease activity probed by RNA plasmon rulers. *Nano Lett.* **8**(1), 214–220.

Smiley, R. D., Zhuang, Z. H., Benkovic, S. J., and Hammes, G. G. (2006). Single-molecule investigation of the T4 bacteriophage DNA polymerase holoenzyme: Multiple pathways of holoenzyme formation. *Biochemistry* **45**(26), 7990–7997.

Sonnichsen, C., Geier, S., Hecker, N. E., von Plessen, G., Feldmann, J., Ditlbacher, H., Lamprecht, B., Krenn, J. R., Aussenegg, F. R., Chan, V. Z. H., Spatz, J. P., and Moller, M. (2000). Spectroscopy of single metallic nanoparticles using total internal reflection microscopy. *Appl. Phys. Lett.* **77**(19), 2949–2951.

Sonnichsen, C., Reinhard, B. M., Liphardt, J., and Alivisatos, A. P. (2005). A molecular ruler based on plasmon coupling of single gold and silver nanoparticles. *Nat. Biotechnol.* **23**(6), 741–745.

Su, K. H., Wei, Q. H., Zhang, X., Mock, J. J., Smith, D. R., and Schultz, S. (2003). Interparticle coupling effects on plasmon resonances of nanogold particles. *Nano Lett.* **3**(8), 1087–1090.

Tabor, C. W., and Tabor, H. (1984). Polyamines. *Annu. Rev. Biochem.* **53**, 749–790.

Tomishige, M., Stuurman, N., and Vale, R. (2006). Single-molecule observations of neck linker conformational changes in the kinesin motor protein. *Nat. Struct. Mol. Biol.* **13**(10), 887–894.

van Oijen, A. M., Blainey, P. C., Crampton, D. J., Richardson, C. C., Ellenberger, T., and Xie, X. S. (2003). Single-molecule kinetics of lambda exonuclease reveal base dependence and dynamic disorder. *Science* **301**, 1235–1238.

Wang, H., and Reinhard, B. M. (2009). Monitoring simultaneous distance and orientation changes in discrete dimers of DNA linked gold nanoparticles. *J. Phys. Chem. C* (in press).

Wei, Q. H., Su, K. H., Durant, S., and Zhang, X. (2004). Plasmon resonance of finite one-dimensional Au nanoparticle chains. *Nano Lett.* **4**(6), 1067–1071.

Xie, X. S., and Lu, H. P. (1999). Single-molecule enzymology. *J. Biol. Chem.* **274,** 15967–15970.

Yang, L., Yan, B., and Reinhard, B. M. (2008). Correlated optical spectroscopy and transmission electron microscopy of individual hollow nanoparticles and their dimers. *J. Phys. Chem. C* **112,** 15989–15996.

Yun, C. S., Javier, A., Jennings, T., Fisher, M., Hira, S., Peterson, S., Hopkins, B., Reich, N. O., and Strouse, G. F. (2005). Nanometal surface energy transfer in optical rulers, breaking the FRET barrier. *J. Am. Chem. Soc.* **127**(9), 3115–3119.

Zhuang, X. W., Kim, H., Pereira, M. J. B., Babcock, H. P., Walter, N. G., and Chu, S. (2002). Correlating structural dynamics and function in single ribozyme molecules. *Science* **296**(5572), 1473–1476.

CHAPTER NINE

QUANTITATIVE ANALYSIS OF DNA-LOOPING KINETICS FROM TETHERED PARTICLE MOTION EXPERIMENTS

Carlo Manzo[1] and Laura Finzi

Contents

1. Introduction	200
2. Change-Point Algorithm	201
3. Data Clustering and Expectation-Maximization Algorithm	203
4. Adaptation of the Method to the Case of TPM Data Analysis	204
5. Performance of the Method	209
6. Comparison with the Threshold Method	212
7. Application to TPM Experiments: CI-Induced Looping in λ-DNA	215
8. Conclusions	217
Acknowledgments	218
References	218

Abstract

In this chapter we show the application of a maximum-likelihood-based method to the reconstruction of DNA-looping single-molecule time traces from tethered particle motion experiments. The method does not require time filtering of the data and improves the time resolution by an order of magnitude with respect to the threshold-crossing approach. Moreover, it is not based on presumed kinetic models, overcoming the limitations of other approaches proposed previously, and allowing its applications to mechanisms with complex kinetic schemes. Numerical simulations have been used to test the performances of this analysis over a wide range of time scales. We have then applied this method to determine the looping kinetics of a well-known DNA-looping protein, the λ-repressor.

Physics Department, Emory University, Atlanta, Georgia, USA
[1] Current address: BioNanoPhotonics Lab, Institut de Bioenginyeria de Catalunya (IBEC), Parc Científic de Barcelona, Barcelona, Spain

1. INTRODUCTION

Transcription regulation of many genes occurs via DNA looping, which brings into close proximity proteins bound at distant sites along the double helix. Looping produces shortening of the DNA molecule and activation or repression (depending on the gene and proteins involved) of promoters adjacent or within the loop region.

By means of the tethered particle motion (TPM) technique, changes in the conformation of DNA molecules can be indirectly monitored *in vitro* at the single-molecule level, by observing the diffusion of a submicron-sized bead tethered to the surface of a microscope coverslip by single DNA molecules (Finzi and Dunlap, 2003; Finzi and Gelles, 1995; Gelles *et al.*, 1995; Pouget *et al.*, 2004; van den Broek *et al.*, 2006; Vanzi *et al.*, 2006; Yin *et al.*, 1994; Zurla *et al.*, 2006, 2007, 2009). In a typical TPM experiment, the molecule is attached with one end to a glass slide and with the other to a submicron-sized particle. As schematically depicted in Fig. 9.1, the measurement of the particle position is used to determine changes in DNA conformation, such as looping, due to specific protein binding.

Due to the Brownian diffusion of the particle and the overlap between the looped and unlooped distributions, the determination of the dynamic changes usually requires time filtering of the raw data, significantly impacting the measurement time resolution and the reliability of the determination of the kinetic constants. Methods have been proposed to either correct for such a drawback (Colquhoun and Sigworth, 1983; van den Broek *et al.*, 2006; Vanzi *et al.*, 2006) or determine the kinetic constants from the raw data (Beausang *et al.*, 2007a,c; Qin *et al.*, 2000). Nevertheless, these approaches require the knowledge of the kinetic mechanism of the reaction being considered and their application is limited to fairly simple reaction schemes.

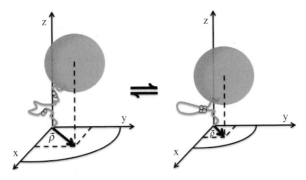

Figure 9.1 Schematic representation of the looping mechanism. Proteins bound at distant sites along the DNA can interact establishing a DNA loop. The loop shortens the DNA, reducing the range of diffusion of the tethered bead.

In this chapter we describe the successful application to TPM experiments of a maximum-likelihood method, previously used for the detection of changes in diffusion coefficients (Montiel et al., 2006) of a freely diffusing particle and in the intensity of fluorescence resonance energy transfer signals (Watkins and Yang, 2005). As tested by analysis of simulated data, this method allows for the reconstruction of the looping kinetics without time filtering at increased time resolution (~200 ms) with respect to the general half-amplitude threshold (HAT) approach (Colquhoun and Sigworth, 1983; van den Broek et al., 2006; Vanzi et al., 2006), which is customarily used to analyze this kind of experimental data.

2. Change-Point Algorithm

The change-point (CP) algorithm is a maximum-likelihood, ratio-type statistical approach for testing a sequence of observations for no change in a given parameter against a possible change, while other parameters remain constant (Gombay and Horvath, 1996; Watkins and Yang, 2005). Here, we discuss its application to the time series of a scalar random variable x_i (which represents the observable of a measurement) with i being a discrete time index. We assume that the probability distribution of the variable is $f(x_i; \sigma)$, where σ is a one-dimensional parameter. The extension of this problem to higher dimensionality cases for both the random variable and the distribution parameter can be found in Gombay and Horvath (1996).

The log-likelihood function for observing a set of N values of x_i is calculated as

$$g(x_i; \sigma) = \log\left(\prod_{i=1}^{N} f(x_i; \sigma)\right) = \sum_{i=1}^{N} \log(f(x_i; \sigma)) \qquad (9.1)$$

By the maximization of the log-likelihood function, the maximum-likelihood estimator $\hat{\sigma}$ can be derived.

To assess the presence of a CP in the parameter σ for a given index k, the null hypothesis,

$$H_0 : \sigma(1) = \cdots = \sigma(k) = \sigma(k+1) = \cdots = \sigma(N), \qquad (9.2)$$

must be compared with the CP hypothesis

$$H_{\mathrm{CP}} : \sigma(1) = \cdots = \sigma(k) = \sigma_1 \neq \sigma_2 = \sigma(k+1) = \cdots = \sigma(N). \qquad (9.3)$$

The test between the two hypotheses can be executed via the calculation of the log-likelihood ratio:

$$R(k) = \log\left(\frac{f(x_{1,\ldots,k};\hat{\sigma}_1)f(x_{k+1,\ldots,N};\hat{\sigma}_2)}{f(x_{1,\ldots,N};\hat{\sigma})}\right) \quad (9.4)$$
$$= g(x_{1,\ldots,k};\hat{\sigma}_1) + g(x_{k+1,\ldots,N};\hat{\sigma}_2) - g(x_{1,\ldots,N};\hat{\sigma}),$$

which represents a measure of the likelihood to have a CP at k.

Since in our problem the location of the CP is unknown, we first need to calculate the position at which the a CP is most likely to occur. This can be easily calculated as the index $i^* = \arg\{\max_{1<k<N}(R(k))\}$.

The problem, then, is the quantification of how large has $R(i^\star)$ to be in order for the hypothesis H_0 to be rejected with a given level of confidence or, in other words, to establish a threshold which allows assessment of the presence of the CP with a known probability.

If we define

$$Z_N = \max_{1<k<N}(2R(k)) = 2R(i^*), \quad (9.5)$$

the problem of the CP identification can be reformulated as

$$\begin{cases} Z_N \geq C_{1-\alpha} \rightarrow H_0 \text{ is rejected,} \\ Z_N < C_{1-\alpha} \rightarrow H_0 \text{ is accepted,} \end{cases} \quad (9.6)$$

where $C_{1-\alpha}$ is intrinsically defined by

$$\text{Probability}(Z_N \geq C_{1-\alpha}) = 1 - \alpha, \quad (9.7)$$

with α being the probability of having a false-positive CP (Type-I error).

The calculation of $C_{1-\alpha}$ requires the knowledge of the distribution of Z_N. Although this distribution has not been calculated in closed form, its approximations and their rate of convergence have been extensively studied (Gombay and Horvath, 1996 and references therein).

In particular, it has been found that the limiting distribution of Z_N can be approximated by the distribution of another random variable, which has been derived in closed form (Gombay and Horvath, 1996). To test for a CP in the value of a one-dimensional parameter, this distribution is given by (Vostrikova, 1981):

$$\frac{\sqrt{C_{1-\alpha}}\exp(-C_{1-\alpha}/2)}{\sqrt{2}\Gamma(1/2)}\left\{T - \frac{1}{C_{1-\alpha}}T + \frac{4}{C_{1-\alpha}}\right\} = 1 - \alpha \quad (9.8)$$

where

$$T = \log\left(\frac{(1-h)^2}{h^2}\right), \quad h = \frac{(\log N)^{3/2}}{N}, \quad \text{and } \Gamma \text{ is the gamma function.}$$

Numerical solution of the equations above allows determination of the asymptotic critical values for type-I error rates. Although these values are a conservative estimate for $C_{1-\alpha}$, in cases where the exact knowledge of the type-I error is not strictly required, their use offers the advantage of a fast calculation, while otherwise the calculation of the critical region needs to be performed numerically (Serge *et al.*, 2008; Watkins and Yang, 2005).

3. DATA CLUSTERING AND EXPECTATION-MAXIMIZATION ALGORITHM

Once the CPs have been determined by the CP algorithm described in Section 2, in the many cases where the changes in the parameter σ occur among several levels, it is necessary to refine the previous analysis by clustering the CP regions in states. Under the hypothesis that the number of states S (i.e., the levels of the parameter σ) is known, it is possible to perform this refinement by means of hierarchical clustering of the data (Fraley and Raftery, 2002).

Such an approach consists in defining a "distance" between the CP regions. This distance is then calculated between every pair of regions and, according to its value, the regions are grouped in a hierarchical tree. After this classification, the data can be partitioned into the S states by means of a grouping criterion. Nevertheless, this kind of clustering is highly sensitive to initial conditions. To overcome this problem, Watkins and Yang (2005), following Fraley and Raftery (1998, 2002), proposed to use the result of this procedure as initial guess for more advanced analysis based on an expectation-maximization (EM) algorithm (Dempster *et al.*, 1977; Fraley and Raftery, 1998, 2002).

This can be understood considering the following. Let us suppose that, from the CP analysis, M CPs are detected. This results in having $M - 1$ CP regions, each with a given maximum-likelihood estimator $\hat{\sigma}_m$. For the hierarchical clustering of these $M - 1$ regions in S state, consistently with Eq. (9.4), a "metric-like" function between two CP regions can be defined as

$$d(m_1, m_2) = g(\{x_{m1}\}; \hat{\sigma}_{m1}) + g(\{x_{m2}\}; \hat{\sigma}_{m2}) - g(\{x_{m1} \cup x_{m2}\}; \hat{\sigma}_{m1 \cup m2}), \tag{9.9}$$

where $m_1, m_2 = 1, \ldots, M - 1$ (Scott and Symons, 1971). The so-defined matrix represents indeed a "distance" between two regions in the maximum-likelihood sense.

Recursively, the two regions having the smallest "distance" are grouped until S clusters are finally formed. According to this procedure, a matrix $p(m, s)$ with $m = 1, \ldots, M - 1$ and $s = 1, \ldots, S$ identifies whether the mth region has been assigned to the sth cluster ($p(m, s) = 1$) or not ($p(m, s) = 0$).

Once the hierarchical clustering has provided the initial conditions, the clustering refinement proceeds through the EM routine. The two steps of this algorithm consist in an expectation step, in which the matrix $p(m, s)$ is updated on the basis of the parameters calculated in the previous step, and a maximization step in which the total likelihood function is maximized. The two steps are repeated until $p(m, s)$ converges.

The total log-likelihood function is given by

$$L = \sum_{m=1}^{M-1} \sum_{s=1}^{S} p(m, s) \log\{p_s g(\{x_m\}; \sigma_s)\}, \qquad (9.10)$$

where p_s represents the relative weight of each state in the mixture.

4. ADAPTATION OF THE METHOD TO THE CASE OF TPM DATA ANALYSIS

The output of a TPM measurement is a time series of the center position ($x(t)$, $y(t)$) of the tethered bead (Zurla et al., 2006). Since the probability distribution of $x(t)$ and $y(t)$ depends on the DNA tether length (Nelson et al., 2006), changes in such distribution observed in the presence of a DNA-looping protein can be indirectly related to changes in the DNA-looping state, providing the possibility to characterize the dynamics of DNA loop formation and breakdown in vitro (Fig. 9.2).

The CP-EM method described in Section 3 was applied to identify changes in the looping state of single DNA molecules. A Gaussian distribution was assumed for the probability density function, pdf, of the time series of $x(t)$ and $y(t)$:

$$f_{TRM}(x_i; \sigma_x) = \frac{1}{\sqrt{2\pi\sigma_x^2}} \exp\left\{-\frac{x_i^2}{2\sigma_x^2}\right\}. \qquad (9.11)$$

It must be noted that the pdf for $x(t)$ and $y(t)$ (Fig. 9.2) deviates from a Gaussian distribution, as expected from the Worm-Like Chain model in the entropic regime (Qian, 2000; Qian and Elson, 1999), because the finite size

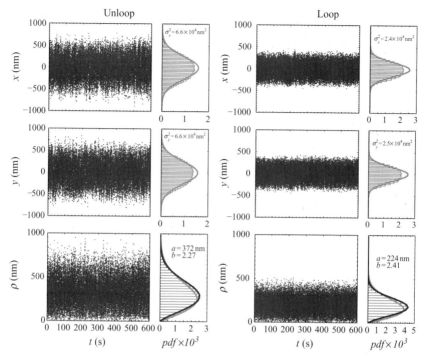

Figure 9.2 Time traces of $x(t)$, $y(t)$, and $\rho(t)$ for a 240-nm-radius bead tethered to an unlooped (length 3477 bp) and looped (effective length 1160 bp) DNA molecule, and corresponding probability distributions. The distribution histogram of $x(t)$ and $y(t)$ deviates from a Gaussian behavior (red curves) due to the excluded-volume effect given by the finite size of the bead. Similarly, the histogram of $\rho(t)$ shows a discrepancy with respect to the Rayleigh distribution (red curve) but is instead well described by a two-parameter Weibull distribution (in black). Histograms have been normalized to the bin width w ($w = 30$ nm) and to the total number of data points n ($n = 30{,}000$).

of the bead causes an excluded-volume effect (Nelson et al., 2006; Segall et al., 2006). The exact distribution of $x(t)$ and $y(t)$ has been calculated through Monte Carlo integration of the Boltzmann distribution (Segall et al., 2006), but an analytical expression for such a distribution has not been derived. Although in principle a parametric expression for the pdf can be calculated from the Pearson system of distributions, based on the knowledge of its first four moments, and the calculation of the maximum-likelihood estimator can be numerically performed via maximization of the likelihood function, this approach critically affects the computational time of the CP determination. Therefore, at this stage, an analytical calculation based on the Gaussian approximation is preferable. By contrast, a numerical approach was used in the EM step, where the shape of the distribution is more critical for the determination of the looping dynamics.

According to the previous equation, the log-likelihood of observing N displacement is given by

$$g_{\text{TPM}}(x_i; \sigma) = \log\left[\prod_{i=1}^{N} \frac{1}{\sqrt{2\pi\sigma_x^2}} \exp\left\{-\frac{x_i^2}{2\sigma_x^2}\right\}\right]$$

$$= -\frac{N}{2}\log(2\pi\sigma_x^2) - \frac{1}{2\sigma_x^2}\sum_{i=1}^{N} x_i^2. \qquad (9.12)$$

Maximization of the latter equation provides the expression for the maximum-likelihood estimator $\hat{\sigma}_x^2 = (1/N)\sum_{i=1}^{N} x_i^2$.

The log-likelihood ratio for the detection of a CP is then calculated as

$$R_{\text{TPM}}^{(x)}(k) = -\frac{k}{2}\log\left(\sum_{i=1}^{k} x_i^2\right) - \frac{N-k}{2}\log\left(\sum_{i=k+1}^{N} x_i^2\right)$$
$$+\frac{N}{2}\log\left(\sum_{i=1}^{N} x_i^2\right) + \cdots + \frac{k}{2}\log(k) + \frac{N-k}{2}\log(N-k) - \frac{N}{2}\log(N). \qquad (9.13)$$

The search for CPs in a TPM measurement is performed using a segmentation algorithm, similarly as described in Watkins and Yang (2005). The time series $x(t)$ and $y(t)$ are segmented in traces of $N = 1000$ points (200 s) each, with a partial overlap of 500 points. As a matter of fact, the error in the detection of a CP depends on its position and has maxima at the edges of the segment and reaches its minimum at the middle point (Watkins and Yang, 2005). The overlap between consecutive fragments thus ensures a nearly constant error rate for all CPs. After the segmentation, the maximum of $R_{\text{TPM}}(k) = R_{\text{TPM}}^{(x)}(k) + R_{\text{TPM}}^{(y)}(k)$ is calculated over the segment and, if it is greater or equal to $C_{1-\alpha}^{\text{sim}}$, the associated index is identified as a CP.

The CP determination routine is schematically represented in Fig. 9.3. This procedure is then repeated over the two distinct fragments divided by the determined CP and recursively applied for each detected CP, until no further CPs are found. After the procedure is completed over a segment, each CP position is refined by the reapplication of the algorithm mentioned above to a segment defined by its two nearest-neighboring CPs.

Calculation of the critical regions $C_{1-\alpha}^{\text{sim}}$ is obtained via numerical simulations. For each k ranging from 2 to 1000, 10,000 traces presenting no CP were simulated for $x(t)$ and $y(t)$ from a Pearson-type distribution having the same first four moments as the experimental data. From the traces, the distribution of $\max(R_{\text{TPM}}(k))$ was then calculated, allowing the determination of the critical region for several levels of confidence (Fig. 9.4).

A similar calculation, based on simulated traces presenting a single CP, allows to obtain the probability of missed events (Fig. 9.4, dot-dashed lines).

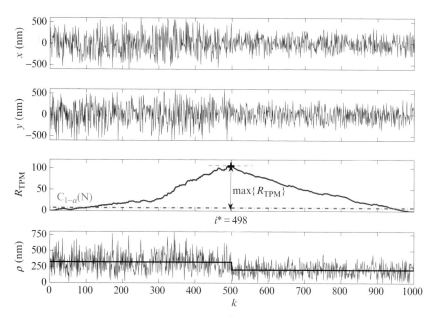

Figure 9.3 Change point determination. Simulated traces for a DNA molecule showing a looping event for $k = 500$ ($t = 10$ s). The time dependence of the bead center position $x(t)$ and $y(t)$ is reported in the two upper panels, respectively. In the third panel, the log-likelihood ratio $R_{TPM}(k)$ is plotted as a function of k. The gray dot-dashed line represents the critical value ($\alpha = 0.99$, $N = 1000$) for the change point detection. The maximum of R_{TPM} is found for $k = i^* = 498$. The bottom panel shows the plot of the radial distance of the bead center $\rho(t) = \sqrt{x^2(t) + y^2(t)}$. The "true" trace is also displayed (black line) for comparison.

The CP determination is then followed by the EM routine, schematically represented in Fig. 9.5. As previously stated, at this step the exact knowledge or a good approximation of the shape of the experimental *pdf* is highly critical for proper clustering of the CP regions. The deviation of the distribution of $x(t)$ and $y(t)$ from a normal *pdf* does not allow us to use the Gaussian approximation. Nevertheless, through the calculation of its first four moments, the *pdf* of $x(t)$ and $y(t)$ can be associated with a Pearson type II distribution and thus approximated to a scaled symmetrical Beta distribution. The maximization of the associated likelihood function can, in principle, be computed numerically. Unfortunately, since the Beta distribution is defined on the interval [0, 1] and the scaling factor is used as one of the parameters of the maximization, the numerical optimization is made nontrivial by the fact that data points can lie outside the range of definition of the function.

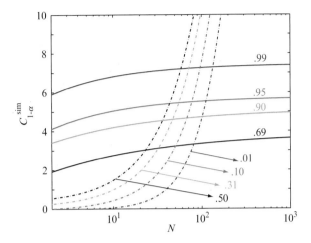

Figure 9.4 Critical regions $C_{1-\alpha}^{sim}$ (continuous lines) obtained by numerical simulations as a function of N for several levels of the confidence interval α for the false-positive event determination. The dependence on N of the critical value for the probability of missed events is reported as dot-dashed lines. Note that at a 31% ($\alpha = 0.69$) confidence level for false-positive event, the probability of missing the detection of a change point occurring in a region of $N \geq 50$ data points ($t \geq 1$ s) is smaller than 10%.

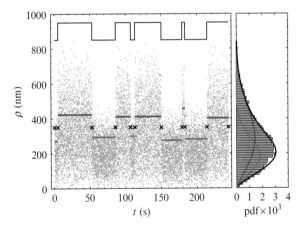

Figure 9.5 Expectation-maximization clustering. *Left panel*: After the change points have been determined (black crosses), the regions between two adjacent change points are clustered into two groups through the maximization of the total log-likelihood function for a Weibull *pdf*. The green and red lines represent the average of the data points in the corresponding change point region. The color identifies the DNA conformation corresponding to the two groups (red unlooped and green looped). The "true" trace (black line) is also reported shifted and scaled for comparison. *Right panel*: Histogram of the time trace and results of the expectation-maximization step. The red and green curves refer to the retrieved Weibull *pdf* for the unlooped and looped state, respectively. The sum of the two *probability density functions* (black line) shows an excellent agreement with the data histogram. (See Color Insert.)

Although the probability distribution of $\rho(t) = \sqrt{x^2(t) + y^2(t)}$ is also a Beta-like distribution (Pearson type I distribution), it can be well approximated by the two-parameter Weibull function:

$$f_{\text{WBL}}(\rho; a, b) = ba^{-b}\rho^{b-1} \exp\left\{-\left(\frac{\rho}{a}\right)^b\right\}, \qquad (9.14)$$

which is defined over the positive real axis and for which the numerical maximization of the likelihood function can be easily performed.

It is worth noting that, for $b = 2$, the Weibull pdf reduces to the well-known Rayleigh function, the distribution function for the modulus of a vector having its two orthogonal components independent and normally distributed. For a Rayleigh pdf the maximum-likelihood estimator is moreover given by $\hat{a} = \sqrt{\sum_{i=1}^{N} \rho_i^2/N}$, which is the quantity usually calculated over a given window size in the threshold method (Nelson et al., 2006; Zurla et al., 2006).

Hierarchical clustering of the CP regions in $S = 2$ states (unlooped and looped) is thus carried out through the recursive calculation of the metric-like matrix:

$$\begin{aligned}d(m_1, m_2) &= g_{\text{WBL}}(\{\rho_{m1}\}; \hat{a}_{m1}, \hat{b}_{m1}) + g_{\text{WBL}}(\{\rho_{m2}\}; \hat{a}_{m2}, \hat{b}_{m2}) \\ &+ \cdots - g_{\text{WBL}}(\{\rho_{m1} \cup \rho_{m2}\}; \hat{a}_{m1\cup m2}, \hat{b}_{m1\cup m2}),\end{aligned} \qquad (9.15)$$

where $g_{\text{WBL}}(\rho; a, b)$ is now defined as

$$g_{\text{WBL}}(\rho; a, b) = \log\left(\prod_{i=1}^{N} ba^{-b}\rho_i^{b-1} \exp\left\{-\left(\frac{\rho_i}{a}\right)^b\right\}\right) \qquad (9.16)$$

and the maximum-likelihood estimators $\hat{a}$ and $\hat{b}$ are obtained through numerical maximization.

The results of the clustering are then used as a first guess for the EM step, which proceeds in this case to the calculation of the matrix $p(m, s)$ via the numerical maximization of the global log-likelihood function:

$$L_{\text{WBL}} = \sum_{m=1}^{M-1} \sum_{s=1}^{S} p(m, s) \log\{p_s g_{\text{WBL}}(\{\rho_m\}; a_s, b_s)\}. \qquad (9.17)$$

5. PERFORMANCE OF THE METHOD

Although the rate of false-positive identification and the probability of missed events for the detection of a single CP can be easily calculated depending on the position of the CP and length of the trajectory, this is not straightforward when dealing with trajectories showing a series of CPs. In this case, the dependence of the probability of detection and the error on the duration of the CP region are also affected by the probability of detection of the neighboring CPs. Intuitively, the power of detection for a CP region depends on its length, that is, on the time the trace dwells in a given state. On the other hand, the error on the calculation of the dwell times also depends on the accuracy at which the contiguous regions are determined and assigned to the "true" state. In order to estimate how the detection probability varies with the dwell time, we used a simplified approach based on numerical simulations. In particular, traces showing transitions between the looped and unlooped states with randomly distributed dwell times were generated and analyzed as described in Section 4. The ratio between the number of detected (n) and simulated (n_s) dwell times with durations lying in windows of exponentially increasing widths was then plotted and considered as an estimate of the power of detection of the method. Figure 9.6 shows the values of n/n_s obtained at a 31% false-positive detection rate ($C^{sim}_{1-0.69}$) for the loop and the unloop state as a function of the dwell time. Although the curves intersect at a dwell time of ~0.8 s with 60% of the events detected, the plot shows an asymmetry between the detection abilities of the loop and the unloop states. This is most likely due to the EM step during which, because of the large overlapping area between the Weibull *pdf*'s, short-lived regions have higher probability to be attributed to the unlooped distribution. It must be noted that these simulations provide just an estimate of the power of detection. In this case no attempt has been made to determine the confidence level for false-positive and missed-event probabilities, as a more rigorous approach would require. Therefore, the results shown in Fig. 9.6

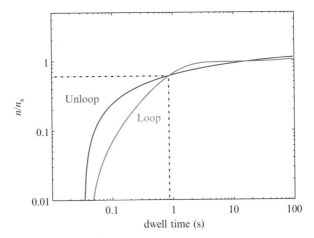

Figure 9.6 Log–log plot of the ratio between the number of detected and the number of simulated change point regions as a function of the dwell time duration.

could be partially biased by some sort of compensation between false-determined and missed events.

To further check the ability of the method to detect DNA transitions between looped and unlooped states in the traces obtained from TPM experiments, time traces were simulated in which the dwell times spent in the two configurations are exponentially distributed with mean lifetimes τ_L and τ_U ranging from 1 s to 1 min. For each case, 20 traces with a duration of 20 min each were generated and analyzed. As already mentioned, actual TPM experimental data show correlation in the bead position that is not present in the simulated trace. The time scale of this correlation is of the order of hundreds of milliseconds for a $\sim$250-nm radius bead tethered to a roughly micron long DNA (Beausang et al., 2007a,b,c). To prevent this correlation from increasing the number of false-positive events shorter than its decay time and inducing a bias in the CP determination, it is necessary to consider only events larger than the correlation time. Although the simulated data do not present this kind of correlation, for sake of consistency only dwell times longer than $T_D = 200$ ms were kept and the mean lifetimes were calculated through the maximization of the likelihood function for the exponential distribution:

$$f = \frac{1}{\tau} \exp\left\{-\frac{t - T_D}{\tau}\right\}.$$

All lifetimes obtained from the fitting are reported on the log–log plot of Fig. 9.7 together with the "true" values τ^{true} as obtained from the simulated traces. The plot shows that although the ability to detect dwell times is affected by the average duration of both states, the method allows for the determination of even very small lifetimes with considerable accuracy. The contour map of the average relative error

$$\Delta(\tau_U^{\text{true}}, \tau_L^{\text{true}}) = \frac{1}{2}\left[\frac{|\tau_U^{\text{true}} - \hat{\tau}_U|}{\tau_U^{\text{true}}} + \frac{|\tau_L^{\text{true}} - \hat{\tau}_L|}{\tau_L^{\text{true}}}\right]$$

shows indeed a flat region with values of $\Delta < 0.2$, with partial degradation of the accuracy at the left and bottom edges. Also in this case an asymmetry between the accuracy in the determination of looped and unlooped lifetimes is observed.

In Fig. 9.8, the histograms of the recovered dwell times for the case $\tau_U = 8$ s, $\tau_L = 4$ s are plotted together with the distribution of the "true" times generated via the simulation (left panels). From the histograms, it is possible to see that the method preserves the exponential shape of the dwell time distribution. As expected, the number of missed events decreased with the dwell times duration, reaching $\sim$10% for values shorter than the mean

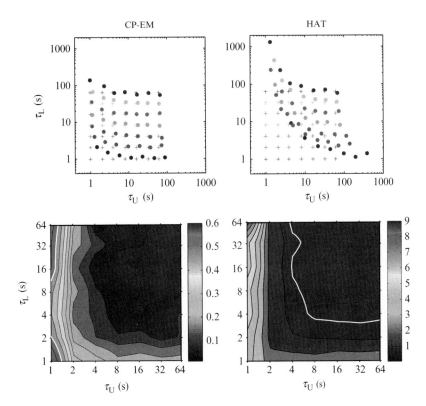

Figure 9.7 Comparison among loop and unloop lifetimes used to generate the simulated time traces (crosses) and the values (filled circles) determined by means of the CP-EM (upper-left panel) and the HAT method (upper-right panel). The lower panels show the contour plots of the average relative error Δ on the loop and unloop lifetimes retrieved using the CP-EM (left) and the HAT (right) method, with respect to the "true" lifetime values used in the simulations. The white line, delimiting the contour region corresponding to $\Delta = 0.5$, is reported in both the plots for comparison.

lifetime and thus allowing a precise determination of the decay constant. Moreover, short-lived missed events do not induce any relevant alteration of the shape of the distribution at longer dwell times through the creation of false long-lived states.

6. COMPARISON WITH THE THRESHOLD METHOD

A common approach to the analysis of TPM data is the HAT method. This approach relies on the time filtering of the ρ^2 time trace with a rectangular or Gaussian filter of given width. This allows the calculation

Quantitative Analysis of DNA-Looping Kinetics

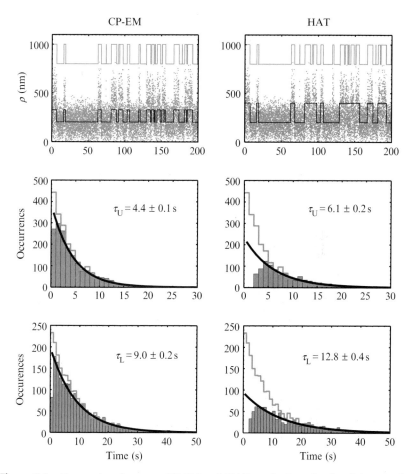

Figure 9.8 Comparison between CP-EM and HAT method on simulated data. In the upper panels a representative part of a simulated trace ($\tau_U = 4$ s, $\tau_L = 8$ s) is shown together with the reconstructed telegraphic-like behavior (black lines) obtained by means of the CP-EM (left) and the HAT (right) method. The "true" signal is also shown (gray lines) scaled and shifted for comparison. The improved CP-EM time resolution allows a better reconstruction of the trajectory, permitting the detection of short dwell times. *Middle panels*: Histograms of the unlooped dwell times as retrieved by the CP-EM (left) and HAT (right) method for a simulated experiment ($\tau_U = 4$ s, $\tau_L = 8$ s). The gray line represents the histogram of the "true" simulated unlooped dwell time and the black line is the exponential fit to the retrieved data. *Bottom panels*: Same plot as above for the looped dwell times. From the histograms it is clear that the CP-EM allows the determination of the mean lifetime for the simulated data with high accuracy with respect to the HAT method, owing to its higher time resolution and the lower number of short-lived missed events.

of the square root of the average signal variance $\sqrt{\langle\rho^2\rangle_W}$ (weighted in the case of Gaussian filter) over the time trace.

In the case of DNA showing looping transitions, this procedure requires that clearly visible steps emerge in the data. The histogram of the filtered data shows indeed a bimodal distribution, which allows defining a threshold as the half amplitude between the locations of the peaks of the distribution. The threshold, as well as the distribution of $\sqrt{\langle\rho^2\rangle_W}$, depends on the filter width. The dwell times are then obtained from the threshold-crossing points of the filtered data. The width of the filter determines the time resolution of the analysis. This is expressed as the dead time, that is, the duration of a loop (or unloop) event that after filtering gives a half-amplitude response (for a rectangular filter, the dead time corresponds to half of the window width) (Colquhoun and Sigworth, 1983). Events with a time duration below the dead time are then neglected in the dwell time calculation. It must be noted that there is no uniform criterion for the choice of an optimal filter. As a rule of thumb, the filter width should be the smallest possible (at least shorter than the minimum lifetime) that still allows for a clear identification of the loop and unloop events from the distribution histogram.

As a consequence, relatively large filter widths must be used ($\sim$4 s), resulting in rather poor time resolution.

Nevertheless, several methods have been proposed to correct for missed events by the use of several filter widths (Vanzi *et al.*, 2006), by bypassing the time-filtering step through the Hidden Markov method (Qin *et al.*, 2000), or by taking into account the diffusion of the bead (Beausang *et al.*, 2007a,c).

Unfortunately, in all of these cases, the knowledge of the kinetic reaction scheme is crucial to obtain the correct value for the rate constant. This makes these methods extremely difficult to apply when dealing with complex kinetic schemes. For the CP-EM method, this information is not required since the dwell time distribution is retrieved without any assumption on the kinetics of the system. Moreover, once the dwell time distribution has been determined with high time resolution, if necessary, other methods (Liao *et al.*, 2007) can be applied to resolve complex kinetics and determine the rate constants.

Our CP-EM is a maximum-likelihood method for the reconstruction of the "true" time traces with high temporal resolution. Its performance was compared with the HAT approach. A direct comparison was performed by applying the HAT method also to the simulated data of Section 5. In the HAT case, we used a rectangular filter of width $W = 4$ s (200 data points) (Nelson *et al.*, 2006; Zurla *et al.*, 2006). In the threshold analysis, only dwell times longer than the dead time (2 s) were retained and the calculation of the mean lifetime was performed through the maximization of the log-likelihood function for an exponential *pdf* on times larger than twice the dead time (Colquhoun and Sakmann, 1981; Colquhoun and Sigworth, 1983).

The results of the analysis performed by the HAT method on the simulated traces are reported in the panels on the right of Figs. 9.7 and 9.8. The calculated lifetimes largely deviate from the "true" values for τs shorter than 10 s and, even for larger τs, the relative error is about one order of magnitude larger than when using the CP-EM (Fig. 9.7). The high cutoff imposed by the time filtering limits the resolution, causing a large number of missed events at short times that in turn induce the detection of false long-lived states (Fig. 9.8), artificially increasing the determined lifetime.

7. Application to TPM Experiments: CI-Induced Looping in λ-DNA

The loop induced in the λ bacteriophage DNA by the λ repressor or CI protein has been recently proposed as the mechanism by which the λ epigenetic switch is regulated (Dodd *et al.*, 2001). Looping strengthens repression of the lytic genes during lysogeny, while simultaneously controlling CI concentration, thus ensuring an efficient switch to lysis, if necessary. The λ loop is achieved via the interaction of up to six CI dimers with two sets of three binding sites separated by 2317 base pairs (bp). According to recent studies (Anderson and Yang, 2008; Dodd *et al.*, 2004; Zurla *et al.*, 2009), the number of CI dimers involved in the loop closure plays a major role in the switch regulation, by affecting the stability of the loop. Therefore, a detailed understanding of the kinetics of the CI-mediated DNA loop formation and breakdown is necessary to understand the mechanism of the switch and its regulatory function.

CP-EM and the HAT analysis was applied to a set of TPM measurements (56 DNA tethers, recording time $\sim$25 min each) performed on fragments of λ-DNA which were 3477 bp long and contained the two triplets of operator sites separated by the wild-type distance of 2317 bp. The tethered beads had a radius of 240 nm, and 20 nM CI protein was used.

In the upper panels of Fig. 9.9, a 100s -long part of a recorded trajectory is reported together with the reconstructed telegraphic-like signal obtained by means of the two methods. Although looping transitions are revealed in both cases, the higher time resolution of the CP-EM method allows for the determination of short-lived events, which are instead missed by the HAT analysis. This obviously affects the determination of the dwell time distribution. The middle panels show the unlooped dwell time histograms. The maximization of the log-likelihood functions for a single exponential

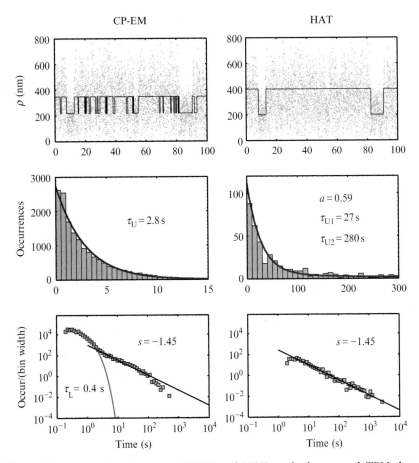

Figure 9.9 Comparison between CP-EM and HAT method on actual TPM data, recorded on λ-DNA in the presence of 20 nM CI protein. In the upper panels a representative part of an experimental time trace is shown together with the reconstructed telegraphic-like behavior (black lines) obtained by means of the CP-EM (left) and the HAT (right) method. *Middle panels*: Histograms of the unlooped dwell times as retrieved by the CP-EM (left) and HAT (right) method for the TPM experiments. The black line is a fit of the retrieved data to a single and a double exponential *pdf* for the CP-EM and the HAT method, respectively. *Bottom panels*: Looped dwell times distribution as retrieved by the CP-EM (left) and HAT (right) method for the TPM experiments. The number of looped dwell times in bin of exponentially increasing size is reported, normalized to the bin width on a log–log scale. The CP-EM method shows a complex kinetics, with a fast decay rate followed by a slow nonexponential tail, whereas the HAT analysis shows a power law distribution.

distribution ($\tau_U = 2.8$ s) fits satisfactorily the data from the CP-EM analysis, whereas for the HAT method, the dwell times are distributed according to a double exponential *pdf* (59% of the events with $\tau_{U1} = 27$ s, 41% with

$\tau_{U2} = 280$ s). The one order of magnitude difference in lifetime is clearly due to the difference in the time resolution of the two methods. Furthermore, the single exponential decay revealed by the CP-EM analysis is consistent with thermodynamic modeling of the λ looping reaction. Since the ΔGs for the binding of CI protein at each of the six sites are known, the probability for a DNA molecule to have at least a pair of cooperatively bound CI dimers for each binding region (minimum requirement for looping) can be calculated (Anderson and Yang, 2008; Dodd *et al.*, 2004; Zurla *et al.*, 2009) to be 98% at 20 nM CI. Since the DNA is almost permanently loaded with at least two dimers per binding region, the occurrence of loop formation is then only regulated by the stochastic rate of encounters between the regions. As a consequence, the loop formation kinetics must be a single exponential with the loop formation rate constant given by the distribution mean lifetime through $k_{LF} = \tau_U^{-1}$.

The situation is quite different for the looped state, in which the determined dwell times span several orders of magnitude. In the bottom panels of Fig. 9.9, the looped dwell time distributions are represented as log–log plots, with the number of occurrences binned with exponentially increasing size and normalized to the bin widths (squares). The HAT analysis shows a power law distribution (straight line on log–log plot), whereas the CP-EM shows a more complex distribution, with a very short-lived population ($\tau_L \approx 0.4$ s) followed by a slow decay, similar to the one observed by the HAT analysis. Analogous distributions have been reported for several ion-channels experiments and many theoretical models associated with complex reaction schemes and high numbers of states (Liebovitch and Sullivan, 1987; Liebovitch *et al.*, 2001; Millhauser *et al.*, 1988; Oswald *et al.*, 1991; Sansom *et al.*, 1989). The quantitative understanding of the observed loop dwell time distribution will be discussed in details in a manuscript in preparation (CM and LF). However, the observed behavior most likely reflects the high complexity of the λ looping kinetics. Such a complexity results from the formation of 32 different looped structures secured by four to six CI dimers (Anderson and Yang, 2008; Zurla *et al.*, 2009), each presumably having a different pathway for loop breakdown.

8. Conclusions

A CP-EM method is presented here for the analysis of DNA-looping time traces and the determination of dwell time distributions from data obtained from TPM measurements. The method was tested on simulated data and its performance is discussed in comparison to more classical analysis methods. Application of the method to experimental TPM data yields information on the looping mechanism with a time resolution an order of magnitude higher than that of the HAT method. Thus, CP-EM improves the performance of TPM as a quantitative single-molecule technique, extending its observable time scale to the lower limit posed by the intrinsic correlation time of the measurement.

ACKNOWLEDGMENTS

The authors are indebted to Dr. Haw Yang for recommending the change-point methods for this problem. The authors also wish to thank Chiara Zurla for providing the TPM data and many useful comments.

This work was supported by the Emory University start-up fund and the NIH (RGM084070A) to L. F.

REFERENCES

Anderson, L. M., and Yang, H. (2008). DNA looping can enhance lysogenic CI transcription in phage lambda. *Proc. Natl. Acad. Sci. USA* **105,** 5827–5832.

Beausang, J. F., Zurla, C., Dunlap, D., Finzi, L., and Nelson, P. C. (2007a). Hidden Markov analysis of tethered particle motion. *Biophys. J.* 417A.

Beausang, J. F., Zurla, C., Finzi, L., Sullivan, L., and Nelson, P. C. (2007b). Elementary simulation of tethered Brownian motion. *Am. J. Phys.* **75,** 520–523.

Beausang, J. F., Zurla, C., Manzo, C., Dunlap, D., Finzi, L., and Nelson, P. C. (2007c). DNA looping kinetics analyzed using diffusive hidden Markov model. *Biophys. J.* **92,** L64–L66.

Colquhoun, D., and Sakmann, B. (1981). Fluctuations in the microsecond time range of the current through single acetylcholine-receptor ion channels. *Nature* **294,** 464–466.

Colquhoun, D., and Sigworth, F. J. (1983). Fitting and Statistical Analysis of Single Channel Recording. (B. Sakmann and E. Neher, eds.), pp. 191–263. Plenum Press, New York.

Dempster, A. P., Laird, N. M., and Rubin, D. B. (1977). Maximum likelihood from incomplete data via EM algorithm. *J.R. Sta. Soc. Series B Method* **39,** 1–38.

Dodd, I. B., Perkins, A. J., Tsemitsidis, D., and Egan, J. B. (2001). Octamerization of lambda CI repressor is needed for effective repression of P-RM and efficient switching from lysogeny. *Genes Dev.* **15,** 3013–3022.

Dodd, I. B., Shearwin, K. E., Perkins, A. J., Burr, T., Hochschild, A., and Egan, J. B. (2004). Cooperativity in long-range gene regulation by the lambda CI repressor. *Genes Dev.* **18,** 344–354.

Finzi, L., and Dunlap, D. (2003). Single-molecule studies of DNA architectural changes induced by regulatory proteins. *Methods Enzymol.* **370,** 369–378.

Finzi, L., and Gelles, J. (1995). Measurement of lactose repressor-mediated loop formation and breakdown in single DNA-molecules. *Science* **267,** 378–380.

Fraley, C., and Raftery, A. E. (1998). How many clusters? Which clustering method? Answer via model-based cluster analysis. *Comp. J.* **41,** 578–588.

Fraley, C., and Raftery, A. E. (2002). Model-based clustering, discriminant analysis, and density estimation. *J. Am. Stat. Assoc.* **97,** 611–631.

Gelles, J., Yin, H., Finzi, L., Wong, O. K., and Landick, R. (1995). Single-molecule kinetic-studies on DNA-transcription and transcriptional regulation. *Biophys. J.* **68,** S73.

Gombay, E., and Horvath, L. J. (1996). On the rate of approximations for maximum likelihood tests in change-point models. *J. Multivar. Anal.* **56**, 120–152.

Liao, J. C., Spudich, J. A., Parker, D., and Delp, S. L. (2007). Extending the absorbing boundary method to fit dwell-time distributions of molecular motors with complex kinetic pathways. *Proc. Natl. Acad. Sci. USA* **104**, 3171–3176.

Liebovitch, L. S., and Sullivan, J. M. (1987). Fractal analysis of a voltage-dependent potassium channel from cultured mouse hippocampal-neurons. *Biophys. J.* **52**, 979–988.

Liebovitch, L. S., Scheurle, D., Rusek, M., and Zochowski, M. (2001). Fractal methods to analyze ion channel kinetics. *Methods* **24**, 359–375.

Millhauser, G. L., Salpeter, E. E., and Oswald, R. E. (1988). Diffusion-models of ion-channel gating and the origin of power law distributions from single-channel recording. *Proc. Natl. Acad. Sci. USA* **85**, 1503–1507.

Montiel, D., Cang, H., and Yang, H. (2006). Quantitative characterization of changes in dynamical behavior for single-particle tracking studies. *J. Phys. Chem. B* **110**, 19763–19770.

Nelson, P. C., Zurla, C., Brogioli, D., Beausang, J. F., Finzi, L., and Dunlap, D. (2006). Tethered particle motion as a diagnostic of DNA tether length. *J. Phys. Chem. B* **110**, 17260–17267.

Oswald, R. E., Millhauser, G. L., and Carter, A. A. (1991). Diffusion model in ion channel gating. *Biophys. J.* **59**, 1136–1142.

Pouget, N., Dennis, C., Turlan, C., Grigoriev, M., Chandler, M., and Salome, L. (2004). Single-particle tracking for DNA tether length monitoring. *Nucleic Acids Res.* **32**, e73.

Qian, H. (2000). A mathematical analysis for the Brownian dynamics of a DNA tether. *J. Math. Biol.* **41**, 331–340.

Qian, H., and Elson, E. L. (1999). Quantitative study of polymer conformation and dynamics by single-particle tracking. *Biophys. J.* **76**, 1598–1605.

Qin, F., Auerbach, A., and Sachs, F. (2000). A direct optimization approach to hidden Markov modeling for single channel kinetics. *Biophys. J.* **79**, 1915–1927.

Sansom, M. S. P., Ball, F. G., Kerry, C. J., McGee, R., Ramsey, R. L., and Usherwood, P. N. R. (1989). Markov, fractal, diffusion, and related models of ion channel gating. *Biophys. J.* **56**, 1229–1243.

Scott, A. J., and Symons, M. J. (1971). Clustering methods based on likelihood ratio criteria. *Biometrics* **27**, 387–397.

Segall, D. E., Nelson, P. C., and Phillips, R. (2006). Volume-exclusion effects in tethered-particle experiments: Bead size matters. *Phys. Rev. Lett.* **96**, 088306.

Serge, A., Bertaux, N., Rigneault, H., and Marguet, D. (2008). Dynamic multiple-target tracing to probe spatiotemporal cartography of cell membranes. *Nat. Methods* **5**, 687–694.

van den Broek, B., Vanzi, F., Normanno, D., Pavone, F. S., and Wuite, G. J. L. (2006). Real-time observation of DNA looping dynamics of type IIE restriction enzymes NaeI and NarI. *Nucleic Acids Res.* **34**, 167–174.

Vanzi, F., Broggio, C., Sacconi, L., and Pavone, F. S. (2006). Lac repressor hinge flexibility and DNA looping: Single molecule kinetics by tethered particle motion. *Nucleic Acids Res.* **34**, 3409–3420.

Vostrikova, L. Y. (1981). Detection of a "disorder" in a Wiener process. *Theory Probab. Appl.* **26**, 356–362.

Watkins, L. P., and Yang, H. (2005). Detection of intensity change points in time-resolved single-molecule measurements. *J. Phys. Chem. B* **109**, 617–628.

Yin, H., Landick, R., and Gelles, J. (1994). Tethered particle motion method for studying transcript elongation by a single RNA-polymerase molecule. *Biophys. J.* **67**, 2468–2478.

Zurla, C., Franzini, A., Galli, G., Dunlap, D. D., Lewis, D. E. A., Adhya, S., and Finzi, L. (2006). Novel tethered particle motion analysis of CI protein-mediated DNA looping in the regulation of bacteriophage lambda. *J. Phys. Condens. Matter* **18,** S225–S234.

Zurla, C., Samuely, T., Bertoni, G., Valle, F., Dietler, G., Finzi, L., and Dunlap, D. D. (2007). Integration host factor alters LacI-induced DNA looping. *Biophys. Chem.* **128,** 245–252.

Zurla, C., Manzo, C., Dunlap, D., Lewis, D. E., Adhya, S., and Finzi, L. (2009). Direct demonstration and quantification of long-range DNA looping by the {lambda} bacteriophage repressor. *Nucleic Acids Res.* **37,** 2789–2795.

CHAPTER TEN

METHODS IN STATISTICAL KINETICS

Jeffrey R. Moffitt,[*] Yann R. Chemla,[†] and Carlos Bustamante[*,‡]

Contents

1. Introduction 222
2. The Formalism of Statistical Kinetics 223
   2.1. From steady-state kinetics to statistical kinetics 223
   2.2. Basic statistics of the cycle completion time 226
   2.3. The "memory-less" enzyme 227
   2.4. Lifetime statistics 229
   2.5. State visitation statistics 230
3. Characterizing Fluctuations 232
   3.1. Fitting distributions 232
   3.2. Calculating moments 235
   3.3. Multiple pathways and multiple steps 236
4. Extracting Mechanistic Constraints from Moments 240
   4.1. The randomness parameter and n_{min} 241
   4.2. Classifying fluctuations 242
   4.3. Mechanistic constraints 244
5. Conclusions and Future Outlook 248
References 255

Abstract

A variety of recent advances in single-molecule methods are now making possible the routine measurement of the distinct catalytic trajectories of individual enzymes. Unlike their bulk counterparts, these measurements directly reveal the fluctuations inherent to enzymatic dynamics, and statistical measures of these fluctuations promise to greatly constrain possible kinetic mechanisms. In this chapter, we discuss a variety of advances, ranging from theoretical to practical, in the new and growing field of statistical kinetics. In particular, we formalize the connection between the hidden fluctuations in the kinetic states that compose a full kinetic cycle and the measured fluctuations in

[*] Department of Physics and Jason L. Choy Laboratory of Single-Molecule Biophysics, University of California, Berkeley, California, USA
[†] Department of Physics and Center for Biophysics and Computational Biology, University of Illinois at Urbana-Champaign, Urbana, Illinois, USA
[‡] Departments of Molecular and Cell Biology, and Chemistry, Howard Hughes Medical Institute, University of California, Berkeley, California, USA

Methods in Enzymology, Volume 475 © 2010 Elsevier Inc.
ISSN 0076-6879, DOI: 10.1016/S0076-6879(10)75010-2 All rights reserved.

the time to complete this cycle. We then discuss the characterization of fluctuations in a fashion that permits kinetic constraints to be easily extracted. When there are multiple observable enzymatic outcomes, we provide the proper way to sort events so as not to bias the final statistics, and we show that these classifications provide a first level of constraint on possible kinetic mechanisms. Finally, we discuss the basic substrate dependence of an important function of the statistical moments. The new kinetic parameters of this expression, akin to the Michaelis–Menten parameters, provide model-independent constraints on the kinetic mechanism.

1. INTRODUCTION

Enzyme dynamics are naturally stochastic. While the directionality of catalyzed reactions is driven by the energy stored in chemical or electrochemical potentials, it is not this energy that drives the internal conformational changes and chemical transformations that compose the kinetic cycle of the enzyme. Rather these transitions are driven by the energy of the surrounding, fluctuating thermal bath. The electrochemical driving potential simply biases these conformational fluctuations along the reaction pathway. As a result, kinetic transitions are stochastic, and the time to complete one full enzymatic cycle is a random quantity. Thus, measures of enzyme dynamics must naturally be statistical.

For much of the twentieth century such fluctuations were ignored due simply to the difficulty in detecting them in large ensembles of unsynchronized copies of enzyme. However, with the recent advances in single-molecule techniques and synchronized ensemble methods, it is now possible to observe these fluctuations directly (Cornish and Ha, 2007; Greenleaf et al., 2007; Moffitt et al., 2008; Sakmann and Neher, 1984). These powerful experimental advances necessarily raise new theoretical questions. In particular, what type of mechanistic information is contained within fluctuations, and how can this kinetic information be extracted in an accurate and unbiased fashion? Moreover, can fluctuations be classified—characterized, perhaps, by quantities in analogy to the Michaelis–Menten constants K_M and k_{cat}? And, if so, what are the implications of such classification, and what do the new kinetic parameters reveal about possible models for the enzymatic reaction?

These are the types of basic questions that face the new field of *statistical kinetics*—the extension of enzyme kinetics from the mean rate of product formation to measures of the inherent fluctuations in this rate. In this chapter, we discuss several recent advances in this field, both theoretical and practical. Our purpose is not to provide a comprehensive discussion of the theoretical foundation of this field nor of the various techniques and methods that are being developed, but rather to complement discussions of a variety of topics, some of them treated extensively in the literature (Charvin et al., 2002; Chemla et al., 2008; Fisher and Kolomeisky, 1999; Kolomeisky and Fisher, 2007; Neuman et al., 2005; Qian, 2008; Schnitzer and Block, 1995; Shaevitz et al., 2005; Svoboda et al., 1994; Xie, 2001). In the first section, we revisit the foundational ideas of statistical kinetics. We adopt a slightly different perspective from other authors, one based on lifetimes rather than kinetic rates, and derive a formal connection between the statistical properties of enzymatic reactions and the statistical properties of their composite kinetic states. In the second section, we turn to more practical matters and discuss different methods of quantifying fluctuations. We explain why, with current methods, fitting the full distribution of lifetimes likely introduces greater risk of bias than extracting kinetic information from statistical moments. We also present simple statistical tests that allow different lifetimes from different kinetic mechanisms to be properly sorted when an enzyme has multiple observable outcomes. In the final section, we discuss methods for extracting kinetic information from the measured moments. In particular, we discuss a newly developed technique for classifying enzymatic fluctuations and the mechanistic constraints provided by the new kinetic parameters introduced in the classification.

2. THE FORMALISM OF STATISTICAL KINETICS

2.1. From steady-state kinetics to statistical kinetics

To illustrate how fluctuations arise in kinetic processes, consider the canonical kinetic model—the Michaelis–Menten mechanism (Michaelis and Menten, 1913):

$$E + S \underset{k_{-1}}{\overset{k_1[S]}{\longleftrightarrow}} ES \overset{k_2}{\longrightarrow} E + P. \qquad (10.1)$$

Here an enzyme, E, binds substrate S to form the bound form ES with the pseudo-first-order rate constant $k_1[S]$. This bound form can then produce product P, or it can unbind the substrate unproductively, returning to E. These processes have rate constants k_2 and k_{-1}, respectively. In general, we will treat the formation of "product" as the only detectable event, though this event could be any kinetic transition that generates an experimentally measurable signal. All other kinetic transitions are hidden from detection. Our measured quantity will be the time between subsequent product formation events or detectable events, the *cycle completion time*. In the case of molecular motors, the formation of "product" is often the generation of a physical motion, such as a discrete step along a periodic track (Fig. 10.1A).

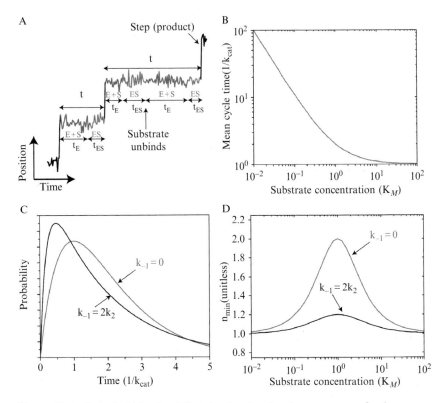

Figure 10.1 Statistical kinetics. (A) A simulated molecular motor trace for the canonical Michaelis–Menten mechanism. Here product formation corresponds to the generation of an observable step—an increase in position. Each full dwell time or cycle completion time, t, can be divided into the lifetimes of the individual kinetic states, t_E and t_{ES}. Fluctuations in t come from both the variations in the individual lifetimes or in different number of visits to each state. (B) Mean cycle completion time versus substrate concentration, measured in units of the maximum rate ($1/k_{cat}$) and the substrate concentration at which the rate is half maximal (K_M). (C) Probability distribution of cycle completion times at a substrate concentration equal to K_M for two different choices of rate constants. Light gray (red online) corresponds to a system in which binding of substrate is irreversible while black correspond to a system in which, on average, two out of three substrate molecules are unbound before catalysis. Despite having the same mean cycle completion time in (B), the fluctuations for these two different systems are distinct. The functional form for these distributions can be found in Chemla et al. (2008) and Xie (2001). (D) A statistical measure of enzymatic fluctuations, n_{min}, as a function of substrate concentration for two different choices of rate constants. Again, these curves are distinct, indicating that while measurements of the mean could not distinguish these mechanisms, measurements of fluctuations can. In contrast to the distributions in (C) simple features of the curves in (D) reveal the difference in binding properties between the two models. The functional form for these curves is described below.

In this case, the cycle completion time is often called a *dwell time* or a *residency time* since during this time the motor resides or dwells at a single place along the molecular track. (Technically speaking the step itself can take some time to complete, just as any kinetic transition requires a small, but finite time to be completed; however, we will ignore this time since it is typically much smaller than the lifetime of the general kinetic state.) This formalism also applies to systems in which there are multiple detectable events per cycle, such as the switching between two enzymatic conformations revealed by fluorescence resonance energy transfer (FRET).

In traditional steady-state kinetics, one would write out the set of coupled, first-order differential equations for the concentration of each of the species, E and ES—equations that describe how the concentration of one species changes continuously into the next species—and then assume that the concentrations of one or various intermediate forms have reached a steady state, that is, are constant in time. This assumption allows differential equations to be changed into algebraic equations, and quantities such as the average rate of product formation to be calculated in terms of substrate concentrations and individual kinetic rates (for a comprehensive discussion, see Segel, 1975). From a single-molecule perspective, however, this picture is flawed. Continuous changes of one species into another are nonphysical since chemical transformations that lead to product formation are discrete, punctate events. Moreover, the system can never be in steady state: at any given time a single enzyme is either in one state or another, not in some constant fraction of both.

Rather than considering a continuous flow of one species to the next, it is more useful, on a single-molecule level, to think of a series of discrete paths through the kinetic cycle—paths that consist of consecutive and discrete transformations of one species into another (i.e., discrete hops between different kinetic states). For example, the above kinetic scheme implies that the following two diagrams represent valid microscopic paths to product formation:

$$E + S \rightarrow ES \rightarrow E + P,$$
$$E + S \rightarrow ES \rightarrow E + S \rightarrow ES \rightarrow E + P. \qquad (10.2)$$

In the first path the enzyme binds substrate and then immediately forms product, whereas in the second path it unbinds this substrate unproductively and has to rebind substrate before making product. Figure 10.1A depicts one possible way in which these two paths might produce an experimental signal. The first dwell represents the first path in Eq. (10.2), while the second represents the second path. It is clear that even for this simple example, there are an infinite number of microscopic paths, representing the infinite number of times, in principle, that the enzyme could release substrate unproductively before completing catalysis.

Now instead of considering the rate at which each species is transformed into the next, we consider the time the enzyme exists as each species. The advantage of this subtle shift in perspective is that the lifetime for each of the above pathways is just the sum of the individual lifetimes of the states visited. For example, the total cycle completion time for the first pathway is $t = t_E + t_{ES}$, where t_E and t_{ES} are the individual lifetimes of the empty and substrate bound states, respectively. The cycle completion time for the second pathway is $t = t_E + t_{ES} + t'_E + t'_{ES}$, where t'_E and t'_{ES} are the distinct lifetimes of these states during the second visit. Despite this shift in perspective, a formal connection can be made between this picture and the first-order differential equations that govern the concentrations of each species by replacing the concentration of any given species with the probability of being in that state (Chemla et al., 2008; Qian, 2008; Schnitzer and Block, 1995).

2.2. Basic statistics of the cycle completion time

In this formalism, it is clear that the statistical nature of the cycle completion time arises in two fashions. First, the individual lifetimes of the states are themselves stochastic; thus, their sum will also be a random variable. Second, the number of times that a given kinetic state is visited in a complete cycle is variable as well. Thus, the number of times that a given lifetime contributes to the total cycle completion time may vary each cycle. Stated simply, the basic statistical problem that arises in enzyme dynamics is one in which the total cycle completion time is a sum of random variables in which the number of terms in the sum is itself variable. Mathematically, if a given kinetic scheme has N states, the cycle completion time, τ, is

$$\tau = \sum_{i=0}^{n_1} t_{1,i} + \sum_{i=0}^{n_2} t_{2,i} + \cdots + \sum_{i=0}^{n_j} t_{j,i} + \cdots + \sum_{i=0}^{n_N} t_{N,i}, \quad (10.3)$$

where n_j is the number of times that kinetic state j was visited during the specific cycle and $t_{j,i}$ represents the stochastic lifetime of state j during the ith visit to that state. While the individual lifetimes are a property of distinct kinetic states, the number of visits to each state is, in general, a function of how all of the kinetic states are interconnected; thus, in general, the lifetimes will be independently distributed random variables while the state visitation numbers will not.

Despite its simplicity, Eq. (10.3) completely determines the relationship between the statistical properties of the cycle completion time and the statistical properties of the hidden kinetic states that compose the cycle. This connection is what forms the basis for the basic premise of statistical kinetics: Measurements of the statistics of the total cycle completion time can provide insight into the properties of the hidden states that compose the cycle. Figure 10.1B–D illustrates this principle. Two different kinetic mechanisms may have the same mean cycle completion time with the same substrate concentration dependence (Fig. 10.1B), yet these two mechanisms are clearly distinguishable via different statistical measures of the fluctuations in the cycle completion time (Fig. 10.1C and D).

Because of its generality, Eq. (10.3) allows the derivation of some basic relationships between the statistics of the cycle completion time and the statistical properties of the individual kinetic states, properties that are independent of the specifics of a given kinetic mechanism. These relations will elucidate how the statistical properties of the individual states generate the statistical properties of the total cycle completion time. In the Appendix, we show that, unsurprisingly, the mean cycle completion time for the arbitrary N state kinetic mechanism is simply the sum of the average amount of time spent in each kinetic state:

$$\langle \tau \rangle = \langle n_1 \rangle \langle t_1 \rangle + \langle n_2 \rangle \langle t_2 \rangle + \cdots + \langle n_N \rangle \langle t_N \rangle. \tag{10.4}$$

We then extend this calculation, in the Appendix, to the next statistical moment, the variance. We find that an N state kinetic cycle has a cycle completion time variance of

$$\begin{aligned}\langle \tau^2 \rangle - \langle \tau \rangle^2 = &(\langle t_1^2 \rangle - \langle t_1 \rangle^2)\langle n_1 \rangle + \cdots + (\langle t_N^2 \rangle - \langle t_N \rangle^2)\langle n_N \rangle \\ &+ (\langle n_1^2 \rangle - \langle n_1 \rangle^2)\langle t_1 \rangle^2 + \cdots + (\langle n_N^2 \rangle - \langle n_N \rangle^2)\langle t_N \rangle^2 \\ &+ 2(\langle n_1 n_2 \rangle - \langle n_1 \rangle\langle n_2 \rangle)\langle t_1 \rangle\langle t_2 \rangle + 2(\langle n_1 n_3 \rangle - \langle n_1 \rangle\langle n_3 \rangle) \\ &\langle t_1 \rangle\langle t_3 \rangle + \cdots + 2(\langle n_{N-1} n_N \rangle - \langle n_{N-1} \rangle\langle n_N \rangle)\langle t_{N-1} \rangle\langle t_N \rangle.\end{aligned}$$

$$(10.5)$$

This expression indicates that the fluctuations in the total cycle completion time arise from both the inherent fluctuations in the individual lifetimes, that is, terms that go as $\langle t_i^2 \rangle - \langle t_i \rangle^2$, and fluctuations in the number of visits to a given kinetic state, that is, terms that go as $\langle n_i^2 \rangle - \langle n_i \rangle^2$. As discussed above, the number of visits to a given state may depend on the number of visits to neighboring kinetic states. The correlation terms in the final two lines capture these interstate relationships. The fact that there are correlation terms in state visitation but not in lifetimes again reflects the fact that lifetimes are properties of individual states while the number of visitations are a function of how states are connected to one another—the topology of the kinetic mechanism.

2.3. The "memory-less" enzyme

At this point, the connection between the statistical properties of the cycle completion time and the statistical properties of the individual kinetic states is independent of the properties of these states. Now we must insert some basic assumptions about enzymatic dynamics in order to determine the statistical properties of the individual kinetic states. First, we require that the properties of a given kinetic state are independent of time—the enzyme has no memory of how long it has lived in a specific state. Second, we require that the properties of a given kinetic state are independent of the past trajectory of the enzyme—the enzyme has no memory of the states from which it came. Thus, we assume that the transitions between kinetic states represent a Markov process, that is, the enzyme is completely "memory-less."

These assumptions arise naturally out of the basic physical properties of the energy landscape that determines the dynamics of enzymes. Protein conformational dynamics can be thought of as diffusion on a complex, multidimensional, energy landscape (Henzler-Wildman and Kern, 2007). In general, the exact position of the system on such a landscape provides a form of memory—the local potential determines the probability of fluctuating in one direction or another, and positions closer to a barrier are more likely to result in fluctuations across that barrier than positions further away. However, it is generally thought that the energy landscapes of most enzymes are characterized by a variety of local minima (Fig. 10.2). If these

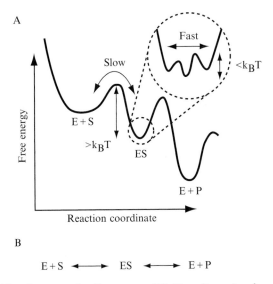

Figure 10.2 The "memory-less" enzyme. (A) One-dimensional projection of the energy landscape of an enzyme. Kinetic states correspond to local minima or energy wells in this landscape. Wells are typically characterized by properties of the enzyme, such as empty (E + S) or substrate bound (ES), that are common to all conformations within a given well. Relatively large barriers between wells ($> k_B T$) creates a separation of timescales in which the enzyme fluctuates many times within a single well before successfully transitioning out of the state. These internal fluctuations "erase" any memory of the amount of time spent in a given well or the past trajectory that brought the system to that well. (B) Because of these physical properties of the energy landscape, the dynamics can often be modeled as discrete hops between distinct kinetic states.

minima are surrounded by energy barriers that are few times the thermal energy available in the bath ($k_B T$), then there is a natural separation of timescales in the diffusional dynamics (Henzler-Wildman and Kern, 2007): Fluctuations can be divided into fast fluctuations within a well and slow fluctuations between wells. Because fluctuations within a well are much faster than fluctuations between wells, the enzyme will fluctuate many times within a well before leaving it. These fluctuations effectively average over differences in transition rates due to the position of the enzyme within a well, erasing any "memory" of its lifetime in the well or the state from which it came. These local minima, or collections of local minima, can be identified as *kinetic states* of the enzyme, and the goal of kinetic modeling is to count the number of these wells, determine their interconnectedness and the height of barriers between them, and identify the chemical state of substrates or products in these wells.

There are a few situations in which these assumptions may not be valid, and discrete hops between kinetic states with no memory may not be a suitable description of enzymatic dynamics (Kolomeisky and Fisher, 2007). For example, if the barriers between local minima are not large, the lifetime of a given kinetic state may not be significantly larger than natural relaxation time of the conformational dynamics within that state; and it may display transition probabilities and lifetimes that depend on the identity of the previous state—assuming distinct conformational states can even be identified. As a rule of thumb, the natural timescales for internal fluctuations within a kinetic state are $\sim$ns–μs (Henzler-Wildman and Kern, 2007); thus, it seems highly unlikely that enzymatic dynamics on the ms and larger timescales will be significantly affected by these complications. However, as experimental techniques push the limit of temporal and spatial resolution, the observation of such transient states may become increasingly common. In such cases, it is likely that more complicated modeling efforts that seek to include this inherent diffusive process will be needed (Karplus and McCammon, 2002; Xing *et al.*, 2005).

2.4. Lifetime statistics

These basic Markovian assumptions completely determine the statistical properties of both the lifetimes and visitation numbers. Since each kinetic state has no memory of when the system arrived in this state, or from where it arrived, the rate at which the system leaves the state (the transition probability per unit time) should be constant and, therefore, the probability of finding the system in that state should decrease exponentially in time. Then, the probability density of observing a lifetime t for that state or, equivalently, its lifetime distribution should be

$$\psi(t) = \langle t \rangle^{-1} e^{-t/\langle t \rangle}. \tag{10.6}$$

Note that the only number that characterizes the distribution of lifetimes, $\psi(t)$, is the mean lifetime, $\langle t \rangle$. Thus, all of the statistical moments of the distribution are completely specified by this mean. For example, the variance in the lifetime is simply the mean squared:

$$\langle t^2 \rangle - \langle t \rangle^2 = \int_0^\infty (t - \langle t \rangle^2) \psi(t) dt = \langle t \rangle^2. \qquad (10.7)$$

This relationship between the variance and the mean has physical implications. States with longer mean lifetimes naturally have larger fluctuations in the lifetime. As we will discuss below, this fundamental relationship between the mean and variance of kinetic lifetimes will prove useful in placing limits on possible mechanisms from statistical measures of fluctuations.

2.5. State visitation statistics

Just as the lack of memory determined the basic statistics of the lifetimes, this assumption also determines the statistics of the number of times an enzyme visits a specific kinetic state during a single cycle. To illustrate this point, consider the options available to an enzyme in a given kinetic state. The enzyme can leave that state and complete the cycle without returning to that state. Let this event occur with a probability, p. Alternatively, the enzyme can return to that state without completing the cycle. Since there are only two options, this event will occur with a probability $1 - p$. With no memory of how it arrived in a given state, once the enzyme returns to this state, the probabilities of completing the cycle must be the same, and completing the cycle must again occur with the same probability p. Thus, each subsequent visit to a given kinetic state occurs with a geometrically decreasing probability.

This argument assumes that the enzyme visits a given kinetic state every cycle, that is, the given state is *on-pathway*. However, there are kinetic pathways in which the enzyme can complete the cycle without visiting a given state—that is, that state is off-pathway or the enzyme has multiple parallel pathways to cycle completion (see Figs. 10.3B or 10.5D and E). Once the system visits this state, however, the above argument applies, so we need only to include a probability, p_0, for visiting the state for the first time. For on-pathway states, $p_0 = 1$.

Combining these arguments yields the visitation statistics for a given kinetic state, that is, the probability that the system completes its cycle with only n visitations to a given kinetic state:

$$P(n) = p_0(1 - p)^{n-1} p. \qquad (10.8)$$

The first term is the probability that the system visits the specific state for the first time. The second term represents the probability of visiting the given state $n-1$ times without completing the cycle while the final term represents the probability of actually completing the cycle from this state. Equation (10.8) applies only for $n \geq 1$, that is, the system visits the state at least once. For no visits to the state, $n = 0$, $P(n = 0) = 1 - p_0$, the probability of not visiting that state for the first time. Summing Eq. (10.8) over all possible visitation numbers and including this $P(0)$ term produces 1, as expected for a normalized distribution. Interestingly, this is the discrete form of the exponential distribution above, and represents the only "memory-less" discrete distribution.

Given the discrete distribution in Eq. (10.8), we can determine the statistics of visiting a given state. The average number of times that a system visits a given kinetic state is

$$\langle n \rangle = \sum_{n=0}^{\infty} n P(n) = \frac{p_0}{p} \qquad (10.9)$$

and the variance is

$$\langle n^2 \rangle - \langle n \rangle^2 = \sum_{n=0}^{\infty} (n - \langle n \rangle^2) P(n) = p_0 \frac{(1 - p_0 + 1 - p)}{p^2}. \qquad (10.10)$$

If a given state is on-pathway, that is, a mandatory state, then $p_0 = 1$ for this state and these expressions simplify. In this case the mean number of visitations completely determines the higher statistical moments of the visitation number. In particular, the variance becomes

$$\langle n^2 \rangle - \langle n \rangle^2 = \langle n \rangle (\langle n \rangle - 1). \qquad (10.11)$$

This expression is almost the mean squared, as is the case with the lifetime statistics. The distinguishing term, $\langle n \rangle - 1$, reflects the discrete nature of the number of visits: If $\langle n \rangle = 1$, the system visits the state once and only once, and the variance in the visitation number must be zero. Again, these statistical properties have clear physical implications. States that are visited on average more frequently will naturally have larger fluctuations in the number of visits.

For off-pathway states or states that compose parallel catalytic pathways, $p_0 < 1$, and the variance in visitation number is not uniquely determined by the mean number of visits. Rather

$$\langle n^2 \rangle - \langle n \rangle^2 = \frac{1}{p_0} \langle n \rangle (\langle n \rangle - p_0 + \langle n \rangle (1 - p_0)). \qquad (10.12)$$

This expression captures the distinction between the visitation statistics of on-pathway and off-pathway states. First, it is important to note that because $0 \leq p \leq 1$, Eq. (10.9) implies that $p_0 \leq \langle n \rangle$, and this expression is always positive, as expected. More interestingly, by comparing Eqs. (10.11) and (10.12), it becomes clear that, for all permissible values of p_0, the variance in the visitation number for an off-pathway state is always *larger* than the visitation number for an on-pathway state with the same average number of visits. Thus, the number of visits to off-pathway states is naturally more stochastic than on-pathway states, and the presence of an off-pathway state or parallel pathways will increase the fluctuations in a system.

3. CHARACTERIZING FLUCTUATIONS

In Section 2, we provide the connection between the statistical properties of the hidden kinetic states and the statistical properties of the total cycle completion time. However, before statistical measures of the cycle completion time can be used to constrain the properties of the kinetic states that compose the kinetic mechanism, care must be taken to characterize these fluctuations in a useful and unbiased way. In particular, not all statistical measures may be as convenient in constraining kinetic mechanisms. Moreover, the characterization of cycle completion or dwell times becomes more subtle when the enzyme has multiple observable outcomes, as is becoming increasingly common. In this section, we discuss these issues.

3.1. Fitting distributions

In the above formalism, we considered only the statistical moments of the cycle completion time; however, one can also compute the full probability distribution for the observation of different cycle completion times. This quantity is often referred to as a *dwell time distribution*, and because all of the statistical moments can be calculated from this distribution, it clearly contains more kinetic information than a subset of the moments. For example, we show in Fig. 10.1 that with the proper choice of rate constants, two different kinetic models (irreversible binding or reversible binding) can have the same mean time to complete the enzymatic cycle (Fig. 10.1B) but very different dwell time distributions (Fig. 10.1C). However, while the differences between these two distributions are clear when they are compared directly, imagine that one has only a single distribution, as would be the case for real data. How would mechanistic information be extracted from this distribution?

One obvious possibility is to derive a function that describes the dwell time distribution and fit this expression to the measured distribution to extract information about the kinetic mechanism. It turns out that, under very general assumptions, it can be shown that a general kinetic model with N states will have a dwell time distribution that is described by a sum of N exponentials with different relative weights and decay rates (Chemla et al., 2008). Thus, any dwell time distribution can be expressed as

$$\psi(t) = a_1 e^{-\lambda_1 t} + a_2 e^{-\lambda_2 t} + \cdots + a_N e^{-\lambda_N t}, \qquad (10.13)$$

where a_i are the weights for each exponential—these can be positive or negative—and λ_i are the eigenvalues of the system—these values set the natural timescales for the cycle completion time and must be positive. If multiple eigenvalues happen to be equal, the m terms in Eq. (10.13) that correspond to these m equivalent eigenvalues are replaced by the term $a_i t^{m-1} e^{-\lambda_i t}$. Distributions of this form are known as *phase-type distributions*, and there is a body of literature on the properties of this class of distribution (cf. Neuts, 1975, 1994; O'Cinneide, 1990, 1999).

As an interesting aside, Eq. (10.13) provides a measure of the maximum kinetic information contained in a dwell time distribution. Since there are N eigenvalues and $N - 1$ free weights (one weight is fixed by the fact that the distribution must be normalized), Eq. (10.13) implies that even the most accurate fit can only extract $2N - 1$ constraints on the system. However, it is easy to imagine an N state kinetic model that has many more than this number of kinetic rates (if every state is connected to every other state—the maximum connectivity—then there are $N(N - 1)$ kinetic rates). Thus, even though much can be determined about these hidden kinetic transitions from enzymatic fluctuations, information is lost in the process of examining only the total cycle completion time. However, some of this information can be restored if there are multiple observables in each cycle. In such a situation, there will be more than one dwell time distribution, and it is possible that additional information can be determined from fits to these multiple distributions.

With Eq. (10.13), we have the appropriate distribution to fit to any measured cycle completion time distribution. However, in practice, there are several problems with using this expression. First, the general dwell time distribution has too many free parameters to be well constrained by typical amounts of data. This is complicated by the fact that the number of terms in the sum, that is, the number of kinetic states in the system, is generally not known *a priori*; thus, one must in principle fit the measured distribution using different numbers of exponentials, comparing the results to determine which number better fits the data. While there are statistically sound methods for performing this comparison (Yamaoka et al., 1978), there is no guarantee that the data will constrain the fits well enough to determine the appropriate number of exponentials. Moreover, if the appropriate number of exponentials is not uniquely determined, then it is not clear how the fit values should be interpreted.

One common solution to this problem is to assume a functional shape for the dwell time distribution that contains far fewer parameters. For example, it is becoming increasingly common to use the gamma distribution to describe such distributions:

$$\varphi(t) = \frac{k^N t^{N-1}}{\Gamma(N)} e^{-kt}, \qquad (10.14)$$

where $\Gamma(N)$ is the gamma function. Since the gamma distribution only has two free parameters—an average rate, k, and the "number" of kinetic states, N—it is often quite well constrained by the data. However, the gamma distribution is the correct functional form for the dwell time distribution only when the underlying kinetic model has N states with equal lifetimes $1/k$ which are connected via irreversible transitions. In other words, the gamma distribution is the correct distribution *only* when the kinetic mechanism is of the form

$$1 \xrightarrow{k} 2 \xrightarrow{k} 3 \xrightarrow{k} 4 \xrightarrow{k} \cdots \xrightarrow{k} N. \qquad (10.15)$$

Unfortunately, it is often not the case that such a mechanism applies to an average kinetic model. First, it is extremely rare that all kinetic rates are identical, and even if this happened to be true for one substrate concentration, it would not be true for arbitrary substrate concentrations. Second, an irreversible transition requires a large energy input, and it is rare for a kinetic mechanism to have only transitions that involve such large energies. Thus, while the gamma distribution is an excellent way to characterize the "shape" of a distribution, it is unlikely that this is often the correct functional form for the distribution, and, thus, it is unclear how the fit values should be interpreted.

Despite these caveats, it is clear that the full cycle completion time distribution contains the most kinetic information available; thus, efforts should be taken to improve methods for extracting this information. There is a recent technique that attempts to extract rate information from dwell time distributions without fitting the distribution. With this method one calculates a rate spectrum—the amplitude of exponentials at each decay rate—via direct numerical manipulation of the distribution (Zhou and Zhuang, 2006). This method is analogous to the Fourier method for disentangling different frequency components from complicated time series. While initial results appear promising, this technique is only reported to perform well when the decay rates are separated by an order of magnitude (Zhou and Zhuang, 2006), which is typically not the case in real experimental data. However, future developments in this direction seem promising.

There is one final issue with fitting distributions. Ignoring the complications with actually extracting the various decay rates, that is, the eigenvalues λ_i and the different weights for each exponential a_i, it turns out that it is difficult, and in some cases impossible, to analytically relate these values to the kinetic rates of a specific kinetic model. The problem is a mathematical one. Calculating analytical expressions for these eigenvalues and weights corresponds to solving for the roots of a polynomial expression with order equal to the number of kinetic states in the model (Chemla et al., 2008). However, Abel's impossibility theorem (Abel, 1826) states that there is no general analytical solution for this problem if the polynomial is of order five or higher. Thus, if there are five or more kinetic states in a kinetic mechanism, the eigenvalues and exponential weights *cannot* in general be expressed in terms of the individual kinetic rates analytically. Of course, this does not mean that a numerical connection cannot be made—there are many techniques capable of calculating these values for any numeric choice of kinetic rates for any mechanism (Liao et al., 2007)—however, since the proper choice of rate constants (and number of rate constants for that matter) are typically not known *a priori*, it is unclear how useful numerical solutions would be.

3.2. Calculating moments

Many of the problems associated with extracting information from the dwell time distribution directly can be relaxed by first calculating properties of this distribution such as its statistical moments or, as we will show below, properties such as its "shape." While it is clear that the moments of the distribution will contain a subset of the information contained in the full distribution, the advantage is that these moments are *model-independent*, that is, a basic kinetic model or form for the distribution does not need to be assumed to calculate the mean dwell time or the variance in the dwell times. Moreover, calculation of these moments is simple and straightforward, and there are well-established techniques to estimate the stochastic uncertainty in these moments directly from the measured data itself (Efron, 1981; Efron and Tibshirani, 1986). Since the uncertainty in the moments can be easily calculated, one can use only the moments that are well constrained by the data, conveniently circumventing issues with unconstrained fits to poorly determined distributions. Finally, methods now exist for calculating analytical expressions for the moments of the cycle completion time for any kinetic mechanism, no matter how complex (Chemla et al., 2008; Shaevitz et al., 2005).

There is one notable disadvantage to characterizing fluctuations via statistical moments. In all measurements, there is a natural dead-time, a time below which events are too quick to observe experimentally. When fitting distributions, one can address this problem simply by fitting only over time durations that are known to be measured accurately. However, a similar method does not exist for calculating moments, and dead-times can introduce bias into the estimation of statistical moments. Fortunately, it is relatively simple to estimate the relative size of this error directly from the data itself with only a few assumptions. In the Appendix, we derive expressions for estimating this systematic bias.

3.3. Multiple pathways and multiple steps

The above discussion makes an implicit assumption: that each random dwell time is derived from the same kinetic mechanism, that is, stochastic passage through the same kinetic states. This is an innocuous assumption when the enzyme takes a single type of step, for example, forward steps of uniform size, since it is likely that identical steps are produced by the same kinetic pathway. However, it is becoming increasingly clear that real enzymes display more complicated behaviors. For example, filament based cargo transport proteins such as kinesin, myosin, and dynein have now been observed to take both forward and *backward* steps and steps of varying size (Cappello *et al.*, 2007; Carter and Cross, 2005; Clemen *et al.*, 2005; Gennerich and Vale, 2009; Gennerich *et al.*, 2007; Mallik *et al.*, 2004; Reck-Peterson *et al.*, 2006; Rief *et al.*, 2000; Yildiz *et al.*, 2008). Moreover, multiple observable events may be produced within a single kinetic cycle, creating in effect multiple classes of dwell times, as is observed for the packaging motor of the bacteriophage φ 29 (Moffitt *et al.*, 2009). Finally, some single-molecule measurements naturally observe multiple events within a single cycle, such as the transitions of a system between multiple FRET states (Cornish and Ha, 2007; Greenleaf *et al.*, 2007) or the folding and refolding of nucleic acids (Li *et al.*, 2008; Woodside *et al.*, 2008) or proteins (Cecconi *et al.*, 2005).

When there are multiple types of steps or observable states, there may also be multiple classes of dwell times, that is, cycle completion times that originate from different kinetic pathways. Clearly, the combined statistical analysis of dwell times derived from different kinetic pathways will provide little insight into each pathway individually. Thus, before one can extract any kind of mechanistic information from fluctuations, one must be sure that dwell times that are generated by the same kinetic mechanism and result in the same basic type of step are properly sorted. However, it is not immediately obvious how this should be done. Does one simply sort dwells based on the type of the step following the dwell? Or is it possible that enzymatic dynamics may have some memory of more distant steps, perhaps the type of step before the dwell as well? Moreover, how does one distinguish between these possibilities; are there simple statistics that can be calculated directly from the data that would allow these different possibilities to be addressed and the appropriate classification scheme to be determined?

It has been recently recognized that there are three basic statistical classes of enzymatic dynamics (Chemla et al., 2008; Linden and Wallin, 2007; Tsygankov et al., 2007) in which there are multiple enzymatic outcomes. For simplicity, we consider different outcomes to represent steps of different sizes, but this discussion can be easily generalized to any measurable enzymatic outcome, for example, off-pathway pauses or dissociation events. In the first statistical class, the size of the step or its direction have no relationship to the hidden kinetic events that must occur before that step (Shaevitz et al., 2005). In this case, the statistics of the dwell times will be uncorrelated to the type of the subsequent step or event. We term such statistics *uncorrelated*. Figure 10.3A contains an example kinetic scheme that would display these statistics.

There are two different classes of correlated statistics. First, the statistics of the dwell time may be related to the type of the subsequent step but not the identity of the previous step. We term such statistics *unconditional* because the statistical properties of the dwells depend only on the identity of the following step and are not conditional on the identity of the previous step. For this statistical class the dwells must be sorted by the identity of the subsequent step. To generate these statistics, each different step must have a distinct kinetic pathway (or subset of a kinetic pathway) in order to generate the distinct dwell statistics, yet all kinetic pathways must start in the same kinetic state. See Fig. 10.3B for an example of a kinetic mechanism which would display such statistics.

Finally, it is possible for an enzyme to actually remember the type of its previous step. This emergent memory arises because different steps need not place the enzyme in the same hidden kinetic state. In this case, the dwell times are *conditional* on both the type of step before and after the dwell, and these durations must be sorted by the identity of both steps. This case is likely for enzymes in which formation of product generates a forward step but the reverse reaction, the recatalysis of substrate from product, corresponds to a backward step (Linden and Wallin, 2007; Tsygankov et al., 2007). See Fig. 10.3C for an example of an enzymatic mechanism that generates correlated, conditional statistics. In this example, a forward step ($+d$) coincides with the formation of product and places the enzyme in the empty state (E) whereas a backward step ($-d$) coincides with the reformation of substrate from product and places the enzyme not in the empty state (E) but in the substrate bound state (ES). Because forward and backward steps place the enzyme into different kinetic states, the microscopic kinetic trajectories (as in Eq. (10.2)) that end in the next step will be different. Thus, the kinetics of a single dwell time will depend on the initial type of step.

Figure 10.3 shows that the individual stepping traces can appear quite similar by eye despite the different mechanisms that produce these traces. Fortunately, there are well-defined statistical properties of such traces that

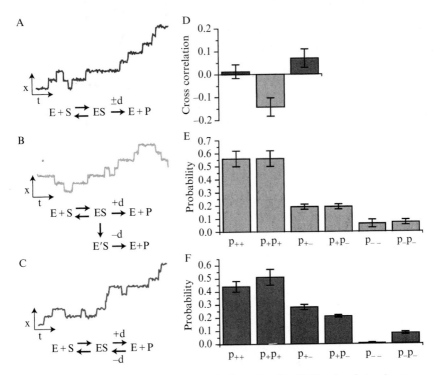

Figure 10.3 Statistical classes of enzymatic dynamics. (A–C) Simulated stepping traces for different mechanisms capable of generating forward and backward steps. (A) The steps and dwell times are uncorrelated because the same kinetic mechanism generates both forward ($+d$) and backward ($-d$) steps. (B) The steps and dwell times are correlated but the dwells are unconditional on the type of the previous step because each step returns the system to the same kinetic state (E). (C) The steps and dwells are correlated and the dwells are conditional on the type of the previous step because forward and backward steps return the enzyme to different kinetic states, E and ES, respectively. (D) Cross-correlation between steps and dwell times (arbitrary units) for the different stepping traces. Correlation values correspond to the different mechanisms in (A) - (C) from left to right, respectively (color online). (E) Single-step probabilities and pair probabilities for the mechanism in (B). (F) Single-step probabilities and pair probabilities for the mechanism in (C). Despite the fact that the individual stepping traces are very similar, the statistical measures in (D)–(F) can clearly distinguish the differences between these mechanisms. Values in (D)–(F) were calculated from kinetic simulations containing 1000 steps with rates set such that the probability of taking a forward step is 0.75 and the average velocity is the same between the different mechanisms. Error bars represent the standard deviation in the various statistics estimated from 100 repetitions of the simulations.

can clearly distinguish the statistical class of the enzyme. For example, if the type of the subsequent step is uncorrelated with the subsequent dwell time, then the appropriate statistical class is *uncorrelated*. In this case, step size is independent of the kinetic pathway, and dwell times can be analyzed together. Mathematically, this can be tested in a variety of ways, but if there is no cross-correlation between step type or size and dwell time, then the steps are uncorrelated, e.g.

$$\langle dt \rangle - \langle d \rangle \langle t \rangle = 0. \tag{10.16}$$

Here d is the step size (or the type of the enzymatic outcome) and t is the dwell or cycle completion time. Figure 10.3D shows that this statistic clearly indicates that the stepping traces for the mechanism in Fig. 10.3A are uncorrelated while the other two traces are correlated.

Often experimental noise broadens step size distributions, and it can be difficult to determine if multiple steps are actually present. In addition, a portion of the steps may have dwells too small for direct detection, giving the appearance that the enzyme can generate steps of multiple sizes. In both cases, the steps will be uncorrelated with the preceding dwell times. Thus, the violation of Eq. (10.16) is a strong criterion for establishing that an enzyme generates multiple types of steps and that such steps are not experimental artifacts.

If the steps are correlated, that is Eq. (10.16) is violated, there is a simple statistical test to determine if the statistical class is conditional or unconditional. Simply compute the probability of observing each pair of events. For the forward and backward step examples considered in Fig. 10.3, this will produce four probabilities, for example, the probability of observing two forward steps in a row, p_{++}, a forward step followed by a backward step, p_{+-}, a backward step followed by a forward step, p_{-+}, and finally two backward steps in a row, p_{--}. These probabilities are computed by simply counting the number of each type of event and dividing by the total number of events, though care must be taken in computing the exact number of events if statistics are small, as pointed out by Tsygankov *et al.* (2007). These quantities can then be compared to the probability of observing a given outcome independent of the previous type of step. For the forward and backward stepping example, this would be the probability of taking a forward step, p_+, and a backward step, p_-, which are again calculated from the number of each step type divided by the total number of steps. If the statistics of the system are unconditional, then the probability for observing two types of events in a row will be equal to the product of the probabilities of observing each of these events individually. If the statistics of the system are conditional, this relation will not be true. For the forward and backward stepping example, there are four equalities to test:

$$p_{++} = p_+p_+, \quad p_{+-} = p_+p_-, \quad p_{-+} = p_-p_+, \quad \text{and} \quad p_{--} = p_-p_-. \tag{10.17}$$

If these equalities are violated, then the statistics of the system are conditional. If these equalities are upheld by the data, then it is likely that the statistics of the system are unconditional. Figure 10.3D and E illustrates the ability of these statistics to distinguish the unconditional and conditional statistics of the mechanisms in Fig. 10.3B and C. These equalities can be easily extended for systems that have more than two outcomes or observable states.

In addition to providing the correct method for sorting cycle completion times, the specific statistical class of an enzyme places clear constraints on the underlying kinetic mechanism (Chemla *et al.*, 2008). For example, a lack of correlation between steps and dwell times can only occur if these processes are determined independently; thus, the observation of uncorrelated statistics implies that the kinetic pathway for each type of step is *identical*. The converse is also true: If the statistics are correlated, then at least a portion of the kinetic pathway that leads to each kinetic event cannot be the same. Moreover, the type of correlated statistics—conditional or unconditional—provides further constraint on the kinetic mechanism. In general, the use of different kinetic pathways to develop different enzymatic outcomes creates an emergent enzymatic "memory." The partial loss of this memory, as is the case in unconditional statistics, requires that all kinetic pathways share at least one common kinetic state after the generation of each step. The memory-less properties of this state is what decouples the statistics of the subsequent dwell time from the identity of the preceding step. While there have been many examples of enzymes that take multiple types and sizes of steps (Cappello *et al.*, 2007; Clemen *et al.*, 2005; Gennerich and Vale, 2009; Gennerich *et al.*, 2007; Kohler *et al.*, 2003; Mallik *et al.*, 2004; Reck-Peterson *et al.*, 2006; Rief *et al.*, 2000; Rock *et al.*, 2001; Sellers and Veigel, 2006; Yildiz *et al.*, 2004), the statistical properties of these steps has so far been underutilized in the analysis of the kinetic mechanism of these enzymes.

4. Extracting Mechanistic Constraints from Moments

Once fluctuations have been properly characterized and measured, the question arises: What can be learned from these statistics? Should candidate models be selected, their properties calculated via the variety of techniques available (Chemla *et al.*, 2008; Derrida, 1983; Fisher and Kolomeisky, 1999; Koza, 1999, 2000; Shaevitz *et al.*, 2005), and then these properties compared to the measured statistics to reject or accept models? This scheme is certainly a viable approach; however, it turns out that, as was seen with the statistical class of an enzyme above, certain properties of the statistics of enzymatic fluctuations can place constraints on candidate models, even before the properties of these models are calculated. In this section, we discuss a method in which statistical measures of fluctuations can constrain kinetic mechanisms without the assumption of candidate models.

4.1. The randomness parameter and n_{min}

In 1994, Schnitzer and Block (Schnitzer and Block, 1995, 1997; Svoboda *et al.*, 1994) introduced a kinetic parameter related to the first and second moments of enzymatic fluctuations: the randomness parameter, $r = 2D/vd$, where D is the effective diffusion constant of the enzyme, v is the average rate of the enzyme, and d is a normalization constant that determines the amount of product each cycle. For molecular motors, where this expression was first introduced, d is the step size. In this context, the diffusion constant, D, is not the rate at which an enzyme diffuses freely through solution, but a measure of how quickly two synchronized enzymes will drift apart from one another. For example, if two identical motors are started at the same location at the same time, they will gradually separate, due to fluctuations, with a squared distance that increases linearly with time. D is a measure of this diffusive-like behavior (Schnitzer and Block, 1995, 1997; Svoboda *et al.*, 1994).

In the limit that the motor takes a single type of step of uniform size and direction (Schnitzer and Block, 1995; Svoboda *et al.*, 1994), the randomness parameter reduces to a quantity that is a function only of the statistics of the dwell times:

$$r = \frac{\langle \tau^2 \rangle - \langle \tau \rangle^2}{\langle \tau \rangle^2} = \frac{1}{n_{min}}, \qquad (10.18)$$

where $\langle \tau \rangle$ is the mean of the cycle completion time distribution and $\langle \tau^2 \rangle - \langle \tau \rangle^2$ is the second moment, the variance. This quantity is known as the squared coefficient of variation, and is used to characterize fluctuations in a wide range of stochastic systems. As we will see below, it is often more convenient to work with the inverse of this parameter—a quantity that we term n_{min}. This new parameter can be thought as a shape parameter for the dwell time distribution (akin to the parameter N of the Gamma distribution). The smaller the variance, the more "sharply peaked" the distribution is, and the larger the value of n_{min}.

We term this parameter, n_{min}, because it has been shown (Aldous and Shepp, 1987) that it provides a strict lower bound on the number of kinetic states that compose the underlying kinetic model, n_{actual}:

$$n_{min} \leq n_{actual}. \qquad (10.19)$$

This inequality is worth a moment's inspection. Equation (10.19) states that a weighted measure of enzymatic fluctuations places a *firm* limit on the minimum number of kinetic states in the underlying kinetic model. The implication is that kinetic schemes with different numbers of kinetic states have fundamentally different statistical properties, and these properties can be used to discriminate between these models. Intuitively, this remarkable property arises from the fact that the variance of an exponentially distributed process is simply the mean squared, as discussed above. Thus, for a kinetic system with a single kinetic state (or dominated by one particularly long-lived state), the ratio of the mean squared to the variance is 1, the number of kinetic states in the system. As additional kinetic states are added (or their lifetimes become comparable), the mean increases more quickly than the variance, and the ratio of the mean squared to the variance increases. While Eq. (10.19) was first introduced as a conjecture in the single-molecule literature (Schnitzer and Block, 1995; Svoboda *et al.*, 1994), it has been formally proven in the context of phase-type distributions (Aldous and Shepp, 1987).

One significant advantage of the randomness parameter is that it can be measured even when the individual steps are obscured by noise, and individual cycle completion times cannot be measured (Schnitzer and Block, 1995; Svoboda *et al.*, 1994). Unfortunately, in recent years, several researchers (Chemla *et al.*, 2008; Shaevitz *et al.*, 2005; Wang, 2007) have shown that the randomness parameter is not always equal to the inverse of n_{min}. In particular, variation in the step size or the stepping pathway will result in correction terms that must be added to r in order to reconstitute n_{min}. Moreover, it does not appear that these terms can be measured from trajectories in which the individual turnovers are obscured by noise, limiting the applicability of the randomness parameter. In addition, in the case of steps of differing size, it is no longer unambiguous what value of d should be used to normalize this parameter (Chemla *et al.*, 2008; Shaevitz *et al.*, 2005; Tsygankov *et al.*, 2007). However, if one can observe the individual cycle completion events, as has been assumed here, different step sizes (or reaction outcomes) can be sorted as above, and it is straightforward to calculate the moments, and, thus, n_{min}, for each of the different classes of dwell times.

4.2. Classifying fluctuations

Under any given experimental conditions, n_{min} places a firm lower limit on the number of kinetic events in the specific kinetic pathway. However, the degree to which n_{min} varies from the actual number of kinetic events depends on the relative lifetimes and visitation statistics of the different kinetic events. If one or more kinetic states tend to produce larger fluctuations, that is, because they have longer lifetimes or smaller probabilities of completing the cycle (see Eqs. (10.6)–(10.12) above), then these states will dominate the measured statistics and will tend to lower $n_{\min}$. However, by changing experimental conditions such as substrate concentration or force, it is possible to change the lifetime and visitation statistics of the different states, making different kinetic states the dominate contributors to fluctuations. In this way, it should be possible to vary the concentration of substrate or force and determine limits on classes of kinetic states, such as the number of substrate binding states or the number of force sensitive kinetic states.

This general dependence of $n_{\min}$ or the related randomness parameter on substrate concentration has been widely recognized, and several different expressions for this dependence have been derived for a variety of specific kinetic mechanisms (Chemla *et al.*, 2008; Garai *et al.*, 2009; Goedecke and Elston, 2005; Kolomeisky and Fisher, 2003; Kou *et al.*, 2005; Moffitt *et al.*, 2009; Schnitzer and Block, 1997; Tinoco and Wen, 2009; Xu *et al.*, 2009). However, we have recently shown (Moffitt *et al.*, 2010) that most if not all of these expressions can be combined into a single expression for the substrate dependence of $n_{\min}$. In particular, it appears that most kinetic mechanisms for which the mean cycle completion time follows the Michaelis–Menten expression:

$$\langle \tau \rangle = \frac{K_M + [S]}{k_{\text{cat}}[S]} \quad (10.20)$$

have a substrate concentration dependence of $n_{\min}$ described by

$$n_{\min} = \frac{N_L N_S \left(1 + \frac{[S]}{K_M}\right)^2}{N_S + 2\alpha \frac{[S]}{K_M} + N_L \left(\frac{[S]}{K_M}\right)^2}. \quad (10.21)$$

Here k_{cat} and K_M are the Michaelis–Menten parameters, which set the maximum rate of a reaction and the substrate concentration at which the rate is half-maximal, respectively. Just as these constants contain the specifics of each kinetic mechanism, the new macroscopic constants for $n_{\min}$, N_L, N_S, and α, contain all of the details of each kinetic mechanism—that is, the specific kinetic rates. Figure 10.4 illustrates the possible shapes permitted by Eq. (10.21) and provides a geometric interpretation of these new kinetic parameters. N_S is the value of $n_{\min}$ at saturating substrate concentrations, N_L is the value of $n_{\min}$ at limiting substrate concentrations, and α controls the height of the peak between these two limits. If $\alpha = 0$, then the peak value is the sum of the two limits whereas if $\alpha > 0$, the peak value is smaller.

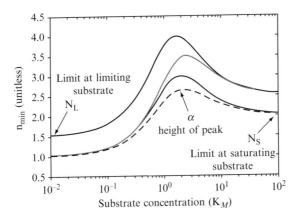

Figure 10.4 n_{min} versus substrate concentration. Different potential curves for n_{min} versus substrate concentration (measured in units of K_M). N_L controls the value of n_{min} at asymptotically low substrate concentration while N_S sets the value of n_{min} at saturating substrate concentration. The value of α controls the height of the peak between these two limits. For $\alpha = 0$, the maximum value of n_{min} is $N_L + N_S$. For $\alpha > 0$, the peak value is less than the sum of the asymptotic limits. The bottom most curves (blue online) correspond to $N_L = 1$, $N_S = 2$, and $\alpha = 0$ (solid) or $\alpha = 0.2$ (dashed). The light gray curve (red online) corresponds to $N_L = 1$, $N_S = 2.5$, and $\alpha = 0$, while the top most curve (green online) corresponds to $N_L = 1.5$, $N_S = 2.5$, and $\alpha = 0$.

4.3. Mechanistic constraints

These new kinetic parameters allow the classification of enzymatic dynamics based on fluctuations, just as the Michaelis–Menten parameters allow such a classification based on measurements of the mean rate versus substrate concentration. However, in contrast to the Michaelis constants, these new parameters also provide clear constraints on the underlying kinetic mechanism. To illustrate these constraints, Fig. 10.5 lists a variety of common kinetic mechanisms and Tables 10.1 and 10.2 list the different kinetic parameters as a function of the individual kinetic rates.

The trends for the kinetic parameters of n_{min} in Table 10.2 can be summarized as follows. First, depending on the kinetic model, each of the different kinetic parameters can be complicated functions of the kinetic rates, just as the Michaelis–Menten parameters (Table 10.1). It is in this fashion that the details of a given kinetic model are "hidden" in these parameters. Moreover, by changing the relative values of the kinetic rates, these parameters can be changed continuously, which implies that N_L and N_S need not be integers. However, this observation does not imply that these parameters can take any value for a specific kinetic mechanism. Rather inspection of these expressions reveals that they have clear upper and lower limits. These limits are included in Table 10.2.

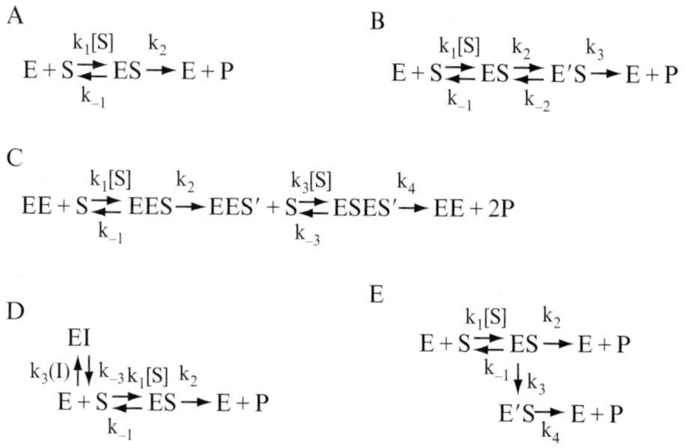

Methods in Statistical Kinetics   245

A
$$E + S \underset{k_{-1}}{\overset{k_1[S]}{\rightleftharpoons}} ES \overset{k_2}{\rightarrow} E + P$$

B
$$E + S \underset{k_{-1}}{\overset{k_1[S]}{\rightleftharpoons}} ES \underset{k_{-2}}{\overset{k_2}{\rightleftharpoons}} E'S \overset{k_3}{\rightarrow} E + P$$

C
$$EE + S \underset{k_{-1}}{\overset{k_1[S]}{\rightleftharpoons}} EES \overset{k_2}{\rightarrow} EES' + S \underset{k_{-3}}{\overset{k_3[S]}{\rightleftharpoons}} ESES' \overset{k_4}{\rightarrow} EE + 2P$$

D
$$\begin{array}{c} EI \\ k_3(I) \updownarrow k_{-3} k_1[S] \;\; k_2 \\ E + S \underset{k_{-1}}{\rightleftharpoons} ES \rightarrow E + P \end{array}$$

E
$$E + S \underset{k_{-1}}{\overset{k_1[S]}{\rightleftharpoons}} ES \overset{k_2}{\rightarrow} E + P$$
$$\downarrow k_3$$
$$E'S \overset{k_4}{\rightarrow} E + P$$

Figure 10.5 Example kinetic mechanisms with a Michaelis–Menten dependence on the substrate concentration. (A) The classic Michaelis–Menten mechanism. (B) One additional intermediate state ($E'S$). (C) Two substrate binding events, separated by irreversible transitions. (D) The Michaelis–Menten mechanism in the presence of a competitive inhibitor (I). (E) A parallel catalytic pathway. The values for the Michaelis–Menten parameters, k_{cat} and K_M, in terms of the pictured rate constants are listed in Table 10.1. Similarly, the values for the new parameters of n_{min}, N_L, N_S, and α, are listed in Table 10.2. The values in these tables can be calculated with a variety of techniques (Chemla et al., 2008; Shaevitz et al., 2005).

Table 10.1 Michaelis-Menten parameters for the example mechanisms in Fig. 10.6

Panel	k_{cat}	K_M
A	k_2	$\frac{k_2+k_{-1}}{k_1}$
B	$\frac{k_2 k_3}{k_2+k_3+k_{-2}}$	$\frac{k_2 k_3 + k_3 k_{-1} + k_{-1} k_{-2}}{k_1(k_2+k_3+k_{-2})}$
C	$\frac{k_2 k_4}{k_2+k_4}$	$\frac{k_1 k_2 k_4 + k_2 k_3 k_4 + k_3 k_4 k_{-1} + k_1 k_2 k_{-3}}{k_1 k_2 k_3 + k_1 k_3 k_4}$
D	k_2	$\frac{k_2+k_{-1}}{k_1}\left(1+\frac{k_3}{k_{-3}}[I]\right)$
E	$k_4 \frac{k_2+k_3}{k_3+k_4}$	$\frac{k_4}{k_1}\frac{k_2+k_3+k_{-1}}{k_3+k_4}$

Investigation of the limits listed in Table 10.2 illustrates the mechanistic constraints provided by each of the different kinetic parameters of n_{min}. Compare, for example, the value of N_S for the two-state mechanism in Fig. 10.5A with the upper limit N_S for the mechanism with one additional state in Fig. 10.5B. This parameter is strictly 1 for the example in Fig. 10.5A, while it is bounded from below by 1 and from above by 2 for the example in Fig. 10.5B. This upper bound is particularly suggestive as it is the number of states that do not involve the binding of substrate. In this sense, the value of

Table 10.2 $n_{\min}$ parameters for the example mechanisms in Fig. 10.6

Panel	N_L	N_S	α
A	1	1	$\dfrac{k_{-1}}{k_{-1}+k_2}$
B	1	$1 \leq \dfrac{(k_2+k_3+k_{-2})^2}{k_2^2+2k_2k_{-2}+(k_3+k_{-2})^2} \leq 2$	$\dfrac{k_{-1}(k_2+k_3+k_{-2})(k_3k_{-2}+(k_3+k_{-2})^2)}{(k_2k_3+k_{-1}k_3+k_{-1}k_{-2})(k_2^2+2k_2k_{-2}+(k_3+k_{-2})^2)}$
C	$1 \leq \dfrac{(k_1k_2(k_{-3}+k_4)+k_3k_4(k_{-1}+k_2))^2}{(k_1k_2(k_{-3}+k_4))^2+(k_3k_4(k_{-1}+k_2))^2} \leq 2$	$1 \leq \dfrac{(k_2+k_4)^2}{k_2^2+k_4^2} \leq 2$	$\dfrac{(k_{-3}k_1k_2^2+k_{-1}k_3k_4^2)}{(k_1k_2(k_{-3}+k_4))^2+(k_3k_4(k_{-1}+k_2))^2}\dfrac{k_2+k_4}{k_2^2+k_4^2} \times$ $\dfrac{k_1k_2(k_{-3}+k_4)+k_3k_4(k_{-1}+k_2)}{(k_1k_2(k_{-3}+k_4))^2+(k_3k_4(k_{-1}+k_2))^2}$
D	1	1	$\dfrac{k_{-1}}{k_{-1}+k_2}+\dfrac{k_3}{k_{-3}}[I]\dfrac{k_2/k_{-3}}{[I]k_3/k_{-3}+1}$
E	1	$0 \leq \dfrac{(k_3+k_4)^2}{k_3^2+k_4^2+2k_2k_3} \leq 2$	$\dfrac{k_{-1}k_4}{k_{-1}+k_2+k_3}\dfrac{k_3+k_4}{k_3^2+k_4^2+2k_2k_3}$

N_S provides a lower limit on the number of nonsubstrate binding states in the kinetic model, and this value can be used to limit possible models. For example, a value larger than 1 immediately rules out the simple Michaelis–Menten mechanism, Fig. 10.5A, among others.

To illustrate the constraints imposed by N_L, note that all mechanisms that have only one substrate binding state have a value of 1 for this parameter. However, this value can be larger than 1 when there are additional substrate binding states in the system (see Fig. 10.5C). In this case, the upper limit is again set by the number of such states in the mechanism; thus, the value of N_L provides a strict lower limit on the number of kinetic states that bind substrate in a given cycle. Remarkably, this statistic can indicate multiple binding events even when the substrate dependence of the mean shows no evidence for cooperativity in binding, as was recently observed for the packaging motor of the bacteriophage φ 29 (Moffitt et al., 2009).

If the kinetic mechanism has no parallel catalytic pathways, then the smallest possible value of N_L and N_S is 1. However, including a parallel catalytic pathway allows this value to be less than 1 as illustrated by the example in Fig. 10.5E. Thus, a measured value of N_L or N_S less than 1 immediately implies that there are multiple catalytic pathways. Multiple catalytic pathways are one possible explanation for enzymes that display dynamics disorder (English et al., 2006; Kou et al., 2005; Min et al., 2006), and it has been predicted that an n_{min} value less than 1 should be possible for these systems.

The constraints placed on the kinetic mechanism by the parameter α are slightly different. Note that in all of the example systems in Fig. 10.5 and Table 10.2, the value of α is proportional to the rate at which substrate unbinds from the enzyme (k_{-1} and k_{-3} for the different examples). Thus, if these rates are zero, $\alpha = 0$. The presence of a competitive inhibitor, Fig. 10.5D, provides a notable exception. In this case, even with irreversible binding $\alpha > 0$ if there is a nonzero concentration of inhibitor. These observations can be summarized in two basic requirements for $\alpha = 0$: (1) the binding of substrate molecules must be irreversible, and (2) the binding competent state cannot be in equilibrium with a nonsubstrate bound state such as an inhibitor-bound state. If these restrictions are met, then $\alpha = 0$; otherwise $\alpha > 0$.

The implications for mechanistic constraints are clear. If the measured value of α is zero, that is, the maximum value of n_{min} is the sum of the two limits $N_L + N_S$, this observation indicates that the binding state has the above properties. This remarkable result stems from the fact that lifetime statistics and visitation statistics have different effects on the statistics of the total cycle completion time, as illustrated above. A binding state not in equilibrium with any other state, that is, with only irreversible transitions out of this state, has only one visit per cycle and, thus, has no fluctuations in the state visitation number. This fact is the reason why such a state has fundamentally different fluctuations—fluctuations that are revealed by

$\alpha = 0$. It is, perhaps, surprising that such clear mechanistic features can be observed from the statistics of fluctuations in which no single binding or unbinding event has been directly observed.

While the constraints described here were developed by considering only a handful of kinetic mechanisms, these constraints have been rigorously proven for all nearest neighbor kinetic models, that is, no off-pathway states or parallel pathways (Moffitt et al., 2010). More importantly, no kinetic model has yet been found which violates these constraints, though a general proof of these properties is still lacking for the arbitrary kinetic model.

5. Conclusions and Future Outlook

Enzymatic dynamics are naturally stochastic, and experimental techniques capable of revealing these natural fluctuations are becoming increasingly commonplace. Thus, it is now time to formalize methods for characterizing these fluctuations and to develop techniques for extracting the full mechanistic information from these statistics. In this chapter, we have contributed to this effort in several ways. First, we have provided an additional theoretical perspective on the relationship between the statistical fluctuations of a total enzymatic cycle and the natural fluctuations of the hidden kinetic states that compose these states. While this connection is of little immediate use to experimentalists, it should help formalize the development of more fundamental connections between the statistical properties that can be measured and the underlying kinetic mechanism of the enzyme, permitting the extraction of more subtle mechanistic details from features of fluctuations. Second, we have discussed methods for characterizing fluctuations. In particular, we have provided statistical quantities that can be calculated for enzymes with multiple measurable outcomes, such as steps of different sizes or directions. These statistics allow the statistical class of the enzymatic dynamics to be identified easily, and from this statistical class, we have demonstrated that powerful constraints can be placed on the underlying mechanism. Finally, we have discussed a novel method for classifying enzymatic fluctuations. We demonstrate that, akin to the Michaelis–Menten expression for the substrate dependence of the mean, there is a general expression for the substrate dependence of a useful measure of enzymatic fluctuations, $n_{\min}$. This expression introduces three new kinetic parameters, and by deriving the expressions for these parameters for a variety of example mechanisms, we illustrate the mechanistic constraints provided by these parameters.

While it is now clear that statistical measures of enzymatic fluctuations cannot uniquely determine a kinetic mechanism, these measures can provide powerful mechanistic constraints, constraints that are not possible from measurements of the mean alone. Given the ease with which statistical moments are calculated and the remarkable additional information provided by the second moment, we expect that generalizations of $n_{\min}$ that include higher statistical moments such as the skew or kurtosis, or more exotic functions of the data (Zhou and Zhuang, 2007), should provide equally powerful constraints. The growing numbers of enzymes for which fluctuations are being accurately measured, not to mention all previously published examples, now await the discovery of these statistics.

APPENDIX

A.1. Calculating the moments of the cycle completion time

Recall that the total cycle completion time is the sum of a random number of random lifetimes for each kinetic state, Eq. (10.3). For convenience, we define the quantity T_i which is the total time spent in the kinetic state i. Thus

$$T_i = \sum_{j=0}^{n_i} t_{i,j} \qquad (10.\text{A}1)$$

and the total cycle completion time is

$$\tau = \sum_{i=1}^{N} T_i. \qquad (10.\text{A}2)$$

To compute the statistical properties of τ, we start by computing the statistical properties of the individual T_i. To compute the mean, we use the tower property of averages. This property states that the mean of a two variable distribution is the average of the mean of the distribution evaluated with a constant value for one parameter averaged again over that parameter. Evaluating the average of $\langle T_i \rangle$ with $n_i = M$, yields:

$$\langle\langle T_i(n_i = M)\rangle\rangle = \left\langle \left\langle \sum_{j=1}^{M} t_{i,j} \right\rangle \right\rangle = \left\langle \sum_{j=1}^{M} \langle t_{i,j} \rangle \right\rangle$$
$$= \left\langle \sum_{j=1}^{M} \langle t_i \rangle \right\rangle = \langle M \langle t_i \rangle \rangle = \langle n_i \rangle \langle t_i \rangle. \qquad (10.\text{A}3)$$

Since the average of a sum of random variables is simply the sum of the averages, this implies that

$$\langle \tau \rangle = \sum_{i=1}^{N} \langle T_i \rangle, \qquad (10.A4)$$

which is the result stated in the main text.

The variance of a sum of random variables is just the sum of the variance of the individual variables and the covariance of all pairs of variables; thus

$$\mathrm{var}(\tau) = \sum_{i=1}^{N} \mathrm{var}(T_i) + 2 \sum_{i=1}^{N} \sum_{j=i+1}^{N} \mathrm{cov}(T_i, T_j). \qquad (10.A5)$$

To compute the variance of the individual T_i, we again use the tower property of averages to calculate the second moment:

$$\langle\langle T_i^2(n_i = M) \rangle\rangle = \left\langle \left\langle \sum_{j=1}^{M} t_{i,j} \sum_{k=1}^{M} t_{i,k} \right\rangle \right\rangle = \left\langle \sum_{j=1}^{M} \langle t_{i,j}^2 \rangle \right\rangle \\ + 2 \left\langle \sum_{j=1}^{M} \sum_{k=j+1}^{M} \langle t_{i,j} t_{i,k} \rangle \right\rangle, \qquad (10.A6)$$

where we have divided the terms into squared terms and cross terms. Since the individual lifetimes are independent and identically distributed, $\langle t_{i,j} t_{i,k} \rangle = \langle t_{i,j} \rangle \langle t_{i,k} \rangle = \langle t_i \rangle^2$. Counting terms in Eq. (10.A6) yields

$$\langle T_i^2(n_i = M) \rangle = M \langle t_i^2 \rangle + M(M-1) \langle t_i \rangle^2 \qquad (10.A7)$$

and averaging over M yields the second moment:

$$\langle T_i^2 \rangle = \langle n_i \rangle \langle t_i^2 \rangle + (\langle n_i^2 \rangle - \langle n_i \rangle) \langle t_i \rangle^2. \qquad (10.A8)$$

Subtracting the mean squared from above yields the variance:

$$\mathrm{var}(T_i) = \langle T_i^2 \rangle - \langle T_i \rangle^2 = \mathrm{var}(n_i) \langle t_i \rangle^2 + \mathrm{var}(t_i) \langle n_i \rangle. \qquad (10.A9)$$

The covariance of the different T_i is defined as

$$\mathrm{cov}(T_i, T_j) = \langle T_i T_j \rangle - \langle T_i \rangle. \qquad (10.A10)$$

The final terms have already been calculated, so we need only the first term T_j.

Again, we calculate this by exploiting the tower property, evaluating the individual $T_{i,j}$ at $n_i = M_i$ and $n_j = M_j$. Thus

$$\left\langle T_i(n_i = M_i) T_j(n_j = M_j) \right\rangle = \left\langle \sum_{k=1}^{M_i} t_{i,k} \sum_{l=1}^{M_j} t_{j,l} \right\rangle = M_i M_j \langle t_i \rangle \langle t_j \rangle.$$

(10.A11)

Averaging over M_i and M_j yields

$$\left\langle T_i T_j \right\rangle = \left\langle n_i n_j \right\rangle \langle t_i \rangle \langle t_j \rangle. \quad (10.\text{A}12)$$

Combining this expression with the averages computed above yields the covariance:

$$\text{cov}(T_i, T_j) = \left\langle n_i n_j \right\rangle \langle t_i \rangle \langle t_j \rangle - \langle n_i \rangle \langle n_j \rangle \langle t_i \rangle \langle t_j \rangle = \text{cov}(n_i n_j) \langle t_i \rangle \langle t_j \rangle.$$

(10.A13)

Combining this result with the variances derived above yields the result in the main text.

A.2. Estimation of systematic errors in statistical moments

Nearly all experimental methods for detecting enzymatic cycle completion times or dwell times of molecular motors have a dead time—a minimum dwell time required for detection of a given event. In the ideal situation, this dead time is much smaller than the typical cycle completion time, and the effects of the dead-time can be ignored. However, in situations in which this is not the case, it would be useful to have quantitative measures of the bias introduced into moments and other statistical properties by the dead-time. Here we provide such an estimate.

If the measurement technique has a finite dead-time t_0, then the cycle completion times that are measured will not follow the actual dwell time distribution, $\varphi(t)$. Rather, these times will be distributed via the modified distribution $\varphi'(t)$:

$$\varphi'(t) = \begin{cases} 0 & t < t_0, \\ \alpha \varphi(t) & t \geq t_0. \end{cases} \quad (10.\text{A}14)$$

Below the dead-time, there is no probability of observing a cycle completion time; thus, the distribution is zero. Above the dead-time, we assume that we can measure all cycle completion times with equal fidelity; thus, the dwell time distribution is the original distribution. An additional factor of α must be included to properly normalize this distribution. A similar technique has been used to produce unbiased estimators of DNA shortening events in single-molecule measurements (Koster *et al.*, 2006).

The measured moments are determined from this distribution via

$$\langle t' \rangle = \int_0^\infty t \varphi'(t) dt \quad \text{and} \quad \langle t'^2 \rangle = \int_0^\infty t^2 \varphi'(t) dt. \tag{10.A15}$$

Thus, the systematic errors arise from both the region of the distribution that is not measured, $t < t_0$, and the fact that not measuring the probability in this region improperly weights the importance of the portion that is measured, that is, $\alpha \neq 1$.

To estimate the systematic errors, we will assume a shape for the distribution for $t < t_0$. As long as the final estimates for the systematic errors are small, then the specific function assumed will not be important. For simplicity, and to allow integrals to be computed directly, we assume that the distribution is well approximated by the gamma distribution during the dead-time:

$$\varphi(t) = \frac{k^n t^{n-1}}{\Gamma(n)} e^{-kt}, \tag{10.A16}$$

where $\Gamma(n)$ is the gamma function. Moreover, since this distribution is only used for a small region of time, its parameters can be estimated from the measured parameters, again, without introducing significant errors into the final estimates of the systematic errors. Namely, the parameters of the gamma distribution are estimated via

$$n \approx \frac{\langle t' \rangle^2}{\langle t'^2 \rangle - \langle t' \rangle^2} \quad \text{and} \quad k \approx \frac{n}{\langle t' \rangle}. \tag{10.A17}$$

The systematic errors in the moments can now be related to integrals of this assumed distribution over the dead-time of the measurement. The error in the mean is

$$\begin{aligned}\langle t \rangle - \langle t' \rangle &= \int_0^\infty t\varphi(t)dt - \int_0^\infty t\varphi'(t)dt = \int_0^{t_0} t\varphi(t)dt \\ &+ (1-\alpha)\int_{t_0}^\infty t\varphi'(t)dt = \int_0^{t_0} t\varphi(t)dt + (1-\alpha)\langle t' \rangle.\end{aligned} \tag{10.A18}$$

In the first line, we took advantage of the fact that the first integral, over $\varphi(t)$, can be split into the region below and above the dead-time, and the region above the dead time can be combined with the same region of the integral over $\varphi'(t)$ as long as the normalization factor α is considered. However, this final integral is by definition the measured moment, $\langle t' \rangle$. Thus, the final expression involves only integrals over the small region of the dead time—the first term and the normalization constant α—and parameters that have already been measured, $\langle t' \rangle$. Since we have assumed a gamma distribution for the distribution during the dead time, the first integral can be performed:

$$\int_0^{t_0} t\varphi(t)dt = \frac{\Gamma(n+1) - \Gamma(n+1, kt_0)}{k\Gamma(n)}. \quad (10.\text{A}19)$$

Similarly the normalization constant, α, can be determined analytically

$$\alpha = \left(\int_{t_0}^{\infty} \varphi(t)dt\right)^{-1} = \left(1 - \int_0^{t_0} \varphi(t)dt\right)^{-1} = \frac{\Gamma(n)}{\Gamma(n, kt_0)}. \quad (10.\text{A}20)$$

Here $\Gamma(n, kt_0)$ is the upper incomplete gamma function. Thus, the final estimate for the systematic error in the mean introduced by a dead-time in the measurement is

$$\langle t \rangle - \langle t' \rangle = \frac{\Gamma(n+1) - \Gamma(n+1, kt_0)}{k\Gamma(n)} + \left(1 - \frac{\Gamma(n)}{\Gamma(n, kt_0)}\right)\langle t' \rangle. \quad (10.\text{A}21)$$

Similar arguments can be applied to determine the error in the second moment. Namely

$$\begin{aligned}\langle t^2 \rangle - \langle t'^2 \rangle &= \int_0^{\infty} t^2\varphi(t)dt - \int_0^{\infty} t^2\varphi'(t)dt \\ &= \int_0^{t_0} t^2\varphi(t)dt + (1-\alpha)\int_{t_0}^{\infty} t^2\varphi'(t)dt \quad (10.\text{A}22) \\ &= \int_0^{t_0} t^2\varphi(t)dt + (1-\alpha)\langle t'^2 \rangle,\end{aligned}$$

where the integral over the dead-time can again be evaluated analytically:

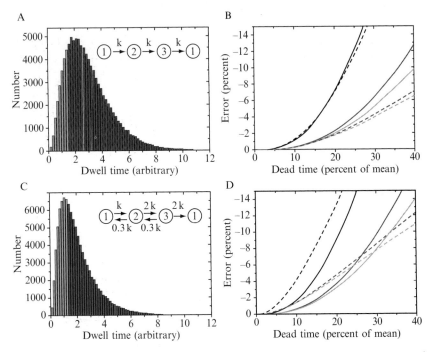

Figure 10.6 Estimating systematic errors from dead-times. (A) Histogram of simulated dwell times for the kinetic mechanism pictured: A three-state irreversible mechanism. The time axis is arbitrary units. Red corresponds to the portion of the distribution which would not be observed if there was a dead-time of 40% of the mean dwell time. (B) Systematic error in the mean (red), second moment (blue), and $n_{\min}$ (black) as a function of dead-time measured in percentage of the mean dwell time. The solid lines correspond to the actual error introduced by the dead-time while the dashed lines correspond to the estimates using the method described here. (C) Histogram of simulated dwell times for the kinetic mechanism pictured: A three-state system with reversible transitions. (B) Systematic errors for this mechanism as a function of dead-time as in panel (B). Data in panels (B) and (D) were calculated from stochastic simulations with 10^6 dwell times. (See Color Insert.)

$$\int_0^{t_0} t^2 \varphi(t) dt = \frac{\Gamma(n+2) - \Gamma(n+2, kt_0)}{k^2 \Gamma(n)}. \quad (10.A23)$$

Thus, the systematic error made in the measurement of the second moment is

$$\langle t^2 \rangle - \langle t'^2 \rangle = \frac{\Gamma(n+2) - \Gamma(n+2, kt_0)}{k^2 \Gamma(n)} + \left(1 - \frac{\Gamma(n)}{\Gamma(n, kt_0)}\right) \langle t'^2 \rangle. \quad (10.A24)$$

The systematic error in functions of the moments, such as the variance or n_{min}, can be calculated by first correcting the measured moments with the estimated systematic errors and then calculating the function of the moments. The systematic error in this function can then be computed by comparing the new estimate to the function evaluated with the measured values. Figure 10.6 shows that these expressions provide reasonable estimates of the systematic errors due to a dead-time as long as the dead-time is relatively small fraction of the mean duration. All statistical properties have errors less than $\sim 10\%$ when the dead-time is less than $\sim 20\%$ of the mean.

In general, it is unlikely that the method used to detect cycle completion events will have a sharp cut-off below which events are never detected and above which events are always detected. If it is necessary to work in conditions in which the dead-time is relatively large, then it may be necessary to include a more representative description of the ability to detect events of different durations. In this case, the general approach described here can be used to estimate systematic errors, and perhaps correct measurements. The discrete distribution in Eq. (10.A14) simply needs to be replaced with the measured response of the detection technique, perhaps estimated from processing of simulated data. The necessary integrals in the subsequent steps can then be performed numerically.

REFERENCES

Abel, N. H. (1826). Beweis der Unmoglichkeit, algebraische Gleichungen von hoheren Graden als dem vierten allgemeinen aufzulosen. *J. fur die reine und Angewandte Math. (Crelle's Journal)* **1**, 65–84.

Aldous, D., and Shepp, L. (1987). The least variable phase type distribution is erlang. *Stochastic Models* **3**, 467–473.

Cappello, G., et al. (2007). Myosin V stepping mechanism. *Proc. Natl. Acad. Sci. USA* **104**, 15328–15333.

Carter, N. J., and Cross, R. A. (2005). Mechanics of the kinesin step. *Nature* **435**, 308.

Cecconi, C., et al. (2005). Direct observation of the three-state folding of a single protein molecule. *Science* **309**, 2057–2060.

Charvin, G., et al. (2002). On the relation between noise spectra and the distribution of time between steps for single molecular motors. *Single Molecules* **3**, 43–48.

Chemla, Y. R., et al. (2008). Exact solutions for kinetic models of macromolecular dynamics. *J. Phys. Chem. B.* **112**, 6025–6044.

Clemen, A. E., et al. (2005). Force-dependent stepping kinetics of myosin-V. *Biophys. J.* **88**, 4402–4410.

Cornish, P. V., and Ha, T. (2007). A survey of single-molecule techniques in chemical biology. *ACS Chem. Biol.* **2**, 53–61.

Derrida, B. (1983). Velocity and diffusion constant of a periodic one-dimensional hopping model. *J. Stat. Phys.* **31**, 433–450.

Efron, B. (1981). Nonparametric estimates of standard error: The jackknife, the bootstrap and other methods. *Biometrika* **68**, 589–599.

Efron, B., and Tibshirani, R. (1986). Bootstrap methods for standard errors, confidence intervals, and other measures of statistical accuracy. *Stat. Sci.* **1**, 54–77.

English, B. P., et al. (2006). Ever-fluctuating single enzyme molecules: Michaelis–Menten equation revisited (vol. 2, p. 87, 2006). *Nat. Chem. Biol.* **2**, 168.
Fisher, M. E., and Kolomeisky, A. B. (1999). Molecular motors and the forces they exert. *Physica a—Stat. Mech. Appl.* **274**, 241–266.
Garai, A., et al. (2009). Stochastic Kinetics of Ribosomes: Single Motor Properties and Collective Behavior. Arxiv preprint arXiv:0903.2608.
Gennerich, A., and Vale, R. D. (2009). Walking the walk: How kinesin and dynein coordinate their steps. *Curr. Opin. Cell Biol.* **21**, 59–67.
Gennerich, A., et al. (2007). Force-induced bidirectional stepping of cytoplasmic dynein. *Cell* **131**, 952–965.
Goedecke, D. M., and Elston, T. C. (2005). A model for the oscillatory motion of single dynein molecules. *J. Theor. Biol.* **232**, 27–39.
Greenleaf, W. J., et al. (2007). High-resolution, single-molecule measurements of biomolecular motion. *Annu. Rev. Biophys. Biomol. Struct.* **36**, 171–190.
Henzler-Wildman, K., and Kern, D. (2007). Dynamic personalities of proteins. *Nature* **450**, 964–972.
Karplus, M., and McCammon, J. A. (2002). Molecular dynamics simulations of biomolecules. *Nat. Struct. Mol. Biol.* **9**, 646–652.
Kohler, D., et al. (2003). Different degrees of-lever arm rotation control myosin step size. *J. Cell Biol.* **161**, 237–241.
Kolomeisky, A. B., and Fisher, M. E. (2003). A simple kinetic model describes the processivity of myosin-v. *Biophys. J.* **84**, 1642–1650.
Kolomeisky, A. B., and Fisher, M. E. (2007). Molecular motors: A theorist's perspective. *Annu. Rev. Phys. Chem.* **58**, 675–695.
Koster, D. A., et al. (2006). Multiple events on single molecules: Unbiased estimation in single-molecule biophysics. *Proc. Natl. Acad. Sci. USA* **103**, 1750–1755.
Kou, S. C., et al. (2005). Single-molecule Michaelis–Menten equations. *J. Phys. Chem. B* **109**, 19068–19081.
Koza, Z. (1999). General technique of calculating the drift velocity and diffusion coefficient in arbitrary periodic systems. *J. Phys. A Math. Gen.* **32**, 7637–7651.
Koza, Z. (2000). Diffusion coefficient and drift velocity in periodic media. *Physica A* **285**, 176–186.
Li, P. T. X., et al. (2008). How RNA unfolds and refolds. *Annu. Rev. Biochem.* **77**, 77–100.
Liao, J.-C., et al. (2007). Extending the absorbing boundary method to fit dwell-time distributions of molecular motors with complex kinetic pathways. *Proc. Natl. Acad. Sci. USA* **104**, 3171–3176.
Linden, M., and Wallin, M. (2007). Dwell time symmetry in random walks and molecular motors. *Biophys. J.* **92**, 3804–3816.
Mallik, R., et al. (2004). Cytoplasmic dynein functions as a gear in response to load. *Nature* **427**, 649–652.
Michaelis, L., and Menten, M. L. (1913). The kinetics of the inversion effect. *Biochem. Z.* **49**, 333–369.
Min, W., et al. (2006). When does the Michaelis–Menten equation hold for fluctuating enzymes? *J. Phys. Chem. B* **110**, 20093–20097.
Moffitt, J. R., et al. (2008). Recent advances in optical tweezers. *Annu. Rev. Biochem.* **77**, 205–228.
Moffitt, J. R., et al. (2009). Intersubunit coordination in a homomeric ring ATPase. *Nature* **457**, 446–450.
Moffitt, J. R., et al. (2010). Mechanistic constraints from the substrate concentration dependence of enzymatic fluctuations. (submitted).
Neuman, K. C., et al. (2005). Statistical determination of the step size of molecular motors. *J. Phys.: Condens. Matter* **17**, S3811.

Neuts, M. F. (1975). Probability distributions of phase type. Liber Amicorum Prof. Emeritus H. Florin, Leuven, pp. 173–206.
Neuts, M. F. (1994). Matrix-geometric solutions in stochastic models: An algorithmic approach. Dover Publications, New York.
O'Cinneide, C. A. (1990). Characterization of phase-type distributions. *Stochastic Models* **6**, 1–57.
O'Cinneide, C. A. (1999). Phase-type distributions: Open problems and a few properties. *Commun. Stat.—Stochastic Models* **15**, 731–758.
Qian, H. (2008). Cooperativity and specificity in enzyme kinetics: A single-molecule time-based perspective. *Biophys. J.* **95**, 10–17.
Reck-Peterson, S. L., et al. (2006). Single-molecule analysis of dynein processivity and stepping behavior. *Cell* **126**, 335–348.
Rief, M., et al. (2000). Myosin-V stepping kinetics: A molecular model for processivity. *Proc. Natl. Acad. Sci. USA* **97**, 9482–9486.
Rock, R. S., et al. (2001). Myosin VI is a processive motor with a large step size. *Proc. Natl. Acad. Sci. USA* **98**, 13655.
Sakmann, B., and Neher, E. (1984). Patch clamp techniques for studying ionic channels in excitable membranes. *Annu. Rev. Physiol.* **46**, 455–472.
Schnitzer, M. J., and Block, S. M. (1995). Statistical kinetics of processive enzymes. *Cold Spring Harb. Symp. Quant. Biol.* **60**, 793–802.
Schnitzer, M. J., and Block, S. M. (1997). Kinesin hydrolyses one ATP per 8-nm step. *Nature* **388**, 386–390.
Segel, I. H. (1975). Enzyme Kinetics. John Wiley & Sons Inc., New Jersey.
Sellers, J. R., and Veigel, C. (2006). Walking with myosin V. *Curr. Opin. Cell Biol.* **18**, 68–73.
Shaevitz, J. W., et al. (2005). Statistical kinetics of macromolecular dynamics. *Biophys. J.* **89**, 2277–2285.
Svoboda, K., et al. (1994). Fluctuation analysis of motor protein movement and single enzyme kinetics. *PNAS* **91**, 11782–11786.
Tinoco, I., and Wen, J. D. (2009). Simulation and analysis of single-ribosome translation. *Phys. Biol.* **6**, 025006.
Tsygankov, D., et al. (2007). Back-stepping, hidden substeps, and conditional dwell times in molecular motors. *Phys. Rev. E Stat. Nonlin. Soft. Matter Phys.* **75**, 021909.
Wang, H. (2007). A new derivation of the randomness parameter. *J. Math. Phys.* **48**, 103301.
Woodside, M. T., et al. (2008). Folding and unfolding single RNA molecules under tension. *Curr. Opin. Chem. Biol.* **12**, 640–646.
Xie, S. N. (2001). Single-molecule approach to enzymology. *Single Molecules* **2**, 229–236.
Xing, J., et al. (2005). From continuum Fokker-Planck models to discrete kinetic models. *Biophys. J.* **89**, 1551–1563.
Xu, W., et al. (2009). Single-molecule kinetic theory of heterogeneous and enzyme catalysis. *J. Phys. Chem. C* **113**, 2393.
Yamaoka, K., et al. (1978). Application of Akaike's information criterion (AIC) in the evaluation of linear pharmacokinetic equations. *J. Pharmacokinet Pharmacodyn* **6**, 165–175.
Yildiz, A., et al. (2004). Myosin VI steps via a hand-over-hand mechanism with its lever arm undergoing fluctuations when attached to actin. *J. Biol. Chem.* **279**, 37223.
Yildiz, A., et al. (2008). Intramolecular strain coordinates kinesin stepping behavior along microtubules. *Cell* **134**, 1030–1041.
Zhou, Y., and Zhuang, X. (2006). Robust reconstruction of the rate constant distribution using the phase function method. *Biophys. J.* **91**, 4045–4053.
Zhou, Y., and Zhuang, X. (2007). Kinetic analysis of sequential multistep reactions. *J. Phys. Chem. B* **111**, 13600–13610.

CHAPTER ELEVEN

VISUALIZING DNA REPLICATION AT THE SINGLE-MOLECULE LEVEL

Nathan A. Tanner[1] and Antoine M. van Oijen[1]

Contents

1. Introduction	260
2. Observing Replication Loops with Tethered Bead Motion	261
2.1. Functionalizing glass coverslips	261
2.2. Functionalizing polystyrene beads	262
2.3. Preparation of forked λ-DNA substrate	263
2.4. Construction of flow chamber	265
2.5. Replication reaction and imaging	268
2.6. Data analysis	269
3. Fluorescence Visualization of DNA Replication	271
3.1. Preparation of rolling-circle template	271
3.2. Construction of flow chamber	272
3.3. Replication reaction and imaging	273
3.4. Data analysis	275
Acknowledgments	277
References	277

Abstract

Recent advances in single-molecule methodology have made it possible to study the dynamic behavior of individual enzymes and their interactions with other proteins in multiprotein complexes. Here, we describe newly developed methods to study the coordination of DNA unwinding, priming, and synthesis at the DNA-replication fork. The length of individual DNA molecules is used to measure the activity of single replisomes engaged in coordinated DNA replication. First, a tethered-particle technique is used to visualize the formation and release of replication loops. Second, a fluorescence imaging method provides a direct readout of replication rates and processivities from individual replisomes. The ability to directly observe transient reaction intermediates and characterize

Department of Biological Chemistry and Molecular Pharmacology, Harvard Medical School, Boston, Massachusetts, USA
[1] Current Address: Single-molecule Biophysics, Zernike Institute for Advanced Materials, Rijksuniversiteit Groningen, Groningen, Netherlands

Methods in Enzymology, Volume 475 © 2010 Elsevier Inc.
ISSN 0076-6879, DOI: 10.1016/S0076-6879(10)75011-4 All rights reserved.

heterogeneous behavior makes these single-molecule approaches important new additions to the tools available to study DNA replication.

1. INTRODUCTION

The duplication of parental DNA into two duplex daughter strands faces an inherent geometric challenge: DNA polymerases catalyze base addition on only the 3' end of a DNA substrate, but the replication template contains two strands in antiparallel orientation. As replication progresses, synthesis of one nascent strand, the leading strand, will follow the direction of DNA unwinding, while the other, the lagging strand, must be synthesized in the opposite direction. Accordingly, polymerization of the leading strand occurs continuously. The orientation of the lagging strand prevents a similar continuity, so the replisome creates transient reversals in lagging-strand directionality in the form of replication loops (Alberts *et al.*, 1983). At each RNA primer synthesized by the primase, a lagging-strand polymerase initiates DNA synthesis in the 5'–3' orientation. As the primer is extended, protein–protein interactions tether the lagging-strand polymerase to the replisome and orient it in the direction of fork progression. The growing lagging strand and helicase-produced single-stranded DNA (ssDNA) form the replication loop, which resolves upon release of the lagging-strand polymerase. This cycle repeats for each new primer, resulting in a series of short double-stranded DNA (dsDNA) regions or Okazaki fragments. The discontinuous production of the lagging strand couples leading- and lagging-strand synthesis to a single replisome as it proceeds through the parental DNA [see recent reviews of coordinated replication and loops (Hamdan and Richardson, 2009; Johnson and O'Donnell, 2005; Yao and O'Donnell, 2008)].

The segmented synthesis of the lagging strand can be demonstrated biochemically, as the short Okazaki fragments can be separated from the long leading strand using reducing agarose gel electrophoresis (Wu *et al.*, 1992a). This approach has been instrumental in understanding the relation between Okazaki fragment length and the frequency of primer synthesis (Tougu and Marians, 1996; Wu *et al.*, 1992b). The ensemble averaging present in such solution-phase approaches, however, makes it challenging to study the kinetics of and the temporal interrelationship between the various enzymatic steps occurring during the production of Okazaki fragments. The use of electron microscopy to image individual replication loops revealed valuable information on the length distributions of Okazaki fragments produced by individual replisomes (Chastain *et al.*, 2003; Park *et al.*, 1998), but the static nature of EM imaging also prevents access to dynamic properties of the loop formation and release process. Recent advances in single-molecule techniques have provided enormous insight into the workings of the individual components of the replication machinery (Ha, 2004; Lee *et al.*, 2006; Pyle, 2008; Seidel and Dekker, 2007; Tanner *et al.*, 2008), but until recently have not been applied to the entire, fully assembled replisome. Here, we describe two methods for real-time, single-molecule observation of coordinated DNA replication by individual replisomes. These techniques utilize length changes in replicating DNA molecules to report on replisomal protein activities and dynamics. First, we explain the observation and measurement of replication loops using beads attached to λ-phage DNA molecules (Hamdan et al., 2009). Second, we describe a simple method of adapting the classic rolling-circle DNA amplification to the single-molecule level. This technique allows for the real-time observation of coupled leading- and lagging-strand synthesis, effectively "watching" high-processivity replication of individual DNA molecules (Tanner et al., 2009).

2. Observing Replication Loops with Tethered Bead Motion

This section details the construction of DNA-bead tethers on functionalized glass coverslips, the stretching of the tethers by laminar flow, and the tracking of the bead-labeled ends of the DNA to visualize replication-loop formation and release.

2.1. Functionalizing glass coverslips

Attachment of the DNA tethers occurs via a biotin–streptavidin–biotin linkage. Glass microscope coverslips are covalently functionalized with biotin-bearing poly(ethylene glycol) (PEG), providing an inert layer between proteins, DNA tethers and the glass to reduce nonspecific interactions (Sofia et al., 1998).

1. Coverslips should be thoroughly cleaned before functionalization reactions. Slips are placed in staining jars containing anhydrous ethanol (EtOH) and sonicated for 30 min. Rinse the coverslips in the jars with deionized water and fill with 1 M potassium hydroxide (KOH). Sonicate again for 30 min. Wash and repeat the EtOH and KOH sonications. The two solvents will remove hydrophobic and hydrophilic contaminants from the glass.
2. For the silanization step, all traces of water need to be removed from the staining jars. After cleaning, fill the jars with acetone and sonicate for 10 min. Empty and refill with acetone. Empty again, pouring the acetone on the lid of the jar and drying the lip of the jar with a paper towel. Fill again with acetone and empty.
3. The silane solution is prepared by making a 2% (v/v) solution of primary amine alkoxy silane (here, 3-aminopropyltriethoxysilane, Sigma) in acetone. Fill the jars with the silane solution and agitate for 2 min. The reaction is quenched with a large excess (>10 jar volumes) of deionized

water poured directly into the container. This reaction is sensitive to water, so it should be performed quickly. Silane stock solution should be stored under dry gas (N_2, Ar) and placed in a desiccator.
4. Dry the silanized coverslips by baking them at 110 °C in an oven for 30 min.
5. While the coverslips are in the oven, the PEG stocks should be taken from the freezer and allowed to warm to room temperature before opening (to avoid condensation, as the PEG succinimidyl esters are labile and will hydrolyze). Once warm, prepare a solution (100 µl per 2 coverslips) of the PEGs. A mixture of biotinylated PEG (here, NHS ester-PEG 5000-Biotin, M-SPA-5000 from Nektar; similarly functionalized PEG can be purchased from Nanocs, CreativePEGWorks etc.) and inert PEG (NHS ester-PEG 5000-Methoxy, Nektar) in 0.1 M $NaHCO_3$, pH 8.2, should be used in order to optimize the biotin surface density. The PEG mixture can be prepared at various ratios to achieve the desired coverage. As described here, 0.2%:15% (w/v) biotin-PEG: methoxy-PEG should give sufficient biotin density for 100–200 λ-DNA-tethered bead constructs at 10× and several hundred tethered M13 substrates at 60× magnification.
6. Pipette 100 µl of PEG solution on the center of a glass coverslip. Place nonfunctionalized spacer coverslips on the edges of the target slip and create a "sandwich" by placing another silanized slip on top of the liquid bubble. This allows good spreading of the PEG solution across the entire surface of the coverslips and eliminates any drying by evaporation. The spacers allow the two coverslips to be separated easily.
7. Incubate the coverslips in the PEG solution for 3 h. Separate the sandwiches and wash each slip thoroughly with deionized water. Dry using compressed air or dry gas and store under vacuum. PEG-functionalized coverslips can be stored under vacuum for at least 1 month without any loss of surface functionality. One preparation of 20–40 coverslips can thus be used repeatedly and remade when depleted.

2.2. Functionalizing polystyrene beads

Beads bind to the ends of surface-attached DNA molecules via a digoxigenin–α-digoxigenin interaction. To do this, paramagnetic polystyrene beads are coated with α-digoxigenin Fab fragments via an amine coupling reaction. This protocol is adapted from the manufacturer's recommendation (Dynal).

1. Resuspend the stock beads (Dynal 142.03, tosyl-activated 2.8 µm diameter) and transfer 0.4 ml to a clean 1.7 ml Eppendorf tube. Buffer is removed by placing the tube in the magnetic separator (Dynal MPC) and separating beads from supernatant. Once the solution clears, remove

buffer with a pipette and remove tube from magnet. Resuspend beads in 1.0 ml 0.1 M H_3BO_3 adjusted to pH 9.5 with NaOH (borate buffer), mixing gently. Remove buffer again using magnet.
2. Antibody solution is prepared by mixing 0.4 ml borate buffer and 0.24 ml Fab solution (Invitrogen, 1.0 mg/ml; gives $\sim$20 μg Fab/mg beads). Add this solution to the bead tube and mix thoroughly. Allowing antibody coupling to proceed for 16–24 h at 37 °C with mild agitation, for example, on a rotator.
3. The next day, pull beads down with the magnet and remove buffer. Resuspend in 1.0 ml 1× PBS (137 mM NaCl, 2.7 mM KCl, 10 mM Na_2HPO_4, 2 mM KH_2PO_4) pH 7.4, 0.1% (w/v) BSA (storage buffer) and incubate at 4 °C for 5 min to wash uncoupled Fab. Repeat this wash step once.
4. Remove storage buffer and resuspend beads in 1.0 ml 0.2 M Tris–HCl, pH 8.5, 0.1% (w/v) BSA to block free tosyl groups. Incubate for 4 h at 37 °C and remove buffer with magnet. Resuspend in 1.0 ml storage buffer and incubate for 5 min at 4 °C.
5. Exchange the storage buffer once more, resuspend thoroughly and aliquot. Beads can be stored at 4 °C for 1 year or longer without substantial loss of quality. As described, this procedure will prepare 1 ml of 1–2 × 10^9 ml^{-1} functionalized beads, enough for >500 experiments.

2.3. Preparation of forked λ-DNA substrate

DNA tethers are constructed using 48.5 kb bacteriophage λ-DNA. The linearized DNA has 12-base single-stranded overhangs at each end, to which modified and unmodified DNA oligonucleotides (in short, oligos) are attached using standard annealing and ligation as described below. A biotinylated oligo is annealed to the 5′ end of the λ-DNA for surface attachment, as the lagging strand needs to be attached in order to visualize length changes during replication. The opposite 3′ end of the λ-DNA is annealed to a short oligo modified with digoxigenin for attachment to the functionalized bead. Preparation of the modified, forked λ-DNA requires four oligonucleotides: (1) λ-complementary fork arm (leading-strand template) 5′-GGGCGGCGACCTGGACAGCAAGTTGGACA ATCTCG-TTCTATCACTAATTCACTAATGCAGGGAGGATTTCAGATATG GCA-3′; (2) leading-strand primer (anneals to 1) 5′-TGCCATATCT-GAAATCCTCCC TGC-3′; (3) biotinylated fork arm (lagging-strand template, anneals to 1) 5′-biotin-A_{16}GAGTACTGTACGATCTAGCAT CAATCACAGGGTCAGGTTCGTTAT TGTCCAACTTGCTGTCC-3′; (4) digoxigenin oligo for bead attachment 5′-AGGTCGCCGCCCA$_{12}$-digoxigenin-3′ (Fig. 11.1).

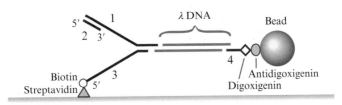

Figure 11.1 Diagram of bead-DNA tether construct. Oligonucleotides 1, 2, and 3 form a primed replication fork with the lagging strand attached to the surface. Oligo 4 anneals to the opposite end of the λ-DNA and binds α-digoxigenin coated beads to complete the tether. Gaps indicate points of ligation as described in Section 2.3.

1. Oligonucleotides 1 and 4 and the λ-DNA need 5′ phosphates for ligation. λ-DNA is purchased in phosphorylated form from supplier (New England BioLabs, NEB), and oligos 1 and 4 can be manufactured with 5′ phosphates (Integrated DNA Technologies, IDT). Alternatively, the 5′ phosphorylation can be performed in house as follows. For each of the two oligos, add 2.0 μl of oligonucleotide (100 μM stock), 15.5 μl of H_2O, 2 μl of 10× ligase/kinase buffer, and 0.5 μl of T4 Polynucleotide Kinase (NEB) and incubate at 37 °C for 1 h. This procedure yields a final concentration of 10 μM phosphorylated oligo for the next steps.
2. Next, fork oligos (1, 2, and 3) are annealed to the λ-DNA. Fork arm 1 anneals directly to its complementary end of the λ-DNA, fork arm 3 anneals to its complementary sequence on arm 1 and primer 2 anneals to the end of arm 1 (Fig. 11.1). Mix 51 μl of 10× ligase/kinase buffer in 400 μl H_2O, then add 56 μl of λ-DNA (14 nM stock), 1.0 μl of oligo 1 (10 μM solution), 2.0 μl of oligo 2 (100 μM solution), and 1.0 μl of oligo 3 (10 μM solution). This provides a ~10× excess of fork oligos (and ~20:1 primer:fork) to ensure that all λ-DNA molecules are annealed to a fork and all forks are primed. To anneal, incubate at 65 °C for 5 min and cool to room temperature slowly by simply turning the power of the heat block off.
3. Ligate the resultant nicks between the oligos and the λ-DNA by adding 2.0 μl of T4 DNA ligase (NEB) and incubating at room temperature for at least 2 h.
4. Last, the digoxigenin end oligo (4) is annealed to the end of the λ-DNA opposite the fork by adding 10 μl of phosphorylated (D) (~100× excess with respect to λ-DNA). Incubate at 45 °C for 30 min and cool to room temperature slowly by turning off heat block.
5. Ligate the digoxigeninated oligo to the λ-DNA by adding 2 μl of T4 Ligase and incubating at room temperature for at least 2 h. This will produce ~500 μl of the double-modified, forked λ-DNA at a concentration of 1.4 nM. Aliquot the sample, pipetting carefully to avoid shearing the long linear DNA and store at 4 or −20 °C.

2.4. Construction of flow chamber

After DNA substrate and beads have been prepared and microscope coverslips functionalized, a flow cell can be assembled and single-molecule experiments performed. Replication experiments are conducted using a simple flow chamber constructed with the functionalized coverslip and a quartz slide. Bead–DNA tethers are constructed *in situ* by flowing λ-DNA forked substrates and functionalized beads into the completed chamber.

1. In preparation for making the flow cell, incubate a functionalized coverslip with streptavidin solution, 25 μl (1 mg/ml stock, Sigma) in 100 μl of 1× PBS, pH 7.3. Spread the solution across the surface (be sure to maintain slide orientation as only one side is functionalized) with a pipette tip and leave at room temperature for 30 min. Incubate the slip in a small box with some water in the bottom to provide high humidity and prevent evaporation of the streptavidin solution.
2. Degas 20 ml of 20 mM Tris–HCl, pH 7.5, 2 mM EDTA, 50 mM NaCl, 0.2 mg/ml BSA, 0.005% (v/v) Tween-20 (blocking buffer) by placing a conical tube in desiccator. Removal of air bubbles is critical for proper flow in the flow chamber.
3. A 1-mm thick, 2-cm × 5-cm sized quartz slide (Technical Glass) will serve as the flow chamber top, and it needs to be drilled for each inlet or outlet tube. Four holes should be made in each slide as two holes ~10 mm apart at each end of the quartz (Fig. 11.2; suitable drill bits can be purchased from DiamondBurs.net). Hole diameter should be

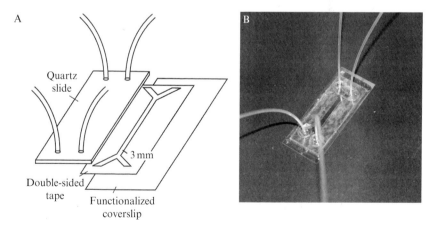

Figure 11.2 Construction of flow chamber. (A) Double-Y channel is cut in double-sided tape, which is affixed to functionalized coverslip and drilled quartz slide. Tubes are inserted into holes in quartz. (B) Example of flow cell for tethered bead experiments. Epoxy around edge of quartz seals chamber to outside air and around tube holes holds tubes in place. Shown channel width is 3 mm.

slightly bigger than the outer diameter of the tubing (Intramedic PE60, 0.76 mm inner diameter, 1.22 mm outer diameter, Becton Dickinson) to hold the tube in place but not block flow. Quartz slides should be cleaned thoroughly with acetone or ethanol before creating a flow cell.

4. The flow channel is cut out of double-sided tape (Grace BioLabs, SecureSeal). Sheets of tape are cut into rectangles to match the quartz slides (2 cm × 5 cm) and the flow channels cut in the desired shape. For experiments with beads, it is best to use two inlet and two outlet channels, as the possibility to switch inlets after flowing beads reduces the amount of washing needed to remove free beads from the tubing. Chamber height is determined by the thickness of the tape (here 100 μm), but the channel width is variable. For coordinated replication experiments, we typically use a 6.0-mm wide channel. Cut the channel outline into a double Y shape, with inlet and outlet channels matching the hole position on the quartz (Fig. 11.2). Align the channel pattern with the holes drilled in the quartz slide by marking the hole position on the tape with a pencil. Once the channel is cut, remove one side of the adhesive backing and carefully place the tape onto the quartz slide. Be sure not to block any of the holes. Apply slight pressure to the tape with plastic forceps to remove any air bubbles from the tape.

5. Wash the streptavidin-coated coverslip thoroughly in water and dry using compressed air, maintaining the orientation of the functionalized side. Once dry, the next steps should be done quickly to minimize air exposure and avoid surface degradation. Remove the backing from the tape/quartz and press onto the functionalized surface of the coverslip and apply slight pressure onto the cover slip with plastic forceps to form a complete seal and remove air bubbles. Apply quick-dry epoxy to the interface between the slide and coverslip, forming a seal around the outer edges of the chamber (Fig. 11.2B).

6. Cut tubing to desired length based on the distance from the microscope stage to the syringe pump. Also, cut the end of the tube that will be inserted through the quartz into the chamber at a $\sim 30°$ angle. This angled cut prevents the face of the tube from sitting too tightly against the flow-cell surface and blocking flow.

7. Insert the four cut tubes into the holes, supporting the tube length so that they are placed stably and vertically into the holes. Apply epoxy to the base of the tube and let dry to seal tube in hole.

8. After the epoxy has dried, the chamber needs to be incubated in blocking buffer to reduce nonspecific interactions. Place the two outlet and one of the inlet tubes into degassed blocking buffer and insert a 21-gauge needle (or other size depending on tube inner diameter) attached to a 1- or 5-ml syringe into the second inlet tube. Slowly draw buffer into the syringe, checking that all tubes permit flow. Expel/draw buffer 2–3 times and leave flow cell full of buffer for 20 min.

9. To connect the flow cell to the syringe pump (Harvard Apparatus 11 Plus) and stabilize any flow irregularities an airspring is used. A 50 ml plastic tube is sealed and the lid affixed using epoxy. Three holes are pierced in the lid, and three lengths of tubing (same tubing as flow cell) are inserted to ~1 cm from the bottom. Using epoxy, the tubes are sealed to the lid, preventing any air from entering or escaping. The tube is filled with 40–45 ml of water and connected to the syringe pump with one of the three tubing pieces. The remaining two connect to the flow-cell tubing using an adaptor piece of slightly larger tubing or a needle. Upon starting the syringe pump, the withdrawal of water from the air spring will result in a pressure drop in the closed air volume. This negative pressure will cause buffer to flow through the flow cell. Any irregularity in the syringe pump motor will not immediately change the negative pressure in the air spring and will be dampened out very effectively.
10. After incubation with blocking buffer, the flow cell is ready for use in experiments. Attach both outlet tubes to the inlet tubes of the airspring. Secure flow cell in place on the stage of an inverted microscope (Olympus IX-51) using stage clamps. Block end of one inlet tube by inserting a needle "plug" filled with epoxy. Any air bubbles should be cleared from tubes by pulling on syringe pump drive arm and flicking the outlet tubes. Switch inlet tubes and repeat. Before flowing DNA, block one outlet tube by kinking to create a flow path from left inlet to right outlet or right inlet to left outlet, facilitating a laminar flow profile.
11. Next, DNA is attached to the surface by flowing into the channel at desired rate and concentration. 1–5 μl of stock DNA construct is diluted in 1 ml of blocking buffer and flowed at 10–40 μl/min (slower flow rates allow for higher surface binding efficiency). This concentration (1–8 pM) can be varied based on observed surface coverage for each experiment or batch of coverslips.
12. Dilute 2–3 μl of stock beads in 1 ml of blocking buffer and mix thoroughly by vortexing, then sonicate for 30 s to disrupt any bead aggregates. Assemble tethers by flowing beads into the channel at 20–50 μl/min.
13. Once beads are added, switch inlet tubes, being sure to close both outlet tubes before any inlet change. Any disturbance to the tubes without the chamber being closed to pressure fluctuations (i.e., if an outlet is open) will exert a strong force on the beads and shear any tethered DNA. Open the other outlet tube and begin to wash flow cell extensively with blocking or replication buffer, manually agitating the microscope stage by tapping to remove any beads nonspecifically stuck to the surface.
14. Tethered DNA can be seen by gentle agitation of the outlet tube or lifting of the airspring to see bead movement based on flow direction. Free beads can be observed visually and will continue to come off the surface for some time. Agitating the stage or outlet tube will speed up the bead removal. Continue until very few free beads are visible.

to base pairs (bp), two parameters need to be determined. First, the pixel size of the camera should be measured by using a 10 µm or similar resolution grating (Edmund Optics) and determining the number of pixels per grid line. Second, the length of a stretched λ-DNA molecule at reaction flow rate should be measured to calibrate the number of bp/pixel at experimental force.

3. For increased accuracy, subtract from the traces of interest a baseline trace of a bead tether that is not enzymatically altered. This should reduce background fluctuation to 100–300 bp.
4. Replication loops will appear as gradual shortening events followed by rapid lengthening (Fig. 11.3). In the T7 system, ssDNA coated with

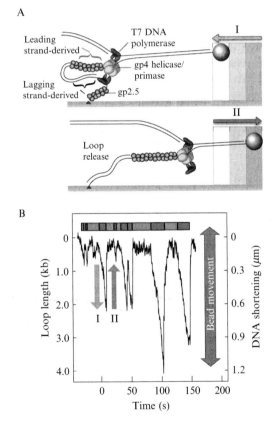

Figure 11.3 Replication loop observation. (A) Loop formation on tethered bead construct. Size of loop is measured by bead displacement from original position. (B) Typical trajectory of replicating molecule. Formation of loop during Okazaki fragment synthesis shortens the overall length of the attached DNA, visible as gradual movement of the bead against flow. Loop release rapidly returns bead to original position as DNA length is restored. Taken with permission from Hamdan et al. (2009).

gp2.5 has the same length as dsDNA, so no length change should be measurable after the loop is released, that is, the position returns to exactly the same level as before the loop was formed (Hamdan et al., 2009). To determine enzymatic processivity, use the Δbp from start to end of a shortening event. For rate calculation, fit the shortening with a linear regression and take the Δy/Δx (slope) of the line as a measure of base pairs/second. Lag time between each loop is measured as the Δx between the complete return of bead to baseline level and the initiation of shortening indicating the formation of a subsequent loop.
5. A successful experiment should yield numerous traces displaying coordinated replication. For each loop, determine rate, loop length, and lag time. Combine all single measurements into distributions to observe population heterogeneity and fit using single-exponential decay (loop length, lag time) or Gaussian distribution (rate).

3. Fluorescence Visualization of DNA Replication

This method utilizes a rolling-circle amplification of dsDNA to observe coordinated replication in real-time. DNA length provides the readout of activity as described above. However, rolling-circle amplification is incompatible with the use of a bead to measure DNA length. Also, the use of a micron-sized bead is not compatible with TIRF imaging, as the large-diameter bead lifts the majority of the stretched DNA out of the 100-nm evanescent wave. Instead, the length of the DNA molecule is measured directly. A rolling-circle substrate is made using a 5′-biotinylated "tail" which anchors the template to the surface. As leading-strand synthesis extends the 3′ end around the circle, the "tail" grows and is converted to dsDNA by lagging-strand synthesis (Fig. 11.4A). The rapidly lengthening dsDNA is stretched with laminar flow and observed with an intercalating dye and TIRF microscopy (Tanner et al., 2009). This method uses the same PEG-biotin coverslips as described above and nearly identical flow chambers.

3.1. Preparation of rolling-circle template

1. Anneal the biotinylated tail oligo to the M13 ssDNA (supplied ~100 nM 40 μl aliquots, NEB) by adding a 10-fold excess of oligo in TBS buffer (final concentrations: ~33 nM M13, 330 nM tail). Heat the mixture to 65 °C in the heat block. Once heated, turn the heat block off and allow slow cooling to properly anneal the primer.
2. To fill in the circle, add the primed M13 to a mixture of 64 nM T7 DNA polymerase (NEB) and T7 replication buffer containing dNTPs and

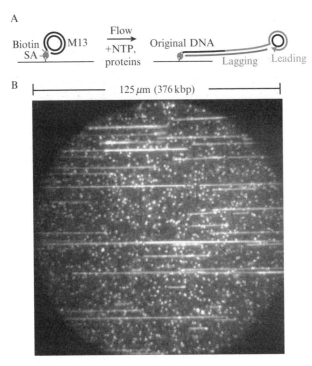

Figure 11.4 Single-molecule rolling-circle experiment. (A) Coordinated leading- and lagging-strand synthesis extends the surface-attached "tail" of the M13 substrate, and the growing DNA is stretched dsDNA products are stained with SYTOX Orange and visible as long lines of DNA (B). Substrates are visible as small bright spots on surface. Each flow cell contains thousands of such fields, each with hundreds of rolling-circle substrates. Taken with permission from Tanner et al. (2009).

MgCl$_2$ (final concentration of primed M13 is 15 nM). Incubate the reaction at 37 °C for 12 min and quench with 100 mM EDTA.
3. Remove replication proteins from the filled-in product DNA solution with phenol/chloroform extraction. Dialyze the DNA into 10 mM Tris–HCl, 1 mM EDTA (TE) or other suitable storage buffer and determine DNA concentration with UV/Vis spectroscopy.
4. One iteration of template preparation can easily make several milliliters of nanomolar concentration template, enough for hundreds to thousands of single-molecule experiments.

3.2. Construction of flow chamber

The flow chamber for the fluorescence experiment is essentially the same as that of the bead experiment described above. The PEG-biotin coverslips are used here with streptavidin solution, double-sided tape, and blocking buffer

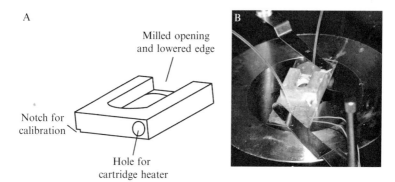

Figure 11.5 Flow cell heater. (A) Schematic of heater. Aluminum block is milled to fit on 2 cm wide quartz slide. A small hole is drilled laterally through the block for insertion of cartridge heater. Though not necessary for fluorescence experiments, a larger hole is drilled down through the center of the block for optical access in bead experiments and for measurement of chamber solution temperature. Milling a small notch alongside edge of heater allows independent temperature readout for checking temperature in each experiment without use of thermocouple in flow channel. (B) Heater is affixed to quartz slide with high thermal conductivity paste and held with epoxy. Cartridge heater is coated with paste and inserted into hole in the block. Leave part of the notched edge unsealed for insertion of thermocouple to measure temperature.

exactly as before. It is desirable to use low-fluorescent background glass, for example, VistaVision (VWR) coverslips. Also, as no bead is present for stretching the DNA in the channel, make the flow channel 1–2 mm wide in order to fully stretch the dsDNA at experimental flow rates (20–50 μl/min). Otherwise high flow rates needed to stretch the DNA either deplete the replication reaction rapidly or require large volumes of proteins to be used. When cutting the channel outline, a single rectangular channel can be used without the forked Y-shape of the bead experiment (Fig. 11.5).

3.3. Replication reaction and imaging

1. After the flow cell has been blocked, everything is ready to begin the single-molecule experiment. Take the flow cell and place it on the microscope stage (Olympus IX-71). Hold the chamber in place with stage clips and be sure that the flow channel is positioned straight so that the flow is aligned with the CCD chip.
2. Connect the flow cell outlet tube to the syringe pump using an airspring as described above. This airspring needs only one inlet tube but still contributes to dampening the syringe pump motor noise. Place the inlet in blocking buffer and pull back on the syringe to remove any air

in the tube. A gentle flick of the outlet tube will help clear any air bubbles trapped in the flow cell.

3. Dilute the stock DNA template to ~25 pM in 1 ml of blocking buffer. Flow into the chamber at moderate flow rate to allow good surface coverage of DNA. This can be changed based on observed or desired surface density, but 20–50 μl/min should provide 100–500 rolling-circle templates in one 60× (125 μm × 125 μm) field of view.
4. Once DNA is flowed into the chamber, wash out excess DNA using blocking buffer. Wash for at least 200 flow-cell volumes to get rid of any free DNA (flow-cell volume should be 4–10 μl depending on channel length and width).
5. Begin degassing the replication buffer. The coordinated T7 reaction is performed exactly as above, but here we will describe an *Escherichia coli* replication experiment. For the *E. coli* reaction, which is performed at 37 °C, the buffer should be degassed at 37 °C to reduce the risk of bubbles in the heated flow chamber.
6. Turn on the laser and camera. The cooled EMCCD (Hamamatsu) needs to reach its operating temperature before it can be used.
7. To achieve 37 °C in the reaction chamber, a small aluminum block can be constructed which is affixed to the flow cell using high thermal conductivity paste and epoxy (Fig. 11.5). The block has a small hole drilled through it along the length of the block, into which is placed a resistive heating element (cartridge heater, Omega Electronics). The cartridge heater is connected to a variable power supply (Elenco) for fine control of voltage and temperature. To calibrate, a flow cell is made as described but with an additional hole drilled in the quartz slide directly above the flow channel. A hole drilled through the top of the block allows access to the flow channel below and permits both placement of the TIRF objective in the flow path and use of the heater in optical microscopy experiments, for example, the bead experiments described above. Insert a thermocouple through the heater opening and into the flow channel hole in the quartz, allowing measurement of the temperature of the flow buffer. Using the flow rate of the replication experiment, the voltage is adjusted to reach 37 °C at the observed region, that is, the middle of the channel. For the TIRF experiment, the oil-immersion objective (Olympus, 60× NA = 1.45) is in contact with the flow cell and acts as a heat sink. This requires the heater to be turned on well ahead of time to equilibrate the temperature. While washing with degassed (37 °C) buffer, begin heating with the objective in contact with the glass slide.
8. While washing and heating, the reaction can be prepared. The *E. coli* replication reaction is performed as: 30 nM DnaB (hexamer), 180 nM DnaC, 30 nM $\alpha\varepsilon\theta$, 30 nM β (dimer), 15 nM $\tau_3\delta\delta'\chi\psi$, 300 n$M$ DnaG, 250 nM SSB (tetramer), 20 nM PriA, 40 nM PriB, 320 nM PriC,

480 nM DnaT in 50 mM HEPES (pH 7.9), 12 mM MgOAc$_2$, 80 mM KCl, 0.1 mg/ml BSA. Replication buffer also contains 40 μM each of dATP, dCTP, dGTP, and dTTP (NEB), 200 μM each of ATP, CTP, GTP, and UTP (Amersham), 10 mM DTT and 15 nM SYTOX Orange (Invitrogen) (Heller and Marians, 2006; Tanner et al., 2009).
9. Flow the reaction mixture into the flow cell. The flow speed for the reaction needs to be sufficient to stretch dsDNA, ~20 μl/min per mm of flow channel width (assuming 100 μm height).
10. Allow time for the mixture to enter the chamber and begin imaging. A good tip is to move the field of view to the side of the channel and focus on the adhesive material. This will ensure that you are near the surface for focusing.
11. Adjust the 532-nm laser power (Coherent Compass 215M-75), microscope focus, and TIRF angle. Use of a dye-appropriate emission filter (here Chroma HQ600/75m) will help eliminate residual laser light. SYTOX and similar DNA stains can cause photocleavage of dsDNA, so it is advisable to perform the experiment at low laser powers (typically 1–5 W/cm^2, should be adjusted based acquisition rate and stain concentration) to minimize photocleavage. Acquire at 1–5 frames/s for several minutes as the rate of replication is ~600 bp/s under described conditions. If desired, repeat to get longer replication trajectories or move to a new field to see more molecules.
12. After the reaction mixture is almost depleted, it is a good idea to flow in more buffer with SYTOX to see other fields of view. Taking multiple images of different replicated molecules provides statistics for processivity determination.

3.4. Data analysis

A typical data set will have several movies of actively replicating molecules. These are easily seen as growing lines of stained DNA. For example, flowing 25 pM of the M13 substrate gives 100–1000 molecules in a 60×, 125 μm × 125 μm field of view. As described, the *E. coli* replisome proteins at 37 °C will yield 10–50 replicating molecules per field of view (Fig. 11.4B). At equivalent concentrations of the T7 replisome, which initiates on the substrate much more efficiently, we observe >70% of the molecules in a field replicate. These conditions yield a product density so high that individual molecules are difficult to resolve and analyze, so the T7 experiments can be performed at lower DNA or protein concentrations.

A quick analysis of the data by eye can give estimates of rates, processivities, and efficiencies from the number of pixels the endpoint moves per frame, the length of the molecule, and the number of products per field, respectively. More accurate data are obtained by fitting the endpoint of the

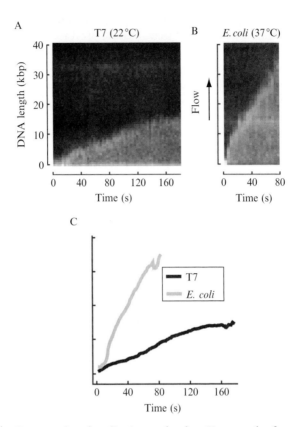

Figure 11.6 Kymographs of replicating molecules. Kymographs from (A) T7 and (B) E. coli replication experiments. (C) Trajectories of endpoints from (A), which are fitted with linear regression to obtain replication rate. Rates of shown examples are 99.4 bp/s (T7) and 467.1 bp/s (E. coli). Taken with permission from Tanner et al. (2009).

DNA molecule over the course of the experiment to obtain rate data. The slope of the endpoint trajectory shows the rate of replication (Fig. 11.6). For processivity, determine the total length of the DNA. Both rate and processivity will need to be converted to base pairs, a calibration done much like that in the bead experiment.

Before performing the experiment, the pixel size of the camera at 60× magnification should be determined. For base pair conversion, a good estimation is simply converting based on the crystallographic length of DNA, 2.9 bp/nm. However, laminar flow will not completely stretch the dsDNA, so a DNA length measurement should be performed by taking a known length DNA, for example, 48,502 bp λ-DNA, attaching it to the flow cell with a biotinylated oligo (the DNA construct from the bead experiment serves this purpose) and measuring its SYTOX-stained length under the experimental flow rate. Determining the number of bp/pixel will allow accurate conversion of the rolling-circle product lengths to base pairs. This experiment allows for visualization of a wider range of DNA product length than an agarose gel, as products in the flow cell can range from 3 to 1000 kbp. A single field of view here is ∼376 kbp wide, but some replicated molecules will stretch across multiple fields (Tanner et al., 2009).

Also, the flow chamber for this experiment is easily modified to have multiple channels. More holes can be drilled in the quartz slide and appropriate channels cut into the double-sided tape to give 1, 2, 3, or more separated reaction channels for screening conditions or inhibitors. PDMS or other microfluidic devices can also be used, increasing the number of channels to dozens if desired.

ACKNOWLEDGMENTS

The development of the bead technique was aided by Dr. Paul Blainey, Dr. Jong-Bong Lee, Candice Etson, and Dr. Samir Hamdan. The fluorescence assay was developed with Drs. Joseph Loparo and Samir Hamdan. T7 replication proteins are received in collaboration with Dr. Charles Richardson, Harvard Medical School. E. coli proteins are received in collaboration with Dr. Nicholas Dixon, University of Wollongong, NSW, Australia. This work is supported by the National Institutes of Health.

REFERENCES

Alberts, B. M., Barry, J., Bedinger, P., Formosa, T., Jongeneel, C. V., and Kreuzer, K. N. (1983). Studies on DNA replication in the bacteriophage T4 in vitro system. *Cold Spring Harb. Symp. Quant. Biol.* **47**(Pt 2), 655–668.

Bustamante, C., Bryant, Z., and Smith, S. B. (2003). Ten years of tension: Single-molecule DNA mechanics. *Nature* **421**, 423–427.

Chastain, 2nd, P. D., Makhov, A. M., Nossal, N. G., and Griffith, J. (2003). Architecture of the replication complex and DNA loops at the fork generated by the bacteriophage t4 proteins. *J. Biol. Chem.* **278**, 21276–21285.

Ha, T. (2004). Structural dynamics and processing of nucleic acids revealed by single-molecule spectroscopy. *Biochemistry* **43**, 4055–4063.

Hamdan, S. M., and Richardson, C. C. (2009). Motors, switches, and contacts in the replisome. *Annu. Rev. Biochem.* **78**, 205–243.

Hamdan, S. M., Johnson, D. E., Tanner, N. A., Lee, J. B., Qimron, U., Tabor, S., van Oijen, A. M., and Richardson, C. C. (2007). Dynamic DNA helicase–DNA polymerase interactions assure processive replication fork movement. *Mol. Cell* **27**, 539–549.

Hamdan, S. M., Loparo, J. J., Takahashi, M., Richardson, C. C., and van Oijen, A. M. (2009). Dynamics of DNA replication loops reveal temporal control of lagging-strand synthesis. *Nature* **457**, 336–339.

Heller, R. C., and Marians, K. J. (2006). Replication fork reactivation downstream of a blocked nascent leading strand. *Nature* **439**, 557–562.

Johnson, A., and O'Donnell, M. (2005). Cellular DNA replicases: Components and dynamics at the replication fork. *Annu. Rev. Biochem.* **74**, 283–315.

Lee, J. B., Hite, R. K., Hamdan, S. M., Xie, X. S., Richardson, C. C., and van Oijen, A. M. (2006). DNA primase acts as a molecular brake in DNA replication. *Nature* **439,** 621–624.
Maier, B., Bensimon, D., and Croquette, V. (2000). Replication by a single DNA polymerase of a stretched single-stranded DNA. *Proc. Natl. Acad. Sci. USA* **97,** 12002–12007.
Park, K., Debyser, Z., Tabor, S., Richardson, C. C., and Griffith, J. D. (1998). Formation of a DNA loop at the replication fork generated by bacteriophage T7 replication proteins. *J. Biol. Chem.* **273,** 5260–5270.
Pyle, A. M. (2008). Translocation and unwinding mechanisms of RNA and DNA helicases. *Annu. Rev. Biophys.* **37,** 317–336.
Seidel, R., and Dekker, C. (2007). Single-molecule studies of nucleic acid motors. *Curr. Opin. Struct. Biol.* **17,** 80–86.
Sofia, S. J., Premnath, V. V., and Merrill, E. W. (1998). Poly(ethylene oxide) grafted to silicon surfaces: Grafting density and protein adsorption. *Macromolecules* **31,** 5059–5070.
Strick, T. R., Allemand, J. F., Bensimon, D., and Croquette, V. (1998). Behavior of supercoiled DNA. *Biophys. J.* **74,** 2016–2028.
Strick, T., Allemand, J., Croquette, V., and Bensimon, D. (2000). Twisting and stretching single DNA molecules. *Prog. Biophys. Mol. Biol.* **74,** 115–140.
Tanner, N. A., Hamdan, S. M., Jergic, S., Loscha, K. V., Schaeffer, P. M., Dixon, N. E., and van Oijen, A. M. (2008). Single-molecule studies of fork dynamics in *Escherichia coli* DNA replication. *Nat. Struct. Mol. Biol.* **15,** 998.
Tanner, N. A., Loparo, J. J., Hamdan, S. M., Jergic, S., Dixon, N. E., and van Oijen, A. M. (2009). Real-time single-molecule observation of rolling-circle DNA replication. *Nucleic Acids Res.* **37,** e27.
Thompson, R. E., Larson, D. R., and Webb, W. W. (2002). Precise nanometer localization analysis for individual fluorescent probes. *Biophys. J.* **82,** 2775–2783.
Tougu, K., and Marians, K. J. (1996). The interaction between helicase and primase sets the replication fork clock. *J. Biol. Chem.* **271,** 21398–21405.
Wu, C. A., Zechner, E. L., and Marians, K. J. (1992a). Coordinated leading- and lagging-strand synthesis at the *Escherichia coli* DNA replication fork. I. Multiple effectors act to modulate Okazaki fragment size. *J. Biol. Chem.* **267,** 4030–4044.
Wu, C. A., Zechner, E. L., Reems, J. A., McHenry, C. S., and Marians, K. J. (1992b). Coordinated leading- and lagging-strand synthesis at the *Escherichia coli* DNA replication fork. V. Primase action regulates the cycle of Okazaki fragment synthesis. *J. Biol. Chem.* **267,** 4074–4083.
Wuite, G. J., Smith, S. B., Young, M., Keller, D., and Bustamante, C. (2000). Single-molecule studies of the effect of template tension on T7 DNA polymerase activity. *Nature* **404,** 103–106.
Yao, N. Y., and O'Donnell, M. (2008). Replisome dynamics and use of DNA trombone loops to bypass replication blocks. *Mol. Biosyst.* **4,** 1075–1084.

CHAPTER TWELVE

MEASUREMENT OF THE CONFORMATIONAL STATE OF F_1-ATPASE BY SINGLE-MOLECULE ROTATION

Daichi Okuno,* Mitsunori Ikeguchi,[†] and Hiroyuki Noji*

Contents

1. Introduction	280
1.1. F_1-Atpase	280
1.2. Rotation of the F_1-Atpase	281
1.3. Conformations of F_1 found in crystal structures and single-molecule studies	281
2. Sample Preparation	282
2.1. Design of a cross-link mutant	282
2.2. Biochemical assay for the β–γ cross-link	284
2.3. Preparation of a hybrid F_1	285
3. Single-Molecule Cross-Link Experiment	287
3.1. Rotation assay	287
3.2. Identification in the rotation assay of the hybrid F_1 with one β(E190D/E391C)	288
3.3. Cross-link formation in the rotation assay	289
3.4. Analysis of pause positions of the cross-linked $\alpha_3\beta_3\gamma$ subcomplex	291
4. Pausing with AMP-PNP or/and N_3^-	293
Acknowledgments	295
References	295

Abstract

F_1-Atpase (F_1) is the water-soluble portion of ATP synthase and a rotary molecular motor in which the rotary shaft, the γ subunit, rotates with 120° steps against the $\alpha_3\beta_3$ stator ring upon ATP hydrolysis. While the crystal structures of F_1 exhibit essentially one stable conformational state of F_1, single-molecule rotation studies revealed that there are two stable conformations of F_1 in each 120° step: the ATP-binding dwell state and the catalytic dwell state.

* The Institute of Scientific and Industrial Research, Osaka University, Osaka, Japan
† Graduate School of Nanobioscience, Yokohama City University, Yokohama, Japan

This chapter provides the experimental procedure for the determination of which catalytic state the crystal structures of F_1 represent, by the use of a cross-linking technique in the single-molecule rotation assay. The β and γ subunits are cross-linked through a disulfide bond between two cysteine residues genetically introduced at the positions where the β and γ subunits have a specific contact in the crystal structures of the ADP-bound form. In the single-molecule rotation assay, the cross-linked F_1 shows a pause at the catalytic dwell state that corresponds to the dwell angle in one turn where the β subunit undergoes ATP hydrolysis. Thus, this experiment reveals not only that the crystal structure represents the catalytic dwell state but also that the ADP-bound β subunit represents the catalytically active state. A protocol for inhibition of the wild-type F_1 with chemical inhibitors such as adenosine-5'-(β,γ-imino)-triphosphate (AMP-PNP) or/and N_3^- under crystallization conditions is also provided.

1. INTRODUCTION

1.1. F_1-ATPase

F_oF_1-ATP synthase is an enzyme that exists in the thylakoid membrane, mitochondrial inner membrane, and bacterial membrane. It consists of a water-soluble portion F_1 and a membrane-embedded portion F_o. F_oF_1-ATP synthase couples the proton translocation across the membranes down the proton motive force (pmf) with the synthesis reaction of adenosine triphosphate (ATP) from adenosine diphosphate (ADP) and inorganic phosphate (P_i) through mechanical rotation (Yoshida et al., 2001). F_1 is an ATP-driven motor subunit, the composition of which in bacteria is $\alpha_3\beta_3\gamma\delta\varepsilon$. The α and β subunits have noncatalytic and catalytic sites, respectively, and form the $\alpha_3\beta_3$ hexameric stator ring. The γ subunit is embedded in the central cavity of the $\alpha_3\beta_3$ ring, and rotates upon ATP hydrolysis against the $\alpha_3\beta_3$ ring in a counterclockwise direction when viewed from the membrane side (Noji et al., 1997). The ε subunit binds to the protruding part of γ and also rotates by being dragged by the γ rotation (Kato-Yamada et al., 1998). F_o is a rotary motor driven by the proton flux down the pmf and has a subunit composition of ab_2c_{10-15}. The c subunits align in a circle to form a ring complex, and the combination of the c-ring with the stator ab_2 subunits forms the proton channel of F_o. Driven by downhill proton flow, the c-ring rotates against the ab_2 subunits in the direction opposite to the γ rotation (Diez et al., 2004). In the F_oF_1 complex, F_o and F_1 are connected through central (γ, ε, and c subunits) and peripheral (δ and b subunits) stalks, and the torque produced by the F_o or F_1 motors is transmitted to the each other through the central stalks. When the pmf is sufficient to overcome the torque of F_1, F_o forcibly rotates the γ subunit in the opposite direction, leading to reversal of the catalytic reaction, that is, to ATP synthesis. When the pmf is insufficient, F_1 hydrolyzes ATP to reverse the c-ring rotation. As a result, F_o is forced to pump protons to build the pmf.

1.2. Rotation of the F_1-ATPase

X-ray crystallographic studies of F_1 provide insights into the rotary mechanism of F_1. The first crystal structure of F_1 from bovine mitochondria (MF_1) (Abrahams *et al.*, 1994) showed that each β subunit has a different conformational state depending on the bound substrate: one bound to adenosine-5'-(β,γ-imino)-triphosphate (AMP-PNP), which is an ATP analogue, another one bound to ADP, and the third one bound to none. These β subunits are denoted as β_{TP}, β_{DP}, and β_{empty}, respectively. While β_{TP} and β_{DP} have a very similar "closed" conformation in which they bend to surround the bound nucleotide in the catalytic site, their α/β interfaces are slightly different, as the α_{DP}/β_{DP} interface is narrower than the α_{TP}/β_{TP} interface. By contrast, β_{empty} has an "open" conformation with lower affinity for a nucleotide. These structural features imply that the cooperative conformational transition of the β subunits would induce a unidirectional rotation of the γ subunit as proposed in the binding change mechanism (Boyer, 1997). Single-molecule rotation assays have revealed that F_1 rotates the γ subunit in 120° steps, each upon a single turnover of ATP hydrolysis (Yasuda *et al.*, 1998), consistent with the expectation from the F_1 crystal structure. These findings implied that the three β subunits change their conformations in a highly cooperative manner to induce unidirectional rotation, which was later confirmed in single-molecule experiments (Masaike *et al.*, 2008).

1.3. Conformations of F_1 found in crystal structures and single-molecule studies

Although the crystal structures of F_1, most of which are of the bovine mitochondrial form, were resolved under various conditions using a wide range of inhibitors, their structural features are essentially the same as evident from the observed small differences in the rotational position of the γ subunit in the $\alpha_3\beta_3$ ring; the variation of the rotary position is mostly within 20°, and the standard deviation is only 8.3° (Fig. 12.1). Thus, the crystal structure of F_1 is thought to represent a specific catalytic state; however, it was unclear which one. In fact, there were several conflicting opinions on the conformational state: some researchers postulated it to be the ATP-waiting state (Gao *et al.*, 2005; Kagawa *et al.*, 2004), whereas others postulated it to be the catalytic-waiting state (Koga and Takada, 2006; Sun *et al.*, 2004). Complementarily, the single-molecule rotation assay of F_1 clearly revealed that F_1 has two stable conformations. Under ATP limiting conditions, F_1 undergoes rotation in 120° steps with intervening pauses to dwell for ATP binding (Yasuda *et al.*, 1998). By high-speed imaging (Yasuda *et al.*, 2001), the 120° step was further resolved into 80° and 40° substeps. Kinetic analysis of the substeps showed that the 80°

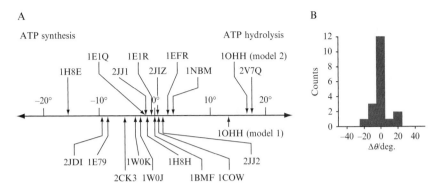

Figure 12.1 Relative orientations of the γ subunit in several crystal structures of bovine MF_1. (A) The relative angular positions of the γ subunit in the crystal structures denoted with Protein Data Bank ID are plotted in the direction of ATP-driven rotation of F_1. The relative angular positions were determined from the center of gravity of the C_α atoms of γ18–25 about an axis of symmetry that was obtained from the Cα atoms of β19 in the three β subunits. 1BMF was used as the reference structure. All structures are derived from the F_1-ATPase of bovine mitochondria (MF_1). (B) Histogram of the angular positions from panel A.

substep is triggered by ATP binding, that is, the dwell before the 80° substep corresponds to the ATP-binding dwell pause. The 40° substep is initiated after two catalytic events, one of which was later revealed to be the ATP hydrolysis step (Shimabukuro et al., 2003). Therefore, the two stable conformational states before the 80° and 40° substeps are referred to as the binding dwell state and the catalytic dwell state, respectively.

In order to reveal the relation between the conformational state of the crystal structures and the two stable states found in single-molecule studies, we carried out an experiment in which the rotary angle of F_1 was fixed during the single-molecule rotation assay by cross-linking, in the position corresponding to that of the F_1 crystal structures (Okuno et al., 2008). This work clarified that the crystal structure represents the catalytic dwell state, consistent with other studies (Masaike et al., 2008; Sielaff et al., 2008). In addition, this experiment revealed that $β_{DP}$ represents the catalytically active state, which promotes hydrolysis of bound ATP. This chapter focuses on the experimental procedures for this cross-link experiment.

2. SAMPLE PREPARATION

2.1. Design of a cross-link mutant

The hybrid F_1 for the cross-link experiment was developed starting from the mutant $α_3β_3γ$ subcomplex from the thermophilic Bacillus PS3 used for the rotation assay, termed $α(His_6$ at the N-terminus$)_3β(His_{10}$ at

N-terminus)$_3\gamma$(S108C/I211C) (Noji et al., 1997; Rondelez et al., 2005). Hereafter, these underlying genetic modifications are omitted for simplicity. For the formation of the β–γ disulfide cross-link, cysteine residues were introduced at positions βE391 and γR84 to generate mutant $\alpha_3\beta$(E391C)$_3\gamma$(R84C) (Fig. 12.2). The arginine at position 84 of the γ subunit is part of an ionic track (Bandyopadhyay and Allison, 2004; Ma et al., 2002) that is composed of positively charged residues around the rotary axis at the basal position of the protruding part of γ. This ionic track is postulated to interact electrostatically with the cluster of negatively charged residues of the β subunit for the mechanical coupling between β and γ; this electrostatic interaction, in combination with a steric interaction, acts as crankshaft to convert the bending and unbending motion of the β subunit into the γ rotation (Ma et al., 2002). The crystal structure of MF$_1$ shows that γR84 is positioned proximally to E391 of the β_{DP} subunit. The distance between the γ carbons of these two residues, which correspond to the sulfur atoms of cysteines, is only 5.9 Å as measured in the crystal structures of bovine mitochondrial F$_1$ (Bowler et al., 2006, 2007; Gibbons et al., 2000; Kagawa et al., 2004; Menz et al., 2001), close enough for a disulfide bond to form. By contrast, the corresponding distance in the β_{TP}–γ and β_{empty}–γ pairs is 11.0 and 22.6 Å, respectively (Table 12.1). Thus, a disulfide bond should be formed predominantly in the β_{DP}–γ pair.

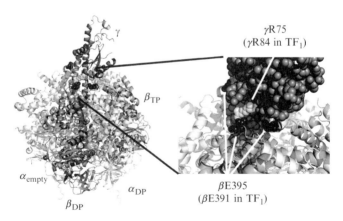

Figure 12.2 Positions of cysteine residues of the β and γ subunits used for cross-linking. βE395 and γR75 are shown in dark gray and light gray, respectively, in a crystal structure (1E79) (Gibbons et al., 2000). These positions correspond to βE391 and γR84, respectively, of F$_1$ from the thermophilic Bacillus PS3, which is used in the present report. The α, β, and γ subunits are shown in light, middle, and middle gray, respectively. The R33, D74, D110, R113, R133, R134, and P135 residues of the γ subunit are removed to show βE395 and γR75. The figure was produced using PyMOL 0.99. (See Color Insert.)

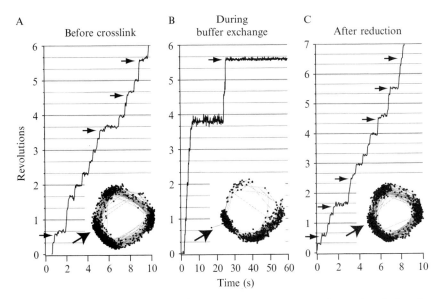

Figure 12.6 Rotation of the $\alpha_3\beta_2\beta$(E190D/E391C)γ(R84C) subcomplex during the cross-link experiment. The inset shows the trajectory of rotation. (A) Rotation before cross-linking. The arrows indicate the catalytic dwell pauses due to the mutant βE190D subunit. (B) Rotation during the buffer exchange with 200 μM DTNB. The rotation stops at $\sim$30 s. The arrows highlight the cross-link-induced pause. (C) Resumption of rotation upon reduction by buffer exchange with 1 mM DTT. The arrows show the catalytic dwell position of due to the mutant βE190D subunit.

Even without being cross-linked, F_1 can spontaneously pause the rotation by lapsing into an inactive state called the ADP-inhibited form, in which F_1 tightly binds to ADP. The ADP-inhibited F_1 pauses at the catalytic dwell angle (Hirono-Hara et al., 2001). Therefore, it is necessary to confirm whether any observed pause is caused by the cross-linking or the ADP-inhibited form of F_1. The ADP-inhibited F_1 can spontaneously resume rotation, or can be reactivated by pushing the γ subunit in the direction of hydrolysis using magnetic tweezers (Hirono-Hara et al., 2005). F_1 is reactivated almost completely when the γ subunit is forcibly rotated over $+80°$ and for >3 s. By contrast, the cross-linked F_1 never resumes rotation. Instead, when released from magnetic tweezers, the γ subunit rapidly returns to the original pause position like a spring (Fig. 12.7A). Thus, one can distinguish a cross-linked F_1 from an ADP-inhibited form from the behavior after releasing from the magnetic tweezers manipulation, exactly whether paused F_1 resumes the spontaneous rotation or not. Other responses to mechanical manipulation are also observed as indicated by the asterisks in Fig. 12.7B–D. For example, when F_1 is rotated with magnetic tweezers by more than $\pm 120°$, it shows a sudden turning in

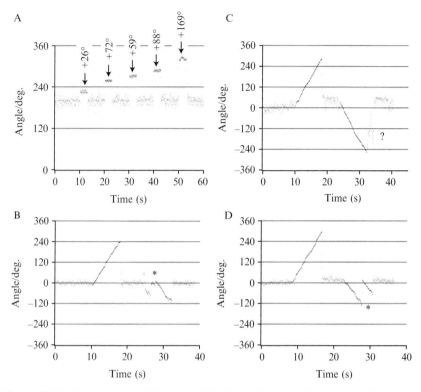

Figure 12.7 Manipulation of the cross-linked F_1 with magnetic tweezers. (A) Time-course of manipulation. The time domains in dark gray each represent a stalling over 3 s with magnetic tweezers. (B–D) Slow forced rotation. Dark gray regions represent the forced rotation. In some cases when F_1 was rotated over 120°, molecules showed irregular responses. Some molecules resisted the external force, and then turned in the angular direction opposite to that of the original magnetic field (asterisks in B and D). This observation is probably due to instantaneous inversion of the magnetic moment of the trapped magnetic bead. Some molecules paused at irregular positions (shown as the question mark in C) after being released from the magnetic tweezers. The reason of this phenomenon is unclear.

some other directions. This observation might be attributed to an aligned magnetic moment transition of the attached magnetic bead because it cannot follow the continuous movement of the external magnetic field.

3.4. Analysis of pause positions of the cross-linked $\alpha_3\beta_3\gamma$ subcomplex

A single dataset of the cross-link experiment and an example of the data analysis are shown in Fig. 12.8. The distance is measured of the cross-link-induced pause angle from the nearest binding angle on the clockwise side or from the

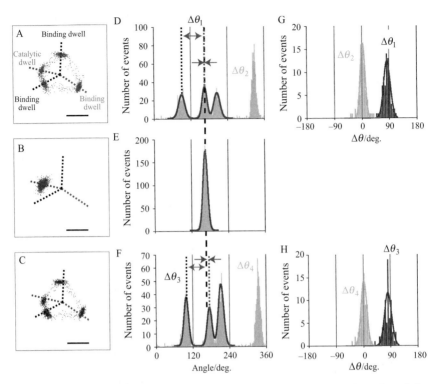

Figure 12.8 Data analysis of the cross-link experiment. (A–C) Trajectories of the centroid of the bead attached to the $\alpha_3\beta_2\beta(\text{E190D/E391C})\gamma(\text{R84C})$ subcomplex. (A) Rotation before cross-link. (B) Pause after cross-link. (C) Rotation after release of the cross-link. Scale bars are 100 nm. (D–F) Angular distributions of the trajectories in panels A, B, and C. The mean angular positions of each ATP-binding dwell and the cross-link pause were determined by fitting with Gaussians. The angle distance of the cross-link-induced angle from the nearest binding angle on the clockwise side ($\Delta\theta_1$) or from the catalytic dwell position of the mutant βE190D subunit ($\Delta\theta_2$) were measured. The same angle distances were also determined as $\Delta\theta_3$ and $\Delta\theta_4$ from the rotation data obtained after reducing the cross-link (C). (G, H) Histograms of $n = 72$ angle distances, $\Delta\theta_1$–$\Delta\theta_4$, obtained from 36 molecules. The following means ($\pm$ S.D.) were determined for $\Delta\theta_1$, $\Delta\theta_2$, $\Delta\theta_3$, and $\Delta\theta_4$ by fitting with Gaussians (dark and light gray): $\Delta\theta_1 = 82.7 \pm 15.9°$, $\Delta\theta_2 = 2.0 \pm 11.0°$, $\Delta\theta_3 = 81.3 \pm 17.0°$, and $\Delta\theta_4 = 1.4 \pm 13.5°$.

catalytic angle observed for βE190D mutation ($\Delta\theta_1$ and $\Delta\theta_2$ in Fig. 12.8D). To ensure that no drift of the attachment site of the molecules occurs, the corresponding angle distances are measured for the datasets from the uncross-linked, reduced F_1 subcomplex ($\Delta\theta_3$ and $\Delta\theta_4$ in Fig. 12.8F). Figure 12.8G shows histograms of the angle distances observed in our experiment; the mean values of $\Delta\theta_1$ and $\Delta\theta_2$ are 82.7 $\pm$ 15.9° and

2.0 ± 11.0°, respectively (Fig. 12.8G; the errors are given as standard deviations). The histograms of $\Delta\theta_3$ and $\Delta\theta_4$ give almost the same values, 81.3 ± 17.0° for $\Delta\theta_3$ and 1.4 ± 13.5° for $\Delta\theta_4$ (Fig. 12.8H). These statistical analyses indicate that the cross-link-induced pause occurs at the catalytic dwell angle of the β(E190D). This observation implies that the conformational state of F_1 in the crystal structures is the catalytic dwell state and that the β_{DP} conformation represents the catalytically active state responsible for ATP hydrolysis.

4. Pausing with AMP-PNP or/and N_3^-

Another method for determining the functional nature of the state observed in the F_1 crystal structure is to measure the pausing position under crystallization condition. Because AMP-PNP and N_3^- are most often used as inhibitors in the crystallization of F_1, it is possible to correlate the crystal structures with a dwell state by the addition of these inhibitors to the single-molecule rotation assay. The pausing positions with inhibitors are measured as the angle distance of the inhibition pause from the ATP-binding position, similar to the cross-link experiment. In the inhibitor experiment, the wild-type of the $\alpha_3\beta_3\gamma$ subcomplex with minimal modifications, α(His$_6$ at the N-terminus)$_3\beta$(His$_{10}$ at N-terminus)$_3\gamma$(S108C/I211C), is used for the rotation assay. Although we use a buffer containing 50 mM Mops–KOH (pH 7.0) instead of HEPES, HEPES-based buffers should be available. After measuring the ATP-binding position at 60 nM ATP, buffer containing 1 μM AMP-PNP and 60 nM ATP or one containing 1 mM N_3^- and 200 μM ATP is infused into the flow cell to inhibit the rotation, similar to the description of the cross-link experiment (Fig. 12.9A and C). To verify that the pause is due to the inhibitor, a pausing F_1 is again manipulated by magnetic tweezers in the manner described for the cross-link experiment. F_1 subcomplexes inhibited with these chemicals do not resume rotation upon manipulation. However, when they are forcibly rotated over +80°, for example +180°, some molecules resume rotation and stop after a few turns. The histograms of each pause position of our experiment are shown in Fig. 12.9B and D. The mean values of the pausing position are 84.2 ± 16.3° for AMP-PNP inhibition and 81.9 ± 16.3° for N_3^- inhibition. The pause position of F_1 inhibited by a combination of AMP-PNP and N_3^- (200 μM AMP-PNP, 5 μM ADP, and 1 mM N_3^-), which mimics the crystallization conditions, is 80.2 ± 14.5° (Fig. 12.9E and F). These results again confirm that the conformational state of the crystal structure of F_1-ATPase is the catalytic dwell state, consistent with the cross-link experiment.

Ma, J., Flynn, T. C., Cui, Q., Leslie, A. G., Walker, J. E., and Karplus, M. (2002). A dynamic analysis of the rotation mechanism for conformational change in F(1)-ATPase. *Structure* **10,** 921–931.

Masaike, T., Koyama-Horibe, F., Oiwa, K., Yoshida, M., and Nishizaka, T. (2008). Cooperative three-step motions in catalytic subunits of F(1)-ATPase correlate with 80 degrees and 40 degrees substep rotations. *Nat. Struct. Mol. Biol.* **15,** 1326–1333.

Menz, R. I., Walker, J. E., and Leslie, A. G. (2001). Structure of bovine mitochondrial F(1)-ATPase with nucleotide bound to all three catalytic sites: Implications for the mechanism of rotary catalysis. *Cell* **106,** 331–341.

Noji, H., Yasuda, R., Yoshida, M., and Kinosita, K., Jr. (1997). Direct observation of the rotation of F1-ATPase. *Nature* **386,** 299–302.

Noji, H., Bald, D., Yasuda, R., Itoh, H., Yoshida, M., and Kinosita, K., Jr. (2001). Purine but not pyrimidine nucleotides support rotation of F(1)-ATPase. *J. Biol. Chem.* **276,** 25480–25486.

Okuno, D., Fujisawa, R., Iino, R., Hirono-Hara, Y., Imamura, H., and Noji, H. (2008). Correlation between the conformational states of F1-ATPase as determined from its crystal structure and single-molecule rotation. *Proc. Natl. Acad. Sci. USA* **105,** 20722–20727.

Rondelez, Y., Tresset, G., Nakashima, T., Kato-Yamada, Y., Fujita, H., Takeuchi, S., and Noji, H. (2005). Highly coupled ATP synthesis by F1-ATPase single molecules. *Nature* **433,** 773–777.

Shimabukuro, K., Yasuda, R., Muneyuki, E., Hara, K. Y., Kinosita, K., Jr., and Yoshida, M. (2003). Catalysis and rotation of F1 motor: Cleavage of ATP at the catalytic site occurs in 1 ms before 40 degree substep rotation. *Proc. Natl. Acad. Sci. USA* **100,** 14731–14736.

Sielaff, H., Rennekamp, H., Engelbrecht, S., and Junge, W. (2008). Functional halt positions of rotary FOF1-ATPase correlated with crystal structures. *Biophys. J.* **95,** 4979–4987.

Sun, S. X., Wang, H., and Oster, G. (2004). Asymmetry in the F1-ATPase and its implications for the rotational cycle. *Biophys. J.* **86,** 1373–1384.

Yasuda, R., Noji, H., Kinosita, K., Jr., and Yoshida, M. (1998). F1-ATPase is a highly efficient molecular motor that rotates with discrete 120 degree steps. *Cell* **93,** 1117–1124.

Yasuda, R., Noji, H., Yoshida, M., Kinosita, K., Jr., and Itoh, H. (2001). Resolution of distinct rotational substeps by submillisecond kinetic analysis of F1-ATPase. *Nature* **410,** 898–904.

Yoshida, M., Muneyuki, E., and Hisabori, T. (2001). ATP synthase—A marvellous rotary engine of the cell. *Nat. Rev. Mol. Cell Biol.* **2,** 669–677.

CHAPTER THIRTEEN

MAGNETIC TWEEZERS FOR THE STUDY OF DNA TRACKING MOTORS

Maria Manosas,[*,†] Adrien Meglio,[*,†] Michelle M. Spiering,[‡] Fangyuan Ding,[*,†] Stephen J. Benkovic,[‡] François-Xavier Barre,[§,¶] Omar A. Saleh,[#] Jean François Allemand,[*,†,||] David Bensimon,[*,†,**] and Vincent Croquette[*,†]

Contents

1. Introduction 298
2. Experimental Setup 299
3. Methods and Protocols 300
   3.1. Surface preparation 300
   3.2. Chamber preparation 301
   3.3. Surface coating 301
   3.4. DNA preparation 301
   3.5. Bead/DNA preparation 306
   3.6. Bead injection 306
   3.7. Selection of beads tethered by a single DNA hairpin 306
   3.8. Selection of beads tethered by a single-nicked DNA molecule 307
   3.9. Selection of beads tethered by a coilable DNA molecule 307
   3.10. Force determination 308
4. Application to the Study of FtsK 308
   4.1. FtsK activity 309
5. Application to the Study of the GP41 Helicase 312
   5.1. Force–extension curve 312
   5.2. Detecting helicase activity 313
   5.3. Optimizing helicase loading 314

[*] Laboratoire de Physique Statistique, Ecole Normale Superieure, Université Paris Diderot, Paris, France
[†] Département de Biologie, Ecole Normale Superieure, Paris, France
[‡] Department of Chemistry, The Pennsylvania State University, University Park, Pennsylvania, USA
[§] CNRS, Centre de Génétique Moléculaire, Gif-sur-Yvette, France
[¶] Université Paris-Sud, Orsay, Paris, France
[#] Materials Department and BMSE Program, University of California, Santa Barbara, California, USA
[||] Institut Universitaire de France, Paris, France
[**] Department of Chemistry and Biochemistry, UCLA, Los Angeles, California, USA

Methods in Enzymology, Volume 475 © 2010 Elsevier Inc.
ISSN 0076-6879, DOI: 10.1016/S0076-6879(10)75013-8 All rights reserved.

297

5.4. Measuring unwinding and ssDNA translocation activities	316
5.5. Using force to investigate the helicase mechanism	318
6. Conclusions	318
Acknowledgments	319
References	319

Abstract

Single-molecule manipulation methods have opened a new vista on the study of molecular motors. Here we describe the use of magnetic traps for the investigation of the mechanism of DNA based motors, in particular helicases and translocases.

1. INTRODUCTION

Single-molecule micromanipulation experiments have brought a new approach to DNA protein interactions (Keller and Bustamante, 2000). They allow one to monitor in real time the activity of single proteins on single DNA molecules and hence reduce the complexity of the system under study. In principle, this leads to a simpler data interpretation. While these experiments were initially challenging because of their novel technical aspects, they are becoming more popular and widespread with the availability of commercial instruments (optical tweezers, atomic force microscopes (AFM), and magnetic traps (MT)).

The principle of magnetic traps is quite simple. It uses the fact that the interaction between a magnetic dipole and a magnetic field produces both a force and a torque on the dipole. Everyone has experienced both effects. A small magnet sticks to the wall of a fridge, because it induces in its magnetic material a dipole of opposite polarity that attracts the magnet to the wall. Similarly, the Earth's magnetic field applies a torque on the magnetized needle of a compass that forces it to align with the field lines and point to the magnetic north pole (which is close to the geographic one). MT uses these common effects to pull on and rotate a micron-sized magnetic bead tethered to a surface by a DNA molecule, through the field of strong and small permanent magnets. The magnetic field applies a force and a torque on the magnetic bead and thus on the DNA molecule. One can thus stretch and coil a DNA molecule (Strick *et al.*, 1996).

Quite often the interaction of proteins with a DNA molecule under tension results in a change of the molecule's extension. This can be due to DNA bending, looping, denaturation, supercoiling, etc. As a consequence, following the molecule's extension as function of time allows for real time monitoring of the DNA/protein interaction dynamics. The applied force is not only a way to stretch the molecule in order to monitor its extension, but it is also a way to affect the DNA/protein interaction. It is thus an alternative tool to temperature, pH, and other standard biochemical parameters that alter chemical equilibria.

In comparison with other popular methods to manipulate single molecules, such as optical tweezers (Svoboda et al., 1994) or AFM cantilevers (Engel et al., 1999), MTs are force clamps. They allow one to set the force applied to a molecule rather than fixing its extension. MT also provide an easy way to fix the orientational angle of the magnetic bead (not the applied torque) and thus the DNA's degree of supercoiling. These properties have a number of experimental consequences. Because MTs are force clamps, their exact positioning is not crucial (the force is not affected significantly if the position of the magnets changes by a few micrometers), which is not the case for optical tweezers or AFM cantilevers. This is quite convenient as it implies that the apparatus does not require particular vibrational isolation or periodic calibrations of the force. It makes an MT setup quite easy to use.

In the present chapter, we are going to first introduce the MT in a more concrete and precise way. We will then give the protocols required to perform an experiment and illustrate some applications in the study of FtsK, a DNA translocase, and gp41, a helicase. These examples will exemplify the use of different DNA substrates in a MT setup.

2. Experimental Setup

A schematic representation of a MT setup is shown in Fig. 13.1A. A DNA molecule (hairpin or double-stranded (ds) DNA) is tethered by its extremities to a glass surface and a magnetic bead. Small magnets positioned above the sample generate a strong magnetic field gradient that pulls the magnetic bead up with constant force. By rotating the magnets one can also coil the molecule. An inverted microscope and a CCD camera are used to image the sample illuminated by the monochromatic focused beam of a LED. The image of the bead displays diffraction rings that are used to estimate its 3-D position as explained elsewhere (Gosse and Croquette, 2002). From the fluctuating positions of the bead both the mean elongation of the molecule and the force applied to it can be deduced (Strick et al., 1996).

Proteins that interact with DNA generally induce distortions in the DNA conformation. These conformational distortions are often translated into changes in the DNA's extension. For example, a helicase that unwinds a DNA hairpin under tension adds two bases of stretched single-stranded (ss) DNA for each unwound base pair (see below). In this way, DNA manipulation techniques allow for the investigation of DNA/protein interactions in real time. In the following, we will demonstrate the use of MT in the study of translocases and helicases.

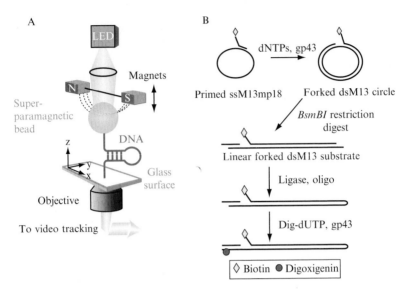

Figure 13.1 Experimental design. (A) Schematic of a magnetic trap setup. (B) DNA hairpin construction used to study gp41 helicase.

3. Methods and Protocols

In order to carry out an MT experiment one needs (in addition to a MT setup) a DNA construct and appropriately functionalized magnetic beads (streptavidin-coated beads are often used) and surfaces. In the following, we present the protocols for preparing surfaces and DNA/bead constructs.

3.1. Surface preparation

Different protocols can be used for preparing the chamber and precoating of the surface. Here we describe one of them (see Lionnet et al., 2008; Revyakin et al., 2003, for alternative protocols).

1. Work in a clean room.
2. Put 60 × 40 mm coverslip on a spin coater.
3. Set acceleration to 4000 rpm^2 and speed to 4000 rpm.
4. Put 200 μL of Teflon AF1600 diluted to 1% in FC72 at the center of the coverslip.
5. Rotate for 10 s.
6. Store Teflon-coated coverslips in a clean container to avoid dust contamination.

Remark: The coated surface is highly hydrophobic.

3.2. Chamber preparation

1. Use a 60 × 40 mm glass coverslip and with a sandblaster make two 1 mm holes in the glass at 50 mm distance.
2. Cut a 60 × 40 mm piece of parafilm. In the center, cut out a band of dimensions 1 × 52 mm.
3. Make a sandwich with the parafilm between a Teflon-coated coverslip (made as above) and the perforated coverslip.
4. Heat the sandwich to 120 °C in order to melt the parafilm and seal the chamber.

Remark: Parafilm can be replaced with double-sided tape.

3.3. Surface coating

1. Fill the chamber ($\sim$10 µL) with a 100 µg/mL solution of polyclonal digoxigenin antibody in phosphate buffer saline (PBS).
2. Incubate for a few hours at 37 °C or overnight at room temperature.
3. Rinse with passivation buffer (PBS + 1 mM EDTA, 1 mg/mL BSA, 0.1% (w/v) pluronic F127, and 10 mM azide).
4. Surfaces can be used after 3 h of incubation at 4 °C.
5. Chambers can be kept for a few weeks in a humid sealed box at 4 °C.

3.4. DNA preparation

In this section, we present two protocols for preparing different DNA constructs: a dsDNA substrate with specific labels at its extremities and a DNA hairpin with labeled tails.

3.4.1. DNA hairpin

Here we present the protocol for preparing a 6.8-kbp-long DNA hairpin that is used to characterize the loading of the gp41 helicase (see below). The protocol for synthesizing the 1.2 kbp hairpin used in the gp41 unwinding experiments (see below) is presented elsewhere (Manosas et al., 2009). A DNA hairpin, 6.8 kbp long, is synthesized by extending a 5′-biotinylated tailed primer annealed to circular single-stranded M13 DNA with a DNA polymerase. The resulting rolling circle product is linearized by digestion with a restriction enzyme and gel purified. A short 5′-phosphorylated oligonucleotide is then ligated to both strands of the 5′-overhang ahead of the fork structure to form a DNA hairpin. The 3′-end of the template strand is labeled in a two-step process using exonuclease digestion followed by 5′-overhang filling with appropriate DNA polymerases and digoxigenin-labeled dUTP. The completed DNA substrate resembles a DNA replication fork with biotin and digoxigenin labels suitable for manipulation using magnetic tweezers (Fig. 13.1).

Procedure

1. The 5′-biotinylated primer (5′-biotin-TTTTTTTTTTTTTTTTT TTTTTTTTTTTTTTTTTTTTGCGCTTAATGCGCCGCTA CAGGGCGCGTAC-3′) may be obtained from numerous companies that provide custom oligonucleotide synthesis. While optional, it is recommended that a primer of this length be purified prior to use. Full-length biotinylated primer may be purified using Pierce Monomeric Avidin UltraLink Resin (Thermo Fisher Scientific) per the manufacturer's instructions.
2. Circular M13 ssDNA is prepared using the large-scale preparation protocol found in Sambrook and Russell (2001).
3. Anneal the 5′-biotinylated tailed primer to positions 5485–5514 of circular M13 ssDNA by heating a mixture of 100 nM M13 ssDNA with 200 nM primer in 1.5 mL of ddH$_2$O to 80 °C for 3 min. Allow the mixture to cool slowly to room temperature over several hours by placing into a water bath that is turned off.
4. To the annealing reaction, add 100 nM T4 DNA polymerase deficient in exonuclease activity (gp43exo$^-$), 250 μM dNTPs, 5% (v/v) DMSO, 20 mM Tris–Ac (pH 7.5), 150 mM KOAc, 10 mM Mg (OAc)$_2$, and ddH$_2$O up to 3 mL (concentrations given are final concentrations). Extend the primer around the M13 ssDNA by incubating the reaction mixture for 1 h at 37 °C. Heat inactivate the polymerase at 65 °C for 20 min.

Note: Another DNA polymerase may be substituted for gp43exo$^-$; however, the substituted DNA polymerase should lack significant strand displacement activity.

Tip: DMSO is included in the synthesis reaction to disrupt any secondary structure of the M13 ssDNA, thereby increasing the complete extension of the primer around the circular M13 single-stranded template. T4 ssDNA-binding protein (gp32) may be used for this purpose instead of DMSO. The efficiency of the extension reaction appears to vary with each batch of purified M13 ssDNA and enzyme. It is recommended that the DNA synthesis be optimized with small test reactions in which the DMSO and gp32 concentrations are varied between 2.5 and 5% (v/v) and 2.5 and 5 μM, respectively. The efficiency of primer extension may be analyzed on a 0.8% agarose gel containing 0.1 μg/mL ethidium bromide run in 0.5× TBE (Tris–Borate–EDTA); the rolling circle product will migrate at an apparently larger size than the M13 ssDNA.

5. Linearize the rolling circle product at position 5976 by adding 60 μL BsmBI restriction enzyme (10,000 U/mL) and incubate for 1.5 h at 55 °C. Quench the digestion reaction with 120 μL of 500 mM EDTA.
6. Purify the linearized forked-DNA product by adding 636 μL of 6× DNA gel loading dye and 0.1% (w/v) SDS (final concentration) to the digestion reaction. Load 18.5 μg DNA (∼100 μL) per well of a 0.8 %

agarose gel containing 0.1 μg/mL ethidium bromide and run in 0.5× TBE until the rolling circle and linearized forked-DNA product bands are separated, approximately 2 h at 100 V for a 12-cm long gel. Cut out the band that migrates at 7.25 kbp (as judged by comparison with a dsDNA ladder) corresponding to the linearized forked-DNA product.

7. The DNA is extracted from the gel slices by electroelution in 1× TBE for 16 h at 50 V. Complete elution of the DNA may be confirmed by restaining the gel slices with ethidium bromide; electroelution is continued on any gel slices still containing DNA. The eluted DNA is precipitated by adding 1/10 volume of 3 M NaOAc (pH 5.2) and 2 volumes of ice cold 95% (v/v) ethanol. Incubate on ice for 1 h. Collect the precipitated DNA by centrifugation at 12,000g for 20 min at 4 °C. Allow the DNA pellet to air dry overnight. Dissolve the DNA in 10 mM Tris–HCl (pH 8.0) with a final DNA concentration around 100 nM. Store at -20 °C.

Tip: Minimize the volume of buffer that the DNA is eluted into to increase the amount of DNA recovered by the ethanol precipitation.

Note: Alternatively, the DNA may be extracted from the gel slices using various gel-extraction kits commercially available per the manufacturer's instructions; however, we find the yield of recovered DNA to be generally lower from these methods.

8. The short oligonucleotide (5′-CCAGGTCAGATGCGTTTTCGCATC TGAC-3′), which forms a hairpin structure with a 4-base loop, 10-base stem, and 4-base overhang complementary to the 5′-cohesive end of the forked-DNA product, may be obtained from numerous companies that provide custom oligonucleotide synthesis. The oligonucleotide may be purchased phosphorylated at the 5′-end or alternatively, may be phosphorylated at the 5′-end with T4 polynucleotide kinase and ATP per the manufacturer's instructions. Following the phosphorylation reaction, the kinase should be heat inactivated at 65 °C for 20 min, but there is no need for further purification.

9. Ligate the 5′-phosphorylated oligonucleotide to the forked-DNA product at 4 °C for a minimum of 24 h in a reaction mixture containing approximately 100 nM forked-DNA, 10-fold excess oligonucleotide, 1× ligation buffer, and T4 ligase per the manufacturer's instructions.

10. Excess oligonucleotide may be removed from the large DNA hairpin product using various PCR cleanup kits commercially available, per the manufacturer's instructions.

Tip: The recovery of the DNA hairpin may often be increased by performing a second DNA elution step and warming the elution buffer to 65 °C.

11. The 5′-overhang of the primer/template end of the forked DNA is increased by exonuclease digestion with T4 DNA polymerase (wild-type gp43). Reactions contain a twofold excess of gp43 over DNA substrate and 1 mM dATP in 20 mM Tris–Ac (pH 7.5), 150 mM KOAc, 10 mM MgOAc$_2$. Incubate the reaction mixture for 20 min at 37 °C. Heat inactivate the polymerase at 65 °C for 20 min.

Note: The extent of exonuclease digestion is limited by the presence of dATP in the reaction causing the polymerase to idle or cycle repeatedly between removing and incorporating dATP when it encounters the first dATP in the template strand.

12. Multiple digoxigenin labels are incorporated as the 5′-overhang is filled-in with T4 DNA polymerase (gp43exo⁻) by adding a twofold excess of gp43exo⁻ over the DNA substrate, 250 μM each dGTP and dCTP, and 50 μM digoxigenin-labeled dUTP directly to the previous reaction. Incubate the reaction mixture for 20 min at 37 °C. Heat inactivate the polymerase at 65 °C for 20 min. Remove protein and excess nucleotides from the DNA hairpin product using various PCR clean-up kits commercially available, per the manufacturer's instructions. Store at -20 °C.

Note: The primer of the forked-DNA substrate is not extended by the gp43exo⁻ polymerase due to its lack of strand displacement activity.

3.4.2. dsDNA construct

The dsDNA construct has been used in FtsK experiments (see below). The protocol must be adapted for the particular DNA sequence being used. Shorter (down to 2 kbps) or longer DNA constructs (typically λ-DNA) may be used.

The guideline is to digest the desired DNA with two restriction enzymes leaving cohesive ends. These ends are then used to bind approximately few hundreds bps PCR products obtained through the incorporation of digoxigenin or biotin modified nucleotides (with a modified/unmodified ratio of about 1/5 or 1/10) as follows (see tables given below)

The PCR products are purified with Microspin SR-400 columns (GE Healthcare): protocol according to the manufacturer's protocol.

Labeling DNA anchoring fragments Reagent	Concentration	Volume (μL)
pBluescriptKS	250 ng/μL	1
Primer A (CTAAATTGTAAGCG TTAATATTTTGTTAAA)	100 μM	1
Primer B, (TATCTTTATAGTCCTGTC GGGTTTCGCCAC)	100 μM	1
dNTPs	10 mM	1.5
Mg^{2+}	25 mM	2
Taq buffer without Mg^{2+}	10×	5
Taq polymerase	Manufacturer stock (NEB)	1
Digoxigenin-11-dUTP or biotin-16-dUTP (Roche)	1 mM	1.5
DI H$_2$O		36.5

PCR program for DNA labeling			
Step	T (°C)	Duration (min)	Number of cycles
1	94	5	1
	94	0.5	30
2	54	1	
	72	1	
3	72	5	1

Central fragment digestion	
Reagent	Volume (μL)
pFX355 (10 kbp at 100 ng/μL)	10
XhoI	1
AatII	1
Eco109I	1
NEB 4	3
H$_2$O	14
37 °C for 1 h + 65 °C for 20 min	

DIG-labeled fragment digestion	
Reagent	Volume (μL)
DIG fragment ($\sim$1 kbp at 50 ng/μL)	10
XhoI	2
NEB buffer 4	4
H$_2$O	4
37 °C for 1 h + 65 °C for 20 min	

Biotin-labeled fragment digestion	
Reagent	Volume (μL)
Biotin fragment ($\sim$1 kbp at 50 ng/μL)	10
AatII	2
NEB buffer 4	4
H$_2$O	4
37 °C for 1 h + 65 °C for 20 min	

Ligation of anchoring fragments to DNA central sequence	
Reagent	Volume (μL)
Digested central fragment	2
Digested DIG fragment	15
Digested Biotin fragment	15
Ligase buffer 10×	10
T4 DNA ligase (Fermentas)	4
H$_2$O	52
16 °C for 2 h then 65 °C for 20 min	

3.5. Bead/DNA preparation

Once the DNA construct has been prepared, mix the DNA with coated magnetic beads at ~1:10 ratio.

1. Pipette 10 μL of MyOne C1 (Invitrogen) streptavidin beads. Clean them according to the manufacturer's protocol. Resuspend in 10 μL PBS.
2. Add typically 1 ng of DNA.
3. After 1 min dilute in 80 μL passivation buffer.
4. After 30 min beads can be used.
5. Beads should be kept on a rotator (10 rpm) at room temperature (to prevent their sedimentation) and can be used for weeks.

3.6. Bead injection

Once the previous steps are complete, inject the DNA/bead construct into the chamber for incubation.

1. Lift the magnets as far away as possible from the sample.
2. With a syringe pump apply a flow and inject 5 μL of the DNA/bead construct. Stop the flow when a large number of beads can be seen.
3. Let the beads sediment for about 5 min.
4. Apply a flow of buffer strong enough to remove the unbound beads, but not so strong as to tear away those that are attached through a DNA molecule to the surface. Stop the flow when no more free beads are passing through the field of view.

3.7. Selection of beads tethered by a single DNA hairpin

Before adding proteins to start the experiment, one needs to identify beads of interest, which are attached to the surface by a single DNA molecule. In order to select suitable beads, we use the previously characterized mechanical properties of the DNA construct.

For the hairpin substrate, the extension remains almost constant below ~15 pN and abruptly increases when the hairpin is unzipped above ~15 pN (see below). If the bead is tethered by two DNA hairpins, the force needed to unzip them is twice as large (~30 pN). Based on these results, we have developed the following protocol for selecting beads with a single DNA hairpin:

1. Change the position of the magnets so that the applied force varies from low force ~1–5 pN to high force ~20 pN.
2. Measure the difference in DNA extension (Δz) between the two applied forces.

3. If Δz is consistent with the length of the unfolded hairpin (typically 1 nm for 1 bp unwound, for example, ~ 1.2 μm for the 1.2 kbp hairpin or 7 μm for the 6.8 kbp hairpin) the bead is selected for the experiment.
4. If Δz is ~ 0, the bead is either nonspecifically bound to the surface or is tethered by two or more DNA molecules. In either case, the bead is ignored.

3.8. Selection of beads tethered by a single-nicked DNA molecule

The simplest way to select a single-nicked dsDNA bead with the magnetic tweezers is to rotate the magnets by a large number of positive turns, thereby strongly supercoiling unnicked DNA molecules or braiding the DNA molecules if a bead is tethered by more than one DNA. These supercoils and braids significantly reduce the extension of the molecules. Therefore, only beads tethered by a single-nicked DNA molecule will remain at a fixed distance from the surface.

3.9. Selection of beads tethered by a coilable DNA molecule

Selection of beads attached with a single-unnicked DNA molecule takes advantage of the fact that negatively supercoiled DNA molecules melt if pulled with a force $F > F_c \sim 0.5$ pN. At forces $F < F_c$ the molecule forms plectonemes that strongly reduce its extension. At larger forces, the molecule does not form plectonemes, but instead denatures (melts) which only slightly affects its extension. To select a bead bound by a single DNA coilable molecule, the idea is to rotate the magnets clockwise by a large number of turns (imposing large negative supercoiling in the DNA). One then tests for a strong change in the bead to surface distance (i.e., the DNA's extension) as the force is varied between values above and below F_c. This type of behavior is not observed if the bead is bound by a nicked DNA or by two or more braided molecules.

Protocol

1. Rotate the magnets clockwise to reach a degree of negative supercoiling of $\sigma \sim -0.1$ (the number of rotations should be about 10% of the number of helical turns in the DNA).
2. Scan the sample while moving the magnets vertically in order to modulate the force around F_c (typically between 0.3 and 1 pN). Beads exhibiting large variations in their distance to the surface are good candidates.
3. At ~ 1 pN (in passivation buffer) force, rotate the magnets counterclockwise (to reach a positive degree of supercoiling of $\sigma \sim 0.1$). The beads of interest should recoil to the surface as positive supercoils are generated.

2. Starting from 10 nM introduce increasing protein concentrations until protein activity is observed. Wait several minutes between protein injections to see if translocation events occur.

Protein activity results in rapid shortening of the DNA extension. Figure 13.3A shows a typical burst of activity. Such events can be recorded for a few hours as long as there are enough active proteins in solution. For data analysis one should make sure that the protein concentration is low enough to observe well separated events to ensure that only one motor is active at any time. Data analysis is relatively simple. Each event (such as the ones shown in

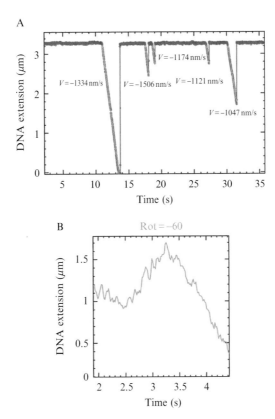

Figure 13.3 Detecting FtsK translocation activity on a DNA molecule by MT (A) FtsK activity events. The DNA molecule shortens as FtsK translocates and forms a DNA loop. The total change in DNA extension indicates the processivity of the enzyme, while the change in extension with time yields the translocation rate for each separate event; the average of these values would be measured in bulk experiments. (B) When negative supercoiling is introduced, one observes an initial increase in the DNA extension due to supercoil removal by FtsK as it twists and translocates on DNA to form a coiled DNA loop.

Fig. 13.3A) is characterized by the slope, the duration, and the extension (height). These values are related to the motor's speed, activity, and processivity, respectively, and may vary with the force, ATP, or salt concentration. In order to relate the enzymatic rate and processivity to the number of base pairs translocated, one must translate the change in the DNA extension at a given force into the DNA contour length. To give an example: at 0.1 pN the DNA extension is only 1/2 of its crystallographic length. Therefore, a decrease in DNA extension at a rate of 1 μm/s corresponds to the motor's actual rate of 2 × 1 μm/s. The relation giving the DNA extension l as a function of the force F is provided by the Worm Like Chain model (Bustamante et al., 1994). A useful approximate formula is (Bouchiat et al., 1999):

$$\frac{F\xi}{k_{\mathrm{B}}T} = \frac{l}{l_0} - \frac{1}{4} + \frac{1}{4(1 - l/l_0)^2} + \sum_{i=2}^{7} a_i (l/l_0)^i$$

where $a_2 = -0.5164228$, $a_3 = -2.737418$, $a_4 = 16.07497$, $a_5 = -38.87607$, $a_6 = 39.49944$, and $a_7 = -14.17718$. Here l_0 denotes the molecular contour length and ξ is the DNA persistence length, which under physiological salt conditions is $\sim$50 nm. This relation gives the relative DNA extension $z = l/l_0$ at each force from which the measured change in DNA extension (δl) can be translated into a change in contour length $\delta l_0 = \delta l/z$.

With MT one also has the ability to investigate the coupling of translocation and rotation as the enzyme moves forming a DNA loop. At forces below 0.5 pN (Strick et al., 1996), supercoils form on unnicked DNA molecules when the bead is rotated. As a result of the formation of plectonemes, the molecule DNA extension decreases by $\sim$40 nm for every turn added. If the motor's step size does not fit the helical pitch of DNA, then it will swivel around the DNA as it proceeds along. If the motor forms a DNA loop, by remaining attached to the DNA molecule at one point while continuing to translocate along the DNA, then the DNA in the loop will be coiled. Since the total linking number of DNA is a topological constant, the amount of coil in the loop has to be compensated for opposite supercoiling in the remaining DNA.

MTs allow one to easily control the degree of DNA supercoiling and as such they are ideally suited to investigate the coupling between translocation and rotation. For that purpose, one studies the motor's activity on a DNA molecule that has been supercoiled through rotation of the magnets. Suppose that the motor's step size is such that it creates supercoils outside the formed DNA loop of opposite sign to the supercoils generated in the stretched DNA. In this case, as the motor proceeds along, the DNA molecule's supercoils are absorbed in twisting of the loop associated with the motor. The observed change in extension of the DNA molecule due to loop formation is therefore smaller than for a nicked (uncoiled) DNA (each adsorbed supercoil lengthens the molecule by about 40 nm). Experimentally, a stretched negatively supercoiled DNA actually lengthens as the FtsK motor translocates, indicating that FtsK generates positive supercoils as it moves along DNA (Fig. 13.3B; Saleh et al., 2005). The experimental protocol is exactly the same as for a nicked DNA except that the molecule is negatively coiled after protein injection.

5. Application to the Study of the GP41 Helicase

DNA helicases are ATP-dependent enzymes capable of unwinding dsDNA to provide the ssDNA template required in many biological processes such as DNA replication, repair, and recombination (Delagoutte and Hippel, 2002, 2003; Hippel and Delagoutte, 2001). Generally, a helicase operating in isolation is difficult to assay as the ssDNA intermediates of the unwinding reaction are transient and may reanneal in the wake of the enzyme. Bulk assays measuring helicase activity use DNA traps such as proteins (e.g., single strand binding (SSB) proteins) or enzymes (e.g., single strand nucleases or the cell replication machinery) that trap or process the ssDNA generated by the helicase activity. In MT experiments, the applied force prevents the DNA from reannealing and allows one to follow the activity of helicases in real time in the absence of DNA trap molecules. Moreover, as discussed below, varying the applied force can help determine the unwinding mechanism of helicases. MT experiments enables one to directly measure the unwinding rate (how many base pairs are opened per second), ssDNA translocation rate (how many nucleotides are translocated per second), and processivity (how many base pairs are unwound before the enzyme dissociates from its substrate) of helicases. In order to illustrate how MT can be used to characterize the behavior of DNA helicases, we next present results on the T4 gp41 replicative helicase working on a DNA hairpin substrate.

5.1. Force–extension curve

First, we characterize the mechanical stability of the DNA hairpin by measuring the extension of the substrate as a function of the pulling force along a force-cycle in which the force is first increased and then relaxed. The force–extension curve for a 1.2 kbp hairpin is shown in Fig. 13.4. As the force is increased, the hairpin remains annealed at a constant extension until the force reaches $F_u = 16 \pm 1$ pN when the extension abruptly increases due to the mechanical unfolding of the hairpin. As the force is decreased below 14 ± 1 pN (F_r), the hairpin reanneals, returning to its initial extension. At forces $F < F_r$, the extension of the DNA molecule remains constant at the level corresponding to the folded hairpin. Thus, in that force range and in the presence of helicase, any unfolding observed results from its unwinding activity.

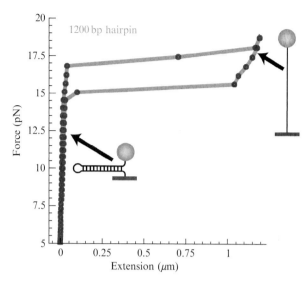

Figure 13.4 Force–extension curve for a 1.2 kbp hairpin. At low forces the hairpin is annealed and displays a constant extension. As the force is increased above ~16 pN, the hairpin is mechanically unzipped and its extension abruptly increases. The force needs to be reduced below 15 pN for the hairpin to fully reanneal; a certain hysteresis is observed (due to the nucleation of a dsDNA seed in the vicinity of the hairpin loop).

Force–extension protocol

1. Move the position of the magnets with respect to the top surface of the flow chamber from $Z_{mag} = -1$ mm ($F \sim 1$ pN) to $Z_{mag} = 0$ mm ($F \sim 17$ pN) and back to $Z_{mag} = -1$ mm in small steps.
2. Use the calibrated force versus magnet position curve to estimate the force (Fig. 13.2).
3. For each position of the magnets measure the DNA extension.

5.2. Detecting helicase activity

When helicase and ATP are added to the chamber, bursts of helicase activity are observed (Lionnet et al., 2007). Unwinding of the hairpin by a single helicase results in an increase in the end-to-end distance of the DNA molecule observed as a change in the distance between the bead and the surface. Complete unwinding is followed by either the instantaneous rehybridization of the hairpin after enzyme dissociation or by hairpin reannealing in the wake of the helicase as it moves along the ssDNA until the extension of the folded hairpin is recovered (Fig. 13.5B).

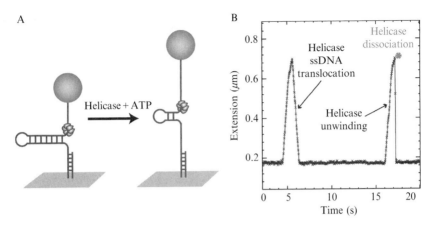

Figure 13.5 Detecting helicase activity on hairpin substrate by MT. (A) Schematic showing increase in DNA extension as a result of helicase activity. (B) Trace showing helicase activity on a 600 bp hairpin (1.2 kbp DNA substrate with a blocking oligo that generate a 600 bp hairpin with 600 nts tails, see below Fig. 13.6B). Two bursts of helicase activity are observed. The first burst corresponds to a full unwinding of the hairpin and the slow hairpin reannealing following the translocation of the helicase on ssDNA, whereas the second burst corresponds to the partial unwinding of the hairpin followed by enzyme dissociation and abrupt hairpin rehybridization.

Materials

1. T4 gp41 helicase (20–100 nM)
2. T4 reaction buffer (25 mM Tris–Ac (pH 7.5), 150 mM KOAc, 10 mM Mg(OAc)$_2$, and 1 mM DTT) and 5 mM ATP.
3. Flow cell with DNA/bead construct.

Protocols

1. Maintain the force constant at the desired value.
2. Wait until an extension increase is detected, indicating the initiation of an helicase burst.
3. Record data during the desired period of time.

5.3. Optimizing helicase loading

The time required for gp41 helicase to load and start unwinding the DNA hairpin decreases as the length of the 5′ ssDNA tail increases (Fig. 13.6A). As the hairpin is unwound by the helicase, the length of the 5′ ssDNA tail increases. Therefore, a second helicase may bind more rapidly as the substrate is unwound, possibly leading to multiple enzymes binding to the substrate. To ensure single-enzyme conditions, the helicase concentration must be well below 100 nM; however, this results in long initial enzyme loading times.

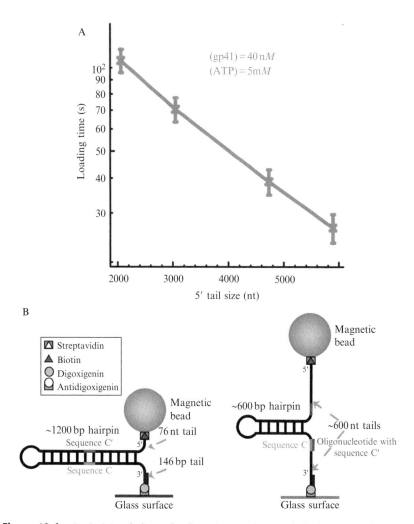

Figure 13.6 Optimizing helicase loading time with annealed oligo. (A) The mean helicase loading time as a function of the length of the 5′ ssDNA tail. Assays measured the loading of 40 nM gp41 at saturating ATP concentration on a 6.8 kbp hairpin substrate. Four oligonucleotides complementary to different sequences along the hairpin were used to obtain 5′ ssDNA tails of approximately 2000, 3000, 4500, and 6000 nt. (B) Schematic representation of the DNA hairpin substrate consisting of a 1239 bp hairpin with a 4-nt loop, a 76-nt 5′-biotinylated ssDNA tail, and a 146-bp 3′-digoxigenin labeled dsDNA tail (Manosas *et al.*, 2009), and the half-hairpin substrate created with a complementary 50-mer oligonucleotide (grey) used to reduce the length of the hairpin and increase the length of the 5′ ssDNA tail.

In order to optimize the conditions for helicase loading on the 1.2 kbp hairpin substrate, we have used a complementary 50-mer oligonucleotide that binds near the middle of the hairpin to increase the length of the 5′ ssDNA tail

(Fig. 13.6B; Manosas et al., 2009). Before starting an experiment, the oligonucleotide is introduced into the chamber at a high concentration, 1 μM. Then a force large enough to unfold the hairpin ($F > 16$ pN) is applied for a few seconds allowing the oligonucleotide to hybridize to its complementary sequence in the hairpin. When the force is reduced to low values, the hairpin reanneals up to the position of the oligonucleotide resulting in a substrate with an ~ 600 bp hairpin, and long 5′ and 3′ ssDNA tails of ~ 600 nucleotides (nt).

Protocol for optimizing helicase loading using a blocking oligonucleotide

1. Inject the 50-mer oligonucleotide at 1 μM diluted in the T4 buffer (25 mM Tris–Ac (pH 7.5), 150 mM KOAc, 10 mM Mg(OAc)$_2$, and 1 mM DTT).
2. Increase the force to ~ 16 pN in order to denaturate the hairpin.
3. Wait for few seconds to allow the oligonucleotide to hybridize to the complementary sequence.
4. Decrease the force to the initial value.

5.4. Measuring unwinding and ssDNA translocation activities

On DNA hairpin substrates, helicase activity is composed of two phases: the unwinding phase (rising edge) corresponding to the release of two nucleotides of ssDNA for each base pair unwound and the rezipping phase (falling edge) corresponding to the slow reannealing of the hairpin following helicase translocation on ssDNA (Lionnet et al., 2007). Conversion of the measured change in DNA extension into the number of base pairs unwound can easily be done by either assigning the maximum DNA extension of the unwinding events to the full length of the unwound hairpin, or by using the previously measured elasticity of ssDNA. The unwinding and ssDNA translocation rates, V_{UN} and V_T, can then be directly computed from the slopes of the unwinding and rezipping phases, respectively (Fig. 13.7A). Alternatively, V_T can be deduced from an experiment where the force is transiently increased to a value of $F > F_u$ in order to denaturate the hairpin during an unwinding event (Fig. 13.7B). After reducing the force to its initial value ($F < F_r$) the hairpin reforms and is shorter by N_t base pairs, corresponding to the distance traveled by the helicase on ssDNA.

Force jump protocol

1. Wait until a burst of helicase activity is observed.
2. Increase the force to ~ 16 pN in order to denaturate the hairpin.
3. Wait for a few seconds (Δt) to allow the helicase to advance along the ssDNA.
4. Decrease the force to the initial value.
5. Calculate the translocation rate by measuring the number of bases the helicase has advanced during Δt.

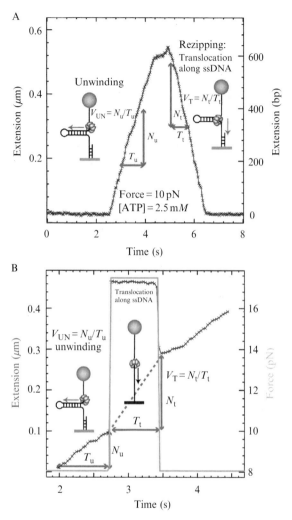

Figure 13.7 Measuring unwinding and ssDNa translocation activities. (A) Experimental trace corresponding to the gp41 helicase activity on the 600 bp hairpin (generated from a 1.2 kbp DNA substrate, Fig. 13.6B). Extension in μm (left axis) is converted to number of base pairs unwound (right axis) by assigning to the maximum length of the unwinding events to the full length of the DNA hairpin. The trace shows the unwinding phase (rising edge) and the rezipping phase (falling edge) in which the enzyme translocates on the ssDNA and the hairpin reanneals in its wake. (B) Experimental trace corresponding to the gp41 helicase activity on the 600 bp hairpin. The applied force (grey) is transiently increased during DNA unwinding by the helicase in order to measure the translocation on ssDNA. (See Color Insert.)

5.5. Using force to investigate the helicase mechanism

From a mechanistic point of view, the most important issue concerning the function of helicases is the coupling between translocation and DNA unwinding. Two mechanisms for helicase unwinding have been proposed. In the passive model the helicase is a ssDNA translocase simply trapping the transient opening fluctuations of the dsDNA; while in the active model the interaction of the helicase with the dsDNA is sufficient to destabilize the double helix removing it as a block to the forward progression of the enzyme. In general, helicases are enzymes that act by lowering the activation barrier of the reaction they catalyze, that is, DNA unwinding. From that point of view, the difference between an active and a passive helicase rests on the size of the activation energy B (Fig. 13.8A). Therefore, an active helicase is one that is able to lower the activation energy significantly below the base pairing energy ($3.4 k_B T$ for a GC base pair, where T is the temperature and k_B the Boltzman's constant), that is $B \ll k_B T$. In contrast, a passive helicase is one that is unable to lower the activation energy resulting in DNA melting being the rate limiting step to DNA unwinding (Fig. 13.8A). In the case of a passive helicase, the application of force on the DNA fork is expected to reduce the activation barrier and result in an increase in the enzyme's unwinding rate at increasing force. In the case of an active helicase, the effect of the force may be negligible as expected for an inch-worm active model (Lohman and Bjornson, 1996) or counterproductive (i.e., slowing down the enzyme) as expected for an active rolling model (Lohman and Bjornson, 1996). Therefore, single-molecule measurements of the rate of a single helicase unwinding a DNA fork under a given tension can yield insight into the mechanism of the studied helicase. Results obtained for gp41 demonstrate that its unwinding rate is extremely sensitive to the applied force, revealing that this helicase is a predominantly passive helicase (Lionnet et al., 2007) (Fig. 13.8B and C). Similar studies have been performed on other helicases (Cheng et al., 2007; Johnson et al., 2007; Sun et al., 2008).

6. Conclusions

In this article we have shown how MTs can be used efficiently to monitor in real time the activity of DNA translocases and helicases. From such data, one can extract enzymatic states (e.g., translocation on dsDNA or ssDNA, forward or backward), unwinding and translocation rates, processivity, and step-size as well as learn about enzymatic mechanisms by studying the variation of these measurables with force and twist on the DNA molecule.

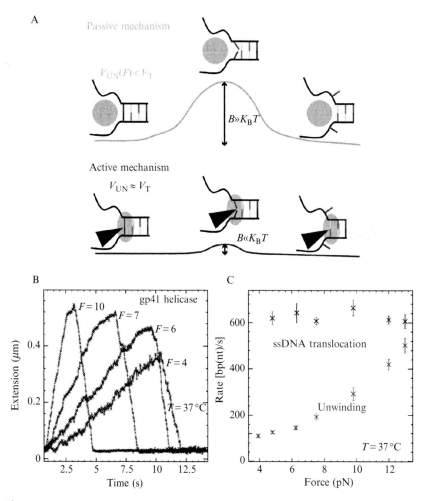

Figure 13.8 Determining helicase mechanism. (A) Schematic representing passive versus active unwinding. (B) Several helicase activity traces at various applied forces. (C) Force dependence of the helicase translocation rate, V_T, and unwinding rate, V_{UN}.

ACKNOWLEDGMENTS

We acknowledge the partial support of grants from ANR, HFSP (RGP0003/2007-C), BioNanoSwitch, IUF and PUF.

REFERENCES

Aussel, L., Barre, F. X., Aroyo, M., Stasiak, A., Stasiak, A. Z., and Sherratt, D. (2002). FtsK is a DNA motor protein that activates chromosome dimer resolution by switching the catalytic state of the XerC and XerD recombinases. *Cell* **108**(2), 195–205.

Bouchiat, C., Wang, M., Block, S. M., Allemand, J.-F., Strick, T., and Croquette, V. (1999). Estimating the persistence length of a worm-like chain molecule from force–extension measurements. *Biophys. J.* **76,** 409–413.
Bustamante, C., Marko, J., Siggia, E., and Smith, S. (1994). Entropic elasticity of λ-phage DNA. *Science* **265,** 1599–1600.
Charvin, G., Allemand, J., Strick, T., Bensimon, D., and Croquette, V. (2004). Twisting DNA: Single molecule studies. *Contemp. Phys.* **45**(5), 383–403.
Cheng, W., Dumont, S., Tinoco, I., and Bustamante, C. (2007). NS3 helicase actively separates RNA strands and senses sequence barriers ahead of the opening fork. *Proc. Natl. Acad. Sci. USA* **104**(35), 13954–13959.
Delagoutte, E., and Hippel, P. H. V. (2002). Helicase mechanisms and the coupling of helicases within macromolecular machines part I: Structures and properties of isolated helicases. *Q. Rev. Biophys.* **35,** 431–478.
Delagoutte, E., and Hippel, P. H. V. (2003). Helicase mechanisms and the coupling of helicases within macromolecular machines part II: Integration of helicases into cellular processes. *Q. Rev. Biophys.* **36,** 1–69.
Engel, A., Gaub, H., and Müller, D. (1999). Atomic force microscopy: A forceful way with single molecules. *Curr. Biol.* **9,** R133–R136.
Gosse, C., and Croquette, V. (2002). Magnetic tweezers: Micromanipulation and force measurement at the molecular level. *Biophys. J.* **82,** 3314–3329.
Hippel, P. H. V., and Delagoutte, E. (2001). A general model for nucleic acid helicases and their "coupling" within macromolecular machines. *Cell* **104,** 177–190.
Johnson, D., Bai, L., Smith, B., Patel, S., and Wang, M. (2007). Asingle-molecule studies reveal dynamics of DNA unwinding by the ring-shaped t7 helicase. *Cell* **129**(7), 1299–1309.
Keller, D., and Bustamante, C. (2000). The mechanochemistry of molecular motors. *Biophys. J.* **78,** 541–556.
Lionnet, T., Spiering, M., Benkovic, S., Bensimon, D., and Croquette, V. (2007). Real-time observation of bacteriophage t4 gp41 helicase reveals an unwinding mechanism. *Proc. Natl. Acad. Sci. USA* **104,** 19790–19795.
Lionnet, T., Allemand, J.-F., Andrey Revyakin, T. R. S., Saleh, O. A., Bensimon, D., and Croquette, V. (2008). Single Molecule Techniques A Laboratory Manual, Chapter 19. Cold Spring Harbor Laboratory Press, Cold Spring Harbor, NY.
Lohman, T. M., and Bjornson, K. P. (1996). Mechanisms of helicase-catalysed unwinding. *Annu. Rev. Biochem.* **65,** 169–214.
Manosas, M., Spiering, M. M., Zhuang, Z., Benkovic, S. J., and Croquette, V. (2009). Coupling DNA unwinding activity with primer synthesis in the bacteriophage T4 primosome. *Nat. Chem. Biol.* **5**(12), 904–912.
Revyakin, A., Allemand, J., Croquette, V., Ebright, R., and Strick, T. (2003). Single-molecule DNA nanomanipulation: Detection of promoter-unwinding events by RNA polymerase. *Methods Enzymol.* **370,** 577–598.
Saleh, O. A., Bigot, S., Barre, F. X., and Allemand, J. F. (2005). Analysis of DNA supercoil induction by FtsK indicates translocation without groove-tracking. *Nat. Struct. Mol. Biol.* **12**(5), 436–440.
Sambrook, J., and Russell, D. W. (2001). Molecular Cloning: A Laboratory Manual, 3rd edn. Cold Spring Harbor Laboratory Press, Cold Spring Harbor, NY.
Strick, T. R., Allemand, J. F., Bensimon, D., Bensimon, A., and Croquette, V. (1996). The elasticity of a single supercoiled DNA molecule. *Science* **271,** 1835–1837.
Sun, B., Wei, K., Zhang, B., Zhang, X., Dou, S., Li, M., and Xi, X. (2008). Impediment of E. coli UvrD by DNA-destabilizing force reveals a strained-inchworm mechanism of DNA unwinding. *EMBO J* **27**(24), 3279–3287.
Svoboda, K., Mitra, P. P., and Block, S. M. (1994). Fluctuation analysis of motor protein movements and single enzyme kinetics. *Proc. Natl. Acad. Sci. USA* **91,** 11782–11786.

CHAPTER FOURTEEN

SINGLE-MOLECULE DUAL-BEAM OPTICAL TRAP ANALYSIS OF PROTEIN STRUCTURE AND FUNCTION

Jongmin Sung,[*,†] Sivaraj Sivaramakrishnan,[*] Alexander R. Dunn,[‡] and James A. Spudich[*]

Contents

1. Introduction	322
2. Insights into Myosin Function Using a Dual-Beam Optical Trap	322
3. Optical Trap Instrumentation	324
3.1. Basic concept of an optical trap	324
3.2. Design and components	325
3.3. Noise and stability considerations	341
3.4. Alignment protocol	344
3.5. Calibration	356
4. Optical Trapping Experiment	358
4.1. Forming an actin dumbbell	358
4.2. Tensing the actin dumbbell	359
4.3. Testing platforms	359
4.4. Identifying binding interactions	361
5. Data Analysis	361
5.1. Measurement of dwell time	361
5.2. Measurement of stroke size: Compliance correction for nonprocessive motors	365
6. Conclusion	369
Acknowledgments	372
References	372

Abstract

Optical trapping is one of the most powerful single-molecule techniques. We provide a practical guide to set up and use an optical trap, applied to the molecular motor myosin as an example. We focus primarily on studies of

[*] Department of Biochemistry, Stanford University School of Medicine, Stanford, California, USA
[†] Department of Applied Physics, Stanford University, Stanford, California, USA
[‡] Department of Chemical Engineering, Stanford University, Stanford, California, USA

myosin function using a dual-beam optical trap, a protocol to build such a trap, and the experimental and data analysis protocols to utilize it.

1. INTRODUCTION

The development of the first dual-beam optical trap for single-molecule analysis was biology driven. The step size of the molecular motor myosin was in great dispute because *in vitro* motility assays to study myosin function examined the collective behavior of multiple motors (Harada *et al.*, 1987; Kron and Spudich, 1986; Uyeda *et al.*, 1990), and the step size of the individual myosin molecule was difficult to be certain of. In the early 1990s, it became clear that one needed to watch a single molecule of myosin go through one chemomechanical cycle and measure the step size directly. Finer *et al.* (1994) "simplified" the Kron and Spudich (1986) *in vitro* motility assay by developing the dual-beam optical trap system. Their study revealed a ~ 10 nm stroke size for muscle myosin II with ~ 5 pN forces generated by it. These results provided strong support for the lever arm model of myosin motion, also referred to as the swinging cross-bridge hypothesis (Huxley, 1969). Since 1994, the dual-beam optical trap has been used to study a variety of proteins at the single-molecule level (Bustamante *et al.*, 2007; Nishikawa *et al.*, 2007; Spudich *et al.*, 2007). Details of the early version of the dual-beam trap in our laboratory have been described elsewhere (Finer *et al.*, 1994; Rice *et al.*, 2003; Spudich *et al.*, 2007). Here, we describe a modern version of the dual-beam trap, which has primarily changed in terms of more highly developed individual components. We use myosin as the prototype molecule in describing the details of how to apply the trap to single-molecule analysis. The reader is referred to review papers for the recent advances in optical traps and their use (Berg-Sørensen and Flyvbjerg, 2004; Moffitt *et al.*, 2008; Neuman and Block, 2004; Perkins, 2009).

2. INSIGHTS INTO MYOSIN FUNCTION USING A DUAL-BEAM OPTICAL TRAP

Since the Finer *et al.* (1994) studies, the dual-beam optical trap has been widely used to study myosins. In translational research, for example, it was used to characterize changes in cardiac myosin II function caused by single-point mutations in the most common inherited heart disease, hypertrophic cardiomyopathy (Palmiter *et al.*, 1999, 2000). Optical trapping showed that the R403Q mutation in β-cardiac myosin, which results in sudden death, decreases myosin dwell time but does not alter stroke size or force-generating ability (Palmiter *et al.*, 2000). The use of the optical trap to dissect single myosin function showed that the functional consequence of the R403Q mutation is a result of reduced coordination between the ATPase cycles of myosins in myosin thick filaments.

Myosin II has a low duty ratio (fraction of the ATPase cycle time spent tightly bound to the actin filament; Marston and Taylor, 1980) with dwell times on the order of 10 ms at saturating ATP concentrations and a stroke size of ~ 10 nm. These small stroke sizes and short dwell times are likely resolvable with the noise and frequency response of the trap described here. To facilitate the separation of such short binding events and to distinguish them from the Brownian motion of the trapped bead, the technique of mean–variance analysis has been used (Guilford et al., 1997).

Unconventional myosins, such as myosin V and VI, by contrast, are high duty ratio motors with significantly larger dwell times (~ 100 ms) and stroke sizes (> 20 nm). Whereas single dimeric molecules of myosin II are nonprocessive, dimeric forms of myosin V and VI are processive motors (Mehta et al., 1999; Rock et al., 2001). A processive motor takes multiple steps along an actin filament without detaching from it. Processivity requires the myosin molecule to be dimeric and have a high duty ratio, such that the leading head remains attached to the actin filament while the trailing head releases and finds its next binding site. Processive motors such as myosin V and VI take several steps along an actin filament in the trap. This feature combined with their long dwell times and large stroke sizes have enabled the use of the optical trap to dissect the workings of the myosin mechanoenzyme in great detail over the last decade (Spudich, 2001; Spudich and Sivaramakrishnan, 2009).

Unlike myosin II, which has two light chains bound to its lever arm (Rayment et al., 1993), myosin V has six (Kad et al., 2008). Purcell et al. (2002) used single-headed truncated versions of myosin V in the dual-beam optical trap to show that the larger stroke of myosin V (~ 25 nm) compared with myosin II (~ 10 nm) is a result of the lever arm that is six light chain-binding domains long. Using a dual-beam optical trap, Veigel et al. (2002) dissected the ~ 36 nm processive step of dimeric myosin V (Mehta et al., 1999) to reveal that it consists of a ~ 25 nm stroke of the lead head, followed by preferred binding of the free head 72 nm from its previous binding site (equivalent to a 36 nm movement of the center of mass of the dimer). The dual-beam optical trap provided adequate time resolution to discern that the 25 nm working stroke consists of two substeps of 20 and 5 nm, respectively. Using the kinetic parameters obtained from analysis of the optical trapping data, the authors suggested that the second 5 nm substep acts to increase processivity by coordinating the release of the rear and lead heads of the motor. This coordination, also referred to as "gating," leads to a hand-over-hand motion of myosin V and has since been observed directly using other single-molecule assays (Churchman et al., 2005; Dunn and Spudich, 2007; Sakamoto et al., 2008). Gating results from intramolecular strain that

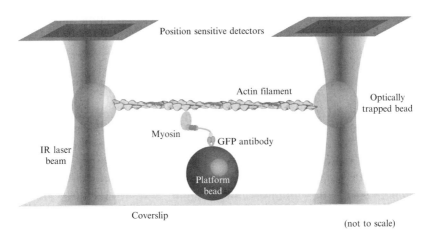

Figure 14.1 Dual-beam, three-bead optical trapping assay for the study of myosins. Schematic of a dual-beam laser optical trap for measuring the lever arm stroke, force generation, and kinetics of single myosin molecules. Two neutravidin-coated polystyrene beads (~ 1 μm, light gray) are optically trapped by focused infrared (IR) laser beams. A single biotinylated actin filament (9 nm diameter) is attached at its ends to the two optically trapped beads. The actin filament is stretched taut (~ 10 μm long) by moving the two trapped beads away from each other by independently controlling each laser beam. Illustrated here is a myosin VI monomer with a C-terminal GFP-tag attached to a ~ 1.5 μm polystyrene platform bead (dark gray) coated with anti-GFP antibody. The position of each of the trapped beads is determined accurately by two position-sensitive detectors (PSD) located in the optical path above the trapped beads. Each PSD is optically conjugated with the back-focal plane of the condenser (not shown). *Note*: For purposes of clarity, the components in the figure are shown at different length scales.

bright-field illumination to image beads, and fluorescence illumination to image actin filament (Figs. 14.2 and 14.3). In this section, we provide an overview of the dual-beam trap setup. Details of the setup are described later (see alignment, Section 3.4). Further description of the setup can also be found in the legends of Figs. 14.2–14.4.

The trapping laser and beam steering optics enable us to trap the two beads and independently control their position, which is essential to make a tight actin dumbbell (Fig. 14.2, panel 1). A pair of polarizing beam splitters (PBS2, PBS3) separate and recombine the p- and s-polarized components of the trapping beam. Piezo-driven mirror mounts (PM1, PM2) are used for independent coarse positioning of the dual trapping beams to enable the formation of the actin dumbbell. We use a pair of acousto-optic deflectors (AOD) for fast (~ 1 μs) and precise (~ 1 nm) control of trap position during the feedback control. This fast and precise feedback control of the trap beam is essential for the study of processive motors such as myosin V. In order to make a stable actin dumbbell, the distance between the two trapped beads must be kept constant. This is achieved by minimizing the path length

Dual-beam optical trap setup

(not to scale)

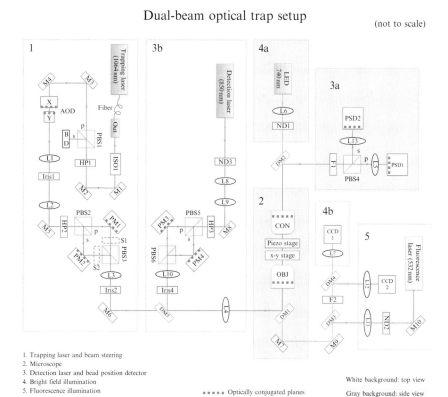

Figure 14.2 Design of a dual-beam optical trap for the three-bead assay. The setup consists of five main parts: (1) trapping laser and beam steering, (2) microscope, (3) detection laser and bead position detector, (4) bright-field illumination, and (5) fluorescence illumination. Note that the white background is the top view while the gray background is the side view. Specific information of each component is found in Table 14.1. (1) Trapping laser and beam steering (panel 1): a collimated 1.6 mm trapping laser beam from a 1064 nm ytterbium-doped fiber laser passes through an isolator (ISO1), which protects the fiber laser from back reflections from any downstream components. Mirrors 1 and 2 (M1, M2) adjust the beam path such that it is parallel to the surface of the optical table. A half-wave plate (HP1) and polarizing beam splitter (PBS1) control the final power of the trapping beam, which is p-polarized. The s-polarized component is dissipated into a beam dump (BD). Alternately, the s-polarized beam can be used for single-beam feedback control (described in Section A.2). There are two different beam steering controls: fast and precise beam steering uses the acousto-optic deflectors (AODs), and slow and coarse beam steering uses the piezo-driven mirror mounts (PMs). The AODs are followed by a telescope with two lenses L1 and L2, which allow optical conjugation of the AODs and PM1/PM2. The PMs are followed by a telescope with two lenses L3 and L4 that optically conjugate the PMs with the back aperture of the objective lens (OBJ). Therefore, both AODs and PMs are conjugated with the OBJ back aperture (gray dotted lines). This allows us to use the AODs for fast and precise control of the optical trap, especially during feedback control of trap position for processive dimeric motors, such as myosin V. PMs, on the other

between PBS2 and PBS3 since it reduces the differential effects of air fluctuation and mechanical vibration on the separated beams. The AODs operate on the combined beam, contrary to previous designs (Rice et al., 2003). In this configuration (Fig. 14.2, panel 1), the AODs cannot be used to steer two trapped beams independently. This feature does not affect studies on myosin function, which we use as an example in this chapter. For studies that require fast feedback control of just one bead, please refer to a subtle modification in this design discussed in Section A.2.

hand, allow the relatively slow and coarse positioning of beads, which is essential for making a taut actin dumbbell. Two sets of telescopic lenses (L1–L4) expand the beam diameter sixfold (from 1.6 to 9.6 mm), so the final beam slightly (10%) overfills the back aperture of OBJ ($D = 9.6$ mm). The second half-wave plate (HP2) followed by two beam-splitters (PBS2/PBS3) create two independent beam paths (s- and p-polarized) that can be steered independently using PM1 and PM2. Two shutters (S1, S2) can be placed between PBS2 and PBS3, such that the two trapping beams can be independently turned on and off. For the distance between each component, see Figs. 14.3 and 14.4. (2) Microscope (panel 2): the steered trapping beam is reflected by the first dichroic mirror (DM1) and passes through the OBJ (NA = 1.45) which focuses it in the sample chamber. The condenser (CON) (NA = 1.35) collects the transmitted beams, which forms interference patterns at the back-focal plane (BFP, gray dotted line) of the CON. The interference patterns are produced from the combination of unscattered incident beams and forward scattered beams from the trapped beads. (3) Detection laser and bead position detector (panel 3): (a) bead position detector—following the CON, the trapping beams are reflected on a dichroic mirror (DM2) followed by two lenses L5/L13 that image the BFP interference patterns at two position-sensitive detectors (PSD1/PSD2) for dual traps. A band-pass filter (F1) is used to attenuate 1064 nm trapping beams. (b) Detection laser—the 850 nm laser beams are used for the detection of the beads. While the 1064 nm trapping laser beams can be used directly to detect trap position, in the three-bead assay discussed here, additional 850 nm laser beams are used for improved accuracy (described in the text). The function of components in the 850 nm laser beam paths is similar to that of the trapping laser. A neutral density filter (ND3) provides optimum power for detection. PM3 and PM4 independently control the position of the detection laser beams relative to the trapped beads. A dichroic mirror (DM5) allows the detection beams to be coupled to the trapping beams. (4) Bright-field illumination (panel 4): we visualize the beads in the sample plane using bright-field illumination with a 740 nm LED light source. A lens (L6) collects a diverging beam from the LED and sends it through a dichroic mirror (DM2) to the CON. The bright-field beam emerges from the OBJ, passes through DM1, and is focused on a charge-coupled display (CCD1) by a tube lens (L7). (5) Fluorescence illumination (panel 5): a 532 nm laser is used to obtain a fluorescence image of the actin filament labeled with fluorescent phalloidin (tetramethylrhodamine (TMR)-phalloidin dye). The fluorescence image is necessary to make a single actin dumbbell between the two trapped beads. DM3 reflects the excitation beam (532 nm), and DM4 reflects the fluorescence (580 nm) of the TMR dye. A tube lens (L12) makes a fluorescent image of the sample plane on CCD2. The focal length of the lenses are L1 = 100 mm, L2 = 200 mm, L3 = 100 mm, L4 = 300 mm, L5 = 75 mm, L7 = 200 mm, L8 = 100 mm, L9 = 200 mm, L10 = 100 mm, L11 = 250 mm, L12 = 200 mm, and L13 = 75 mm.

Single-Molecule Dual-Beam Optical Trap 329

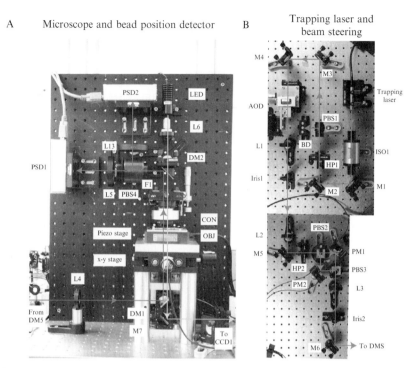

Figure 14.3 Photographs of the actual dual-beam optical trap setup. The primary components of the setup in Fig. 14.2 (microscope, bead position detector, and trapping laser and beam steering) are shown here. (A) Photograph of the microscope and the bead position detector setup. (Green arrow) Beam path from the bright-field light source (LED) to the detector (CCD1)—a diverging beam from the LED light source (740 nm) is collected by a 10× objective lens (L6). The distance between holes in the bread-board is 1 in. Condenser (CON) is mounted on a gimbal optics mount and a z-translation stage. The distance between LED and the front aperture of CON is ~ 11 in. CON collects and focuses the beam into the sample channel which is mounted on the top of the piezo stage during experiments. Objective lens (OBJ) (obscured from view by the piezo stage) collects the beam from the sample plane, and sends it downwards. The mirror turret with dichroic mirror DM1 and reflective mirror M7 is mounted on the optical table, with DM1 and M7, respectively, 3 and 1.5 in. above the table surface. The beam reflected by M7 is collected by a tube lens (L7, $f = 200$ mm, not shown in figure) which images the sample plane on CCD1. (Red arrow) Trapping/detection beam paths from DM5 to the detectors (PSD1, 2)—trapping/detection beams pass through a lens L4 and are reflected on a dichroic mirror DM1. The OBJ focuses the beams in the sample channel and CON collects the transmitted beams. The beams are reflected on a dichroic mirror DM2, pass through F1, PBS4, L5/L13, and finally arrive on the sensors (10 mm^2) of PSD1, 2. We adjust the height of CON such that the transmitted beams ($D \sim 20$ mm at the CON) are collimated. The distance between the BFP of CON and L5 is ~ 230 mm, and the distance between L5/L13 and PSD1/PSD2 is ~ 110 mm. The interference patterns at the BFP of CON are demagnified by a factor of ~ 2 ($M = \frac{110\,\text{mm}}{230\,\text{mm}} \sim \frac{1}{2}$). A filter F1 is mounted on a flipping mount such that it selectively transmits the 850 nm detection laser beams but blocks the 1064 nm beams when it is

The microscope body is set up to accommodate a slide–coverslip sandwich with sample channel (Fig. 14.2, panel 2). The objective lens (OBJ) brings the trapping beam to a focal point in the sample plane. This focused trapping beam generates a restoring force that holds a trapped bead at the center of the trapping beam. The piezo stage (P-731.20, Physik Instrumente) and motorized x–y stage are used for fine and coarse control, respectively, of the position of the platform bead in the sample channel (Fig. 14.2, panel 2). A piezo-based actuator is used to change the distance of the sample plane from the coverslip surface.

For detecting the bead position, we use the back focal plane detection method (Gittes and Schmidt, 1998a; Pralle et al., 1999). Two position-sensitive detectors (PSD1 and PSD2; Fig. 14.2, panel 3a) are used for the detection of each of the two beads. Back focal plane detection could be done using the trapping beam as the detection laser (Gittes and Schmidt, 1998b). However, this is not appropriate for the three-bead assay shown in Fig. 14.1. The trapping beam is not centered on the trapped bead, and this can result in a nonlinear relationship between the back focal detection signal and the bead displacement. To avoid this potential complication, we use a separate detection laser that is centered on the bead (Fig. 14.2, panel 3b) (Neuman and Block, 2004). The positions of the dual detection beams are controlled independently by PM3 and PM4 (see Section 3.2.6). We use PBS5 and PBS6 in conjunction with PM3 and PM4 for independent position control of the detection laser beams (Fig. 14.2, panel 3b).

Imaging the beads and the actin filament is essential for trapping the beads and creating the actin dumbbell. An LED (740 nm) is used to

---

flip-in position. When we use trapping beams for the detection, the F1 is switched to the flip-out position. PSDs are mounted on custom boxes and L-brackets for the stability of the detectors, and the input and output connections of the PSD with power supplies and NI-DAQ are accomplished though the cable mounted on the box. (B) Photograph of the trapping and the beam steering parts setup—the fiber output of the trapping laser (the body is separately mounted off the table) is stably mounted on a V-block mount followed by an ISO1 at 5 in. from the fiber. Mirrors M1 and M2 (4 in. separated) redirect the beams to the pair of HP1 and PBS1 which allows controlling of the trap beam power. The initial beam size is 1.6 mm. Mirrors M3 and M4 redirect the beams to the fast beam steering components (AOD, L1, L2) followed by the slow beam steering components (PM1, L3, L4 (not seen)). Here, the position of AOD, L1, L2, and PM1/PM2 satisfies the *4f arrangement* condition, whose separations are L1, L1 + L2, and L2, respectively. The second pair of HP2, PBS2, and PBS3 separates and recombines the p- and s-polarized beams for independent control of the dual beams. Rotation of the HP2 determines the relative stiffness of the dual beams, and usually we make the trap stiffness of the two beams the same to maximize the stability of the actin dumbbell. The difference of each beam path between PBS2 and PBS3 is 4 in. in our setup, such that noise sources (mechanical vibration and air fluctuation) arising out of them affect the dual beam by the same amount. Note that the distance between the threaded mounting holes on the optical table is 1 in. (See Color Insert.)

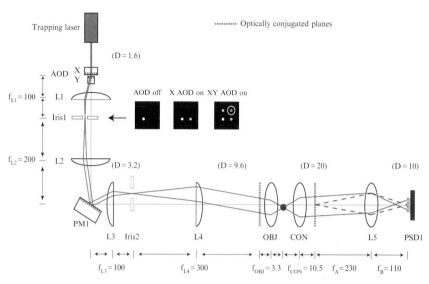

Figure 14.4 Beam path in a single-beam optical trap—the primary components of the dual-trap setup in Fig. 14.2 are shown here for a single-beam trap, with detailed information on the distance between each component and the beam layout. The initial size of the trapping beam is ∼1.6 mm. The combination of AOD, L1 ($f_{L1} = 100$ mm), L2 ($f_{L2} = 200$ mm), and PM1 forms a *4f arrangement* for fast and precise beam steering. Multiple deflected beams (four strong beams) are produced by the AOD, transmitted through L1, and focused in the focal plane of L1. Iris1 transmits only the first-order diffracted beam from the two AODs, as seen in the right panel labeled "XY AOD on." The beam between L2 and L3 is collimated with diameter ∼3.2 mm. Another combination of PM1, L3 ($f_{L3} = 100$ mm), L4 (($f_{L4} = 300$ mm), and back aperture of OBJ forms the second *4f arrangement* for the slow and coarse beam steering. The beam steering of PM1 results in the steering of the beam entrance into the back aperture of OBJ such that the trapped bead translates linearly in the sample plane. The beam is collimated with size ∼9.6 mm, which slightly (10%) overfills the back aperture of OBJ. AODs, PM1, and the back aperture of OBJ are all optically conjugated (dotted gray lines), assuring that the deflected beam is always delivered into the back aperture of OBJ without beam clipping during the beam steering. The trapping beam passing through the trapped bead is collected and collimated by CON (beam size ∼20 mm). L5 (($f_{L5} = 75$ mm) images the back-focal plane (BFP) of CON as drawn in the blue dashed line. Note that the optically conjugated planes are all expressed as dotted green lines shown at the AODs, PM1, BFP of OBJ and CON, and PSD1. The trapping beam is coupled with the detection beam between L4 and OBJ such that the detection beam path is following the same beam path in the detection part.

illuminate the sample plane (Fig. 14.2, panel 4a). A bright-field image of the sample plane is captured by CCD1 (Fig. 14.2, panel 4b). A fluorescence laser (532 nm) is used to excite the fluorescent phalloidin (tetramethylrhodamine (TMR)-phalloidin, Sigma-Aldrich) bound to the actin with the emission being imaged on CCD2 (Fig. 14.2, panel 5). Alternatively, we are exploring the use of another dye (CF 405 phalloidin, Biotium) with a

405 nm laser excitation to leave the green and red portions of the visible spectrum open for future simultaneous optical trapping/fluorescence imaging experiments (Hohng et al., 2007; Ishijima et al., 1998; Lang et al., 2004; van Dijk et al., 2004).

The following sections describe the most important components of the setup in detail. Specific information on all components is found in Table 14.1.

3.2.1. Trapping laser

A typical laser for a dual-beam optical trap is a single mode (Gaussian TEM_{00}), continuous wave (CW) laser of single wavelength ranging from 800 to 1100 nm with a power output ranging from 1 to 5 W (Bustamante et al., 2007; Neuman and Block, 2004). Important features of the laser for a stable trap are its maximum power, power stability, pointing stability, and beam profile. For a detailed description of criteria for selection of a suitable trapping laser, we refer the reader to Neuman and Block (2004) and Bustamante et al. (2007). Our trapping laser is a 5 W, 1064 nm, fiber laser from IPG (YLR-5-1064-LP). It is a diode-pumped, TEM_{00} ($M^2 < 1.1$), linearly polarized CW ytterbium laser with a quoted power instability of $<3\%$ over 4 h ($<2\%$ in the 1 kHz to 20 MHz range) and good pointing stability (~ 1 μrad). We chose a fiber laser, as it provides good pointing and power stability. Fiber coupling allows us to position the laser body and power supply away from the optical table, which reduces vibrations and improves trapping stability. This fiber laser ($\sim\$15,000$) is less expensive than conventional trapping lasers (diode-pumped solid-state lasers $\sim\$30,000$).

3.2.2. Objective (OBJ)

Important features of an OBJ for a good trap are its NA and transmittance. High NA (>1.2, and often >1.4) generates a power gradient sufficient to overcome the radiation force, which results in a stable optical trap. Either an oil or water immersion OBJ can be used for an optical trap. An oil immersion OBJ typically has a higher NA than a water immersion OBJ. One potential problem of using an oil immersion OBJ is the spherical aberration that arises when the beam is focused deep in the solution (more than a few micormeters) due to the index mismatch between the aqueous medium ($n \sim 1.32$) and the immersion oil ($n = 1.512$). The three-bead assay requires focusing the trapping beam close to the coverslip surface (~ 1.5 μm distance) where the spherical aberration is not significant, and an oil immersion OBJ is appropriate for this purpose. Our current OBJ is a CFI Plan Apo oil immersion TIRF objective lens (MRD01602, Nikon) with high NA (1.45) and transmittance (60% at 1064 nm).

Table 14.1 Primary components of the dual-beam optical trap setup

	Company	Model number	Description
Optical table	Technical Manufacturing Corporation	784-35631-01	Top thickness—18 in. (300 mm), top length—8 ft (2.4 m), minimum resonant frequency 200 Hz
Trapping laser	IPG	YLR-10-1064-LP	Diode-pumped CW, ytterbium-doped fiber laser, max power 5 W, wavelength 1064 nm, beam diameter 1.6 mm, TEM00 mode, $M^2 < 1.1$, linearly polarized, power stability <3% (over 4 h) and <2% (1 kH to 20 MHz range), and pointing stability ~1 μrad, fiber length 2 m
Detection laser	World Star Tech	TECIRL-30GC-850	Diode laser, CW, power 30 mW, wavelength 850 nm, beam diameter 1.6 mm, circular, linearly polarized, power stability <0.5%, temperature control, pointing stability <25 μrad (in spec), pointing stability ~1 μrad (measured over 300 s)
Fluorescence laser	Coherent	Compass 215M-20	DPSS laser, CW, power 20 mW, wavelength 532 nm, beam diameter 0.32 mm, TEM00, $M^2 < 1.1$, pointing stability <6 μrad/C, noise (10 Hz to 1 GHz) <0.5% (rms)
Fluorescence laser	World Star Tech	TECBL-10G-405	Diode laser, CW, power 10 mW, wavelength 405 nm, beam diameter 1.6 mm, circular, linearly polarized, temperature control
LED	Mightex	SLS-0111-A	740 nm, 120 mW luminous flux, lens A (narrow beam, 19 mm aperture, half angle 5°, efficiency 85%)
Objective lens	Nikon	MRD01602	TIRF, CFI Plan Apo, magnification 60× VC, NA 1.45, oil immersion, transmittance 60% for 1064 nm

(*continued*)

Table 14.1 (continued)

	Company	Model number	Description
Condenser	Nikon	T-C HNA-oil lens	MEL41410, $f = 10.5$ mm, NA 1.35
Isolator	Thorlabs	IO-5-1064-VHP	Free space isolator, 1064 nm, 4.8 mm aperture, 100 W (max)
Z-translation stage for objective lens	Newport	462-X-SD	462 ULTRAlign™ precision integrated crossed-roller bearing linear stage, 1.0 in. X travel, slide drive
Z translation actuator for objective lens	Newport	PZA12	NanoPZ ultra-high resolution actuator, 12.5 mm travel
	Newport	PZC200	Hand-held controller for PZA12
X–Y translation actuator for sample stage	Newport	406	High-performance large platform two-axis linear stage, aperture platform, 1/4–20 tapped holes
X–Y translation stage for sample stage	Newport	NSA12	11 mm travel motorized miniature linear actuator
	Newport	NSC200	NewStep hand-held motion controller for NSA12
Tube for mounting of OBJ	Thorlabs	SM1L20	SM1 lens tube, 2 in. long, one retaining ring included
Piezo stage	Physik Instrumente	P-731.20	High-resolution XY-positioning stages, range: 100 μm, resonant frequency: 400 Hz, resolution: 1 nm
Breadboard	Thorlabs	MB1824	Aluminum breadboard, 18 in. × 24 in. × 1/2 in., 1/4–20 threaded
Rail	Thorlabs	XT95-750	95 mm construction rail, $L = 750$ mm
Rail base	Thorlabs	XT95P3	Optical table base plate
Microscope body post	Thorlabs	P6	Mounting post, length = 6 in.

Mounting cube for mirror turret (DM1, M7)	Linos	G061-081-000	Cube 40 mm × 40 mm, $D = 30$ mm
	Linos	G061-207-000	Rods 20 mm
	Linos	G065-087-000	Beam steering mirror holder 30
	Linos	G061-011-000	Set of screws M2.3×3 (qty. 150)
	Linos	G061-012-000	Set of threaded pins M2.3×6 (qty. 150)
DM1	Chroma	t800dcspxr	Shortpass, T: 405–740 nm, R: 850, 1064 nm
DM2	Chroma	780dcspxr	Shortpass, T: 740 nm, R: 850, 1064 nm
DM3	Chroma	r532rdc-xt	Longpass, T: 540–1064 nm, R: 532 nm
DM4	Chroma	700dcxr	Longpass, T: 740–1064 nm, R: 460, 580 nm
DM5	Chroma	940dcxr	Longpass, T: 1064 nm, R: 850 nm
Dichroic mirror mount	Newfoucs	9920	45° mirror holder
F1	Thorlabs	FB850-40	Bandpass filter, CWL = 850 nm, FWHM = 40 nm
F2	Thorlabs	FES0800	Shortpass filter, cut-off wavelength: 800 nm
HP1, HP2	Newport	10RP02-34	Zero-order quartz wave plate, 25.4 mm dia, 1064 nm, $\lambda/2$ retardation, 2 MW/cm^2 CW
HP3	Newport	05RP02-30	Zero-order quartz wave plate, 12.7 mm dia, 850 nm, $\lambda/2$ retardation
PBS1	Newport	05BC16PC.9	Laser line polarizing cube beamsplitter, 12.7, 1064 nm
PBS2, 3	Newport	10BC16PC.9	Laser line polarizing cube beamsplitter, 25.4, 1064 nm
PBS4	Newport	10FC16PB.7	Broadband polarizing cube beamsplitter, 25.4 mm, 850–1300 nm, 2 kW/cm^2 CW
PBS5, 6	Newport	05FC16PB.5	Broadband polarizing cube beamsplitter, 12.7 mm, 620–1000 nm
M1–M6, mirrors for PM1, 2	Newport	10D20DM.10	Laser line dielectric mirror, Pyrex, 25.4 dia, 6.0 mm, $\lambda/10$, 1030–1090 nm
M8, mirror for PM3, 4	Thorlabs	BB1-E03	Ø1 in. broadband dielectric mirrors, 750–1100 nm

(*continued*)

Table 14.1 (continued)

	Company	Model number	Description
M7	Linos	G340-523-000	Silver elliptical plane mirrors, laser quality, 22.4 mm × 31.5 mm × 3.5 mm
Mirror mount	Thorlabs	KM100	Kinematic mirror mount for 1 in. optics
L1, L3	Newport	KPX094AR.33	BK 7 plano-convex lens, 25.4 dia, 100 EFL, 1064 nm
L2	Newport	KPX106AR.33	BK 7 plano-convex lens, 25.4 dia, 200 EFL, 1064 nm
L4	Thorlabs	LA1256-B	N-BK7 plano-convex lens, Ø2 in., $f = 300.0$ mm, ARC: 650–1050 nm
L5, L13	Thorlabs	LB1309-B	N-BK7 bi-convex lens, Ø2 in., $f = 75$ mm, ARC: 650–1050 nm
L6			10× objective lens
L7, L9	Thorlabs	LA1708-B	N-BK7 plano-convex lens, Ø1 in., $f = 200.0$ mm, ARC: 650–1050 nm
L8, L10	Thorlabs	LA1509-B	N-BK7 plano-convex lens, Ø1 in., $f = 100.0$ mm, ARC: 650–1050 nm
L11	Thorlabs	LA1461-A	N-BK7 plano-convex lens, Ø1 in., $f = 250.0$ mm, ARC: 350–650 nm
L12	Thorlabs	LA1708-A	N-BK7 plano-convex lens, Ø1 in., $f = 200.0$ mm, ARC: 350–650 nm
Lens mount	Thorlabs	LMR1	Lens mount for Ø1 in. optics, one retaining ring included
Mount for trap beam fiber	Newport	VB-1	V-Block, 3 in. (76.2 mm) length

For most of the mirrors mount, we use 1 in. diameter post RS2P (Thorlabs) with clamping fork CF125 (Thorlabs)
For most of the lens mount, we use 1 in. diameter post extension RS2 (Thorlabs) with adjustable height Post Base RB2 (Thorlabs)

Item	Manufacturer	Model	Description
AOD	IntraAction	DTD-274HA6	2-axis acousto-optic deflection system
AOD RF source	IntraAction	DVE-120	Dual RF frequency source (PCI)
AOD RF amplifier	IntraAction	DPA-502D	Dual RF power amplifier
PM1–4 mount	Newport	AG-M100N	Agilis mount, 1.0 in. (25.4 mm), range ±2 deg, sensitivity 1 μrad, max speed 0.75 deg/s
PM controller	Newport	AG-UC8	Agilis compact 8-axes controller, USB interface
PSD1, 2	Pacific Silicon Sensor	DL100-7PCBA3	X, Y duolateral position sensing photodiode with sum difference circuitry. Includes connector and variable bias option. Active area 10 mm × 10 mm
CCD1	Watec	WAT 902H2 Supreme	Watec WAT 902H2 Supreme B/W 1/2 in. camera low lux
CCD2	Hamamatsu	C2400-08	Real-time fluorescence imaging, 400–850 nm spectral response
NI-DAQ	NI	779068-01	NI PCI-6229
Cable for NI-DAQ	NI	192061-01	SHC68-68-EPM Cable (1 m)
Connector block for NI-DAQ	NI	777960-01	BNC-2120—shielded
Labview	NI	LabVIEW 8.6	
Matlab	Mathworks	Matlab R2009a	
TMR phalloidin	Sigma-Aldrich	P1951	Ex peak: 552 nm, Em peak: 580 nm, phalloidin–tetramethylrhodamine B isothiocyanate conjugated with phalloidin
CF405 dye	Biotium	00034	Ex peak: 405 nm, Em peak: 460 nm, conjugated with phalloidin
Immersion oil	Nikon	MXA22024	50cc NonFluor immersion oil
IR sensor	Newport	F-IRC2-S	IR sensor card, 800–1700 nm, 1 in. × 1.5 in. sensor

3.2.3. Microscope body

The microscope body can either be purchased from commercial manufacturers or custom-made (Fig. 14.3A). The main advantage of using a custom-made body is the enormous flexibility in the design of the objective–condenser assembly, with easy positioning of the PSD and detection units. Furthermore, customized microscope bodies are more cost-effective than commercial bodies. An important measure of a good microscope body is the stable positioning of its components. In the three-bead assay, we quantify the interaction between a myosin molecule immobilized to the coverslip surface and an actin filament suspended between two independently trapped beads. Fluctuations in the relative position of components in the microscope body add to the noise in the bead position measurements. Thus, for an accurate measurement of actin interactions with myosins, which displace the trapped beads by tens of nanometers over millisecond time scales, fluctuations in the microscope body should be an order of magnitude smaller than the myosin-driven actin movements. We use a breadboard mounted on a rail (MB1824, XT95-750, Thorlabs) for the vertical mounting of the microscope, PSD, and LED (Fig. 14.3A). This gives us flexibility in the optical path layout as well as excellent stability. The adaptors that interconnect each component of the microscopy body were machined in house.

3.2.4. Beam steering for slow and coarse positioning

The spatial position of each of the trapped beads in the sample plane needs to be controlled independently in the three-bead assay. This is essential for trapping the two beads and steering them to create a taut actin filament between the beads. The spatial position of the trapped beads can be controlled independently by piezo-driven mirrors PM1 and PM2 (AG-M100N, Newport) (Fig. 14.2, panel 1). Alternate configurations have been used for coarse steering (Rice et al., 2003). The advantages of PMs in the current configuration are their compactness, high sensitivity, improved stability, and low cost ($\sim$$1000 per PM with the controller). The compactness of PMs minimizes the distance over which the two trapped beams pass through separate optical paths. In our current design, this distance reduces to 4 in. Minimizing this distance reduces the contribution of environmental noise arising from mechanical vibration or air flow fluctuations (Bustamante et al., 2007; Moffitt et al., 2006) and improves the stability of the actin dumbbell in the three-bead assay by maintaining the distance between the two traps. Our design provides a $\sim$20 μm range of coarse bead movement and a $\sim$10 μm/s maximum speed in the sample plane. The coarse beam steering of PMs is easily accomplished with either a controller or a custom Labview program (Labview 8.6, National Instruments).

3.2.5. Beam steering for fast and precise feedback control

Fast feedback control of the trapping beam is necessary for either a force clamp or a position clamp of the trapped bead. A force clamp is achieved by maintaining the relative distance between the trapping beam and the trapped bead position. Force clamping has proved to be a very powerful tool in the studies of processive motors, such as a myosin V, to obtain accurate step size and ATPase kinetics in the presence of compliant elements (Rief et al., 2000; Veigel et al., 2002; Visscher et al., 1999). A position clamp is performed by maintaining the bead position as constant as possible (Finer et al., 1994; Simmons et al., 1996). Position clamping is achieved by changing the trap force to prevent bead movement. This technique has been used to measure the maximum force that a single myosin molecule can produce (Finer et al., 1994). Both clamping techniques require fast update of the trap position over submillisecond timescales along with precise positioning over nanometer spatial scales. For detailed information regarding feedback control, see Visscher and Block (1998) and Spudich et al. (2007).

In our design, fast feedback control of the trap is achieved using an AOD. The setup described here is for our AOD model (DTD-274HA6, IntraAction). A PCI board for the dual RF frequency source (DVE-120, IntraAction) generates an RF signal (12 bits amplitude dynamic range, frequency access time ~ 1 μs). The signal is amplified through a dual RF power amplifier (DPA-502D, IntraAction). The full deflection range of our model is 26.9 mrad (16 MHz), which corresponds to ~ 15 μm displacement in the sample plane. With an AOD, only the first-order deflected beam is used, which results in a significant loss of beam power (at least 50%). Use of a high-power laser (5 W) compensates for this lost power in our design. Feedback control of the AOD is achieved with a custom Labview program.

3.2.6. Detection of bead position

Back focal plane detection is one of the most common methods to measure the position of the trapped beads (Gittes and Schmidt, 1998a; Pralle et al., 1999; Visscher et al., 1996) (Figs. 14.4 and 14.5C). At the back focal plane of the condenser, there is interference between the incident trapping beam and the forward scattered beam by the trapped bead. The center of mass of the interference pattern translates linearly with the relative position between the trapping beam and the trapped bead. A PSD or a quadrant photodiode (QPD) can measure the centroid of the interference pattern and thereby tracks the center of the trapped bead. We use a PSD (DL100-7PCBA3, Pacific Silicon Sensor) due to its low noise, linear response on the entire sensor (10 mm^2 area), and ready-to-use preamplifier circuit (Bustamante et al., 2007; Huisstede et al., 2006; Neuman and Block, 2004). Output of the PSD is an x–y voltage signals for the x and y centroids of the beam, as well as the sum signals of them. The bandwidth of this detector, which is an upper

3.3.1. Mechanical vibration of the instrument derived from external vibration sources

(a) Set up the optical trap in a room isolated from vibrations (away from equipment with moving parts such as a centrifuges or an elevator). Optical traps are usually built in the basements of building to further minimize vibrations. The air flow in the room needs to be significantly damped.

(b) Use a heavy floated optical table to isolate the trapping components from all external vibration sources. We are currently using a heavy table (784-35631-01, Technical Manufacturing Corporation; resonance frequency ~ 200 Hz) which is floated on air dampers (Gimbal Piston Isolators, Technical Manufacturing Corporation; resonance frequency below ~ 2 Hz).

(c) Do not mount any equipment with moving parts on the optical table. For example, we mount the body of the trapping laser, piezo stage controller, and other motion controllers away from the table.

(d) Use thick ($D = 1$ in.) and short ($L = 2$ in.) posts for the mounting of the optics including mirrors and lenses. Keep the optical beam close to the optical table (height $= 3$ in.) and the total beam path from the lasers to the detector as short as possible.

(e) Build or purchase a stable microscope body with acceptable vibration amplitude and stage drift in your measurement time. Alternatively, use active feedback control of the sample stage using a piezo stage modulation if necessary (Carter et al., 2009; Nugent-Glandorf and Perkins, 2004).

(f) The effect of pointing fluctuations on the trapping laser can be minimized by using the same laser for both traps and by minimizing the distance over which the dual-beam paths are separated. In our current design, this distance between PBS2 and PBS3 is 4 in.

3.3.2. Pointing and power fluctuation of the trap and detection beam

(a) Use lasers with excellent pointing stability (~ 1 μrad or better), power stability ($< 1–2\%$), and beam profile ($M^2 < 1.1$).

(b) Set an isolator right after the trapping/detection beam that blocks the back-reflected beam to the laser, thereby improving power stability.

(c) Use fiber coupling for improving pointing stability, and use active feedback control of an acousto-optic modulator (AOM) for power stability (Carter et al., 2009; Nugent-Glandorf and Perkins, 2004).

3.3.3. Air fluctuation

(a) Enclose the optical beam path in a plexiglass or a plastic box (~ 0.5 in.) to minimize noise due to the air fluctuations. Minimize the total volume occupied by the beam path, as this also minimizes noise from convection currents.

(b) Have all beam paths pass through plastic tubes, especially the beam-focusing spots (between L1 and L2, L3 and L4) where the contribution of air fluctuation is amplified.
(c) Use lenses with short focal length to minimize the beam path. Larger beam paths have disproportionately higher noise than shorter beam paths.

3.3.4. Thermal expansion

(a) During the experiment, do not touch any optical components. Use controllers which are mounted out of the optical table.
(b) Perform optical trapping in a temperature-controlled room (Bustamante et al., 2007; Carter et al., 2009; Lang et al., 2002; Peterman et al., 2003).

3.3.5. Beam quality in the sample plane

(a) Use lasers with good Gaussian beam profile $M^2 < 1.1$. Test the beam profile of the laser before purchase (see alignment protocol). In our experience, even lasers from reliable vendors have significant variability from unit to unit. Testing this parameter can save significant time and effort in set up and alignment of the instrument.
(b) Prevent beam clipping at any optical components (such as lenses, mirrors, AOD, shutters, or irises).
(c) Expand the beam to a final size that is comparable with the aperture diameter of the OBJ.
(d) Select optics with appropriate antireflection coating.
(e) Do not use lenses with a focal length that is too short ($f < 25$ mm). Spherical aberration increases as $\sim 1/f^3$.
(f) Keep optics free from dirt and dust. If you have to clean the optics, follow the recommended method from the manufacturers.

3.3.6. Electrical noise of data acquisition

(a) Ground the optical table by connecting it to a well-characterized building ground. This minimizes static build-up that could interfere with electrical components.
(b) Avoid ground loop by using a different power supply for each PSD and a floating mode for the input channels of the NI-DAQ connector block.
(c) Minimize ambient light to the detector to avoid 60 Hz noise.

3.4. Alignment protocol

Construction of a dual-beam optical trap for the three-bead assay requires alignment of many components as shown in Fig. 14.2. Here, we describe a detailed procedure for the construction of a dual-beam trap in a step-by-step manner. Modules for alignment are as follows—Step 1: bright-field setup for testing the stability of the microscope; Step 2: single-beam trap; Step 3: dual-beam trap for the three-bead assay. Other useful alignment procedures are found elsewhere (Block, 1998; Bustamante et al., 2007; Lee et al., 2007; Spudich et al., 2007).

Step 1 Bright-field setup for checking the stability of the microscope (follow the green arrow in Fig. 14.3A)

An essential first step is the alignment of the bright-field path, independent of the trapping laser, to test the stability of the microscope body. The stability of the microscope is critical for trapping experiments in general, with specific emphasis on reduced noise for the three-bead assay. A measure of stability is the absence of significant fluctuation or drift effects over the time scale of the experiment. The steps involved in this alignment are (1) set up of components, (2) alignment procedure, and (3) testing. For a conventional microscope whose microscope body is already in place, the reader can skip steps (1) and (2).

(1) Position all the components in the bright-field beam path from the LED light source to the CCD as shown in Fig. 14.3A (see the figure legend for the green arrow path for distance and product specifications).
(2) Alignment:
  (a) Adjust the vertical axis of the LED, L6, and M7 to coarsely coincide with the principal axis of OBJ as shown in Fig. 14.3A. This is accomplished by using the holes in the breadboard as a guide to identify the vertical axis, with components positioned 3.5 in. away from the board.
  (b) Align the vertical axis of the z-translation stage of CON such that it coincides with the fixed principal axis of the OBJ. This is done with the LED turned off. Use a spirit-level for angular adjustment of the CON vertical axis. You have to first check that the optical table and the microscope stage are also leveled. Peer down the CON to see the OBJ with transmitted ambient light. Try to coincide the principal axes of OBJ and CON by lateral translation of the CON. The alignment of CON is completed when the z-translation of the stage produces symmetric changes in the shape of the OBJ front face seen through the back aperture of CON (by eye).
  (c) Turn on the LED. Adjust the position and angle of LED and L6 such that the beam passes through the center of the back aperture of CON. Since 740 nm is visible (far red), the beam path can be

checked using plain paper. For this check, create a film of immersion oil between OBJ and CON. Avoid direct contact between OBJ and CON faces as this could damage the lenses.

(d) The beam should be incident at the center of M7 (Fig. 14.3A). Adjust the angle of M7 and M9 (Fig. 14.2) using the knobs of the mirror mount to make the reflected beam parallel to both the table and a row of threaded holes on the table. Adjust the height of L7 and CCD1 to center them along the beam path. Using an infinity corrected objective, L7 can be positioned anywhere in the beam path followed by the CCD1 at the focal point ($f_{L7} = 200$ mm). Move CON or OBJ along the axial direction (be careful not to damage the lenses by direct contact), and view the beam shape on CCD1. If the beam expands and shrinks symmetrically with a fixed center, you are ready to test the stability of the microscope.

(3) The stability can be tested by recording a bright-field movie of beads adsorbed to a coverslip or a calibration slide (5 μm scale or less).

(a) Place a sample channel that has platform beads (diameter~1.5 μm) on the surface of the coverslip between OBJ and CON. Turn on the LED. The LED output power needs to be adjusted using a variable ND filter (ND1) to avoid saturating CCD1. Adjust the z position of OBJ and CON to visualize the bead images on CCD1. Move the OBJ in the z-direction to focus and defocus the bead image. Check any asymmetry or aberration in the bead image (Fig. 14.5B).

(b) Record the image on the CCD camera with appropriate driver software for ~10 min to test the stability of the microscope system. Measure the fluctuation and drift of the center of the beads. If drift is < 100 nm over 10 min, the microscope setup should be stable enough to start the trapping beam alignment. Otherwise refer back to Section 3.3 on how to improve the stability of the setup.

Step 2 Alignment of a single-beam trap

Alignment of a single-beam trap consists of the following steps: (1) measuring beam profile and stability of the trapping laser; (2) mounting the trapping laser and an isolator; (3) steering the beam parallel to the optical table; (4) setting up a beam dump to control the power of the trapping beam; (5) setting up the overall beam layout; (6) adjusting the beam path through the microscope objective (OBJ) and condenser (CON); (7) setting up lenses L3 and L4 for beam expansion and coarse beam steering; (8) setting up lenses L1 and L2 for beam expansion and fine beam steering; (9) setting up the AOD for feedback control; (10) setting up OBJ and CON; (11) aligning the bright-field light path; (12) testing the single-beam trap; (13) setting up PSD1 for the back focal plane detection.

To visualize the trapping beam during the alignment, we use IR cards (F-IRC2-S, Newport). During the alignment procedure, it is important to attenuate the beam power to <50 mW to prevent damage to the IR cards and avoid accidental laser damage to the user and equipment. Note that aligning M1 and M2 (Fig. 14.3B) is done by attenuating laser output directly from the power source. For subsequent alignment steps the laser output is set to high power (>1 W) for improved beam stability, with a beam dump (HP1 and PBS1 in Fig. 14.3B) to attenuate laser power.

(1) *Measuring beam profile and stability of the trapping laser*
Beam profile, collimation, and laser stability are usually quoted by the manufacturer in the laser specification sheet. Desired specifications include beam profile $M^2 < 1.1$, power fluctuations to <2%, and pointing stability of the laser beam of ~1 μrad or less. It is important to conduct a quick assessment of the laser beam profile, collimation, pointing, and power stability before use to guard against manufacturing defects or damage during the transportation or handling. The beam profile can be characterized using various tools, including a commercial beam profiler, a knife-edge technique, and an iris technique (Bustamante *et al.*, 2007). Laser stability includes power stability and pointing stability. As a first approximation, power stability can be evaluated by monitoring the power output on a power meter over several minutes (short-term measure) to an hour (long-term measure). Pointing stability is assessed by directing the laser output, after attenuation, to a PSD with a path length of a couple of meters from the laser. PSD output is monitored over time and <1 μm change in laser position over several minutes suggests good pointing stability. Please note that these measurements must be made with an attenuated beam (use a polarizing beam splitter followed by a beam dump or a ND filter) with power below the damage threshold of the detector. Collimation of the beam can be tested by measuring the beam diameter at multiple distances from the laser.

(2) *Mounting the trapping laser and an isolator*
Mount the fiber output of the trapping laser (1064 nm, 5 W) followed by an isolator (ISO1) on an optical table (Fig. 14.2, panel 1). The body of the trapping laser is kept off the optical table to minimize vibrations from the power source. Mounting stability of the output fiber of the laser beam affects overall stability of the optical trap. To ensure stable output we use a stable mount (VB-1, Newport) with 1 in. diameter post. The beam path shown in Fig. 14.3B is at a constant height of 3 in. above the optical table to minimize aberrations and simplify the alignment procedure. To minimize beam fluctuations from mechanical vibrations, posts with 2 in. height and 1 in. diameter have been used for most of the mirror mounts. We use an adjustable

tilt mount (KM100, Thorlabs) to run all collimated beams parallel to the surface of the optical table, along the threaded holes. If the beam collimation is not acceptable (>mrad), set two lenses followed by ISO1 to make the beam collimated. Divergence of the laser beam can be identified by measuring the beam diameter several meters away from the beam output.

(3) *Steering the beam parallel to the optical table*
Steering the beam parallel to the optical table is achieved by *walking the beam* using two mirrors M1 and M2 and a set of irises. Set up the two mirrors M1 and M2 (Fig. 14.3B). Prepare the two irises with centers at the same height (3 in. above the optical table). Locate them along the optical table such that they follow the optical path from M2 to M3 (Fig. 14.3B). Locate the first iris close to M2 and the second one >1 m from it. Adjust the position and the horizontal angle of M2 such that the beam passes through the two irises when they are both completely open (0.5 in. diameter). (i) Adjust the angle of M1 making the beam passes through the center of the first iris when it is closed ($\sim$1 mm diameter). (ii) Open the first iris and adjust the angle of M2 to make the beam passes through the center of the second iris when it is closed ($\sim$1 mm diameter). Iterate between (i) and (ii) until the beam passes through both irises when they are closed. When the alignment step is completed, the beam will be parallel to the optical table, 3 in. above it.

(4) *Setting up a beam dump to control the power of the trapping beam*
Mount a half-wave plate (HP1) followed by a PBS1 (Fig. 14.3B). Use HP1 to control the power rather than changing the power of the initial beam to avoid fluctuations in laser output when operating at low power (<1 W). During the alignment, limit the power downstream of the PBS1 to <50 mW for safety. Following alignment, this power can be increased by changing the setting of HP1. The beam diverted to the beam dump can be used for feedback control of a single beam, as described in Section A.2.

(5) *Setting up the overall beam layout*
Set M3–M6 and PM1 using the procedure described in (3) to make the overall beam path parallel to the table (3 in. height) (Fig. 14.3B). The distance between these mirrors is determined by the focal lengths of the lenses L1–L4. We use the *4f arrangement* to create a set of conjugate planes (optical elements highlighted in green dotted lines in Figs. 14.2 and 14.4) to steer the beam efficiently. For the description of the *4f arrangement*, we refer the readers to Bustamante *et al.* (2007) and Lee *et al.* (2007).

For instance, to conjugate the AOD and PM1, the length of the optical path between them is set to twice the sum of the focal lengths of L1 and L2 ($2 \times (f_{L1} + f_{L2}) \sim 24$ in. in our layout),

Look at the beam size of the back-reflected beam (reflection from partial mirror) at a long distance (>1 m). If the beam size is comparable with the initial laser output beam, the trapping beam is sufficiently collimated at the OBJ back aperture. If the beam is much larger than the initial beam, translate L1 in the beam axis such that the back-reflected beam size is minimized. Check whether the final beam size ∼9.6 mm such that the final beam barely overfills the back aperture of the OBJ.

(9) *Setting up the AOD for feedback control*
   (a) Fast (10 kHz) beam steering and feedback control of the trap can be achieved using an AOD. Place an X-AOD (Figs. 14.2, 14.3B, and 14.4) such that its back face is positioned f_{L1} from L1. Please note that the XY-AODs should be mounted on a one-axis translator to change the path length between the AOD and L1 (used for conjugating AOD and PM1). A custom-built Labview program feeds an RF signal to the AOD that changes the angle of the first-order diffraction. Set the frequency of the RF signal to the center frequency recommended by the manufacturer. You will find multiple diffraction spots (only two spots are usually bright enough to be seen on an IR card) at the front focal point of L1. Use Iris1 at this location to transmit only the first-order diffracted beam (Fig. 14.4). Adjust the position and angle of the X-AOD to maximize the power of the first-order diffracted beam (use power meter). Place a Y-AOD such that its front face is positioned at f_{L1} from L1, so the X-AOD and the Y-AOD are very close to each other. Now, you will see four spots clearly before Iris1. Adjust the position of Iris1 to choose only the beam that is first-order diffracted by both AODs. The first-order diffracted beam from both AODs can be identified by turning the X- and Y-AODs on and off. Adjust the position and angle of Y-AOD as done for X-AOD to maximize power output.
   (b) When AOD and PM1 are optically conjugated, the first-order diffracted beam will be incident at the center of PM1. Also, with the Iris1 between AOD and L1 open, all beams transmitted through the AOD should arrive at the same spot at PM1. Use translators to move the AODs along the optical path to alter the position of the AOD along the optical path between M4 and L1 to satisfy these two conditions.
   (c) Setting AOD will change the incident beam (first-order diffracted beam) angle into the OBJ. Rotate the angle of PM1 such that the partially back-reflected beam again overlaps with the incident beam.
(10) *Setting up OBJ and CON*
   (a) Mount the OBJ and CON. Place a film of immersion oil between the OBJ and CON. Adjust the position of the CON such that the

beam transmitted through it is collimated (use an IR card far (>1 m) from the back aperture of the CON). Note that the transmitted beam size from the CON could be different from the incident beam depending on the model of the CON. The aperture size of our CON model is ~ 20 mm. Adjust the position of CON to make the transmitted beam coincide with the mark on the ceiling.

(b) The axes of the trapping beam, OBJ, and CON need to be coincident. To achieve this, we use the procedure described later. Note that with this technique we can also detect aberrations in shape of the beam incident at the back aperture of the OBJ. Move the CON away from the OBJ such that the beam is slightly converging. Place a CCD camera roughly at the focal point of this converging beam (use ND filters to attenuate the beam power below the CCD damage threshold of the CCD). You will notice an airy pattern as shown in Fig. 14.5A. Moving the CON up and down should expand and contract the size of this pattern. Adjust the position of the CON such that the beam expands and contracts symmetrically. Remove the CCD after this alignment step.

(11) *Aligning the bright-field light path*

Make a sample channel that contains both platform beads (diameter ~ 1.5 μm) on the coverslip and freely diffusing beads for trapping (diameter ~ 1 μm). Put immersion oil on both sides of the channel. Mount the channel between OBJ and CON. Turn on the LED (740 nm) and adjust the OBJ height to make a bright-field image of the sample plane on CCD1. Adjust the positions of the LED and L6 such that they are centered on the transmitted trapping beam. Adjust the positions of L7 and CCD1 to obtain a focused bright-field image of the sample channel (Fig. 14.5B). Adjust the height of OBJ such that the beam is focused near the interface of water and glass. Turn off the LED. If the trap is within the field of view of OBJ and the sample plane is close to the glass surface, the back-reflection of the trapping beam at the glass–water interface will be visible on CCD1. Adjust the position of CCD1 and L7 such that this back-reflected beam is at the center of the field of view. The back-reflected beam is visible as an interference pattern on CCD1, which expands and contracts in size with up–down motion of the OBJ. If the trap is correctly aligned, this interference pattern is symmetric in shape (similar to Fig. 14.5A). Note that the back-reflected beam can be used at any stage to locate the position of the trap in the field of view.

(12) *Testing the single-beam trap*

Turn on the LED. Move up the OBJ such that the trapping beam is focused in the aqueous solution. Beads that are located close to the surface appear dark at their center (Fig. 14.5B), whereas beads far away from the surface are light at their center (not shown). Note that only beads between the trap and the coverslip surface (dark center) can be trapped. Beads further in solution than the trap will be pushed away from the trap center by the radiation pressure of the trapping beam.

Increase the power in the trapping beam to ~ 50 mW at the back aperture of OBJ. This can be done by turning up the laser power and/or by using HP1. Try trapping a bead with a dark center by moving the x–y translation stage. Once a bead is trapped, move it using PM1 to test the stability of the trapping beam. If the trap is aligned correctly, the bead should move in response to changes in PM1. An unstable trap is usually a consequence of low power at the back aperture of OBJ, improper alignment of the trap, clipping of the trapping beam at the back aperture of OBJ, or a distorted trapping beam shape. Check all these possibilities if necessary.

(13) *Setting up PSD1 for back focal plane detection*
   (a) PSD1 needs to be optically conjugated with the back focal plane of the CON. To achieve this, place DM2, L5 ($f_{L5} = 75$ mm), and a CCD (in place of PSD1) as shown in Figs. 14.3A and 14.4. Distances between the back focal plane of CON, L5, and CCD are 230 and 110 mm, respectively, which satisfies the condition $1/230 + 1/110 \sim 1/75$. The size of the CON aperture is ~ 20 mm, whereas the size of PSD1 is 10 mm. Hence, we demagnify the beam size twofold ($M = 110/230 \sim 0.5$). Collimate the transmitted beam by adjusting height of the CON. Adjust the position of L5 and the CCD such that the transmitted beam is directed on the center of the CCD. The position of the CCD is optically conjugated with PM1, and the back focal plane of OBJ and CON. Thus, rotation of PM1 should not move the transmitted beam position on the CCD. Test this by steering PM1. If the beam moves with PM1, adjust the position of the CCD along the beam path to minimize this translation.
   (b) Visualize the back-focal plane interference pattern of a bead using the CCD. This can be done using either a platform bead or a trapped bead. Test a platform bead first by moving the trapping beam (use PM1) to the center of a bead. Use the back-reflection of the trapping beam (see (11)). When the beam is focused near the center of the bead, you will see a dark interference pattern at the center of the image as shown in Fig. 14.5C. Changing the relative position of the trapping beam and the platform bead will change the interference pattern, as shown in Fig. 14.5C. Now visualize the interference pattern of a trapped bead. You will see fluctuations in the shape of the interference pattern, which reflects the Brownian motion of the trapped bead.
   (c) Design a stable box for mounting the PSD on the breadboard (Fig. 14.3A). It is important to avoid ground loops with multiple

devices connected to the same ground line. This is avoided by using a different power supply unit for each PSD. The optical table surface must also be grounded to prevent electrostatic problems.

(d) Move the trapping beam to the center of the platform bead as done in (b), such that a dark interference pattern as shown in Fig. 14.5C (center panel) is visible. PSD1 is put in place of the CCD (the detection surface of the PSD should be at the same position as that of the CCD). PSD1 is linked to a National Instruments data acquisition card (NI-DAQ) connected to a personal computer, which runs a custom Labview program and records the four voltage outputs from the PSD (x, y, sum_x, and sum_y). Move the platform bead by known discrete ($\sim$nm) steps using the piezo stage and measure the corresponding normalized voltage change (x/sum_x and y/sum_y). A linear regression fit to these measurements yields the conversion factor (in Δvolts/Δ nm). Record the signal from the platform bead to estimate the noise voltage level. The noise voltage is a measure of the spatial resolution of the trap. Noise voltage can be reduced, if necessary, by techniques discussed in Section 3.3. Power spectrum analysis can be used to pinpoint the main noise source. Once the PSD setup is completed, the single-beam trap is ready for use, following calibration of the AOD, PSD, and trap stiffness (see Section 3.5).

Step 3 Dual-beam trap for the three-bead assay (Fig. 14.2)

Transitioning from a single-beam optical trap to a dual-beam setup is straightforward and is described here. The dual-beam trap requires two independent trapping and detection beams for the two trapped beads. These are obtained by using p- and s-polarized components of the trapping and detection laser beams (Fig. 14.2, panel 1 and 3b). For the second trapping beam path, we add HP2, PBS2, PM2, and PBS3 (Fig. 14.2, panel 1). For detecting the position of the additional bead, we add PBS4, L13, and PSD2 (Fig. 14.2, panel 3a). To illuminate and image the actin filament, we set up a fluorescence laser (532 nm) and CCD2 camera, respectively (Fig. 14.2, panel 5). The laser wavelength is a function of the specific fluorescent phalloidin conjugate used to label the actin filament. The 532 nm laser and corresponding optical components are shown by way of an example for TMR-phalloidin conjugate. When the actin dumbbell is stretched taut, the bead is no longer at the center of the optical trap, and therefore could be outside the linear range of the PSDs. Hence, we recommend the setup of an additional detection laser beam path (Fig. 14.2, panel 3b) for studies that use the dual-beam optical trap for the three-bead myosin assay described here.

(1) *Setting up a second trapping beam path for the dual-beam trap*
   (a) Add HP2, PBS2, PBS3, and PM2 to create the second beam path of the dual-beam optical trap (Fig. 14.2, panel 1). HP2 and PBS2 split the p- and s-polarized components of the trapping beam. PM1 and PM2 steer the two trapping beams independently of each other. PBS3 combines the s- and p-polarized trapping beams. The relative power (trap stiffness) of the two laser traps can be adjusted by rotating the angle of HP2. For the dual-beam assay, the angle of HP2 is adjusted to equalize the trapping power in both trapping beams.
   (b) To align the second trapping beam, roughly adjust the angle of PM2 such that the second beam (s-polarized) overlaps with the first beam (p-polarized). Fine adjustment can then be done as follows. (i) Adjust the angle of PM2 to overlap the two beams at a position between M6 and L4 (use IR card). (ii) Adjust the angle of PBS3 to overlap the two beams at the position between L4 and DM1 (use IR card). (iii) Iterate between (i) and (ii) until both conditions are satisfied.
   (c) Once the two trapping beams are coupled, the dual-beam trap can be tested using a bright-field image on CCD1. Set up a sample channel that has both platform beads and trapping beads. Adjust the height of OBJ such that the platform beads are in focus. Two back-reflected trapping beams (reflected at the interface between the water and the glass coverslip) should be seen using CCD1. PM1 and PM2 can be used to position the trapping beams 5–10 μm from each other, at the center of the CCD1 field. The x–y stage is used to trap the two beads, one in each trap.
(2) *Setting up additional components to detect the position of the second trapped bead*
   Each of the two trapping beams can be used for back focal plane detection of the corresponding trapping bead position. Since the dual trapping beams are orthogonally polarized (s and p), the two transmitted beams can be separated using PBS4. Set PBS4, L13, and the PSD2 as seen in Fig. 14.3A. Adjust the position of PSD2 such that the transmitted beam arrives at the center of the position sensor (use IR card). The distance between PBS4 and PSD2 should be identical with that between PBS4 and PSD1, to ensure that PSD2 is conjugated with the back aperture of CON.
(3) *Setting up a detection laser beam path for the three-bead myosin assay*
   (a) As previously mentioned, an additional detection laser (850 nm) is necessary to ensure linear response of the back focal plane detection technique. The procedure to align the detection beam path is similar to that of the trapping beam path. Place the detection laser, M8, PM3, and DM5 as shown in Fig. 14.2. We need to optically

conjugate PM3 with the back aperture of OBJ. To ensure this, the distance between PM3 and L4 is set equal to the distance between PM1 and L4 (500 mm). We need to couple the detection and trapping beams. This can be done by walking the detection beam by adjusting PM3 and DM5 with the trapping beam as a guide.

(b) Set up L8–L10 as done in the previous trapping beam alignment step (Step 2, lenses L1–L4). In each instance, the detection beam is to be aligned to the trapping beam. To ensure that PM3 is optically conjugated with the back aperture of OBJ, lens L10 is selected to have the same focal length as L3. The focal length of lenses L8 and L9 is selected to expand the detection laser beam to slightly overfill the back aperture of the OBJ (∼9.6 mm). Note that the distance between L10 and L4 is the same as that between L3 and L4 (400 mm). To ensure that the detection beam is collimated, translate L8 or L9 parallel to the beam path. PM3 and PSD1 are optically conjugated if changes in angle of PM3 do not translate the beam at PSD1. To satisfy this, L10 can be translated parallel to the beam path. Subsequently, translate L8 or L9 to ensure that the detection beam is collimated with appropriate beam size.

(c) A second detection beam is generated by adding HP3, PBS5, PBS6, and PM4. The alignment procedure is the same as that for the second trapping beam path (Step 3.1a–b).

(d) Make a sample channel that has both platform and trapping beads. Now, you will see four back-reflected beams (reflected at the interface of water and glass coverslip) on CCD1, two from the trapping beams, and two from the detection beams. By controlling PM3 and PM4, overlap the detection beams with their corresponding trapping beams on CCD1 (PM1–PM3 and PM2–PM4).

(e) Use an ND filter (ND3) at the output of the detection laser (Fig. 14.2, panel 3b) to attenuate the detection beam power at the PSD to a few mW. This is done to optimize the sensitivity of the PSD output. Place a bandpass filter (F1) to block the 1064 nm trapping beam and transmit the 850 nm detection beam. Alternatively, a notch filter that blocks 1064 nm trapping beam also can be used for the filtering. Mount F1 on flipping mirror mounts such that the transition between the trapping beam- and detection beam-based measurement of the bead position is easily achieved.

(f) Trap beads in both traps. The PSD output reflects the Brownian motion of the trapped beads. You can estimate the noise level in the detection beam using a platform bead, as done in Step 2.13d.

(g) (Optional) Set up two shutters between PBS2 and PBS3 for each beam path (shown in Fig. 14.2, not shown in Fig. 14.3) to turn on/off each trapping beam independently. This is useful for making an actin dumbbell. See Section 4.1.

(4) *Set up for visualizing the actin dumbbell*
A 532 nm laser is used to excite the fluorescent phalloidin bound to the actin filament with the emission being imaged on CCD2 (Fig. 14.2, panel 5). Set up the fluorescence laser (532 nm), DM3, DM4, L11-L12, and CCD2 as seen in Fig. 14.2, panel 5. Expand the initial beam size of the 532 nm laser, if necessary, to expand the field of illumination (not shown in Fig. 14.2). DM3 reflects the excitation beam (532 nm) and transmits the emission beam (580 nm peak). DM4 transmits the 740 nm LED beam to CCD1 for the bright-field image, whereas it reflects the 580 nm emission peak of TMR to CCD2. L11 is positioned at its focal length (f_{L11} = 250 mm) from the OBJ. Using an infinity-corrected objective lens, L12 can be positioned anywhere in the beam path, followed by CCD2 at its focal point (f_{L12} = 200 mm) for the epifluorescence illumination. Attenuate the fluorescence beam using ND2 so that the beam does not damage protein molecules or cause rapid photobleaching of the dye. Adjust the angle of DM4 such that the fields of view of CCD1 and CCD2 are similar. Visualizing both the bright field and the fluorescence images simultaneously facilitates the creation of the actin dumbbell (see Section 4.1).

(5) *Cover the overall setup using tubes and a box*
Once all of the alignment procedure is completed, enclose all of the beam paths in a box and tubes. Use small tubes ($\sim$1 in. diameter) to locally enclose all beam paths, and build a thick plastic box ($\sim$0.5 in. thickness) to globally cover all components. This will significantly reduce the effects of air fluctuations, which is essential for the stable trap.

3.5. Calibration

Calibration of AOD, PSD, and trap stiffness are essential procedures for accurate and precise measurement of nanometer displacements and piconewton forces (pN) generated by a single myosin molecule. For this calibration, we need to know the relationships between the displacement of the trapped bead in the sample plane (nm) versus. (1) changes in the frequency input to the AODs, (2) changes in the output voltage of PSD, and (3) the restoring trap force. Here, we introduce simple methods to quantify these conversion factors. Other methods are also found elsewhere (Berg-Sørensen and Flyvbjerg, 2004; Bustamante *et al.*, 2007; Neuman and Block, 2004).

3.5.1. Conversion of changes in the voltage input to the AODs to bead displacement

Apply a voltage input to the AOD and measure the peak-to-peak displacement of a trapped bead on the bright-field image captured by CCD1. Repeat this with stepwise increases in AOD voltage input. The peak-to-peak displacement is read out in CCD1 pixel units, which can be converted to nanometer distance by calibrating the CCD1 field of view using a calibration slide (5 μm scale). Use this procedure to obtain a graph of the peak-to-peak displacement (nanometer) against the voltage input of AOD. Use a linear regression fit to the linear part of this graph (assessed visually) to obtain the nanometer bead displacement per volt input voltage at the AOD.

3.5.2. Conversion of changes in voltage output at the PSD to bead displacement

PSD output is merely a voltage signal, so we need a conversion factor for the real displacement of the bead in the sample plane as a function of PSD voltage output. This can be done using back focal interference detection in conjunction with either a trapped bead or a platform bead.

(1) *Using a trapped bead for calibrating the PSD output*

This method is used when a separate detection beam path is used for measuring bead position. First, trap two beads and separate them by ~5 μm. Ensure that the trapping and detection beams overlap. This is done by adjusting PM1–PM4 to overlap back-reflections of the trapping and detection beams, observed on CCD1 (for details, see Section 3.4, Step 3.3d). Apply a series of voltage inputs to the AOD to raster scan the CCD field of view and collect the normalized PSD voltage outputs (x/sum_x and y/sum_y). The PSD voltage output file can be used to map PSD voltage to AOD voltage, which in turn can be mapped to bead displacement in the sample plane (Section 3.5.1). Since variance in size and shape of the trapped bead can result in changes of the PSD response, make a mapping file for each bead used in an experiment.

(2) *Using a platform bead for calibrating PSD output*

Use the piezo stage to move the platform bead through a series of displacements across the field of view and record changes in the PSD voltage signal. This also yields a 2D mapping file, similar to (1) that directly converts PSD signal to bead displacement at the sample plane. One drawback of this technique is that the z-position of the platform bead relative to the sample plane varies from bead to bead. A combination of (1) and (2), however, is sufficient for accurate calibration of the PSD output.

3.5.3. Measuring trap stiffness

A number of different methods have been developed to measure trap stiffness (Berg-Sørensen and Flyvbjerg, 2004; Neuman and Block, 2004; Tolic-Norrelykke *et al.*, 2004, 2006). Three representative methods are the equipartition theorem method, the power spectrum method, and the hydrodynamic drag force method (for a detailed description, we refer the reader to Neuman and Block (2004) and Perkins (2009)). Since these are complementary approaches, we recommend using all of them for a better estimate of trap stiffness. Power spectrum analysis is a useful method to determine trap stiffness, as it dissects all frequency components of the bead movement. See Berg-Sørensen and Flyvbjerg (2004) for theoretical details and Tolic-Norrelykke et al. (2004) for analysis software. In the case of the three-bead assay, tension in the actin dumbbell displaces the bead from the trap center and alters the effective stiffness experienced by the trapped bead in the presence of the actin dumbbell. The trap stiffness is linearly proportional to the local spatial gradient in trapping power. As the trapped bead moves away from the trap center, this gradient increases, thereby increasing the trap stiffness. We find a ~10–20% increase in trap stiffness following the formation of the actin dumbbell (see Section 5.2.1, Step 1 on compliance correction for details on measuring trap stiffness in the presence of the actin dumbbell).

4. Optical Trapping Experiment

In the three-bead assay, an actin filament is stretched taut between two 1 μm diameter polystyrene beads (Fig. 14.1). The actin filament is biotinylated and the polystyrene beads are coated with neutravidin to facilitate the formation of the actin dumbbell (Rief et al., 2000). The coverslip surface is coated with 1.5 μm diameter glass beads, which serve as platforms for the myosin. The coverslips are dipped in 0.1% solution of nitrocellulose in isoamyl acetate and dried overnight. This sticky surface is coated with antibody against green fluorescent protein (GFP). Myosin chimeras are used that have their C-terminal cargo-binding domains replaced by GFP (Rock et al., 2001). Thus, a single molecule of myosin can be positioned on the top of a glass bead platform with its GFP tail domain bound. Since the myosin is anchored to the surface, as the motor strokes the actin filament slides, thereby pushing against one trapped bead while pulling against the other (Fig. 14.7).

Preparation of reagents, including biotinylated actin filaments, platform bead coated coverslips, and sample cells, is described elsewhere in detail (Spudich et al., 2007). Here, we discuss the procedure to create a taut actin dumbbell and to test platform beads to find a myosin motor that interacts with the actin filament and displaces the trapped beads. Start by creating a sample cell with myosin-coated platform beads, and with biotinylated actin filaments and neutravidin-coated beads to be trapped in trapping buffer (Spudich et al., 2007). Mount the sample channel with immersion oil on both sides, sandwiched between CON and OBJ.

4.1. Forming an actin dumbbell

To avoid photobleaching of actin filaments and beads to be trapped, place a beam block at the fluorescence laser output during the first two steps. Steer traps using PM1 and PM2 such that they are 5–10 μm apart. Trap a single bead in each optical trap by translating the x–y stage.

Sometimes, the trapped beads look asymmetric in shape or are composed of a chunk of multiple beads. Also, some trapped beads are already connected to bundles of actin filaments on the surface. In these cases, close the corresponding shutter and selectively release the bead from its trap.

Remove the beam block and allow the fluorescence laser to illuminate the field of view. Observe actin filaments in solution on CCD2. Translate the x–y stage to bring a single long actin filament (avoid clumps) close to one of the trapped beads. Actin binding to one of the trapped beads can be confirmed by moving the x–y stage. Move the x–y stage parallel to a line connecting the trapped beads, such that the flow caused by this motion brings the actin filament in the vicinity of the second bead. Move the x–y stage perpendicular to the actin filament. If the actin filament is bound to both trapped beads, it will not move away from either trapped bead.

4.2. Tensing the actin dumbbell

Steer the detection beams (PM3 and PM4) so that they overlap with the trapping beams (use back-reflections, see Section 3.4, Step 3.1c). Use PM2 to move the second trapped bead away from the first. Monitor the PSD signals in the direction parallel to the actin filament (for instance, if the actin filament is along the Y-axis, monitor y/sum_y for each PSD). As the actin dumbbell becomes taut, the first trapped bead will begin to move out of the trap center in the bright-field image. At this stage, use small incremental movements to further tense the dumbbell. The PSD outputs reflect the Brownian motion of the two trapped beads. The actin dumbbell is sufficiently taut when the PSD output signals appear correlated in time. It is important to keep repositioning the detection beam during this process to ensure that it coincides with the trapping beam. As the detection beam moves away from the trapping beam, PSD sensitivity, and hence the variance of the PSD output decrease. Reposition the detection beam to maximize variance in the PSD output.

4.3. Testing platforms

Find a single platform bead (avoid doublets or clumps) by steering the x–y stage. The actin dumbbell should be positioned slightly further in solution than the platform beads, to avoid snagging the dumbbell on the surface of the platform bead. Adjust the coverslip height such that the midpoint of the actin dumbbell is at the surface of the platform bead. When an active myosin on the surface of the platform bead interacts with the actin dumbbell, both the variance and mean position of the PSD outputs change (Fig. 14.6).

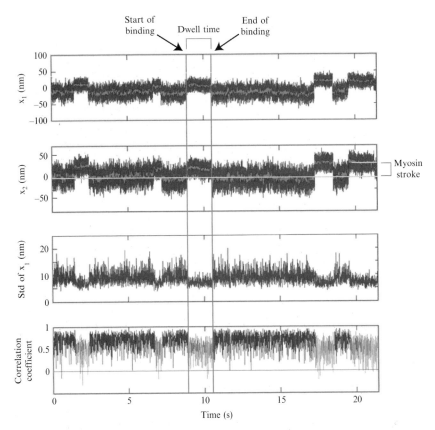

Figure 14.6 Representative experimental result of a nonprocessive motor. The time traces show the interaction of a single-headed myosin VI construct with an actin filament. The first and the second graphs show the time trajectory of the first (x_1) and the second (x_2) bead positions. The blue line shows the raw data with a sampling rate of 10 kHz while the red line is the low-pass filtered result of the raw data (fourth-order low-pass digital Butterworth filter with a cutoff frequency of 50 Hz. The trap stiffness in this particular example is ~ 0.02 pN/nm, and this results in Brownian motion of bead (blue line) with standard deviation (std) of ~ 10 nm. Increasing trap stiffness will reduce the std of Brownian fluctuation. In the figure, there are five very clear binding and unbinding events, showing a sudden jump and drop in the bead positions. The duration between binding and unbinding events is the dwell time. The stroke size refers to the displacement of the bead position when the binding event occurs. This stroke size is before compliance correction which is explained in the legend to Fig. 14.7 and in Section A.2. The third and fourth graphs show the std of the first bead position and the correlation coefficient between the two bead positions, respectively. At the third binding event at ~ 10 s, for example, we observe a sudden drop in the std as well as the correlation between x_1 and x_2. This happens because the myosin binding clamps the actin dumbbell, reducing its Brownian fluctuations, which in turn reduces the Brownian fluctuations of the two trapped beads. We use a custom Matlab program which automatically finds two different states in the correlation graph, so the sudden change of the color (blue and green) in the last graph helps to visualize the binding events. We manually choose the timing of binding and unbinding events based on the std and the correlation. The binding events are chosen only when both the std and the correlation change register a significant change. (See Color Insert.)

4.4. Identifying binding interactions

Lowering the actin dumbbell too far will result in reduced variance in bead movement as a consequence of direct contact with the bead surface. When properly positioned, processive myosins will produce a characteristic staircase pattern. Nonprocessive myosins release the actin before the next binding event. These features can be used to distinguish a properly oriented active platform from a platform that is in direct contact with the actin filament.

Bead displacements that are too close to each other in time or even simultaneous are likely a result of multiple motors on the surface. The motor dilution used should be adjusted so that only one in ten platforms or less interacts with the actin filament. For a detailed discussion of how to avoid multiple motors, see Spudich et al. (2007). Before collecting data on a single motor platform, move the actin dumbbell to the side of the bead and increase tension in the dumbbell to obtain good correlation between the two signals. The piezo stage can be used to optimize the motor position such that myosin binds at well-spaced intervals ($\sim$5 s apart).

5. Data Analysis

The optical trap is a powerful means of discerning the kinetic mechanisms of molecular motors. The processes measured in the optical trap, for instance, the rate of stepping by dimeric track motors, are often difficult to assess in bulk assays. In addition, individual molecules are generally either fully functional or catalytically inactive. Thus, data collected from active molecules represent the behavior of a hypothetical, 100% active preparation. Finally, the measurement of dwell times is often very accurate ($\sim$ms temporal resolution), even in cases where the calculation of absolute displacements is challenging due to the compliant elements. Here, we describe a method in detail for the measurement of dwell time and stepping size in the presence of a compliant element.

5.1. Measurement of dwell time

The kinetic information obtained from an optical trap experiment is the distribution of waiting times between events. This can be the time in between steps for a processive motor-like myosin V (Purcell et al., 2005; Rief et al., 2000), or the length of time that a single-headed, nonprocessive motor remains bound to actin, as does myosin II (Finer et al., 1994) or I (Laakso et al., 2008; Veigel et al., 1999). With the dwell time distribution in hand, the next task is to extract information about the underlying kinetics.

5.1.1. Graphing the data

A common method of portraying dwell time data is as a histogram. This method has several obvious advantages: it is easily interpretable, and is, in principle, readily fit with probability density functions:

$$A \xrightarrow{k} B, \quad f = ke^{-kt}, \tag{14.1}$$

$$A \xrightarrow{k_1} B \xrightarrow{k_2} C, \quad f = \frac{k_1 k_2}{(k_1 - k_2)}(e^{-k_2 t} - e^{-k_1 t}). \tag{14.2}$$

Here, f is the normalized probably density function for a single process (14.1) or sequential exponential processes (14.2). These two distributions are mentioned in particular because they are by far the most common situations encountered in single-molecule data. Single exponential kinetics result when a single process is much slower than all the other steps in the observed portion of the catalytic cycle. An example is ADP release from myosin V at saturating ATP concentrations. Sequential exponential kinetics occurs when two processes have comparable rates. For example, the rates of ATP binding and ADP release are both $\sim 10 \text{ s}^{-1}$ at 10 μM ATP. It is worth noting that the distinction between these two cases is somewhat arbitrary: It is rare in practice to have a dwell time distribution that truly reflects only one underlying rate. However, it is common to observe kinetics where one process is ~ 10 times slower than all the rest.

Unfortunately, histogramming the data suffers from several drawbacks. The choice of bin size is somewhat arbitrary, and can be massaged such that the data coincide with the anticipated model. Furthermore, each binned data point is often erroneously given equal weight during fitting, despite the low statistical significance of the bins containing only a few (<5) observations. Using variable-width bins can of course alleviate this problem by ensuring that each bin contains at least five observations.

The most serious problem from a practical standpoint is that histogramming necessarily greatly reduces the number of data points being fit with the model. For example, ~ 200 dwell time observations, when histogrammed, will result in only ~ 10–15 bins. This makes it difficult to discern the likely underlying mechanism by visual inspection of the data. The reduction in the ratio of data points to parameters also inflates fit errors when performing a straightforward least-squares fit to the data.

The use of cumulative distributions avoids many of the drawbacks associated with histograms. A common form of the cumulative distribution plots the fraction of dwells that are longer than a given time t. As such, the cumulative distribution is 1 minus the integrated probability density function:

$$A \xrightarrow{k} B, \quad f = e^{-kt}, \tag{14.3}$$

$$A \xrightarrow{k_1} B \xrightarrow{k_2} C, \quad f = \frac{k_1}{(k_1 - k_2)}e^{-k_2 t} - \frac{k_2}{(k_1 - k_2)}e^{-k_1 t}. \tag{14.4}$$

Fitting to cumulative distributions removes the complexities associated with properly treating binning artifacts: each observation is given equal weight. Further, the ratio of data points to parameters is much higher than in histogrammed data because the data are not collapsed into bins. Imperfections in the data and differences between the data and model are both often more apparent than in histogrammed data. Cumulative distributions are not intellectually exotic: the integrated rate laws are equally applicable to, for instance, data resulting from an analogous stopped-flow experiment in which a synchronized population evolves with time.

A least-squares fit to the dwell time data using commercial software is often used and has a number of immediate practical advantages. It is very easy to implement in numerous software packages, for instance, Matlab (MathWorks) or Igor (WaveMetrics). The fit is sensitive to outliers. Hence, poor-quality data (for instance, a dataset that is missing observations at short dwell times) will result in obviously poor fits. Finally, commercial packages provide numerous statistical measurements of the goodness of fit, for instance, RMSD, χ^2, reduced χ^2, and confidence intervals.

However, reliance on commercial fitting routines suffers from several important drawbacks. First, as discussed later, it is possible and desirable to find the *most likely* values of the fit parameters given one's data, rather than the parameters that result from a nontransparently implemented nonlinear fit. Second, common commercial fitting software often does not take the covariance of fit parameters into account in computing confidence intervals, which can result in severe underestimates of fit uncertainty. Third, least-squares fits do not take sampling error into account. This is a crucial limitation, given the small sample sizes that often characterize single-molecule experiments.

5.1.2. Maximum likelihood estimation

Here, we give a very brief introduction to the fitting method of maximum likelihood estimation, a better fitting method than using least-square fitting (Rao, 1965). Given a collection of dwell times (t_1, ..., t_n) and a probability density function f with parameters k (k_1, ..., k_m), one may define the likelihood of observing exactly this collection of dwells:

$$P(k, t) \propto \prod_i f(k, t_i). \quad (14.5)$$

In principle, all that is necessary is to find the values of k that maximize P. In practice, it is easier to minimize the negative logarithm:

$$-\log(P(k, t)) \propto \sum_i -\log(f(k, t_i)). \quad (14.6)$$

This minimization can be done using iterative minimization routines that are commonly built into analysis software such as Matlab. However, it is often possible to find the k that minimizes the P analytically by taking the partial derivative with respect to each element in k. Consider the exponential distribution:

$$P(k, t) \propto \prod_i^N k e^{-kt_i}, \qquad (14.7)$$

$$\ln(P(k, t)) \propto \sum_i^N (\ln k - kt_i), \qquad (14.8)$$

$$\frac{\partial}{\partial k} \ln(P(k, t)) \propto \sum_i^N \left(\frac{1}{k} - t_i\right). \qquad (14.9)$$

For convenience, we calculate $\ln(P)$ instead of $-\log(P)$. In order to find the maximum of $\ln(P)$, we set the partial derivative of $\ln(P)$ with respect to k equal to 0:

$$0 = \sum_i^N \left(\frac{1}{k} - t_i\right) = \frac{N}{k} - \sum_i^N t_i,$$

$$\frac{1}{k} = \frac{1}{N} \sum_i^N t_i. \qquad (14.10)$$

Equation (14.10) indicates that the most likely estimate for k is simply the inverse of the average dwell time! An analogous, but algebraically messy derivation of maximally likely values of k_1 and k_2 is likewise possible for the sequential exponential function.

The maximum likelihood estimate is, by definition, the best estimate for the given parameter. Further, its value can often be derived analytically, as earlier. Methods to derive fit errors based on maximum likelihood calculations are well established. However, these methods do not account for fit uncertainty introduced by small sample sizes, as discussed later.

5.1.3. The bootstrap method

This very useful but probably less well-known method of data analysis was introduced and popularized by Efron and Tibshirani (1994). The key underlying assumption is that the observations in the dataset are independent: a given dwell time observation is not influenced by the length of the preceding or following dwell. This is most often, but not always, the case.

The bootstrap method further assumes that the data t constitute a representative sample of the underlying distribution. If one were to repeat the experiment and again make N observations, the resulting observed data and fit parameters would be similar. Any differences would stem principally from the finite sizes of the two datasets. The bootstrap method simulates many repetitions of the experiment: from a dataset t containing N observations, a synthetic dataset b is created by randomly selecting N observations from t with replacement. This means that the same t_i may be chosen more than once when creating b. The dataset b is then fit using the method of one's choice resulting in fit parameters k_b. This process is repeated several thousand times, resulting in a set of synthetic fit parameters (k_b).

The beauty of this procedure is that the error for each element k_i in $\mathbf{k}$ is the standard deviation of corresponding elements in k_b. For example, suppose the data are fit with a single exponential. In order to find the fit uncertainty, the maximum likelihood estimate k_b is calculated for each b. This procedure is then repeated many times. If the data do not suffer from some underlying experimental flaw, the resulting collection of k_b values is approximately normally distributed. The standard deviation of this distribution gives the uncertainty on k.

This approach has several strengths in estimating fit uncertainties. First, it makes no assumptions about the data other than those mentioned earlier. Second, it is easy to implement. Third, it implicitly includes the effect of finite sample size on the resulting fit uncertainty. Fourth, it includes the effect of fit parameter covariance. The bootstrap method has thus become our error estimation method of choice.

5.1.4. A final note: The residual

It is always worthwhile to plot the difference between the model and the data, known as the residual. A residual with obvious, nonrandom features is a sure sign that the model does not fully capture the data. Furthermore, the features of the residual usually stimulate hypotheses about potential deficiencies in either the data collection methodology or the model. The residual is thus often more helpful in the intermediate stages of data interpretation than numerical goodness-of-fit measures, for example, χ^2.

5.2. Measurement of stroke size: Compliance correction for nonprocessive motors

Accurate and precise measurement of stroke sizes for processive and nonprocessive motors is complicated by the presence of compliant connections at the bead–actin interface, compliant elements in the myosin and compliance of the myosin to surface attachment (Fig. 14.7). In the case of force-feedback measurements, all of these springs stay stretched to a constant length, and it is therefore possible to measure displacements directly from

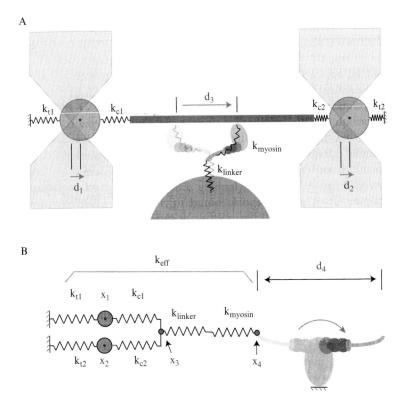

Figure 14.7 Compliance correction in the dual beam optical trap. (A) Cartoon schematic of a myosin in a dual-beam optical trap with compliant elements—a myosin stroke in the presence of compliant elements is depicted. See Fig. 14.1 for the corresponding three-dimensional image. Here, different compliant elements are shown as isotropic linear Hookean springs. The compliant elements include the actin filament to bead connections (k_{c1} and k_{c2}), the $(GSG)_4$ linker (k_{linker}), and the myosin itself (k_{myosin}). In the figure, the stroke displaces the actin filament from left to right, so that the springs on the left side (k_{t1} and k_{c1}) are stretched while those on the right side (k_{t2} and k_{c2}) are compressed. The centers of the trapped beads are indicated by x_1 and x_2. Making a tight dumbbell affects the apparent trap stiffness experienced by the beads. Hence, trap stiffness in the bound state needs to be determined in the compliance correction procedure (see Section 5.2). In the experiment shown in Fig. 14.6, the apparent stroke size before compliance correction is reflected in the displacements d_1 and d_2. The displacement of the position of the myosin–actin binding interface upon going from prestroke to poststroke is d_3. The light and dark myosin images show the prestroke and the poststroke states, respectively. (B) Alternative representation of spring connections in Fig. 14.7A. The springs in Fig. 14.7A can be combined into a serial/parallel combination, such that $k_{eff} = \{(k_{t1}||k_{c1}) + (k_{t2}||k_{c2})\}||(k_{myosin}||k_{linker})$, where $A||B \equiv (A^{-1} + B^{-1})^{-1}$. Despite a change in the position of the second bead, the magnitude and the direction of the force applied by it is the same as in Fig. 14.7A. From the experiment (see first and second graph in Fig. 14.6 as an example), we measure the displacement of the two beads (x_1, x_2). If the spring constant for the compliant elements (k_{c1}, k_{c2}, k_{linker}, and k_{myosin}) are much greater than the trap stiffness (k_{t1}, k_{t2}), the effective stiffness (k_{eff}) can be approximated by $k_{eff} \sim k_{t1} + k_{t2}$. For this case, the bead position displacements (x_1, x_2) reflect the actual stroke size of the myosin (d_4). When the spring constant of the compliant elements is on the order of the trap stiffness, compliance correction is necessary to measure the actual d_4 (compliance correction is explained in Section 5.2).

bead displacements (Visscher et al., 1999). However, stroke-size measurements for nonprocessive motors require compliance correction, especially when the trap stiffness is comparable to or higher than the stiffness of other compliant elements.

5.2.1. Strategy for compliance correction in a dual-beam optical trap

Find interacting platforms with single motors as discussed in Section 4.4. Nonprocessive motors show binding and unbinding events as shown in Fig. 14.6. Binding events result in both a sudden drop in the variance of bead position, as well as in the position correlation between the two beads. These two sources of information, variance and bead-bead correlation, allow us to determine the effective trap stiffness as well as the compliance of individual elements. This technique has been used to measure the stroke contribution of an ER/K α-helix by Sivaramakrishnan et al. (2009).

Figure 14.7 is a schematic of the compliant elements in the three-bead assay. Displacement of positions x_1–x_4 is denoted as d_1–d_4. The PSD output directly measures d_1 and d_2 during the experiment. We want to measure the actual myosin stroke size (d_4) in the presence of many compliant elements, such as the actin filament to bead connections (k_{c1} and k_{c2}) and the (GSG)$_4$ linker between the motor and the bead platform (k_{linker}), as well as the flexibility within the myosin (k_{myosin}).

The effective spring constant for the myosin stroke is a mixed serial and parallel combination of each element, expressed as:

$$k_{\text{eff}} = [(k_{t1}||k_{c1}) + (k_{t2}||k_{c2})]||(k_{\text{myosin}}||k_{\text{linker}}), \qquad (14.11)$$

where $A||B \equiv (A^{-1} + B^{-1})^{-1}$. Here, k_{t1} and k_{t2} are the trap stiffnesses after making a tight dumbbell. Since the formation of the tight dumbbell displaces the trapped beads from the center of the trapping beam, k_{t1} and k_{t2} are usually higher than those for the trapped beads in the absence of a dumbbell (~10% greater by experience). To simplify the mathematical expression, we define ($k_c = k_{c1} || k_{c2}$) for the serial connection of the actin filament to bead connections, and ($k_{xb} = k_{\text{myosin}} || k_{\text{linker}}$) for the stiffness of the crossbridge, which is a serial combination of the myosin and the flexible linker. Simple algebra allows us to derive the diffusion correlation coefficient of the two beads (CorrCoef, see Section A.2) as a function of (k_{t1}, k_{t2}, k_{c1}, k_{c2}, k_{xb}), as first outlined by Mehta et al. (1997). This method is subsequently described in detail. Additionally, we measure the effective spring constant of each bead with the dumbbell present, in both the bound (k_{b1}, k_{b2}) and unbound (k_{u1}, k_{u2}) states. This information is essential to calculate the different compliances.

Step 1 In the unbound state, determine k_{t1}, k_{t2}, and k_c from the measured k_{u1}, k_{u2}, and CorrCoef

We determine the effective trap stiffnesses (k_{t1} and k_{t2}) and the bead–actin connection stiffness (k_c), in the presence of a tight dumbbell, from a combination of analytical expressions and numerical computation.

(a) Select initial values for k_{t1} and k_{t2}. We use values from the calibration of trap stiffnesses without the dumbbell.
(b) Obtain k_c using *CorrCoef* with a given k_{t1} and k_{t2}. We assume $k_{xb} = 0$ (no myosin attachment), and $k_{c1} = k_{c2} = 2k_c$.
(c) Given k_c and the measured k_{u1} and k_{u2}, we can analytically determine the new k_{t1} and k_{t2}:

$$k_{u1} = k_{t1} + (k_{t2} \| k_c), \quad (14.12)$$

$$k_{u2} = k_{t2} + (k_{t1} \| k_c). \quad (14.13)$$

(d) Iterate (b) and (c) until k_{t1}, k_{t2}, and k_c converge. We observe good convergence after a few rounds of iteration.

Step 2 Determine k_{c1} and k_{c2}

We determine k_{c1} and k_{c2} analytically, using the following two expressions, with values for k_{t1}, k_{t2}, and k_c obtained from Step 1, and the measured displacement during the binding event of the first bead (d_1) and the second bead (d_2):

$$k_c = k_{c1} \| k_{c2}, \quad (14.14)$$

$$d_3 = d_1 \frac{k_{t1} + k_{c1}}{k_{c1}} = d_2 \frac{k_{t2} + k_{c2}}{k_{c2}}. \quad (14.15)$$

Thus, we now know the values of k_{t1}, k_{t2}, k_{c1}, and k_{c2}.

Step 3 Determine k_{xb}

We next use the CorrCoef function to fit the data collected during the binding events. We already know four variables of this function, namely k_{t1}, k_{t2}, k_{c1}, and k_{c2}, so we can determine the last unknown k_{xb}.

Step 4 Determine the stroke size of the motor (d_4) from previously determined values for k_{t1}, k_{t2}, k_{c1}, k_{c2}, and k_{xb}

The stroke size of myosin in the presence of compliant elements is given by:

$$d_4 = d_3 \frac{k_{tc} + k_{xb}}{k_{xb}}, \qquad (14.16)$$

where $k_{tc} = (k_{t1}||k_{c1}) + (k_{t2}||k_{c2})$.

6. Conclusion

This chapter has detailed the assembly of a modern dual-beam optical trap, described the experimental protocols involved in analyzing molecular motors using the trap, and provided methods of analysis often missing in similar publications. Even the novice should find it possible to set up the trap and carry out appropriate experiments using this chapter as a guide. We hope we have at least partially achieved our aim.

Appendix

A.1. Alignment for the feedback control of a single beam by the AOD

The dual-beam optical trap setup in Fig. 14.2 is a design optimized for the three-bead assay. However, some dual-beam experiments require feedback control of only one beam. Here, we introduce a simple modification of the current design for the single-beam feedback control. We now use the s-polarized component of PBS1, which is beam-blocked in Fig. 14.2, panel 1. Move BD 5 in. away from PBS such that BD is placed to the left of the beam path from AOD to L1. Insert another PBS (PBS7) between AOD and L1 such that the s-polarized beam from PBS1 is directed to L1. Beam coupling can be done by iteratively adjusting the angle of PBS1 and PBS7. Move the position of HP2 between PBS1 and PBS7. By rotating the angle of HP2, the power of the s-polarized beam can be adjusted and directed to L1. The p-polarized component is passing through the AOD and therefore can be feedback controlled, whereas the s-polarized beam is not controlled by the AOD. In addition, the p- and s-polarized beams are reflected on PM1 and PM2, respectively, so that both of them can be coarsely positioned as described earlier.

A.2. Correlation coefficient between two trapped beads (CorrCoef)

In the three-bead assay, two trapped beads are connected with each other by a tight actin dumbbell. Therefore, the position of each bead in the diffusive motion is correlated with a finite correlation coefficient. We use the diffusion correlation coefficient between two trapped beads to characterize the compliant elements as described in Section 5.2. This method was derived by Mehta *et al.* (1997), and here we show the explicit form of the correlation coefficient function (CorrCoef).

Consider the situation in which two beads are connected to each other through the actin dumbbell and the dumbbell interacts with a motor as depicted in Fig. 14.7A. In this heavily overdamped environment (low Reynolds number), the Langevian equations for the two beads are expressed as

$$F_1(t) - k_{t1}x_1(t) - \gamma \frac{dx_1(t)}{dt} + k_{c1}[x_3(t) - x_1(t)] = 0, \quad (14.A1)$$

$$F_2(t) - k_{t2}x_2(t) - \gamma \frac{dx_2(t)}{dt} + k_{c2}[x_3(t) - x_2(t)] = 0, \quad (14.A2)$$

where $\gamma = 6\pi\eta r$, η is the viscosity of the liquid, and r is the radius of the beads. Note that this description of γ is only strictly true for a sphere far from the coverslip surface. More accurate treatments based on Faxen's law for describing the drag on a sphere near a surface are detailed elsewhere (Neuman and Block, 2004). The four terms in Eqs. (14.A1) and (14.A2) refer to the thermal force, trapping force, drag force, and the force from actin dumbbell, respectively. At any time t, as per Newton's second law, the forces on the actin filament must sum to zero. The equation of motion at x_3 is:

$$k_{c1}[x_1(t) - x_3(t)] + k_{c2}[x_2(t) - x_3(t)] - k_{xb}x_3(t) = 0, \quad (14.A3)$$

where k_{xb} refers to the spring constant for the cross-bridge that is the serial combination of the myosin and anchoring linker on the surface. For example, $k_{xb} = 0$ represents the unbound state, whereas $k_{xb} \to \infty$ indicates the bound state with infinite stiffness of the connection. Solving the three Eqs. (14.A1)–(14.A3), we get

$$F_1(t) - \left[k_{t1} + \frac{k_{c1}(k_{c2} + k_{xb})}{k_{c1} + k_{c2} + k_{xb}}\right]x_1(t) - \gamma\frac{dx_1(t)}{dt} \\ + \left[\frac{k_{c1}k_{c2}}{k_{c1} + k_{c2} + k_{xb}}\right]x_2(t) = 0, \quad (14.A4)$$

$$F_2(t) - \left[k_{t2} + \frac{k_{c2}(k_{c1} + k_{xb})}{k_{c1} + k_{c2} + k_{xb}} \right] x_2(t) - \gamma \frac{dx_2(t)}{dt}$$
$$+ \left[\frac{k_{c1} k_{c2}}{k_{c1} + k_{c2} + k_{xb}} \right] x_1(t) = 0. \qquad (14.\text{A5})$$

Now, we Fourier transform and obtain

$$X_1(f) = \frac{A F_1(f) + C F_2(f)}{AB - C^2}, \qquad (14.\text{A6})$$

$$X_2(f) = \frac{B F_2(f) + C F_1(f)}{AB - C^2}, \qquad (14.\text{A7})$$

where $A = k_{t2} + \frac{k_{c2}(k_{c1}+k_{xb})}{k_{c1}+k_{c2}+k_{xb}} + i2\pi f \gamma$, $B = k_{t1} + \frac{k_{c1}(k_{c2}+k_{xb})}{k_{c1}+k_{c2}+k_{xb}} + i2\pi f \gamma$, $C = \frac{k_{c1} k_{c2}}{k_{c1}+k_{c2}+k_{xb}}$.

Since we are collecting data in a finite measurement time ($T_{\text{msr}} = 1/f_{\text{msr}}$) with a finite sampling rate ($f_{\text{sampling}} = 10\, kHz$), the set of frequency range is given by $f = f_{\text{msr}} \times k$, where $k = [1, 2, \ldots K, N]$ with total sampling number N. Now, we want to calculate the correlation coefficient between two sampled signals $x_1(n)$ and $x_2(n)$ whose discrete Fourier transforms are $X_1(k)$ and $X_2(k)$. Using Parseval's theorem, the correlation coefficient between the two beads is

$$\text{CorrCoef} = \frac{\sum [x_1(n) x_2(n)]}{\sqrt{\sum [x_1(n)]^2 \sum [x_2(n)]^2}} = \frac{\sum [X_1(k) X_2^*(k)]}{\sqrt{\sum [X_1(k)]^2 \sum [X_2(k)]^2}}.$$
$$(14.\text{A8})$$

The thermal forces working on the two beads cannot be correlated with each other, so the cross-term can be dropped as

$$\sum F_1(k) F_2(k) = 0.$$

We assume that $F_1(n)$ and $F_2(n)$ at every frequency are identical on average, since their time scales are much faster than any other related time scales. Thus, we cancel out $|F_1|^2$ and $|F_2|^2$ in the numerator and the denominator in Eq. (14.A8). Therefore, the final expression for the correlation coefficient is

$$\mathrm{CorrCoef} = \frac{\sum \frac{AC^* + B^* C}{|AB - C^2|^2}}{\sqrt{\left[\sum \frac{AA^* + CC^*}{|AB - C^2|^2}\right]\left[\sum \frac{BB^* + CC^*}{|AB - C^2|^2}\right]}}. \quad (14.A9)$$

Since CorrCoef is a function of $[k_{t1}, k_{t2}, k_{c1}, k_{c2}, k_{xb}]$ and also a measurable quantity in our experiment, we can determine one unknown given four other quantities.

ACKNOWLEDGMENTS

This work was supported by a grant to J. A. S. from the National Institutes of Health (GM33289) and Human Frontier Science Program Grant (GP0054/2009). A. R. D. was supported by a Burroughs Wellcome Career Award at the Science Interface, S. S. by an American Cancer Society postdoctoral fellowship, and J. S. by a Stanford BioX fellowship.

REFERENCES

Ashkin, A. (1992). Forces of a single-beam gradient laser trap on a dielectric sphere in the ray optics regime. *Biophys. J.* **61,** 569–582.
Berg-Sørensen, K., and Flyvbjerg, H. (2004). Power spectrum analysis for optical tweezers. *Rev. Sci. Instrum.* **75,** 594–612.
Block, S. M. (1998). Construction of optical tweezers. *In* "Cells: A Laboratory Manual," (D. Spector, R. Goldman, and L. Leinwand, eds.), Cold Spring Harbor Laboratory Press, New York.
Bryant, Z., Altman, D., and Spudich, J. A. (2007). The power stroke of myosin VI and the basis of reverse directionality. *Proc. Natl. Acad. Sci. USA* **104,** 772–777.
Bustamante, C., Chemla, Y. R., and Moffitt, J. R. (2007). High-resolution dual-trap optical tweezers with differential detection. *In* "Single-Molecule Techniques: A Laboratory Manual," (P. R. Selvin and T. Ha, eds.), pp. 297–324. Cold Spring Harbor Laboratory Press, New York.
Carter, A. R., Seol, Y., and Perkins, T. T. (2009). Precision surface-coupled optical-trapping assay with one-basepair resolution. *Biophys. J.* **96,** 2926–2934.
Churchman, L. S., Okten, Z., Rock, R. S., Dawson, J. F., and Spudich, J. A. (2005). Single molecule high-resolution colocalization of Cy3 and Cy5 attached to macromolecules measures intramolecular distances through time. *Proc. Natl. Acad. Sci. USA* **102,** 1419–1423.
Dominguez, R., Freyzon, Y., Trybus, K. M., and Cohen, C. (1998). Crystal structure of a vertebrate smooth muscle myosin motor domain and its complex with the essential light chain: Visualization of the pre-power stroke state. *Cell* **94,** 559–571.
Dunn, A. R., and Spudich, J. A. (2007). Dynamics of the unbound head during myosin V processive translocation. *Nat. Struct. Mol. Biol.* **14,** 246–248.
Efron, B., and Tibshirani, R. J. (1994). An Introduction to the Bootstrap. Chapman & Hall, New York.
Finer, J. T., Simmons, R. M., and Spudich, J. A. (1994). Single myosin molecule mechanics: Piconewton forces and nanometre steps. *Nature* **368,** 113–119.
Gittes, F., and Schmidt, C. F. (1998a). Interference model for back-focal-plane displacement detection in optical tweezers. *Opt. Lett.* **23,** 7–9.

Gittes, F., and Schmidt, C. F. (1998b). Signals and noise in micromechanical measurements. *Method Cell Biol.* **55**, 129–156.
Guilford, W. H., Dupuis, D. E., Kennedy, G., Wu, J., Patlak, J. B., and Warshaw, D. M. (1997). Smooth muscle and skeletal muscle myosins produce similar unitary forces and displacements in the laser trap. *Biophys. J.* **72**, 1006–1021.
Harada, Y., Noguchi, A., Kishino, A., and Yanagida, T. (1987). Sliding movement of single actin filaments on one-headed myosin filaments. *Nature* **326**, 805–808.
Hohng, S., Zhou, R., Nahas, M. K., Yu, J., Schulten, K., Lilley, D. M., and Ha, T. (2007). Fluorescence-force spectroscopy maps two-dimensional reaction landscape of the holliday junction. *Science* **318**, 279–283.
Huisstede, J. H. G., van Rooijen, B. D., van der Werf, K. O., Bennink, M. L., and Subramaniam, V. (2006). Dependence of silicon position-detector bandwidth on wavelength, power, and bias. *Opt. Lett.* **31**, 610–612.
Huxley, H. E. (1969). The mechanism of muscular contraction. *Science* **164**, 1356–1365.
Ishijima, A., Kojima, H., Funatsu, T., Tokunaga, M., Higuchi, H., Tanaka, H., and Yanagida, T. (1998). Simultaneous observation of individual ATPase and mechanical events by a single myosin molecule during interaction with actin. *Cell* **92**, 161–171.
Kad, N. M., Trybus, K. M., and Warshaw, D. M. (2008). Load and Pi control flux through the branched kinetic cycle of myosin V. *J. Biol. Chem.* **283**, 17477–17484.
Kron, S. J., and Spudich, J. A. (1986). Fluorescent actin filaments move on myosin fixed to a glass surface. *Proc. Natl. Acad. Sci. USA* **83**, 6272–6276.
Laakso, J. M., Lewis, J. H., Shuman, H., and Ostap, E. M. (2008). Myosin I can act as a molecular force sensor. *Science* **321**, 133–136.
Lang, M. J., Asbury, C. L., Shaevitz, J. W., and Block, S. M. (2002). An automated two-dimensional optical force clamp for single molecule studies. *Biophys. J.* **83**, 491–501.
Lang, M. J., Fordyce, P. M., Engh, A. M., Neuman, K. C., and Block, S. M. (2004). Simultaneous, coincident optical trapping and single-molecule fluorescence. *Nat. Methods* **1**, 133–139.
Lee, W. M., Reece, P. J., Marchington, R. F., Metzger, N. K., and Dholakia, K. (2007). Construction and calibration of an optical trap on a fluorescence optical microscope. *Nat. Protoc.* **2**, 3226–3238.
Marston, S. B., and Taylor, E. W. (1980). Comparison of the myosin and actomyosin ATPase mechanisms of the four types of vertebrate muscles. *J. Mol. Biol.* **139**, 573–600.
Mehta, A. D., Finer, J. T., and Spudich, J. A. (1997). Detection of single-molecule interactions using correlated thermal diffusion. *Proc. Natl. Acad. Sci. USA* **94**, 7927–7931.
Mehta, A. D., Rock, R. S., Rief, M., Spudich, J. A., Mooseker, M. S., and Cheney, R. E. (1999). Myosin-V is a processive actin-based motor. *Nature* **400**, 590–593.
Moffitt, J. R., Chemla, Y. R., Izhaky, D., and Bustamante, C. (2006). Differential detection of dual traps improves the spatial resolution of optical tweezers. *Proc. Natl. Acad. Sci. USA* **103**, 9006–9011.
Moffitt, J. R., Chemla, Y. R., Smith, S. B., and Bustamante, C. (2008). Recent advances in optical tweezers. *Annu. Rev. Biochem.* **77**, 205–228.
Molloy, J. E., Burns, J. E., Kendrick-Jones, J., Tregear, R. T., and White, D. C. (1995). Movement and force produced by a single myosin head. *Nature* **378**, 209–212.
Neuman, K. C., and Block, S. M. (2004). Optical trapping. *Rev. Sci. Instrum.* **75**, 2787–2809.
Nishikawa, S., Komori, T., Ariga, T., Okada, T., Morimatsu, M., Ishii, Y., and Yanagida, T. (2007). Imaging and nanomanipulation of an actomyosin motor. *In* "Single-Molecule Techniques: A Laboratory Manual," (P. R. Selvin and T. Ha, eds.), pp. 325–346. Cold Spring Harbor Laboratory Press, New York.

Nugent-Glandorf, L., and Perkins, T. T. (2004). Measuring 0.1-nm motion in 1 ms in an optical microscope with differential back-focal-plane detection. *Opt. Lett.* **29,** 2611–2613.

Palmiter, K. A., Tyska, M. J., Dupuis, D. E., Alpert, N. R., and Warshaw, D. M. (1999). Kinetic differences at the single molecule level account for the functional diversity of rabbit cardiac myosin isoforms. *J. Physiol.* **519**(Pt 3), 669–678.

Palmiter, K. A., Tyska, M. J., Haeberle, J. R., Alpert, N. R., Fananapazir, L., and Warshaw, D. M. (2000). R403Q and L908V mutant beta-cardiac myosin from patients with familial hypertrophic cardiomyopathy exhibit enhanced mechanical performance at the single molecule level. *J. Muscle Res. Cell Motil.* **21,** 609–620.

Perkins, T. T. (2009). Optical traps for single molecule biophysics: A primer. *Laser & Photon. Rev.* **3,** 203–220.

Peterman, E. J. G., Gittes, F., and Schmidt, C. F. (2003). Laser-induced heating in optical traps. *Biophys. J.* **84,** 1308–1316.

Pralle, A., Prummer, M., Florin, E. L., Stelzer, E. H., and Horber, J. K. (1999). Three-dimensional high-resolution particle tracking for optical tweezers by forward scattered light. *Microsc. Res. Tech.* **44,** 378–386.

Purcell, T. J., Morris, C., Spudich, J. A., and Sweeney, H. L. (2002). Role of the lever arm in the processive stepping of myosin V. *Proc. Natl. Acad. Sci. USA* **99,** 14159–14164.

Purcell, T. J., Sweeney, H. L., and Spudich, J. A. (2005). A force-dependent state controls the coordination of processive myosin V. *Proc. Natl. Acad. Sci. USA* **102,** 13873–13878.

Rao, C. R. (1965). Linear Statistical Inference and Its Application. John Wiley & Sons, Inc., New York.

Rayment, I., Rypniewski, W. R., Schmidt-Base, K., Smith, R., Tomchick, D. R., Benning, M. M., Winkelmann, D. A., Wesenberg, G., and Holden, H. M. (1993). Three-dimensional structure of myosin subfragment-1: A molecular motor. *Science* **261,** 50–58.

Rice, S. E., Purcell, T. J., and Spudich, J. A. (2003). Building and using optical traps to study properties of molecular motors. *Methods Enzymol.* **361,** 112–133.

Rief, M., Rock, R. S., Mehta, A. D., Mooseker, M. S., Cheney, R. E., and Spudich, J. A. (2000). Myosin-V stepping kinetics: A molecular model for processivity. *Proc. Natl. Acad. Sci. USA* **97,** 9482–9486.

Rock, R. S., Rice, S. E., Wells, A. L., Purcell, T. J., Spudich, J. A., and Sweeney, H. L. (2001). Myosin VI is a processive motor with a large step size. *Proc. Natl. Acad. Sci. USA* **98,** 13655–13659.

Sakamoto, T., Webb, M. R., Forgacs, E., White, H. D., and Sellers, J. R. (2008). Direct observation of the mechanochemical coupling in myosin Va during processive movement. *Nature* **455,** 128–132.

Simmons, R. M., Finer, J. T., Chu, S., and Spudich, J. A. (1996). Quantitative measurements of force and displacement using an optical trap. *Biophys. J.* **70,** 1813–1822.

Sivaramakrishnan, S., Spink, B. J., Sim, A. Y., Doniach, S., and Spudich, J. A. (2008). Dynamic charge interactions create surprising rigidity in the ER/K alpha-helical protein motif. *Proc. Natl. Acad. Sci. USA* **105,** 13356–13361.

Sivaramakrishnan, S., Sung, J., Ali, M., Doniach, S., Flyvbjerg, H., and Spudich, J. A. (2009). Combining single molecule optical trapping and small angle X-ray scattering measurements to compute the persistence length of a protein ER/K alpha-helix. *Biophys. J.* **97,** 2993–2999.

Spudich, J. A. (2001). The myosin swinging cross-bridge model. *Nat. Rev. Mol. Cell Biol.* **2,** 387–392.

Spudich, J. A., and Sivaramakrishnan, S. (2009). Myosin VI: An innovative motor that challenged the swinging lever arm hypothesis. *Nat. Rev. Mol. Cell Biol.* **11,** 128–137.

Spudich, J. A., Rice, S. E., Rock, R. S., Purcell, T. J., and Warrick, H. M. (2007). Optical traps to study properties of molecular motors. *In* "Single-Molecule Techniques: A Laboratory Manual," (P. R. Selvin and T. Ha, eds.), pp. 279–296. Cold Spring Harbor Laboratory Press, New York.

Svoboda, K., and Block, S. M. (1994). Biological applications of optical forces. *Annu. Rev. Biophys. Biomol. Struct.* **23,** 247–285.
Tolic-Norrelykke, I. M., Berg-Sørensen, K., and Flyvbjerg, H. (2004). MatLab program for precision calibration of optical tweezers. *Comput. Phys. Comm.* **159,** 225–240.
Tolic-Norrelykke, S. F., Schaeffer, E., Howard, J., Pavone, F. S., Juelicher, F., and Flyvbjerg, H. (2006). Calibration of optical tweezers with positional detection in the back-focal-plane. *Rev. Sci. Instrum.* **77,** 103101.
Uyeda, T. Q., Kron, S. J., and Spudich, J. A. (1990). Myosin step size. Estimation from slow sliding movement of actin over low densities of heavy meromyosin. *J. Mol. Biol.* **214,** 699–710.
van Dijk, M. A., Kapitein, L. C., van Mameren, J., Schmidt, C. F., and Peterman, E. J. G. (2004). Combining optical trapping and single-molecule fluorescence spectroscopy: Enhanced photobleaching of fluorophores. *J. Phys. Chem. B* **108,** 6479–6484.
Veigel, C., Coluccio, L. M., Jontes, J. D., Sparrow, J. C., Milligan, R. A., and Molloy, J. E. (1999). The motor protein myosin-I produces its working stroke in two steps. *Nature* **398,** 530–533.
Veigel, C., Wang, F., Bartoo, M. L., Sellers, J. R., and Molloy, J. E. (2002). The gated gait of the processive molecular motor, myosin V. *Nat. Cell Biol.* **4,** 59–65.
Veigel, C., Schmitz, S., Wang, F., and Sellers, J. R. (2005). Load-dependent kinetics of myosin-V can explain its high processivity. *Nat. Cell Biol.* **7,** 861–869.
Visscher, K., and Block, S. M. (1998). Versatile optical traps with feedback control. *Methods Enzymol.* **298,** 460–489.
Visscher, K., Gross, S. P., and Block, S. M. (1996). Construction of multiple-beam optical traps with nanometer-resolution position sensing. *IEEE J. Sel. Top. Quantum Electron.* **2,** 1066–1076.
Visscher, K., Schnitzer, M. J., and Block, S. M. (1999). Single kinesin molecules studied with a molecular force clamp. *Nature* **400,** 184–189.

CHAPTER FIFTEEN

AN OPTICAL APPARATUS FOR ROTATION AND TRAPPING

Braulio Gutiérrez-Medina,[*,1,2] Johan O. L. Andreasson,[†,1] William J. Greenleaf,[‡,3] Arthur LaPorta,[§] *and* Steven M. Block[*,‡]

Contents

1. Introduction	378
2. Optical Trapping and Rotation of Microparticles	379
2.1. The principles of optical manipulation	379
2.2. Sources of particle anisotropy	382
3. The Instrument	385
3.1. Overview	385
3.2. The microscope	387
3.3. Signal detection and processing	389
4. Fabrication of Anisotropic Particles	390
4.1. Particles with shape asymmetry	390
4.2. Particles with optical asymmetry	391
5. Instrument Calibration	395
5.1. Standard optical tweezers calibration methods	395
5.2. Force calibration	396
5.3. Torque calibration	398
5.4. Implementation of an optical torque clamp	398
6. Simultaneous Application of Force and Torque Using Optical Tweezers	400
6.1. Twisting single DNA molecules under tension	400
7. Conclusions	402
Acknowledgments	402
References	403

[*] Department of Biology, Stanford University, Stanford, California, USA
[†] Department of Physics, Stanford University, Stanford, California, USA
[‡] Department of Applied Physics, Stanford University, Stanford, California, USA
[§] Department of Physics, Biophysics Program, Institute for Physical Science and Technology, University of Maryland, College Park, Maryland, USA
[1] These authors contributed equally to this work
[2] Present address: Instituto Potosino de Investigación Científica y Tecnológica, San Luis Potosí, Mexico
[3] Present address: Department of Chemistry and Chemical Biology, Harvard University, Cambridge, Massachusetts, USA

Methods in Enzymology, Volume 475   © 2010 Elsevier Inc.
ISSN 0076-6879, DOI: 10.1016/S0076-6879(10)75015-1   All rights reserved.

Abstract

We present details of the design, construction, and testing of a single-beam optical tweezers apparatus capable of measuring and exerting torque, as well as force, on microfabricated, optically anisotropic particles (an "optical torque wrench"). The control of angular orientation is achieved by rotating the linear polarization of a trapping laser with an electro-optic modulator (EOM), which affords improved performance over previous designs. The torque imparted to the trapped particle is assessed by measuring the difference between left- and right-circular components of the transmitted light, and constant torque is maintained by feeding this difference signal back into a custom-designed electronic servo loop. The limited angular range of the EOM ($\pm 180°$) is extended by rapidly reversing the polarization once a threshold angle is reached, enabling the torque clamp to function over unlimited, continuous rotations at high bandwidth. In addition, we developed particles suitable for rotation in this apparatus using microfabrication techniques. Altogether, the system allows for the simultaneous application of forces (~ 0.1–100 pN) and torques (~ 1–10,000 pN nm) in the study of biomolecules. As a proof of principle, we demonstrate how our instrument can be used to study the supercoiling of single DNA molecules.

1. Introduction

Optical tweezers have proved to be an extremely powerful tool in the investigation of biophysical processes, particularly the activity of motor proteins and other processive enzymes, whose biological activity involves converting chemical energy into mechanical work. The ability to measure displacements with subnanometer resolution and to apply piconewton-level forces in a controlled manner makes optical trapping ideal for studying the motion of biological macromolecules (Abbondanzieri et al., 2005; Neuman and Block, 2004; Svoboda and Block, 1994). Through the use of active servo loops or passive optical configurations that can clamp the force, position, loading rate, or other physical quantities, precisely defined perturbations can be applied to identify the kinetic processes that underlie enzyme mechanisms.

Force and displacement, however, represent just one aspect of the physical picture. A host of important biomolecular complexes, including the F_1-ATPase (Noji et al., 1997) and the bacterial flagellar motor (Berg, 2003), generate torque and rotation—rather than force and displacement—as the mechanical output. Moreover, owing to the helical structure of DNA, many processive nucleic-acid-based enzymes undergo rotation and generate torsional strain as a consequence of their translocation. Torsional strain in DNA and chromatin is regulated, directly and indirectly, by a variety of proteins such as topoisomerases, helicases, gyrases, histones, and chromatin remodeling factors (Dong and Berger, 2008), and it is well established that such strain is a major factor in gene expression (Kar et al., 2006). Finally, mechanoenzymes that translocate along linear polymers, such as myosin or kinesin, may also generate (or respond to) torque (Hua et al., 2002). Practical benefit can therefore be gained by generalizing single-molecule methods to include the precise detection of rotation and the application of torque. Currently, the well-established technique of magnetic tweezers permits controlled rotation of magnetic microparticles by adjusting the orientation of an external magnetic field, enabling a variety of single-molecule manipulations. Magnetic tweezers are comparatively simple in their design and operation, and take advantage of the biologically noninvasive character of the magnetic field, typically producing constant forces over distances of several microns. However, conventional magnetic tweezers have certain practical limitations. There is no direct way to measure the torque imparted to the trapped particle, nor to conveniently record its angular displacement with respect to the applied field. Also, most magnetic tweezers determine displacement using frame-by-frame video processing of particle images, and are therefore limited by video acquisition rates, whereas laser-based optical trapping systems tend to use dedicated photosensors (quadrant photodiodes or position-sensitive detectors (PSDs)) with bandwidths exceeding several kilohertz. Finally, the magnitudes and directions of the fields produced by permanent or electromagnets cannot be changed very quickly, making it harder to adjust external control variables, such as the torque and force, as rapidly as one might like.

In the following sections, we describe the latest generation of an optical torque wrench (OTW), based on an improved design. Like conventional optical tweezers, the instrument has the ability to manipulate micron-sized particles, and to monitor both force and displacement. Additionally, it has the capacity to produce rotation, and to measure angle and applied torque with excellent precision (~ 1 pN nm of torque) at high bandwidth (~ 10–100 kHz).

2. Optical Trapping and Rotation of Microparticles

2.1. The principles of optical manipulation

A single beam, gradient-force optical trap is based on the interaction of a micron-sized dielectric particle with a beam of light that is brought to a diffraction-limited focal spot. Under the influence of the electric field (**E**), the particle develops an electric polarization **P** whose magnitude depends on material properties of the particle (Fig. 15.1A). For isotropic materials,

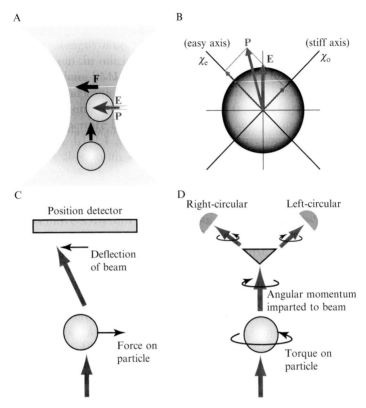

Figure 15.1 Principles of optical manipulation and signal detection. (A) The electric field (E, blue arrow) associated with a tightly focused laser beam induces a collinear electric polarization (**P**, red arrow) in an isotropic dielectric particle. As the particle moves away from the center of the laser focal volume, the induced dipole–electric field interaction produces a net restoring force (F, black arrows) toward the center, confining the particle in 3D. (B) For an optically anisotropic particle, different axes exhibit different polarizabilities. The induced polarization vector, therefore, is not collinear with the external electric field, and inclines toward the most polarizable (easy) axis. This effect gives rise to a net torque on the particle that tends to align the easy axis with the electric field. (C) Force can be detected by measuring trapping beam deflections (corresponding to changes in linear momentum) induced by an off-center, trapped particle, using a position-sensitive detector. (D) Analogously, the torque exerted by a linearly polarized beam on an anisotropic particle can be detected, based on the imbalance in the right- and left-circular polarization components in the beam after scattering by the particle, which can be measured independently. (See Color Insert.)

P = χ**E**, where χ is the polarizability of the material. The interaction of the induced polarization vector with the applied electric field leads to a net force proportional to $\chi \nabla E^2$, where ∇E^2 is the gradient of the electric field intensity (Ashkin, 2006; Ashkin et al., 1986). In the presence of a highly focused laser beam, a particle is therefore drawn toward the focal point, which constitutes the lowest energy state.

An electric dipole, **p**, interacting with an external electric field can also generate a torque, in addition to force, whose magnitude is given by $\tau = |\mathbf{E} \times \mathbf{p}| = Ep \sin\theta$, where θ is the angle between **E** and **p**. To produce nonzero torques, it follows that the net induced dipole moment must have a component perpendicular to the average external field. This condition can be satisfied using anisotropic particles, where either *form* or *material* birefringence (or both) lead to different polarizabilities along perpendicular axes. Form birefringence is a purely geometrical property, arising from the way that the shape of a small particle scatters light (e.g., an oblate ellipsoid), whereas material birefringence is an intrinsic optical property of the material from which the particle is produced (e.g., crystalline quartz). When an object is birefringent, the expression relating the induced dipole to the external field is replaced by the matrix equation $\mathbf{p} = \chi\mathbf{E}$, which reduces to the scalar relations $p_i = \chi_i E_i$ whenever the coordinate system coincides with the principal axes of a birefringent crystal, or with the symmetry axes of an anisotropic shape. Figure 15.1B illustrates an example where the external field makes an angle of 45° with respect to two principal axes. The induced dipole tends toward the easy optical axis, and the resulting torque acts to bring the most polarizable axis into alignment with the external field.

From Newton's laws, the amount of force and torque generated by an optical trap can be computed by considering linear and angular momentum, respectively, transferred from the laser beam to the trapped particle. A single photon carries energy $\varepsilon = hc/\lambda$, linear momentum $p = h/\lambda$, and the angular momentum associated with its spin is $L = \pm h/2\pi$ (for right- and left-circularly polarized light), where h is Planck's constant, c is the speed of light, and λ is the wavelength. The rate of change of linear momentum (equal to the force) is therefore given by $F = \mathrm{d}p/\mathrm{d}t = (1/c)\mathrm{d}\varepsilon/\mathrm{d}t$, or $F = \wp/c$, where $\wp = \mathrm{d}\varepsilon/\mathrm{d}t$ is the optical power. Similarly, the torque, τ, is equal to the rate of change of angular momentum, which leads to $\tau = (\lambda/2\pi c)\wp$. Assuming a laser with $\lambda = 1064$ nm and conversion efficiencies in the range of 1–10% (typical of optical traps, in practice), it follows that an optical trap might generate 0.03–0.3 pN of force and 6–60 pN nm of torque per milliwatt of incident optical power. When these values are compared with the forces needed to extend a coil of double-stranded DNA (dsDNA) (5 pN) (Smith *et al.*, 1992; Wang *et al.*, 1997) or to melt dsDNA (~ 100 pN) (Smith *et al.*, 1996), or with the torque necessary to unwind dsDNA (~ 10–100 pN nm) (Bryant *et al.*, 2003), it becomes clear that the linear and angular momentum carried by a laser beam is adequate to produce biologically relevant forces and torques at modest power.

Torque may also be applied to optically trapped particles by taking advantage of the "orbital" angular momentum of the light, which is associated with the geometry of the laser beam, rather than with the individual photon spin. Practical implementations of this strategy include optical vortices and higher order Gaussian–Laguerre modes (He et al., 1995; Volpe and Petrov, 2006), where the flow of energy (the Poynting vector) in the optical mode carries angular momentum about the beam axis. Orbital angular momentum can also be developed from a pair of ordinary laser beams that propagate along nonintersecting, nonparallel paths, or by using a spatially anisotropic trapping beam in conjunction with an asymmetrically shaped particle. In addition, torque is developed whenever microscopic *chiral* particles scatter light in such a way as to induce orbital angular momentum (Friese et al., 2001). Although the changes in torque from an asymmetric trapping beam might, in principle, be measured (Simpson and Hanna, 2009), doing so would likely involve detecting subtle changes in the phase profile of the outgoing beam, requiring interferometric imaging of the outgoing trap beam and sophisticated real-time image analysis.

By contrast, the OTW represents a conceptually simpler arrangement, based on the "spin" angular momentum of light, which is associated with its polarization. The OTW scheme provides a straightforward way to apply and detect torque (La Porta and Wang, 2004) by monitoring the net change in the polarization of an optical trapping beam as it interacts with a transparent, anisotropic (birefringent) particle. The OTW controls rotation of the specimen by turning the polarization of the incoming beam, and determines the torque by separating the right- and left-circular components of the outgoing beam and measuring their intensities using standard polarization components. It then becomes possible to use a single trapping laser beam for both linear and angular manipulation, and the detection of force (Fig. 15.1C) and torque (Fig. 15.1D). The OTW has the additional advantage that the angular and linear trapping effects are relatively independent, with little crosstalk between them.

2.2. Sources of particle anisotropy

Two types of birefringence are suitable for producing the optical anisotropy necessary for use with an OTW: *form birefringence* and *material birefringence*. In form birefringence, a tiny particle with differing dimensions along perpendicular axes is easier to polarize along the more extended direction, provided that the dimensions are comparable to, or smaller than, the wavelength of the scattered light. Such a small particle will exhibit birefringence even when the material comprising the particle itself is isotropic, resulting in indirect coupling between the particle shape and the polarization of the trap beam. For optical rotation purposes, oblate particle symmetry is desirable, so that the two extended axes lead to stable orientation within the trapping beam and simultaneous alignment of the net polarization vector with the trapping electric field (see Fig. 15.2A). This configuration allows the particle to rotate, as necessary, to present the attachment

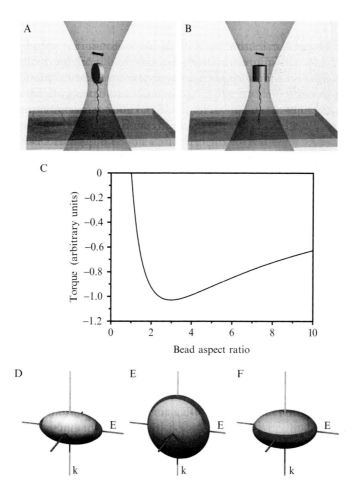

Figure 15.2 Particle anisotropy for optical trapping and rotation. (A) A small, oblate particle made of an optically isotropic material tends to align its long radii with the trapping beam axis (vertical) and the polarization direction (red arrow). (B) A birefringent cylinder tends to align its long axis with the trapping beam axis, allowing the extraordinary optical axis of the crystal to track the trap polarization direction. Possible materials for birefringent particles include quartz and calcite. (C) Theoretical estimate of the torque exerted on a subwavelength, oblate particle subjected to a uniform electric field, shown as a function of its aspect ratio (maximum to minimum radius). Bigger torques correspond to larger negative values; the greatest torque (curve minimum) occurs near an aspect ratio of 3. (D–F) The polarization ellipsoids for quartz (D), and calcite (E), (F) are shown, where red (gray) zones represent regions of maximum (minimum) electric susceptibility. Axes corresponding to red regions tend to align with the direction of the electric field vector **E** (green axis). For optical trapping and rotation, an ideal configuration is obtained when rotation is possible around the direction of the beam propagation vector, **k** (blue axis). This is the case for quartz (D). For calcite, the configuration shown in (E) can be used to exert torque (although alignment of the polarization ellipsoid with respect to the direction of **E** is not unique), whereas in the arrangement shown in (F) the particle exhibits no net birefringence on the plane perpendicular to **k** and no torque can be generated about the beam axial direction. (See Color Insert.)

point of a biological molecule downward in response to upward tension (assuming an inverted microscope arrangement, with the laser beam introduced from below the objective) (Oroszi et al., 2006).

An optimum aspect ratio for form birefringence that maximizes the torque for oblate particles can be derived ($\rho = r_{max}/r_{min}$), assuming that the particle is small compared to the wavelength of the trapping radiation. The torque exerted on an oblate ellipsoid in a uniform electric field is given by:

$$\tau = \frac{(\varepsilon - 1)^2(1 - 3\varepsilon)E^2 V \sin 2\alpha}{8\pi(\pi\varepsilon + 1 - n)([1 - n]\varepsilon + 1 + n)}, \quad (15.1)$$

where V is the volume of the particle, n is the depolarizing factor along the symmetry axis, ε is the dielectric constant, α is the angle between the electric field direction and the polar axis of the particle, and E is the electric field (Landau et al., 1984). Assuming that volume is conserved, the torque asymptotically approaches a constant value as ρ goes to infinity, a limit that is not physically relevant, because the expression for torque is only valid when the major axis remains small compared to the size of the trapping beam. Instead, a meaningful optimization is obtained by varying the aspect ratio under the constraint that the major axis remains constant and on the order of the trap size, in which case the volume of the particle is proportional to ρ^{-1}. The torque, normalized to the incident electric field, E, is plotted in Fig. 15.2C: a maximum value is obtained with an aspect ratio of $\rho \approx 3$. The torque computed from Eq. (15.1) with this aspect ratio for an oblate ellipsoid made of silica is approximately 60% of the corresponding torque for a sphere of equivalent volume made from quartz, which is intrinsically birefringent.

The alternative to form birefringence is material birefringence. Birefringent materials have distinct principal axes exhibiting different polarizabilities. Some substances, such as quartz and calcite, have two (ordinary) axes that are equivalent, and one (extraordinary) axis that is different from the other two. For quartz, the extraordinary axis is the most easily polarized, and so the overall polarizability can be represented by a prolate ellipsoid. In the presence of an external electric field, quartz experiences a torque that tends to align the extraordinary axis with the electric field vector (Fig. 15.2D). For calcite, however, the extraordinary axis is the least polarizable, so the overall polarizability is represented by an oblate ellipsoid. In this case, the extraordinary axis is repelled from, and the two ordinary axes are drawn toward, the electric field vector (Figs. 15.2E and F). While it may be possible to exert torque on calcite particles using the configuration shown in Fig. 15.2E, quartz offers more optimized conditions for combined optical trapping (translation) and rotation. Inside an optical trap, quartz particles with prolate shapes can be stably trapped in a vertical orientation dictated by their shape, then rotated about the optical axis due to the prolate polarizability ellipsoid.

Ideally, particles used in an OTW need to be strongly and stably trapped in all three dimensions, readily rotated, and functionally consistent with any planned biological experiments (e.g., functional attachment of the particles to macromolecules must be possible). One particularly important experimental geometry for single-molecule experiments is the surface-based assay, where the molecule of interest—for example, DNA—is tethered to the coverglass surface by one end and to the trapped particle by the other. In the OTW, force can then be applied upwards to stretch the tether while controlled rotation takes place around the vertical axis. For this purpose, the use of *cylindrical* quartz particles is particularly convenient (Fig. 15.2B). Cylinders whose length exceeds their diameter will naturally tend to align their long axis with the optical axis of the trap, and the extraordinary optical axis of the material can be chosen to lie parallel to the base and top of the cylinder, facilitating rotation in the horizontal plane (Deufel et al., 2007). Micrometer-scale cylinders are also comparatively easy to fabricate, and their flat end-surfaces can be chemically derivatized to facilitate connections to biomolecules. In Section 4, we present detailed protocols to produce both oblate polystyrene ellipsoids and quartz cylinders.

3. THE INSTRUMENT

3.1. Overview

The optical layout for the OTW is shown in Fig. 15.3, highlighting several differences compared with previous implementations (Deufel and Wang, 2006; La Porta and Wang, 2004; Oroszi et al., 2006). A single laser beam is used both for optical trapping and rotation of microparticles, and for the simultaneous detection of force and torque. Rotation of the polarization in the sample plane is achieved by means of an electro-optic modulator (EOM), which replaces the pair of acousto-optic modulators (AOMs) used in a previous apparatus in an interferometer arrangement that shifted the relative phases of the vertical and horizontal polarization components of the input beam (La Porta and Wang, 2004).

Upon passage through the EOM, an incoming polarized light beam with electric field components (E_x, E_y) along the optical axes of the EOM crystal will experience a relative phase retardation (α) between E_x and E_y that is proportional to the applied voltage. The EOM thus behaves as a variable waveplate: it transforms an incoming linear polarization into elliptical polarization, where the degree of ellipticity is controlled by the voltage signal. After the EOM, the laser beam passes through an input $\lambda/4$-waveplate whose main optical axes are aligned at 45° with respect to those of the EOM, thereby

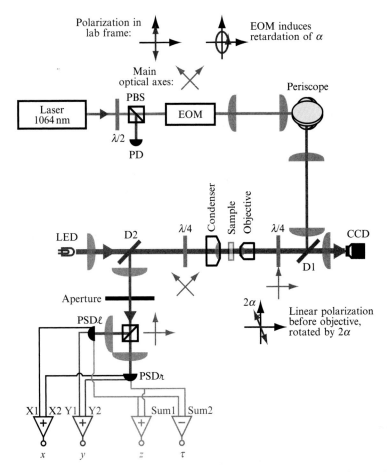

Figure 15.3 Optical schematic of the optical torque wrench. The trapping beam (thick gray line) is produced by a near-infrared laser, and the output power is adjusted by a λ/2-waveplate and polarizing beam splitter (PBS). Rotation of linear polarization in the sample plane is achieved using an EOM and a λ/4-waveplate placed before the microscope objective. The optical axes for components relevant to torque generation and detection are shown (gray axes), along with the polarization state of the trapping beam as it travels along the optical path. To detect torque, a λ/4-waveplate is placed after the condenser to convert the polarization components from circular to linear; these are then separated by a PBS and measured using position-sensitive detectors (PSDℓ, PSDr). Each PSD produces X, Y, and Sum(Z) voltages, which are combined to yield x, y, z, and torque (τ) signals. D1, D2: dichroic mirrors; PD: photodiode; LED: blue light-emitting diode illuminator; CCD: charge-coupled device video camera.

restoring linear polarization, but now rotated by an angle $\theta = 2\alpha$ with respect to its initial orientation. One constraint of the EOM is that its dynamic range is limited, typically corresponding to $\alpha = \pm 90°$ (max). However, this limitation can be circumvented by implementing additional control electronics (see Section 5.4). The use of an EOM offers several distinct advantages over a dual-AOM interferometer: (1) a symmetric beam profile is preserved throughout, which improves optical trap performance and calibration; (2) the polarization is no longer subject to significant long-term drift, as observed in the AOM-based system; and (3) the beam polarization angle is directly proportional to the EOM drive voltage, so there is no longer a need for additional input-angle detection optics. The new design is also considerably simpler to construct and align, involving fewer optical components.

3.2. The microscope

We now describe the instrument in further detail. The trapping beam is produced by a stable, diode-pumped solid-state Nd:YVO$_4$ laser (BL-106C, $\lambda = 1064$ nm, CW, Spectra Physics), operated near its peak power of 5 W to produce a beam with a clean TEM$_{00}$ mode, with typical intensity fluctuations of $<0.2\%$. The optical power used for trapping is computer-controlled by means of a $\lambda/2$-waveplate mounted on a motorized rotary stage (PRM1-Z7E, Thorlabs) followed by a polarizing beam splitter (PBS). An alternative for controlling the power is an AOM; however, we found that the profile of the diffracted beam produced by an AOM was somewhat distorted, leading to an angular asymmetry in the trap that degraded the performance of the instrument. While a motorized stage is comparatively slow, rapid control of the intensity is nonessential, because trapping experiments are typically performed at constant power in the sample plane. After the PBS, the resulting beam has linear polarization of high purity ($>99.5\%$) and a Gaussian profile, providing excellent starting conditions for subsequent manipulation of the polarization.

The next element in the optical pathway is the EOM (360-80, Con-optics), mounted on a V-shaped aluminum block attached to a five-axis alignment mount (9082, New Focus). This scheme allows for manual rotation of the EOM along its longitudinal axis (roll), as well as for fine adjustment of position, pitch, and yaw. Precise control of the orientation is necessary to align the EOM crystal, given its narrow aperture (~ 2 mm) and long length (~ 10 cm). The optical axes of the EOM can be aligned with respect to the incoming polarization by placing a temporary PBS after the EOM, and then oscillating the drive voltage from minimum to maximum range (corresponding to $\alpha = \pm 90°$). As the transmitted intensity is recorded with a photodiode, the EOM housing is rotated until the maximum contrast is observed, signaling polarization changes from linear (vertical) to circular and back to linear (horizontal). While it is possible to rotate the incoming laser polarization instead of rotating the EOM, considerable care must be taken with subsequent polarization alignments.

After the EOM, the beam is expanded to a final waist size of $w \sim 3$ mm and sent to a periscope that elevates the beam height to ~ 20 cm above the optical table. To minimize depolarization effects arising from nonorthogonal reflections, we use silver mirrors in the periscope, rather than dielectric mirrors. Before the beam enters the microscope, two lenses of equal focal length ($f = 75$ mm) are placed in the optical path, forming a 1:1 telescope, with one of the lenses mounted on a x–y–z translation stage and placed in a plane optically conjugate to the back-focal plane of the objective. This telescope provides a means of steering the trap in the sample plane without beam clipping at the back of the objective (Neuman and Block, 2004; Visscher and Block, 1998). Finally, the beam is coupled into the microscope by reflection from a dichroic mirror, passes through the input $\lambda/4$-waveplate, and reaches the entrance pupil of the objective. We confirmed the EOM-controlled rotation of the laser polarization by removing the microscope objective and monitoring the transmitted intensity after an auxiliary polarization analyzer (a PBS).

The uniformity of the beam polarization is the figure of merit in an OTW, and a few precautions were taken to maintain its quality. First, all mirrors transporting the beam after the EOM into the microscope were aligned at 45° angles with respect to the incoming beam direction to avoid depolarization effects. Second, the fast and slow axes of the EOM were matched to the S- and P-axes of any subsequent mirror reflections. In this fashion, any phase retardation between the S and P components is equivalent to an additive constant to the phase retardation generated by the EOM, which can easily be nulled out. Finally, positioning the input $\lambda/4$-waveplate directly below the objective minimizes depolarization effects induced by torque exerted on the mirrors themselves, which would otherwise introduce a spurious torque signature into the output detector.

The apparatus was based on a commercial inverted microscope (Eclipse TE2000-S, Nikon), modified to accommodate the optics needed to couple the trapping beam into the objective and to produce and detect the beam polarization. The vertical arm carrying the microscope condenser was removed and replaced by a structure designed to improve mechanical stability (Fig. 15.4). This structure was formed by two large vertical construction rails (XT95, Thorlabs), cross-linked at the top by a third rail, and further supported by additional beams joining the vertical rails to the optical table. An optical breadboard was suspended vertically from the structure and used to hold the condenser plus all detection optics. We used an oil immersion, high NA objective (100X/1.4NA/PlanApo, part 93110IR, Nikon), which has improved throughput in the near-infrared region and maximizes trapping efficiency while minimizing depolarization effects. The condenser lens (1.4NA, Nikon) was mounted on an x–y–z translation stage, which greatly facilitates alignment. The microscope was set up for brightfield illumination of the sample using a blue LED illumination source (LEDC3, Thorlabs) attached on top of the rail structure. Trapped beads are imaged through the microscope video port using a CCD camera.

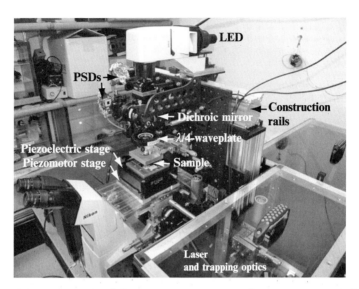

Figure 15.4 Photograph of the OTW instrument, with components as indicated. A commercial Nikon microscope was modified to improve mechanical and optical stability (see main text). The trapping laser, EOM, and associated optics are enclosed in plexiglass boxes (right side) to minimize beam instabilities due to air currents. The sample is held on a 3D piezoelectric stage with a 2D piezomotor substage.

The original Nikon specimen stage was also removed and replaced by a custom-fabricated aluminum mount that supports several items, including the dichroic mirror coupling the laser light into the objective, a precision rotary stage holding the input $\lambda/4$-waveplate below the objective, and piezomotor and piezoelectric stages (M-686.1PM and P-517.3CD, Physik Instrumente). The x–y piezomotor substage is used for coarse positioning of the sample, and features 100-nm step resolution over a 25 mm travel range with enhanced mechanical stability compared to conventional crossed-roller-bearing mechanical stages (Jordan and Anthony, 2009). The x–y–z piezoelectric main stage is used for all fine positioning and has nanometer-level step resolution over $100 \times 100 \times 20$ μm.

3.3. Signal detection and processing

Torque detection from the forward-scattered light exiting the sample chamber is based on an output $\lambda/4$-waveplate placed immediately after the condenser, which maps the right- and left-circular components of the beam polarization into vertical and horizontal linear polarizations, respectively. These polarization components are separated by an analyzer and their intensities measured by separate detectors (Fig. 15.3). In our setup, due to space constrains, an intermediate dichroic mirror placed after the output $\lambda/4$-waveplate directs the beam toward the analyzer. We use two independent duolateral PSDs with built-in preamplifiers (Pacific Silicon Sensors), aligned for back-focal plane detection (Neuman and Block, 2004) to measure bead displacements and the magnitude of polarization components independently. Each detector produces x, y, and sum (z) voltages for its corresponding circular polarization component, either left or right, to generate V_{lx}, V_{ly}, V_{lz}, V_{rx}, V_{ry}, V_{rz}. The net x, y, z, and torque (τ) signals are obtained by combining these voltages using a simple linear analog electronic circuit, with $x = V_{lx} + V_{rx}$, $y = V_{ly} + V_{ry}$, $z = V_{lz} + V_{rz}$, and $\tau = V_{lz} - V_{rz}$. To generate the z signal, the output beam passes through an aperture that allows the intensity of the central portion of the beam (only) to be measured (Pralle et al., 1999). The position and torque voltages go through low-pass multipole filters to remove noise (3988, Krohn-Hite) and are fed directly into a computer-acquisition board (PCI-6052E, National Instruments), where the signals are further processed by custom software written in LabView (Version 7.1, National Instruments).

4. Fabrication of Anisotropic Particles

4.1. Particles with shape asymmetry

One simple way to obtain oblate particles is to mechanically deform polystyrene microspheres by compression (Oroszi et al., 2006). In our procedure, spherical particles (1.1 μm diameter, Bangs Labs) were suspended in water ($\sim$1–3% by volume) and flattened between a pair of glass microscope slides mounted in a simple vise consisting of two machined aluminum blocks. The vise was padded with 3-mm rubber gaskets to seal the sample and maintain uniform pressure, and compression was obtained by gradually tightening the four ¼-20 bolts holding the vise together, producing an estimated pressure of at least 10^7 Pa. After compression at room temperature for $\sim$2 min, the vice was disassembled and the microspheres were washed off the slide surface. A fraction of these pressure-treated beads ($\sim$1–10%) exhibited an aspect ratio of $\sim$3 when examined by electron microscopy from different angles (Fig. 15.5A shows a top view), and were easily distinguished in the light microscope, where they could be trapped and rotated (Fig. 15.5B). Coating of the surface with chemical or antibody labels is possible either before or after compression. Alternative strategies for deforming uniform polystyrene spheres include compression in conjunction with heat treatment to near the glass point of polystyrene ($\sim$90 °C), or flattening between vise faces in the presence of a mixture of smaller, incompressible silica spheres, which act as spacers to limit the compression distance.

We compressed polystyrene spheres initially coated with either biotin or avidin labels. Using either of these approaches, compressed beads were

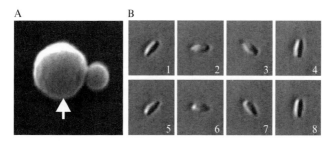

Figure 15.5 Optically anisotropic particles based on form birefringence. (A) Scanning electron microscope picture of an oblate particle (white arrow) created by compressing polystyrene spheres (see main text). The smaller adjacent particle is a 600-nm silica bead. Field of view is 2.5 × 2.5 μm^2. (B) Kymograph showing a sequence of bright-field images corresponding to a trapped, oblate particle being rotated by the OTW. Time between frames, 0.12 s. Field of view is 2 × 2 μm^2.

attached to one end of a single DNA molecule via a biotin–avidin linkage and tethered to a coverslip surface by the opposite end, and torque was exerted on the DNA. While it is possible to perform single-molecule experiments using compressed beads, the uniform labeling of the entire bead surface can make calibration difficult. If the DNA molecule happens to attach to the bead in a nonequatorial position—which, statistically, is the most likely occurrence—the tethered particle will tend to adopt an off-axis orientation, leading to signal crosstalk and thereby to the introduction of uncontrolled forces and torques. This difficulty may be circumvented by carefully selecting for those compressed beads that happen to be tethered by an equatorial point, but these are comparatively rare and may be difficult to identify. However, despite the random variation present in points of attachment, we found compressed beads to be extremely useful during alignment and initial testing phases of the OTW calibration. Because their shape asymmetry is readily seen in the light microscope, it becomes possible to monitor bead rotation using an independent method that does not rely on PSD signals, such as video tracking. This permits measurement of the rotation angle even when the laser trap is off, a helpful trait that we used to confirm the formation of rotationally constrained DNA tethers (see Section 6), by twisting DNA molecules attached to compressed beads and observing unwinding after the trapping beam is blocked.

4.2. Particles with optical asymmetry

Conventional microlithographic techniques can be employed to manufacture birefringent particles of specific shapes in a controlled and reproducible fashion (Deufel et al., 2007). Although chemically produced particles of birefringent materials, such as vaterite, have been used to apply torque in previous applications (Bishop et al., 2004; Funk et al., 2009), lithography offers several advantages. First, the particles can be chemically derivatized for biological labeling on specific surfaces, facilitating their vertical orientation when tethered in surface-based assays, thereby minimizing undesired forces and torques. Second, micro- or nanofabrication methods can yield large numbers of uniform particles, whose sizes can easily be controlled by changing mask features or adjusting etching parameters. Finally, among the many possible birefringent materials that might be used, in principle, quartz is chemically stable, readily available in wafer form at relatively low cost, and suitable for use with a variety of established etching chemistries, making it a natural choice for OTW applications.

Here, we present a protocol to produce cylinder-shaped particles by a single lithographic exposure of a quartz wafer. The particles are designed to have their extraordinary optical axis perpendicular to the axis of cylindrical symmetry, and are chemically functionalized at only one of the bases (Fig. 15.6). Our protocol was developed from a previous implementation (Deufel *et al.*, 2007), but modified to comply with restrictions imposed by the Stanford Nanofabrication Facility, which precluded placing antireflective coatings on the back sides of wafers.

4.2.1. Mask design and wafers

Reticle design typically depends on the stepper system used. To produce arrays of upright cylinders 400–700 nm in diameter, we divide the mask into 4×4 mm^2 areas, at wafer level, each with about 10^7 evenly spaced octagons of a particular size. Lithographic imaging of tiny octagon (or square) shapes in the mask produces nearly circular patterns on wafers after UV exposure, due to light diffraction effects. The exposure to create uniform patterns of small, high aspect ratio cylinders can reach the practical limit of i-line steppers, especially with a transparent substrate. If multiple patterned regions are placed on a single mask, it is therefore advisable to select just one or two patterns and place them in the center of the reticle, thereby reducing possible astigmatism from the outermost part of the stepper lens. We use 4 in. X-cut, single-crystal quartz wafers with double-side polish (University Wafer).

4.2.2. Protocol

1. Clean the wafer in hot piranha solution (9:1 mixture of H_2SO_4 and H_2O_2, 120 °C) for 20 min. Rinse in water followed by a spin rinse dryer cycle.
2. Functionalize surface for biological labeling with 3-aminopropyl-triethoxysilane (APTES) or any other desired organosilane coupling reagent:
   i. Add 0.6 ml of APTES (99%, Sigma-Aldrich) to 30 ml ethanol solution (95% (v/v) ethanol, 5% (v/v) water, pH 5.0 using acetic acid).

An Optical Apparatus for Rotation and Trapping 393

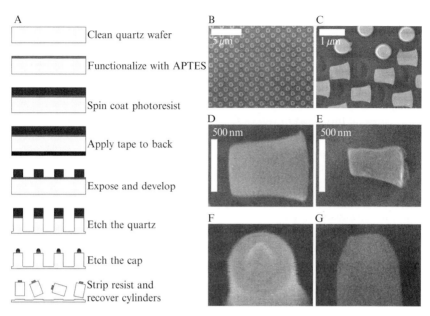

Figure 15.6 Fabrication of birefringent quartz cylinders. (A) Schematic showing the major steps of the fabrication protocol. SEM images of the cylinders during fabrication are shown on the wafer after etching (B) and after cutting (C). (D, E) Examples of particles made with different final sizes. The functionalized area at the top of cylinders may be minimized by shrinking resist (F) or the cylinder itself (G).

    ii. Place wafer in solution and sonicate for 5 min.
    iii. Rinse by sonicating wafer for 30 s in 50 ml methanol, three times.
    iv. Cure in oven for 20 min at 115 °C.
3. Spin coat 1.0 μm photoresist (Megaposit SPR 955-CM, Rohm and Haas Company) onto wafer. We use a Suss MicroTec ACS200 spin coater for these steps:
    i. 1700 rpm, 20 s, followed by edge bead removal (Microposit EC solvent 13) and a final spin at 1200 rpm, 8 s.
    ii. Bake for 90 s on a 90 °C hotplate.
4. Apply dicing tape (Z18551-7.50, Semiconductor Equipment Corporation) to the back side of the wafer and cut along the wafer edge. The use of tape prevents reflections from the stepper exposure chuck and avoids using extra antireflective coatings that may interact with the chuck surface.
5. Expose the pattern. We use an ASML PAS 5500/60 i-line stepper with 5× magnification and a 110 mJ cm^{-2} dose. Adjustments of focus and tilt offsets are necessary for optimal uniformity.
6. Remove tape and bake wafer for 90 s on a 110 °C hot plate.

7. Manually develop the wafer by placing it in the developer (Megaposit MF-26A, Rohm and Haas Company) for 30 s and then gently agitate in solution for an additional 30 s. Rinse in a water beaker and air-blow dry. The manual developing procedure reduces the risk of breaking the high-aspect-ratio resist posts.
8. UV cure for 15 min followed by 1 h bake in 110 °C oven.
9. (Optional) Instead of the previous step, the resist can be cured using a Fusion UV Cure System, which combines high-intensity UV light with a fast temperature ramp (100–200 °C over 45 s). This can improve resist selectivity during the etching process, leading to more vertical cylinder side walls.
10. Etch the wafer. We use an Applied Materials Precision 5000 Etcher at the following settings: power 50 W, pressure 10 mTorr, gas flow 36 sccm CHF_3 and 36 sccm CF_4, magnetic field 30 G, and helium cooling 5 Torr. The resulting etch rate is ~ 150 Å min^{-1}.
11. (Optional) To reduce the APTES-coated area at the top of the cylinder, an additional dry-etching step can be performed using O_2 plasma for 3 min in a Matrix Plasma Asher (3.75 Torr, 450 W, 100 °C, pins down). As the remaining resist cap is etched, the outer rim of the top quartz surface is exposed to the plasma (Fig. 15.6F), removing the APTES in this region. The linking of biological molecules can thereby be concentrated toward the center of the cylinder, reducing any potential wobbling of the particle during rotation in the optical trap. Another way to reduce the top area is to perform Step 10 until the top cylinder diameter shrinks as a result of a diminished resist layer (Fig. 15.6G).
12. The remaining resist is stripped by rinsing and sonicating in acetone for 20 min.
13. Quartz cylinders are recovered by manually scraping the wafer surface with a microtome blade and collecting the material in a test tube. This can be done in the presence of liquid, such as buffer or cross-linking reagents, to maximize yield.
14. The cylinders are functionalized with coupling proteins of interest, such as avidin, by cross-linking to the primary amines in the APTES using a conventional glutaraldehyde kit (Cat. #: 19540, Polysciences).

In the light microscope, nanofabricated quartz cylinders appear as thick, short rods that can be optically trapped with ease. As expected, trapped cylinders align themselves with their long axis along the direction of beam propagation, with the functionalized surface facing toward (or away from) the coverglass surface—an ideal geometry for producing tethers in a surface-based, single-molecule assay.

5. INSTRUMENT CALIBRATION

5.1. Standard optical tweezers calibration methods

A number of well-established methods have been developed to calibrate the stiffness of an optical trap (κ) acting on *spherical* beads. The most common methods are based on analysis of measurements of particle *variance, power spectrum*, or *Stokes' drag*, and have been described in greater detail previously (Neuman and Block, 2004; Svoboda and Block, 1994; Visscher and Block, 1998). Briefly, under low Reynolds number conditions, the thermal motions of a trapped bead in solution depend on the bead's viscous drag coefficient and the trap stiffness. The simplest of all calibration methods is variance-based, and uses the positional variance of a bead $\langle x^2 \rangle$ in combination with the "equipartition theorem" to compute the stiffness from $\kappa_x = \langle x^2 \rangle / k_B T$, where $k_B T$ is the thermal energy. In the power spectrum method, the frequency-dependent amplitude of positional fluctuations is computed, and data are fitted with the behavior of a thermal particle bound in a harmonic potential, which is a Lorentzian function. The spectral roll-off frequency of the fit, $f_c = 1/(2\pi t_0)$, where t_0 is the relaxation time of the bead, can be used to obtain the trap stiffness through the relation $\kappa_x = \beta/t_0$, assuming that the drag coefficient of the spherical particle, $\beta = 6\pi\eta a$, is known, where η is the viscosity and a is the radius. Finally, in the Stokes' drag method, the trapped sphere is subjected to a constant fluid velocity, v_x, and its displacement from the equilibrium position is measured. Flow is typically created by moving the piezoelectric stage holding the sample at constant velocity (e.g., using a triangle wave). The dependence of the bead displacement, x, on v_x has slope $t_0 = \beta/\kappa_x$, from which κ_x can be obtained.

All three methods require a prior calibration of the PSD voltage, V_x, as a function of x, the true displacement from the equilibrium position. In a configuration where the same laser beam is used both for trapping and position detection, the calibration of $V(x)$ is typically achieved by scanning a bead immobilized on the coverglass surface across the laser beam, taking care to perform the scanning directly through the trap center. In 1D, the PSD response is well fit by the derivative of a Gaussian function (Allersma *et al.*, 1998), and from the linear, central, part of the profile the conversion factor from nanometers to volts, ξ, can be obtained. Alternatively, in 2D, the immobilized bead can be raster-scanned throughout the trapping area, and the resulting voltage profile can be fit with a 2D polynomial (Lang *et al.*, 2002).

Although fairly straightforward to implement, these calibration methods generally require knowledge of the particle drag coefficient, which is influenced by its shape and the proximity of any nearby surfaces (Svoboda and Block, 1994). Alternatively, the power spectrum and Stokes' drag calibration methods can be combined to yield experimental estimates for

ξ, κ, and β for trapped particles of any shape. To do so, first the power spectrum of the V_x signal is computed, from which the roll-off frequency $f_c = 1/(2\pi t_0) = \kappa_x/(2\pi\beta)$ and the amplitude at zero frequency, $\widetilde{P}_x = k_B T/(\kappa_x \pi^2 f_c^2 \xi^2)$, are obtained. Next, the Stokes' drag method is used to find V_x versus fluid velocity, v, yielding a linear relationship with slope $s = \beta/\kappa\xi$. The last three equations are combined to yield $\kappa_x = 4k_B T f_c s^2/\widetilde{P}_x$, $\beta = 2k_B T s^2/\pi\widetilde{P}_x$, and $\xi = 1/(2\pi s f_c)$, as functions of the experimentally measured parameters s (V nm^{-1} s), f_c (s^{-1}), and $\widetilde{P}_x$ (V^2 Hz^{-1}).

5.2. Force calibration

The calibration method just discussed may be applied immediately to the case of trapped, nonspherical particles in the OTW along directions transverse to the beam propagation direction (x, y). Because nonspherical particles do not bind to a surface in a unique orientation, the traditional method of scanning the beam diametrically across a stuck bead to obtain the volts-to-nanometers conversion factor poses problems with oblate ellipsoids or quartz cylinders. Instead, we performed x- and y-calibrations using the combined power spectrum Stokes' drag method of the previous section: sample results for quartz cylinders are shown in Fig. 15.7. The Stokes' drag results display the expected linear relationships between $V_{x,y}$ and $v_{x,y}$, and the power spectra calculated from the x- and y-signals are well fit by Lorentzian functions with the parameters shown in the caption of Fig. 15.7.

Careful calibration of the optical trap in the axial direction is, operationally speaking, the most important of all because the OTW apparatus is mainly intended for the simultaneous application of torque and vertical loads. Calibration in the z-direction, however, involves different considerations compared to x–y (Neuman and Block, 2004). First, the finite axial trapping depth restricts the range over which the Stokes' drag calibration method can be performed. Second, when the sample chamber is moved vertically, an intensity modulation at the detector arises from interference between the forward-scattered light and light reflected from the coverslip/solution interface, an effect that must be taken into account. Finally, motions of the sample chamber relative to the microscope objective will induce a focal shift that displaces the trap axial position (Neuman et al., 2005). Recent efforts aimed at addressing some of these limitations include the unzipping of a known DNA template to obtain a calibrated reference (Deufel and Wang, 2006) or performing back-scattered light detection (optionally, with spatial filtering) to reduce systematic errors (Carter et al., 2007; Sischka et al., 2008).

In our instrument, we performed z-calibrations of force and displacement using quartz cylinders by taking advantage of the stable orientation of these nanofabricated particles within the trap. First, we obtained f_c and $\widetilde{P}_z$ from the power spectrum of the z signal (Fig. 15.7). Next, we impaled a trapped, vertically oriented cylinder on the coverglass by adjusting the

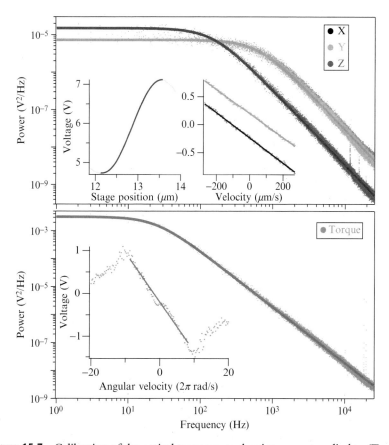

Figure 15.7 Calibration of the optical torque wrench using a quartz cylinder. (Top)- Calibrations of the linear dimensions. First, the power spectra for x, y, and z signals were computed and fit to Lorentzian functions, giving roll-off frequencies $f_{c,x} = 639 \pm 1$ Hz, $f_{c,y} = 631 \pm 1$ Hz, $f_{c,z} = 147.2 \pm 0.2$ Hz, and zero-frequency amplitudes $\widetilde{P}_x = (1.158 \pm 0.003) \times 10^{-6}$ V^2 Hz^{-1}, $\widetilde{P}_y = (1.148 \pm 0.003) \times 10^{-6}$ V^2 Hz^{-1}, and $\widetilde{P}_z = (2.382 \pm 0.005) \times 10^{-6}$ V^2 Hz^{-1}. Next, linefits to Stokes' drag measurements in x and y (Right inset) provided the slopes $s_x = (2.255 \pm 0.008) \times 10^{-6}$ V s nm^{-1} and $s_y = (2.158 \pm 0.008) \times 10^{-6}$ V s nm^{-1}, which were combined with the power spectral results to yield $\xi_x = 1/(2\pi s_x f_{c,x}) = 110$ nm V^{-1} and $\xi_y = 117$ nm V^{-1}. For the z signal, vertical scanning of a fixed cylinder (Left inset) produced a record well fit by the derivative of a Gaussian (amplitude $A = (1.395 \pm 0.006) \times 10^3$ V nm, S.D. $\sigma = 708 \pm 2$ nm), from which we obtain $\xi_z = \sigma^2/A = 359$ nm V^{-1}. These measurements were combined to obtain the trap stiffnesses $\kappa_x = 4.5 \times 10^{-2}$ pN nm^{-1}, $\kappa_y = 4.1 \times 10^{-2}$ pN nm^{-1}, and $\kappa_z = 9.1 \times 10^{-3}$ pN nm^{-1}. (Bottom) Calibration of torque. A procedure analogous to the linear x, y cases was carried out. From the experimentally measured values, $f_{c,\tau} = 24.29 \pm 0.05$ Hz, $\widetilde{P}_\tau = (5.04 \pm 0.01) \times 10^{-4}$ V^2 Hz^{-1}, and $s_\tau = (1.84 \pm 0.03) \times 10^{-2}$ V s rad^{-1}, the volts-to-radians conversion factor $\xi_\tau = 1/(2\pi s_\tau f_{c,\tau}) = 0.37$ rad V^{-1} and the angular trap stiffness $\kappa_\tau = 4k_B T f_{c,\tau} s_\tau^2 / \widetilde{P}_\tau = 264$ pN nm rad^{-1} were obtained. All power spectrum records represent averages from 50 measurements sampled at 66 kHz. For these measurements, the trapping laser power was ~ 20 mW (measured before entry into the objective rear pupil). (See Color Insert.)

piezoelectric stage position until its base bound nonspecifically to the surface. Vertical scanning of the piezoelectric stage while recording V_z yielded the required voltage–displacement calibration factor, ξ_z (nm V^{-1}). Scanning a surface-bound cylinder axially is effective because the trap stabilizes the vertical orientation, contrary to the x–y case, where transverse scanning tends to tilt the particle. The axial stiffness is calculated as before from $\kappa_z = k_B T/(\tilde{P}_z \xi^2 \pi^2 f_c)$. Using spherical test beads, we have compared calibrations obtained by "parking" the particle on the surface with previous methods, and obtained good agreement (data not shown).

5.3. Torque calibration

Torque calibration may be carried out using methods that are entirely analogous to those used for spatial displacement (La Porta and Wang, 2004). In the OTW apparatus described here, the input polarization angle is automatically known (relative to some arbitrary reference), and is proportional to the EOM input voltage. A rotational Stokes' drag method can therefore be implemented by periodically adjusting the EOM voltage such that the input polarization changes with a fixed angular velocity (ω_θ). The power spectrum of the torque signal voltage (V_τ) can also be readily computed. We therefore have used a combination of power spectrum and rotational Stokes' drag techniques to obtain the torque signal volts-to-radians conversion factor, ξ_τ, the trap angular stiffness, κ_τ, and the rotational drag coefficient of the trapped particle, β_τ. The experimental quantities measured were the slope of the V_τ versus ω_θ line, along with the roll-off frequency and zero-frequency amplitude of the angular power spectrum.

Figure 15.7 shows results from torque and force calibrations for a quartz cylinder. Fits of the various calibration signals to the expected functional forms are excellent, and the experimental parameters derived from such fits have uncertainties of less than 1%. For small displacements (< 150 nm) and small angles (< 20°), the detector signals are linear. As anticipated, modest laser power (10–50 mW) is sufficient to provide tight confinement of particles translationally as well as rotationally, making it possible to exert transverse and axial forces in excess of 20 and 5 pN, respectively, and torques of at least 300 pN nm.

5.4. Implementation of an optical torque clamp

An anisotropic particle trapped in a laser beam with fixed linear polarization will undergo rotational thermal motion whose amplitude depends on the angular trapping stiffness. In this "passive" mode, the mean torque exerted by the trap on the particle is zero, and nonzero torques will develop only if the particle is forced to change its angular orientation, for example, when twisted by a molecular motor. While it is possible to study rotary biomolecular processes in the passive mode, data taken under more precisely defined conditions, such as under constant applied torque, can provide more specific information. A torque clamp is the rotary analog of a force clamp, which provides high-resolution data on molecular displacements (Visscher and Block, 1998). We implemented a torque clamp mode by creating a servo loop that feeds the torque signal back into an external electronic circuit driving the EOM (Fig. 15.8A). Although it may be possible to implement

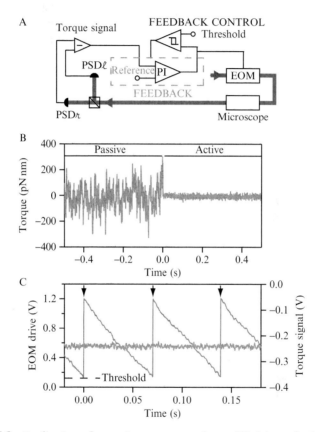

Figure 15.8 Realization of a continuous torque clamp. (A) Schematic showing the signals and feedback loops used in the OTW. Constant torque is maintained by a servo loop that feeds the torque signal into a proportional–integral circuit controlling the EOM. An additional "feedback control" circuit extends the dynamic range of the EOM to maintain constant torque over continuous rotations (see main text). (B) Demonstration of the torque clamp. In passive mode, the torque servo loop is disabled, and the signal reflects the rotational Brownian motion of an oblate trapped particle. At time $t = 0$ s, the servo loop is closed, clamping the torque at $\tau = 0$ pN nm. (C) To clamp torque at nonzero values, the EOM drive signal (gray sawtooth curve) flips by an amount value corresponding to $\pm 180°$ once an angular threshold is reached. These rapid reversals (black arrows) are automatically executed by the feedback control circuit. A nonzero torque signal (light gray curve) remains constant over unlimited rotations.

feedback control in computer software, we decided instead to employ a dedicated proportional–integral (PI) circuit that circumvents delays associated with computer interrupts. The analog PI controller is based on a single operational amplifier (OP27, Analog Devices) that compares the input torque signal with a reference voltage and sends an output voltage proportional to this difference, stabilized by an integral filter that smoothes the response (Gardner, 2005).

The torque clamp works for particles with either form or material birefringence. In the example shown in Fig. 15.8B, an oblate particle is first trapped in the passive mode and then feedback mode is established, keeping the particle at constant (here, zero) torque. As the active mode is enabled, the rms value of the torque signal (τ_{rms}) decreases by sevenfold compared to the rms amplitude of thermal fluctuations in the passive mode, yielding $\tau_{rms} = 14$ pN nm. The torque clamp has high bandwidth (~ 10 kHz) and is limited chiefly by the angular relaxation time of the particle under low Reynolds number conditions.

The restricted dynamic range of an EOM ($\pm 180°$) constrains any simple servo loop in keeping the torque constant over multiple revolutions. To overcome this limitation, we included an additional circuit that monitors the EOM drive signal using a microcontroller (Arduino deicimila, Arduino, Italy). Once an angular threshold is reached, the microcontroller triggers an analog switch (AD7512, Analog Devices) that momentarily disables the feedback, flips the polarization by $\pm 180°$, and reenables the servo loop (Fig. 15.8A). The performance of this circuit is illustrated in Fig. 15.8C, where rapid EOM voltage jumps are evident, but during which the torque signal reflects persistent clamp conditions. Because polarization reversals are completed within 10 μs, the bandwidth of the servo loop is unaffected.

6. SIMULTANEOUS APPLICATION OF FORCE AND TORQUE USING OPTICAL TWEEZERS

6.1. Twisting single DNA molecules under tension

The nanomechanical properties of DNA have been extensively studied at the single-molecule level using magnetic tweezers (Lionnet et al., 2008), but only recently using optical traps (Forth et al., 2008). To demonstrate the capabilities of our calibrated instrument, we studied the supercoiling of single DNA molecules under tension. Nanofabricated quartz cylinders were tethered to a coverglass surface by a 2.1 kb segment of dsDNA using standard protocols (Lang et al., 2004). Briefly, a dsDNA template was constructed with an array of six digoxigenin labels (spaced at 10 bp intervals) located at the 5' end of one strand, and six biotin labels (similarly spaced) at the 5' end of the complementary strand. Neutravidin-labeled quartz cylinders and template DNA molecules were incubated together at ~100 fM in phosphate buffer for several hours at 4 °C. Next, the DNA–cylinder complexes were diluted in PEM80 buffer (80 mM Pipes, pH 6.9, 1 mM EGTA, 4 mM MgCl$_2$) with 20 mg ml^{-1} BSA, introduced into a sample chamber where antibodies against digoxigenin had previously been adsorbed on the coverglass surface, and allowed to bind for 20 min at room temperature. A final wash with PEM80 removed unbound cylinders, and the sample chamber was then sealed and moved to the instrument for measurements. The use of multiple ligands at each end of the DNA molecule hinders free swiveling about the attachment points, creating a rotationally constrained tether, as illustrated in Fig. 15.9. After a tethered bead was identified in the microscope, judging by the restricted Brownian fluctuations, it was captured using the trap and centered with the piezoelectric stage, such that its point of surface attachment was located directly below the trapped cylinder.

The dsDNA tether was then subjected to vertical tension by lowering the piezoelectric stage until the quartz cylinder was displaced below the trap center by a specified distance, Δz, corresponding to a force $F_z = -\kappa_z \cdot \Delta z$, where κ_z is the axial trap stiffness. Then, the bead was rotated by $\pm 180°$ and the corresponding torque signal was recorded. Negligible torque is expected

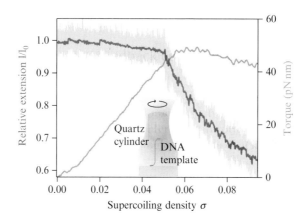

Figure 15.9 Supercoiling of a single 2.1 kbp, dsDNA molecule, tethered to the coverglass surface and a quartz cylinder using multiple dig–antidig and biotin–neutravidin linkages, respectively, to prevent free swiveling at the ends. The DNA molecule was stretched with 3 pN force in the vertical direction while being twisted at a rate of 0.5 turns s^{-1} (inset drawing). The vertical load was kept constant by a software-based PID feedback loop that controlled the piezoelectric stage position with a 20 Hz update rate. The records clearly display the plectonemic transition at $\sigma = 0.05$, and a direct measurement of the imposed torque. Data were collected at 5 kHz (extension, light gray trace) and boxcar averaged to 10 Hz (extension, dark gray trace) and 0.5 Hz (torque, gray thin trace).

to develop from the DNA during these initial half-turns at the forces used (2–10 pN). Therefore, any residual signal variations (arising, e.g., from imperfections in the polarization optics) were taken as an average "background" signal that was subtracted from subsequent data (modulo 2π). After acquisition of this background signal, the dsDNA molecule was twisted, typically at rates of 0.5 turn s^{-1} while F_z was clamped by monitoring the PSDs sum voltage and adjusting the piezoelectric stage position as necessary. (Although this vertical force clamp does not take into account variations in the sum signal arising from interference as the coverglass is displaced, we estimate that these effects introduce errors of less than 10% in F_z.) The effect of twisting dsDNA is shown in Fig. 15.9, where the molecular extension and the torque developed are displayed as functions of the degree of supercoiling, $\sigma = n/Lk_0$, where n is the number of turns and Lk_0 is the number of pitch periods spanned by the dsDNA molecule. Initially, the extension remains nearly constant until a torque $\tau \propto n$ develops. After a characteristic number of turns, n_b (corresponding to a torque for bending, τ_b), the energy required for further twisting of the DNA exceeds that for bending, and the molecule begins to buckle, exhibiting a sharp change in length as n_b is reached. As further twisting proceeds, plectonemes are formed in the DNA, and the torque remains roughly constant as the extension of the molecule decreases linearly with n (Strick et al., 1999). Because supercoiling involves close coupling between torque and force exerted on the DNA, the OTW setup is ideal to monitor and control both of these experimental parameters.

7. CONCLUSIONS

We have constructed an OTW capable of exerting simultaneous torque and force on micron-sized particles exhibiting either form or material birefringence. Compared to previous implementations, the instrument described here features improved mechanical and optical stability along with a simplified design. We presented procedures for the construction and calibration of the new instrument, along with detailed protocols for fabrication of appropriate birefringent particles. We anticipate that the methods presented here will find applications not only in biophysical studies, but in other fields, including colloid- and nano-engineering.

ACKNOWLEDGMENTS

This work was funded by grant GM57035 from the NIH to S. M. B. Fabrication of quartz microparticles was partially supported by a CIS New User Grant from the Stanford Nanofabrication Facility. We thank Kirsten L. Frieda for providing us with the double-stranded DNA construct used in the supercoiling study.

REFERENCES

Abbondanzieri, E. A., Greenleaf, W. J., Shaevitz, J. W., Landick, R., and Block, S. M. (2005). Direct observation of base-pair stepping by RNA polymerase. *Nature* **438**, 460–465.
Allersma, M. W., Gittes, F., deCastro, M. J., Stewart, R. J., and Schmidt, C. F. (1998). Two-dimensional tracking of ncd motility by back focal plane interferometry. *Biophys. J.* **74**, 1074–1085.
Ashkin, A. (2006). Optical Trapping and Manipulation of Neutral Particles Using Lasers: A Reprint Volume with Commentaries. World Scientific Publishing Company, Singapore.
Ashkin, A., Dziedzic, J. M., Bjorkholm, J. E., and Chu, S. (1986). Observation of a single-beam gradient force optical trap for dielectric particles. *Opt. Lett.* **11**, 288–290.
Berg, H. C. (2003). The rotary motor of bacterial flagella. *Annu. Rev. Biochem.* **72**, 19–54.
Bishop, A. I., Nieminen, T. A., Heckenberg, N. R., and Rubinsztein-Dunlop, H. (2004). Optical microrheology using rotating laser-trapped particles. *Phys. Rev. Lett.* **92**, 198104.
Bryant, Z., Stone, M. D., Gore, J., Smith, S. B., Cozzarelli, N. R., and Bustamante, C. (2003). Structural transitions and elasticity from torque measurements on DNA. *Nature* **424**, 338–341.
Carter, A. R., King, G. M., and Perkins, T. T. (2007). Back-scattered detection provides atomic-scale localization precision, stability, and registration in 3D. *Opt. Express* **15**, 13434–13445.
Deufel, C., and Wang, M. D. (2006). Detection of forces and displacements along the axial direction in an optical trap. *Biophys. J.* **90**, 657–667.
Deufel, C., Forth, S., Simmons, C. R., Dejgosha, S., and Wang, M. D. (2007). Nanofabricated quartz cylinders for angular trapping: DNA supercoiling torque detection. *Nat. Methods* **4**, 223–225.
Dong, K., and Berger, J. M. (2008). Structure and function of DNA topoisomerases. In "Protein-Nucleic Acid Interactions: Structural Biology," (P. A. Rice and C. C. Correll, eds.), pp. 234–269. Royal Society of Chemistry, Cambridge.
Forth, S., Deufel, C. Y., Sheinin, M., Daniels, B., Sethna, J. P., and Wang, M. D. (2008). Abrupt buckling transition observed during the plectoneme formation of individual DNA molecules. *Phys. Rev. Lett.* **100**, 148301–148304.
Friese, M. E. J., Rubinsztein-Dunlop, H., Gold, J., Hagberg, P., and Hanstorp, D. (2001). Optically driven micromachine elements. *Appl. Phys. Lett.* **78**, 547–549.
Funk, M., Parkin, S. J., Nieminen, T. A., Heckenberg, N. R., and Rubinsztein-Dunlop, H. (2009). Vaterite twist: Microrheology with AOM controlled optical. In "Proceedings of SPIE. Complex Light and Optical Forces III," (E. J. Galvez, D. L. Andrews, and J. Glückstad, eds.), SPIE, Bellingham, WA, Vol. 7227, p. 72270D.
Gardner, F. M. (2005). Phaselock Techniques. Wiley, Hoboken, NJ.
He, H., Friese, M. E. J., Heckenberg, N. R., and Rubinsztein-Dunlop, H. (1995). Direct observation of transfer of angular momentum to absorptive particles from a laser beam with a phase singularity. *Phys. Rev. Lett.* **75**, 826.
Hua, W., Chung, J., and Gelles, J. (2002). Distinguishing inchworm and hand-over-hand processive kinesin movement by neck rotation measurements. *Science* **295**, 844–848.
Jordan, S. C., and Anthony, P. C. (2009). Design considerations for micro- and nanopositioning: Leveraging the latest for biophysical applications. *Curr. Pharm. Biotechnol.* 10(5), 515–521.
Kar, S., Choi, E. J., Guo, F., Dimitriadis, E. K., Kotova, S. L., and Adhya, S. (2006). Right-handed DNA supercoiling by an octameric form of histone-like protein HU: Modulation of cellular transcription. *J. Biol. Chem.* **281**, 40144–40153.
La Porta, A., and Wang, M. D. (2004). Optical torque wrench: Angular trapping, rotation, and torque detection of quartz microparticles. *Phys. Rev. Lett.* **92**, 190801–190804.

Landau, L. D., Lifshitz, E. M., and Pitaevskii, L. P. (1984). Electrodynamics of Continuous Media. Pergamon Press, Oxford.
Lang, M. J., Asbury, C. L., Shaevitz, J. W., and Block, S. M. (2002). An automated two-dimensional optical force clamp for single molecule studies. *Biophys. J.* **83**, 491–501.
Lang, M. J., Fordyce, P. M., Engh, A. M., Neuman, K. C., and Block, S. M. (2004). Simultaneous, coincident optical trapping and single-molecule fluorescence. *Nat. Methods* **1**, 133.
Lionnet, T., Allemand, J.-F., Revyakin, A., Strick, T. R., Saleh, O. A., Bensimon, D., and Croquette, V. (2008). Single-molecule studies using magnetic traps. *In* "Single-Molecule Techniques: A Laboratory Manual," (P. R. Selvin and T. Ha, eds.), pp. 347–369. Cold Spring Harbor Laboratory Press, Cold Spring Harbor, NY.
Neuman, K. C., and Block, S. M. (2004). Optical trapping. *Rev. Sci. Instrum.* **75**, 2787–2809.
Neuman, K. C., Abbondanzieri, E. A., and Block, S. M. (2005). Measurement of the effective focal shift in an optical trap. *Opt. Lett.* **30**, 1318–1320.
Noji, H., Yasuda, R., Yoshida, M., and Kinosita, K. (1997). Direct observation of the rotation of F1-ATPase. *Nature* **386**, 299–302.
Oroszi, L., Galajda, P., Kirei, H., Bottka, S., and Ormos, P. (2006). Direct measurement of torque in an optical trap and its application to double-strand DNA. *Phys. Rev. Lett.* **97**, 058301–058304.
Pralle, A., Prummer, M., Florin, E. L., Stelzer, E. H. K., and Hörber, J. K. H. (1999). Three-dimensional high-resolution particle tracking for optical tweezers by forward scattered light. *Microsc. Res. Tech.* **44**, 378–386.
Simpson, S. H., and Hanna, S. (2009). Rotation of absorbing spheres in Laguerre–Gaussian beams. *J. Opt. Soc. Am. A* **26**, 173–183.
Sischka, A., Kleimann, C., Hachmann, W., Schafer, M. M., Seuffert, I., Tonsing, K., and Anselmetti, D. (2008). Single beam optical tweezers setup with backscattered light detection for three-dimensional measurements on DNA and nanopores. *Rev. Sci. Instrum.* **79**, 063702.
Smith, S. B., Finzi, L., and Bustamante, C. (1992). Direct mechanical measurements of the elasticity of single DNA molecules by using magnetic beads. *Science* **258**, 1122–1126.
Smith, S. B., Cui, Y., and Bustamante, C. (1996). Overstretching B-DNA: The elastic response of individual double-stranded and single-stranded DNA molecules. *Science* **271**, 795–799.
Strick, T. R., Bensimon, D., and Croquette, V. (1999). Micro-mechanical measurement of the torsional modulus of DNA. *Genetica* **106**, 57–62.
Svoboda, K., and Block, S. M. (1994). Biological applications of optical forces. *Annu. Rev. Biophys. Biom.* **23**, 247–285.
Visscher, K., and Block, S. M. (1998). Versatile optical traps with feedback control. *In* "Methods in Enzymology," (B. V. Richard, ed.), Vol. 298, pp. 460–489. Academic Press, San Diego, CA.
Volpe, G., and Petrov, D. (2006). Torque detection using Brownian fluctuations. *Phys. Rev. Lett.* **97**, 210603.
Wang, M. D., Yin, H., Landick, R., Gelles, J., and Block, S. M. (1997). Stretching DNA with optical tweezers. *Biophys. J.* **72**, 1335–1346.

CHAPTER SIXTEEN

FORCE–FLUORESCENCE SPECTROSCOPY AT THE SINGLE-MOLECULE LEVEL

Ruobo Zhou,[*,1] Michael Schlierf,[*,1] and Taekjip Ha[*,†]

Contents

1. Introduction	406
2. Setup	407
2.1. Optical scheme	407
2.2. Surface-tethered assays	409
3. Optical Trapping	410
3.1. Determination of the trapping height	410
3.2. Position detector calibration	411
3.3. Determination of the trap stiffness	412
4. Fluorescence Detection	413
4.1. Single-molecule confocal microscopy	413
4.2. General methods for data analysis	414
5. Coalignment of Confocal and Optical Trapping	416
5.1. Calibration of piezo-controlled mirror	416
5.2. Setting up the tethered molecule for measurement	418
6. Sample Preparation Protocols	418
6.1. Nucleic acid and protein labeling	418
6.2. Polymer-passivated surface preparation	419
6.3. DNA/RNA sample preparation	420
6.4. Preparation of antidigoxigenin-coated microspheres	421
6.5. Sample assembly	422
7. Applications to Biological Systems	423
8. Outlook	423
Acknowledgments	424
References	424

---

[*] Department of Physics and Center for the Physics of Living Cells, University of Illinois at Urbana-Champaign, Urbana, Illinois, USA
[†] Howard Hughes Medical Institute, Urbana, Illinois, USA
[1] Equal contributions.

Methods in Enzymology, Volume 475
ISSN 0076-6879, DOI: 10.1016/S0076-6879(10)75016-3

© 2010 Elsevier Inc.
All rights reserved.

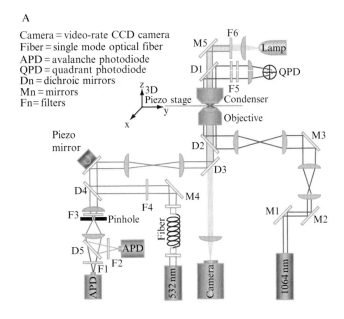

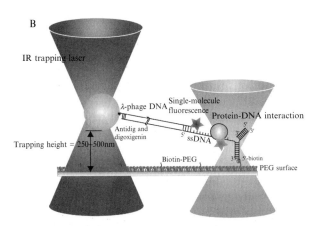

Figure 16.1 Experimental configuration: (A) The combined optical trapping and single-molecule confocal fluorescence instrument is built around a commercial inverted microscope (IX71, Olympus) equipped with a three-dimensional piezo stage (P-527.3CL, Physik Instrumente). The trapping laser beam (1064 nm, 800 mW, Spectra-Physics, Excelsior-1064-800-CDRH) is coupled through the back port of the microscope, while the fluorescence excitation laser beam (532 nm, 30 mW, World StarTech) is directionally controlled by a two-dimensional piezo-controlled steering mirror (S-334K.2SL, Physik Instrumente) and coupled through the right side port. The beams are combined via a dichroic mirror (D2: 780DCSPXR; Chroma) into an oil-immersion objective (UPlanSApo, $100\times$, NA $= 1.4$, Olympus). The intensity profile of the trapping laser in the back focal plane of the condenser (Achromat/Aplanat,

2.2. Surface-tethered assays

Figure 16.1B shows a not to scale cartoon of the surface-tethered assay. The lifetime of the single fluorophore is strongly affected by a nearby IR trapping laser. To overcome this limiting factor, one might either choose alternating excitation and IR trapping beams for a temporal separation or a relatively large spatial separation (Ishijima et al., 1998; Lang et al., 2003). Our setup is built such that we use a long DNA spacer for the large spatial separation of the excitation laser beam and the trapping laser beam. Conveniently, the DNA from the λ-phage (Promega) with a length of 48,502 base pairs (bp) provides a natural long DNA, which has already a double-strand break with a 12 nucleotide (nt) overhang. A complementary short DNA oligonucleotide modified with digoxigenin (Integrated DNA Technologies, Inc., Coralville, IA, USA) is annealed and provides the possibility to specifically attach the λ-phage DNA to antidigoxigenin antibody-coated microspheres. The other lambda DNA overhang is annealed to a freely chosen labeled nucleic acid construct as illustrated in Fig. 16.1B. This construct is tethered to the surface via a specific biotin–neutravidin interaction. Since nucleic acids, proteins and microspheres tend to interact nonspecifically with the coverslip surface, it is crucial to ensure optimal surface passivation, especially if unlabeled and/or labeled proteins are added to an assay. Section 6 gives a detailed description of nucleic acid and protein labeling, surface passivation, antidigoxigenin-coated microsphere preparation, and the final chamber assembly protocols. To avoid nonspecific interactions between the trapped microsphere and the surface, the IR trapping laser focus is chosen to be approximately 250–500 nm above the surface, while the focus of the imaging laser is set to align with the coverslip surface. Using this type of assay our lab was able to study single-molecule kinetics of the Holliday junction using FRET dynamics depending on various external forces (Hohng et al., 2007). However, one could imagine single-molecule fluorescence studies or protein–nucleic acid interaction studies under various forces as illustrated in Fig. 16.1B.

NA $=$ 1.4, Olympus) is imaged onto a quadrant photodiode (UDT SPOT/9DMI) to detect the deviation of the trapped bead position from the trap center. The fluorescence emission is isolated from the reflected infrared light (F3: HNPF-1064.0-1.0, Kaiser) and is band-pass filtered (F1: HQ580/60m, F2: HQ680/60m, Chroma) before imaged onto two avalanche photodiodes, respectively. The bright-field image of the trapped beads is imaged onto a CCD camera (GW-902H, Genwac). (B) Not-to-scale sketch of the combined single-molecule force and fluorescence assay. With the help of λ-phage DNA, a large spatial separation between the trapping laser beam (red) and the excitation laser beam (green) can be achieved. One can probe protein–nucleic acid interactions with single-molecule fluorescence or FRET. The surface passivation is typically achieved with a dense PEG layer, while specific DNA tethering is realized with Biotin–PEG neutravidin interaction. (See Color Insert.)

3. Optical Trapping

3.1. Determination of the trapping height

A critical factor for surface-tethered, combined force–fluorescence assays is the z-height difference between the focus of the confocal imaging beam and the IR trapping laser. One does not want the microsphere trapping height to be too small, since then the microsphere might interact with the coverslip surface. Furthermore, a precise knowledge of the trapping height is important to apply the hydrodynamic drag correction in order to reduce trap stiffness calibration errors. Oil immersion objectives do not allow deep trapping in solution due to spherical aberrations (Neuman et al., 2005). Furthermore, in the case of deep trapping there are necessary corrections on the actual pulling force since the angle between the surface tether point and the tether at the microsphere is no longer close to zero ($\cos(\varphi) \approx 1$). To determine the trapping height, the focus of the confocal beam needs to be preset on the surface of the coverslip such that the diameter of the reflected light spot of the confocal laser detected at the eyepiece CCD camera can be minimized while being focused on the coverslip surface. Taking this focus as a reference height, we use a similar approach described by Lang et al. (2002) in adjusting and finding the actual trapping height. First, a freely floating microsphere in 10 mM Tris–HCl, pH 8.0 (buffer A) is trapped. After finding the focal point of the confocal laser, the fluid cell surface is raised by the piezoelectric stage as illustrated in the inset of Fig. 16.2A, while the back focal plane QPD voltage of the trapping laser is recorded. Similar to Lang et al. (2002), the voltage is initially constant, showing only small oscillations, while when the microsphere touches the surface, the voltage is first rising and then shows a strong drop. The point where the QPD voltage suddenly starts to rise is recorded as the point when the microsphere touches the surface. Since this approach depends on the actual focusing of the surface and the Brownian motion of the microsphere in the trap, this procedure is repeated several times, until a Gaussian distribution of the average microsphere height above the surface is obtained. Figure 16.2B shows such an experimentally obtained Gaussian distribution with an average height of 387 nm. The distribution width originates from two sources, the focal point determination and the Brownian motion of the microsphere. Due to a planar interface between two mismatched indices of refraction, for example, between the coverslip and the aqueous medium, this trapping height has to be corrected following (Neuman et al., 2005):

$$z_{\text{real}} = 0.82 \times z_{\text{measured}} = 0.82 \times 387\,\text{nm} \approx 317\,\text{nm}$$

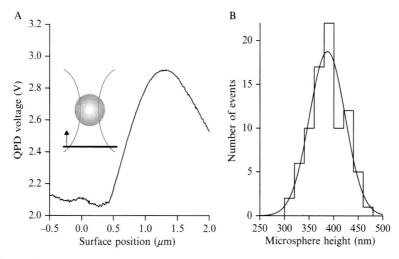

Figure 16.2 Determination of the optical tweezers height. (A) Typical QPD voltage versus surface position curve. A freely floating microsphere is trapped. Then the sample surface is raised to determine the height of the microsphere above the surface. The point (∼0.4 μm) where the curve shows a strong kink and the rises steeply is taken as the microsphere height. Once the microsphere leaves the trapping center the QPD voltage is dropping as shown. (B) Histogram of the microsphere heights. The average height $z_{measured} \approx 387$ nm is found by a Gauss fit to the experimental distribution ($n = 85$).

The real microsphere height is therefore in our case 317 nm and is used in the following calibration procedures.

3.2. Position detector calibration

In the following steps, we describe the calibration of the optical tweezers part. Since the position of the microsphere is imaged in the back focal plane of the condenser on the QPD, a calibration of the QPD signal is required. The microspheres are immobilized on the coverslip nonspecifically by putting the microspheres in buffer B: 10 mM Tris–HCl, pH 8.0, 20–50 mM MgCl$_2$. This ensured that microspheres close to the coverslip surface tend to stick strongly through electrostatic interactions. One such stuck bead is elevated to the trap center. Then the relation of the position of the microsphere to the trapping beam is determined by moving the stuck bead with the piezoelectric stage through the trapping beam in the focal plane of the trapping beam. In detail the QPD signal is calculated in two different directions, V1 and V2. The V1 signal is composed of $[(A + B) - (C + D)]/(A + B + C + D)$, while V2 $= [(B + C) - (A + D)]/(A + B + C + D)$. Similar to Lang et al. (2002), the back focal plane signals V1 and V2 of a stuck bead are taken at various

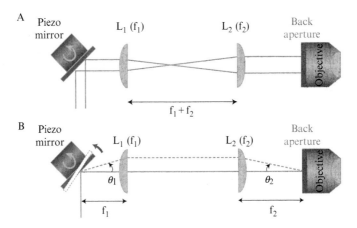

Figure 16.3 Optically conjugated geometry for the piezo-controlled mirror. (A) The distance between L_1 and L_2 is the sum of their focal lengths, to keep the excitation laser beam collimated before and after the telescope system. (B) The piezo-controlled mirror is positioned in the optically conjugated plane of the back aperture of the objective.

A fluorescent microsphere sample is needed for the alignment of the pinhole and the APDs in the confocal light path. The procedure is as follows:

1. Dilute the fluorescently labeled microsphere stock (FluoSpheres® carboxylate-modified microspheres, 0.2 μm, crimson fluorescent, 625/645, 2% solids, Molecular Probes) to an appropriate concentration (typically 200- to 500-fold dilution) in buffer B (10 mM Tris–HCl, pH 8.0, 20–50 mM MgCl$_2$).
2. Inject the diluted microspheres into a sample chamber. Incubate for 10 min.
3. Rinse with buffer B to remove excess microspheres in the solution and use epoxy to seal the chamber.

After focusing the 532-nm confocal excitation laser to the flow chamber surface, the APD and pinhole position can be precisely adjusted with a precision XYZ stage to maximize the photon counting rate of the APDs.

4.2. General methods for data analysis

The donor and acceptor signals are recorded by the two APDs. FRET efficiency, E_{FRET}, is a measure of how much of the energy is transferred from the donor to the acceptor, which can be estimated as $E_{FRET} = I_{acceptor}/(I_{acceptor} + I_{donor})$, where $I_{acceptor}$ and I_{donor} are the emission intensities (APD photon counts) of acceptor and corresponding donor, respectively (Roy et al.,

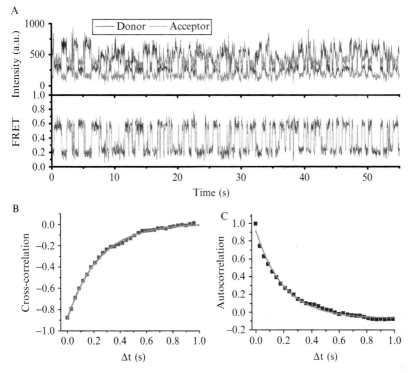

Figure 16.4 General methods to analyze the dynamic FRET trajectories. (A) An example of the donor–acceptor intensity trajectories from a two-state dynamic system. Hidden Markov model (HMM)-derived idealized FRET trajectory (magenta) superimposed on the FRET trajectory (blue). (B) Cross-correlation between the donor and acceptor time trace shown in (A) which is fit to a single-exponential function (solid line). (C) Autocorrelation of the FRET trajectory shown in (A) which is fit to a single-exponential function (solid line). (See Color Insert.)

2008). Some further correction factors, like donor leakage to the acceptor channel should be considered (Selvin and Ha, 2007). Figure 16.4A shows an example of typical donor and acceptor trajectories with two distinct FRET states (time resolution is 30 ms). If the FRET trajectory shows two or more FRET states caused by the dynamics of the molecules, hidden Markov modeling has proved to be extremely useful for determining the total number of FRET states in a trajectory (HaMMy, available for download at http://bio.physics.illinois.edu/HaMMy.html). Using the freely available software, one can generate an idealized trajectory (Fig. 16.4A, magenta line) and obtain the transition rates between two or more FRET states (McKinney et al., 2006). In Fig. 16.4A, we show a FRET time trace with two distinct FRET states. The transition rate from low to high FRET states obtained by HaMMy is $k_{L \to H} = 2.4\,s^{-1}$ and the transition rate from high to low FRET states is

$k_{H\to L} = 2.7\,\mathrm{s}^{-1}$. If the molecule shows too fast dynamics for reliable HaMMy analysis of FRET fluctuations, one would use autocorrelation analysis of the FRET trajectory or cross-correlation analysis of donor and acceptor intensity traces (Kim *et al.*, 2002). Figure 16.4B and C shows the autocorrelation and cross-correlation calculated from the data shown in Fig. 16.4A. By fitting the calculated auto- or cross-correlation functions to a single-exponential function, one obtains two parameters: the characteristic time of the exponential, τ, which reflects the timescale of the FRET fluctuations, and the amplitude of the exponential at $t = 0$. The fitting results in Fig. 16.4B and C yield a characteristic time $\tau = 0.2$ s, which is consistent with the expected relation $\tau^{-1} = k_{L\to H} + k_{H\to L}$ derived for the relaxation kinetics of a reversible two-state system (Joo *et al.*, 2004).

5. Coalignment of Confocal and Optical Trapping

5.1. Calibration of piezo-controlled mirror

To simultaneously operate the single-molecule confocal microscope and optical trap, the confocal excitation beam has to be programmed to follow the motion of the fluorescently labeled molecule when the molecule is moved with the piezoelectric stage for stretching. The deflection angle of the confocal beam can be controlled precisely by the piezo-controlled mirror, but its resulting displacement in the sample plane is unknown and needs to be calibrated. Therefore, mapping is required between the deflection angle of the piezo-controlled mirror (α, β) and the resulting displacement in the sample plane (x, y). In addition, the origin of the piezo-controlled mirror should be preset to a particular position such that the confocal spot is overlapped in the sample plane with that of the trapping laser. The piezo-controlled mirror can be calibrated as follows:

1. Prepare two fluorescent bead samples. One has stuck beads on the coverslip (see protocol in Section 4.1), the other contains free beads in the chamber. To make the free bead sample, the protocol is similar to that for a stuck bead sample, but buffer A (10 mM Tris–HCl, pH 8.0) instead of buffer B is used to dilute and inject the beads.
2. Use the free fluorescent bead sample to reset the mirror origin. Focus the confocal beam to the sample plane and turn on the trapping laser. Some fluorescent beads can be trapped to the center of the laser trap. Steer the mirror to scan the area where the trapping laser spot is located with a step size of 32 nrad. Figure 16.5A (upper panel) shows a typical mirror scan image of trapped beads. Set the origin of the mirror to the center pixel position of the fluorescent spot. Note that the fluorescent spot is ellipse-like

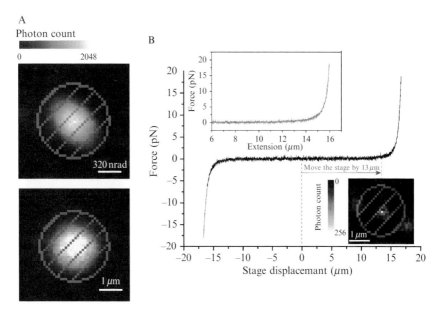

Figure 16.5 Mirror calibration and force–extension curve. (A) The mirror scan image around the area where the fluorescent beads are trapped in the sample plane without (upper) and with (lower) the mirror calibration. (B) A stretching curve and the force–extension curve (upper inset) of the tethered DNA after the origin of the piezo stage is set to the estimated tethered position. A WLC model (red) is used to fit the experimental force–extension curve (blue). The lower inset shows a mirror scan image around the origin of the piezo stage after displacing the stage from its origin by 13 μm. The green dot indicates the center position of the fluorescently labeled molecule that is being stretched. (See Color Insert.)

because the mirror calibration has not been performed yet. Using a calibrated mirror to scan the trapped beads with a step size of 100 nm in the sample plane, a circular fluorescent spot is obtained instead (Fig. 16.5A, lower panel). Calibration is accomplished in the following step.

3. Use the stuck bead sample to calibrate the mirror. The stuck bead sample is imaged sequentially either by scanning with the piezo stage (scan area, 38.4 μm × 38.4 μm) while fixing the mirror or by scanning with the mirror while fixing the piezo stage. Two third-order polynomial fits, $x = \sum_{i,j=0}^{3} m_{ij}\alpha^i\beta^j$, $y = \sum_{i,j=0}^{3} n_{ij}\alpha^i\beta^j$, are used to map angle coordinates into spatial coordinates in the sample plane. Then two mapping files containing the coefficients m_{ij} and n_{ij} are generated, which are later used for steering the confocal beam to any desired position in the sample plane. Typical mapping images were shown previously (Hohng et al., 2007).

With the calibrated piezo-mirror, we can ask the confocal spot to follow the movement of the piezo stage by custom written software such that the confocal spot keeps track of the fluorescently labeled molecule under investigation.

5.2. Setting up the tethered molecule for measurement

Once the calibrations of QPD and piezo-controlled mirror are complete, force–fluorescence measurements can be performed on a sample. After a surface-tethered λ-phage DNA is optically trapped via the attached bead, the stretching curve is obtained by moving the coverslip in x- and y-direction with the piezo stage (Fig. 16.5C). The symmetry of the stretching curves can be used to roughly determine the tethered position by finding the central positions in two orthogonal stretching directions in the sample plane. The origin of the piezo stage can then be reset to this central position. After considering the bead radius and the deviation of the bead from the trap center, force–extension curves of the molecule (blue line, upper inset, Fig. 16.5C) are obtained and can be fitted with the "worm-like chain" (WLC) model (red solid line), yielding a persistence length of about 40–50 nm (Bustamante et al., 1994).

Next the fluorescently labeled molecule is displaced by typically 13 μm for the spatial separation of the trapping and excitation laser beams. The confocal image around the target molecule is taken by scanning the confocal spot in the sample plane (scan area, 3.2 μm × 3.2 μm) using the steering mirror calibration. A more accurate position of the fluorescently labeled molecule and therefore the accurate surface attachment point is then determined from the image (Fig. 16.5C, lower inset). For the force–fluorescence measurement, the piezo stage was moved further away to reach different forces, while both force and the fluorescence signals are recorded.

6. SAMPLE PREPARATION PROTOCOLS

A good sample preparation is essential for force–fluorescence measurement. For example, a high dye-labeling efficiency, appropriate annealing, and sample assembly lead to a successful force–fluorescence measurement.

6.1. Nucleic acid and protein labeling

Since the λ-phage DNA linker is acting as an entropic spring, the force resolution is limited in force–fluorescence spectroscopy. Therefore, the FRET or fluorescence data, rather than the force values, are the read-out of the conformational changes of single molecules. The fluorescent probes (donor and acceptor) need to be engineered to the desired locations on individual nucleic acids or proteins. There are many conjugation strategies for either proteins or nucleic acids as previously described (Roy et al., 2008; Selvin and Ha, 2007), but a high labeling yield is achieved much more easily for nucleic acids than proteins. $3'$ or $5'$ fluorescently end-labeled DNA/RNA oligonucleotides can be ordered from companies (e.g., Integrated DNA Technologies). Fluorescent probes may also be incorporated internally into the nucleic acid chain using phosphoramidite chemistry during oligonucleotide synthesis. This is optimal in cases where the dynamic nature of the biological system is affected little by the internal modification through backbone disruption. In most of the cases, an alternative method is recommended for labeling where the DNA backbone is not broken: an amine-modified base (typically thymine), instead of a fluorescent probe, is inserted into the desired location, which can later react with the N-hydroxysuccinimide (NHS) ester form of the fluorescent probe (GE Healthcare). Purification can be achieved by polyacrylamide gel electrophoresis to separate labeled from unlabeled oligonucleotides. Recombinant engineered cysteine variants of proteins can be easily labeled with maleimide derivatized fluorescent probes, for example, *Escherichia coli* Rep helicase (Myong et al., 2005; Roy et al., 2008). However, there are certain limitations since many proteins carry multiple solvent-exposed cysteines and upon substitution of those cysteines the functionality might be changed or totally lost.

To specifically immobilize the molecules of interest on the glass coverslip for single-molecule experiments, a biotin–neutravidin linkage is commonly used. A biotin modification is easily introduced by commercially ordered DNA oligonucleotides. Furthermore, proteins can be biotinylated using similar conjugation strategies as used for the conjugation of fluorescent probes.

6.2. Polymer-passivated surface preparation

Although biotinylated bovine serum albumin (BSA) can be used to adsorb to the glass surface for immobilization of biotinylated molecules through neutravidin protein sandwiched in the middle, a polymer-passivated surface coated with polyethyleneglycol (PEG) is highly recommended in order to eliminate nonspecific surface adsorption of proteins and efficiently reduce the surface interactions with nucleic acids and beads. The common protocol for preparing the PEGylated surface contains three steps (Selvin and Ha, 2007):

1. Precleaning and surface activation
2. Aminosilanization of the surface
3. PEGylation (coating the amino-modified surface with PEG–NHS esters)

In the third step, a small fraction ($\sim 3\%$) of biotin–PEG–NHS ester (Bio-PEG-SC, Laysan Bio) is mixed with regular PEG–NHS ester (mPEG-SC, Laysan Bio) for the purpose of immobilizing biomolecules. The detailed steps can be inferred from Selvin and Ha (2007). The PEGylation following this protocol on a glass surface is not as good as on a quartz surface. However, dissolving PEG–NHS ester in 50 mM MOPS (pH 7.5) for PEGylation instead of in 0.1 M sodium bicarbonate (pH 8.5) has been found to improve the PEGylation efficiency resulting in further suppression of nonspecific adsorption on a silicate surface (Yokota et al., 2009). A higher concentration of KOH (10 M in Milli-Q water) for the first cleaning step can also be applied to drastically enhance the aminosilanization result and hence improve the PEGylation efficiency.

6.3. DNA/RNA sample preparation

The nucleic acid construct (e.g., a four-way Holliday junction, partial duplex, or forked DNA substrates that interacts with proteins) carrying fluorescence dyes is preannealed from oligonucleotides and contains a 5′ single-stranded tail (5′-GGG CGG CGA CCT) which is complementary to the 12 nt *cos* site of λ-phage DNA. λ-Phage DNA adopts either a circular form or a linear form that has two complementary 12 nt single-stranded overhangs. By heating to above the melting temperature of the *cos* site (~60–70 °C) for approximately 10 min, the circular λ-phage DNA is converted into the linear form with single-stranded 5′ extensions of 12 nt at both ends which are complementary to each other. Thus, we can make the nucleic acid construct annealed with the linear form of a λ-phage DNA using the following protocol:

1. Resuspend and mix the oligonucleotides in a microcentrifuge tube with each final concentration no less than 1 μM in annealing buffer (10 mM Tris–HCl, pH 8.0, 50 mM NaCl). The biotinylated strand should have a slightly lower concentration than the other strands.
2. Put the tube in a heat block at 90–95 °C for 3 min. Remove the heat block from the heater and allow it to slowly cool to room temperature over ~2 h.
3. Dilute the preannealed nucleic acid product to a concentration of 100 nM, make aliquots and store in the freezer.
4. Prepare 40-μl aliquots of λ-phage DNA (~500 μg/ml, Promega) in microcentrifuge tubes. Take one λ-phage DNA aliquot and add 5 μl of 5 M NaCl. Mix very gently (large orifice pipette tips should be used when handling λ-phage DNA to avoid high shearing forces).
5. Place the tube from Step 4 in a heat block at 85–90 °C for 10 min.
6. Bury the tube in ice and incubate for 5 min for fast cooling. Then quickly add 3 μl of 100 nM preannealed mixture from Step 3 and 1 μl of 10 mg/ml BSA (New England Biolabs).
7. Rotate the tube for 1–1.5 h at room temperature.
8. Move the tube to a cold room (4 °C) and keep rotating for another hour.

9. Take out the tube from the cold room and add 1 μl of 10 μM digoxigenin-conjugated DNA oligonucleotide (5′-AGG TCG CCG CCC TTT/digoxigenin/-3′) into the tube and keep rotating the tube in the cold room for 1–1.5 h. This generates the complete DNA/RNA construct with a single digoxigenin-tag on one end of the λ-phage DNA and a biotin-tag on the other end.
10. Add 250 μl of 10 mM Tris–HCl, pH 8.0, 50 mM NaCl into the tube and prepare 10- or 20-μl aliquots of this completed sample (which is now at a concentration of 1 nM). Store aliquots at $-20\,°$C.

6.4. Preparation of antidigoxigenin-coated microspheres

Antidigoxigenin is cross-linked to protein G-coated polystyrene beads following the protocol below so that the beads can be attached to the DNA/RNA template for optical stretching via a digoxigenin–antidigoxigenin interaction.

- Buffer solutions:
  1. *MES buffer*: 100 mM MES–NaOH, pH 6.5 (prepare immediately before use).
  2. *Antibody reconstitution buffer*: 0.019 M NaH$_2$PO$_4$, 0.081 M Na$_2$HPO$_4$, 0.14 M NaCl, 2.7 mM KCl.
  3. *Microsphere storage buffer*: 0.039 M NaH$_2$PO$_4$, 0.061 M Na$_2$HPO$_4$, 0.14 M NaCl, 2.7 mM KCl, 0.1 mg/ml BSA, 0.1% (v/v) Tween-20, 0.02% (w/v) sodium azide.
- Protocol:
  1. Resuspend protein G-coated polystyrene microspheres (1.0 μm, undiluted, 1.25% solids (w/v) aqueous, Polysciences) and take 250 μl of it to a microcentrifuge tube. Exchange the beads into a freshly made MES buffer by repeating buffer wash for three to four times (centrifuge for 4 min at 7000 rpm, carefully pipette off the supernatant, and add 250 μl of MES buffer into the tube).
  2. Dissolve 50 mg of N-(3-dimethylaminopropyl)-N'-ethylcarbodiimide hydrochloride (EDC hydrochloride, Sigma-Aldrich) in 1 ml MES buffer, and dissolve 50 mg of N-hydroxysuccinimide (NHS, Aldrich) in 1 ml MES buffer.
  3. Add 50 μl of EDC hydrochloride and 25 μl of NHS from Step 2 into the tube.
  4. Tumble the tube for 10 min at room temperature.
  5. Dissolve 200 μg of antidigoxigenin antibody (Roche Applied Science) in 200 μl of antibody reconstitution buffer and add 30 μl of antidigoxigenin to the tube. Aliquot and shock freeze the remaining dissolved antidigoxigenin with liquid nitrogen for future use.
  6. Keep tumbling the tube for 2 h at room temperature.

7. Stop the cross-linking reaction by adding 20 μl of Tris–HCl buffer (1 M, pH 6.8) and continue tumbling for 1 h.
8. Wash the beads three times with bead storage buffer by resuspending and centrifuging as in Step 1.
9. Store the beads at 4 °C for future use. This bead solution is 50- to 100-fold diluted in Tris buffer (10 mM Tris–HCl, pH 8.0) for sample assembly in the force–fluorescence experiment as follows.

6.5. Sample assembly

Experimental samples are assembled by sequential infusions of buffer solutions into an open-ended chamber with a volume of approximately 15–20 μl (Roy et al., 2008; Selvin and Ha, 2007). The infusion is a very important step for successful sample preparation, especially after the DNA/RNA samples are immobilized on the coverslip. An overly rapid fluid flow may result in the adsorption of some internal portion of λ-phage DNA onto the coverslip surface. Incubation in a buffer containing tRNA (Ambion) or short double-stranded DNA (20–30 bp) can significantly reduce this nonspecific adsorption. The buffer solutions are delivered drop by drop to one open end of the chamber using a micropipette. Slightly tilt the chamber with a small angle such that the liquid drop infuses into the chamber slowly through gravity and comes out the other end. An alternative way for infusing buffer is to adapt an automated pump (PHD 22/2000 series syringe pump; Harvard Apparatus) by using a sample chamber with two 0.75-mm-diameter inlet/outlet holes (Selvin and Ha, 2007). The incubations given in the following protocol are performed at room temperature by putting the sample chamber in a humid environment (e.g., a water filled pipette tip box) to avoid evaporation.

1. Take out a PEGylated coverslip and a PEGylated microscope slide and assemble a fluid chamber.
2. Infuse 25 μl (slightly larger than the chamber volume) of 0.25 mg/ml neutravidin in T50 buffer (10 mM Tris–HCl, pH 8.0, 50 mM NaCl) and incubate for 5 min.
3. Rinse the chamber with 50 μl of T50 buffer.
4. Infuse 50 μl blocking buffer (10 mM Tris–HCl, pH 8.0, 50 mM NaCl, 1 mg/ml tRNA, 1 mg/ml BSA) and incubate for 1 h.
5. Remove one aliquot of completed DNA/RNA sample and dilute it to a final DNA/RNA concentration of 30–50 pM in 10 mM Tris–HCl, pH 8.0, 50 mM NaCl, 0.1 mg/ml BSA. Infuse the diluted solution into the chamber and incubate for 30 min.
6. Rinse sample chamber with 100–120 μl (more than 5 chamber volumes) of buffer A (10 mM Tris–HCl, pH 8.0).

7. Mix 1 μl of the antidigoxigenin-coated beads as prepared before and 99 μl of buffer A. Complete the buffer exchange from bead storage buffer to the Tris buffer by resuspending and centrifuging twice. Infuse 25 μl of the 100-times diluted beads into the chamber. Incubate for 30 min.
8. Rinse sample chamber with 100–120 μl Tris buffer (10 mM Tris–HCl, pH 8.0).
9. Infuse the final imaging buffer typically containing 20 mM Tris–HCl, pH 8.0, 0.5 mg/ml BSA, 0.01 mg/ml antidigoxigenin, 0.5% (w/v) D-glucose (Sigma), 165 U/ml glucose oxidase (Sigma), 2170 U/ml catalase (Roche), 3 mM Trolox (Sigma), and 0.1% (v/v) Tween 20 (Sigma) as well as appropriate concentrations of NaCl and divalent ions (MgCl$_2$, CaCl$_2$, etc.) for the scientific question at hand. Proteins and other reagents (ATP, DTT, EDTA, glycerol, etc.) can be added to the imaging buffer as needed.

7. Applications to Biological Systems

Force–fluorescence spectroscopy has provided a general method to study nucleic acid conformational dynamics and protein–nucleic acid interactions. This novel approach has made it possible to observe nanometer-scale motion of molecules at sub-piconewton forces, a detection regime previously inaccessible but arguably much more relevant physiologically than the persistent, strong forces used in typical single-molecule force manipulation experiments. A donor–acceptor pair can be positioned to different desired locations on the nucleic acid and protein to probe the dynamics along various vectors, which maximizes the information content collected from the data. For example, we studied the conformational dynamics of the Holliday junction, an important intermediate in genetic recombination. By applying force on three different arms of the junction, the reaction diagram of the Holliday junction dynamics was reconstructed and an intermediate state was revealed (Hohng et al., 2007). We also demonstrated how mechanical force is used to regulate the diffusion of a single-stranded DNA-binding (SSB) protein along DNA and to control the extent of DNA unraveling and ultimate dissociation, with strong implications on how other proteins that directly interact with SSB may gain access to the DNA (Zhou et al., submitted).

8. Outlook

Both single-molecule force and single-molecule fluorescence spectroscopy have proved to be successful and helpful techniques to study the dynamics and mechanics of various biological molecules like DNA, RNA, and proteins. Nevertheless, both methods on their own hold certain limitations that can be resolved with the combination of both techniques. It is important to note that these types of combined experiments still face many challenges and can benefit from potential improvements. For example, the approach of using a long DNA linker for the spatial separation of the trapping and confocal excitation laser beams solves the bleaching problem, but imposes a reduced force resolution due to the long entropic spring that λ-phage DNA represents. As mentioned above a second solution for the bleaching problem of single fluorophores near an IR trapping beam is a temporal separation of both lasers. In principle, this can be achieved by the integration of acousto-optical modulators to rapidly switch (on the order of 10 kHz) between the trapping and excitation lasers (Brau et al., 2006). Using such an approach, one might be able to reduce the tether-bead DNA length and thus increase the force resolution of the optical tweezers. A second limitation in the setup described here is spatial drift of the fluid chamber. Such drift problems, introduced by surface tethers or fixed micropipette tethers, have long been known in the optical tweezers field and a typical solution is a dumbbell trap arrangement (Abbondanzieri et al., 2005; Moffitt et al., 2006). In this experimental approach, instead of trapping a single microsphere a tether is formed between two microspheres that are freely suspended in solution and are thus free of stage and other instrument drift. A stabilization system with feedback loop may also be introduced to improve a surface-coupled optical trapping system (Carter et al., 2009). Following these ideas we anticipate that there are many new opportunities to explore the combined worlds of single-molecule fluorescence and force spectroscopy.

ACKNOWLEDGMENTS

This work was funded by grants from the National Institutes of Health and the National Science Foundation. T. H. is an Investigator of the Howard Hughes Medical Institute. This work was supported by NIH grant GM065367 and NSF grants 0822613 and 0646550 to T.H. M. S. gratefully acknowledges support by the Deutsche Forschungsgemeinschaft (DFG SCHL1896/1-1).

REFERENCES

Abbondanzieri, E. A., Greenleaf, W. J., Shaevitz, J. W., Landick, R., and Block, S. M. (2005). Direct observation of base-pair stepping by RNA polymerase. *Nature* **438**, 460–465.

Berg-Sørensen, K., and Flyvbjerg, H. (2004). Power spectrum analysis for optical tweezers. *Rev. Sci. Instrum.* **75**, 594.

Brau, R. R., Tarsa, P. B., Ferrer, J. M., Lee, P., and Lang, M. J. (2006). Interlaced optical force-fluorescence measurements for single molecule biophysics. *Biophys. J.* **91**, 1069–1077.

Brower-Toland, B. D., Smith, C. L., Yeh, R. C., Lis, J. T., Peterson, C. L., and Wang, M. D. (2002). Mechanical disruption of individual nucleosomes reveals a reversible multistage release of DNA. *Proc. Natl. Acad. Sci. USA* **99**, 1960–1965.

Bustamante, C., Marko, J. F., Siggia, E. D., and Smith, S. (1994). Entropic elasticity of lambda-phage DNA. *Science* **265**, 1599–1600.

Carter, A. R., Seol, Y., and Perkins, T. T. (2009). Precision surface-coupled optical-trapping assay with one-basepair resolution. *Biophys. J.* **96**, 2926–2934.

Cecconi, C., Shank, E. A., Bustamante, C., and Marqusee, S. (2005). Direct observation of the three-state folding of a single protein molecule. *Science* **309**, 2057–2060.

del Rio, A., Perez-Jimenez, R., Liu, R., Roca-Cusachs, P., Fernandez, J. M., and Sheetz, M. P. (2009). Stretching single talin rod molecules activates vinculin binding. *Science* **323**, 638–641.

Ha, T., Chemla, D. S., Enderle, T., and Weiss, S. (1997). Single molecule spectroscopy with automated positioning. *Appl. Phys. Lett.* **70**, 782–784.

Happel, J., and Brenner, H. (1983). Low Reynolds number hydrodynamics: with special applications to particulate media. Kluwer Academic, Dordrecht.

Hohng, S., Joo, C., and Ha, T. (2004). Single-molecule three-color FRET. *Biophys. J.* **87**, 1328–1337.

Hohng, S., Zhou, R., Nahas, M. K., Yu, J., Schulten, K., Lilley, D. M., and Ha, T. (2007). Fluorescence-force spectroscopy maps two-dimensional reaction landscape of the holliday junction. *Science* **318**, 279–283.

Ishijima, A., Kojima, H., Funatsu, T., Tokunaga, M., Higuchi, H., Tanaka, H., and Yanagida, T. (1998). Simultaneous observation of individual ATPase and mechanical events by a single myosin molecule during interaction with actin. *Cell* **92**, 161–171.

Joo, C., McKinney, S. A., Lilley, D. M., and Ha, T. (2004). Exploring rare conformational species and ionic effects in DNA Holliday junctions using single-molecule spectroscopy. *J. Mol. Biol.* **341**, 739–751.

Joo, C., Balci, H., Ishitsuka, Y., Buranachai, C., and Ha, T. (2008). Advances in single-molecule fluorescence methods for molecular biology. *Annu. Rev. Biochem.* **77**, 51–76

Kapanidis, A. N., and Strick, T. (2009). Biology, one molecule at a time. *Trends Biochem. Sci.* **34**, 234–243.

Kim, H. D., Nienhaus, G. U., Ha, T., Orr, J. W., Williamson, J. R., and Chu, S. (2002). Mg^{2+}-dependent conformational change of RNA studied by fluorescence correlation and FRET on immobilized single molecules. *Proc. Natl. Acad. Sci. USA* **99**, 4284–4289.

Lang, M. J., Asbury, C. L., Shaevitz, J. W., and Block, S. M. (2002). An automated two-dimensional optical force clamp for single molecule studies. *Biophys. J.* **83**, 491–501.

Lang, M. J., Fordyce, P. M., and Block, S. M. (2003). Combined optical trapping and single-molecule fluorescence. *J. Biol.* **2**, 6.

McKinney, S. A., Joo, C., and Ha, T. (2006). Analysis of single-molecule FRET trajectories using hidden Markov modeling. *Biophys. J.* **91**, 1941–1951.

Moffitt, J. R., Chemla, Y. R., Izhaky, D., and Bustamante, C. (2006). Differential detection of dual traps improves the spatial resolution of optical tweezers. *Proc. Natl. Acad. Sci. USA* **103**, 9006–9011.

Moffitt, J. R., Chemla, Y. R., Smith, S. B., and Bustamante, C. (2008). Recent advances in optical tweezers. *Annu. Rev. Biochem.* **77**, 205–228.

Myong, S., Rasnik, I., Joo, C., Lohman, T. M., and Ha, T. (2005). Repetitive shuttling of a motor protein on DNA. *Nature* **437**, 1321–1325.

Neuman, K. C., Abbondanzieri, E. A., and Block, S. M. (2005). Measurement of the effective focal shift in an optical trap. *Opt. Lett.* **30**, 1318–1320.

Pralle, A., Florin, E. L., Stelzer, E. H. K., and Horber, J. K. H. (1998). Local viscosity probed by photonic force microscopy. *Appl. Phys. A: Mater. Sci. Process.* **66**, S71–S73.

Puchner, E. M., Alexandrovich, A., Kho, A. L., Hensen, U., Schafer, L. V., Brandmeier, B., Grater, F., Grubmuller, H., Gaub, H. E., and Gautel, M. (2008). Mechanoenzymatics of titin kinase. *Proc. Natl. Acad. Sci. USA* **105**, 13385–13390.

Rief, M., Gautel, M., Oesterhelt, F., Fernandez, J. M., and Gaub, H. E. (1997). Reversible unfolding of individual titin immunoglobulin domains by AFM. *Science* **276**, 1109–1112.

Roy, R., Hohng, S., and Ha, T. (2008). A practical guide to single-molecule FRET. *Nat. Methods* **5**, 507–516.

Schuler, B., and Eaton, W. A. (2008). Protein folding studied by single-molecule FRET. *Curr. Opin. Struct. Biol.* **18**, 16–26.

Selvin, P. R., and Ha, T. (2007). *Single Molecule Techniques: A Laboratory Manual*. Cold Spring Harbor Laboratory Press, Cold Spring Harbor, NY.

Tarsa, P. B., Brau, R. R., Barch, M., Ferrer, J. M., Freyzon, Y., Matsudaira, P., and Lang, M. J. (2007). Detecting force-induced molecular transitions with fluorescence resonant energy transfer. *Angew. Chem. Int. Ed. Engl.* **46**, 1999–2001.

Walter, N. G., Huang, C. Y., Manzo, A. J., and Sobhy, M. A. (2008). Do-it-yourself guide: How to use the modern single-molecule toolkit. *Nat. Methods* **5**, 475–489.

Weiss, S. (2000). Measuring conformational dynamics of biomolecules by single molecule fluorescence spectroscopy. *Nat. Struct. Biol.* **7**, 724–729.

Wiita, A. P., Perez-Jimenez, R., Walther, K. A., Gräter, F., Berne, B. J., Holmgren, A., Sanchez-Ruiz, J. M., and Fernandez, J. M. (2007). Probing the chemistry of thioredoxin catalysis with force. *Nature* **450**, 124–127.

Woodside, M. T., Anthony, P. C., Behnke-Parks, W. M., Larizadeh, K., Herschlag, D., and Block, S. M. (2006). Direct measurement of the full, sequence-dependent folding landscape of a nucleic acid. *Science* **314**, 1001–1004.

Yokota, H., Han, Y. W., Allemand, J. F., Xi, X. G., Bensimon, D., Croquette, V., Ito, Y., and Harada, Y. (2009). Single-molecule visualization of binding modes of helicase to DNA on PEGylated surfaces. *Chem. Lett.* **38**, 308–309.

Zhang, X., Halvorsen, K., Zhang, C. Z., Wong, W. P., and Springer, T. A. (2009). Mechanoenzymatic cleavage of the ultralarge vascular protein von Willebrand factor. *Science* **324**, 1330–1334.

Zhou, R., Roy, R., Kozlov, A., Lohman, T. M., Ha, T. (submitted). Dissecting the mechanisms of protein dissociation and diffusion on single-stranded DNA using tension.

CHAPTER SEVENTEEN

Combining Optical Tweezers, Single-Molecule Fluorescence Microscopy, and Microfluidics for Studies of DNA–Protein Interactions

Peter Gross, Géraldine Farge, Erwin J. G. Peterman, *and* Gijs J. L. Wuite

Contents

1. Introduction 428
2. Instrumentation 430
   2.1. Optical tweezers 430
   2.2. Fluorescence microscopy 437
   2.3. Microfluidics 440
3. Preparation of Reagents 443
   3.1. Terminally labeled DNA molecules 443
   3.2. Fluorescent labeling of DNA 444
   3.3. Fluorescent labeling of proteins 444
4. Combining Optical Trapping, Fluorescence Microscopy, and Microfluidics: Example Protocols 445
   4.1. Example protocol II: Sequential isolation and visualization of a single DNA molecule using YOYO-1 446
   4.2. Example protocol III: Binding of replication protein A to ssDNA 448
   4.3. Example protocol IV: Binding of Rad51 to dsDNA 448
5. Conclusions 450
Acknowledgments 451
References 451

Abstract

The technically challenging field of single-molecule biophysics has established itself in the last decade by granting access to detailed information about the fate of individual biomolecules, unattainable in traditional biochemical assays. The appeal of single-molecule methods lies in the directness of the information

Department of Physics and Astronomy and Laser Centre, VU University, De Boelelaan, Amsterdam, The Netherlands

Methods in Enzymology, Volume 475
ISSN 0076-6879, DOI: 10.1016/S0076-6879(10)75017-5

© 2010 Elsevier Inc.
All rights reserved.

obtained from individual biomolecules. Technological improvements in single-molecule methods have made it possible to combine optical tweezers, fluorescence microscopy, and microfluidic flow systems. Such a combination of techniques has opened new possibilities to study complex biochemical reactions on the single-molecule level. In this chapter, we provide general considerations for the development of a combined optical trapping, fluorescence microscopy, and microfluidics instrument, along with methods to solve technical issues that are critical for designing successful experiments. Finally, we present several experiments to illustrate the power of this combination of techniques.

1. INTRODUCTION

Single-molecule detection methods have made numerous important contributions to biology. Although the range of single-molecule tools is continuously expanding, they can be categorized into two dominant approaches: (i) force detection and manipulation, and (ii) fluorescence imaging and spectroscopy. In this chapter, we will focus on the technical aspects of combining these two approaches for studies of DNA–protein interactions on the single-molecule level.

Single-molecule manipulation techniques, such as magnetic and optical tweezers, are commonly used to study the mechanical properties of biopolymers, in particular DNA, and their interactions with other biomolecules. These interactions lead in many cases to changes in the mechanical properties of DNA (length, flexibility, elasticity) (Neuman and Nagy, 2008; Vladescu et al., 2007; Wen et al., 2008; Wuite et al., 2000b), granting access to detailed information on, for example, the kinetics and mechanochemistry of proteins like the ribosome, RNA polymerase, DNA polymerase, and equilibrium binding properties of DNA intercalators. With fluorescence microscopy, on the other hand, the location and dynamics of fluorescently labeled biomolecules can be measured more directly. In addition, single-molecule fluorescence microscopy techniques can provide information on the dynamics of chemical reactions and local conformational changes of biomolecules like DNAs and proteins (Weiss, 1999). Single-molecule fluorescence microscopy can be classified into two broad categories (Stephens and Allan, 2003): (i) confocal fluorescence microscopy, where an image is generated by scanning a colocalized excitation and detection focal spot over the sample while a pinhole in front of the detector blocks out-of-focus light and (ii) wide-field fluorescence microscopy, where a region of the specimen is excited and observed with an array detector. A successful method to reduce the fluorescence background in wide-field fluorescence microscopy is total internal reflection fluorescence (TIRF) microscopy, where the fluorescence excitation is limited to about 100 nm from a glass–water interface.

Force- and fluorescence-based approaches can provide highly complementary information (van Mameren et al., 2009b). However, the technical challenges involved have limited the number of successful combinations of these single-molecule techniques. Previous studies have incorporated both approaches in several ways. In a first approach, DNA combing (stretching DNA by nonspecific adherence to a surface) and surface-tethering (attaching DNA specifically to a surface by either one or two extremities) have been used to study DNA–protein interaction with TIRF microscopy. This combination has been used to study the diffusion of a variety of fluorescent proteins along the DNA (Gorman and Greene, 2008), including the DNA repair proteins Msh2–Msh6 (Gorman et al., 2007), p53 (Leith et al., 2009), and the DNA polymerase processivity factor PCNA (Kochaniak et al., 2009). In combination with flow stretching, the surface-tethered DNA approach has been used to visualize the dynamics of DNA replication (for phage T7 and *Escherichia coli*) channel without unbound fluorophores (Hamdan et al., 2009; Tanner et al., 2008). In two pioneering studies, it has been demonstrated that a wide-field fluorescence method like TIRF can be successfully be combined with optical trapping (Ishijima et al., 1998, Lang et al., 2003). The challenge of studying the fluorophores close to the surface has been addressed, in both cases, by attaching the molecule of interest to the surface. Lang et al. studied the disruption dynamics of a primer attached to a template that is anchored to the surface by single-molecule FRET, while Ishijima et al. attached myosin to the surface, and used two optical traps and a special design of their flow cell to bring an actin filament into contact with myosin.

In general, however, it is advantageous to position the DNA far away from the surface to avoid interactions of DNA or proteins with the glass and to reduce background fluorescence (van Mameren et al., 2008). One way to achieve this is to attach DNA to one or two optically trapped beads and to position it tens of micrometers away from the surface. Wide-field epi-illumination can then be used for fluorescence excitation. When only one bead is attached to the DNA, buffer flow is required to stretch the DNA. This approach has been pioneered by Kowalczykowski et al. to study RecBCD, a bacterial helicase/nuclease. The translocation of this enzyme was monitored by visualizing the removal of an intercalating dye from the DNA (Bianco et al., 2001) or alternatively by attaching a streptavidin-coated fluorescent bead to RecBCD (Handa et al., 2005). This method, however, requires a continuous buffer flow that can interfere with the dynamics of the protein–DNA interaction (van den Broek et al., 2008). This limitation is not present when both ends of the DNA are attached to beads that are controlled with either two optical traps or with an optical trap and a micropipette (Wuite et al., 2000a). Recent examples of combining fluorescent techniques with optical tweezers are the study of individual quantum-dot-labeled restriction enzymes (such as EcoRV) interacting with double-stranded (ds) DNA

(Biebricher et al., 2009) and an investigation of the interaction and dynamics of fluorescently labeled Rad51 with dsDNA (van Mameren et al., 2009a,b).

In this chapter, we provide background and practical information on how to perform experiments combining single-molecule fluorescence, optical trapping, and microfluidics. In Section 2, we discuss the background of the experimental approach and provide a detailed description of the setup we use. In Section 3 we describe several approaches for the functionalization of DNAs and proteins. In Section 4 we provide several exemplary experimental protocols illustrating how the combination of these techniques allows visualization of DNA and of proteins binding to DNA.

2. INSTRUMENTATION

In this section we focus on the general principle and technical considerations of combining optical trapping, fluorescence microscopy, and microfluidics. In addition, we describe in detail the experimental implementation of such a combined approach based on the instrument we have built in our laboratory (Fig. 17.1). This instrument is constructed around a commercial inverted microscope (Eclipse TE2000-U, Nikon), which provides the mechanical stability for key optical elements, like the microscope objective. Moreover, it allows flexibility in the choice of the dichroic mirrors. The microscope is equipped with a stage riser kit such that two dichroic mirror turrets can be operated independently.

2.1. Optical tweezers

2.1.1. Principles of optical trapping

Optical trapping is based on the transfer of momentum from photons to a transparent object immersed in a medium with a different refractive index. High photon-flux sources like lasers can provide sufficient momentum transfer to manipulate micro-sized polystyrene or silica spheres, as demonstrated by Ashkin in 1970 (Ashkin, 1970; Ashkin et al., 1986). A stable optical trap requires a potential minimum, such that a small excursion results in a restoring force back to the center of the trap. In order to understand this idea, it is illustrative to separate the effect of the momentum transfer of photons into two components: the gradient force, which attracts the microbead toward the focus (Fig. 17.2), and the scattering force, which acts in the direction of light propagation. In most cases the scattering force will dominate over the gradient force. If, however, the laser beam is tightly focused, the intensity gradient around the laser focus becomes steep. In such a situation, close to the tight focus, the gradient force component surpasses the scattering force, which leads to a stable optical trap (Ashkin, 1992) (Fig. 17.2). Using

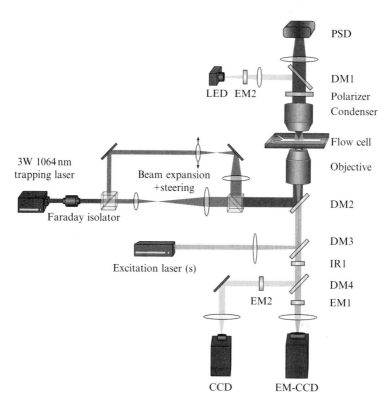

Figure 17.1 Schematic of the combined dual trap and fluorescence microscope with an integrated flow system. Each module (optical trapping, fluorescence microscopy, microfluidics) is described in detail in the designated section. Following abbreviations are used: DM, dichroic mirror; EM, emission filter; IR, infrared short-pass filter; PSD, position-sensitive detector; LED, light-emitting diode; (EM-)CCD, (electron-multiplied)-charged coupled device.

micron-sized beads, optical traps with a spring constant on the order of 250 pN/μm can be generated with a tightly focused laser beam of around 1 W. In this configuration, forces of 100 pN and more can be applied to the trapped bead (Bockelmann et al., 2002). Such forces are sufficient to study unfolding of proteins (Kellermayer et al., 1997), characterize the stall forces of molecular motors, or induce structural rearrangements in DNA.

2.1.2. Light sources for optical tweezers

To create stable optical traps for biological studies, good quality laser sources are required with respect to beam-pointing, wavelength, and power stability. A major cause of beam pointing instabilities is mechanical instability of the laser cavity due to thermal drift. Many modern high-power lasers have

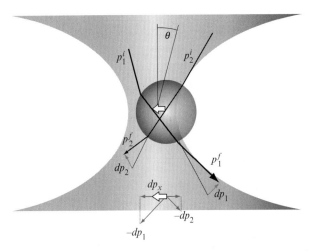

Figure 17.2 Qualitative picture of the origin of the trapping force. The deflection of photons caused by a difference in the index of refraction transfers a net momentum on the microbead, which causes a restoring force back to the focus of the laser beam. Two photon paths, p_1 and p_2, are compared. The momentum balance $dp = dp^i - dp^f$ for two beam paths is depicted in green, and shows that for a microbead that is slightly off the optical axis, a net restoring force, caused by dp_x, back to the center of the trap is generated.

active temperature control to increase stability. Beam pointing instabilities and intensity fluctuations can also be caused by feedback into the laser cavity due to back-reflections. This can be avoided by inclusion of a Faraday isolator in the optical path, immediately after the laser (Fig. 17.1) (IO-5-λ-HP, Optics for Research). A third cause of beam pointing instabilities is fluctuation in air density, caused by air flow. Air flow can result in local refractive index fluctuations, altering the direction of the laser beam. A straightforward way to suppress this unwanted effect is to enclose the laser beam and part of the optics in tubes and boxes.

The choice in wavelength is mainly governed by the inflicted photodamage on biological material, which is minimal in the near-infrared in a spectral window between the absorption of water and proteins ($\sim$800–1200 nm) (Neuman *et al.*, 1999; Peterman *et al.*, 2003a). For optical traps that can generate forces in the range of 100 pN, continuous wave (cw) laser powers in the order of 1 W are required, which can be provided by various laser types. Fiber lasers can produce very high powers (tens of Watts). In addition, the use of a doped single-mode fiber as the active medium/waveguide results in excellent laser beam quality. Another class of lasers are diode pumped solid-state lasers, like Nd:YAG and Nd:YVO$_4$ lasers. These lasers have a compact design, resulting in good (beam pointing) stability and can be obtained with output powers up to ~10 W. In our instrument we use a 3 W Nd:YVO$_4$ laser (Ventus 1064 nm, 3 W cw, Laser Quantum).

2.1.3. Microscope objectives

A stable optical trap requires a tight focus, since photons entering the focus under a high angle θ to the optical axis (Fig. 17.2) contribute the most to the restoring force. Such a tight focus can be generated with a high numerical aperture (NA) objective. The NA of an objective describes the range of angles over which it transmits (or accepts) light and is related to the maximal angle θ_{max} as

$$NA = n \sin(\theta_{max}). \quad (17.1)$$

Here, n is the index of refraction of the immersion medium between the sample and the objective lens. For an oil immersion objective, a typical value of the NA is 1.4, corresponding to a maximum angle of about 130°. Oil immersion objectives suffer, however, from spherical aberrations, resulting in a severely distorted focus when trapping further than 10 μm away from the glass surface (Vermeulen et al., 2006). Water immersion objectives suffer far less from spherical aberrations and are better suited for trapping deep in an aqueous solution than oil immersion objectives, despite their lower NA (typically NA = 1.2). Another important aspect of an objective is its transmission for the infrared light used for trapping. The transmission of some objectives is very low at wavelengths above 950 nm, but many transmit around 70% (Neuman and Block, 2004). In our setup, we generate the optical traps using a 60×, NA = 1.2 water immersion objective (Plan Apo, Nikon).

2.1.4. Generation and steering of a dual trap

In our setup, two optical traps are produced by splitting the 1064 nm laser into two beams with identical power using a polarizing beam splitter cube (10BC16PC.9, Newport). We expand each beam with a 1:2.67 telescope system, which ensures that both laser beams overfill the back-focal plane of the objective. The beams are recombined using another polarizing beam splitter cube and coupled into the objective via a dichroic mirror ("DM2": 950dcsp, Chroma Tech Corp.) (Fig. 17.1).

A dual trap design is capable of reaching very high stability and resolution, as was demonstrated, in both theoretical and experimental studies (Abbondanzieri et al., 2005; Moffitt et al., 2006). The dual trap design is very stable if the differential pathway between the two beams is minimized, reducing differences in air density fluctuations (Bustamante et al., 2008). The effect of beam pointing instabilities in such a design can be suppressed by measuring the difference coordinates of both trapped particles (Moffitt et al., 2006).

In such a dual trap design, care has to be taken that parasitic depolarization effects due to highly curved surfaces in the objective do not lead to interference and crosstalk between the two traps, which would complicate accurate position detection (Mangeol and Bockelmann, 2008). Polarization rectification and frequency shifting of one laser can be used as an effective method to circumvent this problem (Mangeol and Bockelmann, 2008).

In order to manipulate trapped microspheres, a precise and fast method to laterally displace the optical traps is required. This displacement can be achieved by introducing an angular deflection of the direction of the laser beam in the back-focal plane of the objective, which translates into a change in the lateral position in the focal plane of the objective (Fig. 17.3). In our system, using a 60× Nikon objective with focal length 3.33 mm, an angular variation of 1 mrad of one of the two trapping laser beams in the back-focal plane results in a relative displacement of the two traps of around 3 μm.

Several methods can be used to achieve this angular deflection: translation of a moveable lens in a telescope system (Svoboda and Block, 1994), acousto-optic deflectors (AODs) (Noom et al., 2007; Vermeulen et al., 2006), electro-optic deflectors (EODs) (Valentine et al., 2008), or scanning mirrors (Moffitt et al., 2006). All of these methods have their strengths and limitations. In an AOD, a standing ultrasound wave in a crystal generates a modifiable diffraction grating, resulting in a maximal deflection of about 15 mrad (corresponding to a displacement of ∼50 μm in the image plane). The scanning speed of the diffracted laser beam is in principle only restricted to the speed of sound in the crystal. On the downside, the limited diffraction efficiency leads to light losses of around 60%, and nonlinear response of the AOD leads to angular crosstalk (Valentine et al., 2008). EODs offer a fast scanning speed, and suffer less from transmission losses and nonlinear response than AODs. However, their scan range is limited to only ∼1.5 mrad, corresponding to a displacement of ∼5 μm in the image plane. Scanning (piezo or galvano) mirrors offer no light loss and deliver an intermediate scanning range and speed. A final method is to translate a telescope lens. This method is relatively simple to implement, causes no

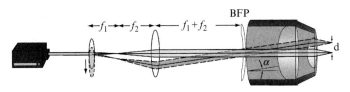

Figure 17.3 Steering of a laser trap using a telescope system with the focal lengths f_1 and f_2. Lateral movement of the first telescope lens introduces an angle α in the beam path. This angular deflection in the back-focal plane (BFP) of the objective translates into the displacement "d" in the focal plane.

light loss and can result in a large scan range. A disadvantage is that scanning speeds are limited to tens of micrometers per second. In our instrument, we have chosen translation of a telescope lens as beam steering method: The first telescope in one beam path can be displaced in two directions using two linear actuators (T-LA28, Zaber Technologies Inc.), allowing positioning of that trap ("moveable trap") in the sample chamber as illustrated in Fig. 17.3.

2.1.5. Laser detection and optical trap calibration

For high-resolution force and distance detection with optical tweezers, accurate position detection of the trapped particles is imperative. The easiest way to achieve this is using back-focal-plane interferometry (Gittes and Schmidt, 1998). In this method, the interference signal of the laser beam with light scattered from the microsphere is measured using a quadrant diode or position-sensitive detector, onto which the back-focal plane of the condenser is imaged (Neuman and Block, 2004). These detectors have a high temporal bandwidth, which is important for detector calibration (see below). In our setup, we use a DL-100-7PCBA (Pacific Silicon Sensor Inc.) position-sensitive detector to measure displacements of a microsphere in the stationary trap, while blocking the light of the moveable trap with a polarizer (03 FPI 003, CVI Melles Griot) (Fig. 17.1). We align the angular orientation of this polarizer by minimizing the total photocurrent of the position-sensitive detector for the moveable trap, while the stationary trap is disabled. Note that nonuniform rotation of the polarization of the laser beam on highly curved surfaces like the objective lens prevents full attenuation. A systematic investigation of this parasitic depolarization can be found in Mangeol and Bockelmann (2008).

The signal from the position-sensitive detector is digitized using a 24-bit A/D converter (NI-PCI-4474, National Instruments). To obtain values for displacement of the bead from the center of the trap and the corresponding force acting on it, the detector needs to be calibrated and the trap stiffness determined. This is achieved by measuring the effect of the optical trap on the Brownian motion of the microbead. The Brownian motion of a microbead in an optical trap can be described analytically and depends on two parameters: the hydrodynamic friction coefficient γ and the trap stiffness k. The equation of motion can be expressed in a Langevin equation:

$$\gamma \frac{dx}{dt} + kx = F(t). \tag{17.2}$$

The external, random force $F(t)$, which generates the Brownian motion, is described by the equipartition theorem, which states that all degrees of freedom receive the same thermal energy $|F(f)| = 4\gamma k_b T$. This implies oxygen scavenging system. It has been demonstrated that photobleaching is substantially enhanced in the proximity of optical traps (Lang et al., 2003). Dyes in their excited state, even with a short excited state lifetime of ∼ns, can absorb an infrared photon due to the enormous photon flux close to the traps. In this higher excitation state, the dye can oxidize (van Dijk et al., 2004). This effect can be partly overcome by the use antioxidants and by the choice of suitable fluorophores (van Dijk et al., 2004). Other approaches include spatial (Harada et al., 1998) or temporal separation of trapping light and fluorescence excitation (Brau et al., 2006).

Typically, fluorophores in aqueous solution photobleach after emitting 10^5–10^6 photons (Schmidt et al., 2002; van Dijk et al., 2004). Consequently, one cannot continuously follow fluorescent dyes over timescales longer than several minutes. A way to circumvent this problem is to use time-lapsed data acquisition, that is, fluorescence excitation and data acquisition are only initiated for a short time interval, τ_1, with a longer waiting interval τ_2, where the sample is not excited (Windoffer and Leube, 1999). The illumination time window τ_1 is governed by illumination conditions, while the choice of τ_2 requires some *a priori* knowledge of the timescale of the molecular process to be investigated, and needs to be iteratively optimized. We, however, present methods to visualize proteins or dyes that nonspecifically bind to DNA. In such a situation, fluorescent images rely on the emission of several fluorophores (typically ∼ 10–100). Consequently, then it is possible to observe the DNA molecule significantly longer than the bleaching time of the individual fluorescent dyes.

2.3. Microfluidics

Single-molecule studies of DNA–protein interactions require precise temporal and spatial control over the chemical environment of the observed DNA molecule. Microfluidic flow cells provide relatively simple and versatile means to this end. Various types of microfluidic flow cells differ by the number of channels (single or multistream flow cell), the material used to construct the flow cell, and the fluid delivery system. The design, building techniques and applications of various flow cells have been reviewed recently by Brewer and Bianco (2008).

For our experiments, we use a microfluidic system with multiple laminar channels. In such a flow chamber, the different input channels converge into a larger common channel where the experiment takes place (Fig. 17.6). In a laminar flow, the adjacent streams of the different fluids only slowly mix upon merging of the channels (Brewer and Bianco, 2008). This feature allows rapid switching of buffer conditions by simply moving the microscope stage in a direction perpendicular to the flow so that the optically trapped molecule(s) move into a different buffer stream. Moreover, it permits exclusion of fluorescent proteins from the detection area.

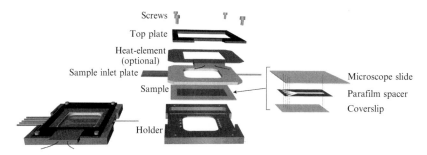

Figure 17.6 Schematic of the laminar multichannel flow system. The disposable sample chamber consists of a channel pattern cut out of parafilm and melted between a glass microscope slide with holes (emphasized by dashed lines) for fluid influx and a thin cover slip (shown on the right). The assembled flow chamber is placed between a holder and a top plate and is connected to incoming and outgoing tubes by a sample inlet plate.

The flow channels in our system are manually cut in a sheet of Parafilm® (Wuite et al., 2000a), serving as a spacer, and sandwiched between a cover slip (24 × 60 mm #1) and a microscope slide (50 × 75 × 1 mm) with drilled holes allowing fluid influx. Before assembly, the glass slides and cover slips are extensively cleaned as follows: (i) acetone washing, (ii) 70% (v/v) ethanol/water washing, (iii) extensive rinsing with deionized water, and (iv) plasma cleaning for 10 min using the plasma cleaner from Harrick Plasma, operating with air of a pressure of about 0.4 mbar. The flow cell is generated by applying heat (~ 130 °C for 15 min) to the glass–parafilm coverslip sandwich by placing a heat block on top of the flow cell, which is situated in a metallic mold. To bring the different liquids into the flow cell, we use a pressurized buffer container, as depicted in Fig. 17.7. To decrease the amount of photobleaching, we keep the pressure chamber in a nitrogen atmosphere, therefore reducing the concentration of dissolved oxygen. We control the pressure in our container by two solenoid valves, where the first is connected to a pressurized nitrogen line, while the other reduces the overpressure by releasing nitrogen out of the container. Generation of flow rates in the order of about 10 μm/s typically requires an overpressure of ~ 100 mbar in our pressure chamber (Fig. 17.7).

In the following we will discuss several complications that can be encountered using such a flow-system.

1. The presence of air bubbles in the flow cell can cause flow perturbation, due to local compression and expansion of the air. Pressure spikes can be high enough to displace the microspheres out of the optical trap. Increasing the flow rate for a short time, or flushing the flow chamber with 70% (v/v) ethanol is usually sufficient to remove air bubbles. In case, higher flow rates are required to flush out air bubbles, the flow

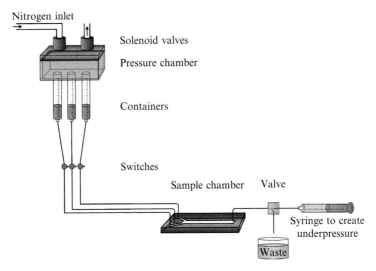

Figure 17.7 Schematic of the flow system. Sample solutions are stored in containers. Solenoid valves control the nitrogen pressure in the pressure chamber. Switches enable quick initiation/stop of each laminar flow channel in the sample chamber. In case of undesired air bubbles getting stuck in the sample chamber, a negative pressure can be manually applied by a syringe, assisting in removing air bubbles in the sample chamber.

rate can be further increased by using a syringe to create a negative pressure at the exit of the flow chamber (Fig. 17.7). In this way, the pressure applied to the flow chamber can be kept low enough to not destroy the flow chamber. Usually, we restrict the maximum overpressure in the pressured container storing the reagents to a maximum of ~ 250 mbar.

2. In general, it is imperative to work with a clean flow cell, tubing and solutions, as small dust particles, bacteria or other contaminations can enter the optical trap and disturb the experiments. Therefore, all solutions need to be filter sterilized (0.22 μm) and flow cells are replaced typically after 1–2 weeks of experiments, or sooner when a new protein with the same fluorescent label is studied. Furthermore, when exchanging the flow cell, the tubing is detached from the flow cell, which enables cleaning of the detached tubing by flushing with 70% (v/v) ethanol under high pressure (typical volumes: 1–5 ml).

3. In general, the high surface-to-volume ratio of the microfluidics system complicates experiments that require low protein concentrations (<1 nM). Under these conditions, adsorption on the surface of the tubing and the flow cell can cause protein depletion. For such low protein concentrations, adsorption to the glass surface should be attenuated by surface passivation using BSA, casein, or other commercially available blocking reagents (for instance, Blocking Agent from Roche).

3. Preparation of Reagents

In this section we describe strategies and protocols for the generation of DNA molecules possessing end labels that enable the linking to optically trapped beads. Furthermore, we give an overview over dyes used to fluorescently stain individual DNA molecules. Finally, we present general considerations and a protocol to fluorescently label proteins.

3.1. Terminally labeled DNA molecules

Choosing the DNA template for a single-molecule experiment depends on various constraints such as the specific nucleotide sequence required for a biological process, the location of the labeling (3'- vs. 5'-end labeling), the shape (DNA forks, overhangs, dsDNA, ssDNA), and the size of the DNA molecule (i.e., number of base pairs). The latter defines the maximum distance between the trapped beads, which is an important parameter in single-molecule experiments. If the DNA molecule is too short, the two traps are in close proximity, which can result in nonnegligible interference and crosstalk between them (Mangeol and Bockelmann, 2008). Moreover, for the combination of trapping and fluorescence detection of proteins on DNA, the high photon flux in the vicinity of the traps can damage biological samples and enhance photobleaching of the fluorophores (van Dijk *et al.*, 2004). DNA molecules used in experiments combining optical trapping and fluorescence are typically between 5 and 50 kilobases (kb) in length.

A dsDNA molecule widely used in single-molecule experiments is bacteriophage lambda DNA, a linear dsDNA of 48,502 base pair (bp) with two complementary 12-nucleotide single-stranded 5'-overhangs. In order to allow specific binding of the DNA to the streptavidin-coated polystyrene beads (1.87 μm diameter, Spherotech), the extremities of the DNA molecules are labeled with biotinylated nucleotides (see protocol below).

We have recently described a protocol to generate ssDNA for optical tweezers by force-induced DNA denaturation (Gross *et al.*, in preparation). For this, we use the linear, terminally biotin-labeled expression vector pTR19-ASDS (10,729 bp) as DNA template. The ends of this vector are labeled with biotinylated nucleotides such that only one of the strands is physically linked to the two microspheres. This dsDNA molecule is denatured by exerting forces that exceed the overstretching force of 65 pN. See Gross *et al.* (in preparation) for a more detailed protocol.

3.1.1. Example protocol I: Biotinylation of lambda DNA
Biotinylated nucleotides are incorporated into the 5'-single-stranded overhangs of lambda DNA by Klenow DNA polymerase exo$^-$:

1. Lambda DNA (4 nM) is incubated with dTTP (100 μM), dGTP (100 μM), biotin-14-dATP (80 μM), biotin-14-dCTP (80 μM), and Klenow DNA polymerase exo$^-$ (0.05 units/μl) at 37 °C for 30 min.
2. The enzymatic reaction is stopped by heating to 70 °C for 15 min.
3. The DNA molecules are purified with a GFX PCR DNA and Gel Band Purification Kit (GE Healthcare).

This labeling reaction results in a DNA molecule containing multiple biotin labels at the 5'-end of each strand.

3.2. Fluorescent labeling of DNA

Fluorescent intercalating dyes are frequently used to directly visualize dsDNA. A number of dimeric cyanine intercalating dyes, covering the entire visible spectrum, are commercially available, for example, TOTO-1, YOYO-1, and POPO-1 (Molecular Probes). These dyes strongly bind to dsDNA and show a 100- to 1000-fold enhancement of their fluorescence quantum yield upon intercalating between the base pairs of nucleic acids. As a result, the background fluorescence from unbound dyes is extremely low, making these dyes ideal for imaging of individual DNA molecules (Netzel et al., 1995). A protocol describing the labeling of a DNA molecule during a single-molecule experiment by the cyanine dye YOYO-1 is provided in Section 4.1. It should be noted that the intercalation of such dye molecules changes the structural properties of the DNA molecule, making it longer and stiffer (Quake et al., 1997), therefore potentially interfering with some biochemical processes. As an alternative to such dyes, fluorophores can be covalently attached to the DNA molecule. For instance, the combined use of Cy5-labeled nucleotides and Cy3-labeled primers in a polymerase chain reaction (PCR) amplification allows introduction of both internal and end-specific fluorescent labels in the DNA molecule (Chan et al., 2006). However, this method can be used only for short and medium length DNA molecule ($\leq$20 kb) due to DNA length limitations of PCR. A broad and detailed description of fluorescent probes for nucleic acids can be found in Wang et al. (2003).

3.3. Fluorescent labeling of proteins

Methods to fluorescently label proteins for optical tweezers experiments can be divided broadly into two categories: (i) the covalent binding of small organic fluorophores to the protein and (ii) the molecular tagging by fusing the gene of a fluorescent protein to the coding sequence of the protein of interest. For both techniques, the ideal fluorophore has high photostability, high fluorescence quantum yield, a large absorption cross section, and shows little intensity fluctuation (Moerner and David, 2003; Peterman et al., 2004).

Most small organic fluorophores with a high photostability, used for covalent labeling of proteins belong to the cyanine and rhodamine families of fluorophores, along with derivatives of fluorescein with an enhanced photostability, like Alexa 488. Among the cyanine dyes, Cy3, Cy5, Alexa 555, and Alexa 647 are extensively used because of their high photostability (reviewed in Berlier et al., 2003). In the rhodamine family of fluorophores, tetramethyl rhodamine (TMR), rhodamine 6G, Texas Red, and some of the Alexa dyes (Alexa 546) are commonly used. The main targets for chemical labeling are the primary amine groups found on lysine residues and at the N-terminus of proteins. However, even if the N-terminus of the protein is preferentially modified by performing the labeling reaction at near neutral pH, such labeling often results in a heterogeneous population of proteins containing a variable number of fluorophores if the target protein contains multiple lysine residues. Alternative targets for labeling are the sulfhydryls of cysteine residues, which are less abundant than primary amines. In case multiple or no sulfhydryls are available on the surface of the protein of interest, cysteine residues may be removed from or added to a protein by point mutation. Hence, cysteine labeling allows for more specific labeling than amine-based labeling.

The use of genetically fused fluorescent proteins such as green fluorescent protein (GFP) ensures a high labeling specificity and avoids protein modification and labeling steps. In addition, a wide spectrum of autofluorescent proteins with high fluorescence quantum yield is now available (for a detailed comparison of fluorescent proteins, see Shaner et al., 2005). However, the fluorescent proteins commonly used are not very photostable and frequently exhibit blinking behavior. Moreover, they are relatively large (~ 27 kDa for GFP). This may cause steric hindrance that can potentially interfere with the function of the target protein. It is, therefore, essential to test for enzymatic activity after labeling the protein of interest.

4. COMBINING OPTICAL TRAPPING, FLUORESCENCE MICROSCOPY, AND MICROFLUIDICS: EXAMPLE PROTOCOLS

Here, we describe a protocol to isolate an individual DNA molecule and successively visualize it using DNA-specific dyes. In addition, we show example experiments focusing on the specific binding of proteins to ssDNA and dsDNA. Furthermore, we demonstrate the power of the combination of optical trapping, fluorescence microscopy, and microfluidics for the reduction of fluorescent background.

4.1. Example protocol II: Sequential isolation and visualization of a single DNA molecule using YOYO-1

In the beginning of an experiment, the different components (beads, DNA, buffer, YOYO-1) are each introduced into a separate channel of the multichannel flow cell and the flow is initiated. In the flow cell the different solutions do not mix because of the laminar flow conditions (Fig. 17.8). A typical buffer for force–extension measurements on DNA contains: 10 mM Tris, pH 7.8, 50–150 mM NaCl. Before inserting the buffer into our flow system, we filter the buffer with a 0.2 μm cutoff filter.

To isolate and visualize a single DNA molecule the following steps are taken:

1. *Trapping of two beads*. This is achieved by turning on the trapping laser in the flow channel containing streptavidin-coated polystyrene beads (stock concentration: 0.5%, w/v) diluted 1/1000 in buffer (a typical buffer for experiments on DNA would be 10 mM Tris–HCl, pH 7.8, and 50–150 mM NaCl). Two beads are usually trapped within a few seconds and their size and shape are visually checked to ensure that only one bead is present in each trap and that no contamination is present.
2. *"Catching" a single DNA molecule*. This is accomplished, in the presence of flow, by moving one bead back and forth around the position where the end of the flow-stretched DNA is anticipated. The presence of a DNA molecule is confirmed when, upon moving the beads with the trap, a correlated movement of the other bead is observed. In a picomolar DNA concentration the DNA–bead attachment usually occurs within a few seconds.

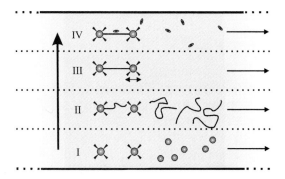

Figure 17.8 Principle of the multiple-channel laminar flow chamber. I. *Bead channel*: two beads are optically trapped; II. *DNA channel*: a single DNA molecule is caught between the two beads; III. *Buffer channel*: a force–extension measurement of the DNA molecule is performed; IV. *Protein channel*: the DNA molecule is incubated with the studied protein or dye.

3. *Force–extension analysis of the trapped DNA.* After "catching" a DNA molecule, the beads are moved into a channel containing only buffer. Using force–extension analysis, we ensure that only one DNA molecule connects the two beads (Fig. 17.9). When dealing with a single DNA molecule, the force–extension curve shows some very distinct features such as the cooperative structural transition at 65 pN, which yields a lengthening of the DNA up to a contour length of 1.7 its unstretched value without significant increase of force (Smith et al., 1996, van Mameren et al., 2009a). Attachment of multiple DNA molecules raises this transition to multiples of 65 pN.
4. In order to visualize the DNA, it is stained with YOYO-1. To this end, the DNA in the buffer channel is first extended to about 90% of its contour length to reduce spatial fluctuations of the DNA molecule and allow for sharp imaging. Next, the DNA is moved to the channel containing 20 nM YOYO-1, 10 mM Tris–HCl, pH 7.8, and 50 mM NaCl. After 5–10 s incubation, the sample is illuminated with the 473 nm excitation laser and a snapshot is taken (Fig. 17.10). The bright fluorescence signal of the dye and the low background allows for a visualization of the DNA molecule, even in a flow channel containing unbound dye in solution. It should be noted that intercalators like YOYO-1 induce photocleavage of DNA. Moderate staining rations ($\sim$1 dye every 500 bp) and low illumination

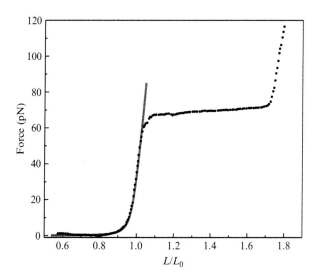

Figure 17.9 Elastic behavior of a single lambda DNA molecule. The gray trace represents the worm-like chain model for the elastic properties of DNA with a persistence length of 55 nm and a stretching modulus of 1350 pN. The distance is normalized to the contour length L_0 of the DNA molecule. At 65 pN the DNA molecule undergoes a structural transition, called overstretching transition.

Figure 17.10 Fluorescent image of the intercalating dye YOYO-1 bound to lambda DNA. The measurement was taken 10 s after the DNA was transferred to a buffer containing 20 nM YOYO-1. A small force (~ 5 pN) suppresses fluctuations during the data acquisition time (0.5 s). A 488 nm laser excites the dye with a power of ~ 3 mW in wide-field excitation, over an area with a diameter of ~ 35 μm.

conditions can reduce this effect. Under such conditions, we can continuously observe DNA on the timescale of a minute, before photocleavage breaks the DNA molecule.

4.2. Example protocol III: Binding of replication protein A to ssDNA

Here, we present an example illustrating the visualization of a DNA–protein interaction: the binding of human replication protein A (RPA) to ssDNA. RPA is a heterotrimeric ssDNA-binding protein that plays essential roles in many aspects of nucleic acid metabolism, including DNA replication and repair. RPA was fused here to the eGFP (kind gift of Mauro Modesti) (van Mameren *et al.*, 2009a). The experiment is conducted as follows:

1. Two beads are optically trapped and a single dsDNA molecule is caught between the beads (see steps 1 and 2, Section 4.1).
2. After "catching" a DNA molecule, the beads are moved into the channel containing only buffer. Next the dsDNA molecule is denatured into ssDNA by putting it under ~ 70 pN tension (see Section 3.1).
3. The traps are moved into the flow channel containing the fluorescent protein (20 nM RPA). After 10 s incubation of the ssDNA molecule with the fluorescent protein, the sample is illuminated with the excitation light (488 nm) and a snapshot is taken (Fig. 17.11A).
4. Alternatively, the traps are moved back to the buffer-only channel and a snapshot is taken in a buffer without fluorescent background (Fig. 17.11B). The comparison of Fig. 17.11A and B clearly illustrates the advantage of a rapid exchange of buffer conditions to drastically reduce the fluorescent background.

4.3. Example protocol IV: Binding of Rad51 to dsDNA

Another example demonstrating the visualization of DNA–protein interactions is the binding of fluorescently labeled Rad51 to dsDNA. Rad51 is an essential component of the eukaryotic homologous recombination system,

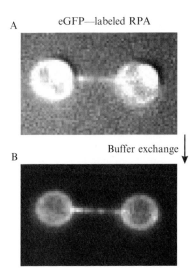

Figure 17.11 Visualization and background comparison of the fluorescent ssDNA-binding protein RPA bound to ssDNA. The GFP-labeled protein RPA, bound to ssDNA under a tension of 6 pN, was observed in (A) the laminar channel containing 20 nM fluorescently labeled RPA or in (B) a buffer containing no unbound fluorophores.

a crucial mechanism for the maintenance of genome integrity. Like other recombinases, Rad51 forms nucleoprotein filaments on both ssDNA and dsDNA. The protein was labeled with Alexa Fluor 555 as described in van Mameren et al. (2006). In this assay, the DNA was preincubated with the fluorescently labeled Rad51 in a buffer containing 1 mM ATP, 30 mM KCl, and 2 mM CaCl$_2$. The latter strongly reduces dissociation of the protein from the DNA. To obtain a patch-wise coverage of the DNA molecule with Rad51 filaments, the stoichiometry of Rad51 proteins versus base is adjusted. The experiment is carried out as follows:

1. Two beads are optically trapped, and a single dsDNA molecule, pre-coated with several Rad51 filaments, is caught between the beads (see steps 1 and 2, Section 4.1).
2. The DNA is brought to the buffer channel and illuminated with the excitation light (532 nm). Here, the fluorescence image and the force exerted on the DNA molecule are recorded simultaneously.
3. The distance between the two traps is then gradually increased, thus exerting a tension on the DNA molecule.

The fluorescence imaging shows an alternation of bare DNA and protein-coated DNA regions (Fig. 17.12), thus providing information on the mechanism of filament formation. At the same time, the force–extension

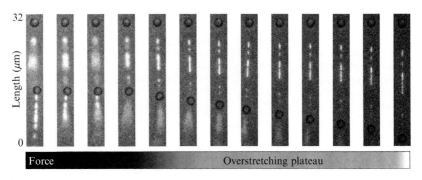

Figure 17.12 Patch-wise formation of fluorescently Rad51 nucleoprotein filaments on dsDNA. One DNA molecule is suspended between two optically trapped microspheres and a second molecule is tethered from the lower bead. By increasing the distance between the traps, tension can be applied to the suspended DNA in a controlled manner (image from van Mameren et al., 2006).

measurements give information on the elastic properties of the heterogeneously coated DNA molecules. In addition, it can be seen that at low tensions, the DNA molecule significantly fluctuates during the camera integration time, causing a blur in the fluorescence image of Fig. 17.12.

5. Conclusions

In this chapter we have presented a detailed description of our instrument that combines optical trapping, fluorescence microscopy, and microfluidics. This instrument grants access to highly complementary information on DNA–protein interactions from optical tweezers and fluorescence microscopy. For example, the fluorescence intensity provides for a direct way to determine the amount of protein bound to a DNA molecule. Spatial information of protein binding on an individual DNA molecule allows for studies of specific binding or cooperatively in binding and unbinding. Moreover, the incorporation of multiple excitation lasers combined with multiband emission filters enables the use of multiple color fluorescence detection, which permits observation of colocalization of different proteins. Additionally, dynamical processes like diffusion, association/dissociation, and directed motion can be directly monitored. Taken together, the incorporation of optical tweezers with fluorescence microscopy and microfluidics combines the most powerful single-molecule manipulation and visualization techniques, and allows for high data throughput due to the ability to rapidly switch the chemical environment. This combination is widely applicable to the study of a vast number of different DNA–protein interactions.

ACKNOWLEDGMENTS

We thank J. van Mameren for kindly providing the data for Figure 17.12. This work is part of the research program of the Stichting voor Fundamenteel Onderzoek der Materie, which is financially supported by the Nederlandse Organisatie voor Wetenschappelijk Onderzoek. P. G. is supported by Atlas, a European Commission-funded Marie Curie early stage training network. This work is supported in part by an Echo and a VICI grant of the Nederlandse Organisatie voor Wetenschappelijk Onderzoek.

REFERENCES

Abbondanzieri, E. A., Greenleaf, W. J., Shaevitz, J. W., Landick, R., and Block, S. M. (2005). Direct observation of base-pair stepping by RNA polymerase. *Nature* **438**, 460–465.

Ashkin, A. (1970). Acceleration and trapping of particles by radiation pressure. *Phys. Rev. Lett.* **24**, 156–159.

Ashkin, A. (1992). Forces of a single-beam gradient laser trap on a dielectric sphere in the ray optics regime. *Biophys. J.* **61**, 569–582.

Ashkin, A., Dziedzic, J. M., Bjorkholm, J. E., and Chu, S. (1986). Observation of a single-beam gradient force optical trap for dielectric particles. *Opt. Lett.* **11**, 288.

Berlier, J. E., Rothe, A., Buller, G., Bradford, J., Gray, D. R., Filanoski, B. J., Telford, W. G., Yue, S., Liu, J. X., Cheung, C. Y., *et al.* (2003). Quantitative comparison of long-wavelength Alexa Fluor dyes to Cy dyes: Fluorescence of the dyes and their bioconjugates. *J. Histochem. Cytochem.* **51**, 1699–1712.

Bianco, P. R., Brewer, L. R., Corzett, M., Balhorn, R., Yeh, Y., Kowalczykowski, S. C., and Baskin, R. J. (2001). Processive translocation and DNA unwinding by individual RecBCD enzyme molecules. *Nature* **409**, 374–378.

Biebricher, A., Wende, W., Escude, C., Pingoud, A., and Desbiolles, P. (2009). Tracking of single quantum dot labeled EcoRV sliding along DNA manipulated by double optical tweezers. *Biophys. J.* **96**, L50–L52.

Bockelmann, U., Thomen, P., Essevaz-Roulet, B., Viasnoff, V., and Heslot, F. (2002). Unzipping DNA with optical tweezers: High sequence sensitivity and force flips. *Biophys. J.* **82**, 1537–1553.

Brau, R. R., Tarsa, P. B., Ferrer, J. M., Lee, P., and Lang, M. J. (2006). Interlaced optical force-fluorescence measurements for single molecule biophysics. *Biophys. J.* **91**, 1069–1077.

Brewer, L. R., and Bianco, P. R. (2008). Laminar flow cells for single-molecule studies of DNA–protein interactions. *Nat. Methods* **5**, 517–525.

Bustamante, C., Chemla, Y. R., and Moffitt, J. R. (2008). High-resolution dual-trap optical tweezers with differential detection. *In* "Single-Molecule Techniques: A Laboratory Manual," (Paul R. Selvin and Taekjip Ha, eds.), pp. 297–324. CSHL Press, New York.

Chan, T.-F., Ha, C., Phong, A., Cai, D., Wan, E., Leung, L., Kwok, P.-Y., and Xiao, M. (2006). A simple DNA stretching method for fluorescence imaging of single DNA molecules. *Nucl. Acids Res.* **34**, e113.

Gittes, F., and Schmidt, C. F. (1998). Interference model for back-focal-plane displacement detection in optical tweezers. *Opt. Lett.* **23**, 7–9.

Gorman, J., and Greene, E. C. (2008). Visualizing one-dimensional diffusion of proteins along DNA. *Nat. Struct. Mol. Biol.* **15**, 768–774.

Gorman, J., Chowdhury, A., Surtees, J. A., Shimada, J., Reichman, D. R., Alani, E., and Greene, E. C. (2007). Dynamic basis for one-dimensional DNA scanning by the mismatch repair complex Msh2–Msh6. *Mol. Cell* **28**, 359–370.

Gross, P., Farge, G., Geertsma, H., Hoekstra, T., Peterman, E. J. G., and Wuite, G. J. L. Generating single-stranded DNA for optical tweezers experiments. Manuscript in preparation.

6. Discussion 503
  6.1. Recommendations for data analysis 503
  6.2. Outlook: Fluorescence image spectroscopy 505
Acknowledgments 505
References 506

Abstract

In the recent decade, single-molecule (sm) spectroscopy has come of age and is providing important insight into how biological molecules function. So far our view of protein function is formed, to a significant extent, by traditional structure determination showing many beautiful static protein structures. Recent experiments by single-molecule and other techniques have questioned the idea that proteins and other biomolecules are static structures. In particular, Förster resonance energy transfer (FRET) studies of single molecules have shown that biomolecules may adopt many conformations as they perform their function. Despite the success of sm-studies, interpretation of smFRET data are challenging since they can be complicated due to many artifacts arising from the complex photophysical behavior of fluorophores, dynamics, and motion of fluorophores, as well as from small amounts of contaminants. We demonstrate that the simultaneous acquisition of a maximum of fluorescence parameters by multiparameter fluorescence detection (MFD) allows for a robust assessment of all possible artifacts arising from smFRET and offers unsurpassed capabilities regarding the identification and analysis of individual species present in a population of molecules. After a short introduction, the data analysis procedure is described in detail together with some experimental considerations. The merits of MFD are highlighted further with the presentation of some applications to proteins and nucleic acids, including accurate structure determination based on FRET. A toolbox is introduced in order to demonstrate how complications originating from orientation, mobility, and position of fluorophores have to be taken into account when determining FRET-related distances with high accuracy. Furthermore, the broad time resolution (picoseconds to hours) of MFD allows for kinetic studies that resolve interconversion events between various subpopulations as a biomolecule of interest explores its structural energy landscape.

1. INTRODUCTION

Over the last decades a vast, increasingly sophisticated, selection of fluorescence-based spectroscopic assays has been developed in order to study biological reactions and the characteristics of freely diffusing or immobilized biomolecules *in vitro* and in living cells (Giepmans *et al.*, 2006; Haustein and Schwille, 2004; Lippincott-Schwartz *et al.*, 2001; Rettig *et al.*, 1999; Valeur and Brochon, 2001). By utilizing the advantages of distinct fluorescence-based methods with respect to selectivity, sensitivity, and high spatial resolution, many questions on supramolecular assemblies and biological systems can be studied: (1) binding equilibria (Al-Soufi et al., 2005; Kask et al., 2000; Wang et al., 2009) and on/off rates (Gansen et al., 2009; Rothwell et al., 2003); (2) complex stoichiometries (Chen and Muller, 2007; Eggeling et al., 2005; Ulbrich and Isacoff, 2007); (3) biomolecular structures with respect to distances, domain orientation, conformational arrangement, dynamics (Doose et al., 2009; Laurence et al., 2005; Margittai et al., 2003; Rosenberg et al., 2005; Schröder et al., 2005), and folding (Michalet et al., 2006; Nettels et al., 2007; Schuler and Eaton, 2008); (4) arrangement and function of complex molecular machines (Blanchard, 2009; Majumdar et al., 2007; Mekler et al., 2002); (5) catalytic reactions (Diez et al., 2004; Joo et al., 2006; Mickler et al., 2009; Pandey et al., 2009); (6) biomolecular concentrations (Haustein and Schwille, 2007; Magde et al., 1972; Muller et al., 2005); and (7) spatially distributed reactions and local mobilities of molecules in complex environments (Dertinger et al., 2007; Digman et al., 2005a,b; Kudryavtsev et al., 2007; Ruan et al., 2004; Weidtkamp-Peters et al., 2009).

To answer these questions, the use of Förster resonance energy transfer (FRET) between two fluorophores has become increasingly popular (Clegg, 1992, 2009; De Angelis, 1999; Szollosi et al., 1998; Vogel et al., 2006; Wallrabe and Periasamy, 2005; Wu and Brand, 1994). FRET is a highly distance-dependent phenomenon that is particularly useful for probing interdye distances between 20 and 100 Å. An excited fluorophore (donor, D) transfers energy to another fluorophore (acceptor, A) when they are in close proximity. As an alternative to native fluorophores, such as tryptophan (Lakowicz, 2006), the green fluorescent protein (Giepmans et al., 2006), a wide range of small and photostable organic fluorophores (Gonçalves, 2009), and chemically activated nanoparticles (Bruchez et al., 1998; Curutchet et al., 2008) have been introduced. These fluorophores can be site-specifically covalently coupled to nucleic acids, proteins, and lipids. As the focus increasingly shifts from protein structure to function, improved labeling strategies have been developed to attach high-performance fluorescent tags to cysteines or to nonnatural amino acids of genetically engineered proteins (Wu and Schultz, 2009). Receptor–ligand reactions or the formation and dissociation of multiprotein or DNA/RNA–protein complexes, for example, can be studied by attaching the donor and the acceptor to different biomolecules (intermolecular FRET) (e.g., Rothwell et al., 2003). For monitoring structural changes that occur, for example, in an enzyme during its functional cycle, both donor and acceptor should be attached to the same biomolecule (intramolecular FRET) (e.g., Margittai et al., 2003).

The advent of optical detection systems for single molecules (Moerner and Kador, 1989; Orrit and Bernard, 1990; Shera et al., 1990) and the successful application of FRET on the ensemble level led to efforts to exploit the same phenomenon on the single-molecule (sm) level in smFRET by Ha et al. (1996).

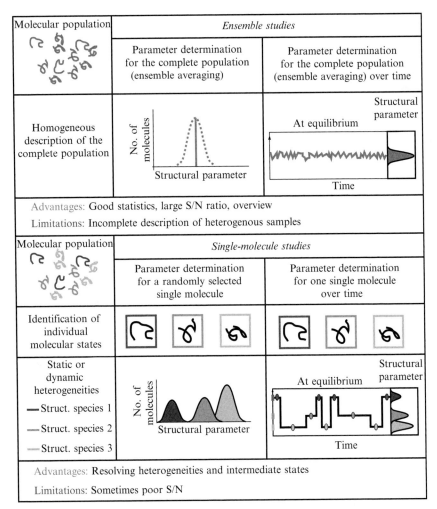

Figure 18.1 Ensemble versus single-molecule studies. The data from fluorescence intensity measurements of an ensemble, although indicative, fail to offer a complete description of molecular behavior due to the averaging inherent in the measurement itself. The revolutionary advantage of single-molecule studies is that they allow one readily to identify heterogeneities. These can be either static, when various subpopulations of molecules remain unchanged over the observation time (left panel) or dynamic (right panel), when there is some interconversion between molecular states during the observation time.

The specific advantages and disadvantages of fluorescence experiments on the ensemble and on the single-molecule level are compared in Fig. 18.1. Two types of experiments must be considered: (i) analysis of all members of an ensemble (left column) and/or (ii) watching the time evolution of a system (right column). If the members of the ensemble exhibit static or dynamic heterogeneity, a single-molecule experiment has clearly several advantages.

Consider for example the case in which a population of enzymes is being observed at equilibrium, while it is performing its function. One could retrieve an average structural parameter (e.g., end-to-end distance) for describing the structure of the whole population. Yet, one could not expect that all the enzymes adopt the same structural conformation at each time point since they do not function in a synchronous manner. This will obscure of the transient intermediate states of the enzymatic cycle since at each time point the average parameter over the various states will be determined. The revolutionary advantage of sm-studies (Bustamante, 2008; Deniz et al., 2008; Kapanidis and Strick, 2009; van Oijen, 2008) is that they offer a way to get access not only to the average value of a parameter describing a molecular population but more importantly to the distribution of the parameter values over the whole population. In this way possible heterogeneities are readily identified. They can be either static, when various subpopulations of molecules remain unchanged over the observation time (left panel) or dynamic (right panel), when there is some interconversion between molecular states during the observation time. The discovery of heterogeneities in biomolecular populations opened the way for answering important biological questions and also ignited the spark for raising new questions about how biological processes are taking place. Nowadays it is well established that heterogeneities in one biomolecular species constitute major aspects of its function and that these states can be only isolated in time (Bahar et al., 2007; Cui and Karplus, 2008; Davydov and Halpert, 2008; Eisenmesser et al., 2005; Frauenfelder et al., 1999; Graham and Duke, 2005; Henzler-Wildman and Kern, 2007; Meier and Ozbek, 2007; Myong et al., 2006; Sytina et al., 2008). Especially, the smFRET approach has proven very fruitful and has provided many new insights regarding a large number of important biological questions, for example: Which structural conformations does a protein adopt during its folding to the native state (Borgia et al., 2008; Haas, 2005; Haran, 2003; Heilemann, 2009; Schuler, 2005; Weiss, 2000), what structural changes are induced in a DNA upon protein binding (Kapanidis et al., 2006), and how is the structure of an enzyme or protein in general related to its functional role (English et al., 2006; Kou et al., 2005; Lu, 2004; Min et al., 2005, 2006; Rothwell et al., 2003; Sugawa et al., 2007; Xie and Lu, 1999)? Results from the application of smFRET have been summarized in several reviews (Ha, 2001; Haas, 2005; Michalet et al., 2006; Schuler and Eaton, 2008; Weiss, 1999).

Despite its widespread use, it is accepted (Berney and Danuser, 2003; Dietrich et al., 2002) that there are some complications in the analysis of data obtained by FRET measurements, especially on the ensemble level. Donor and acceptor photophysics, as well as experimental ambiguities and impurities due to inadequate sample preparation, make the interpretation of FRET results a somewhat delicate task. However, smFRET studies offer unprecedented capabilities to detect molecular subpopulations, and hence allow for the discrimination of heterogeneities arising from experimental artifacts or compromised sample preparation (impurities, aggregates, incomplete labeling, etc.) from those that are biologically relevant. In addition, there is much debate concerning errors in distance measurements by FRET due to unknown angles in the dipole orientation factor κ^2 (Dale et al., 1979; Ivanov et al., 2009; van der Meer, 2002; van der Meer et al., 1994).

To minimize the ambiguity of data interpretation from smFRET studies, we propose in this article the simultaneous determination of a more complete set of parameters describing molecular fluorescence as offered by multiparameter fluorescence detection (MFD). The MFD parameter-space includes (but it is not limited to) the fluorescence lifetimes τ, the fluorescence intensities F, and the steady-state and time-resolved anisotropies r of donor and acceptor (Kuhnemuth and Seidel, 2001; Widengren et al., 2006). Since FRET should be "visible" in all of these parameters, we introduce a multidimensional data histogram analysis. In this way, FRET is monitored with higher specificity by simultaneously inspecting all possible fluorescence dimensions instead of only one. The MFD approach also allows one to combine various fluorescence techniques (see Section 3) and makes use of their complementary advantages to answer a particular question. For example, it is possible to select different single-molecule populations characterized by specific fluorescence properties and combine this subensemble in a single dataset. In this way the statistical noise level of a joint dataset is much reduced compared to individual single-molecule bursts. Thus, it is possible to perform additional, more detailed subensemble analyses to investigate important questions. For example, one may: (i) measure polarization resolved fluorescence decays to describe the fluorophore motion in order to consider possible dye orientation effects, and (ii) study whether an apparently homogeneous burst population contains additional "internal" heterogeneities and dynamics faster than the burst duration. In addition, the temporal evolution of the biological system can be followed easily, because the MFD hardware allows one to register the detected fluorescence photon counts for hours at picosecond resolution.

In summary, while performing only a single experiment the MFD approach allows one to utilize the complete photon information to achieve a more robust quantitative analysis in smFRET studies. The goal of this chapter is to offer the reader insight into the practical application of MFD to smFRET and, perhaps more importantly, to describe a general strategy for data analysis that results in a high degree of confidence in the interpretation of experimental smFRET data.

2. FRET Theory

2.1. Basic FRET theory in a nutshell

Energy transfer from a donor (D) to an acceptor (A) dye may occur via radiative or nonradiative processes, which have different underlying mechanisms. Radiative energy transfer takes place when a photon emitted by the donor is absorbed by the acceptor. Radiative transfer is related to the well-known inner-filter effect in fluorescence (Lakowicz, 2006) and occurs preferentially at higher concentrations, which are irrelevant for single-molecule spectroscopy.

Nonradiative energy transfer is qualitatively a different phenomenon and is strongly distance dependent. It originates from the dipolar coupling of the excited states of D and A. This coupling has quantum-mechanical (for details see literature on Dexter energy transfer or electron exchange excitation transfer) as well as coulombic (electrostatic) components (Andrews and Leeder, 2009; Scholes, 2003; Scholes and Ghiggino, 1994). Förster's theory deals with the case in which only coulombic interactions are considered (Förster, 1948). This is called the "weak coupling regime" since the coulombic coupling does not perturb the energy levels of D and A. Within this regime, energy transfer can be conceptualized with the help of the classical picture of two oscillators (i.e., dipoles) in resonance, hence the term resonance in FRET. Here, only the final equations of Förster's theory are presented, while the reader interested in their exact derivation is referred elsewhere (Andrews, 1989; Andrews and Bradshaw, 2004; Braslavsky et al., 2008; Förster, 1948; van der Meer et al., 1994).

Let us assume that a donor and an acceptor dye are used for labeling two ends of a biomolecule (Fig. 18.2A). The rate of D de-excitation via FRET to A (Fig. 18.2B), k_{FRET}, is given by

$$k_{FRET} = \frac{1}{\tau_{D(0)}} \left(\frac{R_0}{R_{DA}}\right)^6 \qquad (18.1)$$

Here $\tau_{D(0)}$ is the lifetime of D in the absence of A, R_{DA} is the D–A distance (interdye distance), and R_0 is the Förster radius characteristic for the specific dye pair involved in FRET. R_0 depends on the fluorescence quantum yield of D in the absence of FRET, $\Phi_{FD(0)}$, the refractive index of the medium between the dyes, n, the orientation factor, κ^2 (see Section 2.2), and the spectral overlap integral, J:

$$R_0 = \left[\frac{9(\ln 10)}{128\pi^5 N_A} \frac{J\kappa^2 \Phi_{FD(0)}}{n^4}\right]^{1/6} \qquad (18.2)$$

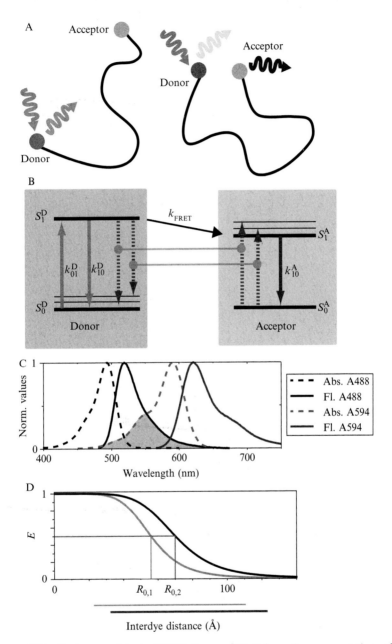

Figure 18.2 Basic essentials in FRET theory. (A) Schematic representation of a biomolecule labeled with donor (D—green sphere) and acceptor (A—red sphere). When the interdye distance is large, only fluorescence signal from D is observed (shown in green while excitation is shown in blue). When an acceptor molecule (A) resides in the vicinity of D, energy transfer takes place. In this case fluorescence signals both from D (dim green) and A (red) will be recorded. The signal of D will be lower

In Eq. (18.2) N_A is the Avogadro constant. Please note that in the literature Eq. (18.2) appears in slightly different forms because different units for the Förster radius R_0 and the variables J and N_A are used (Braslavsky et al., 2008; Lakowicz, 2006; Valeur, 2002). We use a form consistent with the recommended IUPAC units and with Eq. (18.4) below. The spectral overlap integral, $J(\lambda)$, between donor fluorescence (Fl) and acceptor absorbance (Abs) (Andrews and Rodriguez, 2007), is given by Eq. (18.3) and is depicted in Fig. 18.2C (gray area):

$$J(\lambda) = \int_0^\infty F_D(\lambda)\varepsilon_A(\lambda)\lambda^4 \, d\lambda \qquad (18.3)$$

In Eq. (18.3), $F_D(\lambda)$ is the normalized fluorescence of D and $\varepsilon_A(\lambda)$ is the extinction coefficient of A, both at wavelength λ. We explicitly specify the units in Eq. (18.4), which determine the proportionality constant. In most cases it is convenient to report R_0 in Å. Using values of λ in nm and extinction coefficient values in $\text{mol}^{-1} \, \text{dm}^3 \, \text{cm}^{-1}$ in Eq. (18.3), one can derive the following compact form of Eq. (18.2):

$$\frac{R_0}{\text{Å}} = 0.2108 \left[\frac{\kappa^2 \Phi_{FD(0)}}{n^4} \left(\frac{J(\lambda)}{\text{mol}^{-1} \, \text{dm}^3 \, \text{cm}^{-1} \, \text{nm}^4} \right) \right]^{1/6} \qquad (18.4)$$

The efficiency of FRET, E, is defined by the ratio of the number of quanta transferred from D to A over the total number of quanta absorbed by D. From this definition one can derive the kinetically equivalent equations:

(dim green) when energy transfer takes place. (B) Simplified Perrin-Jablonski diagrams of D and A. D is excited at a rate k_{01}^D to the first singlet state S_1^D. In the absence of A, it is depopulated with rate constant k_0^D. Due to the coupling of the possible de-excitation of D and excitation of A, energy transfer can occur at a rate k_{FRET} resulting in the excitation of A from S_0^A to S_1^A which is depopulated with a rate constant k_0^A. (C) The excitation (Abs.) and emission (Fl.) spectra of Alexa488 (A488) and Alexa594 (A594). A488–A594 dyes constitute a commonly used D–A pair in FRET studies. The overlap between the emission of D and excitation of A is highlighted as gray area. The amount of the overlap influences the value of the Förster radius (see text for details about the overlap integral). (D) Plot of FRET efficiency versus the interdye distance for two different values of the Förster radius, $R_{0,1} = 56$ Å (orange), $R_{0,2} = 70$ Å (brown). The value of the Förster radius, R_0, defines the useful dynamic range of distances (orange and brown correspondingly) that can be measured with a specific dye pair as it is illustrated by Eq. (18.6). For ensemble studies this dynamic range is considered to extend from 0.5 to $1.5 R_0$ (Lakowicz, 2006) corresponding to FRET efficiency values from 0.98 to 0.08. For single-molecule studies, FRET efficiencies as low as 0.03 can be discriminated from donor-only populations by PDA (see Sections 5.1 and 5.2; Gansen et al., 2009) and thus longer distances can be measured with the same dye pair. (See Color Insert.)

$$E = \frac{k_{\text{FRET}}}{k_0 + k_{\text{FRET}}} = 1 - \frac{\tau_{D(A)}}{\tau_{D(0)}} = 1 - \frac{F_{D(A)}}{F_{D(0)}} = \frac{F_A/\Phi_{FA}}{F_D/\Phi_{FD(0)} + F_A/\Phi_{FA}} \quad (18.5)$$

where $k_0 = 1/\tau_{D(0)}$ is the overall relaxation rate constant of the excited donor state in the absence of acceptor, and $\tau_{D(A)}$ is the lifetime of D in the presence of A. $F_{D(0)}$ and $F_{D(A)}$ are the fluorescence intensities of D in the absence and presence of acceptor, respectively, F_A represents the corrected fluorescence intensity of A and Φ_{FA} is the fluorescence quantum yield of A. Substituting from Eq. (18.1) to Eq. (18.5) leads to Eq. (18.6), illustrating the strong distance dependence of FRET:

$$E = \frac{1}{1 + (R_{DA}/R_0)^6} \quad (18.6)$$

By using Eqs. (18.5) and (18.6), the interdye distance can be related to observable quantities like the fluorescence lifetime and intensity. In Eqs. (18.7) it becomes obvious how one obtains information on the interdye distance by detecting the fluorescence of D and A.

$$R_{DA} = R_0 \left(\frac{\Phi_{FA}}{\Phi_{FD(0)}} \frac{F_{D(A)}}{F_A} \right)^{1/6} = R_{0r} \left(\Phi_{FA} \frac{F_{D(A)}}{F_A} \right)^{1/6} \quad (18.7a)$$

$$R_{DA} = R_0 \left(\frac{\tau_{D(A)}}{\tau_{D(0)} - \tau_{D(A)}} \right)^{1/6} = R_{0r} \left(\frac{\tau_{D(A)} \Phi_{FD(0)}}{\tau_{D(0)} - \tau_{D(A)}} \right)^{1/6} \quad (18.7b)$$

Note that the R_{DA} is directly related to $F_{D(A)}/F_A$ (without the need for calculating E), which simplifies the D–A distance determination because knowledge of the donor fluorescence quantum yield $\Phi_{FD(0)}$ is not required. This allows us to define a reduced Förster radius R_{0r} in Eq. (18.7c) that is analogous to Eq. (18.4) but lacks $\Phi_{FD(0)}$ (Rothwell et al., 2003):

$$\frac{R_{0r}}{\text{Å}} = 0.2108 \left[\frac{\kappa^2}{n^4} \left(\frac{J(\lambda)}{\text{mol}^{-1} \text{dm}^3 \text{cm}^{-1} \text{nm}^4} \right) \right]^{1/6} \quad (18.7c)$$

Figure 18.2D and Eqs. (18.6) and (18.7) clarify that E is most sensitive to the DA-distance when R_{DA} is in the range of ~ 0.5–$1.5R_0$. Since in single-molecule experiments the typical values of R_0 lie in the region between 40–70 Å, this feature gave rise to the commonly used phrase that FRET provides "a spectroscopic ruler" for measuring distances in the range of 20–100 Å (Schuler et al., 2005; Stryer, 1978). The precision of smFRET is so high that commonly used "rulers" (polyproline and double-stranded (ds) DNA) are observed to be not perfect but partially bent (Best et al., 2007; Doose et al., 2007; Wozniak et al., 2008).

2.2. The orientation factor κ^2

Förster theory treats D and A as point-like oscillating dipoles. The relative orientation of their dipole moments, which is described by the orientation factor κ^2, influences the energy transfer. Naturally, for the case of energy transfer between molecules these dipole moments correspond to the transition dipole moments of D and A, $\vec{\mu}_D$ and $\vec{\mu}_A$ (Fig. 18.3), with corresponding unit vectors, $\hat{\mu}_D$ and $\hat{\mu}_A$, respectively. From Eq. (18.8) it is clear that κ^2 can have values between 0 and 4. In the case where $\kappa^2 = 0$, which for example applies when $\hat{\mu}_D \perp \hat{\mu}_A$ and $\hat{\mu}_D \perp \hat{R}_{DA}$, no energy transfer is possible regardless of the interdye distance.

$$\kappa^2 = [\hat{\mu}_A \cdot \hat{\mu}_D - 3(\hat{\mu}_D \cdot \hat{R}_{DA})(\hat{\mu}_A \cdot \hat{R}_{DA})]^2$$
$$= (\sin\theta_D \sin\theta_A \cos\varphi - 2\cos\theta_D \cos\theta_A)^2 \tag{18.8}$$

Equation (18.8) contains only unit vectors, because solely the angular orientation between the donor and acceptor dipole is relevant. In Fig. 18.3,

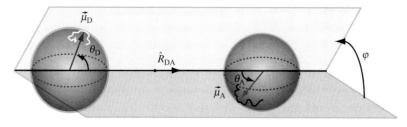

Relative orientation of dipoles	κ^2
↑-----↑	1
→---→	4
↑----→	0
Dynamic averaging (rotational diffusion)	2/3
Static averaging	0.476

Figure 18.3 The orientation factor κ^2. The transition dipole moments of both D and A (respectively $\vec{\mu}_D$, $\vec{\mu}_A$) are changing direction due to rotational diffusion. For the case of totally unrestricted dyes, the rotational diffusion is illustrated by the changes in direction of a vector which start at a center of a sphere and its end is performing translational diffusion on the surface of the sphere. At each time instance the unitary interdye distance vector $\hat{R}_{DA}$ and the unit vectors $\hat{\mu}_D$, $\hat{\mu}_A$ define correspondingly two planes (light gray for $(\hat{\mu}_D, \hat{R}_{DA})$ and dark gray for $(\hat{\mu}_A, \hat{R}_{DA})$). The angles θ_D and θ_A define the dipole orientations with respect to the connecting unit distance vector $\hat{R}_{DA}$ whereas φ is the angle between the two planes. See text for the dependence of the orientation factor κ^2 on the angles and vectors illustrated in the current sketch Eq. (18.8). The table gives κ^2 values for typical dipole orientations.

the angles θ_D and θ_A define the dipole orientations with respect to the connecting unit distance vector $\hat{R}_{DA}$, and φ corresponds to the angle between the two planes defined by $\hat{R}_{DA}$ and the corresponding $\hat{\mu}$ (Fig. 18.3). As the dye molecules rotate (rotational diffusion), $\hat{\mu}_D$ and $\hat{\mu}_A$ change direction resulting in a continuously changing value of κ^2. So, why is it justified to use a constant value of R_0? Under the assumption that the molecular rotational diffusion occurs in a timescale much shorter than the lifetime of the fluorescence emission of D, and dye rotations are unconstrained, all FRET pairs are equivalent and one expects that the dipole moments will take all possible relative orientations during the transfer process (isotropic dynamic averaging regime). Analyses based on rotational diffusion have shown that under these assumptions, the mean value of κ^2 is 2/3. Therefore, unless it is explicitly stated, all R_0 values found in literature are calculated assuming $\kappa^2 = 2/3$. Please note that even if all orientations are considered but the rotational diffusion is slow in comparison with the FRET rate or static, the FRET pairs are not equivalent anymore due to different orientations. Thus a distribution of FRET rates will be measured and a deviation of the mean of this κ^2-distribution from 2/3 is to be expected (for details see van der Meer *et al.*, 1994). Although it is generally impossible to calculate the exact value of κ^2 without *a priori* knowledge of the symmetry of the rotation of the dipole moments, an estimation of the minimum and maximum values of κ^2 can be obtained from anisotropy measurements (van der Meer, 2002). Figure 18.3 gives κ^2 values for typical dipole orientations. In many FRET applications such as kinetic studies, however, only relative changes of R_{DA} are of importance, since absolute values of structural conformations are known beforehand from, for example, NMR and/or crystallographic studies. Hence, the assumption that κ^2 remains constant is sufficient for the interpretation of kinetic FRET studies of biomolecules undergoing conformational changes. In this case, any deviation from $\kappa^2 = 2/3$ will be systematic and will not affect relative distance changes. Nevertheless, there are special cases (i.e., absolute distance determination, experiments on immobilized molecules) for which the validity of the isotropic dynamic averaging hypothesis should be thoroughly tested and κ^2 should be determined precisely. For such cases, FRET theory as presented here should be extended (Hakansson *et al.*, 2004) before it is applied to the experimental data.

3. Fluorescence Properties and Measurement Techniques

3.1. Timescales of biomolecular processes and fluorescence techniques

The beautiful and static insights into protein structure obtained from X-ray diffraction result in the impression that proteins possess absolute functional specificity and a single, fixed structure. This perspective is in conflict with

Accurate Single-Molecule FRET Studies Using Multiparameter Fluorescence Detection   467

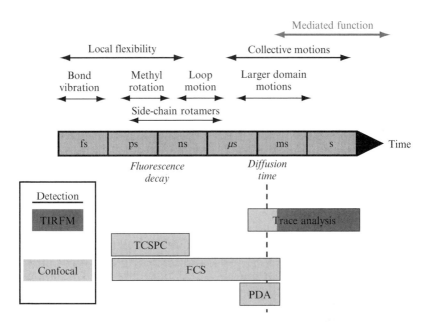

Figure 18.4 Biomolecular and fluorescence timescales. Various distinct fluorescence techniques can be applied to monitor biomolecular processes with different time resolution (see Section 3 for details).

the ability of proteins to adopt different functions and structures and with the intimate relationship between dynamics and molecular function as suggested in Fig. 18.4. Thus, protein function is determined by the thermodynamic stability and kinetic accessibility of the functional states (Henzler-Wildman and Kern, 2007; Henzler-Wildman et al., 2007). In this view, fluorescence spectroscopy is well suited to follow all biomolecular processes slower than the fluorescence lifetime (for organic dyes usually a few nanoseconds). Moreover, fluorescence in combination with single-molecule detection has the advantage of resolving biomolecular heterogeneity over timescales ranging from picoseconds to hours (Fig. 18.1).

Two experimental conditions must be fulfilled for smFRET detection: (i) a limited excitation/detection volume, (ii) low concentrations of fluorescent species. Currently, there are three microscopic modalities in general use for generating a limited excitation volume for single-molecule FRET measurements: total internal reflection microscopy (TIRFM), 2-photon excitation microscopy (2PM) and confocal microscopy (CM), although the emergence of nanoscopic techniques will certainly offer more possibilities in the very near future (Dedecker et al., 2008; Hell, 2007). TIRFM with (electron-multiplied) EMCCD camera detection is used for measuring immobilized molecules, whereas 2PM is mostly for measurements in cells.

CM has been applied to monitor fluorescence of single molecules in any kind of medium (in solution, in cells, on surfaces, etc.). The reason for the broad applicability of CM is its high time resolution in comparison with the other methods and the availability of commercial instruments. Therefore, we focus this review on measurements of smFRET in solution based on the confocal setup described below. More information on TIRFM and 2PM can be found in Thompson and Steele (2007) and So et al. (2000). Nevertheless, it is important to point out that the data analysis scheme presented in the following sections could be adapted to data obtained by using any of the microscopic techniques mentioned above.

3.2. Dimensions of fluorescence

Following the ideas of Förster (Förster, 1951), it is obvious that fluorescence has multiple characteristic parameters that can be exploited to yield-specific information on the environment by multiparameter analysis (Kuhnemuth and Seidel, 2001; Lakowicz, 2006; Valeur, 2002; Widengren et al., 2006). In recent years, the characterization of molecules in single-molecule fluorescence detection measurements has step-by-step been established for the various fluorescence parameters depicted in Fig. 18.5J. Spectral properties of absorption and fluorescence, $F(\lambda_A, \lambda_F)$ (Tamarat et al., 2000), fluorescence brightness and quantum yield, Φ_F (Chen et al., 1999; Fries et al., 1998; Kask et al., 1999), fluorescence lifetime, τ (Tellinghuisen et al., 1994; Zander et al., 1996), and fluorescence anisotropy, r (Ha et al., 1999; Schaffer et al., 1999) are the five intrinsic properties of a fluorophore that are accessible in an MFD experiment and report on its local environment (see Fig. 18.5J).

For molecular systems, changes in fluorescence parameters of a single fluorophore sometimes do not provide enough information for molecular identification or for a detailed investigation of the underlying molecular interactions. Further information can be obtained on the single-molecule level by including more than one fluorophore per particle, thereby increasing the opportunities for determining stoichiometries and interactions of individual particles from photon densities and coincidences (Schmidt et al., 1996; Weston et al., 2002). Moreover, the use of two fluorescing reporters allows for the determination of structural features of particles via FRET. With a donor–acceptor distance approaching or exceeding 100 Å, the FRET efficiency is typically so small that the fluorescence from the acceptor can be detected only by additional direct excitation through, for example, 2PM (Heinze et al., 2000) or two-color excitation in a cw or pulsed mode (Kapanidis et al., 2005; Muller et al., 2005). Such simultaneous excitation of two dyes can be used for monitoring an interaction or binding event by cross-correlation (Schwille et al., 1997), coincidence, colocalization (Lacoste et al., 2000), or two-dimensional (2D) fluorescence intensity distribution analysis (FIDA; Kask et al., 2000) of the resulting emission signals.

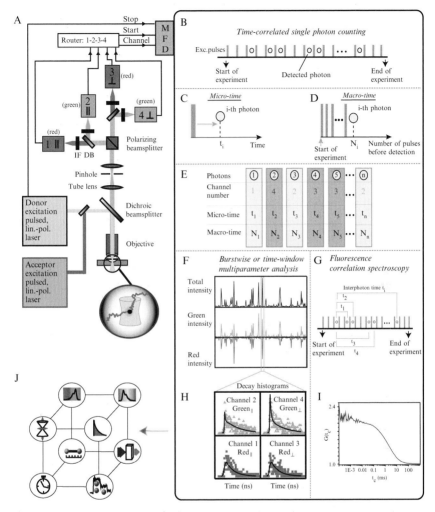

Figure 18.5 Experimental setup and data registration. (A) Typical four-channel confocal microscopy setup (see Section 3.3 for a detailed description). (B–D) Time stamping principle. A cascade of two synchronized counters is used providing a time resolution from ps to s. The laser pulses define the macrotime (D) and the photon arrival times measured by TCSPC correspond to the microtime (C). The time axis of the experiment is constructed by the sequence of excitation laser pulses, whereby no photons are detected for most pulses during a single-molecule experiment. (E) Each detected photon can be uniquely "identified" by the assignment of three tags (channel number, micro- and macrotime). (F–H) Burst-wise or time-window multiparameter analysis. Reconstructed intensity traces (F), where total intensity refers to all four channels. Green intensity corresponds to channels 2 and 4. Red intensity corresponds to channels 1 and 3. From these traces, bursts of fluorescence, due to the passage of a single molecule through the detection volume, are identified. (H) The information on FRET is obtained via lifetime and intensity analysis of each channel (or pairs of channels). (G–I) Fluorescence Correlation Spectroscopy (FCS): By using the interphoton time information the correlation curve is constructed according to the equation

In view of the additional information that can be extracted from a molecule with two or more attached fluorophores, it is reasonable to consider additional fluorescence dimensions. With reference to Fig. 18.5J, we argue that under appropriate conditions the fluorescence parameter space can be extended from the five intrinsic dimensions to at least eight dimensions by adding the three "system/environmental parameters": stoichiometry (via quantized fluorescence intensity), dipolar coupling (FRET), and time (at timescales other than those of excitation and emission).

It is noteworthy that the fluorescence intensity of a dye is influenced by intrinsic transitions to and from photophysical states other than the singlet ground and excited states (e.g., triplet (Widengren et al., 1995) and radical states (Widengren et al., 2007), as well as by extrinsic processes (e.g., quenching or other chemical reactions influencing the fluorescence). Thus, for a given excitation intensity a specific intrinsic blinking pattern of fluorescence is observed, which corresponds to a saturation of the fluorescence intensity (Sanden et al., 2007) and influences also FRET from or to this dye (Hofkens et al., 2003). From the discussion in the previous sections it becomes evident that an accurate monitoring of FRET often requires the investigation of the complete 8D parameter space. Statistical most efficient methods have been developed to determine all fluorescence parameters. The corresponding software is available from our group (see http://www.mpc.uni-duesseldorf.de/seidel/).

3.3. Experimental setup and data registration

The general schematic for a confocal epi-illuminated microscope used for MFD is shown in Fig. 18.5A (Widengren et al., 2006). The fluorescent donor molecules are excited by a linearly polarized, pulsed laser (diode lasers (Picoquant, Germany) or actively mode-locked Ar-ion-laser (Coherent, USA)). The laser light is guided into the microscope (IX81; Olympus, Japan) and is reflected by a dichroic beamsplitter, before it is focused into a dilute (~ 50 pM) solution of labeled molecules by a water immersion lens with high numerical aperture. As a molecule diffuses through the confocal excitation/detection volume, FRET occurs between the dye-pair and a brief burst of fluorescence photons (originating from both D and A) is observed. During a typical dwell time of 1 ms there are $\sim 10^5$ laser pulses and the same molecule is excited $\sim 10^4$ times. The fluorescence photons are

---

$G(t_c) = \langle F(t)F(t+t_c)\rangle / \langle F \rangle^2$, where F is the fluorescence intensity and t_c is the correlation time. (J) 8D fluorescence parameter space starting at the upper left corner with clockwise rotation. (Frontside): Fundamental anisotropy, lifetime, intensity (stoichiometry), detection time; (backside): excitation spectrum, emission spectrum, fluorescence quantum yield, and distance between two fluorophores.

collected by the same objective lens and transmitted through a dichroic beamsplitter that does not allow the reflected laser light to enter the detection pathway. The photons pass through a pinhole that constitutes the central element of any confocal setup (Pawley, 2006). For MFD this photon train is divided initially into its parallel and perpendicular components via a polarizing beamsplitter and then into wavelength ranges that cover the D and A emission spectra. In our work, the channels corresponding to the donor and acceptor fluorescence are usually called the "green" and "red" channels, respectively. High-quality band-pass filters are used in front of the detectors in order to guarantee that only fluorescence photons emitted by either D or A are registered. Single-photon avalanche diodes (SPADs) (single-photon counting module SPCM-AQRH 14 (PerkinElmer Optoelectronics, Canada) or PDM 50CT SPAD (Picoquant, Germany)) are used as detectors with high time resolution, single-photon sensitivity, and low dark counts. The output signals from the SPADs are guided via a router to a time correlated single-photon counting (TCSPC) card (SPC 132 or SPC 832 (Becker&Hickl, Germany) where the data registration takes place. Considering an overall efficiency for photon collection and detection of $\sim 1\%$, on average 100 fluorescence photons are finally registered per millisecond (10^5 Hz).

TCSPC is a versatile mode of data registration that relies on assigning tags on each detected photon (Becker, 2005; Eggeling et al., 1998; Tellinghuisen et al., 1994). Only three tags (channel number, micro- and macrotime) are sufficient for obtaining all the information present in a fluorescence event as described in Fig. 18.5B–E. The channel number provides information for the spectral region and polarization of each detected photon (Fig. 18.5E). The microtime, t, corresponds to the time elapsed between the excitation pulse and the photon detection (TCSPC arrival time). The macrotime information, N, corresponds to the number of excitation pulses that have occurred from the start of the measurement until the detection of a fluorescence photon. As the time interval between two consecutive laser pulses, Δt_L, is known, the exact time of the photon detection can be determined with an accuracy of picosecond over a measurement duration of several hours (Felekyan et al., 2005). The complete time axis can be reconstructed with the help of the equation: (time of detection of the ith photon) $= t_i + N_i \Delta t_L$. Combining this time information with the information regarding the number of detected photons complete time traces of the total fluorescence intensity as well as of the fluorescence intensity for each channel can be generated.

After reconstruction of the full trace photon bursts can be identified by several methods (Eggeling et al., 2001; Fries et al., 1998; Nir et al., 2006) (Fig. 18.5F). For each selected burst a large number of desired parameters can be determined, as discussed above. The five intrinsic properties of the fluorophore are ideally probed during MFD by classical TCSPC, where the reference start time is set by the preceding laser pulse. The resulting spectral and polarization resolved decay histograms (Fig. 18.5H) are, however, time averaged over the burst duration so that information about molecular properties on timescales exceeding the fluorescence lifetime is lost. To preserve this information, analysis of the fluorescence photon train has to be performed prior to averaging (time-window analysis).

In contrast to TCSPC, which is based on the fluorescence lifetime and requires pulsed excitation, a complementary method extracts information contained in intensity fluctuations. This powerful statistical method is based on the calculation of the autocorrelation function of the fluorescence fluctuations (Ehrenberg and Rigler, 1974; Magde et al., 1972) (Fig. 18.5G and I), and is referred to as fluorescence correlation spectroscopy (FCS). The autocorrelation amplitude is proportional to the probability to detect a signal, $F(t + t_c)$, with a time delay t_c after a signal, $F(t)$, that was recorded at time t; that is, the preceding photons set the reference start time (Fig. 18.5G). Only photons emitted by an individual single molecule are correlated. In this way FCS helps identify correlated molecular processes and draw conclusions about chemical reaction equilibria, photophysical processes such as FRET (Margittai et al., 2003; Torres and Levitus, 2007; Widengren et al., 2001), and transport properties (rotation and diffusion) over a time span of several orders of magnitude, ranging from picoseconds to seconds (for a recent review, see Haustein and Schwille, 2007).

The specific photon detection scheme offered by MFD provides unsurpassed flexibility regarding the analysis of data acquired in a single measurement (Neubauer et al., 2007). In addition to the previously mentioned analysis techniques, one can readily employ FIDA (Chen et al., 1999; Fries et al., 1998; Kask et al., 1999) to characterize the brightness Q of molecules, which is determined by taking into account the excitation rate, detection efficiency, and fluorescence quantum yield. FIDA is a well-established analysis scheme based on intensity fluctuations in short time windows (~ 40 μs). For FRET studies that use fluorescence intensities in two spectral detection channels, the information obtained by 2D FIDA is very valuable (Kask et al., 2000).

During the last few years an expansion of MFD was introduced (Fig. 18.5). In this expansion a second laser is used for direct excitation of the acceptor. The pulses of the two lasers must be properly interleaved (Eggeling et al., 2006), exciting D and A in an alternating way; hence, this technique can be found in the literature under interchangeable terms: alternating excitation (nsALEX) (Kapanidis et al., 2005; Lee et al., 2005) or pulsed-interleaved excitation (PIE) (Muller et al., 2005; Ruttinger et al., 2006). For studies of systems with complex and multiple stoichiometries, these methods can have major advantages over the single-color excitation scheme. Using only a single laser for synchronous strong D excitation and weak A excitation is a simple technical alternative to the above techniques (see Section 5.2).

Table 18.1 MFD offers the possibility to test assumptions that are often considered to be valid *a priori* in common FRET spectroscopy

Section	Problems and assumptions	Solutions by the multiparameter approach
4.1. and 4.3.	• Local quenching of D or A? • Problems with the Förster Radius R_0?	Excluded by visual inspection of the 2D histograms of FRET efficiency versus donor lifetime in the presence of the acceptor $(E - \tau_{D(A)})$ and subensemble analysis
4.2.	• Differences in mean mobilities of the dyes? • Orientation factor $\kappa^2 = 2/3$?	Minimize the uncertainty by checking 2D histograms of anisotropy versus lifetime ($r_D - \tau_{D(A)}$) and subensemble analysis
4.4.	• Do not disregard broadening: Does a single peak necessarily correspond to a single species?	• 2D analysis for dynamic FRET • Check also for dynamics by PDA, FCS and fFCS
5.	• Sure about interpretation of parameter distributions?	Remove shot noise with PDA and related approaches
5.7.	• Flexibly linked fluorophores	Check environment and proper averaging of FRET signals: $E(\langle R_{DA}\rangle) \neq \langle E(R_{DA})\rangle$

3.4. Improving accuracy of smFRET measurements

Table 18.1 summarizes general problems and factors complicating the interpretation of data obtained from smFRET studies. In this respect, Table 18.1 will serve as a guide for Sections 4–6 showing the improvements achieved by employing MFD.

Many of these problems, which could result in the misinterpretation of experimental data, are interrelated. For instance, both complex photophysical behavior of a dye and its mobility may influence the Förster radius. Additionally, the linkers used for connecting fluorophores to biomolecules introduce some uncertainty regarding the exact position and orientation of the dipole moment of the dye and the dynamics of its movement. The use of short rigid linkers can produce uncertainties in the generally assumed κ^2 value of 2/3, whereas the use of long flexible linkers allows the dye to "explore" a large diffusion volume (e.g., the radius of the accessible space of a dye attached to DNA via a standard "C6-Linker" is ~20 Å; Sindbert *et al.*, 2010). In kinetic studies in which the identification of short-lived intermediate states as well as the determination of the kinetic rate constants is critical, the analysis of FRET distributions can give rise to artifacts if the aforementioned factors are not properly taken into account (Sections 4.4. and 5). Finally, although a single FRET pair is normally sufficient for monitoring kinetics when the structures are known *a priori*, a network of interrelated FRET pairs should be used to retrieve structural information from smFRET experiments by position triangulation between multiple different labeling sites (without preknowledge at least four DA distances are needed to determine a single dye position in space). As will be illustrated in the following, multiparameter acquisition by MFD offers ways for circumventing and identifying problems, resulting in more reliable data interpretation than possible with single-parameter methods.

4. Qualitative Description of smFRET

4.1. From fluorescence signals to FRET indicators

4.1.1. Intensity measurements

In the previous section, the experimental setup and data registration scheme were described. Here we will explain how, starting from the initial "raw" data, one obtains values of physical quantities that can be related to Förster theory as described in Section 2. The total signal S in the green and red channels that record the fluorescence of the donor and acceptor, respectively (with parallel ($\parallel$) and perpendicular ($\perp$) polarization with respect to the polarization of the excitation light), during a burst (Fig. 18.5) can be written as

$$S_G = S_{G,\parallel} + S_{G,\perp} \quad \text{and} \quad S_R = S_{R,\parallel} + S_{R,\perp} \qquad (18.9)$$

The corresponding fluorescence signal F has to be corrected for the mean background signal $\langle B \rangle$ due to scattered excitation light, background fluorescence, and detector dark counts:

$$F_G = S_G - \langle B_G \rangle \qquad (18.10)$$

$$F_R = S_R - crosstalk - \langle B_R \rangle = S_R - \alpha F_G - \langle B_R \rangle \qquad (18.11)$$

One should also account for the fact that some part of the donor emission may pass the selected filters and beamsplitters and is detected in the red channels. This artifact is called bleed-through or *crosstalk* and is characterized by the *crosstalk* factor α. This factor is determined as the ratio between donor photons detected in the red channels and those detected in the green channels ($\alpha = F_{R(D)}/F_{G(D)}$) for the D-only-labeled sample.

Unfortunately, optical systems neither collect nor detect all photons of interest. Therefore, the fluorescence signals measured for D and A, F_G and F_R, respectively, differ from the true fluorescence, F_D and F_A. The probability of detecting a photon is called detection efficiency. It is wavelength dependent due to the wavelength-dependent properties of the mirrors, filters and detectors of the setup. With knowledge of the detection efficiencies of D (g_G) and A (g_R), one can calculate the corrected fluorescence F_D and F_A:

$$F_D = \frac{F_G}{g_G} = \frac{S_G - \langle B_G \rangle}{g_G} \quad (18.12)$$

$$F_A = \frac{F_R}{g_R} = \frac{S_R - \alpha F_G - \langle B_R \rangle}{g_R} \quad (18.13)$$

Since detection efficiencies are difficult to measure (Fries et al., 1998; Sabanayagam et al., 2005), we prefer to determine the ratio of g_G/g_R, which is what is actually needed for calculating an absolute FRET efficiency (see Eq. (18.14)). The FRET efficiency (Eq. (18.5)) is related to the experimental data through the definitions in Eqs. (18.12) and (18.13):

$$E = \frac{F_R}{\gamma F_G + F_R} = \frac{F_A}{\gamma' F_D + F_A} \quad \text{with} \quad \gamma = \frac{g_R \Phi_{FA}}{g_G \Phi_{FD(0)}} \quad \text{and} \quad \gamma' = \frac{\Phi_{FA}}{\Phi_{FD(0)}}$$
(18.14)

Note that all experimental imperfections are included into a single factor, γ or γ'. Methods to reproducibly determine γ are described in Section 5.6.2. If dyes with well-separated excitation spectra are used, such as Alexa488 as donor and Cy5, Alexa647, or Atto647N as acceptor, and the excitation wavelength is set at 496 nm, no additional effects must be taken into account for a quantitative FRET analysis. If the excitation spectra are less well separated, as is the case for the D–A pair Alexa488 and Alexa594, additional direct acceptor excitation must be taken into account and the equations be modified as described in Section 5.2.

In the literature, not only absolute physical quantities for FRET (the FRET efficiency E and the interdye distance R_{DA}) but also FRET indicators, FIs, such as the proximity ratio, $S_R/(S_G + S_R)$ or $F_G/(F_R + F_G)$ and the ratio between the green and red channels, S_G/S_R or F_G/F_R, respectively, are quite often used to display FRET results. However, one must be aware that FIs cannot be directly linked to FRET efficiencies (and thus to distances) without knowledge of the background intensities and the imperfections of the system (detection efficiencies and fluorescence quantum yields).

4.1.2. Fluorescence lifetime measurements

As indicated in Eq. (18.5), the FRET efficiency can also be determined from fluorescence lifetimes. This approach has the additional advantage that no correction factor (i.e., γ) is necessary. Fluorescence lifetime measurements are easily performed in an MFD setup using a pulsed laser source as shown in Fig. 18.5A. For each selected burst and spectral region (green/red), the fluorescence events are accumulated in decay histograms, which usually have 256 time bins with a width of 65.1 ps (Fig. 18.5H). We have developed a special algorithm to compute a mean fluorescence lifetime τ (Brand et al., 1997; Maus et al., 2001; Zander et al., 1996). Briefly, fitting of the fluorescence lifetimes is performed using an iterative deconvolution approach similar to that applied to ensemble TCSPC data (O'Connor and Phillips, 1984), with several important distinctions: (i) The assumption of Gaussian noise used in the least-square fitting procedure is inappropriate for sm-statistics. Thus, the maximum likelihood estimator (MLE) (Maus et al., 2001; Zander et al., 1996) must be applied instead. (ii) The contribution of background B is rarely negligible and has to be taken into account. (iii) The instrument response function (IRF) is usually obtained by selecting the characteristic narrow peak in a TCSPC histogram measured for water or a buffer sample. (iv) As in MFD, the photons are collected into two polarization channels so that proper weighting (i.e., $S = S_\parallel + 2S_\perp$) should be applied to avoid polarization-related artifacts for the total signal, S (i.e., a fluorescence decay registered with a *single* detector will always contain polarization-related artifacts). (v) A typical burst contains at most a few hundred photons, which limits the complexity of the applied model (fit) function. A typical fit function $M(t, \tau, \beta)$ includes fluorescence $F(t, \tau)$ and a certain fraction β of background contribution $B(t)$ as given by Eq. (18.15a):

$$M(t, \tau, \beta) = (1 - \beta) F(t, \tau) + \beta B(t) \qquad (18.15a)$$

The time dependence of the fluorescence signal resulting from a single laser pulse, $F_{sp}(t, \tau)$, is given as a convolution of the instrumental response function, IRF(t) with a single-exponential decay characterized by the fluorescence lifetime τ, wherein the iterative laser excitation with the frequency ω (in our case 73 MHz) must be taken into account to yield the total observed fluorescence signal, F:

$$F(t, \tau) = \sum_{n=0}^{\infty} F_{sp}(t + n/\omega) \quad \text{where} \quad F_{sp}(t, \tau) = \text{IRF}(t) \otimes \exp(-t/\tau)$$

$$(18.15b)$$

More details can be found in Schaffer et al. (1999) and Maus et al. (2001).

4.1.3. Two-dimensional FRET analysis

All changes observed in the fluorescence intensities of D and A must correlate with the observed changes in the fluorescence lifetime of D, $\tau_{D(A)}$. Thus, any discrepancies between the two methods will indicate deviations from the common description of static FRET states, as will be shown in the next few paragraphs. For the analysis of smFRET experiments, a 2D histogram of $\tau_{D(A)}$ versus E (or equivalently any other FRET indicator) is the most important plot. In a histogram of any FI versus $\tau_{D(A)}$ the theoretically expected relation of FI and $\tau_{D(A)}$ can also be plotted. This curve "connects" all possible FRET populations that originate from the same species of D and A (in terms of fluorescence properties) and therefore constitutes the initial discriminator of FRET-active molecules.

The preceding concepts are clarified by a discussion of typical experimental results depicted in Fig. 18.6. As an example, a mixture of dsDNAs labeled with a FRET pair at different distances was studied (Wozniak et al., 2008). The position of D was fixed, whereas the position of A was varied, resulting in dsDNA populations that exhibit high, middle and low FRET values corresponding to 5, 11, and 19 base pair (bp) D–A distances, respectively. For single-molecule experiments the oligonucleotides were diluted to a final concentration of $<20\ pM$. After mixing, MFD data were taken for >1 h to analyze the sample by counting single molecules. For selected single-molecule bursts (Fig. 18.5F), we determined all possible fluorescence parameters (Fig. 18.5J) including the FRET indicator F_D/F_A (Eq. (18.14)), the donor lifetime in the presence of acceptor $\tau_{D(A)}$ (Eqs. (18.15)), and the donor anisotropy (Eq. (18.17)).

A 2D frequency histogram of F_D/F_A against $\tau_{D(A)}$ is shown in the upper panel of Fig. 18.6A, wherein the number of molecules (or fluorescent bursts) in each bin is colored in gray scale from white (lowest) to black (highest). With decreasing basepair separation, FRET increases as can be seen by a reduction in both F_D/F_A and $\tau_{D(A)}$. Moreover, the changes in F_D/F_A and $\tau_{D(A)}$ have to satisfy Eq. (18.5) if they are related to FRET. For the case of using the ratio F_D/F_A as an FI, the black line in Fig. 18.6 illustrates the relationship between F_D/F_A and $\tau_{D(A)}$, which takes distinct fluorescence quantum yields Φ_F and detection efficiencies into account:

$$\left(\frac{F_D}{F_A}\right)_{static} = \frac{\Phi_{FD(0)}}{\Phi_{FA}} \frac{\tau_{D(A)}}{\tau_{D(0)} - \tau_{D(A)}} \quad (18.16a)$$

$$E_{static} = \frac{F_A/\Phi_{FA}}{F_D/\Phi_{FD(0)} + F_A/\Phi_{FA}} = 1 - \frac{\tau_{D(A)}}{\tau_{D(0)}} \quad (18.16b)$$

All FRET populations arising from different interdye distances should be located along this sigmoidal line, which represents the theoretical relationship between the FRET efficiency and the D lifetime. Any deviation from

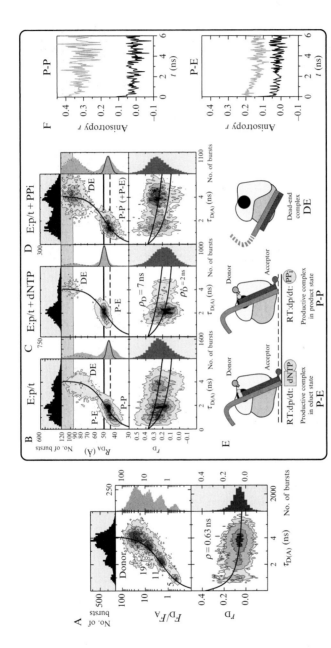

Figure 18.6 Typical 1D and 2D histograms in smFRET for a mixture of quasi-static species. *Left:* (A) Typical results of a smFRET experiment with a mixture of the same dsDNA labeled with the dye pair (Alexa488–Cy5) at three distinct interdye distances (5, 11, and 19 bp separation) and D-only. The parameter histograms for the intensity ratio F_D/F_A, the donor lifetime in presence of acceptor, $\tau_{D(A)}$, and the donor anisotropy, r_D, are arranged in two joint 2D histograms of F_D/F_A versus $\tau_{D(A)}$ and r_D versus $\tau_{D(A)}$ sharing the $\tau_{D(A)}$ axis. The number of bursts corresponds to number of selected single-molecule events. The scaling is from white to black for increasing number of bursts. The parameters for the static FRET line (Eqs. (18.16)) in the upper panel are $\tau_{D(0)} = 4.1$ ns, $\Phi_{FD(0)} = 0.8$, $\Phi_{FA} = 0.4$, and $g_G/g_R = 0.68$. The parameters for the Perrin equation (Eq. (18.18)) in the lower panel are $r_0 = 0.375$ and the estimated mean rotational correlation time of the donor, $\rho = 0.63$ ns. *Right:* MFD analysis of the heterogeneity of HIV-1 (RT):p/t complexes. 2D histograms of R_{DA}

this curve would suggest that additional photophysical processes or interconversions of species (see Sections 4.4 and 5.4) are occurring. Because of these properties, this curve will be referred in the rest of the text as the "static FRET line." For other FIs, analogous dependencies can be derived as evident from Eqs. (18.5), (18.7), and (18.16). Note that the intrinsic lifetime of the donor emission can be estimated from the same measurement since in all sample preparations a small fraction of molecules is labeled only with the donor dye (i.e., we use this population as a valuable internal standard). The importance of these representations of FRET data is highlighted further in the examples below, in which the examination of either the donor lifetime or the F_D/F_A ratio alone would have led to an incorrect interpretation of the data (see Sections 4.3 and 4.4.).

and r_D versus $\tau_{D(A)}$ of different HIV-1 RT:p/t complexes (panels B–D) and subensemble spectroscopy (panel F). For all 2D histograms of R_{DA} versus $\tau_{D(A)}$, the black overlaid line represents the static FRET line (Eq. (18.7b)) of parameters $\tau_{D\,(0)} = 3.1$ ns, $\Phi_{FD(0)} = 0.64$, and $\Phi_{FA} = 0.3$, $R_0 = 53$ Å. *Lower panel*: r_D is plotted versus $\tau_{D(A)}$ together with overlaid Perrin equation (Eq. (18.18)) computed for two rotational correlation times ρ_D (2 and 7 ns). (B) Analysis of the RT:p/t complex showing the presence of three species sketched below: productive complex in educt state (P-E), productive complex in product state (P-P) and dead end complex (DE). (C) Addition of 200 μM dNTPs. The $\tau_{D(A)}$–R_{DA} histogram shows only species P-E and DE. The P-E complex (panel E) interacts with the dp/dt in a state closely resembling the known RT:dp/dt structures. Here, the dNTP is thought to occupy a binding site in the polymerization-active site and is therefore prebound for incorporation into the primer strand (purple). The solid black line in the histogram, and below in the cartoon, indicates the position of the dp/dt bound in the P-E state. In the P-E state, a dNTP occupies the binding pocket. (D) Addition of 200 μM sodium pyrophosphate (NaPP$_i$). The presence of PP$_i$ moves the peak toward shorter distances, indicated by species P-P. The presence of PPi shifts the primer terminus into the binding pocket, forming P-P. The dashed black line in the histogram, and below in the cartoon, indicate the position of the dp/dt bound in the product state, in the presence of the PP$_i$ shifting the dp/dt into the binding cleft. The peak is not Gaussian distributed, and the rotational correlation time, ρ_D, remains high. The position of the dead end (DE) complex remains unchanged. (E) The p66 subunit of RT is colored in light gray, and the p51 subunit is colored dark gray. The polymerase-active site of p66 is colored black. The fluorescence dyes Alexa488 and Cy5 are indicated by a balloon colored in green and red respectively. The FRET data of the p complexes are consistent with the structure obtained by X-crystallography (Kohlstaedt *et al.*, 1992). Preliminary results indicate that the nucleic acid substrate in the DE complex is bound at a site on the p51 subunit, far removed from the nucleic acid binding tract observed by crystallography. (F) Results for subensemble time-resolved anisotropy, r, in the green (light gray) and in the red (dark gray) channels of the species P-P and P-E. Note that the fast decay components are not fully resolved by our setup. Alexa488 has a fundamental anisotropy $r_0 = 0.375$. (See Color Insert.)

4.2. Anisotropy versus donor lifetime

Another useful parameter calculated for each burst is the steady-state polarization anisotropy (Koshioka et al., 1995; Lakowicz, 2006; Schaffer et al., 1999) of the donor, r_D. It is determined by Eq. (18.17):

$$r_D = \frac{GF_{D,\parallel} - F_{D,\perp}}{(1 - 3l_2)GF_{D,\parallel} + (2 - 3l_1)F_{D,\perp}} \qquad (18.17)$$

The fluorescence signal of the donor with parallel and perpendicular polarizations is denoted as $F_{D,\parallel}$, $F_{D,\perp}$, respectively. The ratio of the detection efficiencies of the perpendicular and parallel polarized light is given as $G = g_\perp/g_\parallel$ (for calibration see Section 5.5). The factors l_1 and l_2 above account for polarization mixing due to the objective lens as described in Koshioka et al. (1995). These factors can be determined for a given microscope in a separate calibration measurement (Schaffer et al., 1999).

The anisotropy is linked to the donor lifetime via the Perrin equation:

$$r_D = \frac{r_0}{1 + \tau_{D(A)}/\rho_D} \qquad (18.18)$$

In Eq. (18.18), ρ_D is the rotational correlation time of D, and r_0 is the fundamental anisotropy of D. Strictly speaking, the Perrin equation is valid only if the anisotropy decay is single-exponential, which is usually not the case. Still, the average rotational correlation time in Eq. (18.18) qualitatively reflects the rotational mobility of the dye, which is the collective result of the local rotation of the dye itself and the overall rotation of the biomolecule to which the dye is attached. It is an essential indicator for studying local interactions between the dyes and biomolecule which would lead to sticking of the dye, so that positional averaging is prevented (further details see Section 5.7.2).

Analogously to the relation between F_D/F_A and $\tau_{D(A)}$, a 2D plot for r_D versus $\tau_{D(A)}$ is shown in the lower panel of Fig. 18.6A. Since both 2D diagrams share the $\tau_{D(A)}$ axis, the interrelationship between F_D/F_A and r_D is directly visible, which allows for the assignment of individual anisotropies to each FRET population. The Perrin equation is overlaid as black line, so that the mobility of D can be estimated. Most importantly, one can readily observe whether subpopulations have distinct diffusion properties, which is not the case here (Fig. 18.6A). If there were subpopulations that do not lie on the same Perrin curve, it would suggest differences in the mean correlation time, which may affect κ^2. In this case, individual R_0 values should be used for determining R_{DA} values from the FRET efficiency of each distinct subpopulation. It should be noted that even if all states lie on the same Perrin curve, indicating similar rotational diffusion properties, no exact value of κ^2 can be determined. Any deviation from an assumed value of $\kappa^2 = 2/3$ would still most likely be systematic and not affect any conclusions drawn about relative distance changes. Caution should be used, however, when describing absolute distances. In this case, additional subensemble analysis of time-resolved anisotropy allows one to considerably reduce the uncertainty in κ^2 (see next example).

4.2.1. Example: HIV-1 reverse transcriptase (RT)

One of the earliest examples for the use of MFD to determine the structures of enzyme:substrate complexes was a study of HIV-1 RT (Rothwell et al., 2003). Kinetic studies on this enzyme indicated that the protein:DNA interaction is extremely complex. Although these studies were instructive by uncovering the existence of three distinct HIV-1 RT:primer/template (RT:p/t) complexes, no information could be obtained about their structure. Experiments were performed on donor-labeled HIV-1 RT molecules in combination with acceptor-labeled p/t constructs. Using MFD it was possible to separate Fig. 18.6B three distinct RT:p/t complexes (called productive complex in the educt state (P-E), productive complex in product state (P-P), and dead end complex (DE), see Fig. 18.6E) and to generate a basic structural model from the determined distances (Rothwell et al., 2003).

It is often useful when studying enzymes to "trap" the enzyme in one specific conformation by the addition of substrates, or by the addition of an interaction partner that is known to cause a conformational change. Such experiments give additional confidence in the interpretation of heterogeneity in solution. For the HIV-1 RT:p/t complex this was achieved by termination of the DNA substrate by the incorporation of a dideoxynucleotide, followed by binding of the next base pairing dNTP, resulting in an equilibration toward the P-E complex (Fig. 18.6C). Conversely, binding of pyrophosphate to the terminated HIV-1 RT:p/t complex resulted in an equilibration between the P-E and P-P complexes in which the P-P complex is now predominant (Fig. 18.6D). However, as evident from the 2D r_D versus $\tau_{D(A)}$ plots, there is a conformation-specific anisotropy change. Variations in anisotropy could be indicative of a reduced mobility of the fluorophore, which could affect κ^2 and hence the distances calculated. One of the major advantages of MFD is the ability to perform species-specific analysis. In order to determine the effect of anisotropy on the calculated distances we performed a subensemble analysis by selecting the P-P and P-E populations from the 2D R_{DA} versus $\tau_{D(A)}$ plot. The time-resolved anisotropy decays of both the donor (light gray) and acceptor (dark gray) fluorescence are shown in Fig. 18.6F. It is clearly visible that the P-P and P-E states differ in the amplitude and time of the fast rotational relaxation process, which is followed by the offset due to global macromolecular motion. To estimate the maximum possible distance errors by distinct κ^2-

values, we assume in a worst case scenario that the time-resolved decay of the donor anisotropy remains high, whereas the sensitized acceptor anisotropy rapidly decays to zero indicating a highly mobile acceptor dye. Subensemble analysis thus allows us to estimate maximum errors based on assumption of a worst case scenario (Lakowicz, 2006). The maximum errors of ± 3 Å for the individual P complexes and of ± 16 Å for the DE complex are too small to explain the significant distance differences observed between the species.

4.3. Donor and acceptor quenching

To clearly demonstrate the effects of donor or acceptor quenching on the analysis of smFRET in 2D FI versus $\tau_{D(A)}$ histograms, we simulated a typical FRET experiment using a Brownian dynamics approach (Dix et al., 2006; Enderlein et al., 1997; Kask et al., 1999; Laurence et al., 2004), as described in detail in Felekyan et al. (2009). With this example the power of 2D analysis becomes evident. The results are displayed in Fig. 18.7A–C as 2D F_D/F_A versus $\tau_{D(A)}$ histograms for different conditions. Considering the analysis of static populations in Fig. 18.7A, three populations are clearly visible: a high-FRET (HF) species, a low-FRET (LF) species, and a small fraction of D-only-labeled molecules. All populations reside on the static FRET line (Eq. (18.16a)), which is consistent with a purely FRET-related decrease of the D fluorescence lifetime. In Fig. 18.7B and C, the same states as in Fig. 18.7A were simulated, but this time D or A quenching was added to the simulation, respectively. As can be seen, either effect would lead to a misinterpretation of the data when using single parameter analysis (either F_D/F_A and $\tau_{D(A)}$).

4.3.1. Donor quenching

In the case of donor quenching (Fig. 18.7B), D lifetime analysis alone indicates the presence of three species, an HF species with a low lifetime, an apparent mid-FRET species as well as D-only. Quenching will add another de-excitation pathway (k_Q is the rate constant for the quenching reaction, k_0 is the rate constant for the depopulation of the excited state with $k_0 = 1/\tau_{D(0)}$; see Fig. 18.2B). The decay of D and thus the lifetime $\tau'_{D(A)}$ in the presence of quencher will be shorter than the lifetime $\tau_{D(A)}$ in the absence of quencher.

$$\tau'_{D(A)} = \frac{1}{k_0 + k_Q + k_{FRET}} < \frac{1}{k_0 + k_{FRET}} = \tau_{D(A)} \qquad (18.19)$$

In the 2D plot, however, it can be seen that there is no mid-FRET species, because the mid-FRET population is not located on the static FRET curve (Eq. (18.16a), solid line in Fig. 18.7B). This observation indicates that this D-only species has a lifetime in the absence of the acceptor, $\tau'_{D(0)}$, that is shortened

Accurate Single-Molecule FRET Studies Using Multiparameter Fluorescence Detection   483

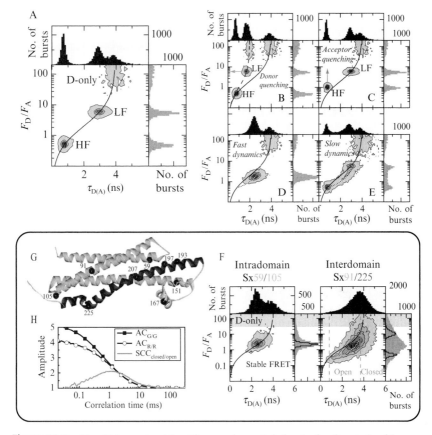

Figure 18.7 Identifying potential artifacts or dynamic behavior by direct visual inspection of the 2D histograms. *Upper panels*: 2D histograms of F_D/F_A versus $\tau_{D(A)}$ for simulations of FRET experiments with two FRET populations and D-only. The effects for the presence and absence of D or A quenching and dynamic interconversion between two different FRET states are considered. (A) Two static FRET populations (LF and HF) and a small percentage of D-only species. All populations are located on the same static FRET curve (full line, Eq. (18.16a)) with the parameters $\Phi_{FD(0)} = 0.8$, $\Phi_{FA} = 0.4$, and $\tau_{D(0)} = 4.0$ ns). (B) Simulation exhibiting the same FRET populations as in panel (A). In the lower FRET population, additional D-quenching reduces $\Phi_{FD(0)}$ from 0.8 to 0.4, so that an additional static FRET curve for D′ ($\Phi'_{FD(0)} = 0.4$, and $\tau'_{D(0)} = 2.0$ ns) must be considered (see Section 4.3 for more details). (C) Same FRET populations as in panel (A). For the HF-FRET population, A-quenching occurs ($\Phi_{FA} = 0.4$ is reduced to 0.2). A deviation of the population from the static FRET line is readily visible (see Section 4.3 for more details). (D) Simulation of dynamic behavior (Case 2, fast equilibrium). The molecules are interconverting between the two FRET states as defined in panel (A) (rate constants for interconversion $k_{1,2} = k_{2,1} = 5$ ms^{-1}). The dynamic population deviates from the static FRET line and must be described by a dynamic FRET line (dashed line, Eq. (18.27a), with the parameters $\Phi_{FD(0)} = 0.8$, $\Phi_{FA} = 0.4$, $\tau_{D(0)} = 4.0$ ns, $\tau_1 = 3.0$ ns, and $\tau_2 = 0.8$ ns). The intersections between the static and dynamic FRET lines give the originating FRET states (1) and (2). (E) Simulation of dynamic behavior (Case 3, slow

by an additional non-FRET-related quenching process. Consequently, an additional static FRET curve is necessary (shown as dashed line in Fig. 18.7B), which was calculated using a different donor lifetime of $\tau'_{D(0)} = 2\,\text{ns}$. As a result, if only the donor lifetime would be used for monitoring FRET then quenching of D could be mistaken as an indication of excessively high-FRET efficiencies. In contrast, data of intensity related FIs from the same measurement lead to an alternative interpretation, in which HF, LF, and D-only species are identified (as in Fig. 18.7A). Using the rate constants defined in Fig. 18.2B, Eq. (18.20) shows that the intensity ratio is not influenced by donor quenching, that is, $(F_D/F_A) = (F_D/F_A)'$:

$$\begin{aligned}\left(\frac{F_D}{F_A}\right)' &= \frac{\Phi'_{FD(0)}}{\Phi_{FA}}\left(\frac{1 - E'_{FRET}}{E'_{FRET}}\right) = \frac{1}{\Phi_{FA}}\frac{k_f}{k_0 + k_Q}\frac{k_0 + k_Q}{k_{FRET}} \\ &= \frac{1}{\Phi_{FA}}\frac{k_f}{k_{FRET}} = \frac{\Phi_{FD(0)}}{\Phi_{FA}}\left(\frac{1 - E_{FRET}}{E_{FRET}}\right) = \frac{F_D}{F_A}\end{aligned} \quad (18.20)$$

---

equilibrium). The molecules are interconverting between the same FRET states as in (A). The interconversion is not as fast as in panel (D) (rate constants for interconversion $k_{1,2} = k_{2,1} = 0.1\,\text{ms}^{-1}$), so that two FRET states are still located on the static FRET line (full line). In addition, a smeared population appears in between them, which is not centered on the static curve. Considering dynamics the whole smeared FRET population now is located on the dynamic FRET line (Eq. (18.27a), same parameters as panel (D), dashed black). *Lower panels*: Analysis of the conformational dynamics of syntaxin-1. (F) 2D histograms of F_D/F_A versus $\tau_{D(A)}$ of the two syntaxin-1 mutants Sx59/105 and Sx91/225, left and right, respectively. The ideal fluorescence was calculated by Eq. (18.32) using following parameters: $\langle B_G \rangle = 7.0$ kHz, $\langle B_R \rangle = 2.5$ kHz, direct acceptor excitation DE = 1.0 kHz, $\alpha = 0.035$ and $g_R/g_G = 1.25$. Two subpopulations are found for the conformationally insensitive mutant Sx59/105, a single static FRET state and a fraction of molecules labeled only with the donor, D-only. In the case of the Sx91-225 mutant, the single static state is described by two slowly interconverting states (open and closed). For both 2D histograms the black overlaid line represents the static FRET line (Eq. (18.16a), with the parameters $\tau_{D(0)} = 4.0$ ns, $\Phi_{FD(0)} = 0.80$, and $\Phi_{FA} = 0.65$). In the right 2D histogram, the dynamic FRET line (dashed line, Eq. (18.27a), with the parameters $\tau_{D(0)} = 4.0$ ns, $\Phi_{FD(0)} = 0.80$, and $\Phi_{FA} = 0.65$, $\tau_1 = 3.7$ ns (open state), $\tau_2 = 0.8$ ns (closed state)) is plotted. (G) Crystal structure of the syntaxin-1 in complex with munc-18 (Misura *et al.*, 2000). The two domains are drawn in light and dark gray respectively. The black numbers represent the positions of the introduced cysteines. The dotted black lines indicate the D–A distances in the intradomain (Sx59/105) and interdomain (Sx91/225) mutants, respectively. (H) FCS curves of the Sx91/225 mutant. Reported are the green, $AC_{G/G}$, and red, $AC_{R/R}$, autocorrelations (black line with black squares and open circles, respectively) and the species cross-correlation closed to open state, $SCC_{closed/open}$ (gray line). As expected the autocorrelations reflect mainly the diffusion properties while the species cross-correlation clearly highlights the presence of a dynamic interconversion of the system with anticorrelation time of 0.6 ms.

4.3.2. Acceptor quenching

The effect of acceptor quenching on the HF species is displayed in Fig. 18.7C. Analysis from the lifetime of D leads to correct FRET efficiencies. In contrast, the intensity-based FIs underestimate E. In the presence of acceptor quenching, only the quantum yield of A will change, $\Phi'_{FA} < \Phi_{FA}$, and as a consequence increases so that $(F_D/F_A)' > F_D/F_A$.

From Fig. 18.7A–C it becomes obvious that the static FRET line offers a guide for direct visual inspection of the data. The fact that one population deviates horizontally from the line offers direct evidence for D quenching, whereas a vertical shift indicates A quenching. In some cases it might be difficult to determine whether a shift is vertical or horizontal (compare Fig. 18.7B and C). In case both dyes might be quenched, additional checks are possible: (i) A good indication for donor quenching is the presence of an additional quenched D-only population (see Fig. 18.7B). (ii) Acceptor quenching can be detected by analyzing the decay of the FRET sensitized A fluorescence (Widengren et al., 2006) or performing additional A lifetime measurements with direct excitation (i.e., by using PIE/ALEX).

4.4. Dynamic behavior

Besides photo-induced dye chemistry and physics, the dynamic interconversion of molecular populations between two or more FRET states results in a systematic deviation of the recorded FRET population from the static FRET line given by Eqs. (18.16). In confocal microscopy the time a molecule stays in the detection volume t_{dwell} (the dwell time is typically a few ms) limits the observation time of dynamic processes (Fig. 18.4). Thereby the dynamics is characterized by the lifetime of a state t_{state}. Depending on the two characteristic times t_{state} and t_{dwell}, three cases must be considered for confocal single-molecule spectroscopy. This consideration is valid for dynamics with state lifetimes larger than the fluorescence lifetime, that is, $t_{state} \gg \tau$.

Case 1 (quasi-static equilibrium, $t_{state} \gg t_{dwell}$): If a dynamic interconversion between states occurs that is much slower than the dwell time, only quasi-static populations will be identified because each molecule will be in the same state while it diffuses through the focus.

Case 2 (fast equilibrium, $t_{state} \ll t_{dwell}$): If a molecule changes its conformation several times during the passage through the detection volume, an average value of the FRET efficiency (weighted by the time the molecule spends in each state and the individual brightnesses) will be recorded. Thus, only the average population will be registered in the 2D histograms for burst-wise analysis (Fig. 18.7D), while individual states will not be "visible." Please note that this population does not fall on the static FRET line (see Eqs. (18.16)). In the next section we will make use of this effect to find the originating states.

Case 3 (slow equilibrium, $t_{state} \approx t_{dwell}$): If the timescale of the dynamics is comparable with the dwell time, then each of the FRET states between which the interconversion occurs is going to be "visible" and the dynamic behavior is detected in the 2D histograms by a "smear" between the two populations (Fig. 18.7E).

In burst-wise analysis, deviations from the static FRET line arise because E values derived from fluorescence intensities are averaged per molecular species fractions, whereas fluorescence lifetimes are averaged per brightness due to the applied MLE (see Section 4.1 and Kalinin et al., 2010b). In the following we make use of this disagreement by employing it as an indicator for the presence of multiple species, which results in a mixed fluorescence signal.

Fluorescence intensities are obtained by single-photon counting and are therefore weighted by the fraction of each species, x_i:

$$\langle F \rangle_{dyn} = \sum_i x_i F_i \qquad (18.21)$$

The species fraction of the ith species is represented by the fraction of time that the molecule spends in that state. The average FRET efficiency and corresponding species weighted average lifetime can be expressed as

$$E_{dyn} = \sum_i x_i E_i \quad \text{and} \quad \langle \tau \rangle_x = \sum_i x_i \tau_i \qquad (18.22)$$

The experimental mean fluorescence lifetime of D is computed by MLE and corresponds to a fluorescence intensity weighted average lifetime $\langle \tau \rangle_f$:

$$\langle \tau \rangle_f = \sum_i f_i \tau_i \quad \text{with} \quad f_i = \frac{x_i \tau_i}{\sum_i x_i \tau_i} \qquad (18.23)$$

To summarize, $\tau_{D(A)}$ in Eqs. (18.16) represents the species-averaged lifetime $\langle \tau \rangle_x$, whereas the MLE provides the intensity-averaged $\langle \tau \rangle_f$. To derive a $(F_D/F_A) - \tau_{D(A)}$ dependence for two interconverting species, one would have to express $\langle \tau \rangle_x$ as a function of $\langle \tau \rangle_f$ and substitute it in the relation between fluorescence intensity ratio (F_D/F_A) and lifetime $\langle \tau \rangle_x$ (see Eqs. (18.16)) (Kalinin et al., 2010b). All FRET species arising from the interconversion between the two FRET states should be located along this modified curve referred to as the dynamic FRET equation:

$$\left(\frac{F_D}{F_A} \right)_{dyn} = \frac{\Phi_{FD(0)}}{\Phi_{FA}} \frac{\langle \tau \rangle_x}{\tau_0 - \langle \tau \rangle_x} \qquad (18.24)$$

For simplicity, the analytical expression of $\langle\tau\rangle_x$ in terms of $\langle\tau\rangle_f$ will be given only for the case of two interconverting FRET species. Let us assume two FRET species characterized by the following set of parameters $(\tau_1, F_{D,1}, F_{A,1})$, $(\tau_2, F_{D,2}, F_{A,2})$. For this case the different lifetime averages are

$$\langle\tau\rangle_x = x_1\tau_1 + (1-x_1)\tau_2 \qquad (18.25a)$$

$$\langle\tau\rangle_f = f_1\tau_1 + (1-f_1)\tau_2 = \frac{x_1\tau_1^2 + (1-x_1)\tau_2^2}{\langle\tau\rangle_x} \qquad (18.25b)$$

which can be rearranged to

$$\langle\tau\rangle_x = \frac{\tau_1\tau_2}{\tau_1 + \tau_2 - \langle\tau\rangle_f} \qquad (18.26)$$

Substituting Eq. (18.26) into Eq. (18.24) yields the corresponding dynamic FRET equations:

$$\left(\frac{F_D}{F_A}\right)_{dyn} = \frac{\Phi_{FD(0)}}{\Phi_{FA}} \frac{\tau_1\tau_2}{\tau_{D(0)}[\tau_1 + \tau_2 - \langle\tau\rangle_f] - \tau_1\tau_2} \qquad (18.27a)$$

$$E_{dyn} = 1 - \frac{\tau_1\tau_2}{\tau_{D(0)}[\tau_1 + \tau_2 - \langle\tau\rangle_f]} \qquad (18.27b)$$

Equations (18.27) require the knowledge of the two lifetimes of the inter-converting states and as such can only be used in combination with other independent measurements of the single states or as an *a posteriori* test using TCSPC in a subpopulation. This method is perfectly suited to detect heterogeneities (here, two FRET states) that are longer lived than the fluorescence lifetime.

4.4.1. Example: Conformational dynamics of syntaxin-1 (Sx)

Syntaxin-1 is a member of the SNARE superfamily and mediates neuronal exocytosis. While performing its function the protein must adopt various conformations (open and closed) depending on its interaction partners (Cypionka et al., 2009; Margittai et al., 2003). This conformational change is due to the motion between two different protein domains colored dark and light gray in Fig. 18.7G. Complex formation between Sx and munc-18 favors the closed conformation, whereas Sx binding to synaptobrevin and SNAP-25 results in the occupation of the open complex. It was previously unknown what conformation the free syntaxin-1 adopts in solution and therefore MFD experiments were performed on a syntaxin-1 mutant containing a FRET pair in the conformationally sensitive area (Sx91/225). From the 2D histograms of F_D/F_A versus $\tau_{D(A)}$ derived from smFRET experiments (Fig. 18.7F), it is evident that both the open and closed conformation coexist in solution. In Fig. 18.7F, a broad smeared peak is visible for Sx91/225, situated between the open and closed states (dashed gray lines). The position of the population in the 2D histogram is not described by the static FRET line (Eq. (18.16a), black line) but rather by the dynamic FRET line (Eq. (18.27a), dotted white line) (Kalinin et al., 2010b).

Next, we compared Sx91/225 with the variant Sx59/105, where the FRET pair is located within a single stable domain (light gray). The labeled Sx59/105 variant should be conformation insensitive and therefore no difference in FRET is expected when syntaxin adopts the open and closed conformations. As expected, MFD FRET analysis shows that Sx59/105 is not sensitive to the dynamics as it has a rather narrow population in the 2D F_D/F_A versus $\tau_{D(A)}$ histogram and its population is located on the static FRET line. To finally confirm the presence of conformational dynamics in free Sx by an alternative analysis method, we calculated the fluorescence species cross-correlation function (SCC) (Felekyan et al., 2009; Gregor and Enderlein, 2007) for the mutant Sx91/225 (Fig. 18.7H, gray line) using the same data. The SCC was computed utilizing the MFD information (in this case $\tau_{D(A)}$) to assign, with a certain probability, each photon to only one of the subpopulations (open or closed), which were then correlated against each other. In this way, any dynamic interconversion between the two subpopulations would give rise to a marked anticorrelation term in the cross-correlation curve that, together with the diffusion term (visible in the green and red autocorrelation functions (AC_{GG} and AC_{RR}), full square and open dots, respectively), produces a characteristic bell shaped curve as seen in the case of Sx91/225 (Fig. 18.7H). These studies are a good example for how MFD makes use of the specific strengths of various distinct analysis techniques. While 2D parameter analysis is essential to analyze the heterogeneity of an ensemble, other techniques such as correlation analysis are required to characterize the time constants for dynamics within the ensemble. Considering the Sx91/225 case, the analysis of the SCC yields a relaxation time of ~ 0.6 ms for the conformational dynamics, which agrees well with the results obtained from other analyses (Margittai et al., 2003).

4.4.2. Possible artifacts

An effect similar to dynamic averaging of FRET states can be observed for a case where the concentration of a heterogeneous sample solution is too high. If more than one molecule is simultaneously present in the focus, the signal of different species can be mixed. The two sources of averaging can be easily discriminated by: (i) the detection of any dynamic term via dynamic PDA (Section 5.4) or FCS measurement, (ii) analysis of the number of recorded bursts per second, which should be less than 2 bursts/s for a characteristic FCS diffusion time of ~ 1 ms to minimize the likelihood of multimolecule events (Fries et al., 1998; Orte et al., 2006).

To summarize, MFD analysis as described in Section 3 is a powerful tool for identifying FRET populations and for qualitatively detecting the presence of dynamic behavior or artifacts induced by photophysics. If, however, a quantitative description of the system (i.e., the molecular fractions of subpopulations or the number and widths of the distributions) is needed, additional analytical tools have to be used. In this context one must consider the fact that the actual width of each distribution of FRET efficiencies is a convolution of many factors, including the photon shot noise (Antonik et al., 2006), actual interdye distance distributions (Sindbert et al., 2010), linker flexibilities (Sindbert et al., 2010), and dye photophysics (Kalinin et al., 2010a). Moreover, in case of dynamic interconversion between two FRET states (Kalinin et al., 2010b), one is interested in obtaining the kinetic rate constants from this measurement. In order to answer these questions, photon distribution analysis (PDA) has been developed and successfully employed as will be described in the next section (Antonik et al., 2006; Kalinin et al., 2007).

5. QUANTITATIVE DESCRIPTION OF SMFRET

5.1. Basic theory of photon distribution analysis (PDA)

The emission of a fluorescence photon is a stochastic process. For this reason the recorded fluorescence signal is naturally distributed around its mean value (shot noise). Structural heterogeneities, dynamic behavior, complex dye photophysics, and numerous artifacts are known to induce an additional broadening of the signal distribution. Traditionally, FRET efficiency distributions derived from single-molecule experiments are fitted with a weighted sum of Gaussian distributions. Although this approach is usually convenient to determine the mean efficiencies for well-resolved states, the information contained in the width is completely ignored. In contrast, PDA makes use of the fact that the shape and width of a FRET distribution can be accurately predicted since they are strictly defined by the underlying stochastic and molecular processes.

PDA is used in fitting experimental FRET efficiency distributions in order to extract the parameters of individual states and to detect and characterize any additional broadening effect. For a single FRET state, the D–A distance R_{DA} or, equivalently, E is the only fitting parameter to define the mean, the width, and the shape of the E distribution. Briefly, the probability of observing a certain combination of green and red photon numbers, $P(S_G, S_R)$, is calculated as

$$P(S_G, S_R) = \sum_{F_G+B_G=S_G; F_R+B_R=S_R} P(F)P(F_G, F_R|F)P(B_G)P(B_R) \quad (18.28)$$

In Eq. (18.28), F and B denote the number of fluorescence and background photons, respectively, S is the total measured signal ($S = F + B$), and $P(F)$ and $P(B)$ represent the fluorescence and background intensity distributions, respectively. In PDA, to simplify the analytical description of the different distributions, the measurement is first divided into time bins (time windows) of fixed length. $P(B)$ obeys a Poissonian distribution with a known mean $\langle B \rangle$ equal to the average recorded background signal. The distribution $P(F)$ can be deconvoluted from the experimental signal intensity distribution $P(S)$ as described (Kalinin et al., 2007). A useful approximation to Eq. (18.30) for $S \gg B$, which avoids deconvolution of $P(F)$, reads (Antonik et al., 2006):

$$P(S_G, S_R) = \sum_{F_G+B_G=S_G; F_R+B_R=S_R} P(S) P(F_G, F_R | S - B_G - B_R) P(B_G) P(B_R)$$

(18.29)

The use of the experimentally measured distribution $P(S)$ in Eqs. (18.28) and (18.29) is a key feature of PDA that distinguishes this method from other approaches to describe the photon statistics of an smFRET experiment (Gopich and Szabo, 2005; Kask et al., 2000). The conditional probability $P(F_G, F_R | F)$ in Eqs. (18.28) and (18.29) can be expressed as a binomial distribution (Antonik et al., 2006):

$$P(F_G, F_R | F) = \frac{F!}{F_G! F_R!} (p_G)^{F_G} (1 - p_G)^{F_R}$$

(18.30)

In Eq. (18.30), the probability to register a "green" photon p_G is determined by the FRET efficiency and fixed experimental parameters as given by

$$p_G = \left(1 + \alpha + \frac{E}{1-E} \frac{g_R \Phi_{FA}}{g_G \Phi_{FD(0)}}\right)^{-1}$$

(18.31)

Equation (18.31) describes the probability to detect a green photon rather than the probability of emitting a donor photon, that is, it takes various experimental imperfections into account. For an "ideal system" ($\gamma = 1$ and $\alpha = 0$), Eq. (18.31) naturally converges to $p_G = 1 - E$.

Experimental $P(S_G, S_R)$ distributions are typically used to generate 1D histograms of E or any other FI (e.g., proximity ratio $S_R/(S_G + S_R)$ or signal ratio S_G/S_R), which are fitted by a theoretical distribution to determine E. The quality of the fit is judged by values for χ_r^2 and analysis of weighted residuals. In this way, PDA is an indispensible tool to judge distributions quantitatively. Any deviation from the shot-noise limit gives us more information on our system as shown in the next sections.

5.2. FRET analysis for acceptors with additional direct excitation

If the absorption and fluorescence spectra of donor and acceptor are less well separated (compare in Fig. 18.2C), one must consider the possibility of additional direct acceptor excitation besides FRET-mediated excitation, that is, even if an excitation wavelength is chosen close to the absorption maximum of the donor dye, there will be a low probability p_{DE} that the acceptor can be directly excited as well. This is important for the FRET DA-pair Alexa488–Alexa594, which is frequently used in single-molecule FRET studies. Thus, the true acceptor fluorescence, F_A, defined in Eq. (18.13) and used for 2D-FRET analysis, must be corrected for additional direct excitation DE as given by Eq. (18.32):

$$F_A = \frac{S_R - (\langle B_R \rangle + DE) - \alpha(S_G - \langle B_G \rangle)}{g_R} = \frac{F_R}{g_R} \qquad (18.32)$$

This weak additional A excitation can also be advantageous (Gansen et al., 2009), since it allows us to distinguish between DA species exhibiting no FRET (i.e., $R_{DA} > 100$ Å) and D-only species without performing an additional and more complex nsALEX or PIE experiment (see Section 3.3).

Moreover, the formalism of PDA has to be adapted for this situation. Let us assume that the direct acceptor excitation probability is p_{DE} as shown in Scheme 18.1.

Scheme 18.1 shows all processes considered in PDA together with their probabilities. It follows directly that the probability for a detected photon to be green is given by Eq. (18.33), which replaces Eq. (18.31):

$$p_G = \frac{(1 - p_{DE})(1 - E)\Phi_{FD(0)}g_G}{(1 + \alpha)(1 - p_{DE})(1 - E)\Phi_{FD(0)}g_G + (p_{DE} + (1 - p_{DE})E)\Phi_{FA}g_R}$$

$$= \left(1 + \alpha + \frac{(p_{DE}(1 - p_{DE})^{-1} + E)\Phi_{FA}}{(1 - E)G\Phi_{FD(0)}}\right)^{-1}$$

$$(18.33)$$

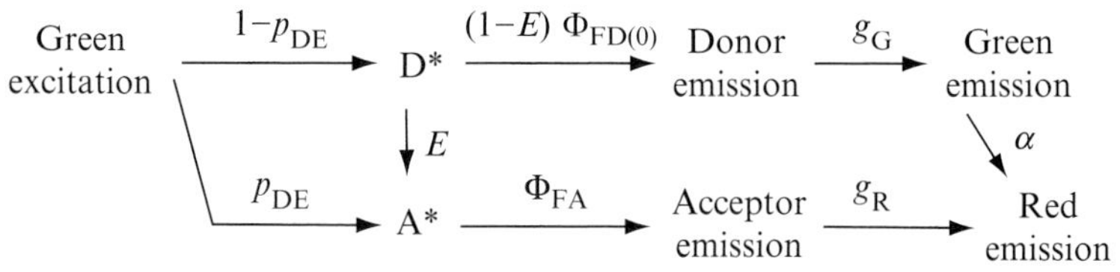

Scheme 18.1 Excitation and emission pathways entering the PDA model function.

For a donor excitation wavelength of 476 nm, p_{DE}(Alexa594) amounts to ~ 0.035. Using the band-pass filters HQ 520/66 nm and HQ 630/60 nm for the donor and acceptor channels, respectively, the spectral donor cross talk amounts to $\alpha = 0.07$.

5.3. Shot-noise limited FRET signal distributions and additional broadening

PDA and related methods (Gopich and Szabo, 2005; Nir et al., 2006) determine the minimum width of a FRET two-color signal distribution. Figure 18.8 (upper panel) shows an example of broadening arising solely from the statistical (Poissonian) nature of the emission of a fluorescence photon (photon shot noise). Experimental conditions such as detection efficiencies, cross talk, and background values strongly influence the shot noise. Taking these parameters into account, PDA predicts the shape of a FRET signal distribution that arises from a single intensity ratio and correspondingly a single fixed distance (Fig. 18.8A).

Conversely, PDA can identify cases where the broadening of the distribution from photon shot noise is larger than expected (Fig. 18.8B). In practice, it frequently happens that FRET populations are best described by Gaussian distributions of donor–acceptor distances with a mean D–A distance $\langle R_{DA} \rangle$ and a half-width, hw. Many reasons can cause the extra broadening, most importantly actual interdye distance distributions, variations in the Förster radius due to heterogeneous populations (in terms of quantum yield and mobility) of the dye species used as a FRET pair, and dynamic interconversion between distinct FRET species (Antonik et al., 2006; Nir et al., 2006; Vogelsang et al., 2007). However, one should note that the millisecond "integration time" of PDA implies that fast processes should cause no extra broadening of E distributions (see Fig. 18.4 and Section 4.4). An important special case is the dye linker dynamics that typically takes place on the sub-μs timescale (Sindbert et al., 2010).

PDA alone cannot identify the underlying cause of additional broadening. Thus, the extraction of biologically relevant information from the width of a FRET distribution becomes a treacherous task since artifacts arising from photophysics can be mistakenly considered to reflect an actual interdye distance distribution of biological importance. For example, we have recently shown that if Cy5 is used as the acceptor dye, the distribution of its fluorescence quantum yields and lifetimes represents the major contribution to the broadening of the E distribution (Kalinin et al., 2010b). We derived a simple theory that predicts that the half-width, hw, of an apparent distance $\tilde{R}$ increases linearly with the mean physical (real) donor–acceptor distance R_{DA} due to variations of the acceptor properties, which results in a slope characterized by the fluorescence quantum yield $\Phi_{F(A)}$ and its variance (var) according to (Kalinin et al., 2010a):

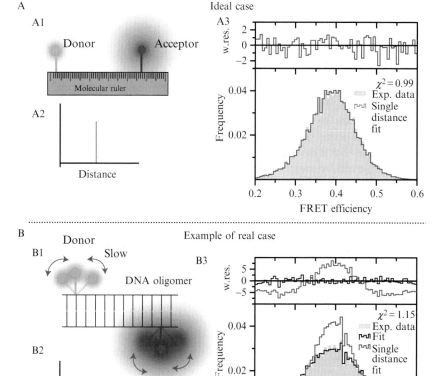

Figure 18.8 Analysis of FRET efficiency distributions by PDA. *Upper panel*: (A1) Ideal case, in which the FRET serves as a "molecular ruler." (A2) A single FRET efficiency value is expected assuming that R_0 is known and the interdye distance is fixed. Nevertheless, a distribution of FRET efficiencies is recorded due to photon shot noise. (A3) This case can be realized experimentally by separating the emission spectrum of a dye in two spectral regions and consider one region as the "green" and the other as the "red" detection channels (pseudo-FRET experiment). In this case a single dye Atto 590 is used and its fluorescence is monitored in two spectral channels. Using the following parameters ($\langle F_G/F_R \rangle = 4.02$, $\langle B_G \rangle = 3.25\,\text{kHz}$, $\langle B_R \rangle = 0.76\,\text{kHz}$, $\alpha = 0$, and $1/\gamma = 0.35$), PDA predicts nicely the distribution of recorded efficiencies using a model function of just a single fixed distance (Antonik *et al.*, 2006). *Lower panel*: (B1) Example of a real case. Assume that one labels a DNA oligonucleotide with a FRET dye pair, whereby the dye linkers are quite rigid. (B2) As a result, there will not be a fixed single D–A distance but a distribution of interdye distances. (B3) A model with a single fixed distance fails to fit the data; however, a model function accounting for a Gaussian distribution of mean distance, $\langle R_{DA} \rangle = 53.9\,\text{Å}$ and $\sigma_{DA} = 2.3\,\text{Å}$ is well suited (see residuals). PDA parameters: $\langle F_G/F_R \rangle = 1.58$, $\langle B_G \rangle = 3.83\,\text{kHz}$, $\langle B_R \rangle = 1.15\,\text{kHz}$, $\alpha = 0.01$, $\Phi_{FA} = 0.32$, $\Phi_{FD(0)} = 0.80$, $g_R/g_G = 2.2$.

with either a static or a dynamic model, using different time-window widths. Clearly, only the dynamic model can produce a reasonable fit to the data for both time windows, $\Delta t = 4$ ms and $\Delta t = 1$ ms, which allows one to rule out a quasi-static equilibrium.

Although the dynamic range of PDA is not comparable to that of FCS (Haustein and Schwille, 2007), it covers a useful range of about 0.1–10 ms and has the unique advantage that the species distributions are resolved directly. In this way, dynamic-PDA complements FCS for cases where the dynamic FRET-anticorrelation term in the cross-correlation function is hidden by the diffusion or any other decay (bunching) term (see Fig. 18.7H and Margittai *et al.*, 2003).

5.6. Methods to calibrate and appropriately display MFD data

For quantitative analysis of FRET efficiencies (Eq. (18.14)) and polarization anisotropies (Eq. (18.17)), ratios of detection efficiencies g_G/g_R and $G = g_\perp/g_\|$ are needed. To determine these parameters the following procedures are employed in our group.

5.6.1. G-factor for polarization

The G-factor is calculated from a measurement of a dye in water (e.g., Rhodamine 110 is used to calibrate the donor channels). It is known that for small molecules the rotational correlation time ρ is about 100–200 ps. Therefore, at a microtime t larger than the rotational correlation time ρ no residual polarization is expected for fluorescence F (i.e., $F_\| = F_\perp$; see Section 3.3). Thus, any observed polarization is due to $g_\perp/g_\| \neq 1$. This feature is used to compute the G-factor as

$$G = \frac{\sum_{t_i \gg \rho}^{t_{\max}} F_\perp(t_i)}{\sum_{t_i \gg \rho}^{t_{\max}} F_\|(t_i)} \quad (18.35)$$

In Eq. (18.35), the fluorescence decay is integrated over several TCSPC histogram channels t_i (corresponding to $\sim$6–13 ns for our setup). Concentrated solutions ($>$2–3 molecules in the detection volume) are used to minimize the background contribution to the signal S (i.e., to assure $S \cong F$). Note that this procedure is sensitive to time shifts between the detection channels that must be taken into account (Felekyan *et al.*, 2005).

5.6.2. g_G/g_R ratio for spectral sensitivity

The g_G/g_R ratio can be determined from FRET data obtained for simple rigid molecules, such as double-labeled dsDNA. For such molecules, Eqs. (18.16) is expected to be fulfilled. For higher reliability we use a mix of at least two different dsDNA oligonucleotides of known properties with low and high E (e.g., Fig. 18.6). A 2D histogram of E versus $\tau_{D(A)}$ is generated from experimental data. By varying g_G/g_R until the static FRET line (Eqs. (18.16)) goes through all observed FRET populations (according to visual inspection), g_G/g_R is accurately determined (± 0.05). Once this procedure is performed for a given setup, a dye with sufficient fluorescence in all detection channels (such as Rhodamine 101) can be used to correct for daily variations of g_G/g_R. Obviously, the g_G/g_R ratio scales linearly with F_G/F_R obtained for such a dye. We have also determined the fluorescence quantum yields of D and A in our dsDNA standard sample to determine the interdye distance R_{DA} from Eqs. (18.7). We tested the precision of the above calibration procedure by determining experimental standard deviations via repeated measurements. To this end, we measured the same sample many times within half a year and analyzed the data by PDA (compare in Fig. 18.8B). The very small experimental standard deviations for the mean distance R_{DA} ($R_{DA} = 58.3 \pm 0.5$ Å) and the broadening characterized by the half-width ($hw = 2.4 \pm 0.5$ Å) indicate a very stable calibration of the setup.

5.6.3. Display of FRET indicators

Several aspects are important for the choice of a FRET indicator FI: (i) discrimination between distinct species; (ii) display of all molecular events; and (iii) effect of the background correction. Each FI has its own advantages and disadvantages, so that there is no unique choice.

In our experience, the representation of ratiometric FIs, such as the ratio of green to red signal, S_G/S_R, and the corrected ratio, F_D/F_A, have the specific advantage that the width of the shot-noise limited distributions in logarithmic scale does not vary drastically over the range of interest (e.g., $0 \leq E \leq 1$) so that complex mixtures with a wide range of parameters can be directly visually analyzed. However, the ratiometric parameters have the disadvantage that they have ill-defined boundary cases: (i) infinity, if the denominator is zero; (ii) no real value if the nominator is zero and a logarithmic x-scale is used. Because of these boundary cases, some data points cannot be displayed on a signal ratio histogram with a logarithmic x-axis (e.g., all points with $S_G = 0$ or $S_R = 0$), which biases the data analysis for small photon numbers.

Normalized functions such as the proximity ratio or FRET efficiency are usually displayed using limited linear scales with a small dynamic range. This approach results in a disadvantage as specified in the following example. On a logarithmic ratio scale, two species with signal ratios of 100 and 10 can be easily distinguished. However, on a linear scale for the proximity ratio these two species fall into a few neighboring bins of the PDA histogram, that is, the shot-noise limited width of normalized parameters, which is proportional to $\langle E \rangle (1 - \langle E \rangle)$ (Gopich and Szabo, 2005), changes drastically on the linear scale of the histogram. This makes a visual interpretation of the histograms very difficult. Normalized FIs have the advantage, however, that they are defined for any total signal $S > 0$. Thus, no data are excluded from the PDA histograms, even those at low photon numbers. Thus, normalized FIs are the preferred display parameters for a quantitative analysis of molecular fractions. If one is interested in a qualitative survey of the different species in a sample, by contrast, a ratiometric FI is the best choice to display the data.

Commonly used spectroscopic parameters, such as the FRET efficiency or F_D/F_A, cannot be directly measured in single-molecule experiments. They can only be calculated, if the contributions of the background signal and (partly unknown) calibration factors are known (Eqs. (18.12)–(18.14)). In contrast to experimental FIs, such as the proximity or signal ratio, spectroscopic parameters have the advantage that their mean values are directly related to molecular properties of interest. In these calculations usually mean background intensities are subtracted, however, so that in some cases E distributions with negative values are obtained, which have no spectroscopic meaning. The reason for why background corrected parameter distributions are partly distorted is the fact that the probability to measure a certain total signal value is a product of the independent probabilities to register fluorescence and background signal values, which corresponds to a convolution of their intensity distributions (Section 5.1 and Kalinin et al., 2007). This result is clearly not equivalent to the subtraction of a mean background value from a parameter distribution.

In summary, we can give the following recommendations. Due to its unsurpassed resolution F_D/F_A is our preferred FI for a qualitative sample survey in a 2D diagram with FRET lines (see Fig. 18.7). Due to its complete coverage of all single-molecule events, E is our preferred FI for a quantitative population analysis with 1D PDA histograms.

5.7. High-resolution FRET structures

FRET-based structure determination will have significant impact in especially two major research areas. Firstly, the complexity and size of multimolecular complexes make any effort to unravel their architecture and function a great challenge. High-resolution crystallographic structural information is often available for individual members of a complex. The overall assembly into a functional machinery, however, is difficult to measure by traditional structure determination. Whether this question can be solved simply by docking the individual components as rigid bodies into lower resolution structures has to be carefully tested for each study (nice examples are: (i) ensemble FRET-study of the structural organization of the bacterial RNA polymerase-promoter open complex (Mekler et al., 2002); (ii) smFRET-derived model of the synaptotagmin 1-SNARE fusion complex (Choi et al., 2010)). Secondly, a quickly growing research area addresses the problem that many proteins or regions of proteins are structurally disordered in their native functional state, that is, they cannot be adequately described by a single equilibrium 3D structure. The existence of so-called intrinsically disordered proteins (IDPs) and intrinsically disordered regions (IDRs) stands in contrast to the traditional view of protein function, which is shaped by well-defined static protein structures. An increasing number of experiments, reviewed for example in Tompa and Fuxreiter (2008) and Henzler-Wildman and Kern (2007), indicate that the function of biomolecules is often governed by their dynamic and/or "fuzzy" character. As a single-molecule technique, smFRET is perfectly suited to study the dynamics and disorder of biomolecules.

5.7.1. Improving accuracy of FRET-based structures

To further increase the degree of acceptance of FRET-based structural studies in the scientific community, routine procedures to establish their accuracy have to be performed. In the literature there are numerous interesting FRET studies for which the anisotropy information is absent and orientational effects are only discussed as possible error sources for distance determination. In these studies any severe influence of orientational effects on FRET is excluded mainly by "biological controls" or separate steady-state anisotropy measurements (Ha, 2001), which is usually sufficient for the study of kinetics with qualitative structural information. If, however, FRET is used to determine biomolecular structures, all possible parameters affecting distance must be taken into account. The observation of a small deviation between the distances measured by FRET and those defined by crystallographic or NMR experiments is not uncommon. Such deviations are usually attributed to the flexible linkers used to attach the dye to the biomolecule. The major problem for obtaining absolute distances from FRET measurements is caused by the uncertainty in fluorophore movement and volume of occupancy because of long flexible linkers. On the one hand, orientational freedom is a prerequisite to safely assume an orientation factor κ^2 of 2/3; in this respect, the use of long flexible linkers is beneficial. On the other hand, flexibility results in an undefined fluorophore position, which prevents fitting FRET data to geometric models based on mean dye positions. In Wozniak et al. (2008), a new conversion function for experimental FRET efficiencies is presented to determine accurate distances by FRET. It considers the dynamics of the movement of the fluorophore-linker system and takes into account the fluorophore volume of occupancy, which is derived from molecular dynamics (MD) simulations (Best et al., 2007; Margittai et al., 2003). Alternatively, if the local structure of the biomolecule in the proximity of the dye with a long linker is known, geometric modeling of the volume accessible to the dye can be employed (Andrecka et al., 2008; Muschielok et al., 2008; Sindbert et al., 2010).

The resolution of smFRET studies can be increased to such an extent that, for example, DNA bending und kinking can be described quantitatively as described next.

5.7.2. Example: DNA structure determination

One of the newest developments involving MFD is a systematic study to improve methodologies for determining fluorophore positions attached to biomolecules, in this case dsDNA (Wozniak et al., 2008). Although the structure of DNA is well known and hence constitutes a "model system" for FRET-based structure determination, the uncertainty in fluorophore positions when coupled to DNA can severely limit the usefulness of the derived distances.

In this study, the starting point was the simulation of the potential dye positions (both D and A) coupled to a dsDNA, taking into account the linker length (Fig. 18.10A). Due to the long linkers used, the dye can explore a large space (± 20 Å). The available space for the dye is determined as the space sterically allowed by the DNA structure and the length of the linker. Unfortunately, the simulated distances between mean D and A positions (R_{mp}) cannot be directly compared with the experimental FRET-weighted mean D–A distance values $\langle k_{DA} \rangle_E$ because FRET efficiencies and distances are averaged differently, that is, $E(\langle R_{DA} \rangle) \neq \langle E(R_{DA}) \rangle$ and $\langle R_{DA} \rangle \neq R_{mp}$. Additionally, the timescale of the translational and orientational diffusion of the dye linker will dictate the proper averaging scheme to be followed.

Considering the rate constant for k_{FRET} (Eq. (18.1)) and the reciprocal diffusion translational and rotational times, giving rate constants k_d and k_R, respectively, three experimentally relevant theoretical boundary cases can be identified: dynamic (k_R, $k_d \gg k_{FRET}$), static (k_R, $k_d \ll k_{FRET}$), and isotropic ($k_R \gg k_{FRET} \gg k_d$) averaging regimes. The mean transfer efficiency calculated from the simulations is different for each of these cases: $\langle E \rangle_{dyn} \neq \langle E \rangle_{stat} \neq \langle E \rangle_{iso}$.

It is then clear that the connection between the transfer efficiency obtained from the measurement and the real interdye distance strongly depends on the averaging model chosen for analysis. It is evident from Fig. 18.10B that the FRET efficiency decreases with the rate of dye dynamics. Thus, the choice of the correct averaging scheme becomes important. For this choice, information about the relation of k_R, k_{FRET}, and k_d is required, and MFD offers the opportunity to accurately determine the averaging regime based on experimental data (anisotropy decay curves) and not on *a priori* assumptions provided (in the best case) from bulk studies. In our study we showed via anisotropy and FCS studies that $k_R > k_{FRET} > k_d$, suggesting that isotropic dynamic averaging should best describe our data. From this result, an empirical relation between the FRET efficiency in the

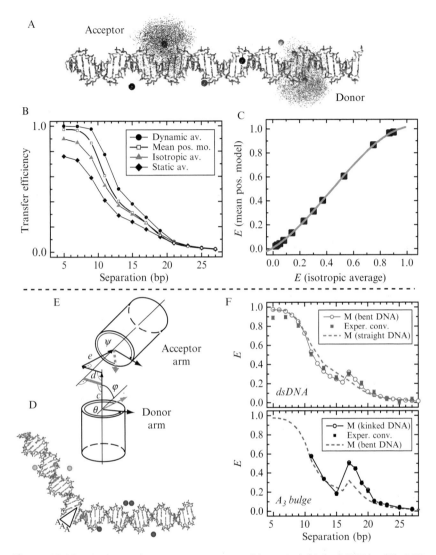

Figure 18.10 Accurate FRET measurement of bent and kinked DNA. (A) MD simulations of the positions of the fluorophores on B-DNA. For the case of 21 bp separation, 5157 possible positions of the reference atoms for the donor (O7—light gray cloud) and the acceptor (C27—dark gray cloud) are illustrated. Here Alexa488 acts as donor and Cy5 as the acceptor. The average positions of the fluorophores for all labeling positions are shown as light and dark gray circles respectively. (B) Transfer efficiency values as predicted from the simulation for the various FRET averaging schemes. (C) Illustration of the empirical relation which connects the transfer efficiencies of the isotropic and the mean position model, $y = 0.008 + 0.679x + 1.470x^2 - 1.141x^3$. (D) Kinked DNA structure induced by the insertion of an A_3 bulge. The various labeling positions for the donor and the acceptor are shown as circles. (E) Geometric model of the kink between the two DNA arms (see text for details).

isotropic averaging scheme and the FRET efficiency that corresponds to the mean position were derived (Fig. 18.10C).

The determination of precise mean dye positions gave us the opportunity to study the implications of sequence-specific bending on the 3D structure of a dsDNA construct. We combined three distinct donor and four acceptor positions to obtain a set of dsDNA molecules with 12 single FRET pairs, where the D–A distance increases mainly in steps of 2 bp to cover a distance range of 30–100 Å. These constructs were used in smFRET measurements (full squares, Fig. 18.10F, upper panel) to detect systematic deviations from a model for a straight helix (dashed line, Fig. 18.10F, upper panel) that manifested in systematic deviations centered around 17 bp, much larger than the experimental error. The high accuracy of our new approach allowed us to detect sequence-dependent dsDNA bending by 16°, in excellent agreement with structure predictions by the program DIAMOD (Dlakic and Harrington, 1998) using NMR-based dsDNA structure parameters (Gabrielian and Pongor, 1996) (open circles, Fig. 18.10F, upper panel). An analogous approach was applied to measure kinked dsDNA of unknown structure. We induced kinks into the above dsDNA sequence by insertion of unpaired adenosines, so-called A_x bulges. Considering a sample containing three adenosines (A_3 bulge) in the donor strand of the dsDNA (Fig. 18.10D), the dependence of E on the interdye separation (full squares, Fig. 18.10F, lower panel) differs strongly from that of bent dsDNA without A_x bulge. In contrast to earlier ensemble FRET studies (Gohlke *et al.*, 1994; Stühmeier *et al.*, 2000), the FRET data of the systematic MFD approach allow us to use a detailed geometric model for kinked dsDNA, as depicted in Fig. 18.10E with three angles and three coordinates for the intersection point of the two main helical axes. To obtain unique structures we performed a quantitative statistical analysis of the conformational search in full space based on triangulation, which uses the known features of helical B-DNA. For the dsDNA with the A_3 bulge we obtained model parameters for the kink angle ($\varphi = 56 \pm 4°$) and two rotation angles ($\theta = 34 \pm 6°$ and $\psi = 38 \pm 9°$) of the helices of the donor and acceptor helix arms,

---

(F) All experimental data were converted by the polynom in panel (C). *Upper graph*: The deviations of the ideal model for straight B-DNA, M(straight DNA), are significantly larger than the experimental error bars. We created a model, M(bent DNA), using the program DIAMOD (Dlakic and Harrington, 1998) and the parameters from Gabrielian and Pongor (1996), which describes the experiment very well without the adjusting any parameter. *Lower graph*: Exceptional agreement between the measured FRET efficiencies for the dsDNA with the A_3 bulge and the transfer efficiencies predicted by the geometric model by applying the experimental FRET constraints (kink angle ($\varphi = 56 \pm 4°$), two rotation angles ($\theta = 34 \pm 6°$ and $\psi = 38 \pm 9°$) of the two helical arms around their individual axes, shifts of the acceptor-carrying arm, one parallel ($c = 3.7 \pm 1.3$ Å) and two perpendicular ($d = -3.2 \pm 2.5$ Å, $e = -0.16 \pm 1.9$ Å) to the axis of the donor-carrying arm (E)).

respectively, that match the experimental FRET data perfectly (open circles, Fig. 18.10F, lower panel).

In conclusion, the structural characterization of the kinked DNA samples in Fig. 18.10 is based on a geometric model, which uses the knowledge of general structural and molecular features. In the future, a combination of FRET-based structure prediction with template-based (or comparative) modeling (Dunbrack, 2006) harbors the potential for a further increase in resolution in order to dissect local molecular features in still greater detail. These studies on a "generally well-behaved" biomolecule, such as dsDNA, have deepened our understanding of how complications originating from orientation, mobility, and position of fluorophores have to be taken into account to determine smFRET-related distances with high accuracy.

6. DISCUSSION

6.1. Recommendations for data analysis

6.1.1. MFD 2D analysis

In order to explore the structure and the conformational landscape of biomolecules by smFRET, the recorded transfer efficiency should only be sensitive to changes of the interdye distance. Is this true in real everyday FRET experiments? In the quest for an ideal system, the experimentalist should ascertain that upon conformational changes two assumptions are fulfilled: (i) the photophysics of the donor and acceptor must not change; and (ii) the mobility of each dye should not change significantly. In most cases these conditions are indeed fulfilled by a careful choice of the dye pair and the labeling positions. It is therefore recommended to create a larger number of molecular constructs with various dye labeling schemes and to proceed to the biologically relevant studies with the construct with the most promising photophysical properties. In addition one should, if possible, verify the biological function (e.g., binding, catalysis, stability) of the labeled biomolecule in order to ensure that labeling does not perturb the system significantly. Nevertheless, such procedures can be painstaking as several independent control measurements at the ensemble level have to be performed to determine all the photophysical characteristics of the donor and acceptor under all biologically relevant conditions. It is therefore not uncommon that the underlying assumptions are not fully validated. The use of the MFD approach in smFRET experiments allows for direct testing of the photophysical assumptions simply by visual inspection of the 2D histograms presented in Section 4 (Fig. 18.7). This feature allows the experimentalist to perform controls using data generated from the same measurement. This is of great importance, especially for the study of biological processes that strongly depend on the concentration used so that the outcome of bulk measurements cannot easily be used as a control for a single-molecule study of the same process.

6.1.2. PDA

With the use of PDA (or related techniques) the contribution of photon shot noise to the apparent width of the FRET populations can be accounted for. In many cases, the width of the FRET distribution has been interpreted as reflecting an actual distribution of distances. In contrast, it is evident by now that a major contribution to the width observed is due to photon shot noise (see Fig. 18.8). Therefore, biological relevance should only be assigned to broadening that cannot be accounted for by photon shot noise or photophysics of the acceptor (Kalinin et al., 2010a; Nir et al., 2006; Vogelsang et al., 2007).

Although PDA has shown its strengths (Gansen et al., 2009), the reader is warned that the fitting procedure is not always straightforward. One usually has to employ a model of several states to fit the data, the choice of the model function may be crucial, and it is not unusual that a decision about the right model cannot be based purely on statistical arguments (χ_r^2 values, F-test). We note that PDA analysis is one-dimensional and depends only on intensity values. To achieve a solid fitting procedure, it is suggested to always thoroughly check the photophysics of the acceptor before choosing the model function and to test for the presence of dynamic terms in the FRET populations by fitting the data with the same model over different time windows (Table 18.1). If the acceptor has multiple brightnesses (see Section 5.3, Eq. (18.34)), the half-width, hw, of an apparent distance distribution increases as a certain fraction of the apparent mean interdye distance $\tilde{R}$ (in case of Cy5 attached to dsDNA this fraction is 7.6 ± 1.2%).

When fitting smFRET data, it is quite common that a number of nonbiologically relevant FRET populations appear. Due to incomplete labeling a population of D-only molecules is always present. This population is usually easy to be excluded from further analysis and does not cause considerable trouble in the analysis except for cases in which a really low FRET population ($E < 10\%$) is present. In addition, due to impurities in the sample preparation it is not uncommon to encounter additional species at lower fractions. Keep in mind that in single-molecule techniques extremely low quantities of impurities (which in bulk would be "invisible" due to averaging and a high concentration of species of interest) are also detected. One must approach the discussion of low fractions (<5%) of apparent FRET states as biologically relevant with caution, especially in cases where this fraction does not change with varying experimental conditions (Roy et al., 2008). Moreover, such contaminants may slightly blur the E distributions of biologically important species. The MFD approach constitutes an unsurpassed tool for the identification of such contaminants and their exclusion from subsequent analysis.

6.2. Outlook: Fluorescence image spectroscopy

We have presented here the use of MFD for monitoring FRET in solution at the single-molecule level. However, it is also a unique feature of fluorescence that it can be used in very complex environments and for imaging purposes. The improvement gained in specificity of smFRET experiments by the application of MFD may be even more significant for the imaging of living cells. Even though monitoring FRET in living cells (Jares-Erijman and Jovin, 2003; Wallrabe and Periasamy, 2005) experienced a boost due to the advent of a palette of fluorescent proteins (Piston and Kremers, 2007), it is well known that FRET in living cells is subject to many artifacts (Berney and Danuser, 2003; Wallrabe *et al.*, 2006). The application of MFD to imaging, although conceptually straightforward, is limited by the number of photons than can be detected from a single dye before it bleaches and by the relatively high background of the cellular environment (i.e., autofluorescence) in comparison with measurements in solutions. What is also critical for MFD is keeping the expression levels of fluorescent proteins quite low so that the "single-molecule" condition is fulfilled. Offering an outlook from the current article, the reader should refer to the initial efforts toward a new modality termed multiparameter fluorescence imaging (MFI) or, interchangeably, multiparameter fluorescence image spectroscopy (MFIS), which have recently been presented (Kudryavtsev *et al.*, 2007; Weidtkamp-Peters *et al.*, 2009). We are quite optimistic about the future of MFIS as a modality for accurate monitoring of smFRET events *in vivo*, encompassing the merits of currently complementary techniques such as ratiometric intensity imaging, fluorescence lifetime imaging spectroscopy (FLIM), anisotropy measurements, and FCS in all its various implementations. With regard to Section 3.2, all eight dimensions of fluorescence can be used in confocal imaging as well. Images obtained with a laser-scanning fluorescence microscope contain a time structure that can be exploited to measure fast dynamics and transport processes of molecules in solution and in cells (Digman *et al.*, 2005a). Moreover, the high time resolution allows one to utilize the presence of photoinduced transient dark states, which are exhibited by practically all common fluorophores. These relatively long-lived states are very sensitive to the local environment and thus highly attractive for microenvironment imaging purposes (transient state (TRAST) imaging; Sanden *et al.*, 2008). MFIS constitutes a versatile modality that allows for the easy integration of these exciting new techniques into the traditional fluorescent toolbox, expanding the multiparameter space of fluorescence measurements.

ACKNOWLEDGMENTS

C. A. M. S. and S. K. thank the German Science Foundation (DFG) in the priority program SPP 1258 "Sensory and regulatory RNAs in prokaryotes" for supporting our work. E. S. was funded by the Swedish foundation STINT. P. J. R. was supported by the EU Marie Curie

Actions Research Training Networks "DNA enzymes." We are grateful to Jerker Widengren for the very fruitful collaboration within the Swedish foundation STINT exchange program. We thank Stefanie Weidtkamp-Peters, Thomas Peulen, Markus Richert, and Joggi Wirz for reading the manuscript and Suren Felekyan for help in preparing figures.

REFERENCES

Al-Soufi, W., Reija, B., Novo, M., Felekyan, S., Kuhnemuth, R., and Seidel, C. A. M. (2005). Fluorescence correlation spectroscopy, a tool to investigate supramolecular dynamics: Inclusion complexes of pyronines with cyclodextrin. *J. Am. Chem. Soc.* **127**, 8775–8784.

Andrecka, J., Lewis, R., Bruckner, F., Lehmann, E., Cramer, P., and Michaelis, J. (2008). Single-molecule tracking of mRNA exiting from RNA polymerase II. *Proc. Natl. Acad. Sci. USA* **105**, 135–140.

Andrews, D. L. (1989). A unified theory of radiative and radiationless molecular-energy transfer. *Chem. Phys.* **135**, 195–201.

Andrews, D. L., and Bradshaw, D. S. (2004). Virtual photons, dipole fields and energy transfer: A quantum electrodynamical approach. *Eur. J. Phys.* **25**, 845–858.

Andrews, D. L., and Leeder, J. M. (2009). Resonance energy transfer: When a dipole fails. *J. Chem. Phys.* **130**, 184504.

Andrews, D. L., and Rodriguez, J. (2007). Resonance energy transfer: Spectral overlap, efficiency, and direction. *J. Chem. Phys.* **127**, 084509.

Antonik, M., Felekyan, S., Gaiduk, A., and Seidel, C. A. M. (2006). Separating structural heterogeneities from stochastic variations in fluorescence resonance energy transfer distributions via photon distribution analysis. *J. Phys. Chem. B* **110**, 6970–6978.

Bahar, I., Chennubhotla, C., and Tobi, D. (2007). Intrinsic dynamics of enzymes in the unbound state and, relation to allosteric regulation. *Curr. Opin. Struct. Biol.* **17**, 633–640.

Becker, W. (2005). Advanced Time-Correlated Single Photon Counting Techniques. Springer, Berlin-Heidelberg.

Berney, C., and Danuser, G. (2003). FRET or no FRET: A quantitative comparison. *Biophys. J.* **84**, 3992–4010.

Best, R. B., Merchant, K. A., Gopich, I. V., Schuler, B., Bax, A., and Eaton, W. A. (2007). Effect of flexibility and cis residues in single-molecule FRET studies of polyproline. *Proc. Natl. Acad. Sci. USA* **104**, 18964–18969.

Blanchard, S. C. (2009). Single-molecule observations of ribosome function. *Curr. Opin. Struct. Biol.* **19**, 103–109.

Borgia, A., Williams, P. M., and Clarke, J. (2008). Single-molecule studies of protein folding. *Annu. Rev. Biochem.* **77**, 101–125.

Brand, L., Eggeling, C., Zander, C., Drexhage, K. H., and Seidel, C. A. M. (1997). Single-molecule identification of Coumarin-120 by time-resolved fluorescence detection: Comparison of one- and two-photon excitation in solution. *J. Phys. Chem. A* **101**, 4313–4321.

Braslavsky, S. E., Fron, E., Rodriguez, H. B., Roman, E. S., Scholes, G. D., Schweitzer, G., Valeur, B., and Wirz, J. (2008). Pitfalls and limitations in the practical use of Förster's theory of resonance energy transfer. *Photochem. Photobiol. Sci.* **7**, 1444–1448.

Bruchez, M., Moronne, M., Gin, P., Weiss, S., and Alivisatos, A. P. (1998). Semiconductor nanocrystals as fluorescent biological labels. *Science* **281**, 2013–2016.

Bustamante, C. (2008). In singulo biochemistry: When less is more. *Annu. Rev. Biochem.* **77**, 45–50.

Chen, Y., and Muller, J. D. (2007). Determining the stoichiometry of protein heterocomplexes in living cells with fluorescence fluctuation spectroscopy. *Proc. Natl. Acad. Sci. USA* **104**, 3147–3152.

Chen, Y., Müller, J. D., So, P. T. C., and Gratton, E. (1999). The photon counting histogram in fluorescence fluctuation spectroscopy. *Biophys. J.* **77**, 553–567.

Choi, U. B., Strop, P., Vrljic, M., Chu, S., Brunger, A. T., and Weninger, K. R. (2010). Single-molecule FRET-derived model of the synaptotagmin 1-SNARE fusion complex. *Nat. Struct. Mol. Biol.* **17**, 318–324.

Clegg, R. M. (1992). Fluorescence resonance energy transfer and nucleic acids. In "Methods in Enzymology (DNA Structures Part A: Synthesis and Physical Analysis of DNA), Vol. 211," (D. M. J. Lilley and J. E. Dahlberg, eds.), pp. 353–388. Academic Press, New York.

Clegg, R. M. (2009). Förster resonance energy transfer—FRET what is it, why do it, and how it's done. In "Laboratory Techniques in Biochemistry and Molecular Biology (Fret and Flim Techniques), Vol. 33," (T. W. J. Gadella, ed.), pp. 1–57. Elsevier, Amsterdam.

Cui, Q., and Karplus, M. (2008). Allostery and cooperativity revisited. *Protein Sci.* **17**, 1295–1307.

Curutchet, C., Franceschetti, A., Zunger, A., and Scholes, G. D. (2008). Examining Forster energy transfer for semiconductor nanocrystalline quantum dot donors and acceptors. *J. Phys. Chem. C* **112**, 13336–13341.

Cypionka, A., Stein, A., Hernandez, J. M., Hippchen, H., Jahn, R., and Walla, P. J. (2009). Discrimination between docking and fusion of liposomes reconstituted with neuronal SNARE-proteins using FCS. *Proc. Natl. Acad. Sci. USA* **106**, 18575–18580.

Dale, R. E., Eisinger, J., and Blumberg, W. E. (1979). Orientational freedom of molecular probes—Orientation factor in intra-molecular energy transfer. *Biophys. J.* **26**, 161–193.

Davydov, D. R., and Halpert, J. R. (2008). Allosteric P450 mechanisms: Multiple binding sites, multiple conformers or both? *Expert Opin. Drug Metab. Toxicol.* **4**, 1523–1535.

De Angelis, D. A. (1999). Why FRET over genomics? *Physiol. Genomics* **1**, 93–99.

Dedecker, P., Hofkens, J., and Hotta, J. I. (2008). Diffraction-unlimited optical microscopy. *Mater. Today* **11**, 12–21.

Deniz, A. A., Mukhopadhyay, S., and Lemke, E. A. (2008). Single-molecule biophysics: At the interface of biology, physics and chemistry. *J. R. Soc. Interface* **5**, 15–45.

Dertinger, T., Pacheco, V., von der Hocht, I., Hartmann, R., Gregor, I., and Enderlein, J. (2007). Two-focus fluorescence correlation spectroscopy: A new tool for accurate and absolute diffusion measurements. *ChemPhysChem* **8**, 433–443.

Dietrich, A., Buschmann, V., Müller, C., and Sauer, M. (2002). Fluorescence resonance energy transfer (FRET) and competing processes in donor-acceptor substituted DNA strands: A comparative study of ensemble and single-molecule data. *Rev. Mol. Biotechnol.* **82**, 211–231.

Diez, M., Zimmermann, B., Borsch, M., Konig, M., Schweinberger, E., Steigmiller, S., Reuter, R., Felekyan, S., Kudryavtsev, V., Seidel, C. A. M., and Graber, P. (2004). Proton-powered subunit rotation in single membrane-bound F0F1-ATP synthase. *Nat. Struct. Mol. Biol.* **11**, 135–141.

Digman, M. A., Brown, C. M., Sengupta, P., Wiseman, P. W., Horwitz, A. R., and Gratton, E. (2005a). Measuring fast dynamics in solutions and cells with a laser scanning microscope. *Biophys. J.* **89**, 1317–1327.

Digman, M. A., Sengupta, P., Wiseman, P. W., Brown, C. M., Horwitz, A. R., and Gratton, E. (2005b). Fluctuation correlation spectroscopy with a laser-scanning microscope: Exploiting the hidden time structure. *Biophys. J.* **88**, L33–L36.

Dix, J. A., Hom, E. F. Y., and Verkman, A. S. (2006). Fluorescence correlation spectroscopy simulations of photophysical phenomena and molecular interactions: A molecular dynamics/Monte Carlo approach. *J. Phys. Chem. B* **110**, 1896–1906.

heterogeneity of rhodamine 6G terminally attached to a DNA helix revealed by NMR and single-molecule fluorescence spectroscopy. *J. Am. Chem. Soc.* **129**, 12746–12755.
Nir, E., Michalet, X., Hamadani, K. M., Laurence, T. A., Neuhauser, D., Kovchegov, Y., and Weiss, S. (2006). Shot-noise limited single-molecule FRET histograms: Comparison between theory and experiments. *J. Phys. Chem. B* **110**, 22103–22124.
O'Connor, D. V., and Phillips, D. (1984). Time-Correlated Single Photon Counting. Academic Press, New York.
Orrit, M., and Bernard, J. (1990). Single pentacene molecules detected by fluorescence excitation in a *p*-terphenyl crystal. *Phys. Rev. Lett.* **65**, 2716–2719.
Orte, A., Clarke, R., Balasubramanian, S., and Klenerman, D. (2006). Determination of the fraction and stoichiometry of femtomolar levels of biomolecular complexes in an excess of monomer using single-molecule, two-color coincidence detection. *Anal. Chem.* **78**, 7707–7715.
Pandey, M., Syed, S., Donmez, I., Patel, G., Ha, T., and Patel, S. S. (2009). Coordinating DNA replication by means of priming loop and differential synthesis rate. *Nature* **462**, 940-U137.
Pawley, J. (2006). Handbook of Biological Confocal Microscopy. Springer, New York.
Piston, D. W., and Kremers, G. J. (2007). Fluorescent protein FRET: The good, the bad and the ugly. *Trends Biochem. Sci.* **32**, 407–414.
Rettig, W., Strehmel, B., Schrader, S., and Seifert, H. (1999). Applied Fluorescence in Chemistry, Biology and Medicine. Springer, Berlin-Heidelberg.
Rosenberg, S. A., Quinlan, M. E., Forkey, J. N., and Goldman, Y. E. (2005). Rotational motions of macromolecules by single-molecule fluorescence microscopy. *Acc. Chem. Res.* **38**, 583–593.
Rothwell, P. J., Berger, S., Kensch, O., Felekyan, S., Antonik, M., Wohrl, B. M., Restle, T., Goody, R. S., and Seidel, C. A. M. (2003). Multiparameter single-molecule fluorescence spectroscopy reveals heterogeneity of HIV-1 reverse transcriptase: Primer/template complexes. *Proc. Natl. Acad. Sci. USA* **100**, 1655–1660.
Roy, R., Hohng, S., and Ha, T. (2008). A practical guide to single-molecule FRET. *Nat. Methods* **5**, 507–516.
Ruan, Q. Q., Cheng, M. A., Levi, M., Gratton, E., and Mantulin, W. W. (2004). Spatial-temporal studies of membrane dynamics: Scanning fluorescence correlation spectroscopy (SFCS). *Biophys. J.* **87**, 1260–1267.
Ruttinger, S., Macdonald, R., Kramer, B., Koberling, F., Roos, M., and Hildt, E. (2006). Accurate single-pair Förster resonant energy transfer through combination of pulsed interleaved excitation, time correlated single-photon counting, and fluorescence correlation spectroscopy. *J. Biomed. Opt.* **11**, 024012.
Sabanayagam, C. R., Eid, J. S., and Meller, A. (2005). Using fluorescence resonance energy transfer to measure distances along individual DNA molecules: Corrections due to nonideal transfer. *J. Chem. Phys.* **122**, 61103–61107.
Sanden, T., Persson, G., Thyberg, P., Blom, H., and Widengren, J. (2007). Monitoring kinetics of highly environment sensitive states of fluorescent molecules by modulated excitation and time-averaged fluorescence intensity recording. *Anal. Chem.* **79**, 3330–3341.
Sanden, T., Persson, G., and Widengren, J. (2008). Transient state imaging for microenvironmental monitoring by laser scanning microscopy. *Anal. Chem.* **80**, 9589–9596.
Schaffer, J., Volkmer, A., Eggeling, C., Subramaniam, V., Striker, G., and Seidel, C. A. M. (1999). Identification of single molecules in aqueous solution by time-resolved fluorescence anisotropy. *J. Phys. Chem. A* **103**, 331–336.
Schmidt, T., Schütz, G. J., Gruber, H. J., and Schindler, H. (1996). Local stoichiometries determined by counting individual molecules. *Anal. Chem.* **68**, 4397–4401.
Scholes, G. D. (2003). Long-range resonance energy transfer in molecular systems. *Annu. Rev. Phys. Chem.* **54**, 57–87.

Scholes, G. D., and Ghiggino, K. P. (1994). Rate expressions for excitation transfer. 1. Radiationless transition theory perspective. *J. Chem. Phys.* **101,** 1251–1261.
Schröder, G. F., Alexiev, U., and Grubmüller, H. (2005). Simulation of fluorescence anisotropy experiments: Probing protein dynamics. *Biophys. J.* **89,** 3757–3770.
Schuler, B. (2005). Single-molecule fluorescence spectroscopy of protein folding. *ChemPhysChem* **6,** 1206–1220.
Schuler, B., and Eaton, W. A. (2008). Protein folding studied by single-molecule FRET. *Curr. Opin. Struct. Biol.* **18,** 16–26.
Schuler, B., Lipman, E. A., Steinbach, P. J., Kumke, M., and Eaton, W. A. (2005). Polyproline and the "spectroscopic ruler" revisited with single-molecule fluorescence. *Proc. Natl. Acad. Sci. USA* **102,** 2754–2759.
Schwille, P., Meyer-Almes, F. J., and Rigler, R. (1997). Dual-color fluorescence cross-correlation spectroscopy for multicomponent diffusional analysis in solution. *Biophys. J.* **72,** 1878–1886.
Shera, E. B., Seitzinger, N. K., Davis, L. M., Keller, R. A., and Soper, S. A. (1990). Detection of single fluorescent molecules. *Chem. Phys. Lett.* **174,** 553–557.
Sindbert, S., Kalinin, S., Kienzler, A., Clima, L., Bannwarth, W., Nguyen, D. T., Appel, B., Muller, S., and Seidel, C. A. M. (2010). Implications of dye linker length and rigidity on accurate distance determination via FRET of labeled DNA and RNA. *J. Am. Chem. Soc.* (submitted).
So, P. T. C., Dong, C. Y., Masters, B. R., and Berland, K. M. (2000). Two-photon excitation fluorescence microscopy. *Annu. Rev. Biomed. Eng.* **2,** 399–429.
Stryer, L. (1978). Fluorescence energy-transfer as a spectroscopic ruler. *Annu. Rev. Biochem.* **47,** 819–846.
Stühmeier, F., Hillisch, A., Clegg, R. M., and Diekmann, S. (2000). Fluorescence energy transfer analysis of DNA structures containing several bulges and their interaction with CAP. *J. Mol. Biol.* **302,** 1081–1100.
Sugawa, M., Arai, Y., Iwane, A. H., Ishii, Y., and Yanagida, T. (2007). Single molecule FRET for the study on structural dynamics of biomolecules. *Biosystems* **88,** 243–250.
Sytina, O. A., Heyes, D. J., Hunter, C. N., Alexandre, M. T., van Stokkum, I. H. M., van Grondelle, R., and Groot, M. L. (2008). Conformational changes in an ultrafast light-driven enzyme determine catalytic activity. *Nature* **456,** 1001–1004.
Szollosi, J., Damjanovich, S., and Matyus, L. (1998). Application of fluorescence resonance energy transfer in the clinical laboratory: Routine and research. *Cytometry* **34,** 159–179.
Tamarat, P., Maali, A., Lounis, B., and Orrit, M. (2000). Ten years of single-molecule spectroscopy. *J. Phys. Chem.* **104,** 1–16.
Tellinghuisen, J., Goodwin, P. M., Ambrose, W. P., Martin, J. C., and Keller, R. A. (1994). Analysis of fluorescence lifetime data for single rhodamine molecules in flowing sample streams. *Anal. Chem.* **66,** 64–72.
Thompson, N. L., and Steele, B. L. (2007). *Total internal reflection with fluorescence correlation spectroscopy* **2,** 878–890.
Tompa, P., and Fuxreiter, M. (2008). Fuzzy complexes: Polymorphism and structural disorder in protein-protein interactions. *Trends Biochem. Sci.* **33,** 2–8.
Torres, T., and Levitus, M. (2007). Measuring conformational dynamics: A new FCS-FRET approach. *J. Phys. Chem. B* **111,** 7392–7400.
Ulbrich, M. H., and Isacoff, E. Y. (2007). Subunit counting in membrane-bound proteins. *Nat. Methods* **4,** 319–321.
Valeur, B. (2002). Molecular Fluorescence: Principles and Applications. Wiley-VCH Verlag, Weinheim.
Valeur, B., and Brochon, J. (2001). New Trends in Fluorescence Spectroscopy: Applications to Chemical and Life Sciences. Springer, Berlin-Heidelberg.

van der Meer, B. W. (2002). Kappa-squared: From nuisance to new sense. *Rev. Mol. Biotechnol.* **82,** 181–196.
van der Meer, B. W., Cooker, G., and Chen, S. Y. (1994). Resonance Energy Transfer: Theory and Data. VCH Publishers, New York.
van Oijen, A. M. (2008). Cutting the forest to see a single tree? *Nat. Chem. Biol.* **4,** 440–443.
Vogel, S. S., Thaler, C., and Koushik, S. V. (2006). Fanciful FRET. *Sci. STKE* **2006**(331), re2.
Vogelsang, J., Doose, S., Sauer, M., and Tinnefeld, P. (2007). Single-molecule fluorescence resonance energy transfer in nanopipets: Improving distance resolution and concentration range. *Anal. Chem.* **79,** 7367–7375.
Wallrabe, H., and Periasamy, A. (2005). Imaging protein molecules using FRET and FLIM microscopy. *Curr. Opin. Biotechnol.* **16,** 19–27.
Wallrabe, H., Chen, Y., Periasamy, A., and Barroso, M. (2006). Issues in confocal microscopy for quantitative FRET analysis. *Microsc. Res. Tech.* **69,** 196–206.
Wang, Y. F., Guo, L., Golding, I., Cox, E. C., and Ong, N. P. (2009). Quantitative transcription factor binding kinetics at the single-molecule level. *Biophys. J.* **96,** 609–620.
Weidtkamp-Peters, S., Felekyan, S., Bleckmann, A., Simon, R., Becker, W., Kuhnemuth, R., and Seidel, C. A. M. (2009). Multiparameter fluorescence image spectroscopy to study molecular interactions. *Photochem. Photobiol. Sci.* **8,** 470–480.
Weiss, S. (1999). Fluorescence spectroscopy of single biomolecules. *Science* **283,** 1676–1683.
Weiss, S. (2000). Measuring conformational dynamics of biomolecules by single molecule fluorescence spectroscopy. *Nat. Struct. Biol.* **7,** 724–729.
Weston, K. D., Dyck, M., Tinnefeld, P., Muller, C., Herten, D. P., and Sauer, M. (2002). Measuring the number of independent emitters in single-molecule fluorescence images and trajectories using coincident photons. *Anal. Chem.* **74,** 5342–5349.
Widengren, J., Mets, Ü., and Rigler, R. (1995). Fluorescence correlation spectroscopy of triplet states in solution: A theoretical and experimental study. *J. Phys. Chem.* **99,** 13368–13379.
Widengren, J., Schweinberger, E., Berger, S., and Seidel, C. A. M. (2001). Two new concepts to measure fluorescence resonance energy transfer via fluorescence correlation spectroscopy: Theory and experimental realizations. *J. Phys. Chem. A* **105,** 6851–6866.
Widengren, J., Kudryavtsev, V., Antonik, M., Berger, S., Gerken, M., and Seidel, C. A. M. (2006). Single-molecule detection and identification of multiple species by multiparameter fluorescence detection. *Anal. Chem.* **78,** 2039–2050.
Widengren, J., Chmyrov, A., Eggeling, C., Lofdahl, P. A., and Seidel, C. A. M. (2007). Strategies to improve photostabilities in ultrasensitive fluorescence spectroscopy. *J. Phys. Chem. A* **111,** 429–440.
Wozniak, A. K., Schroder, G. F., Grubmuller, H., Seidel, C. A. M., and Oesterhelt, F. (2008). Single-molecule FRET measures bends and kinks in DNA. *Proc. Natl. Acad. Sci. USA* **105,** 18337–18342.
Wu, P. G., and Brand, L. (1994). Resonance energy-transfer—Methods and applications. *Anal. Biochem.* **218,** 1–13.
Wu, X., and Schultz, P. G. (2009). Synthesis at the interface of chemistry and biology. *J. Am. Chem. Soc.* **131,** 12497–12515.
Xie, X. S., and Lu, H. P. (1999). Single-molecule enzymology. *J. Biol. Chem.* **274,** 15967–15970.
Zander, C., Sauer, M., Drexhage, K. H., Ko, D. S., Schulz, A., Wolfrum, J., Brand, L., Eggeling, C., and Seidel, C. A. M. (1996). Detection and characterization of single molecules in aqueous solution. *Appl. Phys. B* **63,** 517–523.

CHAPTER NINETEEN

ATOMIC FORCE MICROSCOPY STUDIES OF HUMAN RHINOVIRUS: TOPOLOGY AND MOLECULAR FORCES

Ferry Kienberger,[*,1] Rong Zhu,[†] Christian Rankl,[*,1] Hermann J. Gruber,[*] Dieter Blaas,[‡] *and* Peter Hinterdorfer[*]

Contents

1. Introduction	516
2. Results and Discussion	517
2.1. Rhinovirus production	517
2.2. AFM imaging setting	519
2.3. Immobilization and imaging of HRV2	521
2.4. Force spectroscopy of antibody–virus molecular recognition	528
2.5. RNA release and single molecule unfolding	532
References	536

Abstract

Dynamic force microscopy (DFM) allows for imaging of the structure and assessment of the function of biological specimens in their physiological environment. In DFM, the cantilever is oscillated at a given frequency and touches the sample only at the end of its downward movement. Accordingly, the problem of lateral forces displacing or even destroying biomolecules is virtually inexistent as the contact time and friction forces are greatly reduced. Here, we describe the use of DFM in studies of human rhinovirus serotype 2 (HRV2). The capsid of HRV2 was reproducibly imaged without any displacement of the virus. Release of the genomic RNA from the virions was initiated by exposure to low-pH buffer and snapshots of the extrusion process were obtained. DFM of the single-stranded RNA genome of an HRV showed loops protruding from a condensed RNA core, 20–50 nm in height. The mechanical rigidity of the RNA was determined by single molecule pulling experiments. From fitting RNA stretching curves to the wormlike-chain (WLC) model a persistence length of 1.0 ± 0.17 nm was obtained.

---

[*] Institute for Biophysics, Johannes Kepler University of Linz, Linz, Austria
[†] Christian Doppler Laboratory for Nanoscopic Methods in Biophysics, Johannes Kepler University of Linz, Linz, Austria
[‡] Max F. Perutz Laboratories, Medical University of Vienna, Vienna, Austria
[1] Current address: Agilent Technologies Austria GmbH, Linz, Austria

1. INTRODUCTION

Viruses are multimeric assemblies of nucleic acids and proteins, with many copies of some few proteins constituting the viral shell. They are mainly of globular shape with diameters between 20 nm (parvovirus) and 220 nm (mimivirus) and can also be polymorphic or elongated rods with lengths of up to 300 nm (e.g., tobacco mosaic virus). The capsid encases the RNA or DNA genome and may additionally contain proteins involved in replicative functions. Surface-exposed proteins are targets for neutralizing antibodies. Viral inactivation by antibodies occurs via aggregation (cross-linking of individual virions), competition for receptor-binding sites, and/or by impeding structural changes (via bidentate binding of IgG molecules to symmetry-related sites on the same virion); such conformational modifications are required for release of the genome into the cytosol of the host cell. It is clear that monodentate binding differs from bidentate binding with respect to avidity (Rossmann et al., 1985).

The very first event in viral infection is the recognition of receptors at the plasma membrane. This is eventually followed by attachment to multiple sites and internalization via various endocytotic pathways. In the case of enveloped viruses, cellular uptake is not always required, as some are able to trigger direct fusion of their lipid envelope with the plasma membrane. In most cases, however, membrane fusion occurs within endosomes upon exposure to their acidic environment. As a result, the entire viral core enters the cytosol. The process is much less understood for naked viruses; in their case, structural changes expose hydrophobic stretches of capsid proteins that insert into the endosomal membrane. It is believed that this leads to the formation of pores for the transit of the viral genome. Alternatively, the endosomal membrane is disrupted altogether and the whole virus enters the cytoplasm (Fuchs and Blaas, 2010; Prchla et al., 1994).

Due to the icosahedral geometry of most nonenveloped viruses, interaction with receptors most probably involves initial recognition of single receptor molecules followed by recruitment of more receptors attaching to symmetry-related binding sites on the virion. This process is time-dependent and governed by diffusional motion of the receptors in the plane of the lipid bilayer. Entry might be constitutive or trigged by signals transmitted through the cytoplasmic part of the membrane receptors via conformational changes associated with oligomerization (Hewat et al., 2002).

Genome exit from the protective coat is also poorly understood; at least in the case of some bacterial viruses, this is an active process triggered by opening of a pore through which the highly pressurized DNA is expelled. Very little is known about this process in animal viruses. Interaction with receptors and/or exposure to the acidic environment in endosomes results in conformational alterations concomitant with the formation of channels.

So far, however, the driving force for genome release through these openings in the capsid is enigmatic.

All consecutive steps of infection can be monitored and visualized by atomic force microscopy (AFM), either topographically, or by measuring the unbinding forces between virus and receptors or antibodies. These unbinding forces reflect interaction strengths and are related to the affinity/avidity between the components. In the following we used human rhinovirus type 2 (HRV2) as an example.

HRVs are the main cause of common colds. They are picornaviruses, a large family comprising many animal and human pathogens such as foot-and-mouth disease virus, poliovirus, and hepatitis A virus (Semler and Wimmer, 2002). HRV2 belongs to the minor receptor group that uses members of the low-density lipoprotein receptor family for cell entry. As detailed below, AFM allows for visualization of the virus, for the study of the interaction between virus and neutralizing antibodies, and for single molecule studies of the release of the genomic RNA from single virus particles. Furthermore, topographical AFM imaging of virus particles entering cells allow for detailed mechanistic studies of virus-related diseases under physiological conditions (Kuznetsov et al., 2003).

2. RESULTS AND DISCUSSION

2.1. Rhinovirus production

With few exceptions, HRVs do not replicate in cells other than human. Depending on the virus type, a particular substrain of HeLa cells (HeLa-H1 or HeLa-Ohio) or rhabdomyosarcoma (RD) cells are used for propagation. For mass production, HeLa suspension cultures are most appropriate. However, for poorly understood reasons some HRV types only grow to appreciable titers in monolayer culture. In our case, the protocol given below has turned out to be best suited for the production of HRV2 (American Type Culture Collection).

2.1.1. Protocol: Human rhinovirus production and isolation

1. For virus production, about 1.2×10^8 HeLa-H1 cells (Flow Laboratories, Irvine, Scotland) are seeded per 2 l minimum essential medium modified for suspension (Sigma, Vienna, Austria), containing 7% heat inactivated horse serum, 100 U/ml penicillin and 100 μg/ml streptomycin (all from Gibco, Rockville, MD) and grown at 25 rpm in a 2 l spinner flask (Bellco, Vineland, NJ) at 37 °C for 4 days.
2. Cells, now about 8×10^8, are pelleted at 2000 rpm in a Beckman J6B centrifuge and resuspended in the same medium but containing only 2%

(v/v) horse serum and, in addition, 1 mM MgCl$_2$. For infection, $\sim$8 $\times$ 10^9 TCID50 (tissue infectious dose, i.e., virus concentration that results in infection of 50% of the cells; Blake and O'Connell, 1993) is added and incubation is continued at 34 °C for 16 h.

3. The infected cells are collected in a pellet at 4000 rpm, resuspended in 20 ml of 10 mM EDTA, 10 mM Tris–HCl (pH 7.5) and subjected to three freeze–thaw cycles. Residual cells are broken with a tightly fitting Dounce homogenizer followed by sonication for 3 min at full power in a sonicator bath (Bandelin Sonorex RK31, 35 kHz). This solution is mixed with 20 ml buffer A (20 mM Tris–HCl (pH 7.5), 2 mM MgCl$_2$), and cell debris is removed by low speed centrifugation. Virus in the supernatant is pelleted using a Ti45 fixed angle rotor (Beckman, High Wycombe, UK) for 2 h at 30,000 rpm.

4. Virus pellets obtained from 2 l suspension culture are resuspended overnight in 0.5 ml buffer A and free nucleic acids are digested by addition of RNase I and DNase (Boehringer Mannheim, Vienna, Germany), each to a final concentration of 0.5 mg/ml at room temperature (RT) for 10 min. To also remove contaminating proteins, 0.34 mg/ml trypsin is added and incubation continued at 37 °C for 5 min; native virus is resistant to these hydrolases.

5. Sarcosin is added to 0.1% (w/v) and the mixture is left at 4 °C overnight. Insoluble material is removed by centrifugation in an Eppendorf centrifuge for 15 min and the supernatant is saved. Material stemming from 10 l suspension culture is combined and deposited on six 7.5–45% sucrose density gradients in buffer A, preformed in SW28 Beckman centrifuge tubes. Centrifugation is for 3.5 h at 25,000 rpm at 4 °C. The virus band seen in the middle of the gradient by shining light into the tube from the top is collected with a needle connected to a syringe.

6. Virus from the six gradients is combined, diluted with buffer A, and pelleted in an SW28 Beckman centrifuge tube at 25,000 rpm for 16 h. The pellet is resuspended in 200 μl of 50 mM HEPES–NaOH (pH 7.5), insoluble material is pelleted in an Eppendorf centrifuge and the supernatant saved and stored at -80 °C in aliquots.

7. Purity and concentration are assessed by reducing 15% SDS-polyacrylamide gel electrophoresis (SDS-PAGE) that reveals the capsid proteins VP1, VP2, and VP3 (VP4 is usually not visible) and by capillary electrophoresis (Okun et al., 1999).

2.1.2. Preparation of human His$_6$-tagged MBP-very-low-density lipoprotein receptors

Members of the low-density lipoprotein receptor family act as cellular receptors of the minor receptor group of HRVs, exemplified by HRV2. The ligand-binding domain of these proteins is composed of several ligand-binding modules, about 40 amino acid residues in length that are arranged in tandem. The very-low-density lipoprotein receptor (VLDLR) possesses eight such modules. Each module is stabilized by three disulfide bonds and a Ca^{2+} ion; reduction and/or complexation of the Ca^{2+} ion inactivate the receptor. In the experiments described here, a recombinant soluble VLDLR fused to maltose-binding protein at its C-terminus and carrying a His6-tag at its N-terminus is used. Alternatively, receptor fragments with less than all repeats or concatemers of ligand-binding module 3 (V3) are utilized. They are expressed in *Escherichia coli* from the corresponding plasmids and purified as described (Moser *et al.*, 2005; Ronacher *et al.*, 2000; Wruss *et al.*, 2007). *E. coli* carrying the vector constructs encoding the desired receptor protein is grown in 100 ml LB medium containing 0.1 mg/ml ampicillin at 37 °C overnight. The culture is transferred to 4 l LB medium containing 0.1 mg/ml ampicillin and grown to $A_{600} = 0.7$. Protein expression is induced by addition of IPTG (final concentration 0.3 mM) at 30 °C overnight. The bacteria are pelleted at 5000 rpm (GS3 rotor, Beckman) for 20 min and resuspended in 30 ml Tris-buffered saline containing 2 mM $CaCl_2$ (TBSC). Bacteria are lysed by ultrasonication on ice (six bursts at full power, 10 s each, in a Sonoplus HD200, Bandelin, Berlin). Cell debris is pelleted at 19,000 rpm (SS34 rotor, Beckman) for 20 min. The supernatant is saved and incubated in batch with 5 ml of a slurry of Ni-NTA beads (Quiagen) at 4 °C for 2 h. Unbound proteins are removed by centrifugation at 4000 rpm in a table top centrifuge for 10 min. Ni-NTA beads carrying bound receptor are washed with TBSC followed by 10 mM imidazole (two times for 20 min each). Receptor is eluted in 4 ml aliquots with TBSC containing 250 mM imidazole (20 min each). Samples are dialyzed against TBSC overnight and protein-containing fractions are identified by SDS-PAGE. Refolding is in 1 mM cystamin/ 10 mM cysteamin at 4 °C for 2 days. Virus-binding activity of the receptor is verified by capillary electrophoresis (Konecsni *et al.*, 2004).

2.2. AFM imaging setting

Different AFM modes can be used for imaging biological samples under physiological conditions. Overall, topographical imaging is mainly achieved by contact-mode AFM, in which the cantilever is in continuous touch with the sample during scanning (Muller *et al.*, 1995). However, contact-mode imaging turned out to be less suitable for weakly attached samples, because biomolecules are often pushed away by the AFM stylus during scanning (Karrasch *et al.*, 1993). To overcome this drawback, dynamic force microscopy (DFM) methods such as tapping-mode AFM, and more recently, magnetic AC mode AFM (MACmode AFM) (Han *et al.*, 1996) is often used for imaging soft and weakly attached biological samples (Kienberger *et al.*, 2003; Moller *et al.*, 1999). In DFM the cantilever oscillates and touches the sample only intermittently at the end of its downward movement, which reduces the contact time and minimizes friction forces. An alternating magnetic field drives the oscillation of a magnetically coated cantilever for MACmode AFM, whereas acoustic excitation is used for tapping-mode AFM. MACmode AFM has the advantage over acoustically driven cantilevers that the magnetically coated cantilevers are directly excited by an external magnetic field. This results in a sinusoidal oscillation with a defined resonance frequency. A firm theoretical relationship between oscillation frequency and optimal sensitivity in AFM imaging was established using a modified harmonic oscillator model (Schindler et al., 2000).

2.2.1. Protocol: MACmode AFM imaging

1. A magnetically coated cantilever (Agilent Technologies Inc., Chandler, AZ) is carefully washed with SDS (1% Na-dodecylsulfate), double distilled (dd) water, ethanol, and again with dd water to remove any contamination from the cantilever. The cantilever holder is cleaned by the same procedure. After mounting onto the scanner, it is put into the liquid cell and set to optimal imaging parameters.
2. The laser spot is adjusted on the cantilever and the position of the photodetector is varied to achieve a maximal sum signal (done by μm-screws).
3. To acquire the resonance curve, the cantilever is positioned roughly 100 μm away from the surface; for the cantilevers used in this study (spring constant $\sim$100 pN/nm) the resonance frequency is at about 9 kHz. For imaging, the excitation frequency is set lower by 20% (to 7–8 kHz). The driving signal (coil-current) is adjusted to get free cantilever oscillation amplitudes of about 5 nm.
4. The cantilever is approached to the sample automatically. Since the magnetic field becomes stronger with decreasing distance, the oscillation amplitude of the cantilever increases. To compensate for this, the driving signal is adjusted by reducing the coil-current to maintain the 5 nm amplitude. Automatic approaching stops when the preadjusted amplitude reduction of 20% is reached (this value turned out to work for various cantilevers and samples).
5. Amplitude–distance curves are then measured. The set point (i.e., the amplitude reduction value) used as feedback signal in imaging is preadjusted so that the amplitude reduction is 1 nm (corresponding to roughly 50 pN imaging force).
6. Imaging mode is switched on; the integral and proportional gains are adjusted to obtain a stable time-resolved signal. Imaging is started at larger scan sizes and the scanning size is reduced consecutively (not the other way round). During imaging, the set point is adjusted continuously to minimize the imaging force.

2.3. Immobilization and imaging of HRV2

To obtain high-quality topological images with AFM, a tight immobilization of the biological specimen to a flat surface is pivotal. We use two different virus immobilization schemes for AFM imaging; unspecific electrostatic binding onto mica surface and specific immobilization on a phospholipid bilayer-containing recombinant VLDLR that specifically recognizes HRV2.

2.3.1. HRV2 immobilization onto mica and imaging

HRV2 is noncovalently immobilized in buffer solution onto mica, a highly hydrophilic aluminosilicate (Kienberger et al., 2004a). By using Ni^{2+} ions as electrostatic bridge between the negatively charged mica and the virus, stable binding is obtained. Several other bivalent cations like Mg^{2+} and Co^{2+} were also tested, but in contrast to Ni^{2+}, they did not result in stable imaging of the viruses. Therefore, attachment and measurements is carried out in a 50 mM Tris-buffer containing 5 mM Ni^{2+} (pH 7.6), under which conditions repeated imaging over many hours shows no damage or displacement of the sample. In 50 mM Tris-buffer, no precipitation was observed with 5 mM Ni^{2+} even after prolonged storage. The most likely explanation is weak complexation of Ni^{2+} by Tris base. Nickel is well known to be complexed by primary and secondary amines, thus Tris base is also likely to act as ligand, especially at 50 mM concentration, which serves to permanently keep 5 mM Ni^{2+} in solution at pH 7.6. Physical adsorption from solution onto mica has the advantage that it takes only several minutes to achieve a densely covered surface and that pretreatment of neither virus nor mica surface is necessary. In addition to its simplicity, physical adsorption is easily controlled by variation of concentration and/or time; for instance, layers with different degrees of density are obtained by simply varying the virus concentrations in the adsorption buffer, for example, 1 mg/ml HRV2 gives a densely packed layer, while 20 µg/ml HRV2 results in sparse coverage. The virus coverage on mica scales therefore almost linear with the virus concentration in the adsorption buffer.

2.3.1.1. Protocol: HRV2 immobilization onto mica

1. A stock solution of ~3 mg/ml HRV2 is thawed, 5 µl aliquots at 1 mg/ml HRV2 are prepared by dilution with 50 mM Tris–HCl, pH 7.6, frozen in liquid nitrogen and again stored at −20 °C. For the experiments, always a new sample is prepared; repeated freezing and thawing damages the virus.
2. A buffer containing 5 mM $NiCl_2$, 50 mM Tris–HCl, pH 7.6 (TBN) is prepared in dd water and filtered through a 0.2 µm filter (Schleicher &

Schuell, Dassel, Germany). The solution is stored at 4 °C and should be used within 2 weeks.
3. A vial with 5 μl 1 mg/ml HRV2 is thawed and mixed with 25 μl TBN.
4. A mica sheet (2 cm × 2 cm, Gröpl, Tulln, Austria) is cleaved using Scotch tape. The procedure is repeated two times to ascertain that the removed mica layer is homogenous. The cleaved mica sheet is mounted into the liquid cell of the AFM. The 30 μl virus solution from Step 3 is put onto the mica surface and incubated for 20 min followed by washing five times with TBN. The AFM fluid cell is filled with the same buffer for imaging.
5. The above protocol results in a full coverage of virus on the mica. To obtain a lower coverage, the virus concentration in the final adsorption buffer is reduced to 20 μg/ml, while leaving the other conditions unchanged.

Figure 19.1 shows viruses in a densely packed monolayer, as obtained at high virus concentrations in the adsorption buffer (Fig. 19.1A). The height of the virus particles is about 31 ± 3 nm, as evident from the cross-section profile using an uncovered spot on mica as reference (data not shown); this agrees very well with values derived from X-ray and electron cryo-microscopy measurements of HRV2 (Hewat and Blaas, 1996; Verdaguer et al., 2000). Stable imaging is obtained in buffer solution over many hours by using imaging parameters resulting in low contact forces. In particular, the free oscillation amplitude, the amplitude reduction for the feedback while imaging, and the spring constant of the cantilever are the critical parameters to be adjusted (Kienberger et al., 2004b) (cf. next segment). A careful selection of these parameters results in stable imaging of the weakly adsorbed viruses without mechanical distortion, yielding high topographical resolution (Fig. 19.1B and C). A regularly spaced pattern is observed on almost all viral particles (Fig. 19.1B and C). This might reflect the protrusions on the surface of the viral capsid known from cryo-EM and X-ray diffraction. These substructures are more clearly discernable upon contrast enhancement (Fig. 19.1C). Roughly 20 blobs or protrusions with diameters of 2–4 nm and heights of $\sim$0.3 nm are counted on each viral particle; they arrange in lines of $\sim$5 protrusions per line (see labeling in Fig. 19.1C). Virus capsids do not appear to be strictly globular in shape, but rather the projection of the particles exhibits some straight borders akin to their polygonal surface features that were also observed by X-ray crystallography (Verdaguer et al., 2000). The rhinovirus capsid displays a star-shaped pentameric dome on each of the icosahedral fivefold axes surrounded by a canyon; a triangular plateau is centered on each of the threefold axes. Accordingly, 12 pentameric domes and 20 raised triangular plateaus are present on the full virus particle. Since the AFM tip can effectively scan only the upper half of the virus, 16 protrusions are, in principle, recognizable on a

Figure 19.1 Virus particles adsorbed onto bare mica surfaces. (A) Densely packed HRV2 layer prepared by physical adsorption onto mica by using bivalent cations. The close apposition of the virions results in some deformation. On most of the virus particles substructures can be recognized. Scan size 200 nm. *Inset*: substructures of virus particle. (B) Three-dimensional representation. Roughly 20 protrusions are counted on each single virus particle. Image size 80 nm. (C) Contrast-enhanced image. Singly resolved structures (labeled one to five) with diameters of 3 nm and heights of 0.3 nm are clearly visible. Image size is 80 nm. *Inset*: external surface of HRV2 as derived from X-ray coordinates. All atoms within a distance of 152 Å from the center of the virion are removed in order to emphasize the protrusions at the fivefold and the threefold axes. The view is centered on a threefold symmetry axis. Reprinted with permission from Kienberger et al. (2005).

virus hemisphere; this value is fairly close to the roughly 20 blobs observed. In addition, the center-to-center distance of single protrusions is about 7 nm (Fig. 19.1C). This is in the same range as the distance between the plateau at the threefold axis and the plateau at the fivefold axis seen in the structures solved by cryo-EM image reconstruction and X-ray diffraction.

2.3.2. Crystalline arrangement of HRV2 on model cell membranes

For specific and tight immobilization of virus particles, we use the previously developed method of assembling HRV2 on artificial membrane interfaces that mimic the cell surface (Kienberger et al., 2001) (Fig. 19.2). A supported lipid bilayer containing Ni^{2+}-nitrilotriacetate (NTA) lipids as

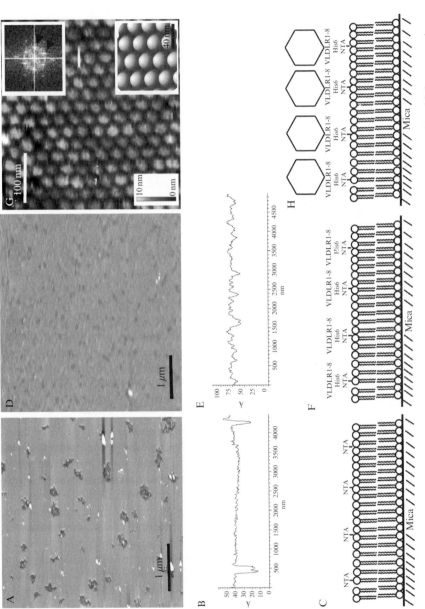

Figure 19.2 Crystalline arrangement of HRV2 on a receptor-containing lipid bilayer. (A) AFM image of lipid bilayer on mica recorded in 10 mM Tris, 150 mM NaCl, pH = 7.5 buffer, showing a flat bilayer surface with some defects. Magnetic AC mode AFM was used with a

functional groups in the outer leaflet at high surface density is assembled onto mica by using the Langmuir–Blodgett technique (Fig. 19.2A). The specific interaction between Ni^{2+}-NTA lipids embedded in the bilayer and soluble His_6-tagged VLDLR1–8 MBP (the entire ligand-binding domain with all eight repeats of the human VLDLR) is used to assemble a planar matrix of laterally diffusing receptor proteins (Fig. 19.2D). A similar strategy has been used previously to obtain two-dimensional crystals of a His_6-tagged HIV reverse transcriptase (Kubalek et al., 1994). This technique appears ideally suited to obtain regular lattices of virus particles under near-physiological conditions with specific binding to the support. However, the soft bilayer most probably impeded high-resolution imaging and better topographical images are obtained when the virus was immobilized directly onto the mica surface (Kienberger et al., 2004a, 2005; Kuznetsov et al., 2001). HRV2 is added to the receptor-carrying membrane by injection into the AFM liquid cell. After incubation for 2 h, a densely packed layer of virus particles covering the surface is observed in the topographical image (Fig. 19.2G). The virions are attached in a tightly packed single layer arrangement. Since Ca^{2+} is required for specific virus–receptor binding (Atkins et al., 1998; Lonberg and Whiteley, 1976), a high coverage results only when Ca^{2+} is present in the buffer solution. In the absence of Ca^{2+} (EDTA contained in the buffer), the virus covers the receptor-modified bilayer only sparsely (Kienberger et al., 2001). The regularly packed layer of

---

spring constant of 0.1 N/m at 1.3 Hz lateral scan velocity. Scan size 5 μm. (B) Cross-section profile along the bilayer. A height of the lipid-bilayer of 3 nm can be derived from the depth of the holes. (C) Sketch of the supported lipid bilayer assembled on mica. The first layer consists of pure POPC and the second layer of 95% POPC and 5% NTA lipid. (D) The lipid bilayer shown in (A) was incubated overnight with hexahistidine-tagged MBP-VLDLR1–8 in 50 mM Na$_2$HPO$_4$, 250 mM NaCl, 5 mM sodium acetate, 10 mM imidazol, at pH = 8.3, washed, and imaged in the same buffer. MacMode topography image using the same conditions as in (A). MBP-VLDLR1–8 covers both bilayer and defect structures. This results in borders appearing smeared out. (E) Cross-section profile. The increase in the height fluctuations indicates receptor binding to the membrane. (F) Sketch illustrating the specific interaction between hexahistidine-tags C-terminally appended to MBP-VLDLR1–8 and the Ni^{2+} NTA-groups in the bilayer. (G) Hexagonal arrangement of the virions on the model cell membrane at a scan size of 400 nm. *Inset* (upper right): the Fourier spectrum exhibits hexagonally arranged spots at frequencies corresponding to the center-to-center distance of HRV2 in the crystal package. *Inset* (lower right): averaged three-dimensional representation of the virus arrangement with a center-to-center distance of 35 nm. (H) Sketch illustrating the specific binding of HRV2 to receptor molecules (MBP-VLDLR1–8) anchored via His6-NTA to a lipid bilayer. A high surface density of receptors on the bilayer leads to full coverage with virus particles. Panels A–F were reprinted with permission from Kienberger et al. (2001) and panels G and H from Kienberger et al. (2005).

virus particles with crystalline patches extends typically over 500 nm patches. The patches appear to be oriented differently to each other (Kienberger et al., 2005), and various defect structures can be observed. A hexagonal arrangement of virus particles on the model cell membrane can be clearly discerned.

2.3.2.1. Protocol: HRV2 immobilization on model cell membranes

1. *Preparation of membranes*
   Freshly cleaved mica is used as solid support for a lipid bilayer. A monolayer of POPC (1-palmitoyl-2-oleoyl-*sn*-glycero-3-phosphocholine; Avanti Polar Lipids Inc., Alabaster, AL) in its liquid state is deposited using the Langmuir–Blodgett technique in a 10 mM Tris–HCl solution (pH = 5.0) at a constant surface pressure of 34 mN/m and RT. The second monolayer, consisting of 95% POPC and 5% of Ni^{2+}-NTA lipid (Ni^{2+} salt of 1,2-dioleoyl-*sn*-glycero-3-{[*N*(5-amino-1-carboxypentyl) iminodiacetic acid] succinyl}; Avanti Polar Lipids Inc.), is transferred by horizontal dipping (Langmuir–Schaefer technique) in 10 mM Tris, 150 mM NaCl, pH 7.5, at the same surface pressure. The bilayer is then moved into the liquid cell of the AFM and imaged in the same buffer.

2. *Receptor (MBP-VLDLR1–8) Immobilization*
   The bilayer is incubated in 10 mM Tris, 150 mM NaCl, 2 mM $NiCl_2$, pH 7.5, for 20 min to load the NTA-groups on the bilayer with Ni^{2+} ions. The buffer in the liquid cell is then changed to 50 mM Na_2HPO_4, 250 mM NaCl, 5 mM sodium acetate, 10 mM imidazol, pH 8.3, and a 13 mM solution of C-terminally His_6-tagged MBP-VLDLR1–8 is added to give a final concentration of 0.3 mM. After overnight incubation, free MBP-VLDLR1–8 is removed by washing with the same buffer and the sample is imaged in 10 mM Tris, 150 mM NaCl, pH 7.5.

3. *HRV2 immobilization*
   The buffer is changed to 10 mM Tris, 150 mM NaCl, 1 mM $CaCl_2$, pH 7.5, prior to addition of a 1 mg/ml HRV2 suspension. After 2 h incubation time, free HRV2 is washed away and the virus-layer is imaged in the same buffer.

Specific immobilization of virus particles to a fluid supported bilayer via their receptor turned out to be very useful for achieving a tight, crystalline arrangement. Specific and strong binding of HRV2 to the model membrane renders the virions resistant to displacement by the movement of the tip.

2.3.3. Imaging of receptor binding to the virus

Binding of receptor molecules to individual virus particles is observed after injection of soluble receptors into the liquid cell. Complexes with zero, one, two, or three receptor molecules attached to the virus are seen

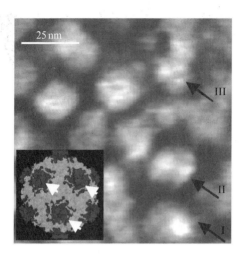

Figure 19.3 Imaging of virus–receptor complexes. Virus was adsorbed to mica as in Fig. 19.1. Then, receptor molecules in solution was injected into the sample chamber. Black arrows indicate receptor–virus complexes with one, two, and three receptor molecules. Scan size 100 nm. The inset shows the X-ray structure of HRV2 carrying 12 receptor molecules, each bound via five modules (PDB accession code 1V9U and 3DPR), displayed with RasWin. White arrows indicate the bound receptor molecules on the fivefold axes. Reprinted with permission from Kienberger et al. (2005).

(Fig. 19.3). The number of receptor molecules associated with virions increases over time. Occasionally, dissociation of single receptor molecules from viral particles is also observed. Note that up to five ligand-binding modules within one single receptor molecule can attach simultaneously around each vertex of the icosahedron. This results in the binding of up to 12 receptors to one virion (Querol-Audi et al., 2009; Verdaguer et al., 2004).

2.3.3.1. Protocol: Binding of receptors to HRV2 immobilized on mica Ten microliters of HRV2 at 1 mg/ml in 50 mM Tris–HCl and 5 mM NiCl$_2$, pH 7.6, is deposited onto freshly cleaved mica for 15 min and washed with the same buffer. Prior to incubation with receptor, the buffer is exchanged to 50 mM Tris–HCl, 5 mM NiCl$_2$, and 2 mM CaCl$_2$. MBP-VLDLR5x3 (a concatemer of five copies of module 3 of VLDLR) is injected into the fluid cell at a final concentration of $\sim$30 μg/ml, incubated for $\sim$1 h, and excess receptor is washed away. In control experiments, we prevented binding of VLDL receptor and HRV2 by doing the experiments with a buffer containing 50 mM Tris–HCl, 5 mM NiCl$_2$, and 1 mM EGTA. The EGTA complexes Ca^{2+} which is required for proper receptor–virus interaction. Therefore, the buffer solution containing EGTA renders the receptor inactive and no binding of receptor and virus was obtained.

2.4. Force spectroscopy of antibody–virus molecular recognition

Using AFM, receptor–ligand interactions (i.e., intermolecular forces) and the unfolding patterns of single biomolecules (i.e., intramolecular forces) can be monitored. For studying specific binding between biomolecules, an increasing force is exerted on a receptor–ligand complex, and the dissociation process is followed over time. Dynamic aspects of recognition are addressed in force spectroscopy experiments in which the time scale is systematically varied in order to analyze the changes in conformations and states during receptor–ligand dissociation (Evans, 2001). Such experiments allow for estimation of affinity, rate constants, and structural data of the binding pocket (Chang et al., 2005; Dobrowsky et al., 2008; Hinterdorfer et al., 1996; Rankl et al., 2008). For studying intramolecular forces, single molecules are stretched between the cantilever tip and the substrate, yielding insight into the molecular determinants of mechanical stability and the role of force-induced conformational changes in physiological function (Carrion-Vazquez et al., 2000).

When ligand–receptor binding is viewed at the single molecule level, the average lifetime of a ligand–receptor bond at zero force, $\tau(0)$, is given by the inverse of the kinetic off-rate constant, $\tau(0) = 1/k_{off}$. Therefore, ligands will dissociate from receptors at times larger than the average lifetime even without any force applied to the bond. In contrast, if molecules are pulled apart very fast, the bond will resist and require a measurable force for detachment (Grubmüller et al., 1996). Accordingly, unbinding forces do not constitute unitary values but depend on the dynamics of the unbinding experiments. On the millisecond to second time scale of AFM experiments, thermal impulses govern the unbinding process. In the thermal activation model, the lifetime of a complex in solution is described by a Boltzmann equation, $\tau(0) = \tau_{osc} \exp(E_{barr}/(k_B T))$ (Bell, 1978), where τ_{osc} is the inverse of the natural oscillation frequency and E_{barr} is the energy barrier for dissociation, yielding a simple Arrhenius dependency of the dissociation rate on barrier height. A force acting on a complex between molecules deforms the interaction energy landscape and lowers the activation energy barrier. The lifetime $\tau(f)$ of a bond loaded with a constant force f is given by $\tau(f) = \tau_{osc} \exp(E_{barr} - fx/(k_B T))$ (Evans and Ritchie, 1997), with x being the distance of the energy barrier from the energy minimum along the direction of the applied force. The lifetime under constant force, therefore, compares to the lifetime at zero force, according to $\tau(f) = \tau(0)\exp(-fx/(k_B T))$.

With the use of AFM, ligand–receptor unbinding is commonly measured in force–distance cycles, where an effective force increase, or the loading rate r, can be deduced from $r = df/dt$, being equal to the product of pulling velocity v and effective spring constant k_{eff}, $r = vk_{eff}$.

The combination of the Boltzmann equation with the stochastic description of the unbinding process predicts the unbinding force distributions at different loading rates r (Strunz et al., 2000). The maximum of each force distribution, $f^*(r)$, thereby reflects the most probable unbinding force at the respective loading rate r. f^* is related to r through $f^*(r) = (k_B T/x) \ln(rx/(k_B T k_{off}))$. Accordingly, the unbinding force f^* scales linearly with the logarithm of the loading rate. For a single barrier, this gives rise to a simple linear dependence of the force on the logarithm of the loading rate, while multiple barriers result in two or more segments of linear curves with different slopes. In the latter case, the various loading rate regimes correspond to different energy barriers in the dissociation pathway of the ligand–receptor interaction.

2.4.1. AFM tip chemistry

Antibodies are covalently coupled to AFM-tips (Fig. 19.4A) using a three-step binding protocol (Hinterdorfer et al., 1998). In the first step, amino groups are generated on Si_3N_4 tips (Veeco Instruments, Santa Barbara, CA) either with ethanolamine hydrochloride or with gas phase silanization (Hinterdorfer et al., 1996, 1998; Lyubchenko et al., 2001). These two procedures do not cause stickiness and yield a very low number of amino groups on the apex of the tip (Riener et al., 2003a) as is desired for single molecule experiments. In the second step of the anchoring protocol, a distensible and flexible linker is often used to space the ligand molecule from the amino-functionalized tip surface by several nanometers. The ligand on the spacer molecule can freely orient and diffuse within a certain volume provided by the length of the tether, thereby achieving unconstrained binding to its receptor. Poly(ethylene glycol) (PEG), an inert water soluble polymer, has often been used as flexible linker (Hinterdorfer et al., 2000). A cross-linker length of 6 nm seems to give a good compromise between high mobility of the ligand and narrow lateral resolution of the target site. The heterobifunctional cross-linkers (termed "PDP–PEG–NHS") used for tip-ligand coupling carry two different reactive groups at their ends (Haselgruebler et al., 1995; Riener et al., 2003b). One is a carboxyl group, activated in the form of an N-hydroxysuccinimide (NHS) ester. This group couples to the amino group on the tip, yielding a stable amide bond. The free-tangling end of the PEG chain carries thiol-reactive group (3-[2-pyridyl]-dithiopropionyl, PDP). Since antibodies do not contain free thiols, they have to be prederivatized with N-succinimidyl-3-(acetylthio)-propionate (SATP), resulting in a protected thiol function. The protecting group is removed with hydroxylamine, leaving a free SH group to react with the PDP group of the PEG linker, resulting in the flexible attachment of the antibody to the tip. The suchlike modified tips are washed with the buffer and stored in the cold room until use.

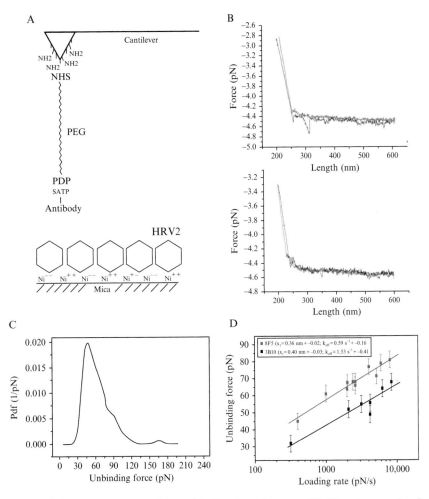

Figure 19.4 Measurements of the unbinding force between HRV2 and monoclonal antibodies. (A) Virus was adsorbed to mica as in Fig. 19.1. Antibody modified with SATP was conjugated to the cantilever tip via an NHS–PEG–PDP linker. (B) Upper panel: typical force–distance curve with an unbinding event. Low panel: after injection of antibody, the force–distance curve shows no binding event. (C) Experimental probability density function of the unbinding force measured via a tip carrying antibody 8F5. (D) Force loading rate dependence of unbinding forces measured with tips carrying antibody 8F5 (bidentate) and 3B10 (monodentate), respectively.

2.4.1.1. Protocol: Tip functionalization with antibodies

1. *Ethanolamine functionalization*

   Ethanolamine hydrochloride (3.3 g) is dissolved in 6.6 ml DMSO by gentle heating to ~70 °C, subsequently 10% (v/v) of molecular sieves (0.4 nm) are added. The solution is allowed to cool to RT. Dissolved air is removed by degassing in a desiccator at aspirator vacuum for 30 min. The cantilevers are incubated in this solution overnight, washed in DMSO (3×) and ethanol (3×), and dried with a gentle stream of nitrogen. When the tips are not used immediately, they are stored in a desiccator under argon atmosphere.

2. *APTES functionalization (gas phase silanization)*
   "APTES tips" are prepared in close analogy to the protocol described in Riener *et al.* (2003a). Before use, APTES is freshly distilled under vacuum. A desiccator (5 l) is flooded with argon gas to remove air and moisture. Then two small plastic trays (e.g., the lids of Eppendorf reaction vials) are placed inside the desiccator, 30 ml of APTES and 10 ml of triethylamine are separately pipetted into two trays, the AFM tips are placed nearby on a clean inert surface (e.g., Teflon), and the desiccator is closed. After 120 min of incubation, APTES and triethylamine are removed, the desiccator is again flooded with argon gas for 5 min, and the tips are left inside for 2 days in order to cure the APTES coating.

3. *PDP–PEG–NHS*
   PDP–PEG–NHS (1 mg) is dissolved in 0.5 ml chloroform and poured into the small glass reaction chamber. Thirty microliters of triethylamine are added, and the ethanolamine-coated AFM tips are immediately immersed for 2 h. The reaction chamber is covered with an upside-down beaker to make sure that the chloroform does not evaporate in a few minutes. After 2 h, the tips are washed with chloroform and dried with N_2.

4. *SATP functionalization of antibodies*
   Antibody–SATP is prepared by reacting 0.5 ml of the antibody (10 mg/ml in a buffer containing 100 mM NaCl and 35 mM boric acid, pH 7.5 adjusted with NaOH) with a 10-fold excess of SATP (5 µl from a 66 mM stock solution in DMSO). The SATP stock solution is slowly added while vortexing, and the mixture is incubated at RT for another 30 min with occasional vortexing. Derivatized antibody is separated from the reagents by gel filtration in buffer B (100 mM NaCl, 50 mM NaH$_2$PO$_4$, 1 mM EDTA, pH 7.5 adjusted with NaOH) using a PD-10 column (Amersham). The column is prewashed with buffer B (25 ml), the outlet is closed, the reaction mixture (0.5 ml) is loaded, and the outlet is opened to allow for elution of 0.5 ml by gravity flow. Two 1 ml portions of buffer B are applied to the column, and after the flow has ceased, one portion of buffer B (1.3 ml) is applied and the eluent (the antibody fraction) is collected. The antibody fraction is split into 40 µl aliquots which are frozen in liquid nitrogen and stored at $-25\,°C$.

5. *Coupling of antibody–SATP to PDP–PEG-tip*
   Sixty microliters of hydroxylamine reagent (500 mM hydroxylamine hydrochloride, 25 mM EDTA, titrated to pH 7.5 by stepwise addition of

solid Na_2CO_3) is added to the antibody–SATP stock in buffer B (100 μl, 0.15 mg/ml) to deprotect the SH-group, vortexed for 2 min, and incubated for 1 h. Cantilevers are placed on a clean dry petri dish covered with parafilm. A small drop of the protein solution is put onto the tips and reacted for 1 h. The tips are subsequently washed three times with phosphate buffered saline (PBS) and stored in PBS at 4 °C.

2.4.2. Force spectroscopy experiments

Conventional force–distance cycles are used to detect single molecule antibody–virus unbinding events for different pulling velocities. A typical force–distance curve with an unbinding event is shown in the upper panel of Fig. 19.4B. For a specific control, the binding sites on the virus are blocked by injection of free antibody. In this case, the force–distance curve usually shows no binding event (lower panel of Fig. 19.4B). Each individual cantilever spring constant is calibrated in solution using the equipartition theorem (Butt and Jaschke, 1995). The recognition forces are quantified using a transition detection algorithm from which empirical probability density functions (pdfs) of forces (Fig. 19.4C) and lengths are constructed (Baumgartner et al., 2000). The most probable unbinding force and respective standard deviations of the mean values were obtained by least square fitting of the force pdfs with Gaussian functions, and plotted in a half logarithmic force spectroscopy plot (Fig. 19.4D). The parameters k_{off} and x are extracted from fitting the experimental plot with the formula $f^*(r) = k_B T / x \, \ln(rx/(k_B T k_{off}))$.

In force spectroscopy experiments, the variation in the pulling speed applied to specific ligand–receptor bonds yields detailed structural and kinetic information of the interaction. Length scales of energy barriers are obtained from the slope of the spectroscopy plot (i.e., force versus loading rate) and extrapolation to zero forces yields the kinetic off-rate for the dissociation of the complex in solution.

Unbinding forces are measured for the interaction between HRV2 and two different monoclonal antibodies; 8F5 binds two sites related by twofold icosahedral symmetry via both of its arms (Hewat and Blaas, 1996) whereas 3B10, because of symmetry constrains, can only bind to one site (Hewat et al., 1998). Experiments with Fab fragments of 8F5 have shown that binding is very weak. Therefore, the higher unbinding force of 8F5, as compared to 3B10 is a result of bidentate binding (Fig. 19.4D).

2.5. RNA release and single molecule unfolding

The release of RNA from the HRV2 capsid is triggered *in vivo* by the low-pH environment (pH of ~5.6) in endosomal carrier vesicles and late endosomes (Prchla et al., 1994). This process was also studied *in vitro* by

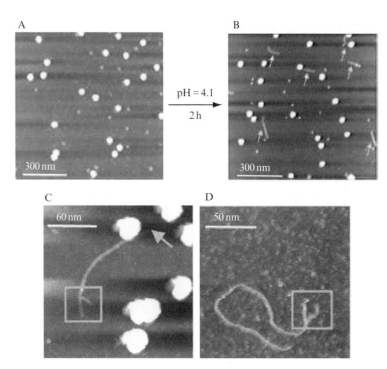

Figure 19.5 Genomic RNA release from the virion at low pH. (A) Single virions adsorbed onto mica and imaged in a pH 7.6 buffer. Scan size 900 nm. (B) The pH of the buffer was changed from neutral to pH 4.1 and maintained for 2 h. Images were acquired after reneutralization to pH 7.6. Virus particles are observed together with RNA molecules (arrows), either separated from the virus capsid or still connected. Scan size 900 nm. (C) RNA molecules released from the virus capsids are occasionally seen to possess a fork-like structure at one end (box). The length of this fork is ∼30 nm. Scan size 200 nm. Note that the RNA molecule is not in the process of extrusion but only touches the virus capsid from the side (arrow). (D) Individual RNA molecule separated from the virus capsid and exhibiting a similar fork-like structure (box). Scan size 150 nm. Reprinted with permission from Kienberger et al. (2004a).

the exposure of isolated virions to low-pH buffer (Kienberger et al., 2004a). In order to induce the process of RNA release, HRV2 bound to mica (Fig. 19.5A) is exposed to a pH of 4.1 for 2 h. Images are then taken again to investigate eventual changes in viral morphology. Indeed, single short RNA molecules are frequently observed (Fig. 19.5B–D). They are either distant from the virus or still connected to it. Presumably, the latter molecules have not entirely left the viral shell and are thus viewed directly during the process of extrusion. No such RNA molecules are observed when the sample is imaged immediately after the cell buffer has been replaced with the low-pH buffer (data not shown). Single virions at different stages of RNA release are observed. However, even after prolonged incubation an entire RNA molecule (of about 2 μm in length) is never seen. This might indicate the degradation of RNA, lack of complete adsorption to the mica, or incomplete releasing of RNA from shell. Virus capsids appear with heights of 25–35 nm, and the heights of the RNA molecules vary between 1 and 1.5 nm, which is in quite good agreement with the reported height of RNA measured by tapping-mode AFM (Hansma et al., 1996).

2.5.1. Protocol: Release of the RNA genome of HRV2

After HRV2 is immobilized on mica at low coverage, the pH in the liquid cell is reduced to 4.1. After 2 h, it is readjusted to pH 7.6 by replacement of the liquid with the original buffer, and images are acquired. Control incubations are done in the presence of RNase A at a final concentration of 0.5 mg/ml, resulting in a complete disappearance of the RNA molecules (Kienberger et al., 2004a).

2.5.2. Protocol: Full genome studies of HRV1A

Topographical imaging (Fig. 19.6) is applied (as described in Section 2.2) to visualize about 7100 nucleotides single-stranded RNA genome of human rhinovirus serotype 1A (HRV1A; Kienberger et al., 2004a, 2007). The viral RNA is synthesized from plasmid pWin1a (kindly donated from D. Pevear, Viropharma) that contains the entire genome of HRV1A, by using standard conditions. Briefly, DNA is linearized with 50 U of SfiI at 50 °C for 2 h. DNA is then extracted with phenol/chloroform, precipitated with Na-acetate/ethanol and resuspended in diethylpyrocarbonate-treated water. For in vitro transcription, 1 μg linearized plasmid is incubated in 50 μl reaction mixture composed of 6 μl T7-polymerase reaction buffer, 200 U T7 RNA-polymerase (both from Promega), 10 mM DTT, 80 U RNasin, and 10 mM NTPs for 90 min at 37 °C. After digestion of the DNA, a 10 μl stock solution of the viral RNA (0.2 mg/ml in 50 mM Tris, pH 7.6) is diluted with 50 μl Tris–Ni^{2+} buffer solution (1 mM Tris, 0.1 mM EDTA, 1 mM NiCl$_2$, pH 7.4) and incubated on freshly cleaved mica for 15 min using a commercially available AFM fluid cell (Agilent Technologies). The sample is thoroughly washed and images are acquired with MAC mode AFM in the same buffer.

Figure 19.6A shows a topographical image of the HRV1A genome. Condensed and aggregated RNA cores roughly 30 nm in height are obtained from the in vitro transcribed RNA. These are similar to those seen for the genomic RNA extracted with phenol/chloroform from poliovirus that belongs to the same virus family (Kuznetsov et al., 2005). The height indicates some expansion of the packed RNA when without the constraint from the HRV capsid. Linearly extended RNA and RNA loops protruding from the globular mass are frequently observed. To probe the

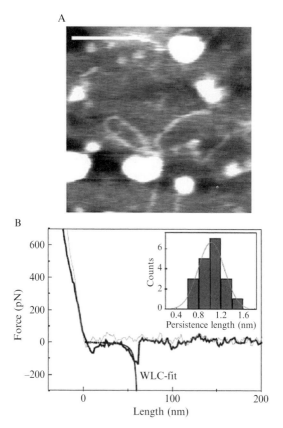

Figure 19.6 Pulling of the viral RNA. (A) Imaging of the single-stranded RNA genome (~7100 nucleotides) from a human rhinovirus in buffer solution. Single RNA cores with protruding RNA loops. Scan size 200 nm, scale bar 70 nm. The z-scale ranges from 0 to 3 nm. (B) Pulling viral RNA in force–distance cycles. After imaging the viral RNA, the tip was lowered to establish a contact for about 1 s and a force versus extension curve was acquired. Approach (red), retraction (black), and WLC-fit (thin black) are shown. The stretching of RNA was observed with a typical nonlinear behavior. *Inset*: histogram of the persistence length obtained from WLC fits of 19 RNA pulling curves. Gaussian fit is in red. Reprinted with permission from Kienberger *et al.* (2007). (See Color Insert)

mechanical rigidity of viral RNA, single molecule pulling experiments are performed (Fig. 19.6B) (Kienberger *et al.*, 2005). After imaging the RNA, the tip is positioned on top of the RNA cores and force–distance cycles are acquired. Depending on the applied force and the encounter time, the tip picks a molecule with limited probability due to unspecific adsorption. Retraction of the tip induces stretching of the suspended molecule. At increased encounter time (1 s) and compression force (1 nN), stretching curves of single-stranded RNA are frequently observed. Using the worm-like chain (WLC) model (Bustamante et al., 1994; Kienberger et al., 2000), an average persistence length of 1.0 ± 0.17 nm is estimated. Such pulling experiments elucidate the molecular determinants of mechanical stability of single molecules (Tinoco, 2004). From the persistence length b, the elastic deformation energy W_b can be estimated to be $W_b = (M_b/2) \int K^2 \, dx$, with M_b the bending module ($M_b = k_b T b$) and K the curvature ($K = 1/R$ for one-dimensional systems; R being the radius of curvature) (Pastushenko, 1991; Pastushenko and Chizmadezhev, 1992).

It is believed that the RNA is released from the virion through a pore at the fivefold icosahedral axis (Hewat et al., 2002). Assuming a pore radius of 1–2 nm (Hewat et al., 2002; Kienberger et al., 2004a), the elastic deformation energy of RNA in the process of extrusion can be estimated to be half of the thermal energy (1.2 kJ/mol; $W_b = k_b T b/R$). Since the mechanics, kinetics, and the energy of the genomic release are largely governed by the viscoelastic properties of the nucleic acids, single molecule pulling experiments (Chemla et al., 2005; Smith et al., 2001) may contribute to the detailed biophysical understanding of viral genomic release.

REFERENCES

Atkins, A. R., Brereton, I. M., Kroon, P. A., Lee, H. T., and Smith, R. (1998). Calcium is essential for the structural integrity of the cysteine-rich, ligand-binding repeat of the low-density lipoprotein receptor. Biochemistry 37, 1662–1670.

Baumgartner, W., Hinterdorfer, P., and Schindler, H. (2000). Data analysis of interaction forces measured with the atomic force microscope. Ultramicroscopy 82, 85–95.

Bell, G. I. (1978). Models for the specific adhesion of cells to cells. Science 200, 618–627.

Blake, K., and O'Connell, S. (1993). Virus culture. In "Virology Labfax," (D. R. Harper, ed.), pp. 81–122. Blackwell Scientific Publications, West Smithfield, London, UK.

Bustamante, C., Marko, J. F., Siggia, E. D., and Smith, S. (1994). Entropic elasticity of X-phage DNA. Science 265, 1599–1600.

Butt, H. J., and Jaschke, M. (1995). Calculation of thermal noise in atomic force microscopy. Nanotechnology 6, 1–7.

Carrion-Vazquez, M., Oberhauser, A. F., Fisher, T. E., Marszalek, P. E., Li, H., and Fernandez, J. M. (2000). Mechanical design of proteins studied by single-molecule force spectroscopy and protein engineering. Prog. Biophys. Biomol. Biol. 74, 63–91.

Chang, M. I., Panorchan, P., Dobrowsky, T. M., Tseng, Y., and Wirtz, D. (2005). Single-molecule analysis of human immunodeficiency virus type 1 gp120-receptor interactions in living cells. J. Virol. 79, 14748–14755.

Chemla, Y. R., Aathavan, K., Michaelis, J., Grimes, S., Jardine, P. J., Anderson, D. L., and Bustamante, C. (2005). Mechanism of force generation of a viral DNA packaging motor. Cell 122, 683–692.

Dobrowsky, T. M., Zhou, Y., Sun, S. X., Siliciano, R. F., and Wirtz, D. (2008). Monitoring early fusion dynamics of human immunodeficiency virus type 1 at single-molecule resolution. J. Virol. 82, 7022–7033.

Evans, E. (2001). Probing the relation between force-lifetime and chemistry in single molecular bonds. Annu. Rev. Biophys. Biomol. Struct. 30, 105–128.

Evans, E., and Ritchie, K. (1997). Dynamic strength of molecular adhesion bonds. *Biophys. J.* **72**, 1541–1555.
Fuchs, R., and Blaas, D. (2010). Uncoating of human rhinoviruses. *Rev. Med. Virol.* DOI: 10.1002/rmv.654.
Grubmüller, H., Heymann, B., and Tavan, P. (1996). Ligand binding: Molecular mechanics calculation of the streptavidin–biotin rupture force. *Science* **271**, 997–999.
Han, W., Lindsay, S. M., and Jing, T. (1996). A magnetically driven oscillating probe microscope for operation in liquids. *Appl. Phys. Lett.* **69**, 4111–4113.
Hansma, H. G., Revenko, I., Kim, K., and Laney, D. E. (1996). Atomic force microscopy of long and short double-stranded, single-stranded and triple-stranded nucleic acids. *Nucleic Acids Res.* **24**, 713–720.
Haselgruebler, T., Amerstorfer, A., Schindler, H., and Gruber, H. J. (1995). Synthesis and applications of a new poly(ethylene glycol) derivative for the crosslinking of amines with thiols. *Bioconjug. Chem.* **6**, 242–248.
Hewat, E. A., and Blaas, D. (1996). Structure of a neutralizing antibody bound bivalently to human rhinovirus 2. *EMBO J.* **15**, 1515–1523.
Hewat, E. A., Marlovits, T. C., and Blaas, D. (1998). Structure of a neutralizing antibody bound monovalently to human rhinovirus 2. *J. Virol.* **72**, 4396–4402.
Hewat, E., Neumann, E., and Blaas, D. (2002). The concerted conformational changes during human rhinovirus 2 uncoating. *Mol. Cell* **10**, 317–326.
Hinterdorfer, P., Baumgartner, W., Gruber, H. J., Schilcher, K., and Schindler, H. (1996). Detection and localization of individual antibody–antigen recognition events by atomic force microscopy. *Proc. Natl. Acad. Sci. USA* **93**, 3477–3481.
Hinterdorfer, P., Schilcher, K., Baumgartner, W., Gruber, H. J., and Schindler, H. (1998). A mechanistic study of the dissociation of individual antibody–antigen pairs by atomic force microscopy. *Nanobiology* **4**, 177–188.
Hinterdorfer, P., Kienberger, F., Raab, A., Gruber, H. J., Baumgartner, W., Kada, G., Riener, C., Wielert-Badt, S., Borken, C., and Schindler, H. (2000). Poly(ethylene glycol): An ideal spacer for molecular recognition force microscopy/spectroscopy. *Single Mol.* **1**, 99–103.
Karrasch, S., Dolder, M., Schabert, F., Ramsden, J., and Engel, A. (1993). Covalent binding of biological samples to solid supports for scanning probe microscopy in buffer solution. *Biophys. J.* **65**, 2437–2446.
Kienberger, F., Pastushenko, V. P., Kada, G., Gruber, H. J., Riener, C., Schindler, H., and Hinterdorfer, P. (2000). Static and dynamical properties of single poly(ethylene glycol) molecules investigated by force spectroscopy. *Single Mol.* **1**, 123–128.
Kienberger, F., Moser, R., Schindler, H., Blaas, D., and Hinterdorfer, P. (2001). Quasi-crystalline arrangement of human rhinovirus 2 on model membranes. *Single Mol.* **2**, 99–103.
Kienberger, F., Stroh, C., Kada, G., Moser, R., Baumgartner, W., Pastushenko, V., Rankl, C., Schmidt, U., Muller, H., Orlova, E., LeGrimellec, C., Drenckhahn, D., *et al.* (2003). Dynamic force microscopy imaging of native membranes. *Ultramicroscopy* **97**, 229–237.
Kienberger, F., Zhu, R., Moser, R., Blaas, D., and Hinterdorfer, P. (2004a). Monitoring RNA release from human rhinovirus by dynamic force microscopy. *J. Virol.* **78**, 3203–3209.
Kienberger, F., Zhu, R., Moser, R., Rankl, C., Blaas, D., and Hinterdorfer, P. (2004b). Dynamic force microscopy for imaging of viruses under physiological conditions. *Biol. Proced. Online* **6**, 120–128.
Kienberger, F., Rankl, C., Pastushenko, V., Zhu, R., Blaas, D., and Hinterdorfer, P. (2005). Visualization of single receptor molecules bound to human rhinovirus under physiological conditions. *Structure* **13**, 1247–1253.
Kienberger, F., Costa, L. T., Zhu, R., Kada, G., Reithmayer, M., Chtcheglova, L., Rankl, C., Pacheco, A. B. F., Thalhammer, S., Pastushenko, V., Heckl, W. M., Blaas, D., *et al.* (2007). Dynamic force microscopy imaging of plasmid DNA and viral RNA. *Biomaterials* **28**, 2403–2411.

Konecsni, T., Kremser, L., Snyers, L., Rankl, C., Kilar, F., Kenndler, E., and Blaas, D. (2004). Twelve receptor molecules attach per viral particle of human rhinovirus serotype 2 via multiple modules. *FEBS Lett.* **568,** 99–104.

Kubalek, E. W., Le Grice, S. F., and Brown, P. O. (1994). Two-dimensional crystallization of histidine-tagged, HIV-1 reverse transcriptase promoted by a novel nickel-chelating lipid. *J. Struct. Biol.* **113,** 117–123.

Kuznetsov, Y. G., Malkin, A. J., Lucas, R. W., Plomp, M., and McPherson, A. (2001). Imaging of viruses by atomic force microscopy. *J. Gen. Virol.* **82,** 2025–2034.

Kuznetsov, Y. G., Victorie, J. G., Robinson, W. E., and McPherson, A. (2003). Atomic force microscopy investigation of human immunodeficiency virus (HIV) and HIV-infected lymphocytes. *J. Virol.* **77,** 11896–11909.

Kuznetsov, Y. G., Daijogo, S., Zhou, J., Semler, B. L., and McPherson, A. (2005). Atomic force microscopy analysis of icosahedral virus RNA. *J. Mol. Biol.* **347,** 41–52.

Lonberg, H. K., and Whiteley, N. M. (1976). Physical and metabolic requirements for early interaction of poliovirus and human rhinovirus with HeLa cells. *J. Virol.* **19,** 857–870.

Lyubchenko, Y. L., Gall, A. A., and Shlyakhtenko, L. S. (2001). Atomic force microscopy of DNA and protein–DNA complexes using functionalized mica substrates. *Methods Mol. Biol.* **148,** 569–578.

Moller, C., Allen, M., Elings, V., Engel, A., and Muller, D. J. (1999). Tapping mode atomic force microscopy produces faithful high-resolution images of protein surfaces. *Biophys. J.* **77,** 1150–1158.

Moser, R., Snyers, L., Wruss, J., Angulo, J., Peters, H., Peters, T., and Blaas, D. (2005). Neutralization of a common cold virus by concatemers of the third ligand binding module of the VLDL-receptor strongly depends on the number of modules. *Virology* **338,** 259–269.

Muller, D. J., Schabert, F. A., Buldt, G., and Engel, A. (1995). Imaging purple membranes in aqueous solutions at sub-nanometer resolution by atomic force microscopy. *Biophys. J.* **68,** 1681–1686.

Okun, V. M., Ronacher, B., Blaas, D., and Kenndler, E. (1999). Analysis of common cold virus (human rhinovirus serotype 2) by capillary zone electrophoresis: The problem of peak identification. *Anal. Chem.* **71,** 2028–2032.

Pastushenko, V. P. (1991). Mobility and electrophoretic mobility of long polymer molecules in gels. *Appl. Theor. Electrophor.* **1,** 313–316.

Pastushenko, V. P., and Chizmadezhev, Y. A. (1992). Energetic estimations of the deformation of translocated DNA and the cell membrane in the course of electrotransformation. *Biol. Membr.* **6,** 287–300.

Prchla, E., Kuechler, E., Blaas, D., and Fuchs, R. (1994). Uncoating of human rhinovirus serotype 2 from late endosomes. *J. Virol.* **68,** 3713–3723.

Querol-Audi, J., Konecsni, T., Pous, J., Carugo, O., Fita, I., Verdaguer, N., and Blaas, D. (2009). Minor group human rhinovirus–receptor interactions: Geometry of multimodular attachment and basis of recognition. *FEBS Lett.* **583,** 235–240.

Rankl, C., Kienberger, F., Wildling, L., Wruss, J., Gruber, H. J., Blaas, D., and Hinterdorfer, P. (2008). Multiple receptors involved in human rhinovirus attachment to live cells. *Proc. Natl. Acad. Sci. USA* **105,** 17778–17783.

Riener, C. K., Stroh, C. M., Ebner, A., Klampfl, C., Gall, A. A., Romanin, C., Lyubchenko, Y. L., Hinterdorfer, P., and Gruber, H. J. (2003a). Simple test system for single molecule recognition force microscopy. *Anal. Chim. Acta* **479,** 59–75.

Riener, C. K., Kienberger, F., Hahn, C. D., Buchinger, G. M., Egwim, I. O. C., Haselgruebler, T., Ebner, A., Romanin, C., Klampfl, C., Lackner, B., Prinz, H., Blaas, D., *et al.* (2003b). Heterobifunctional crosslinkers for tethering single ligand molecules to scanning probes. *Anal. Chim. Acta* **497,** 101–114.

Ronacher, B., Marlovits, T. C., Moser, R., and Blaas, D. (2000). Expression and folding of human very-low-density lipoprotein receptor fragments: Neutralization capacity toward human rhinovirus HRV2. *Virology* **278,** 541–550.

Rossmann, M. G., Arnold, E., Erickson, J. W., Frankenberger, E. A., Griffith, J. P., Hecht, H. J., Johnson, J. E., Kamer, G., Luo, M., Mosser, A. G., *et al.* (1985). Structure of a human common cold virus and functional relationship to other picornaviruses. *Nature* **317,** 145–153.

Schindler, H., Badt, D., Hinterdorfer, P., Kienberger, F., Raab, A., Wielert-Badt, S., and Pastushenko, V. (2000). Optimal sensitivity for molecular recognition MAC-mode AFM. *Ultramicroscopy* **82,** 227–235.

Semler, B. L., and Wimmer, E. (2002). *Molecular Biology of Picornaviruses* ASM Press, Washington, DC20036-2904.

Smith, D. E., Tans, S. J., Smith, S. B., Grimes, S., Anderson, D. L., and Bustamante, C. (2001). The bacteriophage Φ29 portal motor can package DNA against a large internal force. *Nature* **413,** 748–752.

Strunz, T., Oroszlan, K., Schumakovitch, I., Güntherodt, H. J., and Hegner, M. (2000). Model energy landscapes and the force-induced dissociation of ligand–receptor bonds. *Biophys. J.* **79,** 1206–1212.

Tinoco, I. (2004). Force as a useful variable in reactions: Unfolding RNA. *Annu. Rev. Biophys. Biomol. Struct.* **33,** 363–385.

Verdaguer, N., Blaas, D., and Fita, I. (2000). Structure of human rhinovirus serotype 2 (HRV2). *J. Mol. Biol.* **300,** 1179–1194.

Verdaguer, N., Fita, I., Reithmayer, M., Moser, R., and Blaas, D. (2004). X-ray structure of a minor group human rhinovirus bound to a fragment of its cellular receptor protein. *Nat. Struct. Mol. Biol.* **11,** 429–434.

Wruss, J., Runzler, D., Steiger, C., Chiba, P., Kohler, G., and Blaas, D. (2007). Attachment of VLDL receptors to an icosahedral virus along the 5-fold symmetry axis: Multiple binding modes evidenced by fluorescence correlation spectroscopy. *Biochemistry* **46,** 6331–6339.

CHAPTER TWENTY

HIGH-SPEED ATOMIC FORCE MICROSCOPY TECHNIQUES FOR OBSERVING DYNAMIC BIOMOLECULAR PROCESSES

Daisuke Yamamoto,[*,†] Takayuki Uchihashi,[*,†] Noriyuki Kodera,[*,†] Hayato Yamashita,[*] Shingo Nishikori,[‡] Teru Ogura,[†,‡] Mikihiro Shibata,[*] and Toshio Ando[*,†]

Contents

1. Introduction	542
2. Survey of Requirements for High-Speed Bio-AFM Imaging	543
3. Substrate Surfaces	544
3.1. Bare mica surface	544
3.2. Mica surface-supported planar lipid bilayers	546
3.3. Streptavidin 2D crystals as substrates	550
4. Control of Diffusional Mobility	554
5. Protein 2D Crystals as Targets to Study	555
5.1. Streptavidin 2D crystals	555
5.2. p97 2D crystals	557
5.3. Bacteriorhodopsin 2D crystals	558
6. Low-Invasive Imaging	558
7. UV Flash-Photolysis of Caged Compounds	559
8. Cantilever Tip	561
References	562

Abstract

Atomic force microscopy (AFM) enables direct visualization of single-protein molecules in liquids at submolecular resolution. High-speed AFM further makes it possible to visualize dynamic biomolecular processes at subsecond

[*] Department of Physics, Kanazawa University, Kakuma-machi, Kanazawa, Japan
[†] Core Research for Evolutional Science and Technology (CREST) of the Japan Science and Technology Agency, Sanban-cho, Chiyoda-ku, Tokyo, Japan
[‡] Department of Molecular Cell Biology, Institute of Molecular Embryology and Genetics, Kumamoto University, Kumamoto, Japan

resolution. However, dynamic imaging of biomolecular processes imposes various requirements on "wet techniques" and imaging conditions, which are often different from those for static imaging. This chapter first surveys the imposed requirements, then focuses on practical techniques associated with dynamic imaging, highlighting the preparation of substrate surfaces, and presents examples of the use of these techniques.

1. INTRODUCTION

Single-molecule visualization of functional behavior of proteins has been performed using fluorescence microscopy (e.g., Goldman, 2009; Joo et al., 2008; Park et al., 2007). This approach has enriched our knowledge of dynamic processes of proteins. However, even using fluorescence microscopy tools with a spatial resolution beyond the diffraction limit (Huang et al., 2009), it is impossible to observe protein molecules themselves since only single featureless fluorescent spots are observable. Therefore, we need to acquire additional information to bridge the gap between the behavior of fluorescent spots and the actual behavior of labeled protein molecules.

Atomic force microscopy (AFM) allows for the direct observation of individual protein molecules in solutions, at submolecular resolution. However, its imaging rate is too low to trace their dynamic behaviors. Rapidly moving molecules cannot be imaged at all. In the last decade, significant efforts have been carried out to increase the imaging rate of biological AFM (bio-AFM) (e.g., Ando et al., 2001, 2008; Fantner et al., 2006; Viani et al., 1999). Only increasing the scan speed of AFM would not be difficult, if we ignore potential damage to the sample or disturbance of biomolecular processes. Very precise and fast control of the tip-sample distance is mandatory for observing delicate dynamic biomolecular processes. To achieve high-speed and low-invasive AFM imaging, various techniques have been devised (e.g., Kodera et al., 2005, 2006). As a result, some dynamic events and structural changes in proteins have been successfully visualized (Ando et al., 2006; Miyagi et al., 2008; Shibata et al., 2010; Yamamoto et al., 2008; Yamashita et al., 2009). As comprehensive descriptions for realizing the instrumentation are already given elsewhere (Ando et al., 2008), we focus here on other critical aspects of dynamic AFM imaging.

Static AFM imaging of biological samples has been performed by many researchers, and various methods for this imaging have accumulated. On the other hand, dynamic AFM imaging of biological samples is rather new, and hence, methods and scan conditions specifically required for dynamic imaging are not yet found in the literature. Here, we introduce methods useful for dynamic AFM imaging, highlighting the preparation of substrate surfaces.

2. Survey of Requirements for High-Speed Bio-AFM Imaging

A substrate surface, on which a sample is placed, plays an important role in AFM imaging. It should be flat enough so that we can easily identify the molecules of interest deposited on it. To observe dynamically acting protein molecules under physiological conditions, the substrate surface must not firmly bind the molecules so that their physiological function is retained. However, the sample–surface interaction is essential to avoid too fast Brownian motion, particularly for single-protein molecules isolated completely from other molecules. Dynamic AFM imaging requires much more defined conditions for the substrate surface than static AFM imaging does.

We also have to consider requirements for observing dynamic protein–protein interactions. Supposing that one species of protein dynamically interacts with a protein counterpart, only one but not the other can be immobilized to a substrate surface. If both are surface bound, they have almost no chance to interact with each other. Consequently, selective protein attachment to a surface is required. A dynamic protein–protein interaction implies that their association is weak. Tapping mode AFM is suitable for imaging fragile samples because little lateral force is exerted to the sample, while the tip taps the sample. This tapping should not disturb delicate protein–protein interactions. Ideally, we have to achieve a very weak tapping force without slowing the imaging speed.

Protein molecules exhibit Brownian motion alongside physiologically relevant dynamic conformational changes. In many cases, the purpose of observing the dynamic behavior of proteins is to detect physiologically relevant changes in protein conformation. We therefore have to be able to distinguish the two types of movement. Special means sometimes have to be implemented for this distinction, as well as a means to impede Brownian motion.

Biological membranes are very fluidic. When membrane proteins are not assembled into a large ensemble or anchored to an immobile entity, they diffuse rapidly in a membrane, so that there is no way to observe their structure even using high-speed AFM with the capacity to image at 33 frames/s for a scan range of 240×240 nm^2 (Yamashita et al., 2007). In this case, we need some means to slow down diffusion.

In the following sections, several solutions for the difficulties mentioned above are presented and their effects are exemplified.

3. Substrate Surfaces

For AFM imaging, bare surfaces of mica and highly oriented polygraphite (HOPG), as well as gold-coated surfaces of these substrates or silicon have often been used. In aqueous solution, proteins do not attach to HOPG except for rare cases. Gold-coated surfaces are typically used for preparing self-assembled monolayers of molecules containing a thiol. Proteins can be immobilized to these monolayers when the constituting monolayer molecules also possess a reactive or functional group. Since the preparation methods and characterization of these monolayers are well described elsewhere (Hinterdorfer and Dufrêne, 2006; Liu et al., 2008), we here only cover bare mica surfaces, mica surface-supported planar lipid bilayers, and streptavidin 2D crystals formed on lipid bilayers.

3.1. Bare mica surface

Mica (natural muscovite or synthetic fluorophlogopite) has frequently been used as a substrate owing to its surface flatness at the atomic level over a large area. It has a net negative charge and is therefore quite hydrophilic. A bare mica surface adsorbs various proteins by electrostatic interactions. Except in rare cases (e.g., GroEL attachment in an end-up orientation), the orientation of adsorbed proteins is not unique, and the selective attachment of a specific species of protein is not expected. A mica surface is useful for observing the dynamic processes of a single species of isolated protein or membrane proteins in native membrane fragments. We can control the affinity for a specific protein by varying the ionic strength, pH, or by adding divalent cations such as Mg^{2+}. Divalent cations can link negatively charged proteins to a negatively charged mica surface.

Mica disks are prepared from a mica sheet with thickness of <0.1 mm by cutting holes using a sharp puncher. Serrated edge formation, which often accompanies partial cleavage of interlayer contacts in the disk, should be avoided. Hydrodynamic pressure produced by rapid scanning of the sample stage induces vibrations of the disk through movement of the cleaved sites. For high-speed imaging, the disk should be small (1–2 mm in diameter) to avoid generation of too large a hydrodynamic pressure (Ando et al., 2002). The following handling protocol is recommended:

1. Glue a mica disk to a sample stage using epoxy and wait until it has dried (~ 1–2 h).
2. Press a Scotch tape to the surface of the mica disk and then smoothly remove the tape from the mica. The top layer of the mica will be removed with the tape, which can be checked by inspecting the surface of the removed tape.

3. Place a sample solution on the freshly cleaved mica disk surface for 1–3 min, and then rinse with an appropriate buffer solution.
4. Judge the affinity of the sample for the mica surface by AFM imaging.
5. Find an appropriate solution condition by repeated imaging using different solutions (an example is shown in Fig. 20.1 for myosin V). When the affinity is too weak, reduce the ionic strength or pH, or change the concentration of divalent cations.

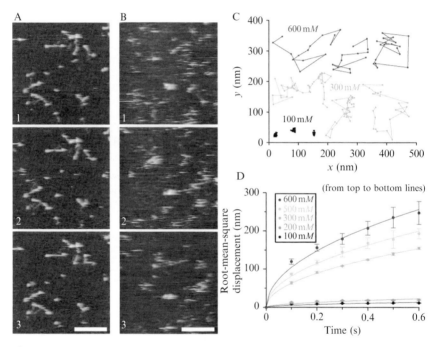

Figure 20.1 Ionic strength dependent diffusion of MyoV-HMM (tail-truncated myosin V) on bare mica surface. (A, B) Typical successive AFM images taken at 100 mM (A) and 600 mM KCl (B). Imaging rate: 102 ms/frame, scan area: 300 × 300 nm^2 with 100 × 100 pixels, scale bar: 100 nm. The number in each frame represents the frame number. The observation buffer is 20 mM imidazole–HCl (pH 7.6), 2 mM MgCl$_2$, 1 mM EGTA, and 100 or 600 mM KCl. (C) Typical trajectories of MyoV-HMM measured in 100, 300, or 600 mM KCl-containing solutions. (D) Root-mean-square displacement (x_{rms}) as a function of time. Each point was calculated from the trajectories of $\sim$100 molecules and well fitted by an equation $x_{rms} = \sqrt{4Dt}$, where D is the diffusion coefficient and t is the time elapse. The values of D (nm^2/s) at respective KCl concentrations are $D_{100\ mM} = 79 \pm 10$, $D_{200\ mM} = 213 \pm 26$, $D_{300\ mM} = (0.99 \pm 0.01) \times 10^4$, $D_{500\ mM} = (1.69 \pm 0.04) \times 10^4$, and $D_{600\ mM} = (2.73 \pm 0.09) \times 10^4$. (See Color Insert.)

3.2. Mica surface-supported planar lipid bilayers

Mica surface-supported planar lipid bilayers are useful for attaching proteins to their surfaces with a controlled affinity or selectivity. A membrane surface with zwitterionic polar head groups such as phosphatidyl choline (PC) and phosphatidyl ethanolamine (PE) is known to resist protein adsorption (Vadgama, 2005; Zhang et al., 1998). Various lipids with functional groups (e.g., biotin and Ni-NTA) on their polar heads are commercially available. They enable specific attachment of biotin- or His-tag-conjugated proteins to planar lipid bilayers. For electrostatic attachment, negatively charged head groups including phosphatidylserine (PS), phosphoric acid (PA), or phosphatidylglycerol (PG) and positively charged head groups such as trimethylammoniumpropane (TAP) and ethylphosphatidylcholine (EPC) are available. Unlike mica surfaces, the surface charge density can easily be controlled by mixing charged and neutral lipids in different ratios as long as no phase separation occurs.

3.2.1. Choice of alkyl chains

Dioleoyl-phosphatidyl-choline (DOPC) is useful for the preparation of 2D crystals of streptavidin and His-tag-labeled proteins when it is used in conjunction with biotinylated lipids and Ni-NTA-containing lipids, respectively (Reviakine and Brisson, 2001; Scheuring et al., 1999). DOPC contains an unsaturated hydrocarbon in each of the two alkyl chains, which causes the bending of the chains and therefore weakens the interaction between neighboring DOPCs. This weak interaction lowers the phase-transition temperature of DOPC to ~ -20 °C, thereby affording considerable fluidity to the planar bilayer at room temperature and facilitating the formation of 2D protein crystals. Such densely packed proteins do not diffuse easily. Dipalmitoyl-phosphatidyl-choline (DPPC) contains no unsaturated hydrocarbons in the alkyl chains. Therefore, its phase-transition temperature is high (~ 41 °C) and it is suitable for preparing planar lipid bilayers with low fluidity. For example, when planar bilayers are formed using DPPC at high temperature (~ 60 °C) together with a suitable fraction of DPPE-biotin, streptavidin sparsely attached to the surface hardly diffuses at room temperature. When DOPE-biotin is used together with DPPC, the sparsely attached streptavidin diffuses at a moderate rate (Fig. 20.2).

3.2.2. Preparation of planar lipid bilayers

For the preparation of planar lipid bilayers on a mica surface, two methods are commonly used; the Langmuir–Blodgett method and the vesicle fusion method. In both cases, uniform and smooth surfaces of lipid bilayers can be obtained. Here, we focus on the latter method, because of its simplicity and lack of a requirement for any expensive instruments and skills. Supported

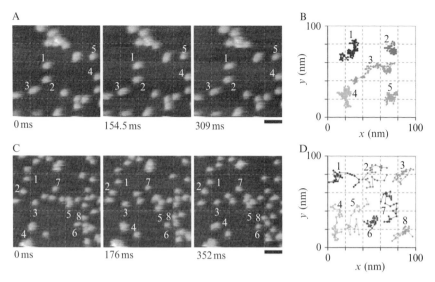

Figure 20.2 Mobility of lipid molecules monitored by diffusion of attached streptavidin molecules and its dependence on lipid compositions. Successive AFM images of streptavidin on lipid bilayers of DPPC + DPPE-biotin (9:1, w/w) and DPPC + DOPE-biotin (9:1, w/w) are shown in (A) and (C), respectively. Imaging rate: 154.5 ms/frame (A), 176 ms/frame (C), scan area, 200 × 200 nm^2 with 100 × 100 pixels, scale bar: 40 nm. The apparent difference in the size of streptavidin molecules between (A) and (C) is due to different cantilever tips. Trajectories of streptavidin molecules observed in (A) for ~15.6 s and (C) for ~4.6 s are plotted in (B) and (D), respectively. The numbers beside trajectories in (B) and (D) correspond to the numbered streptavidin molecules in (A) and (C), respectively.

planar lipid bilayers are formed from small unilamellar vesicles by direct deposition of the solution on a freshly cleaved mica surface. Planar lipid bilayer formation proceeds through: (i) adsorption of small unilamellar vesicles on the mica surface, (ii) rupture of the vesicles, and (iii) fusion of the resulting small membrane patches with juxtaposed lipid membranes. Although protocols for preparing lipid vesicles and supported lipid bilayers can be found elsewhere, we describe here a procedure routinely used in our laboratory.

1. Dissolve each lipid compound in chloroform or in a mixture of chloroform, methanol, and water (follow the instructions provided by the manufacturers).
2. Mix lipid solutions at a desired ratio in a glass test tube.
3. Dry the organic solvent under a stream of N_2 gas. During this procedure, warm up the test tube by gripping it with your hand to facilitate solvent evaporation.

4. To ensure full evaporation, leave the test tube in a desiccator under aspirator vacuum for more than 30 min.
5. Add a buffer solution to the test tube (typical final concentration of lipids, 0.125 mM) and vortex it. If necessary, sonicate the test tube using an ultrasonic bath to disperse the lipids in the medium. At this stage, multilamellar lipid vesicles are formed. The lipid suspension is divided into small aliquots ($\sim$100 μl) and stored at $-80\ ^\circ$C.
6. To obtain small unilamellar vesicles, sonicate the multilamellar vesicle suspension (typically 100 μl) with a tip sonicator at intervals of 1 s with a duty ratio of $\sim$0.5 until the suspension becomes transparent (typically no more than 30 cycles).
7. Place a drop of the small unilamellar vesicle solution on a freshly cleaved mica surface and incubate it for 30 min in a sealed container, while maintaining high humidity in the container to avoid sample drying by stuffing therein a piece of paper damped with water. Subsequently, rinse the sample with an appropriate buffer solution.
8. When the gel–liquid crystal transition temperature of the sample is higher than room temperature, warm up the sample in the sealed container slightly above the transition temperature for $\sim$15 min.
9. Examples for lipid compositions (buffer conditions) are given below. The mixing ratios of lipids should be changed depending on the samples and dynamic events to be visualized.
    - highly fluidic bilayers with biotin; DOPC:DOPS:biotin-cap-DOPE = 7:2:1 (w/w) [10 mM HEPES–NaOH (pH 7.4), 150 mM NaCl and 2 mM CaCl$_2$];
    - low fluidic bilayers with biotin; DPPC:biotin-cap-DPPE = 9:1 (w/w) [10 mM HEPES–NaOH (pH 7.4), 150 mM NaCl and 2 mM CaCl$_2$];
    - slightly fluid bilayers with biotin; DPPC:biotin-cap-DOPE = 9:1 (w/w) [10 mM HEPES–NaOH (pH 7.4), 150 mM NaCl and 2 mM CaCl$_2$];
    - highly fluidic bilayers with Ni-NTA; DOPC:DOPS:DOGS-NTA (Ni) = 7:2:1 (w/w) [10 mM HEPES–NaOH (pH 7.4), 150 mM NaCl and 2 mM CaCl$_2$];
    - low fluidic bilayers with Ni-NTA; DPPC:DOGS-NTA(Ni) = 9:1 (w/w) [50 mM Tris–HCl (pH 8.0), 50 mM KCl and 3 mM MgCl$_2$];
    - low fluidic bilayers with positively charged head groups; DPPC: DPTAP = 7:3 (w/w) or only DPTAP (distilled water);
    - low fluidic bilayers with negatively charged head groups; DPPG 100% [10 mM HEPES–NaOH (pH 7.4), 150 mM NaCl and 2 mM CaCl$_2$];
    - low fluidic bilayers with negatively charged head groups formed on positively charged lipid bilayers; DPPA 100% (distilled water).

3.2.3. Planar lipid bilayers as substrates

3.2.3.1. Electrostatic immobilization Since the mica surface is negatively charged, positively charged lipid bilayers can easily be formed on the surface. Depending on the sample to be imaged, an appropriate positive-charge density has to be found by repeated imaging with various positive to neutral lipid ratios (examples are shown in Fig. 20.3A–C). Negatively charged lipid bilayers can be formed in a wide area on a mica surface

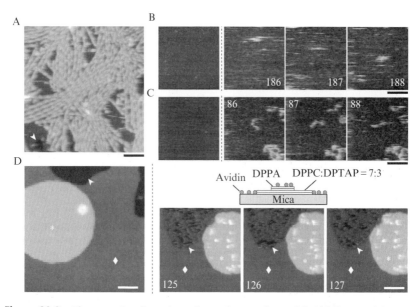

Figure 20.3 Electrostatic adsorption of proteins on charged lipid bilayers. (A) AFM image of actin filament paracrystal formed on positively charged lipid bilayers (DPPC: DPTAP = 7:3, w/w). The white arrowhead indicates an exposed mica surface. Scale bar, 100 nm. (B, C) High-speed AFM images showing the effect of positive-charge surface density on the mobility of attached MyoV-HMM. The lipid compositions are (B) DPPC:DPTAP = 7:3 (w/w) and (C) only DPTAP. The leftmost AFM images show the surface of lipid bilayers before adding MyoV-HMM. Imaging rate, 101 ms/frame; scale bar, 100 nm. The number in each frame represents the frame number. MyoV-HMM was hardly visible (B) due to the large mobility, whereas it was visible (C) due to the low mobility. (D) AFM images of negatively charged lipid bilayers formed on positively charged lipid bilayers. The leftmost AFM image was taken before adding avidin (a positively charged protein). The white arrowhead, diamond, and asterisk represent the surfaces of mica, positively charged lipid bilayers, and negatively charged lipid bilayers, respectively (imaging rate: 418 ms/frame, scale bar: 100 nm). The remaining successive AFM images were taken after adding avidin (imaging rate: 175 ms/frame, scale bar: 100 nm). Avidin molecules are only observed on the negatively charged surfaces of mica and DPPA (see the upper schematic). The observation buffer contains 20 mM imidazole–HCl (pH 7.6), 25 mM KCl, 2 mM MgCl$_2$, and 1 mM EGTA.

using DPPG (transition temperature, T_m, 41 °C). Since the negative charge of DPPG is not positioned at the distal end of the polar head, the bilayer surface weakly adsorbs positively charged proteins. Thus far, negatively charged lipid bilayers have not been formed on a mica surface using DPPA ($T_m = 67$ °C; the negative charge is positioned at the distal end of the polar head) or DPPS ($T_m = 54$ °C). Another means is to form negatively charged lipid bilayers on positively charged bilayers. Although not completely successful yet, at least small patches of DPPA bilayers can be formed by this means (Fig. 20.3D).

3.2.3.2. Immobilization via specific interaction Specific association pairs with strong affinities, such as (biotin and streptavidin) and (His-tag and Ni-NTA), can be used for selective immobilization of proteins. When relatively large molecules mediate the immobilization of a protein on lipid bilayers, the attached mediator would make it difficult to identify the immobilized protein. When streptavidin is used as a mediator, the use of streptavidin 2D crystals is recommended (see Section 3.3). Here, we show only an example of His-tag-conjugated protein immobilization. p97 is a hexameric ATPase of the AAA family and involved in cellular processes (more descriptions of p97 are given later). As shown in Fig. 20.4, recombinant p97 of which each subunit was conjugated with a His-tag was immobilized on a Ni-NTA-containing lipid bilayer. p97 was attached to the bilayer surface in a specific orientation due to the specific-site conjugation of a His-tag to each subunit.

3.3. Streptavidin 2D crystals as substrates

2D crystals of streptavidin formed on biotin-containing lipid bilayers meet various requirements for substrate surfaces used in dynamic AFM imaging. Streptavidin is comprised of four identical subunits, each of which specifically

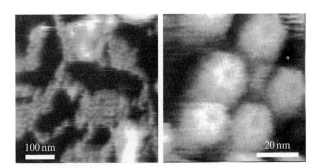

Figure 20.4 p97 attached on Ni-NTA-containing lipid bilayer through His-tag conjugated at the N-terminus of each subunit. (A) Segregation of Ni-NTA lipid in bilayer (p97 are densely packed at bright regions but absent in dark regions). (B) Ring structure of p97 (C-terminus D2 domain surface), z-scales: (A) 20 nm, (B) 14 nm.

binds to one biotin molecule with a strong affinity ($K_a \approx 10^{13}$ M^{-1}) (Green, 1990). In 2D crystals, two biotin-binding sites face the solution, and therefore, can bind to biotinylated samples to be imaged (Darst et al., 1991; Reviakine and Brisson, 2001). Since biotinylated Ni-NTA is commercially available, the surface can also bind to His-tag conjugated recombinant proteins. The surface roughness of a specific type of 2D crystal is very small, as mentioned later. Importantly, streptavidin is resistant to nonspecific binding of many proteins (Green, 1990), which safeguards the surface-bound proteins against dysfunction (Heyes et al., 2004) and allows the selective surface attachment of one species of protein in a multicomponent system. In addition, it allows protein attachment in a controlled orientation through the biotin or His-tag conjugation sites in the protein.

3.3.1. Preparation of streptavidin 2D crystals

On highly fluidic planar lipid layers containing a biotinylated lipid, streptavidin self-assembles into three distinct crystalline arrangements of C222, P2, and P1 symmetries, depending on the crystallization conditions such as pH (Wang et al., 1999) and ionic strength (Ratanabanangkoon and Gast, 2003), as described in the following. Note that all the 2D crystals have P2 symmetry (Yamamoto et al., 2009) but we followed previous reports in the notation for the symmetry groups.

1. Crystallization buffer solutions: (C222 crystal) 10 mM HEPES–NaOH (pH 7.4), 150 mM NaCl, and 2 mM CaCl$_2$; (P2 crystal) 10 mM MES–NaOH (pH 5.8), 450 mM NaCl, and 2 mM CaCl$_2$; (P1 crystal) 10 mM acetate–NaOH (pH 4.0), 450 mM NaCl, and 2 mM CaCl$_2$.
2. Prepare the mica-supported lipid planar bilayers containing biotin as mentioned in Section 3.2.2.
3. After washing the lipid planar bilayers with a crystallization buffer solution, place 0.1–0.2 mg/ml streptavidin in the crystallization buffer solution on the lipid bilayer and incubate at room temperature in a sealed container (keeping 100% relative humidity is a key to the successful crystallization with fewer defects).
4. After incubating at least for 2 h, wash out excess streptavidin with the crystallization buffer solution used.
5. If necessary, chemically fix the streptavidin crystals by applying a 10 mM glutaraldehyde-containing crystallization solution and incubate for 5 min. Then, quench the reaction using 20 mM Tris–HCl mixed in the crystallization buffer.

Figure 20.5 (*top*) shows high-magnification AFM images of the three types of streptavidin crystals. Among the three, the P1 crystal shows the smallest surface roughness (Fig. 20.6C, *bottom*; RMS roughness ~ 0.14 nm) and the highest stability (Yamamoto et al., 2009). Even when tapped with an oscillating cantilever with relatively strong force, the crystal structure is

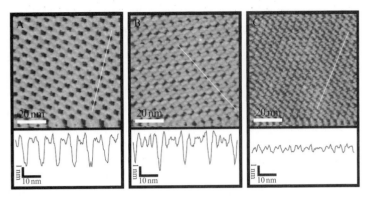

Figure 20.5 High-magnification AFM images of the three types of streptavidin crystals (Yamamoto *et al.*, 2009). The space group symmetries of the crystals are (A) *C*222, (B) *P*2, and (C) *P*1. *Top*: AFM images, according to previous reports (*P*2 is a correct notation for all the crystals). *Bottom*: Surface profile along the lines indicated in the AFM images in *top*. The AFM images were obtained at an imaging rate of 1 s/frame with 256 × 256 pixels.

unchanged, and hence, no chemical fixation is necessary. However, the *P*1 crystal often shows line cracks. The *C*222 crystal shows high uniformity over a wide area but similarly to the *P*2 crystal, its stability is lower than that of the *P*1 crystal, and therefore, chemical fixation is sometimes necessary. Its chemical fixation does not significantly affect the affinity of streptavidin for biotin.

3.3.2. Use of streptavidin 2D crystals as substrates

Supposing that a protein to be imaged is relatively rigid and does not possess flexible chains, its biotinylated species immobilized on a streptavidin 2D crystal surface would not exhibit rapid Brownian motion even when only one biotin per molecule is introduced. If it rapidly rotates around the biotinylated site due to a flexible linker contained in the biotin compound used, we can use reactive dibiotin compounds for less mobility. However, when a protein to be imaged contains a flexible moiety and is biotinylated only at one site, it certainly exhibits too rapid Brownian motion to be imaged (examples are shown in Fig. 20.6 for calmodulin (CaM) and tail-truncated myosin V (MyoV-HMM)). This rapid motion can be stopped by introducing biotin to the protein at more than two sites (Fig. 20.6C). Thus, the streptavidin 2D crystals are excellent substrates for oligomerized proteins as exemplified below but their usefulness is limited for monomeric proteins with flexible polypeptide chains.

We can specifically place a protein on streptavidin 2D crystals in a desired orientation when the protein is biotinylated at designed sites. For example, the D490C GroEL mutant, in which biotin is conjugated to the

High-Speed Atomic Force Microscopy Techniques 553

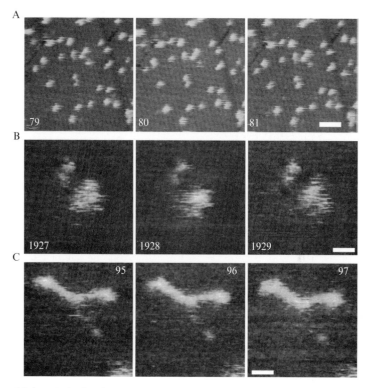

Figure 20.6 Biotinylated proteins attached onto streptavidin 2D crystal surfaces. (A) Successive AFM images of biotinylated calmodulin (CaM) attached on P1 crystal of streptavidin. The images do not reveal a typical dumbbell-like appearance of CaM because one of the two globes connected with a flexible central linker is not biotinylated. Scale bar, 20 nm; imaging rate: 991 ms/frame, scan area: 100×100 nm^2 with 256×256 pixel, z-scale: 4.5 nm. The observation buffer is 20 mM imidazole–HCl (pH 7.6), 100 mM KCl, 2 mM MgCl$_2$, 1 mM EGTA. (B, C) Effect of the number of conjugated biotin per molecule on AFM image of MyoV-HMM immobilized on streptavidin 2D crystal. MyoV-HMM-containing biotinylated CaM was immobilized on the C222 crystal of streptavidin. Scale bar: 20 nm, imaging rate: 487 ms/frame, scan area: 100×100 nm^2 with 128×128 pixels, z-scale: 8.0 nm. The observation buffer was 20 mM Tris–HCl (pH 6.8), 100 mM KCl, 2 mM MgCl$_2$, 1 mM EGTA, and 5 mM DTT. (B) MyoV-HMM having biotinylated CaM only at one of the two necks. It does not show the typical molecular shape because the unbiotinylated head moves rapidly and even the biotinylated head rotates rapidly around the biotin-linker. (C) MyoV-HMM with biotinylated CaM at both necks.

Cys-490 located at the outer surface of the equatorial domain, can attach to the C222 crystal surface in a side-on orientation (Fig. 20.7) (Yamamoto *et al.*, 2009). In this orientation, both rings of GroEL are accessible to GroES floating in bulk solution. GroES never attaches to the crystal surface even

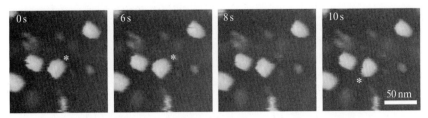

Figure 20.7 AFM images of the GroEL–GroES complex (Yamamoto et al., 2009). The D490C GroEL mutant was biotinylated at the outer surface of the equatorial domain, and was immobilized in a side-on orientation on a streptavidin C222 crystal that is chemically fixed with glutaraldehyde. The GroEL–GroES complex was formed and imaged in the buffer containing 1 μM GroES and 1 mM ADP. The asterisks indicate GroES bound to one of the GroEL rings. The AFM images were obtained at an imaging rate of 2 s/frame with 200 × 200 pixels, z-scale: 19 nm.

when the concentration is increased to the 10 μM range, allowing for the observation of dynamic events of association/dissociation of the GroEL/GroES complex. Figure 20.7 (*asterisks*) shows such a dynamic event in the presence of ADP.

4. Control of Diffusional Mobility

Proteins that are weakly and sparsely attached to a substrate surface often exhibit rapid Brownian motion. Even when densely packed on highly fluidic planar lipid bilayers, some proteins hardly form 2D crystals due to weak association or repulsion between protein molecules. Noncrystalline proteins diffuse due to the fluidity of the lipid bilayers, making it difficult to capture an image. Depending on the situation, we need a specific means to impede the rapid diffusion, as detailed in the following.

1. Brownian motion of a protein that is attached to a surface by electrostatic interaction can be impeded by strengthening the interaction using a low ionic strength solution (Fig. 20.1) or a more charged surface (Fig. 20.3A). This approach is easy and effective for a large number of samples as the stronger electrostatic association usually does not hamper protein function. However, when it does, there are no ideal means to slow down the Brownian motion. One compromise is to place, with an appropriate surface density, small particles on the surface that have a relatively strong affinity for the surface. Diffusion of the protein molecules is slowed down by the contact with the particles. Using this means, we succeeded to visualize the step process of myosin V (MyoV) along actin tracks.

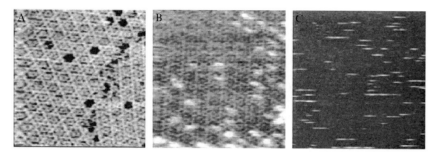

Figure 20.8 (A) An AFM image of the Annexin A-V two-dimensional crystal on lipid bilayer. The lipid bilayer is composed of 65% DOPC, 34% DOPS, and 1% biotin-capped DOPE (w/w). The circular dark areas correspond to the single defects. (B) The streptavidin molecules are trapped in the large defects of the Annexin crystal and observed as bright spots. (C) Streptavidin molecules rapidly diffuse on the lipid bilayer without Annexin and therefore they were observed as spike noises. For all images, the scan area is 200×200 nm and the buffer solution is 10 mM HEPES–NaOH (pH 6.8), 20 mM NaCl, 150 mM KOAc, 2 mM MgCl$_2$, and 2 mM CaCl$_2$.

2. When a protein hardly forms 2D crystals on a highly fluidic lipid bilayer, change the pH to reduce the electrostatic repulsion (an example is shown below), or place the noncrystalline protein on low fluidic lipid bilayers as exemplified in Fig. 20.2.
3. For membrane proteins embedded in native lipid membranes, their rapid diffusion sometimes can be impeded using Annexin (Ichikawa *et al.*, 2006). Annexins are a protein family that bind to negatively charged phospholipids in the presence of calcium and assemble as trimers (Richter *et al.*, 2005). Annexin A-V forms 2D crystals on PS-containing lipid membranes (Fig. 20.8A). Using this property, Annexin A-V can be used to trap membranes' proteins in the defect sites of an Annexin 2D crystal. Although not native membrane proteins, streptavidin molecules are demonstrated to be trapped in the defects of the Annexin A-V crystal formed on very fluidic lipid bilayers (Fig. 20.8B). Without Annexin A-V, no image of streptavidin molecules can be taken due to rapid diffusion (Fig. 20.8C).

5. Protein 2D Crystals as Targets to Study

5.1. Streptavidin 2D crystals

As mentioned in Section 3.2.3, the *C*222 and *P*2 crystals of streptavidin are not stable. Monovacancy defects are formed by increasing the tapping force exerted from an oscillating tip. The defects in the *C*222 crystal move in the

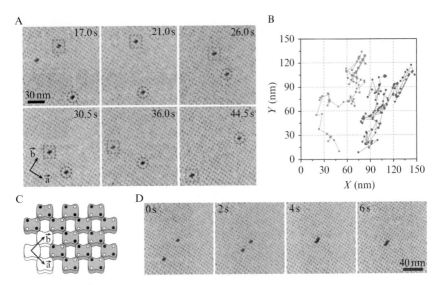

Figure 20.9 Diffusion of monovacancy defects in streptavidin C222 crystal observed by high-speed AFM (Yamamoto et al., 2008). (A) High-speed AFM images of the streptavidin 2D crystal and monovacancy defects migrating in the crystal. The monovacancy defects in the crystal are enclosed by the dashed squares and circles. The arrows indicate the direction of the lattice vectors of the crystal. Successive images were obtained at an imaging rate of 0.5 s/frame with 200 × 200 pixels. (B) Trajectories of the monovacancy defects obtained from the successive images (A). Closed squares and circles correspond to defects enclosed by the open squares and circles in (A), respectively. (C) Schematic of streptavidin C222 crystal. The closed circles indicate biotin-bound subunits of streptavidin in the crystal, while the open circles correspond to biotin-unbound subunits. (D) Fusion of monovacancy defects in the C222 crystal. Two monovacancy defects migrate independently in the crystal at 0–2 s, and then fused to form a divacancy defect at 4 s. The resultant divacancy defect continued to migrate in the crystal (4–6 s). The successive images were obtained at an imaging rate of 0.5 s/frame with 200 × 200 pixels.

plane, and this diffusion exhibits anisotropy correlated with the two crystallographic axes in the orthorhombic C222 crystal (Fig. 20.9A and B) (Yamamoto et al., 2008); one axis (a-axis) is comprised of contiguous biotin-bound subunit pairs whereas the other axis (b-axis) is comprised of contiguous biotin-unbound subunit pairs (Fig. 20.9C). The diffusivity along the b-axis is approximately 2.4 times larger than that along the a-axis. This anisotropy is ascribed to the difference in the free energy of association between the biotin-bound and not biotin-bound subunit–subunit interaction. A preferred intermolecular contact occurs between the not biotin-bound subunits. The difference in the intermolecular-binding free energy between the two types of subunit pairs is estimated from the ratio of diffusion constants to approximately 0.52 kcal/mol. Another dynamic behavior observed for point defects is the fusion of two point defects into a larger defect, which occurs much more frequently than the fission of a point defect into smaller defects (Fig. 20.9D). The diffusivity of point defects increases with increasing defect size. Fusion and higher diffusivity of larger defects are suggested to be involved in the mechanism of defect-free crystal formation.

5.2. p97 2D crystals

A multifunctional AAA ATPase, p97, is involved in cellular processes such as membrane fusion (Hetzer et al., 2001; Rabouille et al., 1998) and extraction of proteins from the endoplasmic reticulum for cytoplasmic degradation (Jarosch et al., 2002; Ye et al., 2001). Each subunit of the hexameric p97 contains two AAA domains (N-terminal D1 and C-terminal D2) and a flexible N-domain (Huyton et al., 2003). Its conformational changes during the ATPase cycle have been studied using cryo-EM and single-particle analysis in the presence of analogues of ATP hydrolysis products (Rouiller et al., 2002). It has been suggested that the relative positioning of the ring-shaped D1 and D2 domains changes and the N-domain moves in a direction perpendicular to the ring planes. We sought to prepare 2D crystals of p97 on Ni-NTA-containing lipid bilayers using a p97 with His-tag at the N-terminus of each subunit. It did not form crystals at neutral pH, but formed a crystal at acidic pH. Interestingly, p97 placed in an end-up orientation showed fluctuations in the height even in the absence of nucleotides (Fig. 20.10), suggesting that the N-domain assumes two conformations in equilibrium. This conformational equilibrium probably shifts to one of the two states depending on the intermediate in the ATPase cycle, which is likely a process needed to extract proteins from the endoplasmic reticulum.

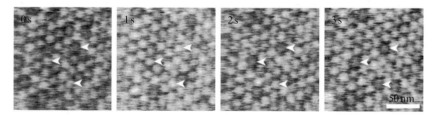

Figure 20.10 High-speed AFM images of two-dimensional crystal of p97 having His-tag at the N-terminus, which was formed on Ni-NTA-containing highly fluidic lipid bilayer. The crystal was formed under acidic pH (pH 5.6). The height of p97 fluctuates relative to other molecules in the crystal (arrowheads). Successive images were obtained at an imaging rate of 0.5 s/frame with 256 × 256 pixels.

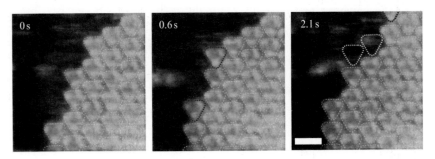

Figure 20.11 Dynamics at the crystal edge of bacteriorhodopsin in purple membrane (Yamashita et al., 2009). The AFM images were taken at 3.3 frames/s (scale bar, 10 nm). The bR molecules encircled by the black dotted lines (at 0.6s) indicate newly bound bR trimer. The white triangles (at 2.1 s) indicate the previously bound trimers.

5.3. Bacteriorhodopsin 2D crystals

In the purple membrane (PM) of *Halobacterium salinarum*, bacteriorhodopsin (bR) monomers associate to form a trimeric structure, and the trimers arrange in a hexagonal lattice (Henderson et al., 1990). The 2D crystals of bR and any crystals in general are in dynamic equilibrium with the constituents at the interface between crystal and liquid. High-speed AFM movies of the crystal edge of PM directly revealed this dynamic equilibrium (Fig. 20.11). Binding of trimeric bR predominates (82%), whereas binding of dimeric and monomeric bR is much less frequent, indicating that trimers are preformed in the noncrystalline region. The residence time of newly bound bR trimers at the crystal edge depends on the number of interaction sites. Within the 2D bR crystal, a trimer can interact with the surrounding trimers through six sites, whereas the number of interaction sites at the crystal edge is reduced to 1, 2, or 3, depending on the binding position of a newly bound trimer. From the ratio of the average residence times (0.19 and 0.85 s) of two bound species containing two or three bonds, the elementary association energy is estimated to about $-1.5k_BT$, which corresponds to -0.9 kcal/mol at 300 K (Yamashita et al., 2009).

6. Low-Invasive Imaging

In tapping mode high-speed AFM, the lateral force exerted from an oscillating cantilever tip to a sample is very small even when an image is taken within 30 ms (scan lines, ~ 100). High-speed AFM for biological samples requires small cantilevers with a high resonant frequency (~ 1 MHz) and a relatively small spring constant (0.1–0.2 N/m). With high-frequency oscillations, the cantilever tip makes contact with a sample for only a very short time (~100 ns). During this virtually instantaneous contact, the sample stage is displaced only by <0.2 nm as long as the scan range is <300 nm, and hence, deformation of the sample in the lateral direction is negligible. On the other hand, the vertical force (tapping force) should be reduced by combining the various protocols and devices described in the following. This is mandatory for visualizing delicate biomolecular processes.

1. The free oscillation amplitude of a cantilever is optimized so that the tapping force is reduced and the feedback operation is not slowed down significantly. For very delicate samples, the amplitude has to be set at <2 nm, sacrificing the feedback speed to some extent.
2. The amplitude set point should be set close to the free oscillation amplitude (>90% of the free oscillation amplitude). Tip parachuting, which often occurs under this condition at the steep downhill regions of the sample, can be avoided using a dynamic proportional-integral-derivative (PID) controller (Kodera et al., 2006). In this controller, the gain parameters can be adjusted automatically and dynamically depending on the cantilever oscillation amplitude.
3. The free oscillation amplitude of a cantilever has to be maintained precisely constant. Otherwise, the tip-sample distance (hence, the tapping force) changes with time even when the cantilever amplitude is maintained during imaging at the set point. Since the difference between the free oscillation amplitude and the set point is less than 0.2–0.3 nm under a very weak tapping force, even small drifts toward smaller free oscillation amplitude easily result in complete tip-sample detachment. This problem can be solved by a compensator for drift in the free oscillation amplitude (Kodera et al., 2006). This compensator controls the AC voltage applied to a cantilever-excitation piezoactuator to maintain a constant second-harmonic amplitude of the cantilever.

7. UV Flash-Photolysis of Caged Compounds

UV flash-photolysis of caged compounds is useful to detect minute structural changes of proteins induced by the interaction with chemicals such as ATP and calcium. Detected structural changes that are synchronized with the flash illumination can be concluded to be caused by the action of the uncaged chemical, facilitating distinction from Brownian motion. After irradiating a small area around the imaging area by a UV flash, the concentration of uncaged compound at the imaging area quickly decreases by rapid diffusion of the chemical. Therefore, repeated observation of structural changes is possible when they are reversible.

For a successful observation, the following precautions are required because UV illumination is prone to disturb the oscillation of a cantilever or to displace the sample stage by thermal expansion.

1. Use high-frequency weak UV pulses (~ 0.1 μJ/pulse) instead of a single shot of strong flash. A Nd:YVO$_4$ 355 nm laser with a repetition frequency >50 kHz is commercially available.
2. Apply laser pulses while the y-scan toward the starting point of imaging is performed (without x-scan) immediately after acquisition of a frame.
3. Withdraw the sample stage from the cantilever tip during the y-scan.

Using a calcium-binding protein, CaM, attached to the neck region of MyoV, we here demonstrate high-speed AFM imaging combined with flash-photolysis of caged calcium. Each neck region of MyoV comprised of 6 IQ motifs to which six light chains (5 CaM and an essential light chain) are attached. It is known that CaM attached to the second IQ dissociates in the presence of Ca^{2+} at micromolar concentrations (Koide et al., 2006). To facilitate the observation of structural changes induced in MyoV by CaM dissociation, MyoV was weakly attached to an aminosilane-coated mica surface on which the neck regions can extend (Fig. 20.12A). On each neck region, three small globes corresponding to three light chains are visible, but the other three light chains face the mica surface. Just after the illumination by ~ 200 UV pulses with 50 kHz, one of the motor domains (a large globe) located at the distal end of the head partially dissociates from the adjacent neck region (Fig. 20.12B), but the three globes on the neck remain visible. Probably one of the three light chains facing the mica surface dissociated from the neck. The partially dissociated motor domain was observed to be linked to the neck region through a thin structure. This behavior was rather rare among many observations. The observations mostly showed shrinkage

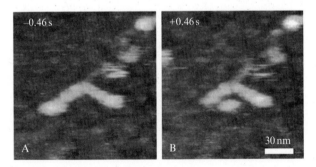

Figure 20.12 Structural change in myosin V upon uncaging calcium from caged-calcium. Calcium was released from caged-calcium at 0 s. (A) Before UV illumination, (B) after UV illumination. The AFM images were taken at 464 ms/frame.

of the neck region after UV illumination, suggesting folding of an exposed IQ motif after dissociation of CaM.

8. Cantilever Tip

Finally, high spatial resolution is also important in high-speed AFM. The resolution is mainly determined by the apex radius and the aspect ratio of a cantilever tip, in addition to the performance of the instrument. Thus far, small cantilevers with a sharp tip are not commercially available. Piece-by-piece attachment of a carbon nanotube to a cantilever tip is possible (Dai et al., 1996). However, this laborious attachment cannot be routinely performed. Instead, we have employed electron beam deposition (EBD) to grow a carbon tip on the original tip (Wendel et al., 1995). A piece of phenol crystal (sublimate) is placed in a small container with small holes (~0.1 mm diameter) in the lid. The container is placed in a scanning electron microscope chamber and small cantilevers are placed above the holes. A spot-mode electron beam is focused onto each original tip, which produces a needle on the original tip at a growth rate of ~17 nm/s. The newly formed carbon tip has an apex radius of 15–25 nm (Fig. 20.13A and B) and is sharpened by plasma etching in argon or oxygen gas, which decreases the apex radius to 4–5 nm (Fig. 20.13C).

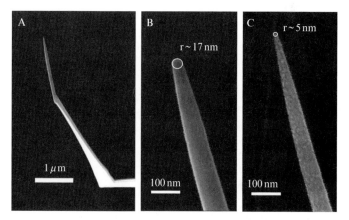

Figure 20.13 (A) An SEM image of an EBD tip grown from phenol as a vapor source. The deposition time was 90 s. The electron energy, working distance, and aperture size of the FE-SEM during the deposition were 20 kV, 4 mm, and 30 μm, respectively. (B) The apex radius of the grown tip is about 17 nm. (C) After etching in Ar gas for 8 min using a plasma etcher with the power of 16 W, the apex radius is reduced to about 5 nm.

REFERENCES

Ando, T., Kodera, N., Takai, E., Maruyama, D., Saito, K., and Toda, A. (2001). A high-speed atomic force microscope for studying biological macromolecules. *Proc. Natl. Acad. Sci. USA* **98**, 12468–12472.
Ando, T., Kodera, N., Maruyama, D., Takai, E., Saito, K., and Toda, A. (2002). A high-speed atomic force microscope for studying biological macromolecules in action. *Jpn. J. Appl. Phys.* **41**, 4851–4856.
Ando, T., Uchihashi, T., Kodera, N., Miyagi, A., Nakakita, R., Yamashita, H., and Sakashita, M. (2006). High-speed atomic force microscopy for studying the dynamic behavior of protein molecules at work. *Jpn. J. Appl. Phys.* **45**, 1897–1903.
Ando, T., Uchihashi, T., and Fukuma, T. (2008). High-speed atomic force microscopy for nano-visualization of dynamic biomolecular processes. *Prog. Surf. Sci.* **83**, 337–437.
Dai, H., Hafner, J. H., Rinzler, A. G., Colbert, D. T. R., and Smalley, E. (1996). Nanotubes as nanoprobes in scanning probe microscopy. *Nature* **384**, 147–150.
Darst, S. A., Ahlers, M., Meller, P. H., Kubalek, E. W., Blankenburg, R., Ribi, H. O., Ringsdorf, H., and Kornberg, R. D. (1991). Two-dimensional crystals of streptavidin on biotinylated lipid layers and their interactions with biotinylated macromolecules. *Biophys. J.* **59**, 387–396.
Fantner, G. E., Schitter, G., Kindt, J. H., Ivanov, T., Ivanova, K., Patel, R., Holten-Andersen, N., Adams, J., Thurner, P. J., Rangelow, I. W., and Hansma, P. K. (2006). Components for high speed atomic force microscopy. *Ultramicroscopy* **106**, 881–887.
Goldman, Y. E. (2009). Imaging and molecular motors. *In* "Single Molecule Dynamics in Life Science," (T. Yanagida and Y. Ishii, eds.), pp. 41–85. Wiley-VCH, Weinheim.
Green, N. M. (1990). Avidin and streptavidin. *Methods Enzymol.* **184**, 51–67.
Henderson, R., Baldwin, J. M., Ceska, T. A., Zemlin, F., Beckmann, E., and Downing, K. H. (1990). Model for the structure of bacteriorhodopsin based on high-resolution electron cryo-microscopy. *J. Mol. Biol.* **213**, 899–929.
Hetzer, M., Meyer, H. H., Walther, T. C., Bilbao-Cortes, D., Warren, G., and Mattaj, I. W. (2001). Distinct AAA-ATPase p97 complexes function in discrete steps of nuclear assembly. *Nat. Cell Biol.* **3**, 1086–1091.
Heyes, C. D., Kobitski, A. Y., Amirgoulova, E. V., and Nienhaus, G. U. (2004). Biocompatible surfaces for specific tethering of individual protein molecules. *J. Phys. Chem. B* **108**, 13387–13394.
Hinterdorfer, P., and Dufrêne, Y. F. (2006). Detection and localization of single molecular recognition events using atomic force microscopy. *Nat. Methods* **3**, 347–355.
Huang, B., Bates, M., and Zhuang, X. (2009). Super-resolution fluorescence microscopy. *Annu. Rev. Biochem.* **78**, 993–1016.
Huyton, T., Pye, V. E., Briggs, L. C., Flynn, T. C., Beuron, F., Kondo, H., Ma, J., Zhang, X., and Freemont, P. S. (2003). The crystal structure of murine p97/VCP at 3.6 Å. *J. Struct. Biol.* **144**, 337–348.
Ichikawa, T., Aoki, T., Takeuchi, Y., Yanagida, T., and Ide, T. (2006). Immobilizing single lipid and channel molecules in artificial lipid bilayers with annexin A5. *Langmuir* **22**, 6302–6307.
Jarosch, E., Taxis, C., Volkwein, C., Bordallo, J., Finley, D., Wolf, D. H., and Sommer, T. (2002). Protein dislocation from the ER requires polyubiquitination and the AAA-ATPase Cdc48. *Nat. Cell Biol.* **4**, 134–139.
Joo, C., Balci, H., Ishitsuka, Y., Buranachai, C., and Ha, T. (2008). Advances in single-molecule fluorescence methods for molecular biology. *Annu. Rev. Biochem.* **77**, 51–76.
Kodera, N., Yamashita, H., and Ando, T. (2005). Active damping of the scanner for high-speed atomic force microscopy. *Rev. Sci. Instrum.* **76**, 053708.

Kodera, N., Sakashita, M., and Ando, T. (2006). Dynamic proportional-integral-differential controller for high-speed atomic force microscopy. *Rev. Sci. Instrum.* **77,** 083704.
Koide, H., Kinoshita, T., Tanaka, Y., Tanaka, S., Nagura, N., zu Hörste, G. M., Miyagi, A., and Ando, T. (2006). Identification of the specific IQ motif of myosin V from which calmodulin dissociates in the presence of Ca^{2+}. *Biochemistry* **45,** 11598–11604.
Liu, M., Amro, N. A., and Liu, G.-Y. (2008). Nanografting for surface: Physical chemistry. *Annu. Rev. Phys. Chem.* **59,** 367–386.
Miyagi, A., Tsunaka, Y., Uchihashi, T., Mayanagi, K., Hirose, S., Morikawa, K., and Ando, T. (2008). Visualization of intrinsically disordered regions of proteins by high-speed atomic force microscopy. *Chem. Phys. Chem.* **9,** 1859–1866.
Park, H., Toprak, E., and Selvin, P. R. (2007). Single-molecule fluorescence to study molecular motors. *Quat. Rev. Biophys.* **40,** 87–111.
Rabouille, C., Kondo, H., Newman, R., Hui, N., Freemont, P., and Warren, G. (1998). Syntaxin 5 is a common component of the NSF- and p97-mediated reassembly pathways of Golgi cisternae from mitotic Golgi fragments in vitro. *Cell* **92,** 603–610.
Ratanabanangkoon, P., and Gast, A. P. (2003). Effect of ionic strength on two-dimensional streptavidin crystallization. *Langmuir* **19,** 1794–1801.
Reviakine, I., and Brisson, A. (2001). Streptavidin 2D crystals on supported phospholipid bilayers: Toward constructing anchored phospholipid bilayers. *Langmuir* **17,** 8293–8299.
Richter, R. P., Him, J. L. K., Tessier, B., Tessier, C., and Brisson, A. R. (2005). On the kinetics of adsorption and two-dimensional self-assembly of annexin A5 on supported lipid bilayers. *Biophys. J.* **89,** 3372–3385.
Rouiller, I., DeLaBarre, B., May, A. P., Weis, W. I., Brunger, A. T., Milligan, R. A., and Wilson-Kubalek, E. M. (2002). Conformational changes of the multifunction p97 AAA ATPase during its ATPase cycle. *Nat. Struct. Biol.* **9,** 950–957.
Scheuring, S., Müller, D. J., Ringler, P., Heymann, J. B., and Engel, A. (1999). Imaging streptavidin 2D crystals on biotinylated lipid monolayers at high resolution with the atomic force microscope. *J. Microsc.* **193,** 28–35.
Shibata, M., Yamashita, H., Uchihashi, T., Kandori, H., and Ando, T. (2010). High-speed atomic force microscopy shows dynamic molecular processes in photo-activated bacteriorhodopsin. *Nat. Nanotech.* **5,** 208–212.
Vadgama, P. (2005). Surface biocompatibility. *Annu. Rep. Prog. Chem., Sect. C: Phys. Chem.* **101,** 14–52.
Viani, M. B., Schäffer, T. E., Paloczi, G. T., Pietrasanta, L. I., Smith, B. L., Thompson, J. B., Richter, M., Rief, M., Gaub, H. E., Plaxco, K. W., Cleland, A. N., Hansma, H. G., and Hansma, P. K. (1999). Fast imaging and fast force spectroscopy of single biopolymers with a new atomic force microscope designed for small cantilevers. *Rev. Sci. Instrum.* **70,** 4300–4303.
Wang, S.-W., Robertson, C. R., and Gast, A. P. (1999). Molecular arrangement in two-dimensional streptavidin crystals. *Langmuir* **15,** 1541–1548.
Wendel, M., Lorenz, H., and Kotthaus, J. P. (1995). Sharpened electron beam deposited tips for high resolution atomic force microscope lithography and imaging. *Appl. Phys. Lett.* **67,** 3732.
Yamamoto, D., Uchihashi, T., Kodera, N., and Ando, T. (2008). Anisotropic diffusion of point defects in two-dimensional crystal of streptavidin observed by high-speed atomic force microscopy. *Nanotechnology* **19,** 384009.
Yamamoto, D., Nagura, N., Omote, S., Taniguchi, M., and Ando, T. (2009). Streptavidin 2D crystal substrates for visualizing biomolecular processes by atomic force microscopy. *Biophys. J.* **97,** 2358–2367.
Yamashita, H., Uchihashi, T., Kodera, N., Miyagi, A., Yamamoto, D., and Ando, T. (2007). Tip-sample distance control using photo-thermal actuation of a small cantilever for high-speed atomic force microscopy. *Rev. Sci. Instrum.* **78,** 083702.

Yamashita, H., Voïtchovsky, K., Uchihashi, T., Antoranz Contera, S., Ryan, J. F., and Ando, T. (2009). Dynamics of bacteriorhodopsin 2D crystal observed by high-speed atomic force microscopy. *J. Struct. Biol.* **167,** 153–158.

Ye, Y., Meyer, H. H., and Rapoport, T. A. (2001). The AAA ATPase Cdc48/p97 and its partners transport proteins from the ER into the cytosol. *Nature* **414,** 652–656.

Zhang, S. F., Rolfe, P., Wright, G., Lian, W., Milling, A. J., Tanaka, S., and Ishihara, K. (1998). Physical and biological properties of compound membranes incorporating a copolymer with a phosphorylcholine head group. *Biomaterials* **19,** 691–700.

CHAPTER TWENTY-ONE

NANOPORE FORCE SPECTROSCOPY TOOLS FOR ANALYZING SINGLE BIOMOLECULAR COMPLEXES

Olga K. Dudko,[*] Jérôme Mathé,[†] and Amit Meller[‡]

Contents

1. Introduction	566
2. The Nanopore Method	567
2.1. Theory of force-driven molecular rupture	571
2.2. Analysis of NFS experiments	575
3. DNA Unzipping Kinetics Studied Using Nanopore Force Spectroscopy	577
3.1. Maximum-likelihood analysis of voltage-ramp data	580
3.2. Histogram transformation method	582
3.3. Temperature rescaling of unzipping data	584
4. Conclusions and Summary	585
Acknowledgments	587
References	587

Abstract

The time-dependent response of individual biomolecular complexes to an applied force can reveal their mechanical properties, interactions with other biomolecules, and self-interactions. In the past decade, a number of single-molecule methods have been developed and applied to a broad range of biological systems, such as nucleic acid complexes, enzymes and proteins in the skeletal and cardiac muscle sarcomere. Nanopore force spectroscopy (NFS) is an emerging single-molecule method, which takes advantage of the native electrical charge of biomolecule to exert a localized bond-rupture force and measure the biomolecule response. Here, we review the basic principles of the method and discuss two bond breakage modes utilizing either a fixed voltage or a steady voltage ramp. We describe a unified theoretical formalism to extract

[*] Department of Physics and Center for Theoretical Biological Physics, University of California, San Diego, La Jolla, California, USA
[†] Laboratoire LAMBE (UMR 8587 —CNRS-CEA-UEVE), Université d'Evry-val d'Essonne, Evry, France
[‡] Department of Biomedical Engineering and Department of Physics, Boston University, Boston, Massachusetts, USA

kinetic information from the NFS data, and illustrate the utility of this formalism by analyzing data from nanopore unzipping of individual DNA hairpin molecules, where the two bond breakage modes were applied.

1. INTRODUCTION

The response of biomolecules to applied force can reveal their most fundamental mechanical properties, their interactions with other biomolecules or self-interactions, and in some cases, their structure. Thus, the trajectory describing the reaction of biomolecules to an applied force contains a wealth of information relevant to their biological function. Advances in single-molecule manipulation have made it possible to measure the forces and strains that develop during these processes with spatial resolution approaching the atomic level (sub-nm), and force sensitivity at the level of thermal fluctuations ($pN \sim k_B T/\text{nm}$ where k_B is the Boltzmann constant that yields the room temperature energy when multiplied with the temperature T). Micromanipulation techniques such as the atomic force microscope (AFM) and optical or magnetic tweezers (described elsewhere in this volume) are the most direct methods of exerting and measuring forces on biomolecules, and thus have been applied to a broad variety of biological systems, ranging from nucleic acids to enzymes to motor proteins (Cecconi et al., 2005; Florin et al., 1994; Gautel et al., 1997; Greenleaf et al., 2008; Kellermayer et al., 1997; Liphardt et al., 2001; Marszalek et al., 1999; Merkel et al., 1999; Schlierf and Rief, 2006; Schlierf et al., 2004).

Nanopores represent a fundamentally different approach for obtaining force spectroscopy data (Wanunu and Meller, 2008). This emerging single-molecule technique utilizes native molecular electric charge to exert force on virtually any biomolecule when it is threaded through a single nanoscale constriction (Akeson et al., 1999; Kasianowicz et al., 1996, 2002; Meller et al., 2000, 2001). In contrast to tweezers and AFM techniques, where force is applied mechanically to one point on the biomolecule by conjugation to a bead or cantilever, the force exerted on biomolecules using nanopore methods is both local to the region inside the pore and applied according to the molecule's effective charge in that region (Q_{eff}). Thus, this force is directly proportional to the electrical voltage drop across the pore (ΔV), $F = (Q_{\text{eff}}/l)\Delta V$, where l is the pore length. To quantify this force, the system's effective charge per unit length ($q_{\text{eff}} = Q_{\text{eff}}/l$) must be determined under the conditions used in each specific experiment. The pore constriction itself then exerts a negative and equal force ($-q_{\text{eff}}\Delta V$) on the molecule. The mechanical force appears as a localized shear force that destabilizes any biomolecular bonds or structures that will not pass easily through the pore, and can lead to their subsequent rupture.

Nanopore force spectroscopy (NFS) takes advantage of our ability to modify the applied voltage (or force) on molecules residing in the pore, in real-time, based on measurements of the ion-current (Bates et al., 2003). The controlled application of local forces on single molecules (or a single-molecular complex) residing in the pore is designed to destabilize and rupture intermolecular bonds, as the response of the molecules is measured. Here we describe two main approaches to force spectroscopy using nanopores, involving either the application of a fixed force level or the application of linearly increasing force, to create mechanical tension. The application of force culminates in a molecular transition (or a "rupture") that is clearly observed in the nanopore system. Some examples for molecular transitions that can be probed are ligand–receptor dissociation, unfolding of a protein, or unzipping of nucleic acids. When performed at constant force (a constant voltage), these experiments directly probe the voltage-dependent lifetime of the system, $\tau(V)$. In contrast, the distribution of rupture voltages, $p(V)$, measured in experiments at a constant voltage-ramp speed needs to be processed to provide information about $\tau(V)$. In this chapter, we discuss these two bond breakage modes in the context of nanopore experiments, and show that the experimental output of these two modes is related quantitatively in an essentially model-free way. We describe a unified theoretical formalism to extract kinetic information from NFS data, and illustrate the utility of this formalism by analyzing data from nanopore unzipping of individual DNA hairpin molecules.

2. THE NANOPORE METHOD

In a nanopore experiment, an electrical force is applied directly to a charged biopolymer threaded through a molecular-sized constriction (a few nanometers), made in an thin insulating membrane separating two reservoirs of buffered salt solution typically 0.2–1 M of monovalent salt (Wanunu and Meller, 2008), as shown schematically in Fig. 21.1. The electric field applied across two electrodes placed on either side of the membrane results in a steady ionic countercurrent of negative and positive ions through the pore. Because the resistivity of the pore is orders of magnitude larger than the resistivity of the bulk solution, the electric field is highly localized to the pore region. When charged biopolymers randomly approach the pore vicinity, they are attracted into the pore region by a residual component of the electrical field, acting to funnel the molecules from bulk into the pore (Wanunu et al., 2010). Once an end of the biopolymer is threaded into the pore, a much stronger force (roughly equal to $F = q_{\text{eff}} \Delta V$) is applied on the biopolymer, causing it to slide from one side of the membrane to the other. This process is usually referred

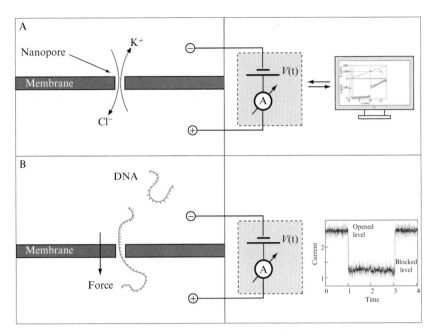

Figure 21.1 The nanopore method. (A) The ion-current flowing through a water-filled single nanoscale pore made in a thin membrane is measured using a pico-ampere electrometer. The voltage applied on the pore is dynamically controlled in real-time by a computer. (B) Insertion and threading of biopolymers cause abrupt blockades in the ion current, from the opened current state to the blocked current state, during the time in which the molecules remain in the pore. An electrical field applied across the membrane results in a strong force, driving the charged biopolymers from the negative to the positive chambers.

to as "translocation" (Meller, 2003). If the biopolymer's cross section is not uniform due to, for example, forming a hairpin or binding proteins, such that a local cross section is larger than the pore diameter, the translocation process will be interrupted until this "obstacle" is cleared. Formally, removal of these obstacles (i.e., unzipping of the hairpin, structures or stripping-off of bound proteins) is described by a crossing of a large energy barrier—much larger than that associated with moving an *unstructured* biopolymer through the pore. Since the waiting time associated with an energy barrier crossing increases exponentially with the barrier height, the characteristic translocation times of unstructured biopolymers (i.e., ssDNA, ssRNA, or polypeptides) are orders of magnitude smaller than the typical time required, for example, to unzip a hairpin of similar length.

In this chapter, we discuss two different kinds of bond-rupture measurements: pulling at a constant force, or with a linearly increasing force (i.e., with a constant force ramp). In nanopore experiments, these two types of measurements can both be realized by dynamically modifying the voltage applied to the biomolecule, after it has been threaded to the pore, using a computerized system (Bates et al., 2003). In the example shown in Fig. 21.2, the alpha-Hemolysin (α-HL) pore is utilized. The narrowest constriction of this protein channel ($\sim$1.4 nm) allows only single-stranded nucleic acids to be translocated, while double-stranded nucleic acids (i.e., hairpin structures) must be unzipped before translocation can proceed. In these experiments the current flowing through the pore is constantly measured, and the computerized data acquisition system is programmed to generate an output voltage signal triggered by an abrupt decrease in the pore current. Panel A depicts a typical unzipping event of a 10 base pair (bp) DNA hairpin where a step in the voltage is applied after molecular capture. The entry of the biomolecule into the pore creates an abrupt decrease in ion current, lowering it from the level of the opened pore to that of the blocked pore, which triggers the dynamic voltage control system. After a brief period of time (sufficient for threading the molecule up to the hairpin) the voltage is set to a constant level, V (90 mV in this case). Bond rupture (designated with an asterisk in Fig. 21.2) is signaled by a jump in the ion current at $t = t_U$. A histogram of hundreds of events collected in this manner is given, showing a peak at $\sim$1 ms and an exponentially decaying tail with a characteristic timescale of $\tau_U \sim 2.7$ ms.

In Panel B of Fig. 21.2, we display bond rupture using the *force-ramp* method. A linearly increasing voltage ("voltage ramp") is applied after the initial threading of a single-stranded overhang into the nanopore. The pore current remains at the blocked level until the moment of rupture (designated with an asterisk), which is easily observed as an abrupt increase in ion current to the opened pore level. The voltage at which final rupture occurs is defined as V_U. Also displayed is a typical distribution of the rupture voltages for $\sim$1000 individual unzipping events similar to the one shown. In a typical experiment, distributions of unzipping events are collected for a wide range of different loading rates (ramp values).

To fully characterize bond-rupture kinetics, the measurements are performed over a broad range of force values (voltages) or force ramps (voltage ramps). The combined measurements represent the response force *spectrum* of the system. While, in principle, the fixed-force and the force-ramp methods formally report the same information on the system, there are practical differences between the two. For example, measuring bond breakage in the limit of small, constant forces could be extremely time-consuming and therefore often impractical. However, the same regime of bond rupture can be effectively "scanned" using the force-ramp method, saving much experimental time. Moreover, we show below that the force-ramp method effectively broadens the range of timescales accessible by the system. The unification of the two methods on a single "master curve," discussed below, reinforces this point.

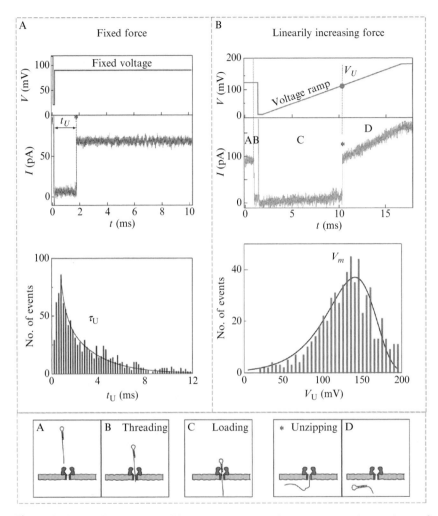

Figure 21.2 Implementation of force spectroscopy using nanopores. Two variants of the method are displayed: (A) Bond rupture using a step in the applied force (or voltage), where rupture is measured at a constant force, shown in blue. (B) Bond rupture under linearly increasing force (voltage ramp) shown in red. In both cases, a typical unzipping event (10 bp hairpin) is shown. Asterisks are used to denote the unzipping moment. In the fixed voltage case, the distribution of the unzipping time is measured for hundreds of events (see bottom histograms) yielding detailed characterization of the system response to force, through the timescale τ_U (fixed voltage) or the maximum voltage V_m (voltage ramp). Lower panel displays schematically the position of the hairpin with respect to the pore at each stage (A–D). Modified from Mathé et al. (2004) with permission.

2.1. Theory of force-driven molecular rupture

The wealth of high-resolution data collected in single-molecule force measurements has to be decoded in order to expose information about the mechanisms that drive biological processes. The interpretation of the experimental output in terms of underlying molecular interactions and structures often turns out to be a challenging task, because not only microscopic dimensions of these systems put the thermal fluctuations from the environment on an equal footing with the applied deterministic force, but also these experiments are often carried out under nonequilibrium conditions. Below we describe a recently introduced unified theoretical approach to analyzing data obtained in single-molecule force experiments.

Because thermal noise is an integral part of the molecular rupture process, biomolecular response to an applied force can only be described in probabilistic terms. Force-driven molecular rupture is viewed here as an irreversible molecular transition induced by a force F, during which the probability distribution of molecular configurations diffuses along the reaction coordinate x on a free-energy surface, $G(x) = G_0(x) - Fx$. The bare free-energy surface $G_0(x)$ is assumed to have a single well, a barrier at a distance $x^{\ddagger}$ from the well center, and an activation free energy, $\Delta G^{\ddagger}$. In the context of DNA unzipping in a nanopore (discussed later), the initial state, in the well of the free-energy surface, represents the single-stranded overhang of DNA threaded into the pore with the hairpin closed. Escape over the barrier involves unzipping of the double-stranded part of the DNA and the pore being cleared (Fig. 21.3A).

Even though this formalism assumes a single barrier and a single bound state, it can be applied to each individual transition in the case of multiple populated states, if the states can be resolved experimentally, for example, based on their molecular extensions along the reaction coordinate. The formalism is applicable to both the forward (e.g., unzipping or dissociation) and the backward (e.g., refolding or rebinding) transitions, as long as these transitions are quasi-irreversible.

We first consider the case of a constant force F accelerating the rate of molecular rupture. The theory (Dudko et al., 2006) based on Kramers' picture of a diffusive barrier crossing (Kramers, 1940), predicts that for a sufficiently high barrier separating the unruptured state from the ruptured state, the lifetime (which is equal to the inverse escape rate $k(F)$) at a constant external force F is

$$\tau(F) = \tau_0 \left(1 - \frac{\nu F x^{\ddagger}}{\Delta G^{\ddagger}}\right)^{1-1/\nu} \exp\left\{-\beta \Delta G^{\ddagger}\left[1 - \left(1 - \frac{\nu F x^{\ddagger}}{\Delta G^{\ddagger}}\right)^{1/\nu}\right]\right\}. \quad (21.1)$$

Equation (21.1) expresses the force-dependent lifetime, which is the output of the constant-force experiment, in terms of three zero-force microscopic

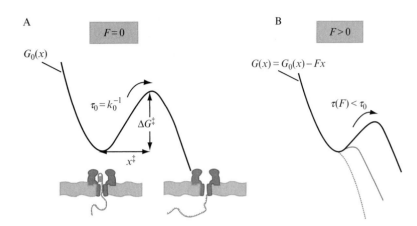

Figure 21.3 Conceptual picture of the molecular rupture under applied force viewed as a diffusive crossing of a barrier on a one-dimensional free-energy surface. (A) Intrinsic (i.e., zero-force) free-energy surface $G_0(x)$ with a well (bound state) and a barrier to quasi-irreversible rupture. In the context of the voltage-driven unzipping of individual DNA hairpin molecules in a nanopore, the well of the free-energy surface corresponds to the single-stranded DNA overhang threaded into the pore constriction with the folded hairpin trapped in the pore vestibule, while escape over the barrier involves the double-stranded part of the DNA being unzipped during its passage through the pore. (B) Free-energy surface $G(x)$ in the presence of an external force F. As the force increases, both the barrier height and the distance to the transition state decrease (gray line), and both eventually vanish (dotted line) when the well and the barrier merge at a critical force.

parameters: intrinsic lifetime τ_0, distance to the transition state $x^{\ddagger}$, and activation free-energy barrier $\Delta G^{\ddagger}$. Throughout this chapter, $\beta = (k_B T)^{-1}$ with k_B being the Boltzmann's constant and T the absolute temperature. Values $v = 2/3$ and $v = 1/2$ of the scaling parameter v correspond to the linear-cubic surface $[G_0(x) = (3/2)\Delta G^{\ddagger} x/x^{\ddagger} - 2\Delta G^{\ddagger}(x/x^{\ddagger})^3]$ and the harmonic-cusp surface $[G_0(x) = \Delta G^{\ddagger}(x/x^{\ddagger})^2$ for $(x < x^{\ddagger})$ and $-\infty$ for $x \geq x^{\ddagger}]$, respectively. For $v = 1$, or for $\Delta G^{\ddagger} \to \infty$ independent of v, the phenomenological expression of Bell (1978) is recovered from Eq. (21.1).

Because Bell's formula, $\tau_{\mathrm{Bell}}(F) = \tau_0 \exp(-\beta F x^{\ddagger})$ (Eq. (21.1) with $v = 1$), predicts that the logarithm of the lifetime is a linear function of the applied force, any deviations from the linear dependence of $\ln \tau(F)$ on F have often been interpreted as a change in rupture mechanism (e.g., switching from one dominant barrier to another, or rebinding). However, Eq. (21.1) shows that simple microscopic models with a *single* barrier and a single bound state (like that depicted in Fig. 21.3) can explain nonlinearity in $\ln \tau$-versus-F plots without introducing additional assumptions of a change in mechanism. The reason for this limitation of the phenomenological Bell formula is the underlying assumption that the distance from the well to the transition state, $x^{\ddagger}$, is independent of force. This assumption cannot be true for all forces: as can be readily seen by examining the behavior of any smooth one-dimensional potential (Fig. 21.3), the barrier and the well must move closer (solid gray line in Fig. 21.3B) as the force increases, because they eventually merge (dotted line in Fig. 21.3B) at a critical force when the barrier to rupture vanishes. As a result, $\ln \tau(F)$ must be a nonlinear function of F.

Equation (21.1) was derived for forces at which a significant barrier (several $k_B T$) still exists, and this equation is thus valid only when the force is below a critical force at which the barrier vanishes, $F < \Delta G^{\ddagger}/v x^{\ddagger}$. To estimate the intrinsic (zero-force) parameters τ_0, $x^{\ddagger}$ and $\Delta G^{\ddagger}$, the logarithm of the lifetimes $\ln \tau(F)$ can be fitted using Eq. (21.1) with several trial values of v. If the resulting estimates for the fitting parameters are relatively insensitive to v over a range of v values that all result in good fits, these estimates can be considered to be independent of the precise nature of the free-energy surface and thus meaningful.

When the force is ramped up linearly with time, such that the force loading rate $\dot{F} \equiv dF/dt = $ const. is constant, the distribution of forces at rupture is (Dudko et al., 2006):

$$p(F|\dot{F}) = \frac{1}{\dot{F}\tau(F)} \exp\left(\frac{1}{\beta x^{\ddagger} \dot{F} \tau_0}\right)$$
$$\exp\left[-\frac{1}{\beta x^{\ddagger} \dot{F} \tau(F)}\left(1 - \frac{v F x^{\ddagger}}{\Delta G^{\ddagger}}\right)^{1-1/v}\right], \quad (21.2)$$

where $\tau(F)$ is the force-dependent lifetime of Eq. (21.1). For intermediate values of the force loading rate $\dot{F}$, the approximate analytical expressions for the mean rupture force $\langle F \rangle = \int F p(F|\dot{F}) dF$ and the variance $\sigma_F^2 = \langle F^2 \rangle - \langle F \rangle^2$ are (Dudko et al., 2006):

$$\langle F \rangle \cong \frac{\Delta G^{\ddagger}}{v x^{\ddagger}} \left\{ 1 - \left[\frac{1}{\beta \Delta G^{\ddagger}} \ln \frac{e^{\beta \Delta G^{\ddagger} + \gamma}}{\beta x^{\ddagger} \dot{F} \tau_0}\right]^v \right\}, \quad (21.3)$$

$$\sigma_F^2 \cong \frac{\pi^2}{6 \beta^2 x^{\ddagger 2}} \left[\frac{1}{\beta \Delta G^{\ddagger}} \ln \frac{e^{\beta \Delta G^{\ddagger} + \tilde{\gamma}}}{\beta x^{\ddagger} \dot{F} \tau_0}\right]^{2v-2}. \quad (21.4)$$

Here $\tilde{\gamma} = \gamma^2 - 3/\pi^2 \psi''(1) \approx 1.064$, $\gamma \approx 0.577$, and $\psi''(1) \approx -2.404$ (Abramowitz and Stegun, 1972). When γ is formally set to zero, Eq. (21.3) is a good approximation for the maximum (mode) of the rupture force distribution.

The microscopic theory based on Kramers' picture of diffusive crossing of a barrier on a class of model potentials ($v = 1/2$ or $v = 2/3$ in Eqs. (21.3) and (21.4)) predicts that the mean rupture force (Eq. (21.3)) depends nonlinearly on the logarithm of the loading rate, $\dot{F}$, and the variance of the rupture force (Eq. (21.4)) is a function of the loading rate. In contrast, the phenomenological theory ($v = 1$ in (Eqs. (21.3) and (21.4)), based on Bell's postulate for the lifetimes, $\tau(F) = \tau_{\text{Bell}}(F)$, leads to a linear dependence of the mean rupture force on the log-loading rate, and to the variance being independent of the loading rate. The predictions of the two approaches are tested against experimental data (Fig. 21.7) below.

Implicit in the derivation of Eqs. (21.2)–(21.4) is the quasi-adiabatic assumption that the force loading rate $\dot{F}$ is not too high so that by the time the barrier is so low that Kramers' theory is invalid, the survival probability (i.e., the probability to find the system unruptured) is effectively zero. If this quasi-adiabatic approximation is indeed valid, the following relation between the constant-force experiments (measuring $\tau(F)$) and constant-speed experiments (measuring $p(F|\dot{F})$) has been established (Dudko et al., 2006, 2008):

$$\tau(F) = \frac{\int_F^\infty p(F')\,dF'}{\dot{F}(F)p(F)}. \tag{21.5}$$

This equation predicts that the rupture force distributions $p(F|\dot{F})$ obtained at different values of the force-ramp speed $\dot{F}$ (right-hand side of Eq. (21.5)) can be directly transformed into the force dependence of the lifetime $\tau(F)$ (left-hand side of Eq. (21.5)) measurable in constant-force experiments. This mapping is independent of the nature of the underlying free-energy surface, and thus it relates the two types of experiments in a model-free way. Equation (21.5) predicts that data obtained at different loading rates must collapse onto a single master curve that yields the force dependence of $\tau(F)$ over a range of forces that may be wider than the range accessible in constant-force measurements. While the analytical solutions in Eqs. (21.2)–(21.4) were derived assuming a constant force loading rate, Eq. (21.5) holds also when the force loading rate, $\dot{F}$, is itself a function of force.

From Eq. (21.5) one can obtain an approximate but quite general relationship between the lifetime at a force equal to the mean rupture force and the variance of the rupture-force distribution (Dudko et al., 2008):

$$\tau(\langle F \rangle) \approx \frac{1}{\dot{F}(F)} \left[\frac{\pi}{2}(\langle F^2 \rangle - \langle F \rangle^2)\right]^{1/2}. \tag{21.6}$$

Whereas Eq. (21.6) gives an estimate for the $\tau(F)$ over a narrower range of F than does Eq. (21.5), the former should prove useful if the data permit estimates of only the mean and variance.

Force-dependent lifetimes $\tau(F)$ can be used to extract microscopic information *independent* of a particular model of the free-energy profile.

The following expression for the force-dependent lifetimes has been derived from Kramers high-barrier theory (Dudko et al., 2008):

$$\tau(F) = \tau_0 \exp\left(-\beta \int_0^F \langle x^{\ddagger}(F')\rangle \, dF'\right), \qquad (21.7)$$

where $\langle x^{\ddagger}(F)\rangle$ denotes the difference in the average positions of the transition state and the bound state along the pulling coordinate as a function of force. A corollary of Eq. (21.7) is that the slope of $\ln \tau(F)$ versus F is a direct measure of how the distance between bound state and transition state changes with the force. Equation (21.7) can be viewed as the generalization of Bell's formula, $\tau_{\text{Bell}}(F) = \tau_0 \exp(-\beta F x^{\ddagger})$, in the framework of Kramers theory. Equation (21.7) shows that Bell's formula is accurate only in the limited range of low forces when the distance from the well to the transition state can be considered to be independent of force (see Fig. 21.3B). Consequently, an uncritical use of the Bell's formula beyond a narrow range of low forces can lead to significant errors in the estimated intrinsic lifetime τ_0 and the distance to the transition state $x^{\ddagger}$.

In nanopore unzipping experiments, the applied voltage V is analogous to the applied mechanical force F in pulling experiments (e.g., those using AFMs or optical tweezers). The voltage drop across the membrane-spanning nanopore results in an electric field that generates a mechanical force on the charged DNA strand threaded into the nanopore (Fig. 21.3A). To adapt the above formalism to nanopore-unzipping experiments, the voltage $V^{\ddagger} = k_B T/Q_{\text{eff}}$ can be defined as the characteristic of the transition state, where Q_{eff} is the effective charge of the DNA inside the pore (Mathé et al., 2004). Equations (21.1)–(21.7) can then be used by making the following change of variables:

$$\beta F x^{\ddagger} \to V/V^{\ddagger}, \qquad (21.8)$$

and with $\dot{F} = dV/dt$ being the voltage-ramp speed.

2.2. Analysis of NFS experiments

2.2.1. Analysis of constant-force experiments

Voltage dependence of the lifetime obtained in constant-voltage measurements can be interpreted in microscopic terms simply by least-squares fitting the data with Eq. (21.1) at several fixed values of v. If the resulting parameters τ_0, $x^{\ddagger}$, and $\Delta G^{\ddagger}$ are relatively insensitive to v in the range of $1/2 \le v \le 2/3$, and thus to the precise shape of the underlying free-energy surface, these parameters may be considered meaningful. Alternatively, Eq. (21.7) that is formally exact within the framework of Kramers' theory can be used to extract information about the transition state as a function of force from the lifetimes, *independent* of the shape of the free-energy surface.

2.2.2. Analysis of force-ramp experiments

Two complementary approaches can be used to extract microscopic information from rupture-voltage distributions obtained in a constant-force-ramp experiment. In the first approach, a maximum-likelihood (ML) formalism is used to fit with Eq. (21.2) all rupture-voltage histograms collected at one or (preferably) several ramp speeds $\dot{F}$ (Dudko *et al.*, 2007). In the second approach, rupture-voltage histograms are transformed according to Eq. (21.5) into the voltage dependence of the lifetime, and the resulting lifetimes can be analyzed as described above in section 2.2.1 (Dudko *et al.*, 2008). Implementation of these two approaches is discussed below.

2.2.2.1. Maximum-likelihood analysis of force-ramp experiments Consider a series of constant force-ramp speed experiments at several ramp speeds $\dot{F}_j$ ($j = 1, \ldots, N$). Molecular rupture will be observed at different forces F_{ij} ($i = 1, \ldots, M_j$). The likelihood function L needs to be maximized with respect to a set of model parameters, $\{\alpha\}$. L can be expressed in terms of the rupture force distribution $p(F|\dot{F})$ at ramp speed $\dot{F}$ as

$$L = \prod_{j=1}^{N} \prod_{i=1}^{M_j} p(F_{ij}|\{\alpha\}; \dot{F}_j). \quad (21.9)$$

To implement this approach, it is convenient to have an analytical expression for $p(F|\dot{F})$. The unified formalism described above provides such an expression in Eq. (21.2) in terms of the model parameters, namely the intrinsic lifetime τ_0, the location of the transition state $x^{\ddagger}$, and the activation free energy $\Delta G^{\ddagger}$. Given experimental measurements, optimal values of τ_0, $x^{\ddagger}$, and $\Delta G^{\ddagger}$ can be found by maximizing L or, equivalently, $\ln[L]$ for different fixed values of ν. As in the case of the constant-voltage data analysis, if the resulting parameters $\{\tau_0, V^{\ddagger}, \Delta G^{\ddagger}\}$ are relatively insensitive to the value of ν in the range $1/2 \leq \nu \leq 2/3$, then they can be considered meaningful.

2.2.2.2. Transformation of rupture-voltage histograms into voltage dependence of lifetime The histogram transformation approach is based on the mapping equation, Eq. (21.5), which transforms rupture-voltage histograms measured at different voltage-ramp speeds directly into voltage dependence of the rupture lifetime. The transformation of histograms using Eq. (21.5) is implemented as follows. Consider a rupture force histogram at the force-ramp speed, $\dot{F}$. The histogram contains N bins of width ΔF, starts at F_0 and ends at $F_N = F_0 + N\Delta F$, and has N_{tot} total number of counts. Let the number of counts in the ith bin be C_i, resulting in a height $p_i = C_i/(N_{\text{tot}}\Delta F)$ in the normalized force distribution. Then the lifetime at the force $F_0 + (k - 1/2)\Delta F$ is

$$\tau[F_0 + (k - 1/2)\Delta F] = \frac{\left(p_k/2 + \sum_{i=k+1}^{N} p_i\right)\Delta F}{\dot{F}(F_0 + (k - 1/2)\Delta F)p_k}, \quad (21.10)$$

where $k = 1, 2, \ldots$ Eq. (21.10) is simply a discrete version of Eq. (21.5).

If the histograms do collapse onto a single master curve, one immediately obtains the force (voltage) dependence of the molecular bond lifetime, $\tau(F)$, or equivalently the rate of rupture, $k(F) = 1/\tau(F)$. The force dependence of the lifetime can now be interpreted in microscopic terms in exactly the same way as the lifetimes that were measured directly in constant-voltage experiments, namely by performing a least-squares fit with Eq. (21.1), or by using a model-independent approach in Eq. (21.7). If the histograms transformed by Eq. (21.5) do not collapse onto a single master curve, then the mechanism of rupture cannot be described as an irreversible, quasi-adiabatic escape over a single barrier. Such behavior may also be evident in non-exponential distributions of the lifetimes in constant-force experiments.

3. DNA Unzipping Kinetics Studied Using Nanopore Force Spectroscopy

The unzipping of double-stranded nucleic acids occurs in a large number of cellular processes, including DNA replication, RNA transcription, translation initiation, and RNA interference. The forces and timescales associated with the breakage of the bonds stabilizing the secondary and tertiary structures of nucleic acids can now be studied at the single-molecule level, revealing information masked heretofore by ensemble averaging, including short-lived intermediate states and multistep kinetic processes. Here we describe the use of nanopores to directly apply and measure unzipping forces on individual DNA and RNA molecules, eliminating the need for molecular linkers, for surface immobilization of the molecules, and for global application of force.

The protein pore α-HL is nearly ideal for nucleic acid unzipping studies. Its heptameric structure (Song *et al.*, 1996), composed of cap and stem portions, has been shown to be highly stable even under high temperatures, voltage gradients (Kang *et al.*, 2006), and a wide range of ionic strengths (i.e., 0.25–2 M KCl) (Jan Bonthuis *et al.*, 2006). The cap portion of α-HL, which is usually assembled on the "*cis*" side of the membrane, contains a wide vestibule-like mouth, which can accommodate double-stranded nucleic acids. The stem portion, which spans the phospholipid membrane, is a nearly cylindrical water-filled channel with an inner diameter ranging from 1.4 to 2.2 nm (Song et al., 1996). It therefore geometrically permits the passage of single-stranded DNA or RNA molecules, but blocks the translocation of double-stranded nucleic acids.

Sauer-Budge et al. (2003) have demonstrated that a DNA duplex molecule (composed of a 100-mer DNA oligonucleotide hybridized to a matching 50-mer oligonucleotide, such that a 50-mer 3′ single-stranded overhang is formed), must be unzipped when the 3′ overhang is threaded through the pore. A quantitative PCR analysis showed that while both the 50-mer and 100-mer oligonucleotides were present in the *cis* chamber, only the 100-mer oligonucleotide was found in the *trans* chamber, after the detection of hundreds of nanopore blockade events. This observation is explained if unzipping occurred at the pore, leaving all 50-mer oligonucleotides in the *cis* chamber while the 100-mer oligonucleotides passed through the pore to the *trans* side (Sauer-Budge et al., 2003). Furthermore, the distribution of nanopore blockade durations (or dwell-times) displayed a characteristic mean time of ~ 435 ms, orders of magnitude longer than the timescale associated with the translocation of single-stranded DNAs of comparable length. Introducing a 6-base mismatch in the duplex region of the hybridized sample resulted in a shortening of the characteristic timescale by more than a factor of two (Sauer-Budge et al., 2003).

Mathé et al. (2004, 2006) employed NFS to study the properties of the unzipping kinetics of DNA hairpin molecules. They systematically probed the unzipping time probabilities of three DNA hairpin molecules composed of a duplex stem region (of 10 bp, 9 bp, which is a 10 bp plus a mismatch at the fifth base pair, or 7 bp, all with a capping 6-base loop) attached to a 3′ poly(dA)$_{50}$ overhang. The poly(dA)$_{50}$ tail is used to insert the molecules into the pore and thus apply the electric force primarily to the duplex region. The unzipping kinetics was probed over a wide range of voltage (30–150 mV) and at several temperatures (from 5 to 20 °C). The temperature dependence will be discussed later on. Figure 21.2 displays a typical unzipping event using this technique at a constant unzipping voltage, where the event's unzipping time is denoted t_U. To obtain the *characteristic* unzipping time (τ_U) approximately 1500 events were accumulated at each voltage leading to the unzipping time distribution. Figure 21.4 displays the time-cumulative distribution for the three hairpin molecules probed at a constant voltage of 120 mV and at 15 °C. These measurements revealed that a single base pair mismatch strongly shifts the unzipping timescale towards shorter times and can therefore be easily detected using unlabeled and unmodified DNA hairpin and DNA hybrids.

The cumulative unzipping time distributions were fitted well with monoexponential functions (shown as dashed line in Fig. 21.4), indicating first-order

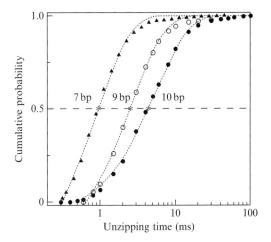

Figure 21.4 Detecting single base mismatches in unlabeled DNA hairpins. The normalized cumulative distributions of the unzipping time measured using DVC at 120 mV and 15 °C, for 10, 9 (10 bp with a mismatch), and 7 bp DNA hairpins (solid circles, empty circles, and triangles respectively). Mono-exponential probability distribution fits (dashed lines) yield characteristic unzipping timescales ($\sim$5, 3, and 1 ms for the 10, 9, and 7 bp hairpins, respectively) used to discriminate between the hairpins.

rate kinetics. The characteristic time of these exponential fits is the mean unzipping time τ_U and its dependence on voltage V is shown in Fig. 21.5 as a semilogarithmic plot. When this characteristic time versus voltage plot was fitted using the phenomenological Bell model, $\tau_U(V) = \tau_0 \exp(-V/V^{\ddagger})$, where $V^{\ddagger}$ contains the effective charge (as explained above) on which the electric force is applied and τ_0 contains the energy barrier height of the duplex region, it was found that $V^{\ddagger}$ does not depend on the duplex region (the slopes are the same) leading to $V^{\ddagger} = 22 \pm 2$ mV and thus an effective charge of $Q_{\text{eff}} = 1.13 \pm 0.1e$, in good agreement with initial experimental determinations by Sauer-Budge et al. (2003).

DNA hairpin unzipping measurements were also performed using voltage-ramp measurements, as explained in Fig. 21.2. Using ramp voltage has several practical advantages over constant-voltage measurement. Most importantly, constant-voltage experiments appear to be much more time consuming as compared to voltage-ramp experiment. This feature is a direct consequence of the roughly logarithmic dependence of V_m (proportional to time with linear ramp voltage) on the ramp as the unzipping timescale in a constant-voltage experiment depends roughly exponentially on the voltage applied. ML analysis and Eq. (21.2) can be used to fit NFS voltage ramp data. Below we also demonstrate how to convert, using Eq. (21.5), the voltage-ramp data directly into the voltage dependence of the unzipping times measured at constant voltage.

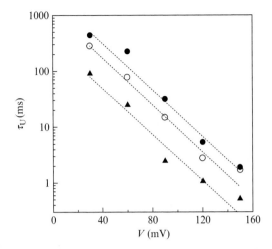

Figure 21.5 Dependence of the characteristic unzipping timescale on the voltage measured using dynamic voltage control, for 10, 9 (10 bp with a mismatch), and 7 bp DNA hairpins (solid circles, empty circles, and triangles respectively). As a first approximation, all molecules follow a mono-exponential dependence on V with the same slope. Reproduced with permission from Mathé et al. (2004).

3.1. Maximum-likelihood analysis of voltage-ramp data

Over 32 independent DNA hairpin unzipping data sets, obtained for different ramp values, were used to construct unzipping voltage histograms and were globally fit using Eq. (21.2) by maximizing the likelihood function, Eq. (21.9). Figure 21.6 presents the results of the fits using Eq. (21.2) with $v = 2/3$ (linear-cubic model) on all ramp speeds of 12 V/s or less. Ramp speeds above 12 V/s were used for further validation of the global fit. Histogram colors represent four voltage-ramp ranges (see caption). As summarized in Table 21.1, the two microscopic theories ($v = 2/3$ and $v = 1/2$ in Eq. (21.2)) produce consistent estimates for the model parameters. ML fitting parameters of the phenomenological theory ($v = 1$ in Eq. (21.2)) are also included for comparison.

Figure 21.7 shows the most probable unzipping voltage V_m as a function of the voltage-ramp speed. The markers represent the experimental data, and the solid line is the theoretical prediction using microscopic theory (Eq. (21.3) with γ set to 0 and $v = 2/3$) with ML parameters as in Table 21.1. The dashed line is a fit with the phenomenological model (Eq. (21.3) with $v = 1$), where V_m is logarithmically dependent on the voltage-ramp speed (when only intermediate and high-voltage ramps are used). In order to account for the observed curvature in the experimental data at low ramp speeds one needs to add to the phenomenological rate model additional molecular processes, such as hairpin-rezipping or switching between multiple states that cannot be

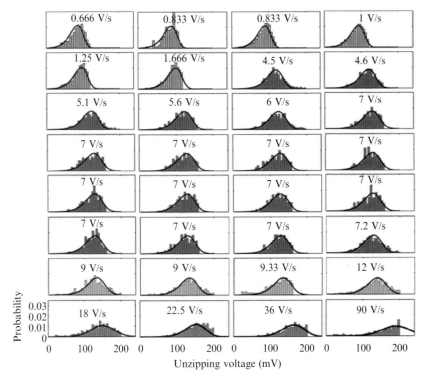

Figure 21.6 Distribution of the unzipping voltage from experiment (histograms) and theory (lines). Theoretical distributions were obtained by maximum-likelihood (ML) global fit of histograms at ramp speed of 12 V/s or less to Eq. (21.2) with $v = 2/3$ (linear-cubic model); the result for $v = 1/2$ (linear-cubic model, not shown) is very similar. The likelihood function, Eq. (21.9), was maximized numerically with respect to the model parameters. Experimental data collected at ramp speeds above 12 V/s (in this regime the DNA hairpin was still intact when the maximum voltage 0.2 V had been reached) were not used in the fit. The microscopic theories reproduce the measured distributions of unzipping voltages very well both in the regime used for the fit and outside that regime. The phenomenological model (fit not shown) with the estimates obtained from the global ML fit was found to be accurate at low ramp speeds, whereas at higher speeds the deviations were substantial. Colors reflect different ranges of the ramp speed and correspond to those in Figs. 21.7 and 21.8. Data reproduced from Dudko et al. (2007) with permission.

represented by a single energy well model (e.g., Fig. 21.3). In contrast, the microscopic models (Eq. (21.3) with $v = 2/3$ and $v = 1/2$) captures the nonlinearity in V_m as a function of $\log \dot{V}$, without the need to make any assumptions beyond a single-well energy landscape. Additionally, it can be verified (Dudko et al., 2007) that the variance, σ_V^2, of the unzipping voltage distributions exhibits a noticeable increase with the ramp speed, in agreement with the microscopic theories [Eq. (21.4) with $v = 1/2$ and $v = 2/3$].

Table 21.1 Maximum-likelihood estimates for the kinetic parameters for nanopore unzipping of DNA

v	$V^{\ddagger}$ (mV)	$\Delta G^{\ddagger}$ ($k_B T$)	τ_0 (s)
1	21.7	–	1.6
2/3	12.7	10.5	8.3
1/2	9.9	11.9	20

Estimates were obtained from data in Fig. 21.6 at ramp speeds 12 V/s or less by maximizing the likelihood function in Eq. (21.9) with the expression in Eq. (21.2) for the unzipping voltage distributions.

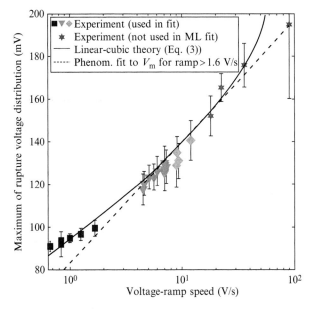

Figure 21.7 Dependence of the maximum V_m of the unzipping voltage distribution on the voltage-ramp speed from experiment (markers) and theory (lines, Eq. (21.3) with γ set to 0). Model parameters for the linear-cubic theory (solid line, Eq. (21.3) with $v = 2/3$) were obtained from global ML fit, see Fig. 21.6 and Table 21.1; result for the harmonic-cusp theory (Eq. (21.3) with $v = 1/2$, not shown) is very similar. Dashed line is the least-squares fit of the maxima of unzipping voltage distributions to the phenomenological model (Eq. (21.3) with $v = 1$) for ramp speeds > 1.6 V/s. Color coding as in Fig. 21.6. Data reproduced from Dudko et al. (2007) with permission.

3.2. Histogram transformation method

Quantitatively relating the constant-voltage data to the voltage-ramp data using Eq. (21.5) and its discrete analog, Eq. (21.10) provides a simple way to obtain the voltage-dependent lifetime $\tau(F)$ directly from the voltage-ramp

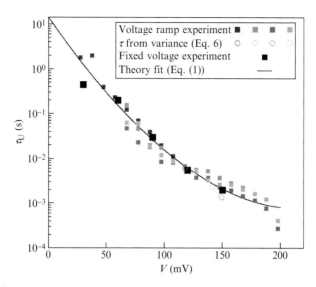

Figure 21.8 Comparison of DNA hairpin unzipping lifetimes τ_U obtained from transformation of the voltage-ramp experiments according to Eq. (21.5) (solid squares) for ramp speeds from 0.83 to 18 V/s, or from the distributions' variance analysis (Eqs. (21.6) and (21.7)). Colors indicate the range of ramp speeds as in Fig. 21.6. Constant voltage measurements of the unzipping lifetime are shown in black. Least-squares fit of the collapsed histograms to Eq. (21.1) with $\nu = 1/2$ displays remarkable agreement with data over a broad timescale range (black line). The fit with $\nu = 2/3$ (not shown) is comparable. Modified from Dudko et al. (2008) with permission.

data. Figure 21.8 is an illustration of the utility of Eq. (21.5) for this purpose. Not only do the lifetimes $\tau(F)$ obtained from histograms at different ramp speeds collapse onto the same curve (colored symbols), but there is also excellent agreement with DNA unzipping lifetimes obtained by an *independent set of measurements* of $\tau(F)$ using constant voltages (open circles). The lifetime obtained from the mean and variance using Eq. (21.6) (filled circles) is also found to agree with the constant-voltage experiments.

Now that the voltage dependence of the lifetime has been obtained from rupture-voltage histograms, the procedure of interpreting this dependence in microscopic terms can be implemented. It is clear from Fig. 21.8 that deviations from the mono-exponential dependence of the lifetime on the voltage are present when the voltage exceeds ~125 mV. The microscopic models ($\nu = 1/2$ and $\nu = 2/3$) allow us to perform a least-squares fit of the data with Eq. (21.1) over the entire range of accessible voltage, producing the following kinetic parameters: $\tau_0 = 14.3$ s, $V^{\ddagger} = 11.1$ mV, and $\Delta G^{\ddagger} = 11.9 k_B T$ for $\nu = 1/2$, and $\tau_0 = 9.6$ s, $V^{\ddagger} = 12.8$ mV, and $\Delta G^{\ddagger} = 10.4 k_B T$ for $\nu = 2/3$, in good agreement with the values of the ML global analysis in Table 21.1. Collapsed data collected at different ramp speeds probe different ranges of the DNA unzipping lifetime. Taken together, they span four orders of magnitude of the bond-rupture lifetime. Furthermore, Fig. 21.8 shows that, although constant-voltage data (open circles) were not used in the fit, the data are accurately predicted by Eq. (21.1).

It can be verified (Dudko *et al.*, 2008) that the microscopic theory in Eq. (21.1) is able to accurately *predict* the original rupture-voltage distributions when the parameters of the least-squares fit of the collapsed distributions with Eq. (21.1) are used. Thus in this case, the histogram transformation procedure gives essentially the same information as the more sophisticated ML method, but is simpler to implement.

3.3. Temperature rescaling of unzipping data

Another way to probe a broader range of unzipping lifetimes is to vary temperature. As shown in Fig. 21.9 (inset) for a constant voltage experiment, changing the temperature by 5 °C is sufficient to shift the characteristic unzipping time by a factor of 2–5. One can consider that changing temperature has a similar effect as changing voltage (both modify the effective energy

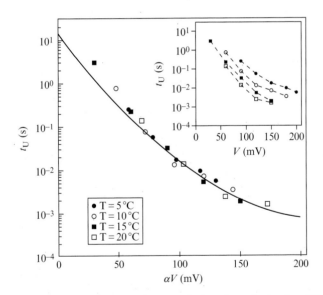

Figure 21.9 Rescaling of constant voltage unzipping time obtained at different temperatures. Lifetime $\tau(V)$ of the DNA hairpin as a function of the rescaled applied voltage V. The rescaling parameter was calculating in order to minimize the distance between points and then collapse all four temperature curves onto a single curve. The line represents Eq. (21.1) with the parameters obtained from Fig. 21.8 data points fit. *Inset*: Unzipping lifetime versus voltage at different temperatures (5, 10, 15, and 20 °C) without rescaling.

barrier), but they are not completely uncoupled since temperature can, in principle, alter the DNA effective charge. However, to first approximation, we can rescale the voltage by a proportionality parameter α as a function of temperature in order to collapse all four temperatures on a single curve while keeping the 15 °C data unchanged ($\alpha = 1$). Since the absolute temperature range is small, it is difficult to extract the dependence of this rescaling parameter on the temperature. We therefore choose to collapse the curves by minimization of the distance between each data point. Figure 21.9 presents the result of this collapse. In addition, we added to this graph the theoretical prediction from Eq. (21.1) using the parameters found previously and displayed in Fig. 21.8. One can see that the rescaled data match the theoretical prediction excepted at low voltages where the timescales are on the order of 1 s and thus large statistics are hard to attain. This consistency shows that the unzipping lifetime range can be broadened by changing temperature and, even better, by performing ramp experiments at various temperatures.

4. Conclusions and Summary

Direct probing of molecular bond strengths reveals biologically relevant information on a molecule's structure and function, and thus has been the objective of multiple single-molecule techniques. NFS utilizes the native electric charge of biomolecules to directly exert forces when a biomolecule is threaded through a nanoscale constriction. Because the method does not involve the formation of a physical attachment between the biomolecule and the pore, biomolecules do not require any modifications, such as attachment of long molecular handles. This highly simplifies experimental procedures and their analysis. In the nanopore method, the pore itself exerts a shear force on those parts of the biomolecular complex that do not fit inside the pore. NFS utilizes these advantages to locally rupture bonds and directly measure $\tau(V)$ and $p(V)$, two important indicators of bond stability.

The process of bond rupture is often described by diffusive crossing of an activation free-energy barrier, where the bound state refers to the unperturbed system, and the unbound state describes the ruptured bond. In many cases, the bound and unbound states are separated by a large energy barrier (many $k_B T$) and the system remains stable over long periods of time. The energy barrier height and thus bond stability are reduced by an external application of force, effectively catalyzing the molecular transition. Force spectroscopy probes a system's response over a broad spectrum of forces by measuring either the distribution of bond times-to-rupture at each given force, or the rupture-force distribution at a given force-ramp speed. Either measurement allows mapping of the system's energy landscape, and extrapolation of the nonequilibrium transition rate to the equilibrium transition rate.

To interpret the experimental output in terms of the underlying molecular properties, we have described here a unified theory for a class of single-barrier free-energy landscape models. While this theory describes a simple case of a single, high energy barrier, it yields closed-form analytical solutions for the experimental observables. The resulting theoretical predictions can be readily used to fit the voltage-dependent bond lifetimes measured for constant force (using Eq. (21.1)), and to employ the powerful ML analysis of the rupture-voltage distributions measured for a constant force ramp (using Eq. (21.2)). The result of these analyses is a set of intrinsic physical parameters characterizing the system, namely the activation free-energy barrier height, its position with respect to the bound state, and the characteristic lifetime (or, equivalently, the inverse intrinsic rate) of the system. It has also been shown that the rupture-voltage histograms obtained at different voltage-ramp speeds can be transformed to determine the voltage-dependence of the bond lifetimes measurable at constant voltage (Eq. (21.5)). This transformation is independent of the functional form of the free-energy landscape, and is valid if the rupture-voltage kinetics at constant voltage is well represented by a single exponential. Although this formalism was developed to describe irreversible rupture, it is applicable to both forward and reverse transitions as long as they can be resolved experimentally.

In this chapter we have illustrated the principles of NFS by focusing on DNA hairpin unzipping kinetics. We show that constant voltage and steady voltage-ramp experiments can be mapped onto the same rupture-time versus voltage curve (Fig. 21.8), which can then be interpreted in microscopic terms using a unified theory (Eq. (21.1)), providing insight into the system's underlying energy landscape and rates. As explained, the single well model applies only to short hairpins for which the unzipping process can be considered to occur in a single prominent step. This should also apply to simple DNA–protein complexes. The unzipping of longer hairpins or duplex regions involves a more complex process, which may entail energy landscapes with several consecutive energy wells. A similar consideration may apply to sequences containing, for example, several GC rich regions. Recently Monte-Carlo simulations of long sequences have been performed (Bockelmann and Viasnoff, 2008) showing that NFS is able to detect the position of a large barrier in a DNA-unzipping free-energy landscape.

The NFS method is quite general and already has been applied to other biomolecular systems. For example, Hornblower et al. (2007) have recently used NFS to study the interactions of Exonuclease I with single-stranded DNA molecules. NFS data can be extrapolated to zero voltage (or zero-force) and be used to measure the dissociation and association constants of the system. While most of the work presented here utilized the membrane channel α-HL, the fragility of the bilayer membrane has limited these measurements to relatively small forces (up to a few tens of pN). A wider range of biomolecular complexes, such as long DNA/RNA molecules or proteins can be studied using NFS if this limitation is removed. The recent progress in fabrication and characterization of synthetic nanopores, specifically nanopores made in thin inorganic membranes (Kim et al., 2006; Li et al., 2001; Storm et al., 2003), has generated vast possibilities for NFS. Solid-state nanopores offer superior mechanical, electrical, and chemical robustness over lipid bilayers used with biological pores. Thus, larger force ranges and a variety of chemical conditions, including extreme pH or denaturants can be used (Wanunu and Meller, 2008). Additionally, a solid-state nanopore can now be tailored to any desired dimension, down to ~1 nm diameter (Kim et al., 2006). Thus a much broader range of bimolecular complexes can be studied, under broader experimental conditions. Specifically, double-stranded DNA (Gershow and Golovchenko, 2007; Heng et al., 2004; Storm et al., 2005; Wanunu et al., 2008), single-stranded nucleic acids and unzipping of duplex regions (Fologea et al., 2005; McNally et al., 2008), DNA–protein interactions as well as DNA–drug interactions can be studied in great detail (Smeets et al., 2009; Wanunu et al., 2009). These developments will undoubtedly be employed for high-throughput NFS analyses of a broad range of biological systems.

ACKNOWLEDGMENTS

A. M. acknowledges fruitful collaboration and discussions with Daniel Branton, Yitzhak Rabin, Mark Akeson, Breton Hornblower, and Anatoly Kolomeisky. O. K. D. is grateful to Attila Szabo and Gerhard Hummer for an exciting collaboration on the theory of single-molecule force spectroscopy. We thank Meni Wanunu for constructive and insightful comments on this manuscript. O. K. D. is supported by the National Science Foundation CAREER Award (MCB-0845099), a Hellman Faculty Fellows Award and the National Science Foundation Grant to the Center for Theoretical Biological Physics (PHY-0822283). J. M. acknowledges financial support from ANR PNANO grant ANR-06-NANO-015-02. A. M. acknowledges grant support from the National Science Foundation (PHY-0646637), the National Institute of Health (HG-004128 and GM-075893), and the Human Frontier Science Program (RGP0036).

REFERENCES

Abramowitz, M., and Stegun, I. A. (1972). Handbook of Mathematical Functions. Dover, New York.
Akeson, M., Branton, D., Kasianowicz, J., Brandin, E., and Deamer, D. (1999). Microsecond time-scale discrimination among polycytidylic acid, polyadenylic acid, and polyuridylic acid as homopolymers or as segments within single RNA molecules. *Biophys. J.* **77**, 3227–3233.
Bates, M., Burns, M., and Meller, A. (2003). Dynamics of single DNA molecules actively controlled inside a membrane channel. *Biophys. J.* **84**, 2366–2372.
Bell, G. I. (1978). Models of the specific adhesion of cells to cells. *Science* **200**, 618–627.

Bockelmann, U., and Viasnoff, V. (2008). Theoretical study of sequence-dependent nanopore unzipping of DNA. *Biophys. J.* **94**, 2716–2724.

Cecconi, C., Shank, E. A., Bustamante, C., and Marqusee, S. (2005). Direct observation of the three-state folding of a single protein molecule. *Science* **309**, 2057–2060.

Dudko, O. K., Hummer, G., and Szabo, A. (2006). Intrinsic rates and activation free energies from single-molecule pulling experiments. *Phys. Rev. Lett.* **96**, 108101.

Dudko, O. K., Mathe, J., Szabo, A., Meller, A., and Hummer, G. (2007). Extracting kinetics from single-molecule force spectroscopy: Nanopore unzipping of DNA hairpins. *Biophys. J.* **92**, 4188–4195.

Dudko, O. K., Hummer, G., and Szabo, A. (2008). Theory, analysis, and interpretation of single-molecule force spectroscopy experiments. *Proc. Natl. Acad. Sci. USA* **105**, 15755–15760.

Florin, E. L., Moy, V. T., and Gaub, H. E. (1994). Adhesion forces between individual ligand–receptor pairs. *Science* **264**, 415–417.

Fologea, D., Gershow, M., Ledden, B., McNabb, D. S., Golovchenko, J. A., and Li, J. L. (2005). Detecting single stranded DNA with a solid state nanopore. *Nano Lett.* **5**, 1905–1909.

Gautel, M., Oesterhelt, F., Fernandez, J. M., and Gaub, H. E. (1997). Reversible unfolding of individual titin immunoglobulin domains by AFM. *Science* **276**, 1109–1112.

Gershow, M., and Golovchenko, J. A. (2007). Recapturing and trapping single molecules with a solid-state nanopore. *Nat. Nanotechnol.* **2**, 775–779.

Greenleaf, W. J., Frieda, K. L., Foster, D., Woodside, M. T., and Block, S. M. (2008). Direct observation of hierarchical folding in single riboswitch aptamers. *Science* **319**, 630–633.

Heng, J. B., Ho, C., Kim, T., Timp, R., Aksimentiev, A., Grinkova, Y. V., Sligar, S., Schulten, K., and Timp, G. (2004). Sizing DNA using a nanometer-diameter pore. *Biophys. J.* **87**, 2905–2911.

Hornblower, B., Coombs, A., Whitaker, R. D., Kolomeisky, A., Picone, S. J., Meller, A., and Akeson, M. (2007). Single-molecule analysis of DNA–protein complexes using nanopores. *Nat. Methods* **4**, 315–317.

Jan Bonthuis, D., Zhang, J., Hornblower, B., Mathe, J., Shklovskii, B. I., and Meller, A. (2006). Self-energy-limited ion transport in subnanometer channels. *Phys. Rev. Lett.* **97**, 128104.

Kang, X. F., Cheley, S., Guan, X. Y., and Bayley, H. (2006). Stochastic detection of enantiomers. *J. Am. Chem. Soc.* **128**, 10684–10685.

Kasianowicz, J., Brandin, E., Branton, D., and Deamer, D. (1996). Characterization of individual polynucleotide molecules using a membrane channel. *Proc. Natl. Acad. Sci. USA* **93**, 13770–13773.

Kasianowicz, J. J., Kellermayer, M., and Deamer, D. W. (eds.) (2002). Structure and Dynamics of Confined Polymers, Springer, Dordrecht.

Kellermayer, M. S. Z., Smith, S. B., Granzier, H. L., and Bustamante, C. (1997). Folding–unfolding transitions in single titin molecules characterized with laser tweezers. *Science* **276**, 1112–1116.

Kim, M.-J., Wanunu, M., Bell, C. D., and Meller, A. (2006). Rapid fabrication of uniform size nanopores and nanopore arrays for parallel DNA analysis. *Adv. Mater.* **18**, 3149–3153.

Kramers, H. A. (1940). Brownian motion in a field of force and the diffusion model of chemical reactions. *Physica* **7**, 284–304.

Li, J., Stein, D., McMullan, C., Branton, D., Aziz, M. J., and Golovchenko, J. A. (2001). Ion-beam sculpting at nanometre length scales. *Nature* **412**, 166–169.

Liphardt, J., Onoa, B., Smith, S. B., Tinoco, I. J., and Bustamante, C. (2001). Reversible unfolding of single RNA molecules by mechanical force. *Science* **292**, 733–737.

Marszalek, P. E., Lu, H., Li, H., Carrion-Vazquez, M., Oberhauser, A. F., Schulten, K., and Fernandez, J. M. (1999). Mechanical unfolding intermediates in titin modules. *Nature* **402**, 100–103.

Mathé, J., Visram, H., Viasnoff, V., Rabin, Y., and Meller, A. (2004). Nanopore unzipping of individual DNA hairpin molecules. *Biophys. J.* **87**, 3205–3212.

Mathé, J., Arinstein, A., Rabin, Y., and Meller, A. (2006). Equilibrium and irreversible unzipping of DNA in a nanopore. *Europhys. Lett.* **73**, 128–134.

McNally, B., Wanunu, M., and Meller, A. (2008). Electro-mechanical unzipping of individual DNA molecules using synthetic sub-2 nm pores. *Nano Lett.* **8**, 3418–3422.

Meller, A. (2003). Dynamics of polynucleotide transport through nanometre-scale pores. *J. Phys.: Condens Matter* **15**, R581–R607.

Meller, A., Nivon, L., Brandin, E., Golovchenko, J., and Branton, D. (2000). Rapid nanopore discrimination between single polynucleotide molecules. *Proc. Natl. Acad. Sci. USA* **97**, 1079–1084.

Meller, A., Nivon, L., and Branton, D. (2001). Voltage-driven DNA translocations through a nanopore. *Phys. Rev. Lett.* **86**, 3435–3438.

Merkel, R., Nassoy, P., Leung, A., Ritchie, K., and Evans, E. (1999). Energy landscapes of receptor–ligand bonds explored with dynamic force spectroscopy. *Nature* **397**, 50–53.

Sauer-Budge, A. F., Nyamwanda, J. A., Lubensky, D. K., and Branton, D. (2003). Unzipping kinetics of double-stranded DNA in a nanopore. *Phys. Rev. Lett.* **90**, 238101.

Schlierf, M., and Rief, M. (2006). Single-molecule unfolding force distributions reveal a funnel-shaped energy landscape. *Biophys. J.* **90**, L33–L35.

Schlierf, M., Li, H. B., and Fernandez, J. M. (2004). The unfolding kinetics of ubiquitin captured with single-molecule force-clamp techniques. *Proc. Natl. Acad. Sci. USA* **101**, 7299–7304.

Smeets, R. M. M., Kowalczyk, S. W., Hall, A. R., Dekker, N. H., and Dekker, C. (2009). Translocation of RecA-coated double-stranded DNA through solid-state nanopores. *Nano Lett.* **9**, 3089–3095.

Song, L., Hobaugh, M. R., Shustak, C., Cheley, S., Bayley, H., and Gouax, J. E. (1996). Structure of staphylococcal α-hemolysin a heptameric transmembrane pore. *Science* **274**, 1859–1865.

Storm, A. J., Chen, J. H., Ling, X. S., Zandbergen, H. W., and Dekker, C. (2003). Fabrication of solid-state nanopores with single-nanometre precision. *Nat. Mater.* **2**, 537–540.

Storm, A. J., Chen, J. H., Zandbergen, H. W., and Dekker, C. (2005). Translocation of double-strand DNA through a silicon oxide nanopore. *Phys. Rev. E* **71**, 051903.

Wanunu, M., and Meller, A. (2008). Single-molecule analysis of nucleic acids and DNA–protein interactions using nanopores. *In* "Single-Molecule Techniques: A Laboratory Manual," (P. Selvin and T. J. Ha, eds.), pp. 395–420. Cold Spring Harbor Laboratory Press, Cold Spring Harbor, NY.

Wanunu, M., Morrison, W., Rabin, Y., Grosberg, A. Y., and Meller, A. (2010). Electrostatic Focusing of Unlabeled DNA into Nanoscale Pores using a Salt Gradient. *Nature Nanotech.* **5**, 160–165.

Wanunu, M., Sutin, J., McNally, B., Chow, A., and Meller, A. (2008). DNA translocation governed by interactions with solid state nanopores. *Biophys. J.* **95**, 4716–4725.

Wanunu, M., Sutin, J., and Meller, A. (2009). DNA profiling using solid-state nanopores: Detection of DNA-binding molecules. *Nano Lett.* **9**, 3498–3502.

CHAPTER TWENTY-TWO

ANALYSIS OF SINGLE NUCLEIC ACID MOLECULES WITH PROTEIN NANOPORES

Giovanni Maglia, Andrew J. Heron, David Stoddart,
Deanpen Japrung, *and* Hagan Bayley

Contents

1. Background: Analysis of Nucleic Acids with Nanopores	593
1.1. Structure of the αHL nanopore	594
1.2. Nucleic acid analysis with αHL nanopores	595
1.3. Homopolymeric strand analysis with protein nanopores	596
1.4. Recognition of specific sequences and single base mismatches through duplex formation	597
1.5. Individual base recognition by αHL nanopores	597
1.6. Control of DNA translocation through nanopores	599
2. Electrical Recording with Planar Lipid Bilayers	601
2.1. Electrical recording equipment	602
2.2. Faraday enclosure	603
2.3. Preparation of electrodes	603
2.4. Chambers	605
2.5. Preparing bilayers	606
2.6. Inserting pores and adding DNA	609
2.7. Perfusion	610
3. Nanopores	610
3.1. Nanopore preparation	610
3.2. Nanopore storage	610
3.3. Measurements with nanopores	611
3.4. Nanopore stability	612
4. Materials	613
4.1. Buffer components	613
4.2. Handling of DNA and RNA	613
4.3. Short single-stranded DNA or RNA	613
4.4. Long single-stranded DNA	614
4.5. Long RNA preparation	615
4.6. Short dsDNA preparation	615

Department of Chemistry, University of Oxford, Oxford, United Kingdom

5. Data Acquisition and Analysis 616
   5.1. Data digitization 616
   5.2. Filtering and sampling 616
   5.3. Acquisition protocols 617
   5.4. Analysis of single DNA/RNA molecules 618
References 619

Abstract

We describe the methods used in our laboratory for the analysis of single nucleic acid molecules with protein nanopores. The technical section is preceded by a review of the variety of experiments that can be done with protein nanopores. The end goal of much of this work is single-molecule DNA sequencing, although sequencing is not discussed explicitly here. The technical section covers the equipment required for nucleic acid analysis, the preparation and storage of the necessary materials, and aspects of signal processing and data analysis.

NOMENCLATURE

ssDNA	single-stranded DNA
dsDNA	double-stranded DNA
ddH$_2$O	distilled, deionized water
f	normalized frequency of DNA interactions with pore (e.g., translocation events) in s^{-1} μM^{-1}
I_O	open pore current
I_B	current through pore partially blocked by DNA
I_F	fractional residual current, that is, I_B/I_O
$I_{\%RES}$	residual current as percentage of I_O, that is, $I_B/I_O \times 100$
$\Delta I_{\%RES}$	difference in $I_{\%RES}$ between residual currents (e.g., for two different DNAs)
t_t	time taken for translocation of an individual nucleic acid molecule
$\bar{t}_t$	mean time taken for translocation of a nucleic acid molecule
t_D	dwell time of a nucleic acid molecule (not necessarily for a translocation event)
$\bar{t}_D$	mean dwell time
t_P	most probable dwell time or translocation time taken from the peak of a dwell time or translocation time histogram
t_{ON}	interevent interval (e.g., time between translocation events)
$\bar{t}_{ON}$	mean interevent interval

1. BACKGROUND: ANALYSIS OF NUCLEIC ACIDS WITH NANOPORES

Nanopores have been used to detect and analyze single molecules (Bayley and Cremer, 2001; Branton *et al.*, 2008; Deamer and Branton, 2002; Dekker, 2007) and to investigate reaction mechanisms at the single-molecule level (Bayley *et al.*, 2008b). In this chapter, we focus on the analysis of nucleic acids with the protein pore formed by staphylococcal α-hemolysin (αHL). First, we present background on the wide variety of experiments that can be performed on nucleic acids. Second, we enlarge on technical aspects of these experiments that have proved useful in our laboratory.

In a nanopore experiment, a single pore is located in a thin barrier that separates two compartments (henceforth called *cis* and *trans*) containing aqueous electrolyte. An electrical potential is applied across the barrier and the flow of ions through the nanopore is monitored. Molecules of interest are investigated by analyzing the associated modulation of the ionic current when the molecules pass through the pore or interact with the lumen (Fig. 22.1A). In this way, information about the length of nucleic acid

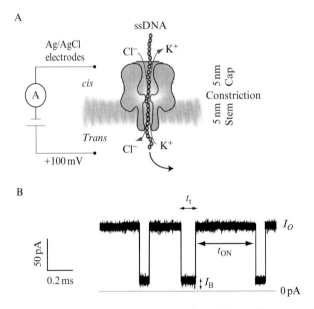

Figure 22.1 Nanopore analysis of DNA. (A) Ag/AgCl electrodes are used to apply a potential (e.g., +100 mV) to drive DNA through an αHL nanopore embedded in a lipid bilayer and to measure the ionic current. (B) Typical trace for DNA translocating through a (wild-type) WT-αHL nanopore showing the translocation times (t_t), the residual current values (I_B), and the interevent intervals (t_{ON}) for individual DNA translocation events.

molecules and their base compositions is gathered. Duplex formation and dissociation, including unzipping in an applied potential, can be examined. The properties of enzymes and binding proteins associated with nucleic acids can also be determined. Stemming from this work, there is the potential to develop a cheap and fast technology to sequence single DNA molecules (Branton et al., 2008), but we do not emphasize this aspect here.

The two most common classes of nanopores are protein pores in lipid bilayers (Fig. 22.1) or synthetic nanopores formed in a variety of thin films, including Si-based materials, PDMS (polydimethylsiloxane), and polycarbonate (Dekker, 2007; Rhee and Burns, 2007; Sexton et al., 2007). Among the advantages of synthetic nanopores are that their dimensions can be adjusted during manufacture and that they can be produced as nanopore arrays. Protein nanopores also have significant advantages; in contrast with synthetic nanopores, their dimensions are reproducible and their structures can be precisely manipulated by site-directed mutagenesis and targeted chemical modification. The stability of solid-state pores in harsh environments is often mentioned as a key advantage, but it should be noted that the αHL pore can withstand denaturants (Japrung et al., 2010), extremes of pH (Maglia et al., 2009b), and temperatures up to $\sim 95\ °C$ (Kang et al., 2005), conditions adequate for the analysis of DNA.

Although we focus on the analysis of nucleic acids with the αHL nanopore, progress on DNA analysis with alternative protein pores is being made, notably with MspA from *Mycobacterium smegmatis* (Butler et al., 2008). The experimental principles developed for αHL are valid in other cases.

1.1. Structure of the αHL nanopore

αHL is a heptameric protein that consists of a stem domain and a cap domain (Fig. 22.1) (Song et al., 1996). The stem domain comprises 14 antiparallel β strands (two per protomer) that form a roughly cylindrical water-filled channel of ~ 2 nm internal diameter. The internal cavity, or vestibule, within the cap domain is roughly spherical with an internal diameter of ~ 4.5 nm. The narrowest part of the interior of the pore is the constriction (1.4 nm in diameter), which is located between the cap and the stem domains. Under an applied potential, ions flow through the pore and the addition of single-stranded (ss)DNA to one side of the bilayer (typically the *cis* side) produces current blockades during the periods when a DNA molecule enters the pore (Fig. 22.1). Entry can result in a brief visit and then exit from the same side or translocation from the *cis* to the *trans* side of the bilayer. The dimensions of the pore are such that double-stranded (ds) DNA cannot pass through it (Kasianowicz et al., 1996), although dsDNA can penetrate as far as the constriction when presented from the *cis* side (Maglia et al., 2009b; Vercoutere et al., 2001). While we most often refer to DNA in this review, the same principles apply to RNA analysis.

1.2. Nucleic acid analysis with αHL nanopores

In electrical recordings from planar bilayers, it is important that sign conventions are followed consistently. We use the following. The *cis* chamber is that to which the protein nanopore is added; after insertion, the cap domain is exposed to the *cis* electrolyte. The *cis* chamber is at ground and the transmembrane potential is given as the potential on the *trans* side (i.e., the *trans* potential minus the *cis* potential; the latter is "zero" in the present case). A positive current is one in which positive charge (e.g., K^+) moves through the pore from the *trans* to the *cis* side, or negative charge (e.g., Cl^-) from the *cis* to the *trans* side.

The information that can be extracted from a typical electrical recording of DNA translocation through an αHL nanopore is the mean translocation time ($\bar{t}_t$), the residual current level during translocation ($I_{\%RES}$), and the mean interevent interval ($\bar{t}_{ON}$), which is the inverse of the frequency of events (f), when $\bar{t}_{ON} \gg \bar{t}_t$ (Fig. 22.1B).

Translocation times of short single-stranded oligonucleotides (<100 nucleotides, nt) are described well by a Gaussian distribution with an exponential tail. The most probable translocation time (t_P) is given by the peak of the distribution (Meller *et al.*, 2000). At +120 mV and 25 °C, the nucleic acid move through the pore at a speed of 1–22 µs per nt (t_P divided by the number of nt). The speed depends on the base composition of the nucleic acid, with purine bases showing the shortest translocation times (Meller, 2003). In addition, t_P decreases exponentially with the temperature or the applied potential (Meller *et al.*, 2000).

The frequency of occurrence of DNA translocation events increases with increased applied potential (Henrickson *et al.*, 2000), temperature (Meller *et al.*, 2000), or salt concentration (Bonthuis *et al.*, 2006). In addition, a threshold potential must be exceeded to observe DNA capture (Henrickson *et al.*, 2000) and even then capture is inefficient. Rough estimates suggest that, at voltages close to the threshold, about one DNA molecule in every 1000 that collide with the wild-type (WT)-αHL pore is captured (Meller, 2003). By contrast, at the highest accessible applied potentials (around +300 mV), ~20% of the collisions result in DNA translocation (Nakane *et al.*, 2002). However, the frequency of DNA capture can be improved by increasing the number of internal positive charges within the pore by site-directed mutagenesis, which also reduces the voltage threshold for DNA translocation through both the αHL (Maglia *et al.*, 2008) and MspA (Butler *et al.*, 2008) nanopores. The most likely mechanisms for the increased frequency of translocation are a strengthened interaction when DNA molecules sample the interior of the pore, increased electroosmotic flow, or a combination of the two that depends on the applied potential (Maglia *et al.*, 2008).

If nanopores are to be used to sequence DNA, single bases will most likely be identified at a recognition point within the nanopore by the extent to which they reduce the ionic current, rather than their dwell times (Bayley, 2006). Therefore, if a large current flows during DNA translocation (I_B), the recognition of the bases is likely to be improved. The percent residual current ($I_{\%RES} = I_B/I_O \times 100$) shows less experiment-to-experiment variation than the residual current (I_B) and therefore it is the descriptor of choice. $\Delta I_{\%RES}$, the difference in $I_{\%RES}$ between residual currents (e.g., for two different DNAs), shows less variation still (Stoddart et al., 2010). WT-αHL nanopores generally show relatively low $I_{\%RES}$ values (typically 10% in 1 M KCl, at +120 mV), which depend on the nucleic acid used, with RNA generally blocking more current than DNA (Meller, 2003). $I_{\%RES}$ increases with increased applied potential (Stoddart et al., 2009) and decreased salt concentration (D. Stoddart, unpublished data). Recent studies have shown that $I_{\%RES}$ also depends on the amino acid side chains that project into the lumen of the pore (Maglia et al., 2008; Stoddart et al., 2009). At +120 mV, additional positively charged residues in the barrel decrease I_B by 60–80% relative to WT-αHL depending on the positioning of the charge (Maglia et al., 2008), while the elimination of ionic interactions and the reduction of side-chain volume at the constriction, in the mutant E111N/K147N, increase I_B by approximately twofold (Stoddart et al., 2009).

1.3. Homopolymeric strand analysis with protein nanopores

Homopolymeric ssDNA or ssRNA blockades can be discriminated with (wild-type) WT-αHL nanopores by means of their residual currents (I_B) and translocation times (t_P). In 1 M KCl and at +120 mV, poly-rA blockades show an $I_{\%RES}$ of 15% and a t_P value of 22 μs per nt, which are readily distinguished from poly-rC blockades ($I_{\%RES} = 9\%$ and 5%, and $t_P = 5.8$ μs per nt) and from poly-rU blockades ($I_{\%RES} = 15\%$ and $t_P = 1.4$ μs per nt) (Akeson et al., 1999). The DNAs, poly-dA and poly-dC, show different translocation times ($t_P = 1.9$ and 0.8 μs per nt, respectively), but similar $I_{\%RES}$ values (12.6% and 13.4%, respectively) (Meller et al., 2000), while poly-dU shows longer blockades than both poly-dC and poly-dA (Butler et al., 2007). Although a fully convincing explanation has yet to be proposed, differences in translocation times and I_B value of homopolymeric strands through αHL nanopores have been associated with differences in base pair-independent secondary structure including base stacking (Akeson et al., 1999; Meller et al., 2000).

The different extents of current blockade associated with homopolymeric RNA strands were exploited to detect the transition from poly-rA to poly-rC segments within the same molecule. When the translocation of poly-rA$_{30}$C$_{70}$ through the WT-αHL nanopore was investigated, ~50% of the events exhibited current steps that reflected the transition (Akeson et al., 1999).

The mean translocation speed is constant for short DNAs (from 12 to 100 nt) (Meller and Branton, 2002; Meller et al., 2001) and for poly-rU from 100 to 500 nt (Kasianowicz et al., 1996). For oligonucleotides shorter than the length of the barrel ($N < 12$), the translocation speed decreases with increasing nucleotide length in a nonlinear fashion. Polynucleotides longer than 500 nt have not yet been carefully characterized.

1.4. Recognition of specific sequences and single base mismatches through duplex formation

αHL pores have been used to recognize specific DNA sequences and single nucleotide mismatches through duplex formation. Howorka et al. (2001) covalently attached an 8-nt oligonucleotide within the vestibule of an αHL pore. The complementary strand was then recognized by measuring the current blockade provoked by strand hybridization (Howorka et al., 2001). Single nucleotide mismatches were identified because they produced shorter current blockades. Based on duplex lifetimes, the DNA-nanopores were able to discriminate mismatches between the oligonucleotide tethered to the pore and complementary 8-base sequences within DNA strands of up to 30 nt. In addition, the last three bases in a 9-nt DNA attached to the pore were sequenced by the sequential addition of known oligonucleotides (Howorka et al., 2001). Vercoutere et al. (2001) detected single-base differences in short, blunt-ended DNA hairpin molecules with up to nine complementary bases. The entry of the hairpins into the vestibule of the αHL pore was accompanied by a characteristic current blockade. Once in the vestibule, the hairpins spontaneously dissociated and a current spike signaled the translocation of the DNA strand through the pore. Single-base mismatches were detected by reduced duplex lifetimes, which reflected the stabilities of the duplexes in solution (Vercoutere et al., 2001). Nakane et al. (2004) used streptavidin to immobilize a biotinylated probe DNA molecule within the αHL pore (Fig. 22.2). The untethered end of the DNA strand entered the *trans* compartment at positive potentials, where it bound a complementary oligonucleotide to form a rotaxane. Reversal of the applied potential exerted a force on the duplex. A shorter mean time for dissociation by comparison with the value for a fully complementary pair was used to recognize single mismatches in probe-oligonucleotide duplexes (Nakane et al., 2004).

1.5. Individual base recognition by αHL nanopores

The recognition of individual DNA bases during translocation is crucial if the sequencing of DNA molecules is to be achieved by using nanopores. However, at present, the translocation of nucleic acids through αHL nanopores is so rapid (1–22 μs per nt) that only artificially bulky bases

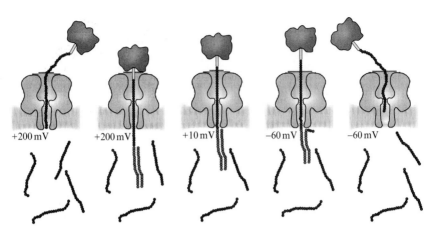

Figure 22.2 Detection of an ssDNA by hybridization. From left to right: A high applied potential (e.g., +200 mV) is used to capture a biotinylated DNA probe (complexed with streptavidin) within the αHL pore. The target DNA strand emerges on the opposite side of the bilayer where it forms a duplex, if it is recognized by a complementary strand. The voltage is lowered to +10 mV. If a duplex has been formed, the probe DNA remains in the pore as a rotaxane. If a duplex has not formed, the probe DNA escapes under the low applied potential. To measure the strength of the duplex, eject the probe DNA and begin a new cycle, the applied potential is stepped to −60 mV.

(Mitchell and Howorka, 2008) or the transition between homopolymer stretches of ∼30 nt can be detected (Akeson et al., 1999). Therefore, to test the feasibility of base detection, DNA strands have been immobilized within the αHL pore by using terminal hairpins (Ashkenasy et al., 2005) (Fig. 22.3C). Because the hairpin might interfere with recognition (Fig. 22.3A), recent work has exploited the interaction between a biotinylated oligonucleotide and streptavidin to immobilize DNA strands (Purnell and Schmidt, 2009; Stoddart et al., 2009) (Fig. 22.3B). In our work, the ability of the αHL pore to discriminate individual bases was investigated by sampling several DNA molecules in which the position of a single base was altered in an otherwise identical homopolymeric background (Fig. 22.3C). While Ghadiri and coworkers were able to identify a recognition point at the *trans* entrance of the αHL nanopore that could discriminate a single dA in a poly-dC background (Ashkenasy et al., 2005), the biotinylated oligonucleotides were used to sample the entire length of the pore (Stoddart et al., 2009). αHL was found to contain three broad recognition sites. When an engineered pore (E111N/K147N) was used, with enhanced current flow when DNA occupies the pore (I_B), all four DNA bases could be distinguished in a poly-dC background. Individual base differences could also be detected in a heteropolymeric DNA strand (Stoddart et al., 2009).

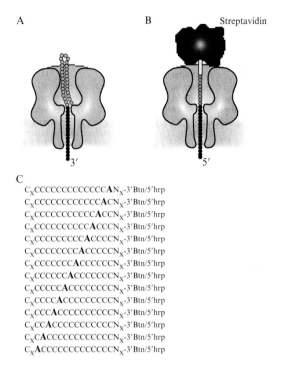

Figure 22.3 Single base recognition in immobilized DNA strands. (A) The translocation of a DNA strand can be arrested by using a terminal DNA hairpin. (B) Alternatively, the translocation of a biotinylated DNA strand can be halted by preincubating the ssDNA with streptavidin. The grey circles indicate the DNA that is in the vestibule and the black circles are the DNA bases that span the barrel of the pore. (C) The barrel of the pore can be sampled by measuring the current blockades provoked by DNA sequences in which the location of a single base is moved within an otherwise identical background. Hrp, hairpin; Btn, biotinylated linker; N_X, nucleotides that are in the vestibule of the pore; C_X, nucleotides that protrude through the *trans* entrance of the pore.

1.6. Control of DNA translocation through nanopores

Controlling the speed at which DNA translocates through a pore is perhaps the single most important challenge that has to be overcome in nanopore sequencing (Branton *et al.*, 2008). At high sampling rates, noise levels are too high for base discrimination. A recent paper estimates that 98% of the small current differences observed for the four DNA nucleotides could be discriminated by an αHL nanopore assuming a mean dwell time of 10 ms (Clarke *et al.*, 2009), but this is at least 3 orders of magnitude longer than the transit time for individual bases in freely translocating ssDNA. Several attempts have been made to reduce the DNA translocation speed. For example, it has been shown recently that increasing the viscosity of the sample solution by adding 63% (v/v) glycerol to the aqueous electrolyte can slow down DNA translocation through αHL nanopores by more than an order of magnitude, but at the expense of a 10-fold decrease in the unitary conductance of the pore (Kawano et al., 2009). In efforts to increase the local viscosity, but preserve the conductance of the pore, a positively charged dendrimer with a hydrodynamic volume of 4.2 nm (Martin et al., 2007) and uncharged PEG molecules of similar size (G. Maglia, unpublished data) were covalently attached inside the vestibule of the αHL pore. The modified nanopores showed a conductance that was roughly half that of the WT-αHL pore. However, both polymers almost completely inhibited the passage of nucleic acids through the pore, rather than slow their movement. Attempts to engineer the internal positive charge of the αHL pore to form "molecular brakes" to reduce the speed of ssDNA translocation have also been made (Rincon-Restrepo, in preparation). Modified versions of the pore were prepared with up to seven internal rings of seven arginine residues. Although the translocation speed was reduced by more than 2 orders of magnitude, the resulting I_B values were too low to allow for discrimination of individual bases. Optical and magnetic traps have also been used to move DNA strands through solid-state nanopores (Keyser et al., 2006; Peng and Ling, 2009), but so far these have been long double-stranded DNAs. Due to the instability of the lipid bilayer and the diffusion of nanopores within it, these techniques cannot be easily implemented with biological nanopores.

Ultimately, DNA-processing enzymes will most likely be used to control DNA translocation through biological pores. ExoI–ssDNA complexes gave current blockades that were approximately fivefold longer and showed a higher I_B value than the current blockades produced by free ssDNA under the same conditions (+180 mV, 1 M KCl) (Hornblower et al., 2007). The disappearance of the long current blockades after the addition of Mg^{2+}, which triggers the activity of the enzyme, confirmed that these blockades were due to DNA in ExoI–ssDNA complexes entering the nanopore. But, the enzymatically driven movement of DNA was not demonstrated in this work. In a similar approach, the interaction of RNA with the αHL pore was monitored in the presence of P4 ATPase, a motor protein from bacteriophage ϕ8, but again the movement of the nucleic acid could not be shown (Astier et al., 2007). Recently, the activity of a DNA polymerase was monitored with single base resolution by using an αHL nanopore (Cockroft et al., 2008). A DNA strand was immobilized within the pore by threading a streptavidin·biotin–phosphoPEG–ssDNA complex through the *trans* entrance under a negative potential (Fig. 22.4). DNA primers (to form a rotaxane) and *TopoTaq* DNA polymerase were then added to the *cis* side and single-nucleotide primer extensions were identified by the current changes caused by the movement of phosphoPEG into the barrel of the pore after the addition of dNTPs (Fig. 22.4). Up to nine successive steps were

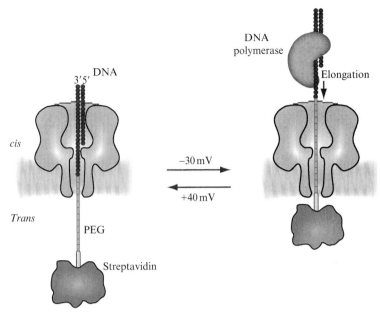

Figure 22.4 Single nucleotide extension monitored with a protein nanopore. (Left) The monitoring configuration. A phosphoPEG sequence occupies a fraction of the barrel of the αHL pore and modulates the ionic current to a level that depends on the extent to which the primer has been elongated. (Right) Elongation configuration. dsDNA is accessible to the DNA polymerase. Elongation by a single nucleotide is controlled by the identity of the dNTPs in solution. The activity of the DNA polymerase can be monitored by stepping the potential between +40 and −30 mV.

observed by switching the potential between −30 mV (a potential at which elongation was possible) and +40 mV (the potential used to monitor the elongation steps). Future efforts will be directed toward combining such an elongation device with an engineered αHL pore capable of base identification.

2. ELECTRICAL RECORDING WITH PLANAR LIPID BILAYERS

Protein nanopores employed for DNA analysis must be reconstituted into artificial bilayers that mimic biological membranes. We carry out our experiments with planar lipid bilayers, examining single pore under *in vitro* conditions where all the constituents of the system are carefully controlled. "Single-channel" experiments require sensitive equipment to measure the small currents passed by individual protein pores. There is a wealth of good literature detailing the various aspects of setting up such equipment and carrying out experiments on single protein channels and pores (see, e.g., The Axon Guide: www.moleculardevices.com/pages/instruments/axon_guide.html) (Hanke and Schlue, 1993; Miller, 1986; Sakmann and Neher, 1995), so here we just cover the specific apparatus and procedures employed in our laboratory for examining the interaction of DNA with the αHL pore.

2.1. Electrical recording equipment

Single-channel electrical measurements can be carried out with a range of equipment, both commercial and custom built. We use Axopatch 200B patch-clamp amplifiers (Molecular Devices) with 16-bit digitizers (132x or 1440A, Molecular Devices). DNA measurements are typically acquired in voltage-clamp mode using resistive WHOLE-CELL ($\beta = 1$) headstage settings. In voltage-clamp mode, the applied voltage is fixed, and the transmembrane current required to maintain that voltage is measured. Voltage clamping does not mimic a process found in nature, but many of the interactions of DNA with the αHL pore are voltage dependent, so it is imperative to control the applied potential. Furthermore, in this mode no capacitive transients are produced (except when changing the applied voltage), and the current flow is proportional to the conductance of the channels.

The Axopatch 200B has a recording bandwidth of 100 kHz, and the digitizers have a maximum sampling rate of 250 kHz per channel, which determines the maximum temporal resolution (4 μs). Amplitude resolution is primarily limited by the noise in the system. Planar bilayers are prone to relatively high background current noise (e.g., >1 pA RMS at 10 kHz) that can obscure small current steps (The Axon Guide) (Mayer et al., 2003; Wonderlin et al., 1990). Signal-to-noise can be improved by applying low-pass filtering, at the expense of a loss of temporal resolution. Therefore, although the electronics can acquire in the microsecond regime (freely translocating nucleic acids pass at ~1–22 μs per nt), in practice there is a balance between acquiring fast enough to observe the events of interest, and low-pass filtering to observe small amplitude current differences.

There are three main sources of internal noise in a planar bilayer system, in increasing order of magnitude (The Axon Guide) (Mayer et al., 2003; Wonderlin et al., 1990): (1) noise originating in the acquisition electronics, (2) noise from clamping the thermal voltage noise in the access resistance (i.e., the resistance of the solutions and electrodes in series with the bilayer) across a large capacitance bilayer, and (3) noise from the nanopore itself. There are various strategies for reducing noise that are well covered in the literature (The Axon Guide) (Mayer et al., 2003; Wonderlin et al., 1990). Access resistance can account for substantial noise in planar bilayer systems in the absence of a pore (Wonderlin et al., 1990). The access resistance can be reduced by: (1) using higher salt solutions in the *cis* and *trans* compartments, (2) using larger area electrodes, and (3) reducing the resistance of the agar bridges (discussed below). The use of suitable aperture materials is also important for reducing dielectric noise (e.g., PTFE films produce lower noise than some other plastics; Mayer et al., 2003; Wonderlin et al., 1990). However, the single most effective means of reducing noise in the absence of a pore is to reduce the size of the bilayer and hence its capacitance (The Axon Guide) (Mayer et al., 2003; Wonderlin et al., 1990). Nevertheless, in practice, the largest source of noise is from the protein pore itself. Protein-derived noise arises from extremely fast fluctuations in conformation and charge states inside a pore (Bezrukov and Kasianowicz, 1993; Korchev et al., 1997). The noise is dependent on physical conditions (pH, temperature, salt concentration, etc.), and for αHL we typically observe RMS noise that is two to three times greater than the noise from the bilayer alone.

2.2. Faraday enclosure

Electrical recording experiments should be carried out in a conducting enclosure (a Faraday cage) to shield the sensitive acquisition electronics from external radiative electrical noise. We typically use steel or aluminum boxes $\sim 50 \times 50 \times 50$ cm in dimensions with rigid side walls >1 mm thick. The boxes are grounded via the common ground in the patch-clamp amplifier (www.moleculardevices.com). Vibration isolation is an important consideration (The Axon Guide). External noise (e.g., from heavy machinery) can be transmitted through a laboratory bench into the box and result in substantial noise in the signal. Therefore, we mount our Faraday cages on pneumatic antivibration tables or soft supports (e.g., partially inflated soft rubber tubing). Unfortunately, antivibration tables do little to prevent sensitivity to environmental noise (i.e., loud noises, or even talking), as the large surfaces of the boxes pick up vibrations in the air. Although the Faraday cage might be housed inside an acoustically isolated enclosure, this source of noise can simply be eliminated by using a Faraday cage with stiff or absorbent walls that are well damped against vibrations. Finally, it is essential that the electrodes within the box are: (1) as short as possible (to reduce access resistance, input capacitance, and the reception of radiated noise) and (2) firmly secured to prevent their vibration.

While a Faraday cage is effective at isolating the system from external noise, care must be taken not to introduce internal sources of noise. As a result, any internal equipment (e.g., stirrers, peltier heaters, etc.) needs to be carefully shielded and should ideally run off a DC battery inside the Faraday cage.

2.3. Preparation of electrodes

Ag/AgCl electrodes are most commonly employed in bilayer recording due to their long-term chemical stability, reversible electrochemical behavior, predictable junction potentials, and superior low-noise electrical performance.

However, it should be noted that Ag/AgCl electrodes only perform well in solutions containing chloride ions. For experiments involving asymmetric solutions (where the chambers contain chloride ions at different concentrations), agar bridges (as discussed later) are used to prevent substantial liquid junction potential offsets (The Axon Guide).

Ag/AgCl electrodes are typically prepared from Ag wire by either: (1) electroplating in HCl solution, (2) dipping in molten AgCl, or (3) treatment with a weak hypochlorite solution (The Axon Guide) (Purves, 1981). Our preferred approach is an overnight treatment with hypochlorite, as this produces long lasting electrodes with a thick coating of AgCl. We make our electrodes from 1.5 mm diameter silver wire (>99.99%, Sigma-Aldrich), using lengths of ~10 mm (Fig. 22.5). These electrodes have a large reactive surface area, which prolongs their working life and minimizes the access resistance. Before treatment, the electrodes are thoroughly cleaned and roughened by abrasion (e.g., with a wire brush or glass paper), and then immersed in 1–5% (w/v) NaClO (Fisher Scientific) in water. After treatment, the electrodes are rinsed with distilled water. The electrodes are prepared and stored in the dark to prevent UV conversion of AgCl back to Ag. In our experience, good electrodes have a rough texture and a dull gray color (Fig. 22.5).

The use of a 3 M KCl/agar bridge between an electrode and the chamber solution is recommended for all experiments, but in particular for experiments carried out under asymmetric salt conditions to reduce junction potential offsets. The 3 M KCl filling solution also reduces the access resistance. Another advantage of the agar bridge is that it prevents Ag^+ ions from the electrode leaching into the solution, which can adversely affect the activity of some protein pores. To create the agar bridges, we incorporate our 1.5 mm Ag/AgCl electrodes into 200 μl pipette tips (Gilson, USA) filled with (initially) molten agar (3% (w/v) low melt agarose

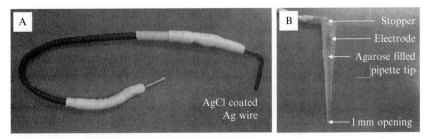

Figure 22.5 Ag/AgCl electrodes. (A) Photograph of an Ag/AgCl electrode, prepared by hypochlorite treatment of a short piece (~10 mm) of 1.5 mm-diameter Ag wire (soldered onto standard electrical wire). (B) To make agar bridges, we incorporate our electrodes into 200 μl pipette tips filled with agar (3%, w/v, 3 M KCl in unbuffered ddH$_2$O), sealing the top with a rubber stopper and parafilm.

in 3 M KCl in pure water), with a rubber stopper as a seal (Fig. 22.5). In addition, the opening of the pipet connecting to the bulk chamber solution should not be too small (we recommend an opening of >1 mm diameter).

Immediately before use, the electrodes are electrically balanced in symmetric solutions of the experimental buffer (e.g., 1 M KCl, 25 mM Tris–HCl, pH 8.0) by using the pipet offset function on the patch-clamp amplifier. Little to no potential adjustment is required for good electrodes. A sign of poor or degraded electrodes is the development of white patches and a requirement to rebalance the potential offset.

2.4. Chambers

We use open planar bilayer chambers for the majority of our DNA experiments. However, in some cases, in particular where limited quantities of reagents are available, we also make use of droplet interface bilayers that have solution volumes of <200 nl (Bayley et al., 2008a). Open chambers come in a wide variety of shapes and formats, with vertical or horizontal bilayers and solution volumes ranging from 50 μl to 1 ml. Although commercial options are available (e.g., Warner Instruments), we use custom vertical bilayer devices machined from Delrin (Fig. 22.6), where the two compartments of 0.5–1.0 ml are separated by a 20 μm-thick PTFE film (Goodfellow, UK). In this format, both the *cis* and *trans* solutions are easily accessed for the purposes of adding reagents or perfusion.

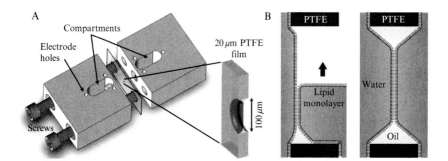

Figure 22.6 Planar bilayer formation. (A) We use custom vertical planar bilayer chambers, with two 1 ml compartments separated by a 20-μm PTFE film, shown here in an exploded view. The two halves are fastened together with screws, clamping the PTFE film between them. The apparatus is sealed with silicone glue (Sylgard 184, Dow Corning) to ensure water-tightness. The PTFE film contains a central aperture ~100 μm diameter, across which the bilayer is formed. (B) The bilayer is formed by the Montal–Mueller method. The aperture is pretreated with an oil (e.g., hexadecane), and the bilayer is then formed by flowing Langmuir–Blodgett lipid monolayers over both sides of the aperture.

The only path between the two compartments is a central aperture in the PTFE film across which the bilayer is formed (Fig. 22.6A, enlargement). The diameter depends on the size of bilayer required, and is typically ~100 μm. While there are many approaches for producing small apertures (Wonderlin et al., 1990), we favor "zapping" the films with a custom high-voltage spark generator (producing a spark ~15 mm long at a frequency of ~1 Hz). Commercial products are available (e.g., Daedelon, USA). To control the location of the aperture, the Teflon film is weakened at the desired point by gently indenting with a sharp needle. The spark passes through this weakness, and the high temperature causes localized melting, which produces a perfectly round aperture. The aperture is sparked several times until the desired diameter is achieved (as measured under a microscope with a graticule). This technique produces high quality apertures with reproducible dimensions (typically >50 and ±10 μm) (Mayer et al., 2003).

Reducing bilayer size, and hence capacitance, reduces noise. However, this must be balanced against the practical necessity of creating bilayers reproducibly (bilayers that are too small are often occluded by enlargement of the annulus) and the difficulty of inserting channels into overly small bilayers. A good compromise is to use bilayers with a capacitance of ~60–100 pF, which corresponds to a diameter of ~105–135 μm (DPhPC/hexadecane, assuming a specific capacitance of 0.70 $\mu F\ cm^{-2}$; Peterman et al., 2002).

2.5. Preparing bilayers

Planar bilayers are typically created across small apertures in a plastic film by variants of two main approaches (White, 1986). In the first, the bilayer is created by directly painting a lipid/oil solution (e.g., 5% (w/v) of DPhPC in hexadecane) around the aperture before adding the aqueous solution. This thick lipid/oil film spontaneously thins and after a few minutes forms a bilayer due to Plateau–Gibbs suction (Niles et al., 1988). In the second approach, the aperture is first pretreated with the oil (e.g., hexadecane), and then the bilayer is formed by flowing Langmuir–Blodgett lipid monolayers across both sides of the aperture (Fig. 22.6B) (Montal and Mueller, 1972; White et al., 1976). In both approaches, the oil (generally an n-alkane such as decane or hexadecane) is extremely important, as it creates an annulus around the rim of the aperture, which allows the bilayer to form an interface with the plastic film (Plateau–Gibbs border). We favor the second "Montal–Mueller" approach. The primary advantages of this technique are that the bilayers are quick to form (with the painting approach it can take tens of minutes for the bilayer to spontaneously thin), and asymmetric bilayers can be created if the two compartments contain lipids of a different composition.

Briefly, we pretreat the aperture with hexadecane by the direct application of a single drop (~5 µl) of 10% (v/v) hexadecane in pentane. After the pentane has evaporated, both compartments of the chamber are filled with the buffered aqueous electrolyte and the electrodes are connected. The aperture should be free of excess oil, so that an electrical current can be measured between the compartments. (At zero applied potential on the amplifier, there remains a very small potential difference between the compartments.) In the case of blockage, excess oil can be removed by gently pipetting solution at the aperture. A small drop (~5 µl) of lipid in pentane solution (e.g., DPhPC, 10 mg ml^{-1}) is then applied directly to the top of the aqueous phase. After waiting ~1 min for the pentane to evaporate fully, solution is pipetted from the bottom of a compartment to lower the lipid monolayer at the water–air interface past the aperture level. The solution level is then slowly raised back above the aperture. After repeating this procedure several times in each compartment, a bilayer forms, and the electrical current drops to zero. We measure the capacitance, as described later, to judge the size and quality of the bilayer.

We use DPhPC bilayers with hexadecane as the oil in most of our studies. DPhPC has ideal bilayer forming properties. It exists in the fluid lamellar phase across a wide temperature range (-120 to $+120$ °C) (Hung et al., 2000; Lindsey et al., 1979), and it is stable to oxidation. Planar bilayers can, however, be prepared from a wide range of phospholipids or phospholipid mixtures. These can have zwitterionic or charged headgroups, and may contain additives such as cholesterol. The correct lipid mixture may be essential for certain proteins to function (e.g., mechano-sensitive channels; Phillips et al., 2009). The phospholipid mixture usually, but not always (Krasne et al., 1971), has fluid lamellar properties under the conditions of the experiment. Many oils can be used to interface the bilayer with the aperture (n-alkanes, squalene, etc.), but experimentalists should be aware that the shorter chain n-alkanes (e.g., decane) interdigitate into the bilayer leading to increased bilayer thickness (White, 1975, 1986). Although this is not important for a robust pore such as αHL, bilayer thickness differences can affect the properties of membrane proteins (Phillips et al., 2009). Hexadecane produces bilayers with little oil content (White, 1975, 1986).

A standard cleaning routine is used between experiments: first rinsing with large amounts of water, then repeated steps of distilled water rinses followed by 100% ethanol rinses. Finally, the chambers are thoroughly dried under a stream of nitrogen. When required, we also employ detergent washing steps, NaOH (1 M) or HCl (1 M) washes, or rinsing with an EDTA solution (e.g., 10 mM) to remove traces of divalent metal ions.

2.5.1. Bilayer stability
The planar bilayers we have described typically have lifetimes of a few hours and can withstand constant applied potentials of 200–300 mV. Smaller bilayers (< 100 µm diameter) are much less susceptible to pressure variations

and mechanical shock, and are therefore longer lived. DPhPC bilayers are stable up to ~100 °C (Kang *et al.*, 2005). In certain cases, alternative oils should be used. For example, hexadecane is frozen below 18 °C, and hexadecene or shorter chains alkanes can be employed at low temperatures. Planar bilayers are also able to withstand substantial osmotic gradients arising from the use of asymmetric salt conditions (we often use asymmetric solutions of 150 mM/1 M salt to measure ionic charge selectivity), although the solutions can be balanced if instability occurs by the addition of an osmolyte (e.g., sucrose) to the low-salt compartment.

2.5.2. Measuring bilayer capacitance

It is important to measure bilayer capacitance, because pores will not insert into misformed bilayers. There are a number of ways of determining capacitance (for reviews see Gillis, 1995; Kado, 1993). One of the most versatile and straightforward is the application of voltage ramps to elicit constant capacitive currents (Fig. 22.7). Since current (I) is proportional to the rate of change of voltage (dV/dt) for a capacitor ($I = C \times dV/dt$), an applied triangular voltage waveform produces a square-wave current output (i.e., dV/dt = constant, for each sweep), which can be used to determine bilayer capacitance (Fig. 22.7). We use commercial analog waveform generators, inputting a triangular wave (~20 Hz) into the patch-clamp amplifier (front-switched input on an Axopatch 200B amplifier). We directly calibrate the input waveform by using a 100 pF bilayer model cell (MCB-1U, Molecular Devices), so that 100 pA = 100 pF. This approach provides a visual measurement of capacitance, which is useful

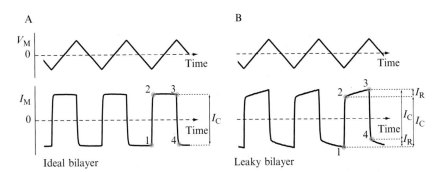

Figure 22.7 Measuring bilayer capacitance. A triangular-wave voltage input applied across a capacitor produces a square-wave current output. (A) Bilayer capacitance can be determined by measuring the peak-to-peak magnitude of the square-wave current output (I_C). (B) A leak in the bilayer is visualized by the triangular ohmic deviation (I_R) from the ideal square-wave.

for determining the quality of a bilayer seal (leaks are readily apparent as a triangular deviation from the ideal square-wave output (I_R in Fig. 22.7)).

2.6. Inserting pores and adding DNA

In the case of αHL, there are several ways to insert the pore into a planar bilayer. αHL is a toxin that assembles spontaneously from water-soluble monomers, so pores can be formed by adding monomeric αHL directly to the chamber. However, it is often difficult to obtain a single channel by using this approach (many pores tend to insert if the monomer remaining after a single pore has formed is not removed by perfusion), and we also find that the αHL pores formed from monomers have substantial pore-to-pore variability (discussed below). We therefore prefer to use preoligomerized protein in a detergent-solubilized form (e.g., after extraction from an SDS–polyacrylamide gel as described below). Typically ~ 0.2 μl of SDS–solubilized heptamer (~ 1 ng μl^{-1}) is manually ejected from a pipette very close to the bilayer beneath the surface of the solution in the *cis* compartment. The solution in the compartment is then stirred gently with a small magnetic stirrer bar until a single insertion event is observed. Electrical recordings are noisy while stirring, as the changing magnetic field produced by the rotating stirrer and stirrer bar induces currents in the electrodes, and stirring is discontinued for the rest of the experiment, unless additions are made to a compartment. Although the protein concentration can be reduced to improve the chances of obtaining a single insertion event, this often requires waiting a considerable length of time. Alternatively, a higher concentration of protein can be added to reduce the time to insertion, followed by prompt perfusion (see below) to remove free protein.

Two other methods of inserting protein employed in our laboratory are vesicle fusion (better suited to proteins that are unstable in detergent) (Morera *et al.*, 2007) and direct introduction with probes (Holden and Bayley, 2005; Holden *et al.*, 2006), which is also suited to unstable proteins or proteins of which only small quantities are available.

DNA, RNA, or other analytes are added directly to the chamber solutions from stocks (typically 100 μM in double-distilled water (ddH$_2$O) or 10 mM Tris-HCl, 100 μM EDTA, pH 8.0) by pipet. Mixing is achieved by stirring the chamber solution for up to 1 min (using a 5 mm stirrer bar in the bottom of the compartment). Depending on the experiment, we typically add oligonucleotides to the 1 ml compartment solution from 100 μM stocks to give final concentrations of 10 nM to 2 μM. The electrode compartments may represent an appreciable dead volume (e.g., 5% of the total volume), which must be taken into account for quantitative work. Better still, the concentration of a DNA or RNA in a compartment should be checked after an experiment, for example, by UV spectrometry.

2.7. Perfusion

We use perfusion to remove free protein from the bulk solution once a single pore has inserted, and to change the solution conditions or analytes in the system while retaining the same pore in the bilayer. We perfuse our devices by manual pipetting (removing some solution and replacing it with fresh solution) or with push–pull syringe drivers (e.g., PHD 2000, Harvard Apparatus) that exchange the solution at rates of ~ 1 ml min^{-1}. It should be noted that solution-filled perfusion tubing entering the Faraday cage acts as an antenna and picks up radiative noise that can prevent useful recording while the perfusion equipment is connected. If this is a problem, drip-feed perfusion can be employed, where the air pocket breaks the electrical continuity in the tubing.

3. NANOPORES

3.1. Nanopore preparation

While monomeric αHL can be obtained commercially, we recommend the use of heptamers carefully prepared in the laboratory. This is of course essential for heteroheptamers with defined subunit combinations (Miles et al., 2002), but we have also found that preassembled homoheptamers give more consistent results than monomers in bilayer experiments.

In our laboratory, αHL homoheptamers used in single-channel recording experiments are usually prepared from αHL monomers expressed by in vitro transcription and translation (IVTT). Heptamers are formed by incubating the monomers with rabbit red blood cell membranes and purified by SDS-PAGE (Cheley et al., 1999). The concentration of heptamer after purification is typically ~ 1 ng μl^{-1}. Alternatively, if a higher concentration of protein is required (e.g., for a protein that inserts poorly; Maglia et al., 2009a), αHL monomers can be expressed in Escherichia coli or Staphylococcus aureus bacterial expression systems (Cheley et al., 1997). In these cases, monomers are purified by ion exchange chromatography before incubation with a surfactant (6.25 mM DOC; Bhakdi et al., 1981; Tobkes et al., 1985) to trigger the heptamerization process. Heptamers are then separated from monomers by size-exclusion chromatography (Cheley et al., 1997). These procedures are more laborious, but αHL heptamers at concentrations of 1–5 mg ml^{-1} can be obtained.

3.2. Nanopore storage

αHL heptamers are usually stored at $-80\ ^\circ$C in 10 mM Tris–HCl buffer at pH 8.0, containing 100 μM EDTA. Heptamers prepared from monomers produced from bacterial expression systems will also contain surfactants (usually 1.25 mM DOC), while heptamers prepared from αHL monomers obtained by IVTT are likely to contain small amounts of polyacrylamide and SDS that are carried over from the SDS-PAGE purification. Polyacrylamide can be removed by using a small size-exclusion column (e.g., P6 Micro Bio-Spin chromatography columns, Bio-Rad), but it usually does not interfere with single-channel recording experiments.

To obtain single channels, protein solutions obtained after IVTT expression, assembly, and electrophoretic purification are usually diluted 10-fold in buffer or ddH$_2$O, and 0.2 μl is added to the buffer in the planar bilayer chamber. At this point, residual SDS is well below its critical micelle concentration (CMC). During an experiment, solutions containing αHL proteins are kept on ice. αHL preparations can be stored at 4 °C for short-term use (less than a week), but the protein slowly loses activity. Multiple freeze/thaw cycles also reduce protein activity. When a reactive group is present (e.g., an oxidation sensitive cysteine residue), the protein must be handled with particular care and long-term storage at −80 °C is usually limited to a few months.

3.3. Measurements with nanopores

Purified WT heptamers form pores that have a high unitary conductance (∼1 nS at +100 mV in 1 M KCl solution). WT-αHL pores show no gating events (openings and closings of the pore), unless the applied potential is high (more than ±180 mV), and even then the events are rare (generally less than 1 min^{-1}). Pore-to-pore variation of the unitary conductance is generally within 5% of the mean. However, very occasionally (<5% of the cases), the pores show a high level of gating or have a dramatically reduced conductance. These cases probably arise from misfolded or misassembled heptamers and are discarded. αHL monomers, rather than purified heptamers, can also be used in single-channel recordings. However, a much greater number of the pores formed from monomers (from 20% to 50%, depending on the batch) show a reduced conductance or gating events. The use of heptamers purified by SDS-PAGE is therefore recommended. Over the course of long experiments (>30 min), the open pore current can drift due to evaporation of the solutions or exhaustion of the Ag/AgCl electrodes. It is recommended, therefore, always to use agarose bridges around the electrodes and to minimize the surface area of the solutions. If lids are used, care must be taken to prevent a short circuit through a film of electrolyte. Layers of oil can be placed on the surfaces of the chamber solutions, but then a complete clean-up will be required if the bilayer is reformed.

The αHL pore can be altered by site-directed mutagenesis to give improvements in f and I_B. Alteration of the lumen of the αHL pore by site-directed mutagenesis generally produces nanopores that have similar unitary conductance (I_O) values to that of WT-αHL (Gu et al., 2001;

Stoddart et al., 2009). Additional charges in the barrel, however, can alter the unitary conductance and rectification properties (Maglia et al., 2008), while the introduction of aromatic residues near the constriction (e.g., M113W and M113Y), or positively charged residues near the *trans* entrance (e.g., T129R and T125R) can produce pores that show a high frequency of gating (M. Rincon-Restrepo, unpublished data). In some cases, the gating events observed with nanopores that have additional positive charge in the barrel might be due to the binding of contaminants to the protein pore (Cheley et al., 2002) as they are more frequent at low salt concentrations when the charges within the pore are less well screened.

3.4. Nanopore stability

αHL heptamers in bilayers are stable under extreme conditions of salt concentration, temperature (Kang et al., 2005), pH (Gu and Bayley, 2000; Maglia et al., 2009b), and in the presence of denaturants (Japrung et al., 2010). Pores can insert into bilayers and remain open over a wide range of ionic strengths (from 0.001 to 4 M KCl). Throughout this range, the conductance of the pore varies linearly with the ionic concentration of the solution, except at very low salt concentrations where there is a small deviation from linearity (Oukhaled et al., 2008).

Electrical recordings of αHL pores have been carried out from 4 to 93 °C (Kang et al., 2005). Over this range, the conductance of the pore increases roughly linearly, largely reflecting the temperature dependence of the conductivity of the electrolyte. DNA translocation experiments are usually carried out at room temperature, but lower temperatures can be used to reduce the speed of DNA translocation through the pore, by an order of magnitude over a range of 25 °C at +120 mV (Meller et al., 2000), which is most likely an effect of the lowered viscosity. High temperatures might be useful for denaturing DNA during translocation experiments.

Current recordings from αHL pores in symmetric solutions have been performed from pH 4 to 13 (Gu and Bayley, 2000; Maglia et al., 2009b). However, the rate of insertion of preassembled heptamers decreases dramatically at pH values greater than 11 and at pH > 11.7 αHL can no longer be transferred into planar lipid bilayers. CD and fluorescence experiments revealed that the cap domain of the protein is unfolded in 0.3% (w/v) SDS at pH 12, suggesting that correct folding of the cap is important for the insertion of αHL heptamers. Once in lipid bilayers, however, αHL pores appear to be more stable than in SDS solution and there is no evidence to suggest that the protein unfolds within the pH range tested. Therefore, to examine αHL at pH > 11, the pores must be reconstituted at lower pH and the pH of the solution brought to the desired pH by adding small portions of 1 M KOH with stirring. The pH of the *cis* compartment (ground) can be monitored by using a pH electrode. However, measurement of the pH in the *trans* compartment provokes breaking of the bilayer, unless the *trans* electrode is removed or a conducting bridge is first formed with a Pt wire between the two compartments.

αHL pores are also stable in high concentrations of denaturants (e.g., up to 8 M urea), which has been used, for example, to investigate the translocation of unfolded polypeptides (Oukhaled *et al.*, 2007) or to reduce the secondary structure of DNA for translocation (Japrung *et al.*, 2010). Since the insertion of αHL occurs only at concentrations of urea lower than 3 M, higher concentrations of denaturants are attained by replacing the buffer after a pore has inserted into a planar bilayer.

4. MATERIALS

4.1. Buffer components

Typically, DNA translocation experiments are carried out in 1 M KCl. Because of possible contaminants, the KCl and other buffer components should be of the highest possible purity. We use KCl from Fluka (catalog no. 05257) which contains less than 0.0005% of trace metals.

4.2. Handling of DNA and RNA

In many experiments, synthetic DNAs and RNAs of 50–100 nt are used. The capture of these nucleic acids is inefficient (e.g., $\sim 4 \text{ s}^{-1} \text{ } \mu M^{-1}$ at $+120$ mV with WT-αHL) (Henrickson *et al.*, 2000; Maglia *et al.*, 2008; Meller, 2003; Nakane *et al.*, 2002). Therefore, DNA or RNA is usually presented at concentrations of > 100 nM and the use of small chambers can help conserve materials (Akeson *et al.*, 1999). Nucleic acids, especially RNA, should be manipulated under nuclease-free conditions. Materials including water, tips, and tubes should be autoclaved prior to use and the experimental area should be cleaned with a product such as RNaseZap (Ambion), to inactivate RNAse. RNaseZap may also be used on the chamber provided it is thoroughly rinsed afterward. Purified RNA can be stored in 50% (v/v) formamide, which is known to protect RNA from degradation (Chomczynski, 1992).

4.3. Short single-stranded DNA or RNA

Several companies, such as ATDBio, Sigma Genosys, RNAeasy kit, Integrated DNA Technologies, and Bio-Synthesis, carry out the custom synthesis of nucleic acids of up to 200 nt in length for ssDNA and up to 50 nt for RNA at the micromole level. An advantage of using synthetic oligonucleotides is that they can be chemically modified at their 5′ or 3′

ends and that additional bases, such as 5′-methylcytosine, can be incorporated at specific sites. Pure oligonucleotides are essential for nanopore investigations and therefore they are usually desalted by the company, purified by HPLC or PAGE as appropriate for the length, and delivered as dried ethanol precipitates. We have also purified desalted short ssDNA (50–100 nt) and ssRNA (<50 nt) from Sigma Genosys by denaturing, 7 M urea, 8% polyacrylamide gel electrophoresis (PAGE). After electrophoresis, the ssDNA and ssRNA are stained with ethidium bromide (10 μg ml^{-1}) and visualized under a UV transilluminator 2000 (Bio-Rad). The desired band of polyacrylamide is cut out and crushed in water (100 μl). The gel is separated from the eluted ssDNA by using P6 Micro-Spin chromatography columns (Bio-Rad). Purity is checked by analytical PAGE.

4.4. Long single-stranded DNA

The preparation of long ssDNA molecules (>200 nt) is usually achieved by using asymmetric PCR. dsDNA templates of the desired length (200–8000 bp) and sequence are prepared by cutting a plasmid DNA with the requisite restriction enzymes. The desired dsDNA is purified from the plasmid DNA by PAGE. The template (e.g., $\sim$0.4 μg of a 1 kb fragment) is then mixed with a 50- to 100-fold excess of a single synthetic primer that is complementary to a specific sequence in the dsDNA template, and 95 asymmetric PCR cycles are carried out (Screaton et al., 1993). Since only one primer is present, ssDNA is generated, which is then separated from the dsDNA template by PAGE. Alternatively, the primer that is used in the asymmetric PCR can be biotinylated, and the ssDNA can be purified by chromatography with immobilized streptavidin (Kai et al., 1997; Pagratis, 1996). Asymmetric PCR produces only one copy of the template in each cycle and the final yield of ssDNA is generally poor (e.g., $\sim$0.2 μg for the 1 kb template). The DNA is purified by using a 0.8% low-melt agarose gel.

In an alternative method, a dsDNA template (0.4 μg) is amplified in 35 PCR cycles by using two primers, one of which is protected at the 5′ end (e.g., with a covalently attached biotin). In this case, amplification is exponential. After 35 cycles, the resulting dsDNA ($\sim$25 μg from a 50 μl reaction) is desalted by using a gel filtration column (Micro Bio-Spin P30 Column, Bio-Rad), and concentrated to 300–500 ng μl^{-1} with a centrifugal vacuum concentrator (SpeedVac, Savant). Then, to obtain ssDNA, the dsDNA is incubated with λ-exonuclease ($\sim$2 U μg^{-1} of DNA) for 3 h at 37 °C, which digests the unprotected DNA strand. After further purification by PAGE, up to 2 μg of ssDNA are obtained from 5 μg of dsDNA.

4.5. Long RNA preparation

Long RNA molecules can be prepared by transcribing a dsDNA template with an RNA polymerase. RNA molecules up to 8000 nt long can be synthesized. Several companies including Promega and Ambion produce *in vitro* transcription (IVT) kits. For example, the T7 RiboMAX™ Express Large Scale RNA Production System kit (Promega) contains materials sufficient for 12 reactions and can synthesize up to 5 μg of RNA per reaction. When using this kit, the dsDNA template must contain the T7 promoter sequence that is recognized by T7 RNA polymerase. A typical reaction contains linear dsDNA template (8 μl, ~4 μg), 2× buffer (20 μl, with rNTPs), RNase-free water (10 μl), and T7 RNA polymerase (2 μl). The mixture is incubated at 37 °C for 1 h to produce the desired RNA and then the dsDNA template is digested with RQ1 RNase-free DNase (4 μl, 1 U μl^{-1}) for 15 min. The digest is brought to 700 μl by the addition of RNase-free water (up to 100 μl, RNAeasy kit), buffer RLT (350 μl, RNAeasy kit) and ethanol (250 μl, 96–100%). The sample is then loaded onto a RNase Mini spin column (RNAeasy kit) and washed twice with buffer RPE (500 μl, RNAeasy kit). The desired RNA is eluted with "double sterile" water (50 μl, RNAeasy kit). The RNA can be purified in a 5% or 8% polyacrylamide gel, or in a 0.8% agarose gel.

4.6. Short dsDNA preparation

dsDNA can be prepared by mixing equimolar amounts of two synthetic complementary ssDNAs in exonuclease I buffer (67 mM glycine–KOH, 6.7 mM MgCl$_2$, 10 mM 2-mercaptoethanol, pH 9.4; New England Bio-Labs). To facilitate the annealing process, the temperature is brought to 95 °C for 1 min and then decreased stepwise to room temperature. At around the calculated annealing temperature (www.basic.northwestern.edu/biotools/oligocalc.html), the temperature is decreased in small steps (e.g., 2 °C, each held for 1 min). Excess ssDNA is digested by incubation with *E. coli* exonuclease I (20 U, New England BioLabs) for 30 min at 37 °C. The dsDNA is ethanol precipitated by first adding 0.1 volumes of 3.0 M sodium acetate at pH 5.2, and then 2.5 volumes of cold 95% (v/v) ethanol. The mixture is left at −20 °C for 1 h and then centrifuged at 25,000 × g for 20 min. The white pellet is washed several times with cold 95% (v/v) ethanol and after drying taken up in 10 mM Tris–HCl 100 μM EDTA, pH 7.5. Residual low molecular weight contaminants can be removed by passing the DNA solution (maximum 75 μl) through a Bio-Spin® 30 column (Bio-Rad). The purity of dsDNA is checked by PAGE by using dyes that stain both single- and double-stranded nucleic acids, for example, Stains-All (Sigma-Aldrich).

5. Data Acquisition and Analysis

5.1. Data digitization

The ionic current that flows through a nanopore can be measured by using commercial instruments such as the Axopatch 200B patch-clamp amplifier. These instruments transform the continuous analog current signal into an analog voltage signal, which is then fed to a digitizer that converts it into binary numbers with finite resolution (digitization). The digitizer is connected to a computer that converts and displays the digitized current information. We use 132x or 1440A 16-bit digitizers (Molecular Devices), with a dynamic input range of ± 10 V, which divides the signal into 2^{16} (65,536) values (bins), providing a resolution of 0.305 mV/bin (20 V/2^{16}). If no output gain is applied, this corresponds to a dynamic range of ± 10 nA and a resolution of 0.305 pA/bin. However, the output gain can be changed (by up to 500-fold) to increase the resolution (decreased bin width), at the expense of dynamic range. Typically, when measuring the translocation of DNA molecules through a single αHL nanopore the output gain is set to at least 10×, which corresponds to a dynamic range of 1 nA and a resolution of 0.0305 pA/bin.

5.2. Filtering and sampling

The signal collected by the amplifier must be filtered to remove high-frequency electrical noise. The most common filtering is low-pass filtering, which limits the bandwidth of the data by eliminating signals above the corner frequency (or cutoff frequency) of the filter. Bessel filters, rather than Butterworth filters, are usually applied for time-domain analyses, as Bessel filters are better at preserving the shape of the single-channel response (The Axon Guide) (Silberberg and Magleby, 1993). It is important to select the correct corner frequency and in order to keep errors in estimates of current levels to <3%, the applied filter cutoff should be greater than five times the inverse of the mean event time (Silberberg and Magleby, 1993). For example, the most probable translocation time for a 92-nt DNA strand through the WT-αHL, at +120 mV in 1 M KCl, 25 mM Tris–HCl at pH 8, is 0.141 ms (Maglia et al., 2008). Therefore, the cutoff frequency should be ~35 kHz or higher (Fig. 22.8A). Over-filtering of the signal can lead to inaccurate estimates of I_B and t_t (Fig. 22.8B). As well as the appropriate filtering rate, it is important to select the correct sampling rate. The Nyquist Sampling Theorem states that the sampling rate should be twice the cutoff frequency of the applied filter (The Axon Guide). In practice, however, it is preferable to oversample (i.e., sample at a rate that exceeds the Nyquist minimum) (Sakmann and Neher, 1995). A sampling rate of 10 times that of

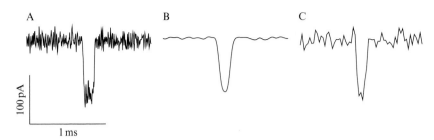

Figure 22.8 The effects of the filter cutoff frequency and sampling rate on the recording of DNA translocation events. The t_t for each event is ~ 0.14 ms suggesting a filter cutoff of ~ 35 kHz (see the text). (A) Signal filtered at 40 kHz with a low-pass 8-pole Bessel filter, and sampled at 250 kHz. The filtering and sampling allow faithful reproduction of the translocation event. (B) The event shown in panel A is filtered at 5 kHz. The sampling rate remains 250 kHz. I_B and t_t can no longer be determined accurately. (C) In this case, although the correct level of filtering has been applied (40 kHz), the signal is sampled at the suboptimal rate of 40 kHz.

the cutoff frequency of the filter is desirable, but five times the cutoff frequency is usually sufficient.

5.3. Acquisition protocols

The software we use to collect and display the current signals is Clampex from Molecular Devices; however, other software such as QuB (www.qub.buffalo.edu) and WinEDR (http://spider.science.strath.ac.uk/sipbs/page.php?show=software_winEDR) can be used. Clampex can also control the 132x or 1440A digitizer to apply a potential through the amplifier (see below). Although Clampex has five data acquisition modes, the most commonly used during DNA translocation experiments, are the "Gap Free," "Variable-Length Events," and "Episodic Stimulation" modes.

In the Gap Free protocol, data are passively and continuously digitized, displayed, and saved. There are no interruptions to the data record, which make it a useful protocol to use when preparing data for publication. However, it should not be used routinely when acquiring with a high sampling rate, or for a long period of time, as the file sizes can become very large (e.g., acquiring at 250 kHz for 1 h will produce a 1.7 GB file).

The Variable-Length Events protocol is ideal for collecting data from a fully characterized system where the frequency of events is very low, and there are long periods of inactivity. Data are visualized as in the gap free protocol, but are acquired only for as long as an input signal has passed a threshold level (which is set manually), so only the "events" and small sections on either side of them are recorded. The file sizes are therefore more manageable, although the data record is interrupted.

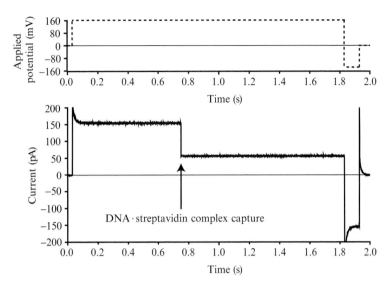

Figure 22.9 Episodic stimulation mode allows both an output voltage (dashed line) to be applied, while simultaneously recording the resulting changes in ionic current. In this example, a positive potential bias of +160 mV is applied to the system, resulting in an open pore current level of ~160 pA. Under the applied potential, a DNA molecule is driven into the pore and becomes immobilized (through a terminal biotin·streptavidin complex), and the current level is reduced. The potential bias is reversed (−140 mV) and the DNA molecule is ejected. The bias is then removed. The protocol can be repeated numerous times.

The Episodic Stimulation protocol can be used to output a voltage signal from the digitizer, which can be applied to the bilayer system, while simultaneously acquiring the resulting current signal. This mode is used to apply voltage steps (e.g., when producing I–V curves) or voltage ramps (Dudko et al., 2007). It can also be used when it is laborious to apply different voltages manually with the patch-clamp amplifier, for example, when studying the repeated immobilization of DNA molecules inside the αHL pore. In this case, a voltage bias can be applied, for a fixed amount of time, to drive a DNA·streptavidin complex into the pore. The bias is then reversed to eject the molecule from the pore (Fig. 22.9) (Purnell et al., 2008; Stoddart et al., 2009, 2010). The changes in the current level that occur as a result of the voltage protocol are simultaneously acquired. The protocol can be repeated hundreds of times per run and therefore many DNA blockades can be recorded. The output waveform can be set manually to apply different voltage steps, for different times, and it does not have to be the same for each sweep during a run.

5.4. Analysis of single DNA/RNA molecules

The current traces recorded by the digitizer can be analyzed by using the "single-channel search" option in the Clampfit software package (Molecular Devices). In doing so, the user must assign the value of the open pore current

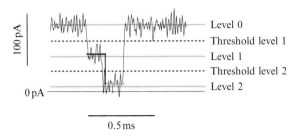

Figure 22.10 Typical DNA translocation event analyzed by the Clampfit software. Levels 0, 1, and 2 are set manually and the software automatically assigns data points to level 1 (red line) or level 2 (green line). (See Color Insert.)

(level 0) and a second level (level 1) that corresponds to the blocked pore level. If a DNA molecule provokes several types of current blockade, of differing amplitude, additional levels can be assigned (levels 2, 3, ..., etc.). Every time the ionic current crosses a threshold (the half distance between two levels), the software recognizes it as an "event"(Fig. 22.10) and records its characteristics such as the mean amplitude, the dwell time, and the interevent interval in a results window. Generally, for short DNA molecules, the average current amplitudes, dwell times, and interevent intervals of numerous events are plotted as histograms. Gaussian fits to the events histogram of the amplitudes and dwell times are used to determine I_B and t_P, respectively; while $\bar{t}_{ON}$ is determined from an exponential fit to the histogram of the interevent intervals. The dwell times (t_D) are described by a Gaussian distribution with an exponential tail; the peak of the Gaussian (t_P) is generally used to describe the most likely dwell time, which often represents the translocation time, t_t (Meller et al., 2000). In the case of DNA strands immobilized within a nanopore, I_B values are the only meaningful features of the events (Ashkenasy et al., 2005; Purnell and Schmidt, 2009; Stoddart et al., 2009).

REFERENCES

Akeson, M., Branton, D., Kasianowicz, J. J., Brandin, E., and Deamer, D. W. (1999). Microsecond time-scale discrimination among polycytidylic acid, polyadenylic acid and polyuridylic acid as homopolymers or as segments within single RNA molecules. *Biophys. J.* **77**, 3227–3233.

Ashkenasy, N., Sánchez-Quesada, J., Bayley, H., and Ghadiri, M. R. (2005). Recognizing a single base in an individual DNA strand: A step toward nanopore DNA sequencing. *Angew. Chem. Int. Ed. Engl.* **44**, 1401–1404.

Astier, Y., Kainov, D. E., Bayley, H., Tuma, R., and Howorka, S. (2007). Stochastic detection of motor protein–RNA complexes by single-channel current recording. *ChemPhysChem* **8**, 2189–2194.

Bayley, H. (2006). Sequencing single molecules of DNA. *Curr. Opin. Chem. Biol.* **10**, 628–637.
Bayley, H., and Cremer, P. S. (2001). Stochastic sensors inspired by biology. *Nature* **413**, 226–230.
Bayley, H., Cronin, B., Heron, A., Holden, M. A., Hwang, W., Syeda, R., Thompson, J., and Wallace, M. (2008a). Droplet interface bilayers. *Mol. BioSyst.* **4**, 1191–1208.
Bayley, H., Luchian, T., Shin, S.-H., and Steffensen, M. B. (2008b). Single-molecule covalent chemistry in a protein nanoreactor. *In* "Single Molecules and Nanotechnology," (R. Rigler and H. Vogel, eds.), pp. 251–277. Springer, Heidelberg.
Bezrukov, S. M., and Kasianowicz, J. J. (1993). Current noise reveals protonation kinetics and number of ionizable sites in an open protein ion channel. *Phys. Rev. Lett.* **70**, 2352–2355.
Bhakdi, S., Füssle, R., and Tranum-Jensen, J. (1981). Staphylococcal α-toxin: Oligomerization of hydrophilic monomers to form amphiphilic hexamers induced through contact with deoxycholate micelles. *Proc. Natl. Acad. Sci. USA* **78**, 5475–5479.
Bonthuis, D. J., Zhang, J., Hornblower, B., Mathe, J., Shklovskii, B. I., and Meller, A. (2006). Self-energy-limited ion transport in subnanometer channels. *Phys. Rev. Lett.* **97**, 128104.
Branton, D., Deamer, D. W., Marziali, A., Bayley, H., Benner, S. A., Butler, T., Di Ventra, M., Garaj, S., Hibbs, A., Huang, X., Jovanovich, S. B., Krstic, P. S., *et al.* (2008). The potential and challenges of nanopore sequencing. *Nat. Biotechnol.* **26**, 1146–1153.
Butler, T. Z., Gundlach, J. H., and Troll, M. (2007). Ionic current blockades from DNA and RNA molecules in the alpha-hemolysin nanopore. *Biophys. J.* **93**, 3229–3240.
Butler, T. Z., Pavlenok, M., Derrington, I. M., Niederweis, M., and Gundlach, J. H. (2008). Single-molecule DNA detection with an engineered MspA protein nanopore. *Proc. Natl. Acad. Sci. USA* **105**, 20647–20652.
Cheley, S., Malghani, M. S., Song, L., Hobaugh, M., Gouaux, J. E., Yang, J., and Bayley, H. (1997). Spontaneous oligomerization of a staphylococcal α-hemolysin conformationally constrained by removal of residues that form the transmembrane β barrel. *Protein Eng.* **10**, 1433–1443.
Cheley, S., Braha, O., Lu, X., Conlan, S., and Bayley, H. (1999). A functional protein pore with a "retro" transmembrane domain. *Protein Sci.* **8**, 1257–1267.
Cheley, S., Gu, L.-Q., and Bayley, H. (2002). Stochastic sensing of nanomolar inositol 1,4,5-trisphosphate with an engineered pore. *Chem. Biol.* **9**, 829–838.
Chomczynski, P. (1992). Solubilization in formamide protects RNA from degradation. *Nucleic Acids Res.* **20**, 3791–3792.
Clarke, J., Wu, H., Jayasinghe, L., Patel, A., Reid, S., and Bayley, H. (2009). Continuous base identification for single-molecule nanopore DNA sequencing. *Nat. Nanotechnol.* **4**, 265–270.
Cockroft, S. L., Chu, J., Amorin, M., and Ghadiri, M. R. (2008). A single-molecule nanopore device detects DNA polymerase activity with single-nucleotide resolution. *J. Am. Chem. Soc.* **130**, 818–820.
Deamer, D. W., and Branton, D. (2002). Characterization of nucleic acids by nanopore analysis. *Acc. Chem. Res.* **35**, 817–825.
Dekker, C. (2007). Solid-state nanopores. *Nat. Nanotechnol.* **2**, 209–215.
Dudko, O. K., Mathe, J., Szabo, A., Meller, A., and Hummer, G. (2007). Extracting kinetics from single-molecule force spectroscopy: Nanopore unzipping of DNA hairpins. *Biophys. J.* **92**, 4188–4195.
Gillis, K. D. (1995) *In* "Techniques for membrane capacitance measurements," (B. Sakmann, and E. Neher, ed), pp. 155, Vol. 2. Plenum, New York.

Gu, L. Q., and Bayley, H. (2000). Interaction of the non-covalent molecular adapter, b-cyclodextrin, with the staphylococcal α-hemolysin pore. *Biophys. J.* **79**, 1967–1975.

Gu, L. Q., Cheley, S., and Bayley, H. (2001). Prolonged residence time of a noncovalent molecular adapter, β-cyclodextrin, within the lumen of mutant α-hemolysin pores. *J. Gen. Physiol.* **118**, 481–494.

Hanke, W., and Schlue, W.-R. (1993). Planar Lipid Bilayers. Academic Press, London.

Henrickson, S. E., Misakian, M., Robertson, B., and Kasianowicz, J. J. (2000). Driven DNA transport into an asymmetric nanometer-scale pore. *Phys. Rev. Lett.* **85**, 3057–3060.

Holden, M. A., and Bayley, H. (2005). Direct introduction of single protein channels and pores into lipid bilayers. *J. Am. Chem. Soc.* **127**, 6502–6503.

Holden, M. A., Jayasinghe, L., Daltrop, O., Mason, A., and Bayley, H. (2006). Direct transfer of membrane proteins from bacteria to planar bilayers for rapid screening by single-channel recording. *Nat. Chem. Biol.* **2**, 314–318.

Hornblower, B., Coombs, A., Whitaker, R. D., Kolomeisky, A., Picone, S. J., Meller, A., and Akeson, M. (2007). Single-molecule analysis of DNA–protein complexes using nanopores. *Nat. Methods* **4**, 315–317.

Howorka, S., Cheley, S., and Bayley, H. (2001). Sequence-specific detection of individual DNA strands using engineered nanopores. *Nat. Biotechnol.* **19**, 636–639.

Hung, W. C., Chen, F. Y., and Huang, H. W. (2000). Order–disorder transition in bilayers of diphytanoyl phosphatidylcholine. *Biochim. Biophys. Acta* **1467**, 198–206.

Japrung, D., Henricus, M., Li, Q., Maglia, G., and Bayley, H. (2010). Urea facilitates the translocation of single-stranded DNA and RNA through the α-hemolysin nanopore. *Biophys. J.* **98**, 1856–1863.

Kado, R. T. (1993). Membrane area and electrical capacitance. *Methods Enzymol.* **221**, 273–299.

Kai, E., Sawata, S., Ikebukuro, K., Iida, T., Honda, T., and Karube, I. (1997). Novel DNA detection system of flow injection analysis (2). The distinctive properties of a novel system employing PNA (peptide nucleic acid) as a probe for specific DNA detection. *Nucleic Acids Symp. Ser.* 321–322.

Kang, X., Gu, L.-Q., Cheley, S., and Bayley, H. (2005). Single protein pores containing molecular adapters at high temperatures. *Angew. Chem. Int. Ed. Engl.* **44**, 1495–1499.

Kasianowicz, J. J., Brandin, E., Branton, D., and Deamer, D. W. (1996). Characterization of individual polynucleotide molecules using a membrane channel. *Proc. Natl. Acad. Sci. USA* **93**, 13770–13773.

Kawano, R., Schibel, A. E., Cauley, C., and White, H. S. (2009). Controlling the translocation of single-stranded DNA through alpha-hemolysin ion channels using viscosity. *Langmuir* **25**, 1233–1237.

Keyser, U. F., Koeleman, B. N., van Dorp, S., Krapf, D., Smeets, R. M. M., Lemay, S. G., Dekker, N. H., and Dekker, C. (2006). Direct force measurements on DNA in a solid-state nanopore. *Nat. Phys.* **2**, 473–477.

Korchev, Y. E., Bashford, C. L., Alder, G. M., Apel, P. Y., Edmonds, D. T., Lev, A. A., Nandi, K., Zima, A. V., and Pasternak, C. A. (1997). A novel explanation for fluctuations of ion current through narrow pores. *FASEB J.* **11**, 600–608.

Krasne, S., Eisenman, G., and Szabo, G. (1971). Freezing and melting of lipid bilayers and the mode of action of nonactin, valinomycin, and gramicidin. *Science* **174**, 412–415.

Lindsey, H., Petersen, N. O., and Chan, S. I. (1979). Physicochemical characterization of 1,2-diphytanoyl-sn-glycero-3-phosphocholine in model membrane systems. *Biochim. Biophys. Acta* **555**, 147–167.

Maglia, G., Rincon Restrepo, M., Mikhailova, E., and Bayley, H. (2008). Enhanced translocation of single DNA molecules through α-hemolysin nanopores by manipulation of internal charge. *Proc. Natl. Acad. Sci. USA* **105**, 19720–19725.

Maglia, G., Heron, A. J., Hwang, W. L., Holden, M. A., Mikhailova, E., Li, Q., Cheley, S., and Bayley, H. (2009a). Droplet networks with incorporated protein diodes show collective properties. *Nat. Nanotechnol.* **4**, 437–440.
Maglia, M., Henricus, M., Wyss, R., Li, Q., Cheley, S., and Bayley, H. (2009b). DNA strands from denatured duplexes are translocated through engineered protein nanopores at alkaline pH. *Nano Lett.* **9**, 3831–3836.
Martin, H., Kinns, H., Mitchell, N., Astier, Y., Madathil, R., and Howorka, S. (2007). Nanoscale protein pores modified with PAMAM dendrimers. *J. Am. Chem. Soc.* **129**, 9640–9649.
Mayer, M., Kriebel, J. K., Tosteson, M. T., and Whitesides, G. M. (2003). Microfabricated teflon membranes for low-noise recordings of ion channels in planar lipid bilayers. *Biophys. J.* **85**, 2684–2695.
Meller, A. (2003). Dynamics of polynucleotide transport through nanometre-scale pores. *J. Phys.: Condens. Matter* **15**, R581–R607.
Meller, A., and Branton, D. (2002). Single molecule measurements of DNA transport through a nanopore. *Electrophoresis* **23**, 2583–2591.
Meller, A., Nivon, L., Brandin, E., Golovchenko, J., and Branton, D. (2000). Rapid nanopore discrimination between single polynucleotide molecules. *Proc. Natl. Acad. Sci. USA* **97**, 1079–1084.
Meller, A., Nivon, L., and Branton, D. (2001). Voltage-driven DNA translocations through a nanopore. *Phys. Rev. Lett.* **86**, 3435–3438.
Miles, G., Bayley, H., and Cheley, S. (2002). Properties of *Bacillus cereus* hemolysin II: A heptameric transmembrane pore. *Protein Sci.* **11**, 1813–1824.
Miller, C. (ed.), (1986). Ion Channel Reconstitution. Plenum, New York.
Mitchell, N., and Howorka, S. (2008). Chemical tags facilitate the sensing of individual DNA strands with nanopores. *Angew. Chem. Int. Ed. Engl.* **47**, 5565–5568.
Montal, M., and Mueller, P. (1972). Formation of bimolecular membranes from lipid monolayers and study of their electrical properties. *Proc. Natl. Acad. Sci. USA* **69**, 3561–3566.
Morera, F. J., Vargas, G., Gonzalez, C., Rosenmann, E., and Latorre, R. (2007). Ion-channel reconstitution. *Methods Mol. Biol.* **400**, 571–585.
Nakane, J., Akeson, M., and Marziali, A. (2002). Evaluation of nanopores as candidates for electronic analyte detection. *Electrophoresis* **23**, 2592–2601.
Nakane, J., Wiggin, M., and Marziali, A. (2004). A nanosensor for transmembrane capture and identification of single nucleic acid molecules. *Biophys. J.* **87**, 615–621.
Niles, W. D., Levis, R. A., and Cohen, F. S. (1988). Planar bilayer membranes made from phospholipid monolayers form by a thinning process. *Biophys. J.* **53**, 327–335.
Oukhaled, G., Mathé, J., Biance, A.-L., Bacri, L., Betton, J.-M., Lairez, D., Pelta, J., and Auvray, L. (2007). Unfolding of proteins and long transient conformations detected by single nanopore recording. *Phys. Rev. Lett.* **98**, 158101.
Oukhaled, G., Bacri, L., Mathé, J., Pelta, J., and Auvray, L. (2008). Effect of screening on the transport of polyelectrolytes through nanopores. *Europhys. Lett.* **82**, 48003(1–5).
Pagratis, N. C. (1996). Rapid preparation of single stranded DNA from PCR products by streptavidin induced electrophoretic mobility shift. *Nucleic Acids Res.* **24**, 3645–3646.
Peng, H., and Ling, X. S. (2009). Reverse DNA translocation through a solid-state nanopore by magnetic tweezers. *Nanotechnology* **20**, 185101.
Peterman, M. C., Ziebarth, J. M., Braha, O., Bayley, H., Fishman, H. A., and Bloom, D. A. (2002). Ion channels and lipid bilayer membranes under high potentials using microfabricated apertures. *Biomed. Microdevices* **4**, 231–236.
Phillips, R., Ursell, T., Wiggins, P., and Sens, P. (2009). Emerging roles for lipids in shaping membrane–protein function. *Nature* **459**, 379–385.

Purnell, R. F., Mehta, K. K., and Schmidt, J. J. (2008). Nucleotide identification and orientation discrimination of DNA homopolymers immobilized in a protein nanopore. *Nano Lett.* **9**, 3029–3034.

Purnell, R. F., and Schmidt, J. J. (2009). Discrimination of single base substitutions in a DNA strand immobilized in a biological nanopore. *ACS Nano* **3**, 2533–2538.

Purves, R. D. (1981). Microelectrode Methods for Intracellular Recording and Ionophoresis. Academic Press.

Rhee, M., and Burns, M. A. (2007). Nanopore sequencing technology: Nanopore preparations. *Trends Biotechnol.* **25**, 174–181.

Rincon-Restrepo, M., Mikhailova, E., Bayley, H., and Maglia, G. Controlled translocation of single DNA molecules through engineered protein nanopores. In preparation.

Sakmann, B., and Neher, B. (1995). Single-Channel Recording. Plenum, New York.

Screaton, G. R., Bangham, C. R., and Bell, J. I. (1993). Direct sequencing of single primer PCR products: A rapid method to achieve short chromosomal walks. *Nucleic Acids Res.* **21**, 2263–2264.

Sexton, L. T., Horne, L. P., and Martin, C. R. (2007). Developing synthetic conical nanopores for biosensing applications. *Mol. BioSyst.* **3**, 667–685.

Silberberg, S. D., and Magleby, K. L. (1993). Preventing errors when estimating single channel properties from the analysis of current fluctuations. *Biophys. J.* **65**, 1570–1584.

Song, L., Hobaugh, M. R., Shustak, C., Cheley, S., Bayley, H., and Gouaux, J. E. (1996). Structure of staphylococcal α-hemolysin, a heptameric transmembrane pore. *Science* **274**, 1859–1865.

Stoddart, D., Heron, A., Mikhailova, E., Maglia, G., and Bayley, H. (2009). Single nucleotide discrimination in immobilized DNA oligonucleotides with a biological nanopore. *Proc. Natl. Acad. Sci. USA* **106**, 7702–7707.

Stoddart, D., Maglia, G., Mikhailova, E., Heron, A., and Bayley, H. (2010). Multiple base-recognition sites in a biological nanopore—Two heads are better than one. *Angew. Chem. Int. Ed.* **49**, 556–559.

Tobkes, N., Wallace, B. A., and Bayley, H. (1985). Secondary structure and assembly mechanism of an oligomeric channel protein. *Biochemistry* **24**, 1915–1920.

Vercoutere, W., Winters-Hilt, S., Olsen, H., Deamer, D., Haussler, D., and Akeson, M. (2001). Rapid discrimination among individual DNA hairpin molecules at single-nucleotide resolution using an ion channel. *Nat. Biotechnol.* **19**, 248–252.

White, S. (1975). Phase transitions in planar bilayer membranes. *Biophys. J.* **15**, 95–117.

White, S. H. (1986). *In* "The Physical Nature of Planar Lipid Membranes Ion Channel Reconstitution," (C. Miller, ed.), pp. 3–35. Plenum, New York.

White, S. H., Petersen, D. C., Simon, S., and Yafuso, M. (1976). Formation of planar bilayer membranes from lipid monolayers. A critique. *Biophys. J.* **16**, 481–489.

Wonderlin, W. F., Finkel, A., and French, R. J. (1990). Optimizing planar lipid bilayer single-channel recordings for high resolution with rapid voltage steps. *Biophys. J.* **58**, 289–294.

Author Index

A

Aathavan, K., 536
Abbondanzieri, E. A., 2, 150, 378, 396, 410, 424, 433
Abel, N. H., 235
Abrahams, J. P., 281
Abràmoff, M. D., 52
Abramowitz, M., 573
Adachi, K., 286, 287
Adams, J., 542
Adams, S. R., 4, 32, 62, 88, 89, 96, 122, 456, 457
Adhya, S., 200, 204, 209, 214, 215, 217, 379
Agard, D. A., 3, 22
Agarwal, H., 93, 178, 182, 184
Ahlers, M., 551
Ahrer, W., 130
Aitken, C. E., 7, 130
Aizpurua, J., 178
Akeson, M., 566, 586, 594–598, 600, 613
Aksimentiev, A., 587
Alani, E., 429
Alberts, B. M., 260
Alder, G. M., 603
Aldous, D., 241, 242
Aleman, E. A., 122
Alexandre, M. T., 459
Alexandrovich, A., 406
Alexiev, U., 457
Al-Hashimi, H. M., 122
Ali, B. M. J., 92
Ali, M. Y., 87, 96, 323, 324, 367
Alivisatos, A. P., 62, 66, 87, 92, 93, 177–179, 181, 182, 184, 186, 195, 457, 468
Allan, V. J., 428
Allemand, J.-F., 268, 269, 297–319, 400, 420
Allen, K. E., 125
Allen, M., 519
Allen, N. S., 85
Allen, R. D., 85
Allersma, M. W., 395
Allison, W. S., 283
Alpert, N. R., 322
Al-Soufi, W., 457
Altan-Bonnet, N., 122
Altman, D., 324
Ambrose, W. P., 28, 468, 471, 482
Amerstorfer, A., 529

Amirgoulova, E. V., 551
Amit, R., 92
Amorin, M., 600
Amro, N. A., 544
Anderson, C. M., 115
Anderson, D. L., 536
Anderson, L. M., 215, 217
Ando, R., 36, 37, 110, 111, 117
Ando, T., 541–561
Andreasson, J. O. L., 158, 377–402, 434
Andrecka, J., 499
Andresen, M., 43, 62
Andrews, D. L., 461, 463
Andrey Revyakin, T. R. S., 300
Angulo, J., 519
Anselmetti, D., 396
Antal, T., 127, 142
Antelman, J., 65, 85
Anthony, P. C., 389, 406
Antonik, M., 457, 459, 460, 464, 468, 470, 481, 485, 486, 488–490, 492–494, 498
Antoranz Contera, S., 542, 558
Aoki, T., 555
Apel, P. Y., 603
Appel, B., 473, 489, 492, 499
Appleyard, S. T., 117
Arai, Y., 459
Ariga, T., 285, 322
Arinstein, A., 578
Armani, M., 157, 161, 169
Arnold, E., 516
Arnold, S., 151
Aroyo, M., 309
Asbury, C. L., 8, 343, 395, 410, 411
Ashkenasy, N., 598, 619
Ashkin, A., 2, 3, 325, 380, 430
Astier, Y., 600
Atkins, A. R., 525
Auerbach, A., 200, 214
Auer, M., 122
Au, L., 195
Aussel, L., 309
Aussenegg, F. R., 178, 179
Austin, R. H., 459
Auvray, L., 612, 613
Averett, L., 3
Axelrod, D., 3, 89
Aziz, M. J., 587

625

B

Babcock, H. P., 122, 176, 444
Bacri, L., 612, 613
Badieirostami, M., 55
Badt, D., 520
Bahar, I., 459
Bai, L., 318
Bakajin, O., 150
Balasubramanian, S., 488
Balci, H., 3, 15, 63, 122, 406, 542
Bald, D., 284
Baldwin, J. M., 558
Balhorn, R., 429
Ball, F. G., 217
Bandyopadhyay, S., 283
Bangham, C. R., 614
Bannwarth, W., 473, 489, 492, 499
Barak, L. S., 84, 122, 123
Barch, M., 406
Barclay, P. E., 178
Barkai, E., 142
Barre, F. X., 297–319
Barroso, M., 505
Barry, J., 260
Barsch, H., 464
Bartoo, M. L., 323, 339
Basché, T., 28
Bashford, C. L., 603
Baskin, R. J., 429
Basu, S. R., 195
Bates, M., 3, 18, 19, 29, 35, 40, 41, 51, 111, 123, 137, 542, 567, 569
Bath, J., 125, 126
Baumgartner, W., 2, 4, 84, 124, 519, 528, 529, 532
Bawendi, M. G., 66
Bax, A., 464, 499
Bayley, H., 577, 578, 591–619
Beane, G. L., 52, 53
Beaufils, F., 114
Beausang, J. F., 200, 204, 205, 209, 211, 214
Becker, W., 457, 471, 496, 505
Beckmann, E., 558
Beck, S., 17
Bedinger, P., 260
Behnke-Parks, W. M., 406
Behr, B., 62
Beljonne, D., 470
Bellaiche, Y., 83, 84, 87, 95
Bell, C. D., 587
Bell, G. I., 528, 572, 574, 575, 579
Bell, J. I., 614
Belov, V., 37
Benesch, R. E., 130
Benkovic, S. J., 176, 297–319
Benner, S. A., 593, 594, 599
Bennett, B. T., 51, 55

Benning, M. M., 323
Bennink, M. L., 339
Bensimon, A., 298, 299, 308, 311
Bensimon, D., 268, 269, 297–319, 400, 402, 420
Bentolila, L. A., 4, 22, 62, 84, 122
Berger, J. M., 379
Berger, S., 457, 459, 460, 464, 468, 470–472, 481, 485
Berg, H. C., 152, 157, 378
Berglund, A. J., 156, 159, 160, 169
Bergman, K., 432
Berg-Sørensen, K., 322, 356–358, 412
Berland, K. M., 468
Berlier, J. E., 445
Bernard, J., 3, 457
Berne, B. J., 406
Berney, C., 459, 505
Bertaux, N., 203
Bertoni, G., 200
Bertozzi, C. R., 62
Best, R. B., 464, 499
Betton, J.-M., 613
Betzig, E., 3, 18, 28–31, 33, 43, 62, 93, 110, 112–114, 117, 118, 123, 124
Beuron, F., 557
Bewersdorf, J., 51–53, 55
Bezrukov, S. M., 603
Bhakdi, S., 610
Bhunia, A. K., 62
Biance, A.-L., 613
Bianco, P. R., 429, 440
Biebricher, A., 430
Bigot, S., 312
Bilbao-Cortes, D., 557
Bishop, A. I., 391
Biteen, J. S., 27–55
Bjorkholm, J. E., 2, 3, 380, 430
Bjornson, K. P., 318
Blaas, D., 515–536
Blab, G. A., 35, 37
Blainey, P. C., 176
Blake, K., 518
Blanchard, S. C., 457
Blankenburg, R., 551
Blass, D., 522
Bleckmann, A., 457, 505
Blehm, B., 83, 86
Block, S. M., 2, 3, 8, 10, 82, 83, 103, 110, 158, 170, 223, 226, 241–243, 299, 311, 322, 325, 330, 332, 339, 343, 344, 356, 357, 367, 370, 377–402, 406, 409–411, 424, 429, 432–436, 440, 566
Blom, H., 470
Bloodgood, R. A., 83
Bloom, D. A., 606
Blosser, T. R., 18, 35
Blumberg, W. E., 460
Bobroff, N., 2, 124

Bockelmann, U., 431, 434, 435, 443, 586
Bock, H., 43
Boder, E. T., 64, 66, 71, 72
Bokinsky, G., 150
Bomar, B., 156, 159, 161
Bonifacino, J. S., 3, 18, 29, 43, 62, 93, 110
Bonthuis, D. J., 595
Bordallo, J., 557
Borgia, A., 459
Borken, C., 529
Borsch, M., 280, 457
Bosco, D. A., 459
Bossi, M., 37
Bottka, S., 384, 385, 390
Bouchiat, C., 311
Boukobza, E., 150
Bouwer, J., 89
Bowler, M. W., 283
Bowman, G. R., 31, 35, 43, 45–47, 49
Boyer, P. D., 281
Bradford, J., 445
Bradshaw, D. S., 461
Brady, S. T., 85
Braff, R. A., 151
Braha, O., 606, 610
Braig, K., 281, 283
Brakenhoff, G. J., 28
Brandenburg, B., 20, 35, 37, 43, 84
Brandin, E., 566, 594–598, 612, 613, 619
Brand, L., 457, 468, 471, 472, 475, 476, 488, 490
Brandl, D. W., 178
Brandmeier, B., 406
Brandt, A., 97
Branton, D., 566, 578, 579, 587, 593–599, 612, 613, 619
Braslavsky, I., 92
Braslavsky, S. E., 461, 463
Brauchle, C., 457, 468, 472
Brau, R. R., 406, 424, 440
Bredas, J. L., 470
Brenner, H., 412
Brereton, I. M., 525
Brewer, L. R., 429, 440
Brichta, J., 65
Briggs, L. C., 557
Brisson, A. R., 546, 551, 555
Brocchieri, L., 170
Brochon, J., 456
Broggio, C., 200, 201, 214
Brogioli, D., 204, 205, 209, 214
Brooks, R. A., 127
Brower-Toland, B. D., 406
Brown, A. E. X., 83, 86
Brown, C. M., 457, 505
Brown, P. O., 525
Bruchez, M., 62, 457
Bruchez, M. P., 22
Bruckner, F., 499

Brune, D., 155
Brunger, A. T., 498, 557
Bruno, M. M., 8
Bryant, Z., 151, 170, 269, 324, 381
Buchinger, G. M., 529
Buchner, J., 457
Buldt, G., 519
Buller, G., 445
Bünemann, M., 62
Buranachai, C., 63, 122, 406, 542
Burns, J. E., 325
Burns, M. A., 567, 569, 594
Burov, S., 142
Burr, T., 215, 217
Buschkamp, P., 90, 92
Busch, M., 97
Buschmann, V., 459
Bustamante, C., 3, 151, 170, 221–255, 268, 269, 298, 311, 318, 322, 332, 338, 339, 341, 343, 344, 346, 347, 356, 381, 406, 418, 424, 428, 429, 433, 434, 447, 459, 536, 566
Butler, T. Z., 593–596, 599
Butt, H. J., 532

C

Cai, D. W., 4, 81–103, 444
Cai, L., 150
Cai, Y., 62, 114
Caldwell, W. B., 150, 176
Campbell, R. E., 37, 87
Cande, W. Z., 3
Canfield, B., 156, 159, 161
Cang, H., 156, 157, 160, 169, 201
Cappello, G., 83, 84, 87, 95, 236, 240
Carballido-López, R., 45
Carlson, B. L., 62
Carlton, P. M., 3
Carrico, I. S., 62
Carrion-Vazquez, M., 528, 566
Carter, A. A., 217
Carter, A. P., 8
Carter, A. R., 342, 343, 396, 424
Carter, N. J., 236
Carugo, O., 527
Cauley, C., 600
Cecconi, C., 236, 406, 566
Cermelli, S., 83
Ceska, T. A., 558
Chadd, E. H., 432
Chandler, M., 200
Chang, D. J., 429
Chang, M. I., 528
Chang, Y-P., 61–76, 85, 88
Chan, S. I., 607
Chan, T.-F., 444
Chan, V. Z. H., 179
Chao, G., 71, 72

Charvin, G., 223, 308
Chastain, II, P. D. 260
Chaudhary, S., 157, 161, 169
Cheezum, M. K., 8, 103, 111, 115
Cheley, S., 577, 578, 594, 597, 608, 610–612
Chemla, D. S., 16, 122, 150, 176, 413, 457, 468, 482
Chemla, Y. R., 3, 221–255, 322, 332, 338, 339, 341, 343, 344, 346, 347, 356, 406, 424, 433, 434, 536
Cheney, R. E., 323, 339, 358, 361
Chen, F. Y., 607
Cheng, M. A., 457
Cheng, W., 318
Chen, J. H., 587
Chen, J. Y., 195
Chen, L. Q., 8
Chennubhotla, C., 459
Chen, P., 83, 84, 86
Chen, S. Y., 126, 127, 150, 457, 460, 461, 468, 472, 505
Chepurnykh, T. V., 113
Cherayil, B. J., 122, 459
Cherry, R. J., 84, 115
Cheung, C. Y., 445
Chhun, B. B., 3
Chiba, P., 519
Chichester, R. J., 3, 123, 124
Chizmadezhev, Y. A., 536
Chmyrov, A., 470
Choi, E. J., 379
Choi, U. B., 498
Chomczynski, P., 613
Chou, Y. H., 2, 11, 12, 84, 85, 95
Chow, A., 587
Chowdhury, A., 429
Chtcheglova, L., 534, 535
Chudakov, D. M., 36, 37, 62
Chu, J., 600
Chung, J., 8, 379
Churchman, L. S., 3, 15–17, 95, 122, 125, 137, 139, 323
Chu, S., 2, 3, 122, 150, 176, 339, 380, 416, 430, 444, 498
Cimprich, K. A., 429
Cisse, I., 150
Claridge, S. A., 195
Clarke, J., 459, 599
Clarke, R., 488
Clarke, S., 64
Clark, S. W., 22
Clegg, R. M., 457, 502
Cleland, A. N., 542
Clemen, A. E., 236, 240
Clima, L., 473, 489, 492, 499
Cockroft, S. L., 600
Cognet, L., 35, 37
Cohen, A. E., 149–171

Cohen, C., 324
Cohen, F. S., 606
Colbert, D. T. R., 561
Colquhoun, D., 200, 201, 214, 215
Coluccio, L. M., 361
Conlan, S., 610
Conley, N. R., 27–55
Cooker, G., 460, 461
Coombs, A., 586, 600
Cornish, P. V., 222, 236
Cornish, V. W., 62, 114
Coronado, E., 177, 195
Correa, I. R., Jr., 114
Corzett, M., 429
Costa, L. T., 534, 535
Cotlet, M., 470, 476
Cotton, S. L., 83
Courty, S., 83, 84, 87, 95, 96
Cox, E. C., 457
Cozzarelli, N. R., 170, 381
Craik, C. S., 92
Cramer, P., 499
Crampton, D. J., 176
Cremer, C., 28, 123
Cremer, P. S., 593
Crocker, J. C., 115, 116
Cronin, B., 605
Croquette, V., 151, 268, 269, 297–319, 400, 402, 420
Cross, R. A., 236
Cubitt, A. B., 4, 35, 43
Cui, Q., 283, 459
Cui, Y., 381, 447
Curutchet, C., 457
Cypionka, A., 487

D

Dabbousi, R. O., 66
Dabby, N., 123, 125, 127–129, 133, 136–140, 142, 143
Dahan, M., 63, 64, 83, 84, 87, 95, 122
Dai, H., 561
Daijogo, S., 534
Dalc, R. E., 460
Daltrop, O., 609
Damjanovich, S., 457
Daniel, C., 161
Daniell, X. G., 125
Daniels, B., 400
Danuser, G., 115, 116, 459, 505
Darst, S. A., 551
Das, R., 177
Davenport, R. J., 429
David, P. F., 444
Davidson, M. W., 3, 4, 18, 29–31, 33, 36, 37, 39, 43, 51, 62, 85, 87, 93, 110, 113, 118
Davis, I., 10

Davis, L., 156, 159, 161
Davis, L. M., 37, 457
Davydov, D. R., 459
Dawson, J. F., 3, 15–17, 95, 122, 125, 137, 139, 323
Dax, T. G., 130
Day, R. N., 35, 37
Deacon, S. W., 85
Deamer, D. W., 566, 593, 594, 596–599, 613
De Angelis, D. A., 457
Debyser, Z., 260
deCastro, M. J., 395
Dedecker, P., 37, 117
Deerinck, T. J., 89
Dejgosha, S., 385, 391, 392
Dekker, C., 260, 587, 593, 594, 600
Dekker, N. H., 587, 600
DeLaBarre, B., 557
Delagoutte, E., 312
Delp, S. L., 214
del Rio, A., 406
Dempsey, G. T., 3, 18, 19, 41
Dempster, A. P., 203
Deng, Y., 53
Deniz, A. A., 122, 150, 176, 459
Denk, W., 22
Dennis, C., 200
Derrida, B., 240
Derrington, I. M., 594, 595
Dertinger, T., 167, 457
Dertouzos, J., 122
Desbiolles, P., 430
De Schryver, F. C., 476
de Silva, C., 121–144
Deufel, C. Y., 385, 391, 392, 396, 400
De Vos, A., 95
Dholakia, K., 344, 347
Dickson, R. M., 35, 43, 150
Diekmann, S., 502
Diesselhoff-Den Dulk, M. M., 95
Dieterich, P., 142
Dieterlen, M. T., 97
Dietrich, A., 459
Dietz, H., 125
Diez, M., 280, 457
Digman, M. A., 457, 505
Dimitriadis, E. K., 379
Ding, F., 297–319
Distefano, M. D., 62
Ditlbacher, H., 179
Ditzler, M. A., 122
Divakaruni, A. V., 45
Di Ventra, M., 593, 594, 599
Dixit, R., 10
Dix, J. A., 142, 482
Dixon, N. E., 260, 261, 268, 271, 272, 275–277, 429
Dlakic, M., 502

Dobrowsky, T. M., 528
Dodd, I. B., 215, 217
Dolder, M., 519
Dominguez, R., 324
Dong, C. Y., 468
Dong, K., 379
Doniach, S., 323, 324, 367
Donmez, I., 457
Donnermeyer, A., 90, 92
Doose, S., 4, 22, 62, 84, 122, 457, 464, 492, 504
Doub, L., 38
Douglas, N., 159, 161, 169
Douglas, S. M., 125
Dou, S., 318
Downing, K. H., 558
Drenckhahn, D., 519
Drexhage, K. H., 468, 476
Duckworth, B. P., 62
Dudko, O. K., 571, 574–576, 581–584, 618
Dudko, O. L., 565–587
Duff, S. R., 4, 89
Dufrene, Y. F., 544
Duke, T., 459
Dumont, S., 318
Dunbrack, R. L. Jr., 503
Dunlap, D. D., 200, 204, 205, 209, 211, 214, 215, 217
Dunn, A. R., 92, 321–372
Dupuis, D. E., 322, 323
Durant, S., 178, 183, 184
Dyba, M., 3
Dyck, M., 468
Dye, N., 45
Dziedzic, J. M., 2, 3, 380, 430

E

Eaton, W. A., 122, 150, 406, 457, 459, 464, 499
Ebner, A., 529, 531
Ebright, R. H., 300, 457, 459, 472, 498
Ebright, Y. W., 457, 498
Edmonds, D. T., 603
Efron, B., 235, 364
Egan, J. B., 215, 217
Eggeling, C., 37, 43, 122, 123, 457, 468, 470–472, 475, 476, 480, 488
Egner, A., 3, 21, 37, 43
Ego, C., 470
Egwim, I. O. C., 529
Ehrenberg, M., 472
Eid, J. S., 475
Einstein, A., 152
Eisenman, G., 607
Eisenmesser, E. Z., 459
Eisinger, J., 460
Elghanian, R., 179
Elings, V., 519
Ellenberger, T., 176

Ellisman, M. H., 4, 62, 89, 96, 122, 456, 457
El-Sayed, M. A., 178, 184, 194
Elson, E. L., 117, 141, 142, 204, 457, 472
Elston, T. C., 243
Encell, L. P., 43, 62
Enderlein, J., 5, 158, 160, 167, 457, 482, 488
Enderle, T., 16, 122, 176, 413, 457
Endesfelder, U., 19
Engel, A., 299, 519, 546
Engelbrecht, S., 282
Engelhardt, J., 21
Engh, A. M., 332, 400
English, B. P., 122, 247, 459
English, D. S., 176
Erickson, J. W., 516
Errington, J., 45
Escude, C., 430
Essevaz-Roulet, B., 431
Euteneuer, U., 2, 3
Evans, E., 528, 566
Everett, J., 122

F

Falkenberg, M., 429, 430, 447, 448
Fananapazir, L., 322
Fantner, G. E., 542
Farge, G., 427–450
Fasshauer, D., 457, 472, 487, 488, 496, 499
Faulhaber, A. E., 122
Fay, N., 457, 468, 472, 490
Fehr, A. N., 8, 158, 434
Feldmann, J., 179
Felekyan, S., 280, 457, 459, 463, 464, 471–473, 481, 482, 486–494, 496, 498, 499, 504, 505
Fenn, T., 467
Fernandez, J. M., 406, 528, 566
Fernandez-Suarez, M., 114
Ferrer, J. M., 406, 424, 440
Fersht, A. R., 429
Fetter, R. D., 51, 118
Fetting, D., 90, 92
Fields, A. P., 149–171
Figge, R. M., 45
Filanoski, B. J., 445
Finer, J. T., 322, 325, 339, 361, 367, 370
Finkel, A., 602, 603, 606
Finley, D., 557
Finzi, L., 199–218, 381
Fisher, M. E., 177, 223, 229, 240, 243
Fisher, T. E., 528
Fishman, H. A., 606
Fita, I., 522, 527
Florin, E. L., 330, 339, 390, 412, 566
Flynn, T. C., 283, 557
Flyvbjerg, H., 322–324, 356–358, 367, 412
Fölling, J., 37
Fologea, D., 587

Foote, R. S., 168
Forde, N. R., 127
Fordyce, P. M., 332, 400, 406, 409, 429, 440
Forgacs, E., 323
Forkey, J. N., 2, 7, 8, 17, 86, 93, 95, 110, 122, 124, 125, 137, 457
Formosa, T., 260
Fornasiero, E. F., 116
Förster, T., 461, 468
Forstner, M. B., 142
Forth, S., 385, 391, 392, 400
Foster, D., 566
Fradkov, A. F., 113
Fraley, C., 203
Franceschetti, A., 457
Frankenberger, E. A., 516
Franzini, A., 200, 204, 209, 214
Frauenfelder, H., 459
Frechet, J. M. J., 195
Freemont, P. S., 557
French, R. J., 602, 603, 606
Freyzon, Y., 324, 406
Frieda, K. L., 566
Friedman, J. M., 479
Friedman, M. D., 178
Friedman, N., 150
Friese, M. E. J., 382
Fries, J. R., 468, 471, 472, 475, 488
Fromm, D. P., 34, 47, 52, 179
Fron, E., 461, 463
Frydman, J., 151, 159, 161, 169
Fuchs, J., 116
Fuchs, R., 516, 532
Fujisawa, R., 282, 286
Fujita, H., 283
Fujiwara, T., 84, 142
Fukami, K., 62
Fukuma, T., 542
Funatsu, T., 3, 89, 332, 406, 409, 429, 440
Funk, M., 391
Furuike, S., 286
Füssle, R., 610
Fuxreiter, M., 499

G

Gabdoulline, R. R., 62
Gabrielian, A., 502
Gafni, A., 122
Gaiduk, A., 489, 490, 492, 493
Gaietta, G., 62, 89
Gaiko, N., 472
Galajda, P., 384, 385, 390
Galbraith, C. G., 3, 18, 30, 31, 33, 36, 51, 114, 118
Galbraith, J. A., 3, 18, 30, 31, 33, 36, 51, 114, 118
Gall, A. A., 529, 531
Galli, G., 200, 204, 209, 214

Gall, K., 457, 468, 472, 482, 490
Gambhir, S. S., 4, 22, 62, 84, 122
Gansen, A., 457, 463, 491, 494, 504
Gao, Y. Q., 281
Garai, A., 243
Garaj, S., 593, 594, 599
Gardner, F. M., 400
Gast, A. P., 551
Gaub, H. E., 299, 406, 542, 566
Gautel, M., 406, 566
Gautier, A., 114
Geertsma, H., 443
Geier, S., 179
Geisler, C., 43
Geissler, P. L., 177
Gelfand, V. I., 2, 9–12, 83–86, 88, 95, 125
Gelles, J., 2, 8, 28, 84, 94, 200, 379, 381
Gendreizig, S., 88, 114
Gennerich, A., 8, 82, 236, 240
Gensch, T., 476
George, N., 62
Georgiou, G. N., 115
Gerken, M., 460, 468, 470, 485
Germann, J., 156, 159, 161
Gershow, M., 587
Ghadiri, M. R., 598, 600, 619
Ghiggino, K. P., 461
Ghosh, I., 63
Ghosh, R. N., 2, 115
Gibbons, C., 283
Giepmans, B. N. G., 4, 37, 62, 89, 96, 122, 456, 457
Gileadi, O., 92
Gillette, J. M., 3, 18, 36, 43, 51, 109–118
Gillis, K. D., 608
Gin, P., 62, 457
Girirajan, T. P. K., 3, 18, 29, 123
Gitai, Z., 29, 45, 46, 50, 150
Gittes, F., 330, 339, 343, 395, 432, 435, 436
Gober, J. W., 45
Goedecke, D. M., 243
Goelet, P., 62
Gohlke, C., 502
Golding, I., 457
Gold, J., 382
Goldman, R. D., 2, 11, 12, 84, 85, 95
Goldman, Y. E., 2, 7, 8, 10, 17, 86, 93, 95, 110, 122, 124, 125, 137, 457, 542
Goldstein, L. S., 3, 8
Golovchenko, J. A., 566, 587, 595, 612, 619
Golubovskaya, I. N., 3
Gombay, E., 201, 202
Gonçalves, M. S. T., 457
Gonzalez, C., 609
Goodrich, G. P., 177
Goodwin, P. M., 156, 158, 160, 468, 471, 482
Goody, R. S., 457, 459, 464, 481
Gopich, I. V., 457, 459, 464, 490, 492, 497, 499

Gordon, M. P., 3, 15, 16, 122, 137
Gore, J., 170, 381
Gorman, J., 429
Goshima, G., 2, 9–11, 83, 84, 86, 88, 95, 125
Gosse, C., 151, 299
Gouaux, J. E., 594, 610
Gouax, J. E., 577, 578
Gould, T. J., 51, 55
Gouzer, G. G., 64
Graber, P., 280, 457
Grady, N. K., 177
Graf, F., 125
Graham, I., 459
Granzier, H. L., 566
Gräter, F., 406
Gratton, E., 84, 150, 156, 159, 160, 169, 457, 468, 472, 505
Gray, D. R., 445
Gray, M. L., 151
Grecco, H. E., 195
Greene, E. C., 429
Greenleaf, W. J., 2, 103, 222, 236, 377–402, 424, 433, 566
Green, N. M., 551
Green, S. J., 125, 126
Gregor, I., 167, 457, 488
Gribbon, P., 122
Grier, D. G., 115, 116
Griesinger, C., 472
Griffin, B. A., 4, 88
Griffis, E. R., 3
Griffith, J. D., 260
Griffith, J. P., 516
Grigoriev, M., 200
Grimes, S., 536
Grimley, R., 122
Grimsdale, A., 470
Grinkova, Y. V., 587
Grinstein, S., 115
Gronemeyer, T., 88
Groot, M. L., 459
Grosberg, A. Y., 567
Gross, D. J., 142
Gross, L. A., 4
Gross, P., 427–450
Gross, S. P., 83, 85, 339
Grubera, H. J., 528
Gruber, H. J., 2, 4, 84, 124, 130, 468, 515–536
Grubmüller, H., 406, 457, 464, 472, 477, 487, 488, 496, 499, 500, 528
Grunwell, J. R., 122
Guan, X. Y., 577
Guffey, M. J., 151
Guilford, W. H., 8, 83, 103, 111, 115, 323
Guilfoyle, R. A., 133
Gu, L.-Q., 594, 608, 611, 612
Gundlach, J. H., 594–596
Gunewardene, M. S., 36, 37

Gunnarsson, L., 178
Güntherodt, H. J., 529
Gunther, R., 471
Guo, F., 379
Guo, L., 457
Guo, W., 70
Guo, Z., 133
Gurskaya, N. G., 113
Gusek, T. W., 155
Gustafsson, M. G. L., 3, 22, 28
Gutiérrez-Medina, B., 158, 377–402, 434
Güttler, F., 28
Gu, W. W., 87
Guydosh, N. R., 158, 434
Guyot-Sionnest, P., 151

H

Haas, E., 459
Habuchi, S., 37, 117, 429
Ha, C., 444
Hachmann, W., 396
Hackel, B. J., 71, 72
Hadjantonakis, A.-K., 62
Haeberle, J. R., 322
Hafner, J. H., 178, 561
Hagberg, P., 382
Hahn, C. D., 130, 529
Hahn, H., 3
Hakansson, P., 466
Halas, N. J., 177, 178, 195
Hall, A. R., 587
Halpert, J. R., 459
Halvorsen, K., 406
Hamadani, K. M., 471, 492, 494, 504
Hama, H., 36, 62, 110, 111
Hamdan, S. M., 260, 261, 268, 270–272, 275–277, 429
Hamilton, J. D., 166
Hamilton, R. S., 10
Hammann, C., 194
Hammes, G. G., 176
Handa, N., 429
Hanke, W., 602
Hanna, S., 382
Hansch, C., 38
Hansma, H. G., 534, 542
Hansma, P. K., 542
Hanson, G. T., 4, 89
Hanstorp, D., 382
Han, W., 519
Hanyu, A., 62
Han, Y. W., 420
Happel, J., 412
Harada, Y., 3, 89, 322, 420, 440
Hara, K. Y., 285, 290
Haran, G., 150, 459
Harbury, P. A., 177

Harms, G. S., 35, 37, 130
Harrington, R. E., 502
Hartland, G. V., 195
Hartmann, R., 167, 457
Hartzell, D. D., 43, 62
Haselgruebler, T., 529
Ha, T. J., 2, 3, 6–8, 15, 16, 63, 86, 89, 93, 95, 110, 122, 124, 125, 130, 137, 150, 176, 222, 236, 260, 332, 405–424, 457, 459, 468, 499, 504, 542
Hauger, F., 457, 463, 491, 494, 504
Haugland, R. P., 122
Haussler, D., 594, 597
Haustein, E., 456, 457, 472, 487, 488, 496, 499
Hayakawa, T., 94
Hayashi, Y., 62
Haynes, C. L., 178
Hazelwood, K. L., 4, 36, 37, 39, 85, 87
Hecht, E., 123
Hecht, H. J., 516
Heckenberg, N. R., 382, 391
Hecker, N. E., 179
Heckl, W. M., 534–535
Hedde, P. N., 116
Hegner, M., 529
He, H., 382
Heilemann, M., 19, 459
Heine, J. R., 66
Heinis, C., 114
Heintzmann, R., 3, 17, 28, 123
Heinze, K. G., 468
Heller, R. C., 275
Hell, S. W., 3, 4, 21, 22, 28, 37, 43, 62, 123, 467
He, M., 129
Henderson, R., 558
Hendrix, G. M., 116
Heng, J. B., 587
Henrickson, S. E., 595, 613
Henricus, M., 594, 612, 613
Henriques, R., 116
Hensen, U., 406
Henzler-Wildman, K. A., 228, 229, 459, 467, 499
Hernandez, J. M., 487
Hernandez, L. I., 178
Heron, A. J., 591–619
Herschlag, D., 406
Herten, D. P., 468
Heslot, F., 431
Hessel, N., 151
Hess, H. F., 3, 18, 29, 43, 51, 62, 93, 110, 112, 113, 117, 118
Hessling, M., 457
Hess, S. T., 3, 18, 29, 36, 37, 51, 55, 123
Hetzer, M., 557
Hewat, E. A., 516, 522, 532, 536
He, Y., 126, 127, 142

Author Index

Heyes, C. D., 551
Heyes, D. J., 459
Heymann, B., 528
Heymann, J. B., 546
Hibbs, A., 593, 594, 599
Hibino, K., 63
Higuchi, H., 63, 92, 332, 406, 409, 429
Hildt, E., 472
Hillbert, M., 43
Hillisch, A., 502
Him, J. L. K., 555
Hinterdorfer, P., 515–536, 544
Hintersteiner, M., 122
Hippchen, H., 487
Hippel, P. H. V., 312
Hirano, K., 94, 150
Hira, S., 177
Hirono-Hara, Y., 282, 286, 290
Hirose, S., 542
Hiroshima, M., 63
Hisabori, T., 280
Hite, R. K., 260, 268
Hnilova, M., 65
Hobaugh, M. R., 577, 578, 594, 610
Ho, C., 587
Hochschild, A., 215, 217
Hodges, C., 170, 428
Hoekstra, T., 443
Hoffman, M. T., 9
Hoffmann, A., 457
Hoffmann, C., 62
Hofkens, J., 37, 117, 470, 476
Hofmann, M., 123
Hofmeister, W., 156, 159, 161
Hogberg, B., 125
Hohenau, A., 178
Hohng, S., 89, 332, 406, 407, 409, 413–415, 417, 419, 422, 423, 504
Holden, H. M., 323
Holden, M. A., 605, 609, 610
Holmgren, A., 406
Holten-Andersen, N., 542
Holzbaur, E. L., 10
Hom-Booher, N., 4
Hom, E. F. Y., 482
Honda, T., 614
Hooijman, P., 429, 430, 447, 448
Hopkins, B., 177
Hörber, J. K. H., 330, 339, 390, 412
Hormes, R., 194
Hornblower, B., 577, 586, 595, 600
Horne, L. P., 594
Horvath, L. J., 201, 202
Horwich, A. L., 151
Horwitz, A. R., 457, 505
Ho, S. O., 459
Hosokawa, A., 62
Hostetter, D. R., 92

Howard, J., 82, 83, 357
Howarth, M., 62, 64
Howorka, S., 597, 598, 600
Huang, B., 3, 18–21, 35, 37, 41, 43, 51, 53, 84, 93, 123, 542
Huang, C. Y., 89, 122, 406, 413
Huang, F., 429
Huang, H. W., 607
Huang, W. Y., 178, 184, 194
Huang, X., 593, 594, 599
Hua, W., 8, 379
Hubner, C. G., 467
Hudspeth, A. J., 82
Hugel, T., 457
Hui, N., 557
Huisstede, J. H. G., 339
Hu, M., 195
Hummer, G., 571, 573–576, 581–584, 618
Hung, W. C., 607
Hunter, C. N., 459
Huxley, H. E., 322
Huyton, T., 557
Hwang, W. L., 605, 610

I

Ichikawa, T., 555
Ide, T., 555
Iida, T., 614
Iino, R., 282, 286
Ikebukuro, K., 614
Ikeguchi, M., 279–294
Ikura, M., 122
Imamoto, N., 89
Imamura, H., 282, 286
Imamura, T., 62
Inami, H., 150
Inoue, Y., 92
Irngartinger, T., 28
Isacoff, E. Y., 457
Isaksson, M., 466
Ishihama, A., 440
Ishihara, K., 546
Ishii, Y., 322, 459
Ishijima, A., 332, 406, 409, 429
Ishitsuka, Y., 63, 122, 406, 542
Ishiwata, S., 94
Ishizuka, K., 290
Itoh, H., 94, 281, 282, 284, 286
Ito, Y., 420
Ivanchenko, S., 18, 36, 37, 111
Ivanova, K., 542
Ivanov, T., 542
Ivanov, V., 460
Iwane, A. H., 459
Iwasawa, K., 142
Iyer, G., 61–76
Izhaky, D., 3, 338, 341, 424, 433, 434

J

Jacobson, K., 3, 84, 141, 142
Jacobson, S. C., 168
Jager, M., 457, 459
Jager, S., 457
Jahn, R., 3, 4, 62, 457, 472, 487, 488, 496, 499
Jain, P. K., 178, 184, 194
Jakobs, S., 3, 21, 43, 62, 123
Jan Bonthuis, D., 577
Janesick, J., 136
Japrung, D., 591–619
Jaqaman, K., 115, 116
Jardine, P. J., 536
Jares-Erijman, E. A., 505
Jarosch, E., 557
Jaschke, M., 532
Javier, A., 177
Jawhari, A., 499
Jayasinghe, L., 599, 609
Jennings, T., 177
Jensen, K. F., 66
Jergic, S., 260, 261, 268, 271, 272, 275–277, 429
Jett, J. H., 3
Jiang, Y., 159, 161, 169
Jing, T., 519
Johansson, L. B. A., 466
Johnson, A., 260
Johnson-Buck, A., 123, 125, 127–129, 133, 136–140, 142, 143
Johnson-Buck, A. E., 121–144
Johnson, D. E., 268, 318
Johnson, I., 444
Johnson, J. E., 516
Johnsson, K., 62, 85, 88, 114
Johnsson, N., 62
Jones, S. A., 20, 35, 37, 43, 84
Jongeneel, C. V., 260
Jontes, J. D., 361
Joo, C., 63, 122, 150, 406, 413, 415, 416, 419, 457, 542
Jordan, S. C., 389
Jordens, S., 470
Jovanovich, S. B., 593, 594, 599
Jovin, T. M., 3, 17, 28, 123, 505
Joyce, G. F., 127
Juelicher, F., 357
Juette, M. F., 51–53, 55
Juillerat, A., 114
Junge, W., 282
Jungmann, J., 457, 468, 472, 490
Jun, Y. W., 92

K

Kaback, H. R., 457
Kada, G., 130, 519, 529, 534–536
Kadavanich, A. V., 66
Kad, N. M., 323, 324

Kador, L., 3, 123, 457
Kado, R. T., 608
Kagawa, R., 281, 283
Kahn, J. D., 176
Kai, E., 614
Kainov, D. E., 600
Kalaidzidis, Y., 115
Kalinin, S., 455–505
Kalisch, S. M. J., 429
Kallio, K., 4
Kall, M., 178
Kamer, G., 516
Kamin, D., 4, 62
Kanaar, R., 429, 430, 447–450
Kanchanawong, P., 51, 118
Kandori, H., 542
Kang, X. F., 577, 594, 608, 612
Kao, J. P. Y., 32
Kapanidis, A. N., 62, 413, 457, 459, 468, 472, 482, 498
Kapitein, L. C., 332, 432, 436, 440, 443
Karassina, N., 43, 62
Karlin, S., 170
Karplus, M., 229, 281, 283, 459, 467
Karrasch, S., 519
Kar, S., 379
Karube, I., 614
Kasai, R. S., 84
Kasemo, B., 178
Kashiwagi, S., 62
Kasho, V., 457
Kasianowicz, J. J., 566, 594–598, 603, 613
Kas, J. A., 142
Kask, P., 457, 468, 472, 482, 490
Kasper, R., 19
Kato-Yamada, Y., 280, 283
Katsura, S., 150
Kaul, N., 81–103
Kawano, R., 600
Kay, L. E., 459
Keighley, W. W., 122
Keller, D., 268, 298, 428
Kellermayer, M. S. Z., 431, 566
Keller, R. A., 3, 37, 457, 468, 471, 482
Kelly, K. L., 177, 195
Kendrick-Jones, J., 325
Kenndler, E., 518, 519
Kennedy, G. G., 17, 323
Kensch, O., 457, 459, 464, 481
Kenworthy, A. K., 116, 456
Keppler, A., 88, 114
Kern, D., 228, 229, 459, 467, 499
Kerry, C. J., 217
Keyser, U. F., 600
Kho, A. L., 406
Kienberger, F., 515–536
Kienzler, A., 473, 489, 492, 499
Kilar, F., 519

Kim, H. D., 2, 9–11, 83, 84, 86, 88, 95, 122, 125, 150, 176, 416
Kim, K., 534
Kim, M.-J., 587
Kim, S. Y., 29, 45, 46, 50, 150, 151, 155
Kim, T., 587
Kindermann, M., 114
Kindt, J. H., 542
King, D., 63
King, G. M., 396
King, J., 156, 159, 161
King, S. J., 125
Kinkhabwala, A., 29, 45, 46, 50, 150
Kinns, H., 600
Kino, G., 179
Kinoshita, T., 560
Kinosita, K., 94, 378
Kinosita, K. Jr., 280–287, 290
Kirei, H., 384, 385, 390
Kishino, A., 322
Kis-Petikova, K., 159
Klafter, J., 142
Klages, R., 142
Klampfl, C., 529, 531
Klar, T. A., 3
Klar, T. W., 28
Kleimann, C., 396
Klenerman, D., 488
Knaus, H. G., 130
Kner, P., 3
Knight, J., 457, 498
Knöchel, W., 62
Koberling, F., 472
Kobitski, A. Y., 551
Kochaniak, A. B., 429
Koculi, E., 194
Kodera, N., 541–561
Ko, D. S., 468, 476
Koeleman, B. N., 600
Koga, N., 281
Kohler, D., 240
Kohler, G., 519
Köhler, J., 28
Kohlstaedt, L. A., 479
Koide, H., 560
Kojima, H., 332, 406, 409, 429
Köllner, M., 468, 471, 472, 475, 488
Kolomeisky, A. B., 223, 229, 240, 243, 586, 600
Koltermann, A., 468
Komori, T., 322
Kondo, H., 557
Kondo, J., 84, 142
Konecsni, T., 519, 527
Kong, X. X., 457, 468, 472, 482
Konig, M., 280, 457, 472, 487, 488, 496, 499, 505
Kon, T., 125
Korchev, Y. E., 603
Kornberg, R. D., 551

Kortkhonjia, E., 457, 459, 498
Korzhnev, D. M., 459
Koshioka, M., 480
Koster, D. A., 252
Kotova, S. L., 379
Kotthaus, J. P., 561
Kou, S. C., 122, 243, 247, 459
Koushik, S. V., 457
Kovchegov, Y., 471, 492, 494, 504
Kowalczykowski, S. C., 429
Kowalczyk, S. W., 587
Koyama-Horibe, F., 281, 282
Koza, Z., 240
Kozlov, A., 405–424
Kraemer, J., 122
Kramer, B., 472
Kramers, H. A., 571, 574, 575
Krapf, D., 600
Krapivsky, P. L., 127, 142
Krasne, S., 607
Kreibig, U., 177
Krementsova, E. B., 17
Kremers, G. J., 36, 37, 39, 505
Kremser, L., 519
Krenn, J. R., 178, 179
Kreuzer, K. N., 260
Kriebel, J. K., 602, 603, 606
Kron, S. J., 322
Kroon, P. A., 525
Krstic, P. S., 593, 594, 599
Kubalek, E. W., 525, 551
Kubitscheck, U., 35, 37, 440
Kudryavtsev, V., 280, 457, 460, 468, 470, 471, 485, 496, 505
Kuechler, E., 516, 532
Kuhnemuth, R., 457, 460, 468, 471, 473, 489, 492, 496, 499, 505
Kulic, I. M., 83, 86
Kumar, M., 116
Kumke, M., 464
Kunetsky, R., 37
Kural, C., 1–22, 83–86, 88, 95, 125
Kurokawa, H., 62
Kusumi, A., 84, 141, 142
Kuwata, H., 115
Kuznetsov, Y. G., 517, 525, 534
Kwok, P.-Y., 444
Kwon, A. H., 127
Kwon, G. S., 62

L

Laakso, J. M., 361
Labeikovsky, W., 459
Lackner, B., 529
Lacoste, T. D., 3, 468
Lagerholm, B. C., 3
Laib, J. A., 83

Laird, D. W., 89
Laird, N. M., 203
Lairez, D., 613
Lakowicz, J. R., 31, 457, 461, 463, 468, 480, 482
Lal, S., 177, 178
Lamb, D. C., 457, 468, 472
Lamprecht, B., 178, 179
Lancaster, L., 170, 428
Landau, L. D., 384
Landick, R., 2, 200, 378, 381, 424, 433
Laney, D. E., 534
Lang, M. J., 332, 343, 395, 400, 406, 409–411, 424, 429, 440
Langowski, J., 457, 463, 491, 494, 504
La Porta, A., 377–402
Larabell, C., 87
Larizadeh, K., 406
Larson, D. R., 5, 7, 22, 29–31, 48, 53, 83, 103, 114, 115, 124, 137, 269
Lassiter, J. B., 178
Latorre, R., 609
Laurence, T. A., 62, 457, 468, 471, 472, 482, 492, 494, 504
Lauterbach, M. A., 4, 62
Lau, W. L., 71, 72
Learish, R., 43, 62
Ledden, B., 587
Leeder, J. M., 461
Lee, H. D., 35, 37–41
Lee, H. T., 525
Lee, J. B., 260, 268
Lee, K. T., 459
Lee, N. K., 194, 468, 472
Lee, P., 424, 440
Lee, W. M., 344, 347
Le Grice, S. F. J., 150, 525
LeGrimellec, C., 519
Lehmann, E., 499
Lei, M., 467
Leith, J. S., 429
Leitner, A., 178
Lelek, M., 116
Lemay, S. G., 600
Lemke, E. A., 459
Lenhard, J. R., 444
Leo, A., 38
Lescano, I., 156, 159, 161
Leslie, A. G., 281, 283
Lessard, G. A., 156, 158, 160
Lessard, M. D., 51, 55
Letsinger, R. L., 179
Leube, R. E., 440
Leung, A., 566
Leung, L., 444
Lev, A. A., 603
Levi, M., 457
Levi, S., 63
Levis, R. A., 606

Levitus, M., 472
Levi, V., 84, 85, 156, 159, 160, 169
Levy, R., 457, 498
Lewis, D. E. A., 200, 204, 209, 214, 215, 217
Lewis, J. H., 361
Lewis, R., 499
Lew, M. D., 55
Liang, H. Y. W., 195
Liang, J., 150, 176
Lian, W., 546
Liao, J.-C., 214, 235
Lidke, K., 3, 17
Liebovitch, L. S., 217
Liedl, T., 125
Lifshitz, E. M., 384
Li, H., 528, 566
Li, H. B., 566
Li, J. J., 4, 22, 62, 84, 122, 127, 587
Li, J. L., 587
Li, L., 444
Lilley, D. M. J., 332, 407, 409, 416, 417, 423, 502
Li, M., 318, 460
Lin, C. W., 62
Linden, M., 237
Lindsay, S. M., 519
Lindsey, H., 607
Lindwasser, O. W., 3, 18, 29, 43, 62, 93, 110
Ling, X. S., 587, 600
Linke, H., 127
Lin, M. Z., 4, 85, 87
Lionberger, T. A., 81–103
Lionnet, T., 300, 313, 316, 318, 400
Liou, G. F., 432
Liphardt, J., 93, 175–195, 566
Lipman, E. A., 150, 464
Lippincott-Schwartz, J., 3, 18, 29, 36, 43, 51, 62, 93, 109–118, 122, 456
Lippow, S. M., 71, 72
Li, P. T. X., 236
Li, Q., 594, 610, 612, 613
Lis, J. T., 406
Liu, G.-Y., 544
Liu, J. X., 445
Liu, M., 151, 544
Liu, N., 35, 37–41, 51, 53–55
Liu, R., 406
Li, X. D., 156, 159, 161, 195
Li, Z. Y., 195
Loerke, D., 115
Lofdahl, P. A., 470
Lohman, T. M., 318, 405–424
Lohse, M. J., 62
Lomholt, M. A., 429
Lommerse, P. H. M., 35, 37
Lonberg, H. K., 525
Loparo, J. J., 261, 268, 271, 272, 275–277, 429
Lord, S. J., 27–55
Lorenz, H., 561

Loscha, K. V., 260, 268, 429
Los, G. V., 43, 62
Lounis, B., 468
Lubelski, A., 142
Lubensky, D. K., 578, 579
Lucas, R. W., 525
Luccardini, C., 63, 83, 84, 87, 95
Luchian, T., 593
Lu, H. P., 150, 176, 459, 566
Lukyanov, K. A., 36, 37, 62, 113
Lukyanov, S., 36, 37, 62, 113
Lund, K., 123, 125, 127–129, 133, 136–140, 142, 143
Luo, G., 122
Luo, G. B., 459
Luo, M., 516
Lu, P. J., 156, 157, 160
Lutter, R., 281
Lu, X., 610
Lu, Y., 127
Lu, Z., 37, 40
Lyubchenko, Y. L., 529, 531

M

Maali, A., 468
Mabuchi, H., 156, 159, 160, 169
Macarthur, J. B., 156, 157, 160
Macdonald, R., 472
Madathil, R., 600
Magde, D., 457, 472
Magennis, S. W., 489, 492
Magleby, K. L., 616
Maglia, G., 591–619
Maglia, M., 594, 612
Maier, B., 268
Maier, S. A., 178
Ma, J., 283, 557
Majumdar, D. S., 457
Makhov, A. M., 260
Malghani, M. S., 610
Malkin, A. J., 525
Mallick, K., 142
Mallik, R., 236, 240
Mangeol, P., 434–435
Mangoel, P., 443
Manley, S., 36, 43, 51, 109–118
Manosas, M., 297–319
Mantulin, W. W., 457
Manzo, A. J., 89, 121–144, 406, 413
Manzo, C., 199–218
Mao, C. D., 126, 127
Mao, Y., 10
Marchi-Artzner, V., 64
Marchington, R. F., 344, 347
Margeat, E., 64, 74, 75, 459, 468, 472
Margittai, M., 457, 472, 487, 488, 496, 499
Marguet, D., 203
Marians, K. J., 260, 275

Marin, J. A., 83
Marino, J. P., 122
Marko, J. F., 3, 311, 418, 536
Marlovits, T. C., 519, 532
Marquez, M., 195
Marqusee, S., 406, 566
Marshall, R. A., 7, 130
Marston, S. B., 323
Marszalek, P. E., 528, 566
Martens, J. R., 83, 88, 96
Martin, C. R., 594
Martin, D. S., 142
Martinez, O. E., 195
Martin, H., 600
Martin, J. C., 3, 468, 471
Maruyama, D., 542, 544
Marziali, A., 593–595, 597, 599, 613
Masaike, T., 281, 282
Mason, A., 609
Mason, M. D., 3, 18, 29, 123
Masters, B. R., 468
Mastroianni, A. J., 177, 181, 186
Masuhara, H., 480
Mathé, J., 565–587, 595, 612, 613, 618
Mathew-Fenn, R. S., 177
Matsudaira, P., 406
Matsuura, S., 150
Mattaj, I. W., 557
Mattoussi, H., 62, 66
Matyas, S. E., 65, 71–73
Matyus, L., 457
Maus, M., 476
Mayanagi, K., 542
May, A. P., 557
Mayer, M., 602, 603, 606
McCammon, J. A., 229
McCauley, M. J., 428
McDougall, M. G., 43, 62
McEwen, D. P., 83, 88, 96
McFarland, A. D., 178
McGee, R., 217
McHale, K., 156, 159, 169
McHenry, C. S., 260
McKeown, M. R., 4, 85, 87
McKinney, S. A., 2, 7, 8, 86, 93, 95, 110, 122, 124, 125, 130, 137, 415, 416, 457
McMullan, C., 587
McNabb, D. S., 587
McNally, B., 587
McPherson, A., 517, 525, 534
Medda, R., 37
Meglio, A., 297–319
Mehta, A. D., 3, 323, 339, 358, 361, 367, 370
Mehta, K. K., 618
Meier, S., 459
Mekler, V., 457, 498
Meller, A., 475, 565–587, 595–597, 600, 612, 613, 618, 619

Meller, P. H., 551
Mendez, J., 43, 62
Menten, M. L., 223
Menz, R. I., 283
Merchant, K. A., 464, 499
Merkel, R., 566
Merrill, E. W., 261
Meseth, U., 150, 155, 167
Mets, L., 15, 123
Mets, Ü., 457, 468, 470, 472, 490
Mettlen, M., 115
Metzger, N. K., 344, 347
Metzler, R., 142, 429
Meyer-Almes, F. J., 468
Meyer, H. H., 557
Meyhofer, E., 4, 81–103
Mhlanga, M. M., 116
Michaelis, J., 499, 536
Michaelis, L., 223
Michalet, X., 3, 4, 22, 29, 61–76, 84, 87, 122, 457, 459, 468, 471, 472, 492, 494, 504
Michelotti, N., 121–144
Mickler, M., 457
Mikhailova, E., 595, 596, 598, 610, 612, 613, 616, 618, 619
Mikulec, F. V., 66
Miles, G., 610
Millar, D. P., 122
Miller, C., 602
Miller, L. W., 62, 114
Miller, S. C., 62
Millet, O., 459
Millhauser, G. L., 217
Milligan, R. A., 83, 361, 557
Milling, A. J., 546
Mills, R., 155
Milosevic, J., 97
Min, W., 122, 247, 459
Mirkin, C. A., 179
Mirny, L. A., 429
Misakian, M., 595, 613
Misura, K. M. S., 484
Mitchell, N., 598, 600
Mitchell, T. E., 127, 129
Mitchison, T. J., 97
Mitra, P. P., 299
Miyagi, A., 542, 543, 560
Miyata, T., 62
Miyawaki, A., 36, 37, 110, 111, 117
Miyoshi, H., 62
Mizuno, A., 150
Mizuno, H., 36, 37, 110, 111, 117
Mizuuchi, K., 460
Mlodzianoski, M. J., 51–53, 55
Mock, J. J., 178
Modesti, M., 429–430, 447–450
Moerner, W. E., 3, 27–55, 122, 123, 129, 150, 151, 156, 157, 159, 161, 162, 167, 169–171, 179, 444, 457

Moffitt, J. R., 3, 221–255, 322, 332, 338, 339, 341, 343, 344, 346, 347, 356, 406, 424, 433–434
Moller, C., 519
Moller, M., 179
Molloy, J. E., 323, 325, 339, 361
Molski, A., 34
Montal, M., 606
Montgomery, M. G., 281, 283
Montiel, D., 156, 157, 169, 201
Moore, H. P., 63, 64, 74, 75
Mooseker, M. S., 323, 339, 358, 361
Morera, F. J., 609
Morgan, M. A., 176
Morikawa, K., 542
Morimatsu, M., 322
Morimura, T., 62
Moronne, M., 62, 457
Morris, C., 323
Morrisoni, I. E. G., 115
Morrison, W., 567
Moser, R., 519, 521–523, 525, 527, 533–534, 536
Mosser, A. G., 516
Moy, V. T., 566
Mucic, R. C., 179
Mueller, P., 606
Mukherjee, A., 19
Mukhopadhyay, J., 457, 472, 498
Mukhopadhyay, S., 459
Müllen, K., 470
Müller, B. K., 457, 468, 472
Müller, C., 459, 468
Müller, D. J., 299, 519, 546
Müller, H., 519
Müller, J. D., 150, 457, 468, 472
Müller, M., 28
Müller, S., 473, 489, 492, 499
Muneyuki, E., 280, 285, 290
Murakami, K., 440
Murakoshi, H., 84
Murase, K., 84
Murchie, A. I. H., 502
Murphy, C. S., 36, 37, 39
Murray, C. B., 66
Muschielok, A., 499
Muto, E., 92
Myong, S., 8, 176, 419, 457, 459

N

Nafisi, K., 444
Nagpure, B. S., 51, 55
Nagura, N., 551–554, 560
Nagy, A., 428
Nahas, M. K., 332, 407, 409, 417, 423
Nakada, C., 84
Nakakita, R., 542
Nakamura, M., 457
Nakane, J., 595, 597, 613

Author Index

Nakashima, T., 283
Nandi, K., 603
Nangreave, J., 123, 125, 127–129, 133, 136–140, 142, 143
Nan, X. L., 83, 84, 86, 92, 93
Nassoy, P., 566
Neher, B., 602, 616
Neher, E., 2, 222
Nelson, P. C., 83, 86, 200, 204, 205, 209, 211, 214
Nelson, S. R., 87, 96
Nettels, D., 457
Netzel, T. L., 444
Neubauer, H., 472
Neuhauser, D., 471, 492, 494, 504
Neuman, K. C., 223, 322, 325, 330, 332, 339, 356, 357, 370, 378, 388, 390, 395, 396, 400, 410, 428, 432, 433, 435, 436
Neumann, E., 516, 536
Neuts, M. F., 233
Neuweiler, H., 457, 464
Newman, R., 557
Nguyen, D. T., 473, 489, 492, 499
Nichols, B. J., 116
Niederweis, M., 594, 595
Nieminen, T. A., 391
Nienhaus, G. U., 18, 36, 37, 111, 116, 122, 416, 551
Nienhaus, U., 35, 37, 440
Nie, S., 22
Niles, W. D., 606
Nir, E., 457, 471, 492, 494, 504
Nishikawa, S., 322
Nishikori, S., 541–561
Nishimura, S. Y., 37, 40
Nishiura, M., 125, 290
Nishiyama, M., 92
Nishizaka, T., 94, 281, 282, 286
Niu, L., 43, 113, 118
Niu, W., 457, 498
Nivon, L., 566, 595, 597, 612, 619
Noguchi, A., 322
Noji, H., 279–294, 378
Noller, H. F., 170, 428
Nollmann, M. N., 170
Nonet, M. L., 2, 12, 14, 95
Nonoyama, Y., 440
Nordlander, H. P., 178
Nordlander, P., 178
Normanno, D., 200, 201
Norris, D. J., 66, 150
Nossal, N. G., 260
Novo, M., 457
Nugent-Glandorf, L., 342
Nunez, M. E., 428
Nutter, H. L., 37
Nutter, L., 3
Nyamwanda, J. A., 578, 579
Nyquist, H., 29–31, 33

O

Oberdorff-Maass, S., 62
Oberhauser, A. F., 528, 566
Ober, R. J., 29–31, 66
O'Cinneide, C. A., 233
O'Connell, S., 518
O'Connor, D. V., 476
O'Donnell, M., 260
Oesterhelt, F., 406, 457, 464, 477, 499, 500, 505, 566
Ogawa, M., 62
Ogletree, D. F., 16, 122, 176, 457
Ogura, T., 541–561
Ohana, R. F., 43, 62
Ohkura, R., 125
Ohlendorf, D. H., 62
Oiwa, K., 281, 282, 286
Okada, C. Y., 96
Okada, T., 322
Okamoto, K., 176
Oki, H., 156, 157, 160
Okten, Z., 3, 15–17, 95, 122, 125, 137, 139, 323
Okumus, B., 150
Okuno, D., 279–294
Okun, V. M., 518
Olenych, S., 3, 18, 29–31, 33, 43, 62, 93, 110
Olsen, H., 594, 597
Omabegho, T., 125, 129
Omote, S., 551–554
Ong, N. P., 457
Onoa, B., 566
Oppenheim, A., 92
Orlova, E., 519
Ormo, M., 4
Ormos, P., 384, 385, 390
Oroszi, L., 384, 385, 390
Oroszlan, K., 529
Orrit, M., 3, 457, 468
Orr, J. W., 122, 150, 416
Orte, A., 488
Osawa, H., 62
Ostap, E. M., 361
Oster, G., 281
Ostroverkhova, O., 129
Oswald, F., 18, 36, 37, 62, 111
Oswald, R. E., 217
Ott, M., 467
Oukhaled, G., 612, 613
Ozbek, S., 459
Ozwald, F., 116

P

Pacheco, A. B. F., 534–535
Pacheco, V., 167, 457
Pagratis, N. C., 614
Painter, O., 178
Palmer, A. E., 37

Palmiter, K. A., 322
Paloczi, G. T., 542
Palo, K., 457, 468, 472, 482, 490
Pandey, M., 457
Panorchan, P., 528
Papermaster, B. W., 4
Parker, D., 214
Park, H., 4, 8, 89, 542
Parkin, S. J., 391
Pasternak, C. A., 603
Pastushenko, V. P., 519, 520, 523, 525, 527, 534–536
Patel, A., 599
Patel, G., 457
Patel, R., 542
Patel, S. S., 318, 457
Patlak, J. B., 323
Patterson, G. H., 3, 18, 29, 35–37, 43, 62, 93, 110–113, 117, 118, 122
Pavani, S. R. P., 35, 38, 40, 51, 53–55
Pavlenok, M., 594, 595
Pavone, F. S., 200, 201, 214, 357
Pawley, J., 471
Payne, G., 65, 71–73
Pei, R., 123, 125, 127–129, 133, 136–140, 142, 143
Pelta, J., 612, 613
Pelton, M., 151
Peng, H., 600
Peng, X., 66, 70
Pereira, M. J. B., 122, 176
Perez-Jimenez, R., 406
Periasamy, A., 457, 505
Perkins, A. J., 215, 217
Perkins, T. T., 322, 342, 343, 357, 396, 424
Persson, G., 470, 505
Pesic, J., 151
Peterman, E. J. G., 332, 343, 427–450
Peterman, M. C., 606
Petersen, D. C., 606
Petersen, N. O., 607
Peters, H., 519
Peterson, C. L., 406
Peterson, S., 177
Peters, T., 519
Petrov, D., 382
Petsko, G. A., 467
Pflughoefft, M., 62
Phillips, D., 476
Phillips, R., 205, 607
Phong, A., 444
Pick, H., 62, 88, 114
Picone, S. J., 586, 600
Piehler, J., 64
Pierce, D. W., 3, 4
Pierce, N. A., 125
Piestun, R., 35, 38, 40, 51, 53–55
Pietrasanta, L. I., 542

Pinaud, F., 64
Pinaud, F. F., 4, 22, 62–65, 69, 71–75, 84, 85, 122, 468
Pingoud, A., 430
Piston, D. W., 35–37, 39, 505
Pitaevskii, L. P., 384
Plakhotnik, T., 28
Platz, M. S., 38
Plaxco, K. W., 542
Plenert, M. L., 155
Pljevaljcic, G., 122
Plomp, M., 525
Pomerantz, A. K., 37, 40
Pongor, S., 502
Pons, T., 62
Pouget, N., 200
Pous, J., 527
Pozharski, E., 467
Pralle, A., 330, 339, 390, 412
Prchla, E., 516, 532
Preckel, H., 122
Premnath, V. V., 261
Preuss, R., 142
Prikulis, J., 178
Prinz, H., 529
Probst, R., 157, 161, 169
Prodan, E., 178
Prummer, M., 330, 339, 390
Pschorr, J., 457, 468, 472, 490
Puchner, E. M., 406
Puglisi, J. D., 7, 130
Purcell, T. J., 322–325, 328, 338, 339, 344, 358, 361
Purnell, R. F., 598, 618, 619
Purves, R. D., 604
Pye, V. E., 557
Pyle, A. M., 8, 260

Q

Qian, H., 117, 141, 142, 204, 223, 226
Qimron, U., 268
Qin, F., 200, 214
Quake, S. R., 444
Querol-Audi, J., 527
Quinlan, M. E., 17, 457
Qu, L., 70
Qu, X. H., 15, 123

R

Raab, A., 520, 529
Rabin, Y., 567, 570, 575, 578, 580
Rabouille, C., 557
Radloff, C., 178
Raftery, A. E., 203
Ram, S., 29–31
Ramsden, J., 519
Ramsey, J. M., 168

Ramsey, R. L., 217
Rangelow, I. W., 542
Rankl, C., 515–536
Rao, C. R., 363
Rapoport, T. A., 557
Rappaport, A., 429
Rasnik, I., 130, 176, 419, 457
Ratanabanangkoon, P., 551
Ratzke, C., 457
Rausch, J. W., 150
Rayment, I., 323
Ray, S., 83
Rechberger, W., 178
Rechsteiner, M., 96
Reck-Peterson, S. L., 8, 236, 240
Reece, P. J., 344, 347
Reems, J. A., 260
Reese, T. S., 3, 8, 82, 85
Reichman, D. R., 429
Reich, N. O., 177
Reid, S., 599
Reif, J. H., 125
Reija, B., 457
Reilein, A. R., 85
Reinhard, B. M., 93, 175–195
Reithmayer, M., 527, 534–535
Remington, S. J., 4
Remmert, C. L., 116
Renn, A., 28
Rennekamp, H., 282
Restle, T., 457, 459, 464, 481
Rettig, W., 456
Reuter, R., 280, 457
Revenko, I., 534
Reviakine, I., 546, 551
Revyakin, A., 300, 400, 457, 498
Rhee, M., 594
Ribi, H. O., 551
Rice, P. A., 479
Rice, S. E., 322, 323, 325, 328, 338, 339, 344, 358, 361
Richardson, C. C., 176, 260, 261, 268, 270, 271, 429
Richter, M., 542
Richter, R. P., 555
Rief, M., 3, 236, 240, 323, 339, 358, 361, 406, 542, 566
Rieger, B., 3, 17
Riener, C. K., 130, 529, 531, 536
Rigler, R., 150, 155, 167, 468, 470, 472
Rigneault, H., 203
Rincon-Restrepo, M., 595, 596, 600, 612, 613, 616
Ringler, P., 546
Ringsdorf, H., 551
Rinzler, A. G., 561
Ripley, B. D., 117

Rist, M. J., 122
Ritchie, K., 84, 142, 528, 566
Riveau, B., 63
Rizvi, A. H., 157, 160, 169
Rizzoli, S. O., 3, 4, 62
Robbins, D. L., 482
Robertson, B., 595, 613
Robertson, C. R., 551
Robinson, W., 156, 159, 161
Robinson, W. E., 517
Roca-Cusachs, P., 406
Rocker, C., 18, 111
Röcker, C., 36, 37
Rock, R. S., 3, 15–17, 95, 122, 125, 137, 139, 240, 322, 323, 325, 339, 344, 358, 361
Rodriguez, H. B., 461, 463
Rodriguez, J., 463
Rodriguez-Viejo, J., 66
Rohler, D., 35, 37, 440
Rolfe, P., 546
Roman, E. S., 461, 463
Romanin, C., 529, 531
Romberg, L., 3
Romero, I., 178
Romuss, U., 97
Ronacher, B., 518, 519
Rondelez, Y., 283
Rong, G. X., 93, 181, 195
Roos, M., 472
Rosenberg, S. A., 17, 457
Rosenfeld, S. S., 9, 10
Rosenmann, E., 609
Ross, J., 90, 92
Ross, J. L., 10
Rossmann, M. G., 516
Rostaing, P., 63
Roth, C. M., 90, 92
Rothe, A., 445
Rothemund, P. W. K., 125, 128
Rothwell, P. J., 455–505
Rotman, B., 4
Rouiller, I., 557
Roullier, V., 64
Rouzina, I., 428
Roy, R., 89, 405–424, 504
Ruan, Q. Q., 156, 159, 160, 457
Rubin, D. B., 203
Rubinsztein-Dunlop, H., 382, 391
Rudchenko, S., 127, 129
Rueda, D., 122
Runzler, D., 519
Rusek, M., 217
Russell, D. W., 302
Rust, M. J., 29, 40, 111, 123, 137
Ruttinger, S., 472
Ryan, J. F., 542, 558
Rypniewski, W. R., 323

S

Sabanayagam, C. R., 475
Sable, J., 62
Sacconi, L., 200, 201, 214
Sachs, F., 200, 214
Safer, D., 8
Saito, K., 3, 89, 542, 544
Sakamoto, T., 323
Sakashita, M., 542, 559
Sakata-Sogawa, K., 89
Sakaue-Sawano, A., 62
Sakmann, B., 2, 215, 222, 602, 616
Sako, Y., 63, 84, 141
Saleh, O. A., 297–319, 400
Salemme, F. R., 62
Salih, A., 18, 36, 37, 111
Salmon, E. D., 85
Salome, L., 200
Salpeter, E. E., 217
Samaii, L., 127
Sambrook, J., 302
Samuel, R., 35, 37–41
Samuely, T., 200
Sánchez-Quesada, J., 598, 619
Sanchez-Ruiz, J. M., 406
Sanden, T., 470, 505
Sandhagen, C., 457, 471, 496, 505
Sansom, M. S. P., 217
Santoro, S. W., 127
Sasaki, K., 480
Sauer-Budge, A. F., 578, 579
Sauer, M., 19, 457, 459, 464, 468, 476, 492, 504
Sawata, S., 614
Saxton, M. J., 84, 116, 141, 142
Sazinsky, S. L., 71, 72
Schabert, F. A., 519
Schaeffer, E., 357
Schaeffer, P. M., 260, 268, 429
Schaertl, S., 122
Schafer, L. V., 406
Schafer, M. M., 396
Schaffer, J., 468, 471, 472, 476, 480
Schäffer, T. E., 542
Schaible, D., 64
Schatz, G. C., 177, 178, 195
Scheller, R. H., 484
Scherer, N. F., 15, 123, 151
Scheuring, S., 546
Scheurle, D., 217
Schibel, A. E., 600
Schilcher, K., 528, 529
Schilde, J., 43
Schindler, H., 2, 4, 84, 124, 468, 520, 523, 525, 528, 529, 532, 536
Schitter, G., 542
Schlamp, M. C., 66
Schlierf, M., 405–424, 566

Schliwa, M., 2, 3, 7, 82
Schlue, W.-R., 602
Schmid, S. L., 115
Schmidt-Base, K., 323
Schmidt, C. F., 8, 10, 83, 110, 170, 330, 332, 339, 343, 395, 432, 433, 435, 436, 440, 443
Schmidt, J. J., 28, 598, 618, 619
Schmidt, M. A., 151
Schmidt, R., 21
Schmidt, T., 2, 4, 35, 37, 84, 124, 440, 468
Schmidt, U., 519
Schmitt, F., 18, 36, 37, 111
Schmitz, S., 324
Schmitz-Salue, R., 62
Schnapp, B. J., 2, 3, 8, 10, 28, 83, 84, 94, 110, 170
Schneibel, J. H., 168
Schneider, B., 97
Schnitzer, M. J., 83, 223, 226, 241–243, 339, 367
Scholes, G. D., 457, 461, 463
Schönle, A., 37, 43
Schrader, S., 456
Schriven, E. F. V., 38
Schröder, G. F., 457, 464, 472, 477, 487, 488, 496, 499, 500
Schroer, T. A., 125
Schuck, P. J., 37, 40, 179
Schuler, B., 122, 150, 406, 457, 459, 464, 499
Schulten, K., 332, 407, 409, 417, 423, 566, 587
Schultz, P. G., 122, 150, 176, 457
Schultz, S., 178
Schulz, A., 468, 476
Schumakovitch, I., 529
Schuttpelz, M., 19
Schutze, K., 2, 3
Schütz, G. J., 2, 4, 84, 124, 468
Schwab, A., 142
Schwarz, J., 97
Schwarz, S. C., 97
Schweinberger, E., 280, 457, 472, 487, 488, 496, 499
Schweitzer, G., 461, 463, 470
Schwille, P., 456, 457, 468, 472, 496
Scott, A. J., 204
Screaton, G. R., 614
Sczakiel, G., 194
Sedat, J. W., 3, 22
Seeman, N. C., 125, 129
Segall, D. E., 205
Segel, I. H., 225
Seidel, C. A. M., 280, 455–505
Seidel, R., 260
Seifert, H., 456
Seitzinger, N. K., 457
Sellers, J. R., 240, 323, 324, 339
Selvin, P. R., 1–22, 83–86, 88, 89, 93, 95, 110, 122, 124, 125, 137, 176, 415, 419, 422, 457, 542
Semler, B. L., 517, 534

Semyonov, A. N., 37, 40, 151
Sengupta, P., 457, 505
Sens, P., 607
Seol, Y., 342, 343, 424
Serge, A., 203
Serpinskaya, A. S., 2, 11, 12, 84, 85, 95
Sethna, J. P., 400
Seuffert, I., 396
Sewing, A., 122
Sexton, L. T., 594
Shaevitz, J. W., 2, 53, 223, 235, 237, 240, 242, 245, 343, 378, 395, 410, 411, 424, 433
Shaner, N. C., 4, 37, 85, 87, 97, 113, 445
Shank, C. V., 118
Shank, E. A., 406, 566
Shannon, C. E., 29–31, 33
Shan, X. Y., 142
Shao, L., 3
Shapiro, B., 157, 161, 169
Shapiro, L., 29, 31, 35, 43, 45–50, 150
Sha, R., 125, 129
Shcheglov, A. S., 113
Shear, J. B., 155
Shearwin, K. E., 215, 217
Sheetz, M. P., 2, 3, 8, 28, 62, 82, 84, 85, 94, 114, 117, 141, 142, 406
Sheikholeslami, S., 92, 181, 186
Sheinin, M., 400
Shen, G., 156, 159, 161
Shepp, L., 241, 242
Shera, E. B., 3, 37, 457
Sherman, W. B., 125
Sherratt, D., 309
Shibata, M., 541–561
Shih, W. M., 125
Shi, J., 122
Shimabukuro, K., 285
Shimada, J., 429
Shima, T., 125
Shin, J. S., 125
Shin, S.-H., 593
Shiroguchi, K., 125
Shklovskii, B. I., 577, 595
Shlyakhtenko, L. S., 529
Shroff, H., 3, 18, 30, 31, 33, 36, 110, 112–114, 117, 118
Shtengel, G., 51, 118
Shubeita, G. T., 83
Shuman, H., 361
Shustak, C., 577, 578, 594
Sielaff, H., 282
Siggia, E. D., 3, 311, 418, 536
Sigworth, F. J., 200, 201, 214, 215
Sikorski, Z., 156, 159, 161
Silberberg, S. D., 616
Siliciano, R. F., 528
Sim, A. Y., 324
Simmons, C. R., 385, 391, 392

Simmons, R. M., 3, 322, 325, 339, 361
Simon, R., 457, 505
Simon, S. M., 63, 89, 606
Simpson, D., 43, 62
Simpson, S. H., 382
Sims, P. A., 83, 84, 86, 92, 93, 156, 157, 160
Sindbert, S., 473, 489, 492, 499
Sisamakis, E., 455–505
Sischka, A., 396
Siu, M., 93, 178, 182, 184
Sivak, D. A., 177
Sivaramakrishnan, S., 321–372
Skalicky, J. J., 459
Skewis, L. R., 93, 181, 186, 195
Sligar, S., 587
Smalley, E., 561
Smeets, R. M. M., 587, 600
Smiley, R. D., 176
Smirnova, I., 457
Smith, A. M., 22
Smith, B. L., 318, 542
Smith, C. L., 406
Smith, D. A., 3
Smith, D. E., 536
Smith, D. R., 178
Smith, L. M., 133
Smith, R., 323, 525
Smith, S. B., 3, 151, 268, 269, 311, 322, 341, 381, 406, 418, 428, 431, 447, 536, 566
Snapp, E. L., 87, 456
Snyers, L., 519
Sobhy, M. A., 89, 122, 406, 413
Sofia, S. J., 261
Sokolov, I. M., 142
Sommer, T., 557
Song, L., 577, 578, 594, 610
Sonnenfeld, A., 150
Sonnichsen, C., 179
Soper, S. A., 3, 37, 457
So, P. T. C., 150, 468, 472
Sosa, H., 444
Sosinsky, G. E., 89
Sougrat, R., 3, 18, 29, 43, 51, 62, 93, 110, 118
Soundararajan, N., 38
Sparrow, J. C., 361
Spatz, J. P., 179
Spiering, M. M., 297–319
Spindler, K. D., 18, 36, 37, 111
Spink, B. J., 324
Springer, T. A., 406
Spudich, J. A., 3, 15–17, 92, 95, 122, 125, 137, 139, 214, 321–372
Staroverov, D. B., 113
Stasiak, A. Z., 309
Stavans, J., 92
Steel, D., 122
Steele, B. L., 468
Stefanovic, D., 127, 129, 144

Steffensen, M. B., 593
Stegun, I. A., 573
Steiger, C., 519
Steigmiller, S., 280, 457
Stein, A., 487
Steinbach, P. A., 4, 37, 85, 87, 97, 445
Steinbach, P. J., 464
Stein, D., 587
Steitz, T. A., 479
Stelzer, E. H. K., 22, 330, 339, 390, 412
Stephens, D. J., 428
Stevens, B. C., 459
Stevenson, G. V. W., 115
Stevenson, P. E., 38
Stewart, R. J., 395
Stiel, A. C., 43
Stienen, G. J. M., 433
Stoddart, D., 591–619
Stojanovic, M. N., 123, 125, 127–129, 133, 136–140, 142–144
Stone, M. D., 170, 381
Storch, A., 97
Storhoff, J. J., 179
Storm, A. J., 587
Strehmel, B., 456
Strickler, J. H., 22
Strick, T. R., 268, 269, 298–300, 308, 311, 400, 402, 413, 459
Striker, G., 468, 476, 480
Stroh, C. M., 519, 529, 531
Strop, P., 498
Strouse, G. F., 177
Strunz, T., 529
Stryer, L., 122, 464
Stühmeier, F., 122, 502
Stuurman, N., 176
Subach, F. V., 36, 43, 113, 117, 118
Subramaniam, V., 339, 468, 476, 480
Su, B. Y., 444
Sugawa, M., 459
Su, K. H., 178, 183, 184
Sullivan, J. M., 217
Sullivan, L., 211
Sun, B., 318
Sundaramurthy, A., 179
Sundaresan, G., 4, 22, 62, 84, 122
Sung, J., 321–372
Sun, H. Y., 459
Sun, S. X., 281, 444, 528
Surtees, J. A., 429
Sutin, J., 587
Sutoh, K., 125
Suzuki, K., 84
Svoboda, K., 8, 10, 83, 110, 170, 223, 241, 242, 299, 325, 378, 395, 434
Sweeney, H. L., 3, 8, 15, 323, 324, 358, 361
Syeda, R., 605
Syed, S., 2, 9–11, 83, 84, 86, 88, 95, 125, 457

Symons, M. J., 204
Sytina, O. A., 459
Szabo, A., 459, 490, 492, 497, 571, 573–576, 581–584, 618
Szabo, G., 607
Szollosi, J., 457

T

Tabler, M., 194
Tabor, C. W., 194
Tabor, H., 194
Tabor, S., 260, 268
Taft, R. W., 38
Tafvizi, A., 429
Tajon, C., 92
Takada, S., 281
Takahashi, M., 63, 261, 268, 271, 429
Takai, E., 542, 544
Takao, K., 62
Takeuchi, S., 283
Takeuchi, Y., 555
Tamarat, P., 468
Tanaka, H., 332, 406, 409, 429
Tanaka, S., 546, 560
Tanaka, Y., 560
Tang, J., 118
Taniguchi, M., 551–554
Tanner, N. A., 259–277, 429
Tans, S. J., 536
Tarsa, P. B., 406, 424, 440
Tavan, P., 528
Taxis, C., 557
Taylor, E. W., 323
Taylor, S. K., 123, 125, 127–129, 133, 136–140, 142, 143
Telford, W. G., 445
Tellinghuisen, J., 468, 471
Terekhov, A., 156, 159, 161
Tessier, B., 555
Tessier, C., 555
Thai, V., 467
Thaler, C., 457
Thalhammer, S., 534, 535
Thiel, A. J., 133
Thirumalai, D., 194
Thomen, P., 431
Thompson, J. B., 542, 605
Thompson, M. A., 27–55
Thompson, N. L., 3, 468
Thompson, R. E., 5, 7, 29–31, 48, 53, 83, 103, 114, 115, 124, 137, 269
Thurner, P. J., 542
Thyberg, P., 470
Tian, Y., 126, 127
Tibshirani, R. J., 235, 364
Timp, G., 587
Timp, R., 587

Ting, A. Y., 62, 64, 114, 150, 176
Tinnefeld, P., 90, 92, 468, 470, 492, 504
Tinoco, I. J., 170, 243, 318, 428, 536, 566
Tobi, D., 459
Tobkes, N., 610
Toda, A., 542, 544
Tokunaga, M., 3, 89, 332, 406, 409, 429
Tolic-Norrelykke, I. M., 358
Tolic-Norrelykke, S. F., 357
Tomchick, D. R., 323
Tomic, S., 62
Tomishige, M., 8, 83, 95, 176
Tompa, P., 499
Toner, M., 151
Tonsing, K., 396
Toprak, E., 1–22, 542
Torres, T., 472
Tosteson, M. T., 602, 603, 606
Tougu, K., 260
Tour, O., 89
Toussaint, K. C., 151
Toyoshima, Y. Y., 125
Tran, P., 85
Tran, S. L., 83
Tranum-Jensen, J., 610
Trautman, J. K., 3
Travis, J. L., 85
Tregear, R. T., 325
Tresset, G., 283
Triller, A., 63
Troll, M., 596
Truong, K., 122
Trybus, K. M., 17, 87, 96, 323, 324
Tsay, J. M., 4, 22, 62, 63, 69, 84, 122
Tselentis, N. K., 31, 35, 43, 45–50
Tsemitsidis, D., 215
Tseng, Y., 528
Tsien, R. Y., 4, 32, 35, 37, 43, 62, 85, 87, 88, 96, 97, 122, 445, 456, 457
Tsunaka, Y., 542
Tsygankov, D., 237, 239, 242
Tuma, J., 472
Tuma, M. C., 85
Tuma, R., 600
Turberfield, A. J., 125, 126
Turlan, C., 200
Twieg, R. J., 35, 37–41, 51, 53–55, 129, 151
Tyn, M. T., 155
Tyska, M. J., 322

U

Uchihashi, T., 541–561
Ulbrich, M. H., 457
Ullmann, D., 457, 468, 472, 482, 490
Urh, M., 43, 62
Ursell, T., 607
Usherwood, P. N. R., 217
Uyeda, T. Q., 322

V

Vach, P., 175–195
Vadgama, P., 546
Valentine, M. T., 158, 434
Vale, R. D., 3, 4, 8, 82, 83, 85, 95, 176, 236, 240
Valeri, A., 455–505
Valeur, B., 456, 461, 463, 468
Valle, F., 200
Valtorta, F., 116
van de Linde, S., 19
Vandenbelt, J. M., 38
van den Broek, B., 200, 201, 429
van der Meer, B. W., 460, 461
van der Werf, K. O., 339
van Dijk, M. A., 332, 432, 436, 440, 443
van Dorp, S., 600
van Duffelen, M., 10
Van Duyne, R. P., 178
Van Furth, R., 95
van Grondelle, R., 459
van Mameren, J., 332, 429–430, 440, 443, 447–450
van Oijen, A. M., 28, 176, 259–277, 429, 459
van Rooijen, B. D., 339
van Stokkum, I. H. M., 459
Vanzi, F., 200, 201, 214
Vargas, G., 609
Vaziri, A., 118
Veigel, C., 240, 323, 324, 339, 361
Vendra, G., 10
Vercoutere, W., 594, 597
Verdaguer, N., 522, 527
Verdier, L., 472
Verheijen, W., 37, 117
Verhey, K. J., 4, 81–103
Verkhusha, V. V., 36, 43, 113, 117, 118
Verkman, A. S., 142, 482
Vermeulen, K. C., 433
Verschuren, H., 117
Vershinin, M., 83
Viani, M. B., 542
Viasnoff, V., 431, 570, 575, 576, 578, 580
Victorie, J. G., 517
Viergever, M. A., 52
Vilardaga, J. P., 62
Villalobos, V., 62
Viskovic, T., 65
Visram, H., 570, 575, 578, 580
Visscher, K., 83, 339, 367, 388, 395, 399
Visser, J. A., 95
Vladescu, I. D., 428
Vogel, H., 62, 88, 114, 150, 155, 167
Vogelsang, J., 492, 504
Vogel, S. S., 457
Voïtchovsky, K., 542, 558
Voldman, J., 151
Volkmer, A., 468, 471, 472, 476, 480
Volkwein, C., 557

Vollmer, M., 177
Volpe, G., 382
von der Hocht, I., 167, 457
von Middendorff, C., 43
von Plessen, G., 179
Vosch, T., 470
Vostrikova, L. Y., 202
Vrljic, M., 498

W

Wacker, S. A., 62
Wade, R. C., 62
Walker, J. E., 281, 283
Walker, S., 157, 161
Walker, W. F., 8, 103, 111, 115
Wallace, B. A., 610
Wallace, M., 605
Walla, P. J., 487
Wallin, M., 237
Wallrabe, H., 457, 505
Walter, J. C., 429
Walter, N. G., 89, 121–144, 176, 406, 413
Walther, K. A., 406
Walther, T. C., 557
Wan, E., 444
Wang, C. J., 3
Wang, F., 323, 324, 339
Wang, H. Y., 37, 40, 93, 168, 181, 195, 242, 281
Wang, J., 479
Wang, M. D., 311, 318, 381, 382, 385, 391, 392, 396, 398, 400, 406
Wang, Q., 159, 161, 169
Wang, R. F., 133
Wang, S.-W., 551
Wang, T., 62
Wang, W., 3, 18, 19, 51, 53
Wang, Y. B., 444
Wang, Y. F., 457, 472
Wanunu, M., 566, 567, 587
Ward, E. S., 29–31
Warren, G., 557
Warrick, H. M., 322, 323, 325, 339, 344, 358, 361
Warshaw, D. M., 17, 87, 96, 322–324
Watanabe, N., 97
Watanabe, T. M., 63
Waterman, C. M., 51, 118
Watkins, L. P., 201, 203, 206
Webb, M. R., 323
Webb, W. W., 2, 5, 7, 22, 29–31, 48, 53, 83, 84, 103, 114, 115, 122–124, 137, 142, 269, 457, 472
Weber, P., 62
Weber, R., 37, 40
Wegner, F., 97
Weidtkamp-Peters, S., 457, 505
Wei, K., 318

Weinreb, G. E., 3
Wei, Q. H., 178, 183, 184
Weiss, S., 3, 4, 16, 22, 29, 61–76, 84, 85, 122, 150, 176, 406, 413, 428, 457, 459, 468, 471, 472, 482, 492, 494, 504
Weis, W. I., 484, 557
Weitz, D. A., 156, 157, 160
Wells, A. L., 323, 358
Welte, M. A., 83
Wendel, M., 561
Wende, W., 430
Wendoloski, J. J., 62
Weninger, K. R., 498
Wen, J.-D., 170, 243, 428
Wennmalm, S., 63
Werner, J. H., 156, 158, 160
Wesenberg, G., 323
Westlund, P. O., 466
Weston, K. D., 468, 470
Westphal, V., 3, 4
Whitaker, R. D., 586, 600
White, D. C., 325
White, H. D., 3, 18, 36, 323
White, H. S., 600
Whiteley, N. M., 525
White, S. H., 606, 607
Whitesides, G. M., 602, 603, 606
White, Y., 156, 159, 161
Wichmann, J., 3, 28
Wickenden, M., 122
Widengren, J., 457, 460, 468, 470, 472, 485, 487, 488, 496, 499, 505
Wiebusch, G., 19
Wiedenmann, J., 18, 36, 37, 62, 111, 116
Wielert-Badt, S., 520, 529
Wiener, D. M., 81–103
Wiggin, M., 597
Wiggins, P., 607
Wiita, A. P., 406
Wildinga, L., 528
Wild, U. P., 28
Willets, K. A., 37, 40, 129
Williams, G., 122
Williams, M. C., 428
Williamson, J. R., 122, 150, 416
Williams, P. M., 459
Williams, R. M., 22
Willig, K. I., 3
Wilson-Kubalek, E. M., 557
Wilson, M. A., 467
Wimmer, E., 517
Windoffer, R., 440
Winfree, E., 123, 125, 127–129, 133, 136–140, 142, 143
Winkelmann, D. A., 323
Winkler, D., 457
Winoto, L., 3
Winters-Hilt, S., 594, 597

Author Index

Wirtz, D., 528
Wirz, J., 461, 463
Wise, F. W., 22
Wiseman, P. W., 457, 505
Witkowski, J. A., 117
Wittrup, K. D., 64, 66, 71, 72
Woehlke, G., 7, 82
Wohland, T., 150, 155, 167
Wohrl, B. M., 457, 459, 464, 481
Wolf, D. H., 557
Wolfrum, J., 468, 476
Wolf-Watz, M., 459, 467
Wolter, S., 19
Wolynes, P. G., 459
Wonderlin, W. F., 602, 603, 606
Wong, C. M., 157, 160, 169
Wong, O. K., 200
Wong, W. P., 406
Wood, K., 43
Wood, M. G., 43, 62
Woodside, M. T., 103, 236, 406, 566
Woodson, S. A., 194
Work, S. S., 17
Wozniak, A. K., 457, 464, 473, 477, 489, 492, 499, 500, 505
Wright, G., 546
Wruss, J., 519, 528
Wu, A. M., 4, 22, 62, 84, 122
Wu, C. A., 260
Wu, D., 15, 123
Wu, H., 599
Wuite, G. J. L., 200, 201, 268, 427–450
Wu, J., 323
Wu, P. G., 457
Wurm, C. A., 21
Wu, X., 444, 457
Wyman, C., 429, 430, 447–450
Wyss, R., 594, 612

X

Xiao, M., 444
Xia, Y. N., 195
Xie, S. N., 223, 224
Xie, X. S., 83, 84, 86, 92, 93, 122, 150, 176, 260, 268, 459
Xing, J., 229
Xi, X. G., 318, 420
Xu, C. S., 156, 157, 160, 169
Xu, J., 83
Xun, L., 150
Xu, W., 243

Y

Yafuso, M., 606
Yamaguchi, A., 150
Yamamoto, D., 541–561
Yamamoto-Hino, M., 36, 110, 111

Yamamoto, M., 141
Yamaoka, K., 233
Yamashita, H., 541–561
Yanagida, T., 3, 84, 89, 92, 322, 332, 406, 409, 429, 440, 459, 555
Yan, B., 178
Yang, H., 156, 157, 160, 169, 201, 203, 206, 215, 217
Yang, J. H., 444, 610
Yang, L., 178
Yang, W., 281
Yang, Z., 8
Yan, H., 123, 125, 127–129, 133, 136–140, 142, 143
Yao, N. Y., 260
Yardimci, H., 10
Yassif, J. M., 175–195
Yasuda, R., 280–285, 290, 378
Yeh, R. C., 406
Yeh, Y., 429
Ye, Y., 557
Yildiz, A., 2, 7–9, 83, 86, 93, 95, 110, 122, 124, 125, 137, 236, 240
Yin, H., 200, 381
Yin, P., 125–127
Yokota, H., 420
Yoshida, M., 280–286, 290, 378
Yoshimura, S. H., 170, 428
You, C., 64
Young, M., 268, 428
Ypey, D. L., 95
Yue, S., 445
Yu, J., 113, 118, 332, 407, 409, 417, 423
Yun, C. S., 177
Yu, P., 43
Yu, W. W., 70

Z

Zandbergen, H. W., 587
Zander, C., 468, 476
Zaychikov, E., 457, 468, 472
Zechner, E. L., 260
Zemlin, F., 558
Zeri, A.-C., 170, 428
Zhang, B., 318
Zhang, C. Z., 406
Zhang, J. X., 150, 577, 595
Zhang, N., 8
Zhang, S. F., 546
Zhang, X., 178. 183, 184, 318, 406, 557
Zhang, Z., 62
Zhao, L. L., 177, 178, 195
Zhao, M., 444
Zhao, R., 122
Zhao, Z. S., 444
Zheng, W., 127
Zhou, J., 534
Zhou, M., 63

Zhou, R., 332, 405–424
Zhou, Y., 234, 249, 528
Zhuang, X. W., 3, 18–20, 29, 35, 37, 40, 41, 43, 51, 53, 84, 111, 122, 123, 137, 150, 176, 234, 249, 542
Zhuang, Z. H., 176, 301, 315, 316
Zhu, R., 515–536
Ziebarth, J. M., 606
Zima, A. V., 603
Zimmerberg, J., 116

Zimmer, C., 116
Zimmermann, B., 280, 457
Zimprich, C., 43, 62
Zipfel, W. R., 22
Zochowski, M., 217
Zuckermann, M. J., 127
zu Hörste, G. M., 560
Zunger, A., 457
Zurla, C., 200, 204, 205, 209, 211, 214, 215, 217

Subject Index

A

ABEL traps. *See* Anti-Brownian electrokinetic traps
Accuracy, 2
Adenosine triphosphate (ATP), 280–281
AFM. *See* Atomic force microscopy
AFM cantilevers, 299
Alignment, dual-beam optical trap
  bright-field setup, microscope stability, 344–345
  single-beam trap
    acousto-optic deflectors (AOD), 350
    bright-field light path, 351
    dump setting, 347
    expansion and fine steering, L1 and L2 lense, 349–350
    L3 and L4 lense, expansion and coarse beam steering, 348–349
    laser, profile and stability, 346
    layout setting, 347–348
    mounting, trapping laser and isolator, 346–347
    OBJ and CON setting, 350–351
    path adjustment, OBJ and CON, 348
    PSD1 back focal plane detection, 352–353
    steering, optical table, 347
    steps, 345
    testing, 351–352
  three-bead assay
    actin dumbbell visualization, 356
    description, 353
    myosin, laser beam path, 354–355
    position detecting components, 354
    second trapping beam path, 354
    tubes and box, 356
Anisotropic particle fabrication
  asymmetric shape
    polystyrene sphere compression, 390–391
    single-molecule experiment, 391
  birefringent quartz cylinders, 393
  mask design and wafers, 392
  optical asymmetry
    birefringent materials, 391–392
    cylinder-shaped particles, 392
  protocol, 392–394
Annexins, 555
Anti-Brownian electrokinetic traps (ABEL traps)
  illumination system
    confocal scan pattern, 165
    EODs and optical system, 164
    pixel spacing, 165
  microfluidics
    cell, 161–162
    polydimethylsiloxane (PDMS), 168–169
    sample holder, 168
  optical layout, 163
  photon-by-photon feedback, 163–164
  polydimethylsiloxane (PDMS) trapping chamber, 162–163
  tracking and feedback system
    field-programmable gate array, 165, 167
    Kalman filter signal-processing, 166–167
    single axis algorithm, 165–166
  trapping, 162
Anti-Brownian traps
  ABEL trap
    description, 161–162
    feedback system, 162
    illumination system, 164–165
    microfluidics, 168–169
    photon-by-photon feedback, 163–164
    tracking and feedback system, 165–167
    trapping, 162
  applications, 169–170
  confinement, factors, 152–153
  diffusion and diffraction balance, 154
  feedback systems
    electrokinetic, 160–161
    forces, 155–156
    laser tracking, 160
    parameters, 155
    particle position adjustment, 159
    photon diffraction, 153–154
    stage, 160
    velocity, 154–155
  general scheme, 153
  literature and performance, 156
  single fluorophores, en route
    confinement, 170
    Cy3, 171
  tracking systems
    camera, 156–157
    multifocus, 159
    multiphotodiode, 157–158
    scanning, 158–159
APDs. *See* Avalanche photodiodes

649

Subject Index

Atomic force microscopy (AFM). *See also*
High-speed AFM techniques; Human
rhinovirus; Mica surface, AFM imaging
antibodies, tip functionalization
APTES, PDP–PEG–NHS and SATP, 531
ethanolamine, 530–531
SATP to PDP–PEG-tip coupling, 531–532
cantilever, 299
spiders, 129
tip unbinding force, HRV2 and monoclonal
antibodies, 530
virus visualization, 517
Avalanche photodiodes (APDs). *See also*
Single-molecule confocal microscopy
Geiger mode, 157
PMTs, 157–158
pulse, 165–166
silicon, 407
Azido DCDHF fluorogen, 39

B

Bacteriorhodopsin 2D crystals, 558
Bell's formula, 572
Bifunctional rhodamine dyes, 8–9
Biomolecular motors
cellular functions, 82
in living cells, 83–84
mechanism, 82–83
Bright-field imaging with one nanometer
accuracy (bFIONA), 11–12, 86
Brownian motion. *See also* Optical tweezer;
DNA-protein interactions; Streptavidin 2D
crystals, dynamic AFM imaging
features, force field, 152
instrumentation, optical trap, 341
optical trapping, 410
single-molecule spectroscopy
diffusing molecules, 150–151
feedback traps scales, 151
immobilization, 150
laser tweezers, 151
optical trapping, dielectrophoresis and
magnetic tweezers, 151

C

Cell culture, 97–99
Change-point (CP) algorithm. *See also*
DNA-looping kinetics, TPM
determination, 207
false-positive, 202
log-likelihood function, 201–202
one-dimensional parameter, 202–204
Clampfit software package, 618–619
Clamp implementation, OTW
EOM, 400
form/material birefringence, 400
passive mode, 398–399

realization, 399
servo loop, 399–400
Confocal fluorescence microscopy, 428
Conventional microlithographic techniques, 391
β–γ Cross-link biochemical assay
ATP hydrolysis activity, 284
subunit formation, 285
Cross-link mutant design
β and γ subunits cysteine residue position, 283
βE395 and γR75, γ-carbon distance, 284
α_3 $\beta_3\gamma$ subcomplex, 282–283
p97 2D Crystals, 557–558
Cy3–Cy5 covalent heterodimers
photophysical properties, 42–43
photoswitching system, 40–41
single-molecule behavior, 43
synthesis, 41–42

D

Data acquisition, single nucleic acid molecules
analysis
digitization, 616
DNA/RNA
amplitude, dwell times and interevent
intervals, 619
single-channel search, 618
filtering, 616
protocols
Clampex, 617
episodic stimulation, 618
Gap Free, 617
variable-length event, 617
sampling, 616–617
Data analysis, smFERT
MFD 2D
biomolecule biological function, 503
photophysical assumption, 503–504
transfer efficiency, 503
PDA, 504
DH-PALM microscope, 52
Digital CCD cameras, 93
DNA-based nanowalkers
behavior-based molecular robots, 127–128
characterizing methods
processive movement mechanism, 126
spider
biological molecular proteins, 125
deoxyribozymes catalytic power
(*See* DNAzymes)
description, 127
DNA-looping kinetics, TPM
analysis, data
Boltzmann distribution, 205
CP-EM method, 204
expectation-maximization (EM) routine,
207–208
Gaussian distribution, 204–205
log-likelihood, 206

Subject Index

Rayleigh function, 209
segmentation algorithm, CPs, 206–208
Weibull function, 209
CI-induced looping, λ-DNA
 CP-EM and HAT analysis, 215–217
 description, 215
 dwell times, 217
CP algorithm
 false-positive, 202
 log-likelihood function, 201–202
 one-dimensional parameter, 202–203
data clustering and expectation-maximization
 algorithm
 CP, distance, 203
 M CPs, 203–204
 steps, 204
mechanism, 200
performance
 CPs trajectories, 209–210
 dwell time, 210–211
 looped and unlooped lifetimes, 211–212
vs. threshold method
 CP-EM, 214–215
 dwell times, 214
 HAT method, 212, 214
DNA replication, single-molecule level
fluorescence visualization
 data analysis, 275–277
 flow chamber construction, 272–273
 reaction and imaging, 273–275
 rolling-circle template preparation, 271–272
loops, tethered bead motion
 data analysis, 269–271
 flow chamber construction, 265–267
 forked λ-DNA substrate preparation, 263–264
 glass coverslip functionalization, 261–262
 polystyrene beads functionalization, 262–263
 reaction and imaging, 268–269
DNA unzipping kinetics, NFS
duplex molecules, 578
hairpin measurements, 579
histogram transformation method
 lifetime, 582–584
 microscopic theory, 584
 properties, 578
protein pore α-HL, 577–578
single base mismatches, 579
temperature rescaling
 data, 585
 vs. voltage, 584–585
time distribution, 578–579
voltage-ramp data (See Voltage-ramp data, DNA unzipping)
DNAzymes, 125–127
Double-helix point spread function (DH-PSF)
 generation, 51

single-molecule superlocalization, 54
SLM, 51–52
Double-stranded DNA (dsDNA) molecule
 supercoiling, 401
 template construction, 400–401
 tethering, 401
 twisting, 402
 vertical tension, 401–402
Rad51 binding, DNA-protein interactions
 description, 448–449
 experiment, 449
 fluorescence imaging, 449–450
 replication protein A binding, DNA-protein interactions, 448
 short, preparation, 615
dsDNA. See Double-stranded DNA
Dual-beam optical trap
 assay, 326
 compliance correction
 binding and unbinding events, 367
 myosin, stroke size, 369
 stroke sizes, nonprocessive motors, 365–367
 three-bead assay, 367
 trap stiffnesses, 368
 setup, 329
 three-bead assay (See also Alignment, dual-beam optical trap)
 design, 327
Dual channel image recording, 94
Dwell time, optical trap experiment
 bootstrap method, 364–365
 data graphing
 commercial software, 363
 cumulative distributions, 362–363
 histogram, 362
 maximum likelihood estimation
 definition, 363–364
 fitting method, 363
 motor-like myosin V, 361
 residual, 365

E

Electrical recording, single nucleic acid molecules analysis
bilayers preparation
 capacitance measurement, 608–609
 DPhPC, 607
 hexadecane penetration, 607
 Langmuir–Blodgett lipid monolayers, 606
 Plateau–Gibbs suction, 606
 stability, 607–608
chambers
 open planar bilayer, 605
 planar bilayer formation, 605
 zapping, 606
electrodes preparation

Electrical recording, single nucleic acid
  molecules analysis (cont.)
  Ag/AgCl, 603–604
  3 M KCl/agar, 604–605
  equipments
    Axopatch 200B patch-clamp amplifiers, 602
    internal noise, planar bilayer system,
      602–603
  Faraday enclosure, 603
  perfusion, 610
  pore insertion and DNA adding, 609
  single-channel experiments, 601–602
Electrokinetic feedback, 160–161
Electroosmosis, 160
Electrophoresis
  ABEL traps, 160
  biomolecules, 151
Enhanced yellow fluorescent protein (EYFP)
  EYFP-MreB, *C. crescentus* cells, 46
  fluorescent protein fusions, 43, 45
  photoreactivation, live cells, 45–46
  photoswitchable emitter, 43
EYFP-MreB protein
  fusion, live *C. crescentus* cells, 45
  super-resolution images, 48–49

F

F_1-ATPase
  AMP-PNP and N_3 pausing
    inhibitors, 293
  $\alpha_3\beta_3\gamma$ subcomplex, 294
  crystal structures conformation
    ATP limiting conditions, 281–282
    and single-molecule two stable state, 282
  rotation, 281
  sample preparation
    β–γ cross-link biochemical assay, 284–285
    cross-link mutant design, 282–284
    hybrid F_1, 285–287
  single-molecule cross-link experiment
    formation, 289–291
    hybrid F_1, one β(E190D/394C), 288–289
    rotation assay, 287
  $\alpha_3\beta_3\gamma$ subcomplex pause position, 291–293
FIONA. *See* Fluorescence imaging with one
  nano-meter accuracy
Flow cell heater, 273
Fluctuation, cycle completion time
  fitting distribution
    decay rates, 235
    dwell time, 232–233
    gamma, 234
    kinetic information, 234
    moment calculation, 235–236
  pathway and steps, multiple
    correlated statistics, 237
    dwell time, 236

enzymatic dynamic classes, 237
kinetic mechanism, 240
statistical test, 239–240
stepping traces, 237–239
Fluorescence image spectroscopy, 505
Fluorescence imaging with one nano-meter
  accuracy (FIONA). *See also* Kinesin
  accuracy, 102
  description, 95
  fluorescent marker, 102
  limitation, 103
  live organism
    ELKS, 13
    fluorescently labeled protein complexes, 14
    genetic engineering, 12–13
    and SHREC, 125
  singly fluorophore-labeled myosin V motor
    proteins, 124
  without fluorescence
    bFIONA, 10–12
    Xenopus melanosomes tracking, 10
Fluorescence microscopy. *See also* Protein
  interactions, DNA
  cellular processes, 63
  photobleaching
    cause, 439–440
    fluorophores, 440
    proteins and nucleic acids, 413
  single molecule
    design, 437–438
    epi-fluorescence illumination, 438–439
    nano-technology, 123
    optical trapping, 438
    trapping laser, 439
  Stokes' shift, 179
  super-resolution, 128
Fluorescence photoactivated localization
  microscopy (FPALM), 3, 18, 112
Fluorescence resonance energy transfer (FRET).
  See Förster resonance energy transfer
Fluorescence, smFRET
  accuracy, 473–474
  biomolecular processes timescale
    confocal microscopy (CM), 468
    detection, 467
    2-photon excitation microscopy (2PM),
      467–468
    protein structure, 466–467
  dimension
    characteristics, 468
    dye intensity, 470
    molecular systems, 468
  experimental setup and data registration
    confocal epi-illuminated microscope, 470
    FIDA, 472
    FRET, 470–471
    intensity fluctuation calculation, 472
    MFD, 472

Subject Index

TCSPC, 471
  trace photon burst, 471–472
Fluorescence visualization, DNA replication
  data ananlysis, 275–277
  flow chamber construction, 272–273
  reaction and imaging, 273–275
  rolling-circle template preparation, 271–272
Fluorescent labeling
  DNA, 444
  proteins
    GFP, 445
  techniques, 444–445
Fluorescent proteins (FPs). See also Fluorescent labeling; Single protein tracking, live cells
  fusion plasmids, 96–97
  genetic encoding, 62
  labeled motors, 96
  microtubule markers and labeled kinesin motors, 99
Force calibration, OTW
  axial direction, 396
  parking, 398
  quartz cylinders, 396–398
Force-fluorescence spectroscopy, single-molecule
  biological system application, 423
  confocal and optical trapping
    piezo-controlled mirror calibration, 416–418
    tethered molecule, setting up, 418
  fluorescence detection
    data analysis methods, 414–416
    single-molecule confocal microscopy, 413–414
  optical trapping
    calibration, 411–412
    height determination, 410–411
    stiffness determination, 412–413
  sample preparation
    antidigoxigenin-coated microspheres, 421–422
    assembly, 422–423
    DNA/RNA, 420–421
    nucleic and protein labeling, 418–419
    polymer-passivated surface, 419–420
  setup
    optical scheme, 407–408
    surface-tethered assays, 409
Force-ramp method, 569
Force spectroscopy
  antibody-virus molecular recognition
    AFM tip chemistry, 529–532
    biomolecules binding, 528
    experiments, 532
    ligand-receptor binding, 528
    unbinding, ligand-receptor, 528–529
    experiments, 532
  nanopore (See Nanopore force spectroscopy)

Förster resonance energy transfer (FRET). See also Qualitative description, smFRET
  biological system, 176
  Cy3 optical excitation, 42
  DNA structure determination
    3D, dsDNA, 502–503
    dsDNA, 500
    dye positions, dsDNA, 502
    kinked, characterization, 503
    measurement, bent and kinked, 501
    rate constants, 500
    transfer efficiency, 500–502
  efficiency, 414–415
  fluctuations, 416
  limitations, 176
  Raman based molecular rulers, 177
  states, 415
  time trace, 415–416
  trajectories dynamic analysis, 415
  two-dimensional analysis, smFRET
    fluorescence intensities, 477
    indicator dependencies, 479
    interdye distances, 477–479
FRET. See Förster resonance energy transfer
FtsK
  activity
    MT, DNA loop, 311–312
    protocol, 309–310
    translocation, DNA molecule, 310–311
  description, 308–309
Functionalized-peptide-coated-QDs (FL-pc-QD). See also Single protein tracking, live cells
  advantage, 66
  stoichiometry variation
    affinity pair, 68–69
    hapten labeling density, 69
Functionalized-QDs (FL-QDs)
  binding, anti-scFv fusion, 71–73
  imaging, DNA construct, 73–74

G

GP41 helicase
  activity detection
    hairpin unwinding, 313
    materials and protocols, 314
  experimental design, 300
  force extension curve
    DNA hairpin, 312–313
    protocol, 313
  force, mechanism, 318
  loading, optimization
    protocol, oligonucleotide, 316
    5′ ssDNA tail, 314–316
  unwinding and ssDNA translocation activities
    force jump protocol, 316
    measurement, 317
    V_{UN} and V_1, 316

Green fluorescent proteins (GFPs)
  Aequorea victoria, 96
  description, 4
  ELKS proteins, 13, 14
  enhanced (eGFP), 439, 448
  PALM and FPALM
    photoactivated, 29
    photoswitchable, 18
  peroxisomes, 11

H

HAT method, 212–215
α Hemolysin (HL) nanopores. *See also* Nanopore method; Single nucleic acid molecules analysis, nanopores
  DNA bases recognition
    feasibility, 598
    immobilized strands, 599
    nucleic acids, 597–598
  nucleic acid analysis
    cis and *trans* potential, 595
    DNA translocation, 595–596
    structure, 594
Hidden Markov method, 214
High-speed AFM techniques
  bio-imaging, 543
  cantilever tip, 561
  diffusional mobility
    Brownian motion, 554
    2D crystals, 555
    low-invasive imaging, 558–559
    protein 2D crystals
      bacteriorhodopsin, 558
      p97, 557
      streptavidin, 555–557
    substrate surfaces
      bare mica, 544–545
      mica, planar lipid bilayers, 546–550
      streptavidin 2D crystals, 550–554
  UV flash-photolysis, caged compounds, 559–561
HRV2. *See* Human rhinovirus serotype 2
HRV1A. *See* Human rhinovirus serotype 1A
HRVs. *See* Human rhinovirus
Human rhinovirus (HRVs)
  AFM imaging setting
    contact-mode, 519
    magnetic AC (MAC) mode, 519–520
    topographical, 519
  antibody-virus molecular recognition, force spectroscopy
    AFM tip, 529–532
    binding, ligand-receptor, 528
    biomolecule binding, 528
    force spectroscopy experiments, 532
    ligand-receptor unbinding, 528–529
  immobilization and imaging, HRV2

  crystalline arrangement, model cell membrane, 523–526
  mica, 521–523
  receptor binding, virus, 526–527
  production
    His$_6$-tagged MBP-very-low-density lipoprotein receptors, 518–519
    and isolation protocol, 517–518
  RNA release and single molecule unfolding
    genome release, HRV2, 534
    HRV1A full genome, 534–536
    pH, 532–534
Human rhinovirus serotype 2 (HRV2)
  imaging, receptor binding
    immobilization, mica, 527
    liquid cell and soluble, 526–527
  immobilization, mica
    crystalline arrangement, model cell membranes, 523–526
    particles, 523
    protocol, 521–523
  RNA release
    genome, 534
    HRV1A, 534–536
    pH, 532–534
Human rhinovirus serotype 1A (HRV1A)
  RNA
    cores, 534–536
    release, 536
  topographical imaging, 534
Hybrid F$_1$ preparation
  α$_3$β$_2$β(E190D/E391C)γ(R84C), 286–287
  E391C and E190D mutation, 285
  image, 286

I

Imaging, super-resolution
  molecules
    Cy3/Cy5 heterodimers synthesis and characterization, 40–43
    EYFP, 43–46
    photoactivatable fluorophores, 30–34
    photoswitchable fluorophores, acido-DCDHF, 34–40
  MreB, live *C. crescentus* cells, 49
  PALM, live *C. crescentus* bacterial cells
    image reconstruction, 47–49
    sample preparation, 46–47
    time-lapse imaging, 49–50
  reconstruction
    EYFP-MreB protein, 48–49
    localizations, 47–48
    putative single-molecule emitters, 47
  three-dimensional single-molecule imaging
    DH-PALM, 53–55
    DH-PSF, 51
    estimator, 53–54

Subject Index

localization precision, 53
SLM, 51–52
standard point spread function, 50–51

K

Kalman filter signal-processing
field-programmable gate array, 167
tasks, 166–167
Kinesin
axonal transport, 85
description, 8
and dynein *in vivo*, 10, 11
FIONA time resolution, 8–9
fluorescent protein-labeled, 99
kinesin-1 motor, 83–84, 87, 97
kinesin-2 motors, 86
molecules, stepping dynamics, 9
QDs, 9–10, 87
tracking, 98
Kramers' picture, 571, 574

M

Magnetic trap (MT)
DNA strands movement, 600
principle, 298
setup, 300
Magnetic tweezers, DNA tracking motors
bead/DNA preparation, 306
chamber preparation, 301
coating, surface, 301
DNA preparation
dsDNA construct, 304–305
hairpin, 301–304
experimental setup, 299–300
force determination, 308, 309
FtsK application, 308–312
GP41 helicase
activity detection, 313–314
force-extention curve, 312–313
loading, optimization, 314–316
mechanism, 318
unwinding and ssDNA translocation activities, 316–317
injection, bead, 306
surface preparation, 300
tethered beads selection, 306–308
MATLAB function nlinfit, 54
Melanosomes, 85–86
MFD. *See* Multiparameter fluorescence detection
Mica surface, AFM imaging
disk preparation, 544
ionic strength dependent diffusion, 545
planar lipid bilayers
adsorption, 549
alkyl chains, 546
electrostatic immobilization, 549–550
immobilization, specific interaction, 550

lipid molecules mobility, 547
preparation, 546–548
protocol, 544–545
Michaelis–Menten mechanism
kinetic, 245
molecular motor trace, simulated, 224
parameter, 243, 245
Microparticles optical trapping
anisotropy sources
cylindrical quartz particles, 385
form birefringence, 382–384
material birefringence, 384–385
manipulation principles
form/material birefringence, 381
isotropic materials, 379–381
OTW, 382
and signal detection, 380
torque, 381–382
Moment mechanistic constraints extraction
candidate model, 240–241
fluctuation classification
kinetic states, 242–243
Michaelis–Menten expression, 243–244
n_{min}, 242
kinetic mechanisms, 247–248
Michaelis–Menten parameters, 244–245
n_{min}
kinetic parameters, 244
limits, 244–247
model, kinetic, 241–242
N_S, two-state mechanism, 245, 247
randomness parameter
advantage, 242
molecular motors, 241
Montal Mueller approach, 605, 606
Multicolor TIRF microscopy
laser beam coupling, 90
limitation, 90, 92
microscope setup, 91
Multiparameter fluorescence detection (MFD). *See also* Quantitative description, smFRET
acquisition, 474
2D analysis, 503–504
enzyme structure, 481–482
epi-illuminated microscope, 469, 470
fluorescence SCC computation, 488
fluorophore intrinsic properties, 471–472
FRET spectroscopy, 473
photon
detection scheme, 472
information, 460
train, 471
setup, pulsed laser source, 469, 476
MyoV-HMM
biotinylated CaM, 553
ionic strength dependent diffusion, 545
positive-charge surface density, 549

N

Nanopore force spectroscopy (NFS)
  constant force analysis, 575–576
  DNA unzipping kinetics
    duplex molecule, 578
    hairpin measurement, 579
    histogram transformation method, 582–584
    properties, 578
    protein pore α-HL, 577–578
    temperature rescaling, 584–585
    time distribution, 578–579
    voltage-ramp data maximum-likelihood analysis, 580–582
  force-ramp experiment
    analysis, 576
    maximum-likelihood analysis, 576
    rupture-voltage histograms, 576–577
  implementation, 570
Nanopore method. *See also* Single nucleic acid molecules analysis, nanopores
  bond rupture
    force ramps, 569
    voltage ramp, 569
  DNA unzipping kinetics, NFS (*See* DNA unzipping kinetics, NFS)
  electrical force, 567
  force-driven molecular rupture
    Kramers' picture, diffusive crossing, 574
    lifetime, 571–573
    loading rate, 573
    one dimensional free-energy surface, 572
    quasi-adiabatic approximation, 574–575
    thermal noise, 571
    unzipping experiment, 575
  α-HL pore, 568–569
  measurements
    αHL pore lumen alteration, 611–612
    WT-αHL pores, 611
    measurement types, 568–569
  NFS experiments
    constant-force, 575–576
    force-ramp, 576–577
  preparation, 610
  stability, αHL pore
    denaturants, 613
    electrical recordings, 612–613
  storage, 610–611
  translocation, 567–568
Nanowalkers, single-molecule fluorescence tracking
  experimental procedure, 133–134
  imaging instrumentation
    Newport ST-UT2 vibration isolation table, 134, 136
    polarized light, 136
    single-molecule based TIRFM setup, 135
  slide preparation, 131–133

Near-field scanning optical microscopy (NSOM), 124
NFS. *See* Nanopore force spectroscopy
Nonradiative energy transfer, 461
Nyquist sampling theorem, 616
Nyquist–Shannon theorem, 29, 33

O

Okazaki fragments, 260–261
Optical apparatus, rotation and trapping
  anisotropic particles fabrication
    asymmetric shape, 390–391
    optical asymmetry, 391–394
  calibration
    clamp implementation, torque, 398–400
    force, 396–398
    optical tweezers, methods, 395–396
    torque, 398
  instrument (*See* Optical torque wrench)
  microparticles
    anisotropy sources, 382–385
    manipulation principles, 379–382
  optical tweezers, force and torque, 400–402
Optical torque wrench (OTW)
  anisotropy sources
    form birefringence, 382, 384
    material birefringence, 384–385
    surface-based assay, 385
  calibration
    force, 396–398
    implementation, clamp, 398–400
    optical tweezer method (*See* Optical tweezers)
    quartz cylinder, 397
    torque, 398
  cylindrical quartz particles, 385
  electro-optic modulator (EOM), 385–387
    layout, 385–386
  microscope
    acousto-optic modulator, 387
    beam polarization uniformity, 388
    commercial inverted, 388
    electro-optic modulator, 387–388
    LED illumination source, 388
    Nikon specimen stage, 389
    photograph, 389
  signal detection and processing
    back-focal plane detection, 390
    polarization components, 389–390
    torque, 389
Optical trap analysis
  data, measurements
    dwell time (*See* Dwell time, optical trap experiment)
    stroke size, 365–369
  experiment
    actin dumbbell forming, 358–359

Subject Index 657

binding interactions, 361
platform testing, 359–360
tensing, actin dumbbell, 359
instrumentation
acousto-optic deflectors (AOD), 326, 328
air fluctuation, 342–343
alignment protocol (See Alignment, dual-beam optical trap)
bead position detection, 330, 339–341
beam path, dual-trap setup, 331
beam steering, fast and precise feedback control, 339
calibration, 356–358
components, 333–337
concept, 324–325
coupling, 325
data acquisition electrical noise, 343
dichroic mirrors, 341
dual beam setup, 329
imaging, beads and actin filament, 330–332
laser and beam steering optics, 326
mechanical vibration, 342
microscope body, 338
objective (OBJ), 332
pointing and power fluctuation, 342
quality, beam, 343
slow and coarse positioning, beam steering, 338
thermal expansion, 343
three-bead assay, 325–327
trapping, laser, 332
trapping (See Trapping, optical)
Optical tweezer, DNA-protein interactions
distance determination, 437
dual trap generation and steering
angular deflection, 434–435
design, 433–434
microspheres, 434
telescope system, 434
light sources
beam pointing instabilities, 431–432
lasers, 432–433
microscope objectives, 433
trap calibration and laser detection
back-focal-plane interferometry, 435
Brownian motion, microbead, 435–436
power spectrum, 437
trapping principle
gradient/scattering force, 430–431
high photon-flux source, 430
Optical tweezers. See also Optical apparatus, rotation and trapping
See also Optical tweezer, DNA-protein interactions
calibration
equipartition theorem, 395
PSD voltage, 395–396
QPD signal, 411–412

single DNA molecules twisting, tension
dsDNA (See Double-stranded DNA)
supercoiling, 400
Organic fluorophores
chemically synthesized, 4
genetic tags
advantage, 87–88
labeling methods, 88–89
molecular mechanism and functional regulation, 88
STORM, 18

P

PALM. See Photoactivated localization microscopy
Peptide-coated-QDs (pc-QDs)
absorbance spectra, normalized, 69
advantages, 63
biotinylated, 64
functionalized (See Functionalized-peptide-coated-QDs)
Photoactivatable fluorescent proteins (PA-FPs), 111, 113
Photoactivatable fluorophores
effective turn-on ratio
activated molecule and preactivated fluorogen, 32
localization, 33
on state molecules, 33–34
labeling density, 31
photobleaching quantum yield, 31–32
photoconversion, 32
photon-count analysis, single-molecule, 34
and photoswitchable fluorophores, 35–37
single-molecule, 30–31
Photoactivated localization microscopy (PALM). See also Imaging, super-resolution
analysis, 111
live *C. crescentus* bacterial cells
image reconstruction, 47–49
sample preparation, 46–47
time-lapse imaging, 49–50
and STROM, 17–19
Photon-by-photon feedback, 163–164
Photon distribution analysis (PDA). See also Data analysis, smFERT
analysis, FRET efficiency distributions, 493
dyanamic system
advantage, 496
determination, 495
use, 494, 496
FRET efficiency distributions, 489
green photon detection, 490
model function pathways, 491
multiple species, 494
Poissonian distribution, 490

Photon distribution analysis (PDA). *See also* Data analysis, smFERT (*cont.*)
  shot noise, 504
Photoswitchable fluorophores, azido-DCDHF class
  design
    aryl azides photochemistry, 38
    photocaged and push–pull, 38
    fluorogenic molecules, 38
  photophysical properties, 35–37
  photoreaction characterization
    effective turn-on ratio, 40
    photoactivation, 40, 41
    push–pull fluorogens, 39
  push–pull, 34, 38
Piezo-controlled mirror calibration
  fluorescent bead preparation, 416
  and force-extension curve, 417
  origin reset, 416–417
  stuck bead sample, 417–418
Piezo-stepper experiments, 10
Plasmon rulers, enzymology
  annealing, ssDNA/RNA tethered nanoparticles, 186
  assembly, 184–185
  calibration
    dsDNA tethered gold nanoparticle dimers, 183
    heterogeneity, colloidal nanoparticle, 182, 184
    interparticle separation and resonance wavelength, 184
  color, 181
  coupling, distance dependence
    charge distributions and dipole orientations, 178
    feasibility, 179
    fixed interparticle distances, 178–179
    optical properties, 177
    resonance wavelength, 178
  DNA bending and cleavage, EcoRV
    global kinetics, 187
    high-intensity state duration, 191–192
    proteins, 186–187
    restriction nuclease, 187–188
    scattering trajectories, 188–190
    signal change, 188
    spacers intensity, 189–190
  intensity, 181–182
  parallel single-molecule assays, 188
  polarization, 182
  purification, 186
  RNA, spermidine modulated ribonuclease activity
    advantages, 192
    cleavage time distributions, 192–194
    RNA-DNA hybridization, 192
    subpopulations, 194

single particle Rayleigh scattering spectroscopy
  dark-field microscope, 180
  inverted microscopes, 180
  microscopy techniques, 179
Postimage processing, 94–95
Protein interactions, DNA
  instrumentation
    fluorescence microscopy, 437–440
    microfluidics, 440–442
    optical tweezer, 430–437
  optical trapping, fluorescence microscopy and microfluidics
    Rad51 and dsDNA binding, 448–450
    replication protein A and ssDNA binding, 448
    single DNA molecule isolation and visualization, 446–448
  reagents preparation
    fluorescent labeling, 444
    molecules, terminally labeled, 443–444
    proteins fluorescent labeling, 444–445
Protein structure and function
  data analysis
    dwell time, 361–365
    stroke size, 365–369
  dual-beam optical trap, myosin function
    myosin II and V, 323–324
    myosin VI, 324
    R403Q mutation, β-cardiac, 322–323
  optical trap instrumentation
    acousto-optic deflectors (AOD), 326, 328
    air fluctuation, 342–343
    alignment protocol, 344–356
    bead position detection, 330, 339–341
    beam steering, fast and precise feedback control, 339
    calibration, 356–358
    components, 333–337
    concept, 324–325
    coupling, 325
    data acquisition electrical noise, 343
    dichroic mirrors, 341
    dual beam setup, 329
    imaging, beads and actin filament, 330–332
    laser and beam steering optics, 326
    laser trapping, 332
    mechanical vibration, 342
    microscope body, 338
    objective (OBJ), 332
    pointing and power fluctuation, 342
    quality, beam, 343
    setup, dual-trap beam path, 331
    slow and coarse positioning, beam steering, 338
    thermal expansion, 343
    three-bead assay, 325–327
  optical trapping experiment

Subject Index

actin dumbbell forming, 358–359
binding interactions, 361
platform testing, 359–360
tensing, actin dumbbell, 359
Proton motive force (pmf), 280

Q

Qualitative description, smFRET
  acceptor quenching, 483, 485
  anisotropy *vs.* donor lifetime
    diffusion properties, 480–481
    HIV-1 reverse transcriptase
      (RT), 481–482
    Perrin equation, 480
    polarization, 480
  donor quenching
    intensity, 484
    lifetime analysis, 482–484
  dynamic behavior
    artifacts, 488–489
    burst-wise analysis, 486
    fluorescence intensities, 486–487
    lifetime, 485–486
    syntaxin-1 (Sx), 487–488
  fluorescence signals and FRET indicators,
    474–475
  intensity measurement, 474–475
  lifetime measurement, 476
  quasi-static species 1D and 2D
    histogram, 478
  two-dimensional analysis, 477–479
Quantitative description, smFRET
  acceptor FRET analysis, 491–492
  MFD data
    FRET indicators display, 497–498
    G-factor, polarization, 496
    spectral sensitivity, g_G/g_R ratio,
      496–497
  PDA
    dynamic system, 494–496
    FRET efficiency distribution, 489
    intensity distribution, 490
    multiple species, 494
    shot-noise limited FRET signal distribution
      and broadening
    biological phenomena, 494
    Cy5, 492, 494
    PDA, 491–493
    structure, FRET
      accuracy, 499–500
      DNA determination, 500–503
      multimolecular complexes, 498–499
Quantum dots (QDs). *See also* Single protein
    tracking, live cells
  description, 87
  Drosophila kinesin-1 motor domains, 87
  functionalization, hapten, 65
  pc-QDs (*See* Peptide-coated-QDs)

R

Radiative energy transfer, 461
Resolution, 2
Rotation assay
  basal buffer contents, 287
  cross-link formation
    ADP-inhibited F_1, 290–291
    $\alpha_3\beta_2\beta(E190D/E391C)\gamma(R84C)$ rotation,
      290
  F_1 manipulation magnetic tweezers, 291
  protocol, 289
  flow cell construction, 287
  hybrid F_1, one $\beta(E190D/394C)$
  ATP conditions, 289
  stepping time course, 288

S

Sample preparation, force-fluorescence
  spectroscopy
  antidigoxigenin-coated microspheres
    buffer solution, 421
    protocol, 421–422
  assembly
    incubation protocol, 422–423
    infusion, 422
  DNA/RNA
    nucleic acid, 420
    protocol, 420–421
  nucleic acid and protein labeling
    biotin-neutravidin linkage, 419
    fluorescent probes, 418–419
  polymer-passivated surface, 419–420
Single-chain variable fragment (scFv). *See also*
  Single protein tracking, live cells
  anti-scFv construction, 73
  α-FL and FL-QD binding affinity
    measurement and determination, 72
  optimization, 73
  preparation, 71
Single DNA molecule, YOYO-1
  isolation and visualization steps
    beads trapping, 446
    catching, 446
    force–extension analysis, 447
    multiple-channel laminar flow chamber,
      446
    spatial fluctuation, 447–448
  lambda DNA
    bound, 448
    elastic behavior, 447
Single EcoRV restriction enzymes
  cleavage, DNA
    intensity distribution, 191–192
    scattering trajectories, 188–189
    spacer intensity, 189–190
  DNA bending
    binding enzymes, 187

Single EcoRV restriction enzymes (cont.)
  cleavage kinetics, 187–188
  proteins, 186–187
Single molecular nano-assemblies, nanometer-scale movements
  DNA-based nanowalkers
    behavior-based molecular robots, 127–128
    characterizing methods, 129
    spider, 125–127
  fluorescence imaging
    fluorophore lifetime, 129–130
    stage and focal drift control, 130–131
  fluorescent single-particle tracking
    advances, 124
    FIONA and SHREC, 124–125
    position measurement statistical error, 124
    single chromophores, 123–124
  nanowalkers, fluorescence tracking
    experimental procedure, 133–134
    imaging instrumentation, 134–136
    slide preparation, 131–133
  super-resolution position information
    emission channels mapping, 136–137
    error analysis, 143
    presentation, 139–142
    quality control, 137–139
Single-molecule confocal microscopy
  excitation beam, 413
  piezo-controlled mirror, 414
  pinhole and APDs alignment, 414
  use, 413
Single-molecule cross-link experiment
  rotation assay
    basal buffer contents, 287
    flow cell construction, 287
    formation, 289–291
    hybrid F_1, one β(E190D/394C), 288–289
  $\alpha_3\beta_3\gamma$ subcomplex pause position analysis
    data, 292
    distance measurement, 291–293
Single-molecule Förster resonance energy transfer (smFRET)
  advantages and disadvantages, 458–459
  complication, 459–460
  data analysis
    MFD 2D, 503–504
    PDA, 504
  ensemble vs. single-molecule, 458
  enzyme population, 459
  essentials, 462
  fluorescence image spectroscopy, 505
  MFD, 460
  nutshell, FRET theory
    distance dependence, 464
    efficiency, 463–464
    nonradioactive energy transfer, 461, 463
    radioactive energy transfer, 461
  optical detection system, 457

  orientation factor κ^2, FRET theory
    donor and acceptor, 465–466
    energy transfer, 465
    molecular rotational diffusion, 466
    transition dipole moment, 465
  properties and measurement techniques, fluorescence
    accuracy measurements, 473–474
    biomolecular processes timescale, 466–468
    dimensions, 468–470
    experimental setup and data, 470–473
  qualitative description
    anisotropy vs. donor lifetime, 480–482
    donor and acceptor quenching, 482–485
    dynamic behavior, 485–489
    fluorescence signals and FRET indicator, 474–479
  quantitative description
    acceptors and direct excitation, 491–492
    dynamic systems, PDA, 494–496
    high-resolution structures, 498–503
    MFD data, 496–498
    multiple species, PDA, 494
    PDA theory, 489–490
    shot-noise limited FRET signal, 492–494
    two fluorophores, 457
Single-molecule high-resolution colocalization (SHREC)
  CaMs, 17
  description, 95
  FIONA, 125
  fluorophores, 124–125
  FRET, 15–16
  short DNA segment measurement, 16–17
Single molecule high-resolution imaging with photobleaching (SHRIMP)
  description, 14–15
  and SHREC, 17–18
  technique, 16
Single molecule tracking
  algorithms
    diffusing, 116
    sptPALM, 115
  diffusion, scan pattern, 165
  identification
    image, 114–115
    intensities and shapes, 115
  intact cells, 84
  QDs, 64
Single motor proteins tracking, mammalian cell cytoplasm
  experimental procedures
    cell culture, 97–99
    events identification, 100–101
    fluorescent protein-labeled motors, 96
    high-resolution tracking, 102–103
    image sequence recording, 100
    molecule recording and analyzing, 99–100

Subject Index

plasmids, fluorescent protein fusion, 96–97
SD map, 101–102
transient expression, 99
instrumentation, *in vivo*
dark-field excitation and nanoparticles, 92–93
dual channel image recording, 94
microscope accessories, 95
multicolor TIRF illumination, 90–92
postimage processing, 94–95
single-molecule events recording, 93–94
TIRFM, 89
molecular motors labeling, *in vivo*
organelles, cytoplasmic, 85–86
organic fluorophores and genetic tags, 87–89
quantum dots, 87
principles, 84
Single nucleic acid molecules analysis, nanopores
cis and *trans* potential, 595
classes, 594
control, DNA translocation
αHL nanopores, 599–600
noise levels, 599
processing enzymes, 600–601
single nucleotide extension, 601
data acquisition and
acquisition protocols, 617–618
digitization, 616
filtering and sampling, 616–617
single DNA/RNA molecules, 618–619
DNA, 593
duplex formation, 597
electrical recording, planar lipid bilayers
chambers, 605–606
electrode preparation, 603–605
equipment, 602–603
Faraday enclosure, 603
preparation, 606–609
single-channel experiments, 601–602
αHL
individual base recognition, 597–599
structure, 594
inserting pores and DNA addition, 609
materials
buffer components, 613
DNA and RNA handling, 613
long RNA preparation, 615
long single-stranded DNA, 614
short dsDNA preparation, 615
short single-standard DNA/RNA, 613–614
measurements, 611–612
perfusion, 610
preparation, 610
protein, homopolymeric strand, 596–597
stability, 612–613
storage, 610–611

translocation, DNA, 595–596
use, 593
Single-particle tracking photoactivated localization microscopy (sptPALM)
acquisition and analysis, 112
description
living cell proteins, 112–113
PA-FPs, 111–112
labeling
PA-FPs, 113
photocaged dyes, 114
membrane protein
diffusion coefficients, 117
motion, 116
single molecule tracking
algorithms, 115–116
identification, 114–115
trajectories and diffusion coefficient, 117
Single protein tracking, live cells
anti-scFv fusion constructs, FL-QDs binding
quantification, α-FL scFv, 71–73
yeast cell, growth and induction, 71
FL molecule quantification, FL-pc-QD
absorbance spectra, 69
coating computation, 70–71
QDs functionalization
FL-pc-QDs, stoichiometry, 68–69
peptide coating, 66–68
reagents, peptide coating, 66
scFv-hapten pair, QD targeting
affinity pair, 65–66
antibodies and molecules, 64
anti-FITC, 64–65
and FP-protein, 65
single FL-QD imaging, DNA constructs
anti-scFv 4M5.3, 73
labeling, mammalian cells, 74
single molecule imaging, mammals
fluorophore imaging, 74–75
scFv-PrP imaging, 76
TIRF and numerical aperture (NA), 74
smFRET. *See* Single-molecule Förster resonance energy transfer
Spatial light (phase) modulator (SLM), 51–52
Species cross-correlation function (SCC), 488
Spectroscopic ruler, 464
Spider
behavior-based molecular robots
capability assessment, 128
influence of, 127–128
motion initiation, 128
biological molecular proteins, 125
Cy3-labeled, 133, 137, 138
deoxyribozymes catalytic power (*See* DNAzymes)
description, 127
origami applications, 130, 131
PAGE and AFM, 129

Spider (cont.)
  position value, 139–140
  SHREC analysis, 139
  walking mechanism, 141
Stage feedback, 160
Standard deviation map (SD map), 101–102
Statistical kinetics
  cycle completion time
    fluctuations, 227
    and hidden kinetic states properties, 226–227
  fluctuations
    fitting distributions, 233–235
    moment calculation, 235–236
    pathways and steps, multiple, 236–240
  lifetime
    distribution, 230
    probability density, 229
  memory-less enzyme
    energy landscape physical properties, 228–229
    fluctuations and lifetime, 229
    individual kinetic state properties, 227–228
  moment mechanistic constraints extraction
    candidate models, 240–241
    fluctuation classification, 242–244
    mechanistic constraints, 244–248
    randomness parameter and n_{min}, 241–242
  physical motion generation, 224
  state visitation
    enzymatic dynamics classes, 238
    kinetic, 230–231
    off-pathway, 231–232
    on-pathway, 230, 232
    single cycle, enzyme, 230
  steady-state kinetics
    cycle paths, 225
    dwell time/residency time, 225
    fluctuation, 223
    species enzyme time, 226
Stimulated emission depletion microscopy (STED)
  fluorescence process, 19–20
  microscopy, 21
  twofold resolution, 20, 22
  variant, 21
Stochastic optical reconstruction microscopy (STORM)
  application, 111
  Cy3 and Cy5 molecules, 29, 40–41
  2-D and 3-D, 20
  and PALM
    activator and reporter, 18–19
    colors, organic fluorophores, 18
    2-D and 3-D, 20
    direct STORM (dSTROM), 19

SHRIMP and SHREC, 17–18
Stokes' drag calibration methods, 395
Streptavidin 2D crystals
  biotin-bound and-unbound subunit pairs, 556
  dynamic AFM imaging
    preparation, 551–552
    subunits, 550–551
    use, substrates, 552–554
  monovacancy defects, 555–556
  nonspecific binding, 551
  point defects, 557
Super-accuracy
  background noise, 5–6
  calculation
    dynein and kinesin in vivo, 10
    FIONA without fluorescence, 10–12
    kinesin, 8–10
    live organism, FIONA, 12–13
    translationally immobile dye, 7–8
  description, 2
  determination, 4–5
  fluorescence, 7
  fluorescent dye CCD image, 5
  pixel size, 5
  TIRF, 6
  total internal reflection, 2
Super-resolution
  description, 13
  emission channel mapping, 136–137
  ensemble-averaged mean-square displacement plots, 141–142
  error analysis, 143
  PALM and STROM
    activator and reporter, 18–19
    colors, organic fluorophores, 18
    2-D and 3-D, 20
    direct STORM (dSTROM), 19
    SHRIMP and SHREC, 17–18
  presentation
    displacement plots, 139, 140
    nanowalker movement characterization, 138
    trajectory plots, 139
  quality control
    colocalization, 139
    displacement, 137, 139
    intensity and ellipticity, 137
    nanowalker trajectories extraction, 138
  SHREC (See Single-molecule high-resolution colocalization)
  SHRIMP
    fluorescent probes, 14–15
    NALMS, 15
  STED (See Stimulated emission depletion microscopy)
Surface-tethered assays, 409
Surface tethered optical tweezers assay, 407

Subject Index

T

Tapping mode AFM, 543
Tethered bead motion, replication loops observation
   data analysis, 269–271
   flow chamber construction
      description, 265–267
      single-molecule experiments, 265
   forked λ-DNA substrate preparation
      description, 264
      oligonucleotides, 263
      glass coverslip functionalized, 261–262
      polystyrene beads functionalization description, 262–263
      α-digoxigenin Fab fragments, 262
   reaction and imaging
      description, 268–269
      proteins and nucleotides, 268
Tethered beads selection
   coilable DNA molecule, 307–308
   force determination, 308, 309
   single DNA hairpin
      identification, 306
      protocol, 306–307
   single-nicked DNA molecule, 307
Tethered particle motion (TPM) technique. *See also* DNA-looping kinetics, TPM
   CI-induced looping, λ-DNA
      CP-EM and HAT analyses, 215–217
      description, 215
      dwell time, 217
      DNA molecules, 200
Three-bead assay, dual beam optical trap. *See also* Alignment, dual-beam optical trap
   design, 327
   myosins, 326
Time correlated single-photon counting (TCSPC)
   description, 471
   instrument response function (IRF), 476
   MFD, 471–472
Time-lapse imaging, 49–50
TopoTaq DNA polymerase, 600
Total internal reflection fluorescence (TIRF)
   DNA-protein interaction, 429
   imaging, 18, 114–115
   microscopy (TIRFM) (*See* Total internal reflection fluorescence microscopy)
   setup, single molecule prism based, 135
   single-molecule imaging experiment, 74–75
   two-color experiment, 96, 97
Total internal reflection fluorescence microscopy (TIRFM)
   description, 89
   EMCCD camera, 467–468
   multicolor
      fluorophores, 90
      multiwavelength excitation, 90–92
      setup, 91
Trapping, optical. *See also* Optical tweezer, DNA-protein interactions
   calibration
      optical tweezers, 411
      quadrant photodiode (QPD) signals, 411–412
   and confocal trapping coalingment, 416–418
   experiment
      actin dumbbell, 358–359
      binding interaction identification, 361
      coverslip, 358
      nonprocessive motor, 360
      platform testing, 359
   height determination
      Brownian motion, 410–411
      force-fluorescence assays, 410
   monomeric myosin VI, 324
   principles, 430
   and single-molecule confocal fluorescence instrument, 408
   stiffness determination
      equipartition theorem, 412
      microsphere, Brownian motion, 412–413
      spring constant, 413
Trapping systems, anti-Brownian. *See* Anti-Brownian traps
Turning on process, 29

U

UV flash-photolysis, caged compounds
   CaM calcium-binding protein, 560–561
   precautions, 560
   structural changes, proteins, 559

V

Vesicular stomatitis virus G (VSVG) proteins, 117
Virus
   description, 516
   genome exit, 516–517
   HRVs (*See* Human rhinovirus)
   inactivation, 516
   infection, 516
VistaVision (VWR) coverslips, 273
Voltage-ramp data, DNA unzipping, 580–582

W

Wide-field fluorescence microscopy, 428, 437
Worm-like chain (WLC) model, 204, 269, 311, 418, 447

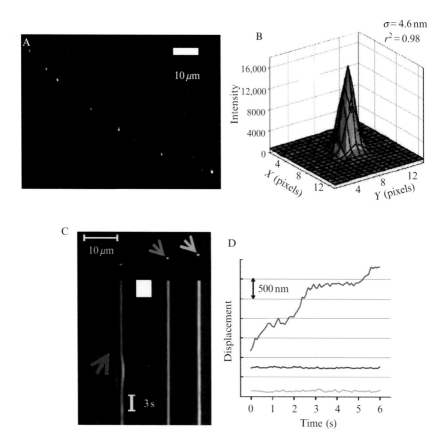

Erdal Toprak et al., Figure 1.6 Fluorescently labeled protein complexes can be accurately localized within a live *C. elegans*. (A) DENDRA2::ELKS spots decorate the mechanosensory neurons along the *C. elegans* body. (B) The peak of the two dimensional Gaussian fit to the emission of a DENDRA::ELKS puncta can be localized within less than 5 nm in 4 ms. (C) A kymograph showing the moving and stationary GFP::ELKS spots in neurons. The displacement of spots pointed with red, blue and green arrows are shown in (D) with red, blue and green colors respectively. Adapted from Kural *et al.* (2009).

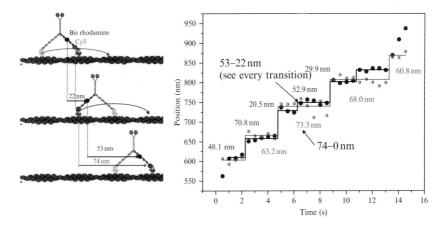

Erdal Toprak et al., Figure 1.9 SHREC. Two fluorophores which emit at different wavelength can be imaged separately and the position of each dye fit with FIONA-like accuracy. The minimum distance between the two dyes, that is, the resolution, is approximately 10 nm. In this particular case, a myosin V, which moves with 37 nm center-of-mass, is labeled with a Bis-rhodamine (blue) and a Cy5 (red) (left). With the Cy5 labeled very close to the motor domain, a "hand-over-hand" motion causes the Cy5 to move 0-74-0-74 nm. The other dye, bis-rhodamine, labeled on the "leg," translocates by $(37-x)$ and $(37+x)$ nm, where x is the distance from the center-of-mass. Adapted from Toprak and Selvin (2007).

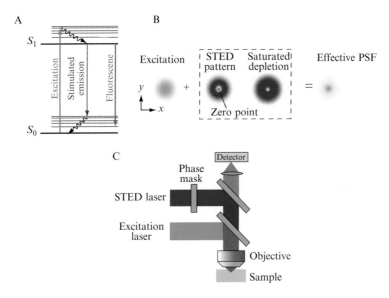

Erdal Toprak et al., Figure 1.12 STED microscopy. STED consists of two laser beams: one which forms the normal fluorescence excitation beam (green); the other, called the STED laser, which causes relaxation of select fluorophores (red). These get forced to relax by stimulated emission. By putting in a phase mask in the STED laser, it can be made to form a doughnut-like shape. When combined with the excitation pulse, the result is a very narrow PSF. Adapted from Huang et al. (2009).

A

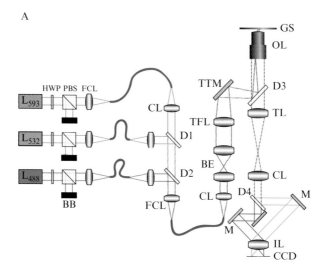

B

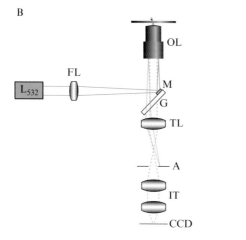

Dawen Cai et al., Figure 4.1 Microscope setups for tracking single motor proteins. (A) Tri-color TIRFM. Multicolor excitation is provided by three laser sources with wavelengths of 488 nm (Melles Griot Ar$^+$, solid blue path), 532 nm (CrystaLaser DPSS, solid green path), and 593 nm (CrystaLaser DPSS, solid orange path). The output power of each laser is independently modulated by rotating a $\lambda/2$-waveplate (HWP, Tower Optics) to adjust the polarization axis with respect to a polarizing beam splitter (PBS, Linos Photonics) which divides the light between a beam block (BB) and a lens (FCL) which in turn couples the light into a 3-μm single-mode fiber (Oz Optics). After exiting the fiber, the laser light is collimated (CL) and the three wavelengths are colocalized using dichroic mirrors (D1, D2, Chroma Technology and Semrock). This combined, multiwavelength beam is coupled into a 3-μm single-mode fiber (Oz Optics) to easily interface with the TIRF microscope. A collimation lens (CL) and a telescope (BE) expand the beam to the appropriate diameter such that it can be focused with the TIR focusing lens (TFL) and directed with a kinematic tip-tilt mirror (TTM, Newport

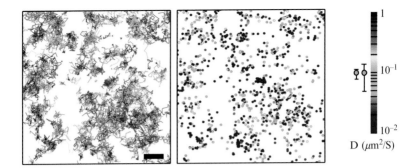

Suliana Manley et al., Figure 5.3 Example of sptPALM trajectories (left) and diffusion coefficient map (right) for the membrane protein VSVG. Color bar provides a scale for diffusion coefficients. Scale bar is 1 μm.

Ultima U100) and an appropriate dichroic mirror (D3, multiband laser dichroic from Chroma Technology) into the back focal plane of the TIRF microscope objective. The TTM allows precise alignment of the laser excitation sources for TIRF. Fluorescence emission signals (here shown as green and red paths) are transmitted through the dichroic mirror (D3) and imaged by the tube lens (TL). A custom dual-view system splits (via dichroic mirror D4) the image into two-color components and projects them side-by-side (with mirrors (M) and lens (IL) onto the sensor of a single EMCCD camera (Photometrics 512B Cascade). (B) Objective Dark-field. A small mirror (M) attached to an antireflection-coated glass substrate (G) reflects the illumination laser beam (solid line) into the back-focal plane of the microscope objective lens (OL). The scattered light from gold nanoparticles (dashed line) is collected by the objective, passed through the glass substrate, and is imaged by the TL and a magnifying imaging telescope (IT) onto a CCD. Back-reflected illumination light from the glass coverslip is blocked by an aperture (A) to increase the signal-to-background ratio.

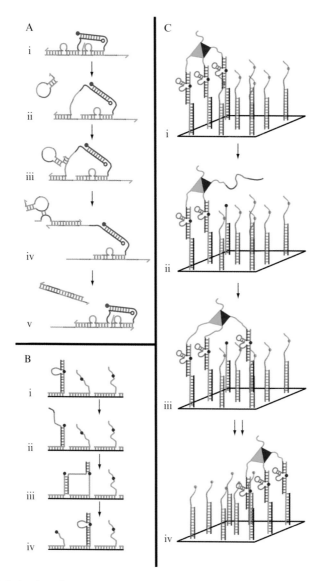

Nicole Michelotti et al., Figure 6.1 Example mechanisms for processive movement of DNA-based nanowalkers. (A) Biped nanowalker from Green et al. (2008) that utilizes fuel consisting of two complementary DNA hairpins. Colors (gray scales) represent complementary sequences. (i) The competition between identical feet to bind to the track permits the exposure of a toehold region in the left foot (ii). A hairpin hybridizes to the toehold region (iii), displacing the left foot from the track. A second, complementary hairpin hybridizes to the first hairpin (iv) to form a waste product, allowing the foot to rebind to the track with equal probability to the left or right. (B) Single-stranded deoxyribozyme-based nanowalker from Tian et al. (2005). (i) The

10–23 deoxyribozyme (red with orange active site) is able to cleave its substrate (green) at a specific site (purple) in the presence of Mg^{2+}. The shorter end dissociates from its product (ii) and hybridizes to a neighboring strand (iii). Via displacement of the cleavage product by neighboring substrates, the deoxyribozyme progresses along the track (iv). (C) Spider moving along a three-substrate-wide origami track (Lund et al., 2010). The spider is composed of a streptavidin body, a capture leg, spacers, all shown in blue, and three 8–17 deoxyribozyme legs (red binding arms with orange active site). (i) The deoxyribozyme spider legs hybridize to substrates (green) that are attached to the origami scaffold via hybridization to staple overhangs (black). In the presence of Zn^{2+}, each leg of the spider cleaves its substrate, dissociates from its products (ii) and hybridizes to a neighboring strand (iii). The greater affinity of the leg for the substrate than the cleavage products makes it energetically favorable for the strand to bind to the full substrate, generating a biased-random walk from the cleaved toward the uncleaved substrate (iv).

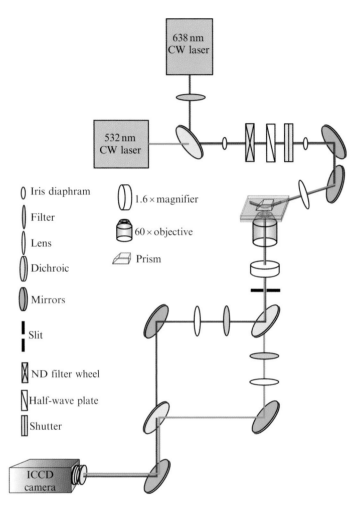

Nicole Michelotti et al., Figure 6.3 Single-molecule prism-based TIRF setup. Linearly polarized light from the 638-nm red diode laser passes through a 638 ± 10 nm clean-up filter and is reflected off a 610-nm cutoff dichroic. 532-nm light from the Nd: YAG passes through same dichroic to join the same beam path. Light from both lasers then passes through the following components, in order: an iris diaphragm, a neutral density filter, a $\lambda/2$-wave plate for 532-nm light, a shutter, an iris diaphragm, a series of mirrors (two are shown for ease of representation; three were used during our experiments), a focusing lens, the TIR prism, the microscope slide containing the sample, the 60× objective, a 1.6× magnifier (along with additional filters, mirrors, and image transferring lenses contained within the microscope), a slit, and a 610-nm cutoff dichroic mirror. The emission from Cy3 passes through a band-pass filter, while that of Cy5 passes through a long pass filter. These separated images are projected side-by-side onto an intensified charge-coupled device (ICCD) camera.

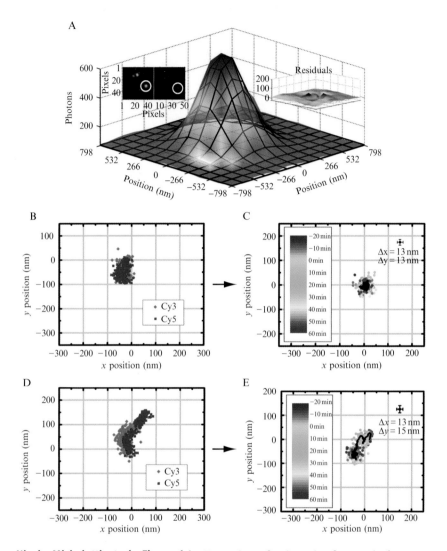

Nicole Michelotti et al., Figure 6.4 Extraction of trajectories from a single nanowalker on a linear origami track imaged by TIRF microscopy (Lund et al., 2010). (A) Point spread functions (PSFs) of Cy3-labeled spiders and Cy5-labeled origami are colocalized (inset, upper left) and fit separately to two-dimensional Gaussian functions in each movie frame to determine their coordinates over time. The fit has low residuals (inset, upper right). (B–E) Centroids from Gaussian fitting are plotted as a function of time to yield nanowalker trajectories. Even in the absence of the divalent metal ion cofactor, Cy3 and Cy5 coordinates show considerable drift (B), but when the trajectory of Cy5 is subtracted from that of Cy3 it becomes clear that the nanowalker is stationary on its track (C). In contrast, in the presence of 5 mM ZnSO$_4$, subtraction of the raw Cy5 and Cy3 trajectories (D) yields net movement of about 90 nm. In (C) and (E), HBS buffer containing either 0 or 5 mM ZnSO$_4$ was added to the sample at time $t = 0$ min.

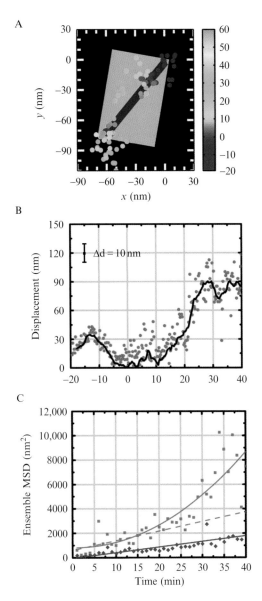

Nicole Michelotti et al., Figure 6.5 Characterization of nanowalker movement (Lund et al., 2010). (A) The two-dimensional trajectory of an individual spider is plotted as a function of time and compares reasonably well to a scale-drawn schematic of the origami track design (gray rectangle). HBS buffer containing 5 mM ZnSO$_4$ was added at time $t = 0$ min. (B) Displacement of the spider in panel (A) measured with respect to its position at the time of adding 5 mM ZnSO$_4$. Displacement plots calculated from both the raw trajectory (green dots) and a 4-min rolling average of the raw trajectory (black line) are shown. (C) Ensemble mean-square displacement (MSD) calculated from 16 spiders observed in HBS buffer containing 5 mM (green squares) or 0 mM (red diamonds) ZnSO$_4$ added at time $t = 0$ min. The former plot is fit with a power law function (green solid line), and the latter with a straight line (red line). To clearly illustrate the concave-up shape of the MSD in 5 mM ZnSO$_4$, a straight line is also fit to the first 15 min of this plot (green dotted line).

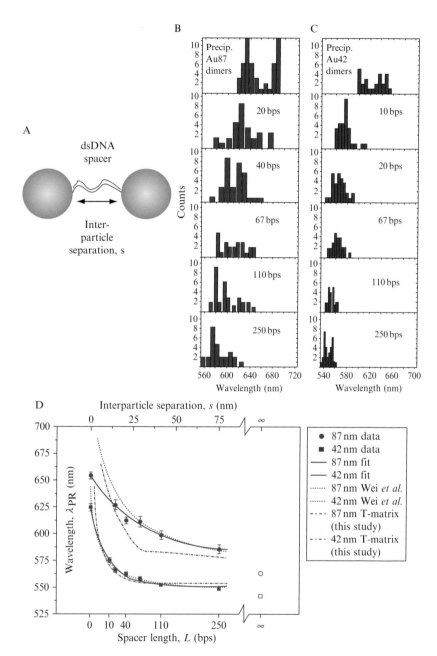

Björn M. Reinhard et al., Figure 8.3 Plasmon resonance versus interparticle separation in dsDNA tethered gold nanoparticle dimers. (A) Schematic drawing of the plasmon rulers. Distributions of measured plasmon resonance wavelengths for selected dsDNA spacer lengths are plotted in (B) for 87-nm Au plasmon rulers and in (C) for 42-nm Au plasmon rulers. Data points for salt-precipitated dimers were included. (D) Plot

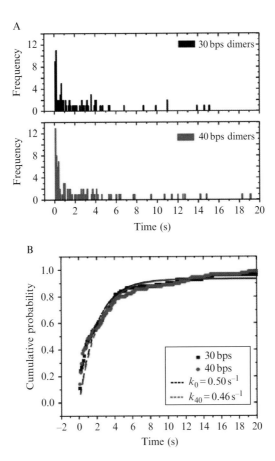

Björn M. Reinhard et al., Figure 8.8 Kinetic and thermodynamic analysis of single-molecule data, in particular dwell-time analysis of plasmon rulers in high-intensity state. (A) Histograms with a bin size of 100 ms for 30- (upper) and 40-bp (lower) plasmon rulers. (B) Cumulative probability of plasmon ruler dissociation. Plots show the percentage of dimers with dwell times less than the indicated time. First-order kinetic fits are included as dashed lines ($k_{30} = 0.50$ s^{-1}, $k_{40} = 0.46$ s^{-1}).

---

of the average plasmon resonance as a function of spacer length, L (bottom axis) and approximated interparticle distance x (top axis) for 42 nm (red squares) and 87 nm (blue circles) plasmon rulers. The plasmon resonance wavelengths for dimers with infinite separations (monomers) are included as open symbols. The reported errors are the standard errors of the mean. The continuous lines show fits (single exponentials $y(x) = A_0 \exp(-x/D_0) + C$) to the experimental data. Best-fit parameters for 42-nm Au particles: $C = 550.87$ nm, $A_0 = 73.48$ nm, $D_0 = 10.24$ nm; for 87-nm Au particles: $C = 579.66$ nm, $A_0 = 74.42$ nm, $D_0 = 30.23$ nm. The dotted lines represent T-matrix simulations by Wei et al. (2004) assuming illumination with light polarized along the interparticle axis and dot-dashed lines are T-matrix simulations assuming nonpolarized illumination.

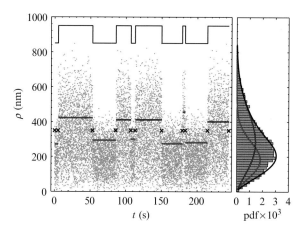

Carlo Manzo and Laura Finzi, Figure 9.5 Expectation-maximization clustering. *Left panel*: After the change points have been determined (black crosses), the regions between two adjacent change points are clustered into two groups through the maximization of the total log-likelihood function for a Weibull *pdf*. The green and red lines represent the average of the data points in the corresponding change point region. The color identifies the DNA conformation corresponding to the two groups (red unlooped and green looped). The "true" trace (black line) is also reported shifted and scaled for comparison. *Right panel*: Histogram of the time trace and results of the expectation-maximization step. The red and green curves refer to the retrieved Weibull *pdf* for the unlooped and looped state, respectively. The sum of the two *probability density functions* (black line) shows an excellent agreement with the data histogram.

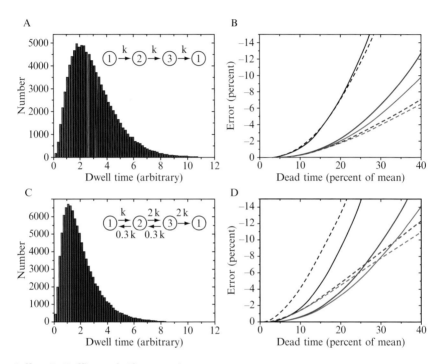

Jeffrey R. Moffitt et al., Figure 10.6 Estimating systematic errors from dead-times. (A) Histogram of simulated dwell times for the kinetic mechanism pictured: A three-state irreversible mechanism. The time axis is arbitrary units. Red corresponds to the portion of the distribution which would not be observed if there was a dead-time of 40% of the mean dwell time. (B) Systematic error in the mean (red), second moment (blue), and $n_{\min}$ (black) as a function of dead-time measured in percentage of the mean dwell time. The solid lines correspond to the actual error introduced by the dead-time while the dashed lines correspond to the estimates using the method described here. (C) Histogram of simulated dwell times for the kinetic mechanism pictured: A three-state system with reversible transitions. (B) Systematic errors for this mechanism as a function of dead-time as in panel (B). Data in panels (B) and (D) were calculated from stochastic simulations with 10^6 dwell times.

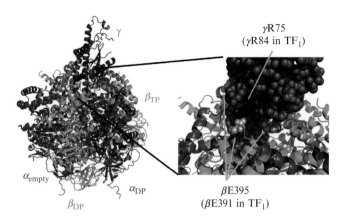

Daichi Okuno et al., Figure 12.2 Positions of cysteine residues of the β and γ subunits used for cross-linking. βE395 and γR75 are shown in dark gray and light gray, respectively, in a crystal structure (1E79) (Gibbons et al., 2000). These positions correspond to βE391 and γR84, respectively, of F_1 from the thermophilic *Bacillus* PS3, which is used in the present report. The α, β, and γ subunits are shown in blue, green, and red, respectively. The R33, D74, D110, R113, R133, R134, and P135 residues of the γ subunit are removed to show βE395 and γR75. The figure was produced using PyMOL 0.99.

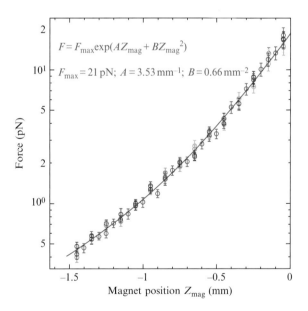

Maria Manosas et al., Figure 13.2 Force calibration. Force as a function of the magnets position (Z_{mag}) measured for several micron-sized beads (Myone Invitrogen) attached by a single λ-DNA molecule (each color corresponds to a separate bead). Note that the position Z_{mag} is measured with respect to the sample chamber (e.g., $Z_{mag} = 0$ when the magnets touch the upper surface of the chamber and when the magnets are 1 mm above the chamber $Z_{mag} = -1$). The blue line corresponds to a fit of the data to the function $F = F_{max} \exp(AZ_{mag} + BZ_{mag}^2)$, which yields $F_{max} = 21$ pN, $A = 3.53$ mm^{-1} and $B = 0.66$ mm^{-2}. This curve can be used to estimate the force on the bead at a known position of the magnets.

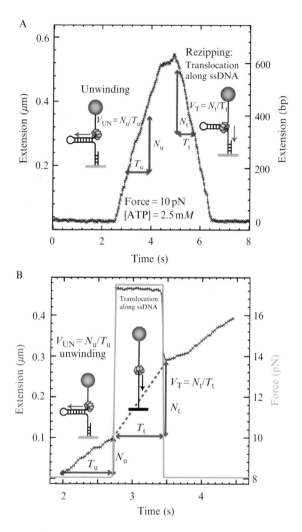

Maria Manosas et al., Figure 13.7 Measuring unwinding and ssDNa translocation activities. (A) Experimental trace corresponding to the gp41 helicase activity on the 600 bp hairpin (generated from a 1.2 kbp DNA substrate, Fig. 13.6B). Extension in μm (left axis) is converted to number of base pairs unwound (right axis) by assigning to the maximum length of the unwinding events to the full length of the DNA hairpin. The trace shows the unwinding phase (rising edge) and the rezipping phase (falling edge) in which the enzyme translocates on the ssDNA and the hairpin reanneals in its wake. (B) Experimental trace corresponding to the gp41 helicase activity on the 600 bp hairpin. The applied force (grey) is transiently increased during DNA unwinding by the helicase in order to measure the translocation on ssDNA.

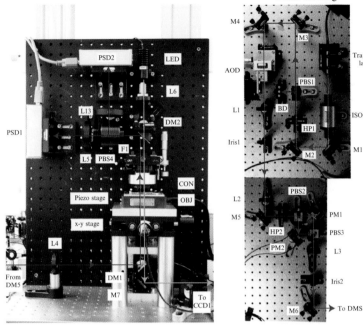

Jongmin Sung et al., Figure 14.3 Photographs of the actual dual-beam optical trap setup. The primary components of the setup in Fig. 14.2 (microscope, bead position detector, and trapping laser and beam steering) are shown here. (A) Photograph of the microscope and the bead position detector setup. (Green arrow) Beam path from the bright-field light source (LED) to the detector (CCD1)—a diverging beam from the LED light source (740 nm) is collected by a 10× objective lens (L6). The distance between holes in the bread-board is 1 in. Condenser (CON) is mounted on a gimbal optics mount and a z-translation stage. The distance between LED and the front aperture of CON is ~11 in. CON collects and focuses the beam into the sample channel which is mounted on the top of the piezo stage during experiments. Objective lens (OBJ) (obscured from view by the piezo stage) collects the beam from the sample plane, and sends it downwards. The mirror turret with dichroic mirror DM1 and reflective mirror M7 is mounted on the optical table, with DM1 and M7, respectively, 3 and 1.5 in. above the table surface. The beam reflected by M7 is collected by a tube lens (L7, $f = 200$ mm, not shown in figure) which images the sample plane on CCD1. (Red arrow) Trapping/detection beam paths from DM5 to the detectors (PSD1, 2)—trapping/detection beams pass through a lens L4 and are reflected on a dichroic mirror DM1. The OBJ focuses the beams in the sample channel and CON collects the transmitted beams. The beams are reflected on a dichroic mirror DM2, pass through F1, PBS4, L5/L13, and finally arrive on the sensors (10 mm^2) of PSD1, 2. We adjust the height of CON such that the transmitted beams ($D \sim 20$ mm at the CON) are collimated. The distance between the BFP of CON and L5 is ~230 mm, and the distance between L5/L13 and PSD1/PSD2 is ~110 mm. The interference patterns at the BFP of CON are demagnified by a factor of ~2 ($M = \frac{110\,\text{mm}}{230\,\text{mm}} \sim \frac{1}{2}$). A filter F1 is mounted on a flipping mount such that it selectively transmits the 850 nm detection laser beams but blocks the 1064 nm beams when it is flip-in position. When we use trapping beams for the detection, the F1 is switched to the flip-out position. PSDs are mounted on custom boxes and L-brackets for the stability of the detectors, and the input and output connections of the PSD with power supplies and NI-DAQ are accomplished though the cable mounted on the box. (B) Photograph of the trapping and the beam steering parts setup—the fiber output of the trapping laser (the body is separately mounted off the table) is stably mounted on a V-block mount followed by an ISO1 at 5 in. from the fiber. Mirrors M1 and M2 (4 in. separated) redirect the beams to the pair of HP1 and PBS1 which allows controlling of the trap beam power. The initial beam size is 1.6 mm. Mirrors M3 and M4 redirect the beams to the fast beam steering components (AOD, L1, L2) followed by the slow beam steering components (PM1, L3, L4 (not seen)). Here, the position of AOD, L1, L2, and PM1/PM2 satisfies the *4f arrangement* condition, whose separations are L1, L1 + L2, and L2, respectively. The second pair of HP2, PBS2, and PBS3 separates and recombines the p- and s-polarized beams for independent control of the dual beams. Rotation of the HP2 determines the relative stiffness of the dual beams, and usually we make the trap stiffness of the two beams the same to maximize the stability of the actin dumbbell. The difference of each beam path between PBS2 and PBS3 is 4 in. in our setup, such that noise sources (mechanical vibration and air fluctuation) arising out of them affect the dual beam by the same amount. Note that the distance between the threaded mounting holes on the optical table is 1 in.

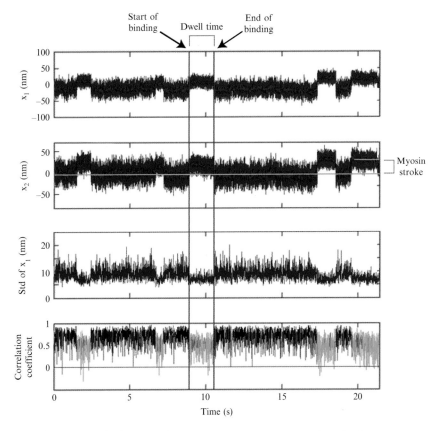

Jongmin Sung et al., Figure 14.6 Representative experimental result of a nonprocessive motor. The time traces show the interaction of a single-headed myosin VI construct with an actin filament. The first and the second graphs show the time trajectory of the first (x_1) and the second (x_2) bead positions. The blue line shows the raw data with a sampling rate of 10 kHz while the red line is the low-pass filtered result of the raw data (fourth-order low-pass digital Butterworth filter with a cutoff frequency of 50 Hz. The trap stiffness in this particular example is ~ 0.02 pN/nm, and this results in Brownian motion of bead (blue line) with standard deviation (std) of ~ 10 nm. Increasing trap stiffness will reduce the std of Brownian fluctuation. In the figure, there are five very clear binding and unbinding events, showing a sudden jump and drop in the bead positions. The duration between binding and unbinding events is the dwell time. The stroke size refers to the displacement of the bead position when the binding event occurs. This stroke size is before compliance correction which is explained in the legend to Fig. 14.7 and in Section A.2. The third and fourth graphs show the std of the first bead position and the correlation coefficient between the two bead positions, respectively. At the third binding event at ~ 10 s, for example, we observe a sudden drop in the std as well as the correlation between x_1 and x_2. This happens because the myosin binding clamps the actin dumbbell, reducing its Brownian fluctuations, which in turn reduces the Brownian fluctuations of the two trapped beads. We use a custom Matlab program which automatically finds two different states in the correlation graph, so the sudden change of the color (blue and green) in the last graph helps to visualize the binding events. We manually choose the timing of binding and unbinding events based on the std and the correlation. The binding events are chosen only when both the std and the correlation change register a significant change.

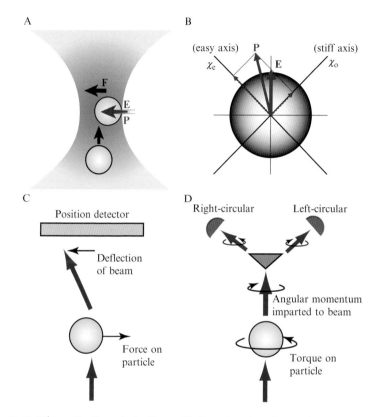

Braulio Gutiérrez-Medina et al., Figure 15.1 Principles of optical manipulation and signal detection. (A) The electric field (E, blue arrow) associated with a tightly focused laser beam induces a collinear electric polarization (P, red arrow) in an isotropic dielectric particle. As the particle moves away from the center of the laser focal volume, the induced dipole–electric field interaction produces a net restoring force (F, black arrows) toward the center, confining the particle in 3D. (B) For an optically anisotropic particle, different axes exhibit different polarizabilities. The induced polarization vector, therefore, is not collinear with the external electric field, and inclines toward the most polarizable (easy) axis. This effect gives rise to a net torque on the particle that tends to align the easy axis with the electric field. (C) Force can be detected by measuring trapping beam deflections (corresponding to changes in linear momentum) induced by an off-center, trapped particle, using a position-sensitive detector. (D) Analogously, the torque exerted by a linearly polarized beam on an anisotropic particle can be detected, based on the imbalance in the right- and left-circular polarization components in the beam after scattering by the particle, which can be measured independently.

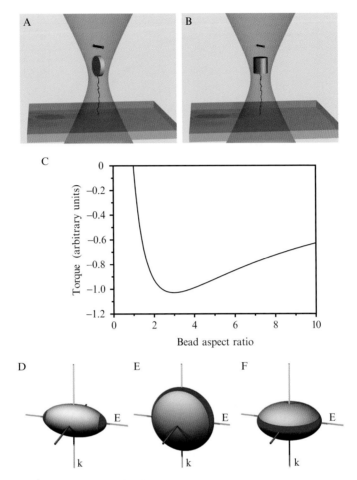

Braulio Gutiérrez-Medina et al., Figure 15.2 Particle anisotropy for optical trapping and rotation. (A) A small, oblate particle made of an optically isotropic material tends to align its long radii with the trapping beam axis (vertical) and the polarization direction (red arrow). (B) A birefringent cylinder tends to align its long axis with the trapping beam axis, allowing the extraordinary optical axis of the crystal to track the trap polarization direction. Possible materials for birefringent particles include quartz and calcite. (C) Theoretical estimate of the torque exerted on a subwavelength, oblate particle subjected to a uniform electric field, shown as a function of its aspect ratio (maximum to minimum radius). Bigger torques correspond to larger negative values; the greatest torque (curve minimum) occurs near an aspect ratio of 3. (D–F) The polarization ellipsoids for quartz (D), and calcite (E), (F) are shown, where red (gray) zones represent regions of maximum (minimum) electric susceptibility. Axes corresponding to red regions tend to align with the direction of the electric field vector **E** (green axis). For optical trapping and rotation, an ideal configuration is obtained when rotation is possible around the direction of the beam propagation vector, **k** (blue axis). This is the case for quartz (D). For calcite, the configuration shown in (E) can be used to exert torque (although alignment of the polarization ellipsoid with respect to the direction of **E** is not unique), whereas in the arrangement shown in (F) the particle exhibits no net birefringence on the plane perpendicular to **k** and no torque can be generated about the beam axial direction.

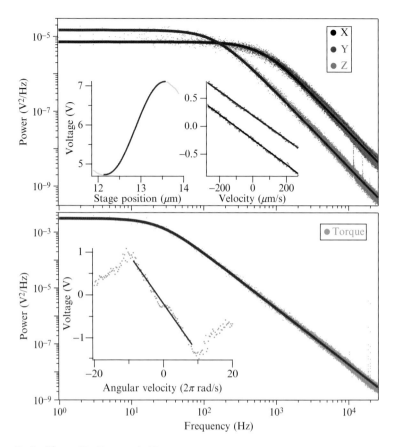

Braulio Gutiérrez-Medina et al., Figure 15.7 Calibration of the optical torque wrench using a quartz cylinder. (Top) Calibrations of the linear dimensions. First, the power spectra for x, y, and z signals were computed and fit to Lorentzian functions, giving roll-off frequencies $f_{c,x} = 639 \pm 1$ Hz, $f_{c,y} = 631 \pm 1$ Hz, $f_{c,z} = 147.2 \pm 0.2$ Hz, and zero-frequency amplitudes $\widetilde{P}_x = (1.158 \pm 0.003) \times 10^{-6}$ V^2 Hz^{-1}, $\widetilde{P}_y = (1.148 \pm 0.003) \times 10^{-6}$ V^2 Hz^{-1}, and $\widetilde{P}_z = (2.382 \pm 0.005) \times 10^{-6}$ V^2 Hz^{-1}. Next, linefits to Stokes' drag measurements in x and y (Right inset) provided the slopes $s_x = (2.255 \pm 0.008) \times 10^{-6}$ V s nm^{-1} and $s_y = (2.158 \pm 0.008) \times 10^{-6}$ V s nm^{-1}, which were combined with the power spectral results to yield $\xi_x = 1/(2\pi s_x f_{c,x}) = 110$ nm V^{-1} and $\xi_y = 117$ nm V^{-1}. For the z signal, vertical scanning of a fixed cylinder (Left inset) produced a record well fit by the derivative of a Gaussian (amplitude $A = (1.395 \pm 0.006) \times 10^3$ V nm, S.D. $\sigma = 708 \pm 2$ nm), from which we obtain $\xi_z = \sigma^2/A = 359$ nm V^{-1}. These measurements were combined to obtain the trap stiffnesses $\kappa_x = 4.5 \times 10^{-2}$ pN nm^{-1}, $\kappa_y = 4.1 \times 10^{-2}$ pN nm^{-1}, and $\kappa_z = 9.1 \times 10^{-3}$ pN nm^{-1}. (Bottom) Calibration of torque. A procedure analogous to the linear x, y cases was carried out. From the experimentally measured values, $f_{c,\tau} = 24.29 \pm 0.05$ Hz, $\widetilde{P}_\tau = (5.04 \pm 0.01) \times 10^{-4}$ V^2 Hz^{-1}, and $s_\tau = (1.84 \pm 0.03) \times 10^{-2}$ V s rad^{-1}, the volts-to-radians conversion factor $\xi_\tau = 1/(2\pi s_\tau f_{c,\tau}) = 0.37$ rad V^{-1} and the angular trap stiffness $\kappa_\tau = 4k_B T f_{c,\tau} s_\tau^2 / \widetilde{P}_\tau = 264$ pN nm rad^{-1} were obtained. All power spectrum records represent averages from 50 measurements sampled at 66 kHz. For these measurements, the trapping laser power was ~ 20 mW (measured before entry into the objective rear pupil).

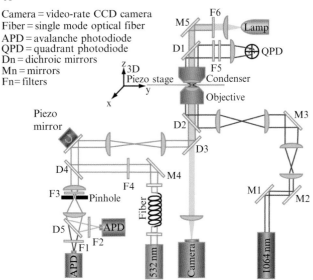

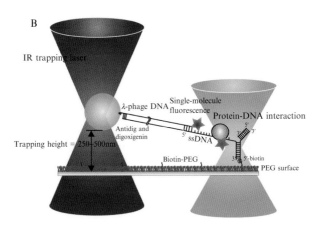

Ruobo Zhou et al., Figure 16.1 Experimental configuration: (A) The combined optical trapping and single-molecule confocal fluorescence instrument is built around a commercial inverted microscope (IX71, Olympus) equipped with a three-dimensional piezo stage (P-527.3CL, Physik Instrumente). The trapping laser beam (1064 nm, 800 mW, Spectra-Physics, Excelsior-1064-800-CDRH) is coupled through the back port of the microscope, while the fluorescence excitation laser beam (532 nm, 30 mW, World StarTech) is directionally controlled by a two-dimensional piezo-controlled steering mirror (S-334K.2SL, Physik Instrumente) and coupled through the right side port. The beams are combined via a dichroic mirror (D2: 780DCSPXR; Chroma) into an oil-immersion objective (UPlanSApo, 100×, NA = 1.4, Olympus). The intensity profile of the trapping laser in the back focal plane of the condenser

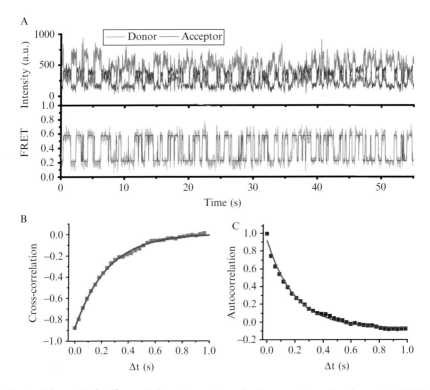

Ruobo Zhou et al., Figure 16.4 General methods to analyze the dynamic FRET trajectories. (A) An example of the donor–acceptor intensity trajectories from a two-state dynamic system. Hidden Markov model (HMM)-derived idealized FRET trajectory (magenta) superimposed on the FRET trajectory (blue). (B) Cross-correlation between the donor and acceptor time trace shown in (A) which is fit to a single-exponential function (solid line). (C) Autocorrelation of the FRET trajectory shown in (A) which is fit to a single-exponential function (solid line).

(Achromat/Aplanat, NA = 1.4, Olympus) is imaged onto a quadrant photodiode (UDT SPOT/9DMI) to detect the deviation of the trapped bead position from the trap center. The fluorescence emission is isolated from the reflected infrared light (F3: HNPF-1064.0-1.0, Kaiser) and is band-pass filtered (F1: HQ580/60m, F2: HQ680/60m, Chroma) before imaged onto two avalanche photodiodes, respectively. The bright-field image of the trapped beads is imaged onto a CCD camera (GW-902H, Genwac). (B) Not-to-scale sketch of the combined single-molecule force and fluorescence assay. With the help of λ-phage DNA, a large spatial separation between the trapping laser beam (red) and the excitation laser beam (green) can be achieved. One can probe protein–nucleic acid interactions with single-molecule fluorescence or FRET. The surface passivation is typically achieved with a dense PEG layer, while specific DNA tethering is realized with Biotin–PEG neutravidin interaction.

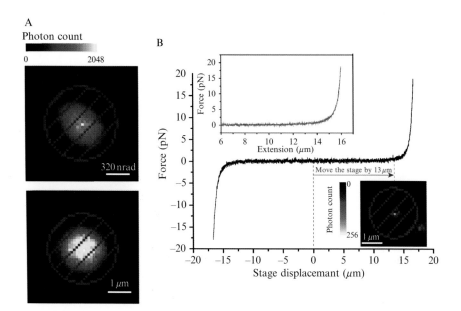

Ruobo Zhou et al., Figure 16.5 Mirror calibration and force–extension curve. (A) The mirror scan image around the area where the fluorescent beads are trapped in the sample plane without (upper) and with (lower) the mirror calibration. (B) A stretching curve and the force–extension curve (upper inset) of the tethered DNA after the origin of the piezo stage is set to the estimated tethered position. A WLC model (red) is used to fit the experimental force–extension curve (blue). The lower inset shows a mirror scan image around the origin of the piezo stage after displacing the stage from its origin by 13 μm. The green dot indicates the center position of the fluorescently labeled molecule that is being stretched.

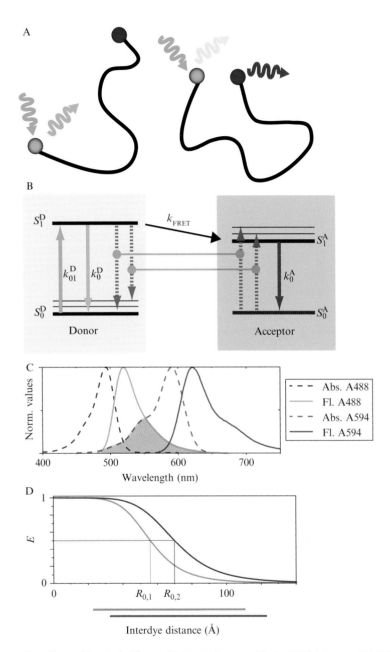

Evangelos Sisamakis et al., Figure 18.2 Basic essentials in FRET theory. (A) Schematic representation of a biomolecule labeled with donor (D—green sphere) and acceptor (A—red sphere). When the interdye distance is large, only fluorescence signal from D is observed (shown in green while excitation is shown in blue). When an acceptor molecule (A) resides in the vicinity of D, energy transfer takes place. In this case fluorescence signals both from D (dim green) and A (red) will be recorded. The signal of D will be lower (dim green) when energy transfer takes place. (B) Simplified Perrin-Jablonski diagrams of D and A. D is excited at a rate k^D_{01} to the first singlet state S^D_1. In the absence of A, it is depopulated with rate constant k^D_0. Due to the coupling of the possible de-excitation of D and excitation of A, energy transfer can occur at a rate k_{FRET} resulting in the excitation of A from S^A_0 to S^A_1 which is depopulated with a rate constant k^A_0. (C) The excitation (Abs.) and emission (Fl.) spectra of Alexa488 (A488) and Alexa594 (A594). A488–A594 dyes constitute a commonly used D–A pair in FRET studies. The overlap between the emission of D and excitation of A is highlighted as gray area. The amount of the overlap influences the value of the Förster radius (see text for details about the overlap integral). (D) Plot of FRET efficiency versus the interdye distance for two different values of the Förster radius, $R_{0,1} = 56$ Å (orange), $R_{0,2} = 70$ Å (brown). The value of the Förster radius, R_0, defines the useful dynamic range of distances (orange and brown correspondingly) that can be measured with a specific dye pair as it is illustrated by Eq. (18.6). For ensemble studies this dynamic range is considered to extend from 0.5 to 1.5R_0 (Lakowicz, 2006) corresponding to FRET efficiency values from 0.98 to 0.08. For single-molecule studies, FRET efficiencies as low as 0.03 can be discriminated from donor-only populations by PDA (see Sections 5.1 and 5.2; Gansen et al., 2009) and thus longer distances can be measured with the same dye pair.

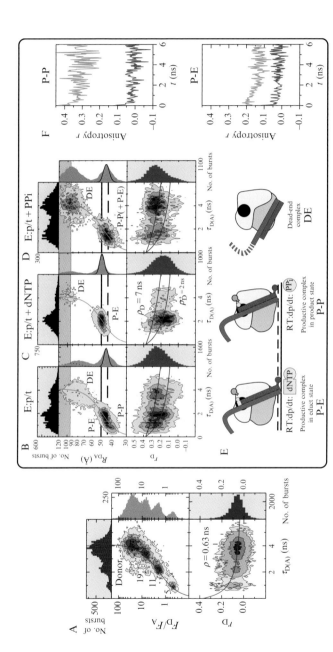

Evangelos Sisamakis et al., Figure 18.6 Typical 1D and 2D histograms in smFRET for a mixture of quasi-static species. *Left:* (A) Typical results of a smFRET experiment with a mixture of the same dsDNA labeled with the dye pair (Alexa488–Cy5) at three distinct interdye distances (5, 11, and 19 bp separation) and D-only. The parameter histograms for the intensity ratio F_D/F_A, the donor lifetime in presence of acceptor, $\tau_{D(A)}$, and the donor anisotropy, r_D, are arranged in two joint 2D histograms of F_D/F_A versus $\tau_{D(A)}$ and r_D versus $\tau_{D(A)}$ sharing the $\tau_{D(A)}$ axis. The number of bursts corresponds to number of selected single-molecule events. The scaling is from white to black for increasing number of bursts. The parameters for the static FRET line (Eqs. (18.16)) in the upper panel are $\tau_{D(0)} = 4.1$ ns, $\Phi_{FD(0)} = 0.8$, $\Phi_{FA} = 0.4$, and $g_G/g_R = 0.68$. The parameters for the Perrin equation (Eq. (18.18)) in the lower panel are $r_0 = 0.375$ and the estimated mean rotational correlation time of the donor, $\rho = 0.63$ ns. *Right*: MFD analysis of the heterogeneity of HIV-1 (RT):p/t complexes. 2D histograms of R_{DA} and r_D versus $\tau_{D(A)}$ of different HIV-1 RT:p/t complexes (panels B–D) and subensemble spectroscopy (panel F). For all 2D histograms of R_{DA} versus $\tau_{D(A)}$, the magenta overlaid line represents the static FRET line (Eq. (18.7b)) of parameters $\tau_{D(0)} = 3.1$ ns, $\Phi_{FD(0)} = 0.64$, and $\Phi_{FA} = 0.3$, $R_0 = 53$ Å. *Lower panel*: r_D is plotted versus $\tau_{D(A)}$ together with overlaid Perrin equation (Eq. (18.18)) computed for two rotational correlation times ρ_D (2 and 7 ns). (B) Analysis of the RT:p/t complex showing the presence of three species sketched below: productive complex in educt state (P-E), productive complex in product state (P-P) and dead end complex (DE). (C) Addition of 200 μM dNTPs. The $\tau_{D(A)}$–R_{DA} histogram shows only species P-E and DE. The P-E complex (panel E) interacts with the dp/dt in a state closely resembling the known RT:dp/dt structures. Here, the dNTP is thought to occupy a binding site in the polymerization-active site and is therefore prebound for incorporation into the primer strand (purple). The solid black line in the histogram, and below in the cartoon, indicates the position of the dp/dt bound in the P-E state. In the P-E state, a dNTP occupies the binding pocket. (D) Addition of 200 μM sodium pyrophosphate (NaPP$_i$). The presence of PP$_i$ moves the peak toward shorter distances, indicated by species P-P. The presence of PPi shifts the primer terminus into the binding pocket, forming P-P. The dashed black line in the histogram, and below in the cartoon, indicate the position of the dp/dt bound in the product state, in the presence of the PP$_i$ shifting the dp/dt into the binding cleft. The peak is not Gaussian distributed, and the rotational correlation time, ρ_D, remains high. The position of the dead end (DE) complex remains unchanged. (E) The p66 subunit of RT is colored in light gray, and the p51 subunit is colored dark gray. The polymerase-active site of p66 is colored black. The fluorescence dyes Alexa488 and Cy5 are indicated by a balloon colored in green and red respectively. The FRET data of the p complexes are consistent with the structure obtained by X-crystallography (Kohlstaedt *et al.*, 1992). Preliminary results indicate that the nucleic acid substrate in the DE complex is bound at a site on the p51 subunit, far removed from the nucleic acid binding tract observed by crystallography. (F) Results for subensemble time-resolved anisotropy, *r*, in the green and in the red channels of the species P-P and P-E. Note that the fast decay components are not fully resolved by our setup. Alexa488 has a fundamental anisotropy $r_0 = 0.375$.

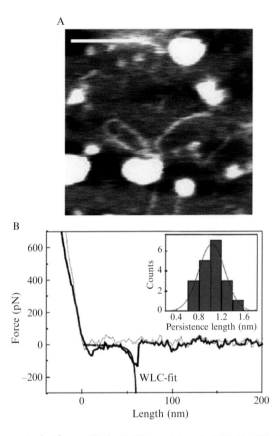

Ferry Kienberger et al., Figure 19.6 Pulling of the viral RNA. (A) Imaging of the single-stranded RNA genome (∼7100 nucleotides) from a human rhinovirus in buffer solution. Single RNA cores with protruding RNA loops. Scan size 200 nm, scale bar 70 nm. The z-scale ranges from 0 to 3 nm. (B) Pulling viral RNA in force–distance cycles. After imaging the viral RNA, the tip was lowered to establish a contact for about 1 s and a force versus extension curve was acquired. Approach (red), retraction (black), and WLC-fit (thin black) are shown. The stretching of RNA was observed with a typical nonlinear behavior. *Inset*: histogram of the persistence length obtained from WLC fits of 19 RNA pulling curves. Gaussian fit is in red. Reprinted with permission from Kienberger *et al.* (2007).

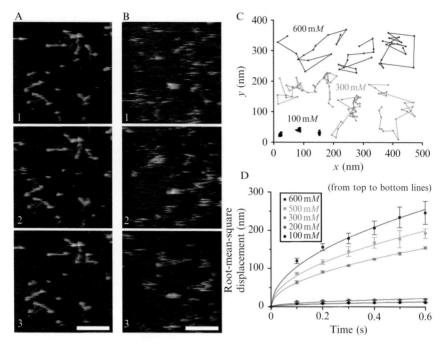

Daisuke Yamamoto et al., Figure 20.1 Ionic strength dependent diffusion of MyoV-HMM (tail-truncated myosin V) on bare mica surface. (A, B) Typical successive AFM images taken at 100 mM (A) and 600 mM KCl (B). Imaging rate: 102 ms/frame, scan area: 300 × 300 nm² with 100 × 100 pixels, scale bar: 100 nm. The number in each frame represents the frame number. The observation buffer is 20 mM imidazole–HCl (pH 7.6), 2 mM MgCl$_2$, 1 mM EGTA, and 100 or 600 mM KCl. (C) Typical trajectories of MyoV-HMM measured in 100, 300, or 600 mM KCl-containing solutions. (D) Root-mean-square displacement (x_{rms}) as a function of time. Each point was calculated from the trajectories of ∼100 molecules and well fitted by an equation $x_{rms} = \sqrt{4Dt}$, where D is the diffusion coefficient and t is the time elapse. The values of D (nm²/s) at respective KCl concentrations are $D_{100\ mM} = 79 \pm 10$, $D_{200\ mM} = 213 \pm 26$, $D_{300\ mM} = (0.99 \pm 0.01) \times 10^4$, $D_{500\ mM} = (1.69 \pm 0.04) \times 10^4$, and $D_{600\ mM} = (2.73 \pm 0.09) \times 10^4$.

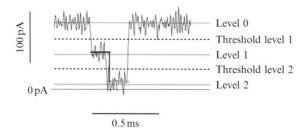

Giovanni Maglia et al., Figure 22.10 Typical DNA translocation event analyzed by the Clampfit software. Levels 0, 1, and 2 are set manually and the software automatically assigns data points to level 1 (red line) or level 2 (green line).